Collective Excitations in Unconventional Superconductors and Superfluids

Collective Excitations in Unconventional Superconductors and Superfluids

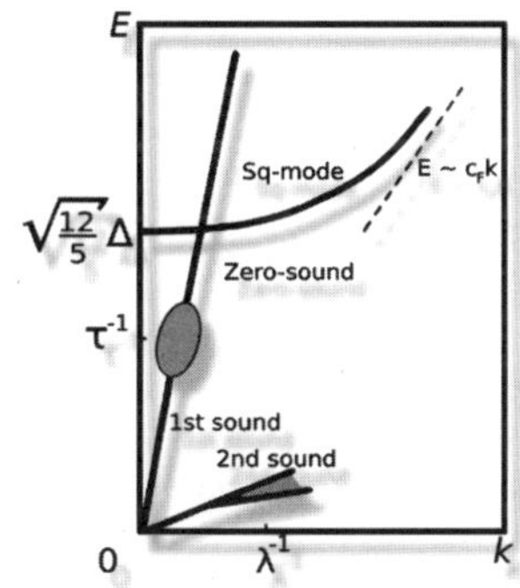

Peter Brusov
Pavel Brusov

World Scientific
NEW JERSEY · LONDON · SINGAPORE · BEIJING · SHANGHAI · HONG KONG · TAIPEI · CHENNAI

Published by

World Scientific Publishing Co. Pte. Ltd.

5 Toh Tuck Link, Singapore 596224

USA office: 27 Warren Street, Suite 401-402, Hackensack, NJ 07601

UK office: 57 Shelton Street, Covent Garden, London WC2H 9HE

British Library Cataloguing-in-Publication Data
A catalogue record for this book is available from the British Library.

COLLECTIVE EXCITATIONS IN UNCONVENTIONAL SUPERCONDUCTORS AND SUPERFLUIDS

ISBN-13 978-981-277-123-0
ISBN-10 981-277-123-9

Printed in Singapore by World Scientific Printers

To our Teacher
Professor Victor Popov,
whose name should have been in the author's list.

Preface

The monograph is devoted to the theoretical investigation of the collective excitations in superfluid quantum liquids and superconductors with nontrivial pairing. It based mostly on the original papers by the authors (with coauthors) and generalizes their study during last 33 years in Leningrad branch of Mathematical Institute named by V. A. Steklov, Low Temperature Laboratory of Helsinki University of Technology, Rostov State University (Rostov–on–Don), Northwestern University (Evanston), Cornell University (Ithaca), Texas Center for Superconductivity of University of Houston, Osaka City University, The Harish–Chandra Research Institute (Allahabad, India), S. N. Bose National Centre for Basic research (Kolkata, India).

Among superfluid quantum liquids we consider superfluid ^{3}He, superfluid ^{4}He, superfluid ^{3}He–^{4}He mixtures, superfluid ^{3}He–films, superfluid ^{3}He and superfluid ^{3}He–^{4}He mixtures in aerogel. Among superconductors we consider high temperature superconductors, heavy–fermion superconductors and superconducting films. We describe shortly the relativistic analogs of ^{3}He and discuss the connections between ^{3}He theory and gauge field theory.

Important feature of this manuscript is the use of one method through the whole book – the path (functional) integral technique. The required background on the functional integration method in statistical physics is presented in Chapter I. By this method we build the three– and two–dimensional models for *s*–, *p*– and *d*–wave pairing in neutral as well as in charged Fermi– and Bose–systems and their mixtures. The use of path integral technique has allowed to describe the collective modes of order parameter in such complex systems with great details and create "the spectroscopy of collective modes in unconventional superfluids and superconductors".

Monograph will be useful for theorists as well as for experimentalists, studying superfluids and superconductors, low temperature physics, condensed matter physics, solid state physics, quantum liquids. It could be used by graduate students specializing in the same areas. Because description in the monograph includes the basic principles and is self–consistent and detailed the advanced undergraduate students can use it as a textbook on low temperature physics, quantum liquids and superconductors as well as on condensed matter physics.

Contents

Introduction

The collective properties of superfluid quantum liquids and superconductors with nontrivial pairing are studied in the monograph. In quantum liquids the De Broigle wave length corresponding to the thermal movement of atoms becomes comparable to the average distance between atoms at sufficiently low temperatures. Thus quantum effects play a leading role and determine the main properties of quantum liquids. Nowadays we know such quantum liquids as the helium isotopes (^{3}He, ^{4}He, ^{6}He) and their mixtures, electron liquids in metals, semimetals and semiconductors, nuclear matter and also the spin oriented hydrogen isoropes ($H\uparrow, D\uparrow, T\uparrow$). Last 20–30 years there is significant progress in study of unconventional superconductors, like heavy–fermion superconductors and especially high temperature superconductors.

There are both Bose– and Fermi–superfluid systems. The superfluidity phenomena in Bose–systems is connected with the Bose–condensate of bosons, while in Fermi–systems it is due to the formation of Cooper pairs of fermions and their subsequent Bose–condensation.

The level of understanding of the collective properties varies for different quantum liquids. For example, superfluidity in ^{4}He and superconductivity of electrons have been studied in details, but investigations of ^{6}He and of spin – oriented hydrogen are in a rudimentary state. The small lifetime of ^{6}He and necessity for strong magnetic fields demand special experimental conditions. During last 38 years the superfluid phases of ^{3}He (which were discovered in 1971) have been intensively studied. In the monograph we also examine different systems with a different level of detail. The major part of the book is devoted to the investigation of the collective excitations in superfluid ^{3}He, ^{4}He and ^{3}He–^{4}He solutions. We also discuss Fermi–systems with *s*–pairing, which can be considered as a model for ordinary superconductors and as a model of superfluid nuclear matter.

The high temperature superconductors, studying very intensively during 25 years, are far from complete understanding. It is enough to say that even the pairing mechanism in most of the high temperature superconductors is unknown.

Some original results by the authors, which were obtained within the framework of a unique approach, the method of functional (continuum) integration (path integral technique), are presented. The required background on the functional integration method in statistical physics is presented in Chapter I.

The application of the functional integration methods allows one to derive many interesting results. The theory of phase transitions of the second kind, superfluidity, superconductivity, lasers, plasma, Kondo effect, Ising model – this is an incomplete list of problems, for which the application of the functional integration method appears to be very useful. In Chapter I we consider the functional integrals and diagram techniques for Bose– and Fermi–particles as well as the method of successive integration over fast and slow Fermi–fields.

In Chapter II we investigate superfluid Fermi–systems with *s*–pairing. As examples we consider the Fermi–gas model with an attractive interaction and also the generalization of this model in which short–range attractive and long–range Coulomb repulsive potentials act. The application of the functional integration method allows the construction of an effective (or "hydrodynamical") action functional which is specially adapted for the investigation of collective excitations in superfluid Fermi–systems.

A more complicated and interesting superfluid Fermi–system is ^{3}He, with *p*–wave pairing, which is discussed in Chapters III–XVIII. The complexity arises from pairing of ^{3}He atoms in the *p*–state which results in an order parameter with 18 degrees of freedom. This complexity leads to difficulties in building up a microscopic theory. ^{3}He–type systems have superfluid, magnetic, and liquid–crystal properties simultaneously. Investigation of the spectrum of collective excitations has been the most important tool for studying both the microscopic and macroscopic properties of ^{3}He along with thermodynamic and nuclear magnetic resonance (NMR) experiments.

More traditional methods of investigation of the collective excitations, such as the kinetic equation method and using the Bethe–Solpiter equation, had solved a number of problems. However, the functional integration method turns out to be much more effective tool for treating collective excitations. The subsequent functional integration with respect to "fast" and "slow" Fermi–fields along with the introduction of auxiliary Bose–field (corresponding to Cooper pairs of quasi–fermions instead of the initial Fermi–field) allows us to construct a model for Fermi–systems with *p*–pairing. In the framework of this model we obtain an effective (hydrodynamical) action functional, which describes all the physical properties of the system, particularly its collective excitation spectrum. This approach allows us to investigate both Goldstone–modes of the Bose–spectrum (arising as a result of the symmetry properties of the system) and also nonphonon–modes connected with various oscillations of the self–consistent field).

Sound experiments play the most important role in study of collective modes in unconventional superfluids and superconductors. In Chapter III we make general remarks concerning sound propagation in superfluid ^{3}He and superconductors.

The experimental properties of the superfluid ^{3}He are described in Chapter IV. We discuss the phase diagram of the system, its superfluid and magnetic properties. Special attention is devoted to experiments on ultrasound propagation connected with the excitation of collective modes.

In Chapter V we build the *p*–pairing model of the superfluid phases of ^{3}He by path integral technique. Here we discuss the alternative kinetic equation method.

Different forms of the self–consistent field oscillations in the *A*– and *B*–phases are investigated in Chapters VI and VII. In Chapter VI we calculate the whole collective mode spectrum in ^{3}He–*B* with dispersion corrections. Knowledge of the dispersion corrections is important when comparing the theory with experiments (sound waves generate excitations (collective modes) with nonzero momenta). Comparison with experiment as well allows to obtain the temperature dependence of the energy gap, which is the most important parameter of a superfluid Fermi–system.

A microscopic theory of the collective excitations (CE) in the *A*–phase of ^{3}He is constructed in Chapter VII with the use of a path–integration method. The whole collective mode spectrum, taking damping into account, is calculated. The cause of additional Goldstone modes in the weak–coupling approximation and their analogy to *W*–bosons in the weak–interaction theory are discussed. The whole set of equations, which describe the collective excitations in arbitrary magnetic fields, is obtained. They are solved for small magnetic fields and the linear Zeeman effect for clapping and pair–breaking modes is obtained.

In Chapter VIII we discuss the method of the identification of ^{3}He–*A* by ultrasound experiments. On the basis of available experimental data it has been suggested that ^{3}He–*A*, conventionally identify as the axial–phase, may actually be an axi–planar phase. We investigate this problem within a simple, time–dependent Ginsburg–Landau model as well as by studying the second variation of the free energy functional. Both methods show that the spectrum in the axial–phase is degenerate, while it is split in the axi–planar phase. This fact may serve as a sensitive test of the existence of the latter phase and appropriate measurement of the collective mode spectrum could resolve the issue.

Chapter IX is devoted to the problem of the stability of the Goldstone–modes in *A*–, *B*– and 2*D*–phases of ^{3}He with respect to various decay mechanisms. It is shown, that all the Goldstone branches in the *B*–phase are stable, while in the *A*– and 2*D*–phases the stability depends on the angle between the momentum of a collective excitation and the preferred direction of the phase: $\boldsymbol{l}$–vector in the *A*–phase and the direction of the magnetic field $\boldsymbol{H}$ in the 2*D*–phase. The Goldstone–modes turn out to be stable inside some cones around the preferred direction. Outside the stability regions, the energy of the excitation becomes complex, that physically means the possibility of the decay of the excitation into constituted fermions whose momenta are close to the preferred direction. In this Chapter the dispersion coefficients for all Goldstone modes in above three phases are calculated.

In Chapter X the ^{3}He–model is generalized in order to take into account a magnetic dipole interaction, which is responsible for the phenomena connected with the NMR. The generalized effective action functional which allows us describe all the NMR phenomena have been built. The influence of the dipole interaction on velocities of collective excitations and their stability in the *B*–phase of ^{3}He is investigated, as well as the magnetic field influence on the Bose–spectrum in the *A*– and *B*–phases. The problem of the magnetic field influence on the number of Goldstone modes is considered.

The influence of the outer electric field on the collective modes in the superfluid ^{3}He and ^{4}He is investigated in Chapter XI. Interaction of induced electric dipole moments of atoms lead to anisotropy of the sound velocity which can decrease and vanish at some critical field. This leads also to disappearance of the superfluidity in directions orthogonal to the electric field.

In Chapters X and XI we consider the influence of a magnetic field, dipole interaction and electric field on the spectrum of the collective modes due to some supplementary terms in the free energy connected with these perturbations. In these Chapters it was supposed that the order parameter remains unperturbed. But it is clear that the outer perturbation will deform the order parameter, and it can change the spectrum of collective excitations significantly. In Chapter XII the influence of one kind of order parameter deformation on the collective modes is studied, namely the deformation of the gap in the Fermi–spectrum induced by dipole interaction, electric and magnetic fields, superfluid flows and rotational effects (due to vortices and gyromagnetism). Some new effects are obtained such as the splitting of the spectrum of the nonphonon modes under the perturbation, the intersection of the branches of modes with $J = 2$ at nonzero momentum of collective modes, the existence of the so–called pairbreaking modes (with $E \approx 2\Delta$) as resonances and some other effects. The existence of pairbreaking modes leads to new very interesting features in sound experiments at the absorption edge.

In Chapter XIII we discuss the structure of the spectrum of the ultrasound absorption into the squashing–mode in superfluid ^{3}He–*B*, where a doublet splitting of this mode has been observed. We show that this phenomenon is induced by superflow and based on these experiments we suggest a method of determining the superfluid velocity by ultrasound experiments: we show that the order parameter collective mode splitting may be a measure of the superfluid velocity, $\boldsymbol{v}_S$, which is not easy to obtain.

In Chapter XIV we analyze the old and recent transverse sound experiments in superfluid ^{3}He–*B* and solve the old problem of superfluid quantum liquids in confined geometry: what is the boundary state of ^{3}He–*B*. We pay special attention to difference of transverse sound experiments data from ones of longitudinal sound experiments. We consider a few possible explanations of above experimental data: existence of a new superfluid phase in the vicinity of a boundary, excitation of different branches of squashing–mode by longitudinal and transverse sounds and deformation of *B*–phasc near the boundary. We come to conclusion that last possibility seems the most likely and boundary state of ^{3}He–*B* is deformed *B*–phase, predicted by Brusov and Popov more then twenty five years ago for case of presence of external perturbations like magnetic and electric fields. Results of Chapter XIV mean, that influence of wall or, generally speaking, of confined geometry does not lead to existence of a new phase near the boundary, as it was supposed many years ago and seemed up to now, but like other external perturbations (magnetic and electric fields, etc.) wall deforms the order parameter of *B*–phase and this deformation leads to very important consequences. In particular, frequencies of collective modes in the vicinity of boundary are shifted up to 20%.

The superfluid phases of ^{3}He in addition to the isotropic *B*–phase, the anisotropic *A*–phase and the A_1–phase also include a 2*D*–phase. This phase has not yet been observed, but its existence under certain conditions was deduced by many researchers. Chapter XV is devoted to the study of the collective excitations in the planar 2*D*–phase of superfluid ^{3}He. We have obtained the whole collective mode spectrum in ^{3}He–2*D* at zero momenta of collective excitations. We show that, the spectrum of a planar 2*D*–phase in a magnetic field contains modes, which are similar to those in the *A*–phase without a magnetic field, as well as a number of new modes. The former consist of six Goldstone–modes, four clapping–modes, and two pairbreaking–modes. Two quasi–Goldstone–modes and two quasi–pairbreaking–modes are obtained from the Goldstone– and the pairbreaking–modes respectively by substituting

$E^2 \to E^2 - 4\mu^2 H^2$. The gap in the quasi Goldstone–mode spectrum is $\propto 2\mu H$. Finally, we obtained two new modes having no analogs in the A–phase. They correspond to the variables u_{33} and v_{33}, and they are not degenerated, as well the difference between their frequencies is small. Interestingly, whereas for the clapping– and pairbreaking–modes there exists in the A–phase a linear Zeeman effect (threefold splitting in a magnetic field), the frequencies of this modes in the 2D–phase are independent of the magnetic field, while the energies of the quasi–pairbreaking modes and of the two "new" modes are quadratic in the field. Note also that the energies of all the nonphonon modes, except the two "new" ones, have imaginary parts due, just as in A–phase, to the vanishing of a Fermi–spectrum gap in a special direction (that of the magnetic field). The frequencies of all the nonphonon modes of the spectrum turn out to be complex, in view of the possible decay of the collective excitations into the initial fermions (owing to the vanishing of the Fermi–spectrum gap along the field direction). Just as in the A– and B–phases, collective modes can be excited in the 2D–phase in ultrasound and NMR experiments.

Note that notwithstanding some similarity between the spectra of the A– and 2D–phases, they also have substantial differences, that can possibly help identify the 2D–phase. Just as in the latter, there exist some nonphonon modes absent from the A–phase (and also from the B–phase), and the behavior of the spectrum (and even of the analog modes) in the 2D–phase and in the A–phase is quite difference: in the A–phase we have a linear splitting of the pairbreaking– and clapping–modes, while in the 2D–phase one part of the spectrum is independent of the field, whereas the other part has a quadratic field dependence.

In Chapter XVI the whole collective mode spectrum in axial– and planar–phases of superfluid ^{3}He with dispersion corrections is calculated. In axial A–phase the degeneracy of clapping–modes depends on the direction of the collective mode momentum $\boldsymbol{k}$ with respect to the vector $\boldsymbol{l}$ (mutual orbital moment of Cooper pairs), namely: the mode degeneracy remains the same as in case of zero momentum $\boldsymbol{k}$ for $\boldsymbol{k} \| \boldsymbol{l}$

only. For any other directions there is a three–fold splitting of these modes, which reaches maximum for $\boldsymbol{k} \perp \boldsymbol{l}$.

In planar 2*D*–phase, which exists in magnetic field (at $H > H_C$) we find that for clapping–modes the degeneracy depends on the direction of the collective mode momentum $\boldsymbol{k}$ with respect to the external magnetic field $\boldsymbol{H}$, namely: the mode degeneracy remains the same as in case of zero momentum $\boldsymbol{k}$ for $\boldsymbol{k} \| \boldsymbol{H}$ only. For any other directions different from this one (for example, for $\boldsymbol{k} \perp \boldsymbol{H}$) there is twofold splitting of these modes.

The obtained results means that new interesting features can be observed in ultrasound experiments in axial– and planar–phases: the change of the number of peaks in ultrasound absorption into clapping–mode. One peak, observed for these modes by Ling et al. (J. Low Temp. Phys. **78**, 187 (1990)) will split into two peaks in planar–phase and into three peaks in axial–phase under change the ultrasound direction with respect to the external magnetic field $\boldsymbol{H}$ in planar–phase and with respect to the vector $\boldsymbol{l}$ in axial–phase.

In planar–phase some Goldstone–modes in magnetic field become massive (quasi–Goldstone) and have similar twofold splitting under change the ultrasound direction with respect to the external magnetic field $\boldsymbol{H}$.

The obtained results will be useful under interpretation of the ultrasound experiments in axial– and planar–phases of superfluid ^{3}He.

In Chapter XVII we study one more superfluid phase of ^{3}He – the polar phase – and calculate the collective mode spectrum in this phase. It turns out that six high–frequency–modes $E = \Delta_0(1.20 - i \cdot 1.75)$, corresponding to variables $v_{13}, v_{23}, u_{11}, u_{21}, u_{22}, u_{12}$, have a large enough imaginary part, of the same order as the real one. This relates to the fact that in the polar–phase the gap disappears at the equator line which is opposite to the case of the axial– and planar–phases, where the gap has point nodes, and the imaginary parts of the mode energies are small compared with the real parts.

Three superfluid phases have been discovered in ^{3}He: the isotropic *B*–phase, the anisotropic *A*–phase and the A_1–phase. The latter phase exists due to the fact that the superconducting transition temperature for fermions with spins oriented along the magnetic field ($T_{C\uparrow}$) and against the field ($T_{C\downarrow}$) are different: $T_{C\downarrow} < T_{C\uparrow}$. For this reason, the particles (^{3}He atoms) with spins directed along the field are the first to undergo a superconducting transition in a magnetic field upon decreasing of the temperature at $T = T_{C\uparrow}$, which is followed by a superconducting transition in the particles with spins directed against the field at $T = T_{C\downarrow}$. The A_1–phase exists in the temperature region confined between $T_{C\uparrow}(H)$ and $T_{C\downarrow}(H)$. We study the A_1–phase in Chapter XVIII and obtained the whole collective mode spectrum in it. The spectrum contains the modes determined earlier for the *A*–phase of ^{3}He: Goldstone ($E = 0$), clapping ($E = (1.17 - i \cdot 0.13)\Delta$) and pairbreaking ($E = (1.96 - i \cdot 0.31)\Delta$) modes as well as the modes determined in the planar 2*D*–phase ($E = 2\mu H$). The frequencies of six modes are found to be imaginary, which is apparently due to the fact that the A_1–phase turns out to be unstable with respect to small perturbations in the given model. Nevertheless the results obtained for other modes can be useful for interpreting experimental results on NMR and absorption of ultrasound, since the modes with a real spectrum actually exist in the A_1–phase.

The interest to the physics of thin helium films absorbed on different sublayers remains significant during long time. Thus the investigation of the two–dimensional ^{3}He–systems, the possible phase transitions, superfluid phases and collective excitations in such systems becomes very important. These problems are discussed in Chapter XIX. Brusov and Popov have predicted the possibility of existence of two superfluid phases (*a*– and *b*–) in ^{3}He–films and have proved their stability. All the branches of the Bose–spectrum (12 branches in each phase) are obtained. Stability of Goldstone–modes is proved and dispersion laws for all the nonphonon modes were obtained. The magnetic field influence on the number of Goldstone modes in the *a*– and *b*–phases is also discussed. The long wavelength behavior of the correlators of Bose–fields which is responsible for superfluid properties of the system is considered. It is shown that such correlators decrease according to the power law in the presence of the magnetic field that proves possibility of superfluidity in ^{3}He–films. Experiments on the search of the two–dimensional superfluidity are discussed and the new possibility for the search of such 2D–superfluidity in ^{3}He–films are suggested.

Chapter XX is devoted to the problem of the collective excitation spectrum in superfluid solutions ^{3}He–^{4}He. Both cases of *s*– and *p*–pairings in Fermi–systems are considered. In the bulk solutions we consider a system of the type ^{3}He*A*–^{4}He as well as ^{3}He*B*–^{4}He and calculate the collective mode spectrum in them. We study the two–dimensional films of the types ^{3}He*a*–^{4}He and ^{3}He*b*–^{4}He. It was shown that in case of *p*–pairing the interaction between Fermi– and Bose–subsystems lead to renormalization of sound velocities in both subsystems.

Last decade new techniques for producing impure superfluids with unique properties have been developed. This new class of systems includes superfluid helium confined to aerogel, HeII with different impurities (D_2, N_2, Ne, Kr), superfluids in Vycor glasses, and watergel. These systems exhibit very unusual properties including unexpected acoustic features. In Chapter XXI we discuss the sound properties of these systems and show that sound phenomena in impure superfluids are modified from those in pure superfluids.

We calculate the coupling between temperature and pressure oscillations for impure superfluids and for superfluid He in aerogel. We show that the coupling between these two kinds of oscillations is governed by terms proportional either to impurity or to aerogel density rather than by thermal expansion coefficient, which is enormously small in pure superfluids. This replacement plays a fundamental role in all sound phenomena in impure superfluids. It enhances the coupling between the two sound modes (first and second sounds) that leads to the existence of such phenomena as the slow mode and heat pulse propagation with the velocity of first sound observed in superfluids in aerogel. This means that it is possible to observe in impure superfluids such unusual sound phenomena as slow "pressure" (density) waves and fast "temperature" (entropy) waves.

The enhancement of the coupling between the two sound modes decreases the threshold values for nonlinear processes as compared to pure superfluids. Sound conversion, which has been observed in pure superfluids only by shock waves should be observed at moderate sound amplitude in impure superfluids. Cerenkov emission of second sound by first sound (which never been observed in pure as well as in impure superfluids) could be observed in impure superfluids. We have shown that the enhanced coupling between first and second sound changes even the nature of the sound modes in impure superfluids. It leads as well to significant shift in fast mode frequency at transition temperature. We also have derived for the first time the nonlinear hydrodynamic equations for superfluid helium in aerogel. These equations are generalizations of McKenna et al. equations for the case of nonlinear hydrodynamics and could be used to study sound propagation phenomena in aerogel–superfluid system, in particular – to study sound conversion phenomena. We get two alternative sets of equations, one of which is a generalization of a traditional set of nonlinear hydrodynamics equations for the case of an aerogel–superfluid system, while second one represents equations which are the analogy of Putterman's hydrodynamic equations for superfluids.

Coming to study the solid state properties (meaning study of superconductors) we describe in Chapter XXII an approach to the microscopic theory of periodic structures developed by Popov et al. in the framework of the path integral technique. The starting point is a system of electrons and ions with the Coulomb interaction. The properties of crystals are determined by the collective excitations (phonons). Clearly, a microscopic theory must describe phonons and their interactions starting from the system of electrons and ions. The functional integral method allows us to realize this aim. The main idea is to go from the initial action of electrons and ions to the effective action functional in terms of the electric potential field $\varphi(x,\tau)$. This field has an immediate physical meaning and provides the collective variable we need. These variables correspond to normal oscillations of the crystalline lattice.

In the Chapter XXIII this approach is applied for construction of the effective interaction between electrons near the Fermi–surface. It is not difficult to show that this interaction for the model considered has an attractive character and may lead to superconductivity at sufficiently low temperatures.

Chapters XXIV–XXXI are devoted to study unconventional superconductors: high temperature superconductors (HTSC) and heavy fermion superconductors (HFSC).

Up to now study of the collective excitations in unconventional superconductors (USC) carries exotic character via a few reasons. First of all while there were some evidences of nontrivial type of pairing in some superconductors (HFSC, HTSC etc.) there was not superconductor in which unconventional pairing has been established exactly. Secondly, there was not found strong evidences of existing of the collective excitations in superconductors.

The situation has changed drastically within last few years removing study of the collective excitations in USC into real plane. In light of recent experiments this topic becomes very important. First of all an amplitude mode (with frequency of order 2Δ) has been observed in films of ordinary superconductors. Secondly, now the type of pairing is established for many superconductors. We have ***s*–pairing** in ordinary superconductors and electron–type HTSC; ***p*–pairing** in pure ^3He; ^3He in aerogel, Sr_2RuO_4 (HTSC), UPt_3 (HFSC) and ***d*–pairing** in hole–type HTSC, organic superconductors, some HFSC (UPd_2Al_3, $CePd_2Si_2$, $CeIn_3$, $CeNi_2Ge_2$ etc.).

Recently Northwestern University (John Ketterson's group) has presented results of a microwave surface impedance study of the heavy fermion superconductor UBe_{13}. They clearly have observed an absorption peak whose frequency and temperature–dependence scales with the BCS gap function $\Delta(T)$. This was the first direct observation of the resonant absorption into a collective mode, with energy approximately proportional to the superconducting gap. This discovery opens a new page in study of the collective excitations in unconventional superconductors.

The significance of studying of collective modes connects to the fact that they exhibit themselves in ultrasound attenuation and microwave absorption experiments, neutron scattering, photoemission and Raman scattering. The large peak in the dynamical spin susceptibility in HTSC arises from a weakly damped spin density–wave collective modes. This gives rise to a dip between the sharp low energy peak and the higher binding energy hump in the ARPES spectrum. Also, the collective modes of amplitude fluctuation of the d–wave gap yields a broad peak above the pair–breaking threshold in the B_{1g} Raman spectrum. The contribution of collective modes to microwave absorption and ultrasound attenuation maybe substantial.

In Chapter XXIV we build the path integral models of *p*– and *d*–pairing for bulk superconductors. While model of *p*–pairing is similar to the model obtained by us for unconventional superfluids, *d*–pairing model is quite different from the *p*–pairing one. The first version of the model of *d*–pairing in superconductors constructed by the method of functional integration was proposed by Brusov and Brusova in 1994 (Physica B **194–196**, 1479) when the idea of *d*–pairing in HTSC compounds was just put forth. We describe in Chapter XXIV an improved self–consistent model of superconductors with *d*–pairing and apply it in following Chapters for analyzing the collective mode spectrum in high–temperature and heavy–fermion superconductors.

In Chapter XXV we give general information concerning high–temperature superconductors and describe their physical properties.

Chapter XXVI is devoted to discussion of the symmetry of order parameter in high temperature superconductors. We provide the symmetry classification of HTSC states for square lattice, tetragonal one as well as orthorhombic lattices. Different methods of experimental identification of the type of pairing and the order parameter, such as Josephson effects, transition splitting, spontaneous strain and magnetism, collective modes are discussed.

In Chapter XXVII we calculate the collective mode spectrum for superconducting phases appearing in the symmetry classification of high–temperature superconductors under *d*–pairing. We considered the following states: $d_{x^2-y^2}, d_{xy}, d_{xz}, d_{yz}, d_{3z^2-r^2}$. For each superconducting phase, five high–frequency modes were determined as well as five Goldstone (quasi–Goldstone)–modes whose energies are either equal to zero or small ($\leq 0.1\,\Delta_0$).

The results on high–frequency modes can be useful in determining the order parameter and the type of pairing in HTSC as well as for interpreting the ultrasound and microwave absorption experiments with these systems. It turns out that collective modes are damped much strongly in case of *d*–pairing than in case of *p*–pairing. This fact is connected with the nodal structure of energy gap. As a rule one has points of nodes under *p*–pairing and lines of nodes under *d*–pairing.

The most scientists believe that in oxide superconductors a *d*–wave pairing takes place. At the same time the different ideas concerning extended s–wave pairing, mixture of *s*– and *d*–states, as well as of different *d*–states still discuss actively. One of the cause of such a situation is the uncertainty in answer the question: do we have exact zero gap along some chosen lines in momentum space (like the case of $d_{x^2-y^2}$) or gap is anisotropic but nonzero everywhere (except maybe some points). Existing experiments (tunneling etc.) do not give the certain answer this question while the answer is quite principle. From other side there are some experiments which could be explained under suggestion about realization in high–temperature superconductors of a mixed states, like $d_{x^2-y^2}+id_{xy.}$ Annett et al. considered the possibility of mixture of different d–wave states in high–temperature superconductors and came to conclusion that $d_{x^2-y^2}+id_{xy}$ is the most likely state. Pavel Brusov and Peter Brusov (Physica B, **281&282**, 949 (2000)) suggested one of the possible ways to distinguish the mixture of two *d*–states from pure *d*–states. For this they considered the mixed $d_{x^2-y^2}+id_{xy}$ state and calculate the spectrum of collective modes in this state. The comparison of this spectrum with the spectrum of a pure d–wave states of high–temperature superconductors shows that they are significantly different and could be the probe of the symmetry of the order parameter in high–temperature superconductors. We describe the results obtained by Pavel Brusov and Peter Brusov in Chapter XXVIII.

While these results for equal admixtures of $d_{x^2-y^2}$ and id_{xy} states shows the principle possibility to use the collective modes as the probe of the symmetry of the order parameter, the most interesting case is the case of small γ (γ is the admixture of d_{xy}–state). We consider this case in Chapter XXVIII. We suppose that dominant state is $d_{x^2-y^2}$–state and admixture of d_{xy}–state is small, say 3–10% and expand all expressions in powers of small γ. This allows obtain the corrections to the spectrum of pure $d_{x^2-y^2}$–state, which has been found before.

Obtained results could be useful for identification of the type of pairing and determination of the exact form of the order parameter in unconventional superconductors. In particular, they allow to estimate the extent of admixture of a d_{xy}–state in a possible mixed state. Derived equations allow to calculate the whole collective mode spectrum, which could be used for interpretation of the sound attenuation and microwave absorption data.

Obtained results could allow answer three very important questions:

1) does the gap disappear along some chosen lines?
2) do we have a pure or mixed *d*–wave state in high–temperature superconductors ?
3) how large is the admixture of d_{xy}–state in a possible mixed state?

P–pairing is realized not only in pure superfluid ^{3}He, superfluid ^{3}He in aerogel, but as well in unconventional superconductors: Sr_2RuO_4 (HTSC) and UPt_3 (HFSC). In Chapter XXIX we study the collective mode spectrum in bulk superconducting phases which are realized in HTSC as well as in HFSC under *p*–pairing.

The existence of CuO_2 planes – the common structural factor of high–temperature superconductors – suggests we consider two–dimensional (2D)–models. In Chapter XXX we develop a 2D–model of *p*– and *d*–pairing, using a path integration technique. The models involve a hydrodynamic action functional, obtained by path integration over "fast" and "slow" Fermi–fields. These functionals determine all properties of two–dimensional superconductors (for example, of CuO_2 planes of HTSC) and, in particular, the spectrum of collective excitations. We calculate the collective mode spectrum for both type of pairing.

Heavy–fermion superconductors were the first superconductors, which have demonstrated the unconventional pairing in charged systems. In Chapter XXXII we study the collective mode spectrum in heavy–fermion superconductors. We consider three superconducting states: $d\gamma$, Y_{2-1} and $\sin^2\theta$ and calculated for each state the complete spectrum consisting of eighteen modes. We find ten collective modes in each phase: five high–frequency–modes as well as five Goldstone–(quasi–Goldstone) modes with vanishing energies (of order $(0.03 \div 0.08)\Delta_0(T)$).

In Chapter XXXII we consider the problem of relativistic analogs of ^{3}He and connections between ^{3}He–theory and relativistic quantum field theory.

Among previous literature we can mentioned only several books, where the problem of the collective modes in superfluid ^{3}He is discussed in some details. These are:

1. P. Wolfle, D. Volhardt "Superfluid ^{3}He" (Taylor&Francis, London, 1990).
2. W. P. Halperin, L. P. Pitaevskii (Eds.) Helium Three, North–Holland, Amsterdam, 1990.
3. E. R. Dobbs, Helium three, Oxford University Press, 2001, 1088 p.

Note, however, that in each of this book only one Chapter is devoted to discussion of the collective mode problem, while our monograph is completely devoted to investigation of the collective modes. And the number of the objects, which collective properties we study is much larger in current monograph: in addition to superfluid ^{3}He, we consider superfluid ^{4}He, superfluid ^{3}He–^{4}He mixtures, superfluid ^{3}He–films, superfluid ^{3}He and superfluid ^{3}He–^{4}He mixtures in aerogel. As well we consider the influence of different perturbations on the collective modes. The problem of collective modes in unconventional superconductors is discussed in this monograph for the first time: we consider such systems as high temperature superconductors, heavy–fermion superconductors and superconducting films.

The authors are grateful to Victor Popov, O. Lounasmaa, Matti Krusius, John Ketterson, Mikko Paalanen, Alex Balatsky, James Annett, J. Pekola, Paul Chu, Tony Leggett, Bill Halperin, D. Vollhardt, V. Dmitriev, D. Tsakadze, Z. Nadirashvili, G. Kharadze, Z. Zhao, E. Dobbs, K. Kitazawa, K. Nagai, O. Ishikawa, Y. Okuda for many helpful discussions on the problem considered in this monograph.

The following scientists have made the significant contribution to the study of collective excitations in unconventional superfluids and superconductors:

Theory:

Peter Brusov, Victor Popov, Peter Wolfle, James Sauls, R. Combescot, G. Volovik, Tony Leggett, K. Nagai, Alex Balatsky, James Annett, K. Maki, Natali Orehova (Brusova), K. Nagai, L. Tewordt, D. Hirashima, Yu. Vdovin, V. Mineev, V. Koch, Pavel Brusov, N. Schopohl, D. Vollhardt, J. Serene.

Experiment:

John Ketterson, Matti Krusius, Bill Halperin, R. Movshovich, E. Dobbs, B. Shivaram, D. Paulson, O. Avenel, V. Dmitriev.

Chapter I

Functional Integration Method

1.1. Functional Integrals in Statistical Physics

Functional integration is one of the most powerful methods of contemporary theoretical physics, enabling us to simplify, accelerate and make more clear the process of the theoretician's analytical work. Application of the functional integration methods to systems with an infinite number of degrees of freedom allows us to introduce and formulate the diagram perturbation theory in quantum field theory and statistical physics. This approach is significantly simpler as compare with the operator approach.

The application of the functional integration methods allows one to derive many interesting results. The theory of phase transitions of the second kind, superfluidity, superconductivity, lasers, plasma, Kondo effect, Ising model – this is an incomplete list of problems, for which the application of the functional integration method appears to be very useful. In some of the problems, it allows us to prove results, obtained by other methods, clarify the possibilities of their applicability. If an exact solution is possible, the functional integration method gives a simple way to obtain it. In problems far from being exactly solvable (general theory of phase transitions), the application of functional integrals helps to understand the qualitative picture of the phenomenon and to develop the approximate methods of calculations.

Functional integrals are especially useful for the description of collective excitations such as plasma oscillations in the theory of the systems of particles with Coulomb interaction, quantum vortices and long–wave phonons in the theory of superfluidity and superconductivity, collective excitations in the systems like ^{3}He and ^{3}He–^{4}He solutions, superconductors with unconventional pairing. That is the case when standard perturbation theory should be modified. Functional integrals represent a sufficiently flexible mathematical apparatus adjusted for such a modification and suggesting the method of its concrete realization.

In this Chapter one can find the basic information on the functional integrals in quantum statistical physics describing systems of interacting particles at the finite temperatures T. The method of functional integration is useful for obtaining a diagram perturbation techniques and also for a modification of the perturbation theory in application to the superfluid and superconducting systems.

There exist several modification of Green's functions for quantum mechanical systems at temperatures different from zero: temperature, temperature–time and so on. The Green's functions method in statistical physics was firstly derived in the pioneer works by Bogoliubov and his school[1-8] and than it was successfully aprobated in application to the problems of superconductivity, plasma, magnetism, the theory of model Hamiltonians. The diagram techniques for perturbation theory can be directly obtained only for temperature Green's functions. They are convenient for calculation of the thermodynamical characteristics of the system, and they also contain information on the quasiparticle spectrum and weakly nonequilibrium kinetical phenomena. The theory of Green's functions in the operator approach can be found in the excellent book by Abrikosov, Gor'kov and Dzyaloshinski[9].

1.2. Functional Integrals and Diagram Techniques for Bose–particles

Let us consider the functional integral formalism for the system of Bose–particles placed into a cubic volume $V = L^3$ with periodic boundary conditions. The functional integral for such a case, is an integral over the space of complex–valued functions ("fields") $\psi(\mathbf{x},\tau), \bar{\psi}(\mathbf{x},\tau)$, where $\mathbf{x} \in V$, periodical in the parameter τ ("time") with period $\beta = kT^{-1}$, where k is the Boltzmann constant and T is the absolute temperature. In the following we shall use the system of units with $\hbar = k = 1$, where $\hbar$ is the Planck constant and k is the Boltzmann constant.

We define the Green's functions as expectation values in the above defined space of the products of several fields $\psi(\mathbf{x},\tau)$, $\overline{\psi}(\mathbf{x},\tau)$ with different arguments weighted with $\exp S$, where S is the functional of having the meaning of action

$$S = \int_0^\beta d\tau d^3x \sum_s \overline{\psi}(\mathbf{x},\tau)\partial_\tau \psi(\mathbf{x},\tau) - \int_0^\beta H'(\tau)d\tau . \tag{1.1}$$

Here, $H'(\tau)$ is the Hamiltonian of the form

$$H'(\tau) = \int d^3x (2m)^{-1} \nabla \overline{\psi}(\mathbf{x},\tau) \nabla \psi(\mathbf{x},\tau) - \lambda \overline{\psi}(\mathbf{x},\tau)\psi(\mathbf{x},\tau)$$
$$+ \frac{1}{2} \int d^3x d^3y U(\mathbf{x}-\mathbf{y}) \overline{\psi}(\mathbf{x},\tau)\overline{\psi}(\mathbf{y},\tau)\psi(\mathbf{y},\tau)\psi(\mathbf{x},\tau).$$

where λ is the chemical potential of the system and $U(\mathbf{x}-\mathbf{y})$ is the pair interaction potential between two particles.

For example, we can define the one–particle Green's function

$$G(\mathbf{x},\tau;\mathbf{x}',\tau') = -<\psi(\mathbf{x},\tau)\overline{\psi}(\mathbf{x}',\tau')> = -\frac{\int \psi(\mathbf{x},\tau)\overline{\psi}(\mathbf{x}',\tau') \exp S D\overline{\psi} D\psi}{\int \exp S D\overline{\psi} D\psi} \tag{1.2}$$

as a (formal) ratio of two functional integrals over a space of complex–valued functions. The measure of integration is denoted by $D\overline{\psi}D\psi$.

Due to periodicity, the functions $\psi(\mathbf{x},\tau)$, $\overline{\psi}(\mathbf{x},\tau)$ can be decomposed into the Fourier series

$$\psi(\mathbf{x},\tau)=(\beta V)^{-1/2}\sum_{k,\omega}a(k,\omega)\exp(\omega\tau+k\mathbf{x}),$$

$$\overline{\psi}(\mathbf{x},\tau)=(\beta V)^{-1/2}\sum_{k,\omega}a^{+}(k,\omega)\exp[-(\omega\tau+k\mathbf{x})], \tag{1.3}$$

where $\omega=\omega_n=2\pi n/\beta$, $k_i=2\pi n_i/L$, n, n_i are integers.

After substitution (1.3) into the action (1.1) we get $S=S_0+S_1$, where

$$S_0=\sum_p\left(i\omega-\frac{k^2}{2m}+\lambda\right)a^{+}(p)\,a(p),$$

$$S_1=-(2\beta V)^{-1}\sum_{p_1+p_2=p_3+p_4}V(k_1-k_3)a^{+}(p_1)a^{+}(p_2)a\,(p_3)a\,(p_4).$$

Here, p denotes the set ($\boldsymbol{k}$, ω) and $V(k)$ is the Fourier transform of $U(x)$ defined by

$$U(x)=V^{-1}\sum_k V(k)\exp(k\mathbf{x}).$$

The use of Fourier coefficients $a(p),a^{+}(p)$ gives an exceptional simple form to the functional integral as well as to diagram perturbation theory. The measure in functional integral denoted symbolically by $D\overline{\psi}D\psi$ can be written as

$$D\overline{\psi}D\psi=\prod_p da^{+}(p)da(p).$$

The functional integral can be regarded as a limit of the finite–dimensional integral over the variables $a(p), a^{+}(p)$ with

$|k| < k_0, |\omega| < \omega_0$, where $k_0, \omega_0 \to \infty$.

One–particle Green's functions (1.2) depends, evidently, on differences $\boldsymbol{x} - \boldsymbol{x}'$, $\tau - \tau'$. Substituting the Fourier expansions (1.3) for $\psi(\mathbf{x},\tau)$, $\overline{\psi}(\mathbf{x},\tau)$, we come to the conclusion that the function (1.2) can be expressed in terms of averages

$$G(p) = -< a(p)\, a^{+}(p) > = -\frac{\int a(p)\, a^{+}(p) \exp S \prod_p da^{+}(p) da(p)}{\int \exp S \prod_p da^{+}(p) da(p)}. \qquad (1.4)$$

Now we can build up the perturbative scheme for Green's function such as (1.2) and (1.4). It is founded on the representation of the action S, as a sum of an unperturbed part S_0 and a perturbation S_1

$S = S_0 + S_1$.

Then we use the expansion

$$\exp S = \exp S_0 \cdot \exp S_1 = \exp S_0 \sum_{n=0}^{\infty} \frac{S_1^n}{n!}, \qquad (1.5)$$

and integrate term by term these series. We obtain the standard perturbation theory, if we take the quadratic form in $\psi(\mathbf{x},\tau)$, $\overline{\psi}(\mathbf{x},\tau)$ (or in $a(p), a^+(p)$) as S_0:

$$S_0 = \sum_p \left(i\omega - \frac{k^2}{2m} + \lambda \right) a^+(p)\, a(p),$$

and considering the fourth–order form in S as S_1

$$S_1 = -(2\beta V)^{-1} \sum_{p_1+p_2=p_3+p_4} [V(k_1-k_3)+V(k_1-k_4)] a^+(p_1) a^+(p_2) a\,(p_3) a\,(p_4)$$

Here, we have changed the potential function $V(k_1-k_3)$ by the symmetrized potential

$$V(k_1-k_3)+V(k_1-k_4).$$

In order to build up the perturbation theory we need the formula for the generating functional for unperturbed Green's functions. This functional is defined as follows

$$Z_0(\eta,\eta^+) = \frac{\int \left(\prod_p da^+(p) da(p) \right) \exp\left[S_0 + \sum_p \eta^+(p) a(p) + a^+(p) \eta(p) \right]}{\int \left(\prod_p da^+(p) da(p) \right) \exp S_0}.$$

This expression is equal to the product over p of the ratios of two–dimensional integrals. It is not hard to calculate each of them.

The result is

$$Z_0(\eta,\eta^+)=\exp\left[-\sum_p \eta^+(p)G_0(p)\eta(p)\right], \tag{1.6}$$

where

$$G_0(p)=\left(i\omega-\frac{k^2}{2m}+\lambda\right)^{-1}.$$

Differentiating (1.6) first with respect to $\eta^+(p)$, then with respect to $\eta(p)$ and putting $\eta^+(p)=\eta(p)=0$, we receive an expression for the unperturbed Green's function

$$G_0(p)=-\frac{\int a(p)\; a^+(p)\exp S_0\prod_p da^+(p)da(p)}{\int \exp S_0\prod_p da^+(p)da(p)}. \tag{1.7}$$

Moreover differentiating (1.6) *n* times with respect to $\eta(p_i)$ and *n* times with respect to $\eta^+(p_j)$ (*i*, *j* =1, 2, … *n*) and then putting $\eta^+(p)=\eta(p)=0$, we obtain the expression for the average

$$\left\langle \prod_{i=1}^n a(p_i)\prod_{j=1}^n a^+(p_j)\right\rangle_0, \tag{1.8}$$

where

$$\left\langle f\left(a,a^{+}\right)\right\rangle_{0}=\frac{\int f\left(a,a^{+}\right)\exp S_{0}\prod_{p}da^{+}(p)da(p)}{\int\exp S_{0}\prod_{p}da^{+}(p)da(p)}.$$

From the right–hand side of (1.6) it is clear that the average (1.8) is equal to the sum of products of all possible pair averages

$$\left\langle a(p_i)a^{+}(p_j)\right\rangle_0=-\delta_{p_ip_j}G_0(p_i),$$

(the sum over all possible ways to select n pairs $a(p_i),a^{+}(p_j)$ out of n objects $a(p_i)$ and n objects $a^{+}(p_j)$.

This statement is known as Wick's theorem. For instance, if $n = 2$, we have

$$\begin{aligned}\left\langle a(p_1)a(p_2)a^{+}(p_3)a^{+}(p_4)\right\rangle_0&=\left\langle a(p_1)a^{+}(p_3)\right\rangle_0\left\langle a(p_2)a^{+}(p_4)\right\rangle_0+\\&+\left\langle a(p_1)a^{+}(p_4)\right\rangle_0\left\langle a(p_2)a^{+}(p_3)\right\rangle_0\end{aligned}$$

(the averages $\langle aa\rangle_0$ and $\langle a^{+}a^{+}\rangle_0$ vanish).

Now we can develop the perturbation theory. The starting point is the expansion

$$\exp S=\exp S_0\cdot\exp S_1=\exp S_0\sum_{n=0}^{\infty}\frac{S_1^n}{n!}.$$

We express Green's function (1.4) as a ratio of two series

$$G_0(p) = -\frac{\sum_{n=0}^{\infty}\frac{1}{n!}\left\langle a(p)\ a^+(p)S_1^n\right\rangle_0}{\sum_{n=0}^{\infty}\frac{1}{n!}\left\langle S_1^n\right\rangle_0}. \tag{1.9}$$

Each of the S_1 inside $\langle\ldots\rangle_0$ is a fourth–order form in the integration variables a, a^+. Thus, we deal with the expectation values

$$\left\langle \prod_{i=1}^{n} a^+(p_{1i})a^+(p_{2i})a(p_{3i})a(p_{4i})\right\rangle_0 \tag{1.10}$$

in the denominator of (1.9). The following expectation values

$$a(p)a^+(p)\left\langle \prod_{i=1}^{n} a^+(p_{1i})a^+(p_{2i})a(p_{3i})a(p_{4i})\right\rangle_0 \tag{1.11}$$

appear in the numerator.

Now we need Week's theorem, in order to write down the averages $\langle\ldots\rangle_0$ such as (1.10) or (1.11) as sums of products of all possible pair averages. This allows us to calculate every term in the series in the numerator and the denominator. Feynman suggested to assign a picture diagram to each term of the series analogous to (1.9) in quantum field theory. A perturbation theory in which every term corresponds to a diagram is called the diagram technique. It is not difficult to build up the diagram technique in statistical physics as well. For the case of a system of Bose–particles with pair interaction we arrive to diagrams as follows. To the average (1.10) we assign a diagram made of n vertices of fourth–

order (points with two incoming and two outgoing arrows). For $n=2$ the diagram has the form

(1.12)

The arrows incoming into the vertex correspond to the variable a, the outgoing ones to a^{+}. To the average (1.11) we assign the diagram

(1.13)

Diagrams introduced in such a way will be called prediagrams in order to distinguish them from those which will arise later.

In accordance with Wick's theorem, the expectation values (1.10), (1.11) are sums of all possible pair averages. To each fixed way of composing the pair average we assign a diagram by connecting each pair of verices i, j with a line if the average

$$\left\langle a(p_i)a^{+}(p_j)\right\rangle_0$$

is present among the pair expectation values. For $n = 2$ we can list all the diagrams arising from the prediagram (1.12)

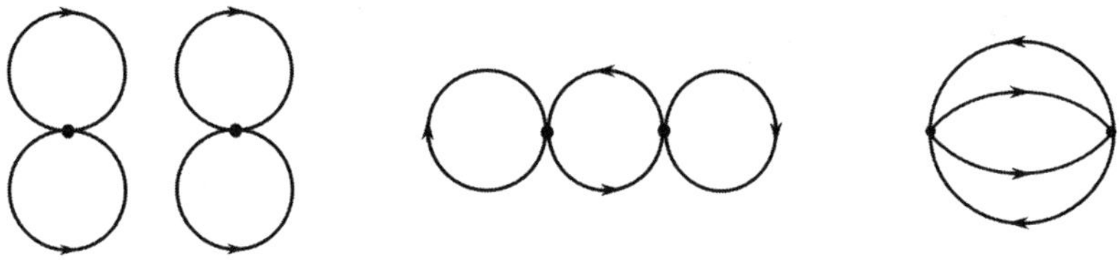

and also all diagrams arising from prediagram (1.13)

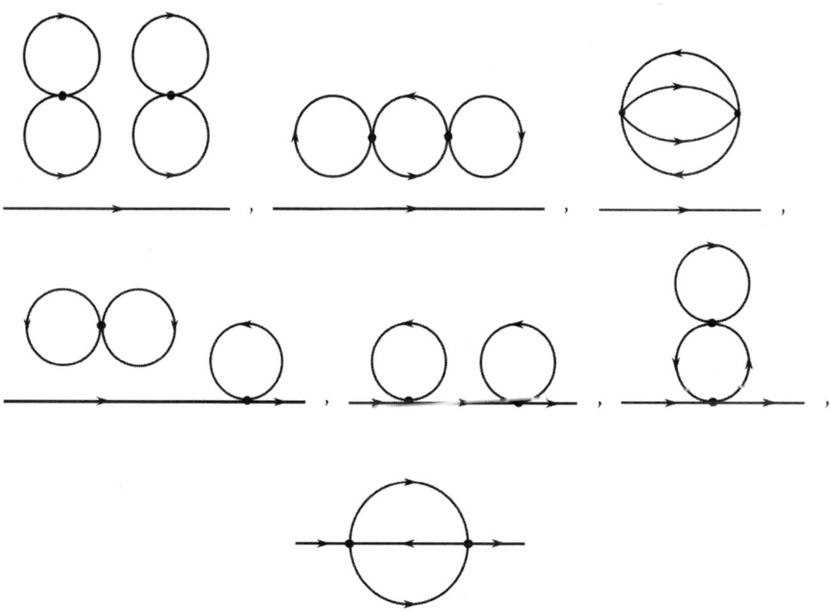

We get the expression corresponding to a given diagram if the product of the pair expectation values is multiplied by

$$\frac{(-1)^n}{n!}\prod_{i=1}^{n}\frac{V(k_{1i}-k_{3i})+V(k_{1i}-k_{4i})}{4\beta V} \tag{1.14}$$

and also by the number of ways through which a given diagram can be obtained from a prediagram. Then we have to perform the summation over all independent four–momenta p_i of internal lines.

It is convenient to reformulate slightly the resulting rules of correspondence. The coefficient of $a^+(p_1)a^+(p_2)a(p_3)a(p_4)$ in S_1 is symmetric with respect to the permutations $p_1 \leftrightarrow p_2$ and $p_3 \leftrightarrow p_4$. The demonstrated symmetry allows us to speak about the prediagram symmetry group of the order $R = 4^n \cdot n!$. The factor 4^n corresponds to symmetry groups at each fourth–order vertex; $n!$ corresponds to the group of vertex permutations. Let us note that in (1.14) we just have exactly R^{-1}. It is easily seen that $n = R/r$, where N is the number of ways to construct a given diagram from a prediagram, r is the order of the diagram symmetry group. The presence of the symmetry factor r^{-1} is a common feature in diagram perturbation technique in quantum field theory and statistical physics.

The above arguments allow us to formulate the rules of correspondence as follows. We shall assign Green's function $G_0(p)$ to a line of the diagram and the symmetrized potential to a vertex

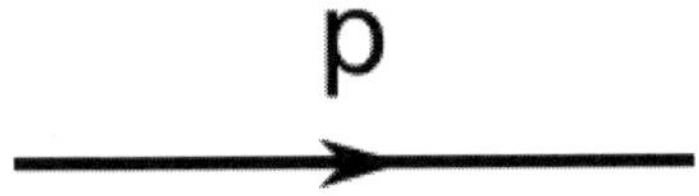

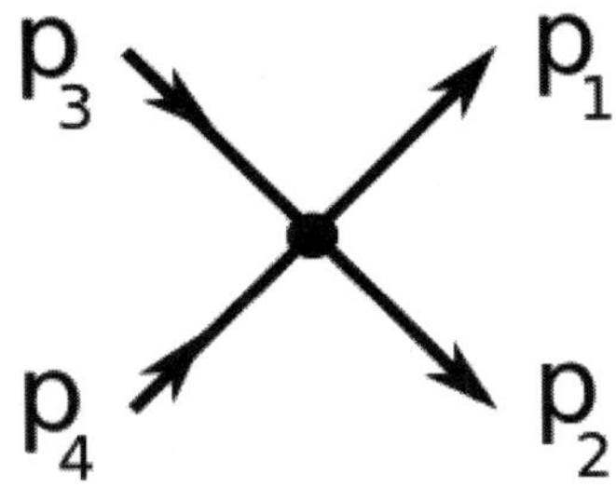

$$G_0(p)=\left(i\omega-\frac{k^2}{2m}+\lambda\right)^{-1},\quad V(k_1-k_3)+V(k_1-k_4).$$

The expression corresponding to a given diagram for the denominator in (1.9) can be obtained by summing over independent four–momenta of the product of expression which correspond to lines and vertices of the diagram and then multiplying the result by

$$r^{-1}\left(\frac{-1}{\beta V}\right)^{l-n} \tag{1.15}$$

where l is the number of lines, n is the number of vertices, r is the order of the diagram symmetry group. For the diagram corresponding to the numerator the factor in front of the diagram is equal to

$$r^{-1}\left(\frac{-1}{\beta V}\right)^{l-n-1} \tag{1.16}$$

where $l-n-1=C$ is the number of independent loops of the diagram.

Now let us note that the denominator in (1.9) can be written in exponential form:

$$\sum_{n=0}^{\infty} \frac{\left\langle S_1^n \right\rangle_0}{n!} = \exp\left(\sum_i D_i^{vac} \right), \tag{1.17}$$

where $\sum_i D_i^{vac}$ is the sum of contributions from all connected vacuum diagrams (i.e. diagrams without ingoing and outgoing lines). The exponential is due to the symmetry factor

$$r^{-1} = \prod_i \left(n_i ! r_i^{n_i} \right)^{-1}$$

for any diagram which consists of n_1 connected components of the first kind, n_2 components of the second kind and so on. The numerator of (1.9) is equal to the factor (1.17) multiplied by the sum of contributions from all connected diagrams without vacuum components. This is why the factor (1.17) in the numerator cancels that in the denominator, so we can take into account only connected diagrams without vacuum components. The transition to connected diagrams is a common fact as well as the above–mentioned appearance of the factor r.

The diagram techniques developed above in the functional integral approach coincide with the well–known Matsubara–Abrikosov–Gor'kov–Dzyaloshinski perturbation theory for the temperature Green functions[9]. Its derivation in functional integral formalism is simpler than in operator formalism. Moreover, functional methods are very useful for a reformulation of perturbation theory in many cases when the perturbative scheme is not applicable in its standard form, for instance for superfluid Bose– or Fermi–systems. From the functional integration point of view such a reformulation is an alternative method of asymptotic evaluation of the functional integral.

The perturbation theory for Green's functions, constructed here, allows modifications connected with different methods of partial summation of diagrams. It has been shown above that vacuum diagrams give no contribution to the expressions for the Green's functions, so that when dealing with Green functions we may only take into account connected diagrams, which can be passed from the entry to exit moving along diagram lines.

It is well–known that the Green functions $G(p)$ can be expressed in terms of its irreducible self energy part $\sum_p$ ($\sum_p$ is the sum of all diagrams with two tails

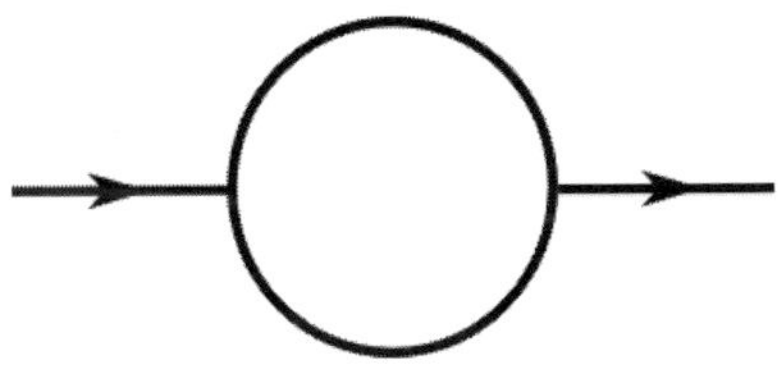

such that is impossible to cut the diagram into two without cutting one line of the diagram).

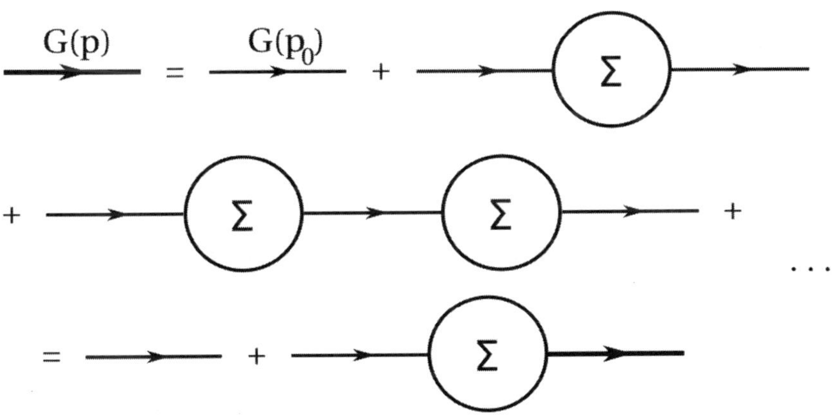

This coincides with the well–known Dyson equation. Its analytical form is as follows:

$$G(p) = G_0(p) + G_0(p)\sum(p)G_0(p) + G_0(p)\sum(p)G_0(p)\sum(p)G_0(p) + \\ + \ldots = G_0(p) + G_0(p)\sum(p)G(p). \tag{1.18}$$

Its solution is

$$G(p) = \left(G_0^{-1}(p) - \sum(p)\right)^{-1}. \tag{1.19}$$

Thus in order to evaluate the full Green function it is sufficient to find its irreducible self–energy part.

Now we shall consider the modification of perturbation theory called the skeleton diagram techniques. We come to this form of perturbative scheme if we perform partial summation of diagrams which is equivalent to replacing the bare Green function $G_0(p)$ by the full Green function for each inner line of the diagram. So the elements of the skeleton diagram techniques are the full Green functions and bare vertices

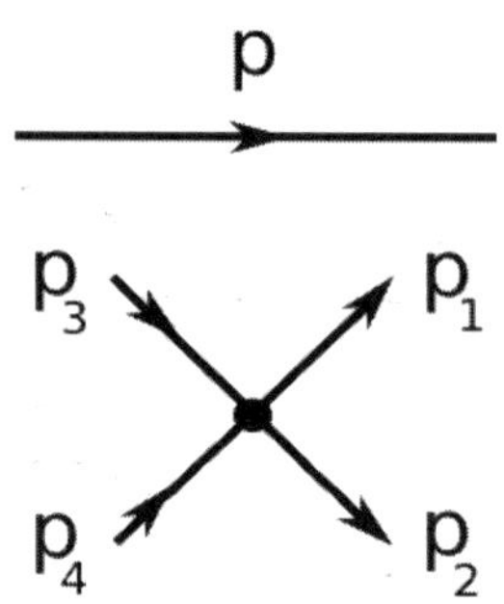

$$G(p), \quad V(k_1 - k_3) + V(k_1 - k_4)$$

In contrast with the ordinary diagram technique however, there is no need to take into account diagrams with self–energy part insertions into inner lines. Of course, we do not know $G(p)$ from the very beginning, so we have to deal with a system of equations which would allow us to find $G(p)$. The first of them is the Dyson equation (1.19), and the second one is the equation for the self energy part

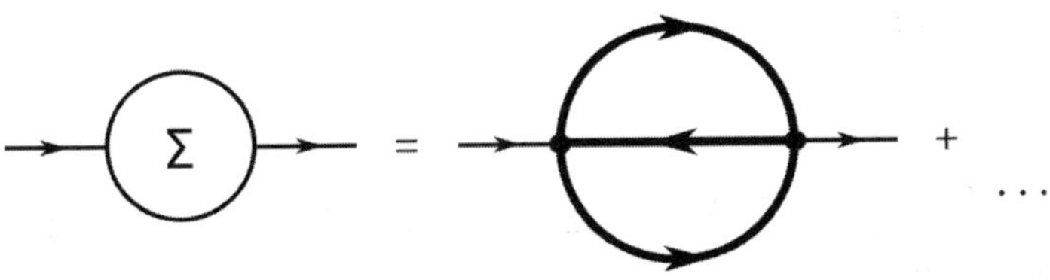

(1.20)

representing it as a sum of an infinite number of diagrams of the skeleton diagram technique, the elements of which are full Green's functions and bare vertices.

The skeleton diagram technique appears to be especially suitable in theories with anomalous Green's functions which are identically equal to zero when calculated by the standard scheme of perturbation theory. Such a situation takes places in the theory of superfluidity and superconductivity, where there exist anomalous Green's functions below the phase transition temperature. Equations for such anomalous functions have, besides trivial solutions, also nontrivial ones, emerging below the phase transition point. The possibility of partial summation is a common feature of the diagram perturbation technique, valid for statistical physics as well as for relativistic quantum field theory.

To conclude this section we shall show how an information about physical properties of the system can be obtained by using the Green's functions.

We can obtain the average number of particles N in the Bose–system by averaging the functional

$$\int \overline{\psi}(\mathbf{x},\tau)\psi(\mathbf{x},\tau)\, d^3x\,.$$

This implies the formula

$$N = \frac{\int \exp S \int \overline{\psi}(\mathbf{x},\tau)\, \psi(\mathbf{x},\tau) D\overline{\psi} D\psi}{\int \exp S D\overline{\psi} D\psi} = \int < \overline{\psi}(\mathbf{x}',\tau')\, \psi(\mathbf{x},\tau) > d^3x. \quad (1.21)$$

If our system is a translation–invariant one, the Green's function (1.2) can only depend on $\mathbf{x}-\mathbf{x}', \tau-\tau'$. So we come to the following result:

$$N = V \lim_{\substack{\mathbf{x}'\to\mathbf{x} \\ \tau'\to\tau}} < \psi(\mathbf{x},\tau)\overline{\psi}(\mathbf{x}',\tau') > .$$

It turns out that the result depends on the way of taking the limit. Let us demonstrate this for the ideal Bose–system. Here we have

$$< \psi(\mathbf{x},\tau)\, \overline{\psi}(\mathbf{x}',\tau') >_0 = (\beta V)^{-1} \sum_{k,\omega} G_0(p) \exp i(\omega(\tau-\tau') + k(\mathbf{x}-\mathbf{x}')).$$

Putting here $\mathbf{x}'=\mathbf{x},\ \tau'=\tau-\varepsilon\ (\varepsilon>0)$ and summing over the frequencies ω we find

$$\beta^{-1} \sum_{\omega} \frac{\exp(i\omega\varepsilon)}{i\omega - \varepsilon(k)} = \frac{\exp(\varepsilon\varepsilon(k))}{\exp(\beta\varepsilon(k)) - 1} \quad ,$$

where $\varepsilon(k)=\dfrac{k^2}{2m}-\lambda$.

The limit $\varepsilon\to+0$ gives

$$n\ (k)=\frac{1}{\exp(\beta\varepsilon(k))-1}.$$

If we take the limit $\varepsilon\to-0$, we come to a different result

$$\frac{1}{\exp(\beta\varepsilon(k))-1}+1.$$

The correct answer is received for $\tau-\tau'\to+0$. In that case we find the well–known Bose–partition function

$$\rho=N/V=V^{-1}\sum_k\left(\exp(\beta\varepsilon(k))-1\right)^{-1}. \tag{1.22}$$

for the density $\rho=N/V$. The approaching $\tau-\tau'\to+0$ gives the correct answer for nonideal system, too. As a result, the following formula is obtained

$$\rho=N/V=-\lim_{\varepsilon\to+0}(\beta V)^{-1}\sum_p\exp(i\omega\varepsilon)G(p). \tag{1.23}$$

It expressed the density through the Green's function $G(p)$ in p– representation.

The expectation values for the momentum and kinetic energy can be obtained by averaging the following functionals

$$\frac{i}{2}\int d^3x\left(\nabla\bar{\psi}\psi-\bar{\psi}\nabla\psi\right), \qquad (2m)^{-1}\int d^3x\left(\nabla\bar{\psi}\nabla\psi\right).$$

This leads to the following formulae

$$\boldsymbol{K}/V=-(\beta V)^{-1}\sum_p \exp(i\omega\varepsilon)\cdot\boldsymbol{k}\cdot G(p), \tag{1.24}$$

$$H_{kin}/V=-(\beta V)^{-1}\sum_p \exp(i\omega\varepsilon)\cdot\frac{k^2}{2m}\cdot G(p), \tag{1.25}$$

in which $\varepsilon\to+0$.

We shall derive further the expression for pressure starting from a formula for the ratio of partition functions of nonideal and ideal systems:

$$\frac{\int \exp S D\bar{\psi}D\psi}{\int \exp S_0 D\bar{\psi}D\psi}=\frac{Z}{Z_0}=\exp\beta(\Omega-\Omega_0) \tag{1.26}$$

Here, Z, Z_0 are partition functions of ideal and nonideal systems, $\Omega_0=-p_0V, \Omega=-pV$, where p_0 is the pressure of the ideal system, p is the pressure of the nonideal one. The left–hand side of (1.26) is equal to

$$\exp\left(\sum_i D_i^{vac}\right). \tag{1.27}$$

where $\sum_i D_i^{vac}$ is the sum of all connected vacuum diagrams. We thus have

$$p = p_0 + (\beta V)^{-1} \sum_i D_i^{vac} . \tag{1.28}$$

This formula expresses the pressure p in terms of p_0 and the sum of contribution of all vacuum diagrams.

1.3. Functional Integrals and Diagram Techniques for Fermi–particles

In the previous section we discussed a quantization scheme for Bose–fields in the functional integral approach. Quantization of Fermi–fields may be performed using the functional integral over anticommuting variables (for details see Berezin's book[10]). Here we need the following basic facts.

We can define the integral over Fermi–fields as a limit of the integral over the Grassmann algebra with a finite (even) numbers of generators $x_i, x_i^+ \, (i = 1,\ldots,n)$, which anticommute with each other

$$x_i x_j + x_j x_i = 0, \quad x_i^+ x_j^+ + x_j^+ x_i^+ = 0, \qquad x_i x_j^+ + x_j^+ x_i = 0. \tag{1.29}$$

According to (1.29) we have $x_i^2 = 0$, $\left(x_i^+\right)^2 = 0$ and each element of the algebra can be written in the following form

$$f(x,x^+) = \sum_{a_i,b_i=0,1} C(a_1,...,a_n,b_1,...,b_n)\ x_1^{a_1}...x_n^{a_n}(x_n^+)^{b_n}...(x_1^+)^{b_1}\,. \tag{1.30}$$

Let us define the involution operation in the algebra by the formula

$$f \to f^+ = \sum_{a_i,b_i=0,1} \overline{C}(a_1,...,a_n,b_1,...,b_n)\ x_1^{b_1}...x_n^{b_n}(x_n^+)^{a_n}...(x_1^+)^{a_1}\ .$$

We can now introduce the functional integral over the algebra

$$\int f(x,x^+)dx^+dx = \int f(x_1,...,x_n,x_1^+,...,x_n^+)dx_1^+dx_1\,...dx_n^+dx_n\,.$$

This integral is defined through the relations

$$\int dx_i = 0,\ \int dx_i^+ = 0,\ \int x_i\,dx_i = 1,\ \int x_i^+dx_i^+ = 1$$

The symbols $dx_i\,,dx_i^+$ must anticommute with each other and with the generators of the algebra. We demand also that the natural condition of linearity is fulfilled:

$$\int (c_1 f_1 + c_2 f_2)dx^+dx = c_1\int f_1 dx^+dx + c_2\int f_2 dx^+dx\,,$$

where coefficients $c_1\,,c_2$ are complex numbers.

So integrating the polynomial function (1.30), we obtain

$$\int f(x,x^+)dx^+dx = C(1,...,1,...1)\,.$$

The following two formulae will be important for future applications

$$\int \exp(-x^+ Ax) dx^+ dx = \det A. \tag{1.31}$$

$$\frac{\int \exp(-x^+ Ax + \eta^+ x + x^+ \eta) dx^+ dx}{\int \exp(-x^+ Ax) dx^+ dx} = \exp(\eta^+ A \eta). \tag{1.32}$$

Here,

$$x^+ Ax = \sum_{i,k} a_{ik} x_i^+ x_k$$

is a quadratic form of the generators x_i, x_i^+ corresponding to the matrix A. The expressions $\eta^+ x = \sum_i \eta_i^+ x_i, x^+ \eta = \sum_i x_i^+ \eta_i$ are linear forms of the generators x_i, x_i^+, whose coefficients η_i, η_i^+ anticommute with each other and with the generators. The elements η_i, η_i^+ together with the generators x_i, x_i^+ can be regarded as generators of a larger algebra. The expression $\eta^+ A^{-1} \eta$ in (1.32) is the quadratic form of the matrix A^{-1} inverse to A.

The exponentials in the integrals (1.31), (1.32) can be expressed through the expansion into series, in which, due to anticommutation relations (1.29), only several first terms are nonzero. We can prove (1.31) by expanding the exponential function and then noticing that only the n–th term gives a contribution to the integral. As for (1.31) we can prove it by using the shift transformation $x \to x + \tilde{\eta}$, $x^+ \to x^+ + \tilde{\eta}^+$, which cancels the linear terms in the exponent of the integrand.

Now we shall briefly discuss the functional integral and diagram technique for Fermi–systems. The quantization of a Fermi–system can be obtained as a result of integration over the space of anticommuting functions $\psi(\mathbf{x},\tau)$, $\overline{\psi}(\mathbf{x},\tau)$ (the elements of an infinite Grassman algebra), where $\mathbf{x}\in V, \tau\in[0,\beta]$. To obtain the correct result it is necessary to impose on $\psi(\mathbf{x},\tau)$, $\overline{\psi}(\mathbf{x},\tau)$ the antiperiodicity conditions in the variable τ:

$$\psi(\mathbf{x},\beta) = -\psi(\mathbf{x},0),\ \overline{\psi}(\mathbf{x},\beta) = -\overline{\psi}(\mathbf{x},0).$$

As a result, we have the following Fourier series for $\psi(\mathbf{x},\tau)$, $\overline{\psi}(\mathbf{x},\tau)$ in the Fermi case:

$$\psi(\mathbf{x},\tau) = (\beta V)^{-1/2}\sum_{k,\omega} a(k,\omega)\exp[i(\omega\tau + k\mathbf{x})],$$

$$\overline{\psi}(\mathbf{x},\tau) = (\beta V)^{-1/2}\sum_{k,\omega} a^{+}(k,\omega)\exp[-i(\omega\tau + k\mathbf{x})]. \tag{1.33}$$

Here, $p = (\mathbf{k},\omega)$ and $\omega = \omega_n = (2n+1)\pi/\beta$ are the Fermi–frequencies.

In contrast to the case of Bose–system, the Fermi–frequencies are proportional to the half–integers $(n+1/2)$. Let us notice, that the Fourier coefficients $a(p), a^{+}(p)$ in (1.33) may be considered as generators of an infinitely dimensional Grassmann algebra.

Green's functions for Fermi–systems are defined by the same equations (1.2), (1.4), as for Bose–systems, namely as the ratio of two functional integrals. Such a ratio can be understood as a limit of the ratio of two finite–dimensional functional integrals arising when only the coefficients with $|\mathbf{k}| < k_0$ and $|\omega| < \omega_0$ are taken. These integrals coincide with the previously defined integrals over the Grassmann algebra with generators $a(p), a^{+}(p)$. It then remains to take the limit $\omega_0, k_0 \to \infty$.

The action functional S and the Hamiltonian have the same form as those for the Bose–system. For instance, the unperturbed action has the form

$$S_0 = \sum_p \left(i\omega - \frac{k^2}{2m} + \lambda \right) a^+(p)\, a(p),$$

where $\omega = \omega_n = (2n+1)\pi / \beta$ is the Fermi–frequency. It is convenient to write the perturbative term S_1 as

$$S_1 = -(4\beta V)^{-1} \sum_{p_1+p_2=p_3+p_4} [V(k_1 - k_3) - V(k_1 - k_4)] a^+(p_1) a^+(p_2) a(p_4) a(p_3)$$

with the antisymmetrized potential $V(k_1 - k_3) - V(k_1 - k_4)$.

The derivation of the diagram technique is completely analogous to that performed above for Bose–systems. We come to the diagram technique with the elements

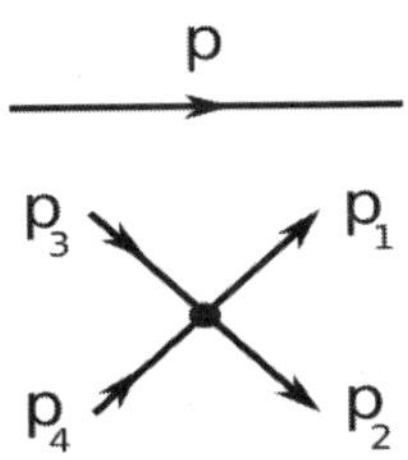

$$G_0(p) = \left(i\omega - \frac{k^2}{2m} + \lambda \right)^{-1}, \quad \omega = (2n+1)\pi / \beta,$$
$$V(k_1 - k_3) - V(k_1 - k_4).$$

The "Fermi"–diagram techniques differ from the Bose case in the following points:

1) The Fermi frequences $\omega = \omega_n = (2n+1)\pi / \beta$ are multiplies of half–integers, whereas the Bose–frequencies $\omega = \omega_n = 2\pi n / \beta$ are multiplies of integers.

2) We have an antisymmetrized potential $V(k_1 - k_3) - V(k_1 - k_4)$ instead of a symmetrized one in the Bose case.

3) The sum over independent momenta is multiplied by the factor

$$(-1)^F r^{-1} \left(\frac{-1}{\beta V} \right)^{l-n} \tag{1.34}$$

for vacuum diagrams and by

$$(-1)^F r^{-1} \left(\frac{-1}{\beta V} \right)^{l-n-1} \tag{1.35}$$

for diagrams corresponding to a one–particle Green function. Expressions (1.34), (1.35) differ from (1.15), (1.16) for Bose–systems by the factor $(-1)^F$, where F is the number of independent closed Fermi–loops of the diagram. The presence of the factor $(-1)^F$ is a consequence of the fact that the fields $\psi(\mathbf{x}, \tau)$, $\bar{\psi}(\mathbf{x}, \tau)$ anticommute.

The formula (1.28) for the pressure is valid for both Bose– and Fermi–systems. The equations for the average number of particles, momentum and kinetic energy per unit volume for the Fermi–system differ from the corresponding equations (1.23), (1.24), (1.25) for the system by replacement the Bose frequencies $\omega = \omega_n = 2\pi n / \beta$ with Fermi ones $\omega = \omega_n = (2n+1)\pi / \beta$ and have opposite signs.

1.4. Method of Successive Integration over "Fast" and "Slow" Variables

Many problems of quantum field theory and statistical physics are connected with the evaluation of the Green's functions at small energies and momenta (the infrared asymptotics). If quanta with arbitrary small energies are present in the system, the standard perturbation theory leads to difficulties. Such a situation takes place in quantum electrodynamics and in various branches of statistical physics: superfluidity, superconductivity, plasma theory, the theory of ^{3}He and the general theory of phase transitions. The difficulties mentioned come from the fact that every graph of the ordinary perturbation theory has a so–called infrared singularity. It means that the expression assigned to a graph that is an integral over inner momentum variables is singular when the external momenta tend to zero. In such cases a modification of the perturbation theory is desirable. One possible modification may be called the method of subsequent integration, first over "fast" and then over "slow" variables.

We shall represent every field that is integrated over as a sum of two terms. One of them will be called the slowly oscillating field, $\psi_0(\mathbf{x},\tau)$, and the other the fast oscillating field $\psi_1(\mathbf{x},\tau)$

$$\psi(\mathbf{x},\tau)=\psi_0(\mathbf{x},\tau)+\psi_1(\mathbf{x},\tau).$$

In the theory of Bose–systems the function $\psi(\mathbf{x},\tau)$ is supposed to be periodic in its arguments and it is decomposed into the Fourier series

$$\psi(\mathbf{x},\tau)=(\beta V)^{-1/2}\sum_{k,\omega}a(k,\omega)\exp[i(\omega\tau+k\mathbf{x})]. \tag{1.36}$$

The sum of terms in (1.36) with $|\mathbf{k}|<k_0$ and $|\omega|<\omega_0$ will be called the slowly oscillating part $\psi_0(\mathbf{x},\tau)$ of $\psi(\mathbf{x},\tau)$. The difference $\psi(\mathbf{x},\tau)-\psi_0(\mathbf{x},\tau)$ will be called the fast oscillating part $\psi_1(\mathbf{x},\tau)$. This difference is, of course, the sum of terms in (1.36) with $|\mathbf{k}| > k_0$ or $|\omega| < \omega_0$.

It is worth noting that the boundary between the slow and fast fields in the concrete problems of statistical physics is to some extent conditional. It reflects in the fact that the parameters distinguishing between the "slow" and "fast" variables are not determined exactly but only in order of value.

In the theory of Fermi–systems the decomposition of Fermi–fields has the same form (1.36), as that for Bose–systems, but with Fermi– frequencies $\omega = \omega_n = (2n+1)\pi / \beta$ replacing the Bose ones $\omega = \omega_n = 2\pi n / \beta$. In many cases it is natural to call the sum of terms in (1.36) with $|\mathbf{k} - \mathbf{k}_F| < k_0$, $|\omega| < \omega_0$ the "slow" part of the Fermi field $\psi(\mathbf{x}, \tau)$. Here the momenta belong to the narrow shell around the Fermi sphere $k = k_F$. Such a definition of "slow" and "fast" variables turns out to be very useful in the theory of superfluid Fermi–systems.

The functional integral in statistical physics is the integral with the measure

$$\prod_{k,\omega} da^+(k,\omega) da(k,\omega),$$

as it has been shown in 1.1.

The fundamental idea of the modification of the perturbation theory consists in the successive integration, at first over the "fast" field and the over the "slow" one, using different schemes of perturbation theory at the two different stages of integration. At the first stage we integrate over the Fourier coefficients $a(\boldsymbol{k},\omega), a^+(\boldsymbol{k},\omega)$, whose indices $(\boldsymbol{k},\omega)$ lie in the region of fast variables. At the second stage integration over slow variables is carried out.

When integration over the fast fields we can use the perturbation theory and corresponding diagram techniques that differ from the standard ones explained in the previous sections in two points:

(1) The integrals (sums) over the four–momenta are cut off at a low limit.

(2) Supplementary vertices will emerge describing the interaction of the fast field $\psi_1(\mathbf{x},\tau)$ with the slow field $\psi_0(\mathbf{x},\tau)$.

The first point is obvious because the variables (k,ω) of the fast field lie in the "fast" region. The supplementary vertices appear because the starting action S is expressed via $\psi(\mathbf{x},\tau)=\psi_0(\mathbf{x},\tau)+\psi_1(\mathbf{x},\tau)$, so that the crossing terms that may not appear in the quadratic form do contribute to the terms of the third and higher degrees.

Cutting of integrals at a low limit prevents the emergence of infrared divergences at the first stage of integration. At the second stage (integration over the slow fields) we may achieve the vanishing of infrared divergences if a nontrivial perturbative scheme exploiting the special features of the system is adopted. In the superfluidity theory, for example, when integrating over the slow Bose–fields $\psi_0(\mathbf{x},\tau)$, $\overline{\psi}_0(\mathbf{x},\tau)$, it proves convenient to pass to the polar coordinates

$$\psi_0(\mathbf{x},\tau)=(\rho(\mathbf{x},\tau))^{1/2}\cdot\exp[i\varphi(\mathbf{x},\tau)],$$

$$\overline{\psi}_0(\mathbf{x},\tau)=(\rho(\mathbf{x},\tau))^{1/2}\cdot\exp[-i\varphi(\mathbf{x},\tau)]$$

and integrate over fields $\rho(\mathbf{x},\tau)$, $\varphi(\mathbf{x},\tau)$. The perturbation theory for the integral over slow fields can be formulated in terms of Green's functions of the fields $\rho(\mathbf{x},\tau)$, $\varphi(\mathbf{x},\tau)$. Such a perturbative scheme is free from infrared divergences. In the case of superfluid Fermi–systems it turns out to be convenient to pass from the integral over slow Fermi–fields to the integral over some auxiliary Bose–field corresponding to the "collective" degrees of freedom of the system considered.

In what follows we give concrete illustrations of how one can imply the idea of the successive integration first over fast and then over slow variables. This approach proves to be very effective for the description of collective excitations in superfluid systems.

Chapter II

Collective Excitations in Superfluid Fermi–Systems with *s*–Pairing

2.1. Effective Action Functional of the Superfluid Fermi–gas

The effective Bose–excitations in superfluid Fermi–systems were investigated in the work by Bogoliubov, Tolmachev and Shirkov[1] simultaneously with the creation of the microscopic theory of superconductivity. Particularly in this work the excitations of the Bose type were firstly obtained, which were called later "Bogoliubov sound" or "Bogoliubov–Anderson sound".

In this Chapter the functional integration method is applied for description of the collective excitations in the model of Fermi–gas with attractive interaction. We shall use the idea of the subsequent integration over the fast and then over the slow variables which was outlined in the section 1.4. After the integration over the fast Fermi–fields we shall obtain the functional integral which can be transfored into the integral over some auxiliary Bose–field corresponding to Cooper pairs of fermions. The quanta of this new field are just collective excitations of Fermi–systems. This Chapter follows the paper by Andrianov and Popov[2].

A nonrelativistic system of Fermi–particles with spin is described by the anticommuting functions $\chi_s(\mathbf{x},\tau)$, $\overline{\chi}_s(\mathbf{x},\tau)$, defined in the cubic volume $V = L^3$ and antiperiodic in "time" τ with period $\beta = T^{-1}$ ($s = \pm$ is the spin index). Such functions maybe expanded into the Fourier series

$$\chi_s(\mathbf{x},\tau) = (\beta V)^{-1/2} \sum_{k,\omega} a_s(k,\omega) \exp[i(\omega\tau + k\mathbf{x})], \tag{2.1}$$

where $\omega = \omega_n = (2n+1)\pi/\beta$.

Green's functions can be defined as averages of products $\chi_S(\mathbf{x},\tau)$, $\bar{\chi}_S(\mathbf{x},\tau)$ with different space–time arguments with the weight factor $\exp S$ over the space of these anticommuting functions. The action functional has the following form

$$S = \int_0^\beta d\tau d^3x \sum_s \bar{\chi}_s(\mathbf{x},\tau)\partial_\tau \chi_s(\mathbf{x},\tau) - \int_0^\beta H'(\tau) d\tau, \tag{2.2}$$

where

$$H'(\tau) = \int d^3x \sum_s (2m)^{-1} \nabla\bar{\chi}_s(\mathbf{x},\tau)\nabla\chi_s(\mathbf{x},\tau)$$

$$-(\lambda + s\mu_0 H)\bar{\chi}_s(\mathbf{x},\tau)\chi_s(\mathbf{x},\tau)$$

$$+\frac{1}{2}\int d^3x d^3y U(\mathbf{x}-\mathbf{y})\sum_{ss'}\bar{\chi}_s(\mathbf{x},\tau)\bar{\chi}_{s'}(\mathbf{y},\tau)\chi_{s'}(\mathbf{y},\tau)\chi_s(\mathbf{x},\tau) \tag{2.3}$$

has a meaning of the Hamiltonian.

It was shown in the section 1.3 of Chapter II, how the standard temperature diagram techniques[3] can be obtained in the functional integration approach, if we regard the quadratic form S_0 in (2.3) as an unperturbed action, and the quadratic form S_1 as the perturbative term.

The standard perturbation theory does not converge in the vicinity of the Fermi–surface and it is not applicable for the superfluid state. We are going to demonstrate that the idea of integration over fast and slow fields, together with the introduction of new Bose–field variable instead of Fermi ones, provides a method for describing collective Bose–excitations in Fermi–superfluids. The phase transition of a Fermi–system into the superfluid state will be described as the Bose–condensation of some Bose–system in this formalism. The method may apply to the case of a pairing in the s –state superconductor model as well as to the case of the p–pairing (^{3}He – model).

First of all we integrate $\exp S$ over the fast Fermi–fields $\chi_{1S}(\mathbf{x},\tau)$, $\overline{\chi}_{1S}(\mathbf{x},\tau)$ such that their Fourier components $a_S(\mathbf{x},\tau)$, $a_S^*(\mathbf{x},\tau)$ have $|k-k_F|>k_0$ or $|\omega|>\omega_0$. The parameters k_0 and ω_0 are only defined up to their order of magnitude. All the physical results must not depend on the specific choice of k_0, ω_0. So we have

$$\int \exp S D\overline{\chi}_{1s} D\chi_{1s} = \exp\tilde{S}\left(\chi_{0s},\tilde{\chi}_{0s}\right), \tag{2.4}$$

where

$$\int \exp S D\overline{\chi}_{1s} D\chi_{1s} = \exp\tilde{S}\left(\chi_{0s},\tilde{\chi}_{0s}\right)$$

is the effective action functional, depending on the slow fields $\chi_{0s},\tilde{\chi}_{0s}$, which only have the Fourier components for which $|k-k_F|<k_0$ and $|\omega|<\omega_0$.

The most general form of $\tilde{S}$ is

$$\tilde{S}=\sum_{n=0}^{\infty}\tilde{S}_{2n}, \tag{2.5}$$

where $\tilde{S}_{2n}$ is the form of order $2n$ in the fields $\chi_{0s},\tilde{\chi}_{0s}$.

The constant term $\tilde{S}_0$ in (2.5) will be irrelevant in the further considerations. The higher–order terms $\tilde{S}_6$, $\tilde{S}_8$ and so on in (2.5) maybe neglected when the “low–energy shell” $|k-k_F|<k_0$ is thin . The form $\tilde{S}_2$ corresponds to noninteracting quasi–particles near the Fermi surface: The most general form of $\tilde{S}_2$ is

$$\tilde{S}_2 = \sum_{\substack{S,\ |\omega|<\omega \\ |k-k_F<k_F|}} \varepsilon_S(k,\omega,H) a_S^+(p)_S a(p), \tag{2.6}$$

Supposing that $\varepsilon_S(k = k_F, \omega = 0, H = 0)$, we can use the following form of ε_S:

$$\varepsilon_S(k,\omega,H) \approx Z^{-1}(i\omega - c_F(k-k_F) + s\mu H), \tag{2.7}$$

Here we expand ε_S into a series in $\omega, k-k_F, H$ and take into account only the linear terms. Here c_F is the velocity on the Fermi–sphere, μ is the magnetic momentum of a Fermi–quasiparticle and Z is the normalization constant.

The form $\tilde{S}_4$ describes the pair interaction of quasiparticles and is equal to

$$\begin{aligned}\tilde{S}_4 = &-(\beta V)^{-1} \sum_{p_1+p_2=p_3+p_4} t_0(p_1,p_2,p_3,p_4) a_+^+(p_1) a_-^+(p_2) a_-(p_4) a_+(p_3) - \\ &-(2\beta V)^{-1} \sum_{p_1+p_2=p_3+p_4} t_1(p_1,p_2,p_3,p_4)(2a_+^+(p_1) a_-^+(p_2) a_-(p_4) a_+(p_3) + \\ &+ a_+^+(p_1) a_+^+(p_2) a_+(p_4) a_+(p_3) + a_-^+(p_1) a_-^+(p_2) a_-(p_4) a_-(p_3)).\end{aligned} \tag{2.8}$$

Here $t_0(p_i)$ is a scattering amplitude symmetric under the permutations $p_1 \leftrightarrow p_2$; $p_3 \leftrightarrow p_4$; $t_1(p_i)$ is antisymmetric under the these permutations.

Near the Fermi–surface we may put $\omega_i = 0, \boldsymbol{k}_i = k_F \boldsymbol{n}_i$ in $t_0(p_i)$ and $t_1(p_i)$, where $\boldsymbol{n}_i$ are unit vectors obeying the following condition

$\boldsymbol{n}_1+\boldsymbol{n}_2=\boldsymbol{n}_3+\boldsymbol{n}_4$.

The scattering amplitudes $t_0(p_i)$ and $t_1(p_i)$, must depend only on two invariants, for instance on $(\boldsymbol{n}_1,\boldsymbol{n}_2)$ and $(\boldsymbol{n}_1-\boldsymbol{n}_2,\boldsymbol{n}_3-\boldsymbol{n}_4)$. So we have

$$t_0=f((\boldsymbol{n}_1,\boldsymbol{n}_2),(\boldsymbol{n}_1-\boldsymbol{n}_2,\boldsymbol{n}_3-\boldsymbol{n}_4));$$

$$t_1=(\boldsymbol{n}_1-\boldsymbol{n}_2,\boldsymbol{n}_3-\boldsymbol{n}_4)\,\mathrm{g}((\boldsymbol{n}_1,\boldsymbol{n}_2),(\boldsymbol{n}_1-\boldsymbol{n}_2,\boldsymbol{n}_3-\boldsymbol{n}_4)). \qquad (2.9)$$

Both f and g are even functions of the second variable.

The functional $\tilde{S}_2+\tilde{S}_4$ is the most general expression which describes Fermi–quasiparticles near the Fermi–surface and their pairing interaction.

The functions f and g can be calculated for low–density systems. For realistic systems they must be defined experimentally. The approach outlined for obtaining $\tilde{S}_2+\tilde{S}_4$ in the functional integral formalism is an alternative to the Landau theory of the Fermi–liquid[4,5].

We will consider below the two simplest cases

$f=f_0>0,\ g=0$ "superconductor"

$f=0,\ g=g_0>0,$ "^{3}He–model" (2.10)

In both cases we can go from the integral over Fermi–fields to the integral over some auxiliary Bose–field. In order to do so, we insert a Gaussian integral of $\exp(\bar{c}Ac)$, (A is some operator) over the fields c, $\bar{c}$ into the integral over the Fermi field and than perform the shift transformation which cancels the $\tilde{S}_4$ functional. After this transformation we obtain a Gaussian integral over the Fermi–fields which can be evaluated in closed form. We obtain the following effective action functional

$$S_{eff}(c,\bar{c})(=)\bar{c}Ac + \ln\det\frac{\hat{M}(c,\bar{c})}{\hat{M}(0,0)}. \tag{2.11}$$

This functional contains all the information about collective Bose–excitations in a given Fermi–system.

The Bose–system described by functional (2.11) may undergo a phase transition into the superfluid state due to the Bose–condensation. We can find the phase transition temperature T_C for both cases (2.10). Then we shall look for the Bose–spectrum of quasiparticles for $T > T_C$ as well as for $T < T_C$.

First of all let us consider the simplest case $t_0 = f_0 < 0, t_1 = 0$.

We introduce the following Gaussian integral:

$$\int \exp\left(f_0^{-1} \sum_P c^+(p)c(p) \right) \prod_P dc^+(p)dc(p) \tag{2.12}$$

into the integral over Fermi–fields. This is an integral over the complex Bose–field $c(\mathbf{x},\tau)$ with Fourier coefficients $c(p)$.

If we now make the shift transformation

$$c(p) \to c(p) + f_0(\beta V)^{-1/2} \sum_{P_1+P_2=P} a_+(p_1)a_-(p_2), \tag{2.13}$$

which reduces out the interacting term $\tilde{S}_4$ in (2.5), we obtain the following action which depends on both Fermi– and Bose–fields:

$$f_0^{-1}\sum_P c^+(p)c(p)+\sum_{P,S} Z^{-1}(i\omega-c_F(k-k_F))a_S^+(p)a_S(p)+$$
$$+(\beta V)^{-1/2}\sum_{P_1+P_2=P}\left(c^+(p_3)a_+(p_1)a_-(p_2)+a_-^+(p_2)a_+^+(p_1)c(p_3)\right).$$
$$(2.14)$$

The quadratic form of the Fermi–variables maybe written down as

$$c(p)\to c(p)+f_0(\beta V)^{-1/2}\sum_{P_1,P_2,a,b}\chi_a^+(p_1)M_{ab}(p_1,p_2)\chi_b(p_2), \qquad (2.15)$$

where

$$\chi_1(p)=a_+^+(p),\chi_2(p)=a_-(p) \qquad (2.16)$$

can be regarded as entries of the column on which the operator of the form

$$\widehat{M}(p_1,p_2)=\begin{pmatrix} Z^{-1}(i\omega-\xi)\delta_{p1p2} & (\beta V)^{-1/2}c(p_1+p_2) \\ -(\beta V)^{-1/2}c^+(p_1+p_2) & Z^{-1}(-i\omega+\xi)\delta_{p1p2} \end{pmatrix} \qquad (2.17)$$

acts. Here, $\xi=c_F(k-k_F)$.

It is now possible to evaluate the integral over the Fermi–fields in closed form. We obtain the following effective action functional

$$S_{\text{eff}}=f_0^{-1}\sum_p c^+(p)c(p)+\ln\det\frac{\hat{M}(c,c^+)}{\hat{M}(0,0)}, \qquad (2.18)$$

where $\widehat{M}$ defined by (2.17).

2.2. Bose–spectrum of Superfluid Fermi–gas

The effective action (2.18) describes collective excitations in Fermi–gas with short–range attractive interaction. This functional was investigated by Andrianov and Popov[2]. The problem of Bose–excitations in the BCS (Bardeen–Cooper–Schriffer) model was discussed by Bogoliubov, Tolmachev and Shirkov[1]. The functional integral approach to Bose–excitations in Fermi–gas near the phase transition temperature was considered by Svidzinski[6].

First of all we shall find the phase transition point. Near the phase transition (in the Ginsburg–Landau region $|\Delta T| << T_C$) we may expand $\ln\det$ as a power series in $c,\overline{c}$. Denoting

$$\hat{M}(0,0)=\widehat{G}^{-1},\ \ \hat{M}\left(c,c^{+}\right)=\widehat{G}^{-1}+\widehat{u}\,, \tag{2.19}$$

we have

$$\ln\det\frac{\hat{M}\left(c,c^{+}\right)}{\hat{M}(0,0)}=\ln\det\left(I+\widehat{G}\widehat{u}\right)=Tr\ln\left(I+\widehat{G}\widehat{u}\right)=-\sum_{n=0}^{\infty}\frac{1}{2n}Tr\left(\widehat{G}\widehat{u}\right)^{2n}. \tag{2.20}$$

Only even powers of the ln–expansion contribute to $Trln$.

Near the phase transition point we will only take into account the first two terms in (2.20), so that

$$\begin{aligned} S_{\text{eff}} &= -\sum_{p} A(p)c^{+}(p)c(p)- \\ &-(2\beta V)^{-1}\sum B(p_i)c^{+}(p_1)c^{+}(p_2)c(p_3)c(p_4). \end{aligned} \tag{2.21}$$

Here,

$$A_0 = -f_0^{-1} - (\beta V)^{-1/2} \sum_{P1} Z^2 (i\omega_1 - \xi_1)^{-1} (i\omega + i\omega_1 - \xi_2)^{-1} =$$

$$= \frac{Z^2}{2V} \sum_{|k-k_F|<k_0} \frac{th\frac{\beta\xi_1}{2} + th\frac{\beta\xi_2}{2}}{i\omega - \xi_1 - \xi_2} - f_0^{-1} \tag{2.22}$$

In the second term in (2.21) we may substitute $B(0)$ instead of $B(p_i)$ and write down this term as

$$-\frac{g}{2} \int d\tau\, d\mathbf{x} |c(\mathbf{x}, \tau)|^4 , \tag{2.23}$$

where

$$g = B(0) = \frac{Z^4}{\beta V} \sum_P \left(\omega^2 + \xi^2\right)^{-2} . \tag{2.24}$$

We can evaluate sums over momenta in the vicinity of the Fermi–surface ($|k - k_F| < k_0$) according to the following rule

$$V^{-1} \sum_{|k-k_F|<k_0} \to \frac{k_F^2}{(2\pi)^3 c_F} \int_{|\xi|<c_F k_0} d\Omega d\xi . \tag{2.25}$$

where $\xi = c_F(k - k_0)$, $d\Omega$ is an element of a solid angle.

Using this prescription, we obtain

$$g = \frac{k_F^2 Z^4}{(2\pi)^2 c_F \beta} \sum_\omega \int \frac{d\xi}{\left(\xi^2 + \omega^2\right)^2} = \frac{k_F^2 Z^4}{8\pi c_F \beta} \sum_\omega |\omega|^{-3} = \frac{7\zeta(3) k_F^2 Z^4}{16\pi^4 c_F T^2} . \tag{2.26}$$

Now we can find the phase transition temperature from the equation

$$A_0(0) = 0 \tag{2.27}$$

or

$$\frac{Z^2}{(2\pi)^3} \int\limits_{|k-k_F|<k_0} d^3k \frac{th\dfrac{\beta\xi}{2}}{2\xi} + f_0^{-1} = \frac{Z^2 k_F^2}{2\pi^2 c_F} \int\limits_0^{c_F k_0} \frac{d\xi}{\xi} th\frac{\beta\xi}{2} + f_0^{-1} = 0 . \tag{2.28}$$

The integral in (2.28) depends on k_0 logarithmically. In order that $T_C = \beta_C^{-1}$ be independent of. k_0 , we have to suppose that f_0^{-1} depends on k_0 in the following way:

$$f_0^{-1} = f^{-1} - \frac{Z^2 k_F^2}{2\pi^2 c_F} \ln\frac{k_0}{k_F}, \tag{2.29}$$

where f does not depend on k_0. Substituting (2.29) into (2.28) and using the formula

$$\int\limits_0^{c_F k_0} \frac{d\xi}{\xi} th\frac{\beta\xi}{2} = \int\limits_0^{c_F k_{/0}/2T} \frac{dx}{x} thx \approx \ln\frac{2c_F k_0 \gamma}{\pi T} . \tag{2.30}$$

where $\ln\gamma = C$ is the Euler constant, we obtain

$$T_C = \frac{2c_F k_F \gamma}{\pi} \exp\left(-\frac{2\pi^2 c_F}{|f| Z^2 k_F} \right). \tag{2.31}$$

Now we are able to obtain the Bose–spectrum near T_C.

1). $|\Delta T| << T_C$

a) If $T > T_C$ the equation for the spectrum is

$$A_0(p) = A_0(0) + \left(A_0(p) - A_0(0)\right) = 0 \tag{2.32}$$

where the analytic continuation $i\omega \to E$ is to be made. Then

$$A_0(0) = \frac{Z^2 k_F^2}{2\pi^2 c_F} \int_0^\infty \frac{d\xi}{\xi} \left(th\frac{\beta_C \xi}{2} - th\frac{\beta\xi}{2} \right) = \frac{Z^2 k_F^2}{2\pi^2 c_F} \ln\frac{\beta_C}{\beta} \approx$$
$$\approx \frac{Z^2 k_F^2}{2\pi^2 c_F} \frac{T - T_C}{T_C}, \tag{2.33}$$

$$A_0(p) - A_0(0) = \frac{Z^2}{2(2\pi)^3} \int d^3k_1 \left(\frac{th\frac{\beta\xi_1}{2}}{\xi_1} + \frac{th\frac{\beta\xi_1}{2} + th\frac{\beta\xi_2}{2}}{i\omega - \xi_1 - \xi_2} \right) \approx$$
$$\approx -\frac{Z^2 k_F^2 |\omega|}{16\pi c_F T} - \frac{7\zeta(3) Z^2 k_F^2 c_F^2}{96\pi^4 T^2}. \tag{2.34}$$

Substituting (2.33) and (2.34) into (2.32) and changing $|\omega| \to -iE$ we obtain the following result for the Bose–spectrum for $T > T_C$

$$E(\boldsymbol{k}) = -i\frac{7\zeta(3)c_F^2k^2}{6\pi^3 T_C} - i\frac{8}{\pi}(T - T_C). \tag{2.35}$$

This spectrum is purely imaginary. The value $E(0)$ goes to zero as $T \to T_C$ (the system becomes a superconductor).

b) $T < T_C$.

If $T < T_C$, there exists a Bose–condensate density. We can find it by substituting $c(p), c^+(p) \to (\beta V)^{1/2} c\delta_{p0}$ into (2.21):

$$-\beta V\left[A_0(0)|c|^2 + \frac{g}{2}|c|^4\right]. \tag{2.36}$$

Looking for a maximum of (2.36) we obtain the condensate density

$$\rho_0 = |c|^2 = -\frac{A_0(0)}{g}E(\boldsymbol{k}) = \frac{k_F^2Z^2}{2\pi^2 g c_F}\frac{T - T_C}{T_C} = \frac{8\pi^2 T_C(T_C - T)\zeta(3)k_F^2c_F^2}{7\zeta(3)Z^2}. \tag{2.37}$$

Now let us perform the shift transformation

$$c(p) = b(p) + (\rho_0\beta V)^{1/2}\delta_{p0},\ c^+(p) \to b^+(p) + (\rho_0\beta V)^{1/2}\delta_{p0} \tag{2.38}$$

and consider the quadratic form of the new variables

$$-\frac{1}{2}\sum_p \left[2A(p)b^+(p)b(p)+B(p)\left(b(p)b(-p)+b^+(p)b^+(-p)\right)\right], \quad (2.39)$$

where

$$A(p)=A_0(p)+2g\rho_0\ ,\quad B(p)=g\rho_0\,. \quad (2.40)$$

The shift transformation (2.38) will be sufficient if we confine ourselves to considering only the quadratic form (2.39). In order to construct the perturbation theory (without singularities) it would be more suitable to go to the density–phase variables.

The Bose–spectrum corresponding to the form (2.39) can be found from the equation

$$\det\begin{pmatrix} A(p) & B(p) \\ B(p) & A(-p)\end{pmatrix}=A(p)A(-p)-B^2(p)=0\,. \quad (2.41)$$

According to (2.34) and (2.40) both $A(p)$ and $B(p)$ are even functions of p, and (2.41) splits into two equations

$$A(p)-B(p)=0,\ A(p)+B(p)=0, \quad (2.42)$$

which give two branches of the spectrum

$$E_1(\boldsymbol{k})=-i\frac{7\zeta(3)c_F^2k^2}{6\pi^3T_C},$$

$$E(\boldsymbol{k}) = -i\frac{7\zeta(3)c_F^2k^2}{6\pi^3T_C} - i\frac{16}{\pi}(T - T_C). \tag{2.43}$$

The first branch is defined by the equation $A(p) - B(p) = 0$. It begins from zero ($E_1(\boldsymbol{k}) \to 0$ if $\boldsymbol{k} \to 0$). The second one is defined by the equation $A(p) + B(p) = 0$. It differs from the branch (2.35) existing for $T > T_C$ by the change $T - T_C \to 2(T_C - T)$.

2). $T << T_C$

Here we have to take into account all the terms of expansion (2.20). It is convenient to make the shift transformation (2.38), where ρ_0 is the condensate density. We can find ρ_0 demanding S_{eff} to have a maximum after the substitution $c(p) = c^+(p) = (\rho_0\beta V)^{1/2}\delta_{p0}$.

This maximum condition can be written in the form

$$\int_0^\infty d\xi \left(\frac{th\frac{\beta}{2}\left(\xi^2 + \Delta^2\right)^{1/2}}{\left(\xi^2 + \Delta^2\right)^{1/2}} - \frac{th\frac{\beta\xi}{2}}{\xi} \right) = 0, \tag{2.44}$$

where $\Delta = Z\rho_0^{1/2}$. For instance, if $T = 0$ $(\beta = \infty)$ (2.44) implies

$$\Delta(0) = \frac{\pi}{\gamma}T_C = 2c_Fk_F \exp\left(-\frac{2\pi^2c_F}{|f|Z^2k_F^2} \right). \tag{2.45}$$

The Bose–spectrum is defined by (2.41), where $A(p), B(p)$ are the coefficients of the quadratic form in the variable $b(p), b^{+}(p)$. They are defined by the following equations:

$$A(p) = (\beta V)^{-1} \sum_{p1} G_{+}(p_1) G_{-}(p + p_1) - f_0^{-1},$$
$$B(p) = (\beta V)^{-1} \sum_{p1} G_{1}(p_1) G_{1}(p + p_1), \tag{2.46}$$

where

$$G_{\pm}(p) = Z \frac{-i\omega \pm \xi(k)}{\omega^2 + \xi^2(k) + \Delta^2}, \quad G_1(p) = \frac{Z\Delta}{\omega^2 + \xi^2(k) + \Delta^2} . \tag{2.47}$$

We can split (2.41) into two equations, $A(p) - B(p) = 0$ and $A(p) + B(p) = 0$, as we did for the case $|\Delta T| << T_C$. The first of them ($A(p) - B(p) = 0$) has a solution $p = 0$ owing to the equation determining the gap

$$A(0) - B(0) = (\beta V)^{-1} \sum_{p1} \left[G_{+}(p) G_{-}(p) - G_1^2(p) \right] - f_0^{-1} . \tag{2.48}$$

So we can rewrite the equation $A(p) - B(p) = 0$ in the form

$$A(p) - B(p) - A(0) + B(0) = 0 \tag{2.49}$$

or

$$\frac{Z^2}{\beta V}\sum_{p1+p2=p}\left[\frac{1}{\omega_1^2+\xi_1^2+\Delta^2}-\frac{(i\omega_1+\xi_1)(i\omega_2+\xi_2)+\Delta^2}{(\omega_1^2+\xi_1^2+\Delta^2)(\omega_2^2+\xi_2^2+\Delta^2)}\right]=0, \quad (2.50)$$

where $\xi_i=\xi(k_i),\ i=1,2.$

For $T\to 0$ the sum over $\boldsymbol{k}$ can be replaced by the integral and then we can go to the integral over the neighborhood of the Fermi–sphere according to the rule

$$(\beta V)^{-1}\sum_{p_0}\ \to(2\pi)^{-4}\int d\omega\, d^3k \to(2\pi)^{-4}\int d\omega\, k^2 dk\ d\Omega\to$$
$$\to\frac{k_F^2}{(2\pi)^4 c_F}\int d\omega\, d\xi\ d\Omega.$$

In order to evaluate the integral with respect to ω_1,ξ_1 it is useful to apply the Feynman trick standard in relativistic quantum theory and based on the identity

$$(ab)^{-1}=\int_0^1 d\alpha[\alpha a+(1-\alpha)b]^{-2}.$$

We substitute

$$a=\omega_1^2+\xi_1^2+\Delta^2,\ b=\omega_2^2+\xi_2^2+\Delta^2\ .$$

and make replacements

$$\omega_1 \to \omega_1 + \alpha\omega,\ \ \omega_2 \to -\omega_1 + (1-\alpha)\omega,$$
$$\xi_1 \to \xi_1 + \alpha c_F(\boldsymbol{nk}),\ \ \xi_2 \to \xi_1 - (1-\alpha)c_F(\boldsymbol{nk}),$$

where c_F is a velocity on the Fermi surface, $\boldsymbol{n}$ is a unit vector orthogonal to the Fermi surface and $(\boldsymbol{k},\omega) = p$ is an external four–momentum. After these substitutions the integral with respect to ω_1, ξ_1 can easily be evaluated and the left–hand side of (2.50) takes a form

$$\frac{\pi k_F^2 Z^2}{(2\pi)^4 c_F}\int_0^1 d\alpha \int d\omega \Big[\ln\big(1+\Delta^{-2}\alpha(1-\alpha)(\omega^2 + c_F^2(\boldsymbol{nk})^2)\big)+ \\ + \frac{2\alpha(1-\alpha)(\omega^2 + c_F^2(\boldsymbol{nk})^2)}{\Delta^2 + \alpha(1-\alpha)(\omega^2 + c_F^2(\boldsymbol{nk})^2)}\Big] \tag{2.51}$$

For small ω, k we can expand the integrand function up to the second power of $q^2 = \omega^2 + c_F^2(\boldsymbol{nk})$ and obtain

$$\frac{k_F^2 Z^2}{8\pi^2 c_F \Delta^2}\left[\omega^2 + \frac{1}{3}c_F^2 k^2 - \frac{1}{6\Delta^2}\left(\omega^4 + \frac{2}{3}\omega^2 c_F^2 k^2 + \frac{1}{5}c_F^4 k^4\right)\right]. \tag{2.52}$$

Replacing $i\omega \to E$ and putting (2.52) equal to zero we find the spectrum

$$E = uk\left(1 - \gamma k^2\right),\ \ u = \frac{c_F}{\sqrt{3}},\ \ ,\gamma = \frac{c_F^2}{45\Delta^2}. \tag{2.53}$$

This branch of the spectrum (the Bogoliubov sound) is linear in k for small k. Positivity of the coefficient γ ("dispersion") means a stability of an excitation with respect to a decay into two (or more) excitations of the same type.

Now we consider excitations described by the equation.

Let us write the equation of the form

$$\frac{Z^2}{\beta V}\sum_{p1+p2=p}\left[\frac{1}{\omega_1^2+\xi_1^2+\Delta^2}-\frac{(i\omega_1+\xi_1)(i\omega_2+\xi_2)-\Delta^2}{(\omega_1^2+\xi_1^2+\Delta^2)(\omega_2^2+\xi_2^2+\Delta^2)}\right]=0 \ . \tag{2.54}$$

By using the Feynman method and by integrating with respect to ω_1, ξ_1 we write the left–hand side of (2.54) in the form

$$\frac{\pi\, k_F^2 Z^2}{(2\pi)^4 c_F}\int_0^1 d\alpha \int d\Omega \left[2+\ln\left(1+\Delta^{-2}\alpha(1-\alpha)(\omega^2+c_F^2(\boldsymbol{nk})^2)\right)\right]=$$
$$\frac{\pi\, k_F^2 Z^2}{(2\pi)^4 c_F}\int d\Omega\ \left[\omega^2+4\Delta^2+c_F^2(\boldsymbol{nk})^2\right]\int_0^1 d\alpha\left[\Delta^2+\alpha(1-\alpha)(\omega^2+c_F^2(\boldsymbol{nk})^2\right]^{-1} \tag{2.55}$$

Expression (2.55) tends to zero when $\omega^2 \to -4\Delta^2$ and $k \to 0$. For $E^2=-\omega^2$ near to $4\Delta^2$ and small k the internal integral in (2.55) is reciprocal to the root $\left(4\Delta^2-E^2+c_F^2(nk)^2\right)^{1/2}$. As a result (2.55) goes to

$$\int_{-1}^{1} dx\left(x^2+z^2\right)^{1/2}=0. \tag{2.56}$$

or

$$\left(1+z^2\right)^{1/2}+\frac{z^2}{2}\ln\frac{\left(1+z^2\right)^{1/2}+1}{\left(1+z^2\right)^{1/2}-1}=0. \tag{2.57}$$

By introducing the new variable

$$t=\ln\frac{\left(1+z^2\right)^{1/2}+1}{\left(1+z^2\right)^{1/2}-1}. \tag{2.58}$$

we can reduce the solution of equation (2.57) to finding nontrivial $(t\neq 0)$ roots of the equation

$$sht+t=0. \tag{2.59}$$

The roots group to "quartets" $\pm a\pm ib$. For the smallest nontrivial root with respect to its modulus in the first quadrant t_1 and for asymptotics t_n for $n\rightarrow\infty$ we have

$$t_1\approx 2.251+i\ 4.212$$

$$t_n\approx \ln\pi(4n-1)+i\ (2\pi n-\frac{\pi}{2})+o(1). \tag{2.60}$$

The sequence t_n determines the sequence of the roots of the equation $A(p)+B(p)=0$

$$E_n=2\Delta-\frac{c_F^2k^2}{4\Delta sh^2(t_n/2)}. \tag{2.61}$$

These roots lie on different sheets of a Riemann surface and concentrate at $E = 2\Delta$ for $n \to \infty$. The immediate physical meaning has root E_1, the first one appearing when analytic continuation from the upper to the lower halfplane is performed. It corresponds to oscillations of the densitywhich may be excited by acting on the system with the frequency near to 2Δ.

We found the Bose–spectrum of a superfluid Fermi–gas in the approximation corresponding to noninteracting quasiparticles of the fields c, c^+. Nevertheless, in general, the spectrum turns out to be complex.

In order to take into account interaction between quasiparticles and, in particular, to construct the kinetic theory, the higher terms in the expansion S_{eff} of fields b, b^+ (fluctuations of c, c^+ around their condensate values) are necessary. The summation of diagrams which reduces to solution of the kinetic equations must lead to the branch of the second sound for $0 < T < T_C$. For $T \to 0$ its velocity is

$$u_2 = \frac{c_F}{3}. \tag{2.62}$$

Let us still note, that in order to build up a perturbation theory without infrared singularities it is more suitable to use density–phase variables instead of c, c^+ according to the equations

$$c(\boldsymbol{x},\tau) = \left(\rho(\boldsymbol{x},\tau)\right)^{1/2} \exp(i\phi(\boldsymbol{x},\tau)), \quad c^+(\boldsymbol{x},\tau) = \left(\rho(\boldsymbol{x},\tau)\right)^{1/2} \exp(-i\phi(\boldsymbol{x},\tau)). \tag{2.63}$$

In the new variables the branch of the Bogoliubov sound (2.53) corresponds to oscillations of the phase, and equations (2.61) correspond to oscillations of the density.

2.3. Fermi–gas with Coulomb Interaction

Let us now dwell on the Fermi–system in which the short–range attractive interaction between Fermi–particles is supplemented by the long–range Coulomb interaction. This is a more realistic model of superconductor than the one considered above. In this case the starting action functional differs from (2.2) by the addend

$$-\frac{e^2}{2}\int_0^\beta d\tau\, d^3x d^3y |x-y|^{-1} \sum_{s,s'} \bar{\chi}_s(\boldsymbol{x},\tau)\bar{\chi}_{s'}(\boldsymbol{y},\tau)\chi_{s'}(\boldsymbol{y},\tau)\chi_s(\boldsymbol{x},\tau). \tag{2.64}$$

We can go from the long–range Coulomb interaction to the interaction via the electric potential field. We can do it, introducing the integral over a new variable $\varphi(\boldsymbol{x},\tau)$ (it is just the field of electric potential) into the integral over anticommuting Fermi–variables $\chi_S, \bar{\chi}_S$

$$\int D\varphi \, \exp\left(-(8\pi)^{-1}\int d\tau\, d^3x (\nabla\varphi(\boldsymbol{x},\tau))^2\right). \tag{2.65}$$

The Coulomb interacting term (2.64) can be eliminated by making a shift transformation

$$\varphi(\boldsymbol{x},\tau) \to \varphi(\boldsymbol{x},\tau) + ie\int d^3y |x-y|^{-1} \sum_{s'} \bar{\chi}_{s'}(\boldsymbol{y},\tau)\chi_{s'}(\boldsymbol{y},\tau). \tag{2.66}$$

Such an approach turns to be useful in plasma theory[7]. Introducing the Gaussian integral (2.12) and eliminating the short–range interaction by means of the shift transformation (2.13), we obtain the action depending on the Fermi–fields $a_S(p), a_S^+(p)$ and the Bose–fields $c(p), c^+(p), \varphi(p)$ (Fourie coefficients of the field $\varphi(\boldsymbol{x},\tau)$):

$$f_0^{-1}\sum_p c^+(p)c(p)-(8\pi)^{-1}\sum_p k^2\varphi(p)\varphi(-p)+$$
$$+Z^{-1}\sum_{p,s}\left(i\omega-c_F(k-k_F)\right)a_S^+(p)a_S(p)+$$
$$+(\beta V)^{-1/2}\sum_{p1+p2=p}\left(c^+(p_3)a_+(p_1)a_-(p_2)+a_-^+(p_2)a_+^+(p_1)c(p_3)\right)- \tag{2.67}$$
$$-ie(\beta V)^{-1/2}\sum_{S,p1+p2=p}\varphi(p_3)a_S^+(p_1)a_S(p_2)$$

After the integration over Fermi–fields we obtain the following effective action

$$S_{eff}=f_0^{-1}\sum_p c^+(p)c(p)-(8\pi)^{-1}\sum_p k^2\varphi(p)\varphi(-p)+$$
$$+\ln\det\hat{M}\left[c,c^+,\varphi\right]/\hat{M}\left[0,0,0\right] \tag{2.68}$$

where $\hat{M}$ is equal to

$$\hat{M}=\begin{pmatrix} Z^{-1}(i\omega_1-\xi(k_1))\delta_{p1p2}- & \\ -ie(\beta V)^{-1/2}\varphi(p_1-p_2)\,; & (\beta V)^{-1/2}c(p_1+p_2) \\ & Z^{-1}(-i\omega_1+\xi(k_1))\delta_{p1p2}+ \\ -(\beta V)^{-1/2}c^+(p_1+p_2)\,; & +ie(\beta V)^{-1/2}\varphi(p_1-p_2) \end{pmatrix} \tag{2. 69}$$

The spectrum of collective excitations is defined by the quadratic part of the functional S_{eff}. At $T>T_C$ this quadratic form is equal to

$$-\sum_p A(p)c^+(p)c(p) - \frac{1}{2}\sum_p C(p)\varphi(p)\varphi(-p). \tag{2.70}$$

Here, $A(p)$ coincides with the quadratic part of S_{eff} in the model without Coulomb interaction. This part leads to the branch of the collective excitations (2.35). The coefficient $C(p)$ in the second term in (2.70) has the following form

$$C(p) = \frac{k^2}{4\pi} - \frac{e^2}{\beta V}\sum_{p1}\left[G_+(p_1)G_+(p+p_1) + G_-(p_1)G_-(p+p_1)\right] =$$

$$= \frac{k^2}{4\pi} + \frac{e^2}{(2\pi)^3}\int_{|k-k_F|<k_0} d^3k_1 \frac{th\frac{\beta}{2}\xi\left(k_1+\frac{k}{2}\right) - th\frac{\beta}{2}\xi\left(k_1-\frac{k}{2}\right)}{\xi\left(k_1+\frac{k}{2}\right) - \xi\left(k_1-\frac{k}{2}\right) + i\omega}. \tag{2.71}$$

The equation $C(p)=0$ (after analytic continuation $i\omega \to E$) defines the plasmon spectrum

$$E^2 = \omega_{pl}^2 = \frac{4\pi e^2 \rho}{m} \tag{2.72}$$

in the limit $k \to 0$. Here $\rho = k_F^3 / 3\pi^2$ is a density of the Fermi–gas.

At $T > T_C$ we can make the shift transformation (2.38) in order to take the condensate into account. In the first approximation the condensate density can be defined from Eq. (2.44). Here we obtain the quadratic form with the matrix

$$\hat{K} = \begin{pmatrix} A(p) & B(p) & D(p) \\ B(p) & A(-p) & D(-p) \\ D(p) & D(-p) & (p) \end{pmatrix} . \tag{2.73}$$

instead of (2.70). The coefficient function $D(p)$ is odd. One can show that the Bose–spectrum can be obtained from the equation

$$\det \hat{K} = 0 \tag{2.74}$$

At $T > T_C$ and small p we have $D(\omega) \propto \omega$. It leads to two branches of Bose–spectrum (2.43) and also to plasmon branch (2.72).

At $T > T_C$ we have the functions $C(p)$, $D(p)$ look as follows

$$\begin{aligned} C(p) &= \frac{k^2}{4\pi} + \frac{e^2}{\beta V} \sum_{p1} [2G_1(p_1)G_1(p+p_1) - \\ &- G_+(p_1)G_+(p+p_1) - G_-(p_1)G_-(p+p_1) , \\ D(p) &= \frac{ie}{\beta V} \sum_{p1} [G_+(p_1)G_1(p+p_1) - G_-(p_1)G_-(p+p_1)] . \end{aligned} \tag{2.75}$$

$A(p)$, $B(p)$, $C(p)$ are even functions of p, $D(p)$ is an odd function of p. As a result we can split (2.75) into two equations

$$(A-B)C - 2D^2 = 0, \quad A+B=0 . \tag{2.76}$$

The second equation is the same as for the system without Coulomb interaction. It was investigated in 2.2 and defines at small k the branch of the spectrum $E^2 = 4\Delta^2 + c_F^2 k^2 (0.237 - i0.295)$. The analysis of the first equation in (2.76) shows that this equations has at $E \le \Delta$ only a trivial root $k = 0$. It means that the second mode disappears. The phonon Bogoliubov spectrum turns into the plasma oscillation mode.

It is possible also to include the interaction with a quantized electromagnetic field described by the scalar potential $\varphi(\boldsymbol{x}, \tau)$ and the vector potential $A(\boldsymbol{x}, \tau)$ into this scheme. It is appropriate to impose the Coulomb gauge condition $div\mathbf{A} = 0$. The Bose–spectrum in this case is defined by the fourth order matrix which turns out to be quasidiagonal. The spectrum equation splits in two ones. The first one coincides with (2.74). The second equation describes propagation of electromagnetic waves in superconductors.

Chapter III

Sound Propagation in Superfluid ^{3}He and Superconductors

3.1. Sound Propagation in Superfluid ^{3}He

Sound experiments play the most important role in study of collective modes in unconventional superfluids and superconductors. In this Chapter we make general remarks concerning sound propagation in superfluid ^{3}He and superconductors, following to Ref.1.

There are order parameter collective mode in superfluid ^{3}He, whose energies lie between 0 and 2Δ above the Fermi–surface. Since T_C is between 0.9 mK and 2.5 mK, according to the BCS model, the maximum value of $2\Delta(0) = 3.5k_B T_C$ is about 180 MHz, which is easily accessible by sound.

In a typical experiment the transducer would be pulsed or cw resonated at an odd harmonic of its fundamental frequency. Consider the attenuation of sound at a fixed frequency, *f*, as the temperature is reduced to T_C and below. Since gap $\Delta(T)$ is temperature–dependent, this is analogous to keeping the temperature fixed and varying the frequency. Above T_C the attenuation is relatively small (and frequency–independent), being in the collisionless limit. At T_C the liquid goes superfluid, and Cooper pairs are formed. However, since the sound frequency (energy) is larger than $2\Delta(T)$, the sound quanta can break these pairs, giving rise to an absorption or attenuation increase. As the temperature is lowered the number of pairs increased, and the attenuation is increased. This pair–breaking mechanism turns off when the sound quantum becomes smaller than the pair–breaking energy, $2\Delta(T)$, there is relatively abrupt drop (edge) in the attenuation at this point, designated the pair–breaking. At a somewhat lower temperature the sound can excite the collective modes, resulting in large attenuation peaks. Associated with these peaks are large dispersions of the phase velocity and a large dip in the group velocity. The amplitude of these peaks and their widths depend on the coupling of the sound to the collective modes and the (temperature–dependent) relaxation time. Any fine structure in these collective modes – for example, a splitting in an applied magnetic field, a dispersion induced splitting, a superlow induced splitting etc. – results in a splitting in the attenuation peak as the temperature is varied (provided this splitting is of the order of or larger than the width of the attenuation peak). A nicc example of such a splitting is the three–fold dispersion induced splitting of the real–squashing mode, observed by Ketterson group at Northwestern[2]. First only one peak has been observed in absorption into this mode, but after theoretical predictions of three–fold splitting by Brusov et al.[3], the experiment has been done more careful and three–fold splitting has been observed. The distances between peaks were in excellent agreement with Brusov et al. theory[3].

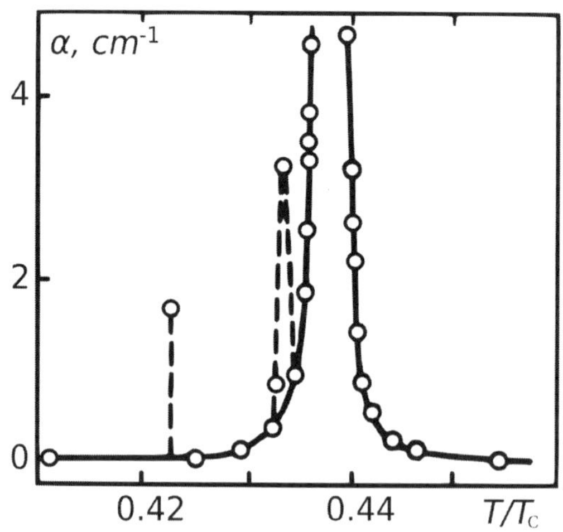

FIG. 3.1. Dispersion induced splitting of the real squashing (*rsq*) – mode[2].

Some technical difficulties can arise if the attenuation from some collective mode turns out to be too high. In this case it is impossible to completely resolve in the transmission experiments (Northwestern experiments could measure up to $10cm^{-1}$). Such situation took place in B–phase of superfluid ^{3}He for pair–breaking and the real–squashing mode. However, in the cw impedance technique (see below) these high attenuation features could be completely resolved.

Another way to sweeping through the collective mode resonances is to depressurize continuously. This technique was developed simultaneously by the Northwestern and Cornell groups. Since T_C depends on the pressure, changing the pressure at the fixed temperature changes the gap and the value of all the collective modes. Using this technique the group velocity of zero sound near the squashing–mode was measured by Movshovich et al.[4,5] at very low temperatures (where the energy gap varies very slowly with temperature and it is awkward to sweep through the spectrum using temperature as the variable) while the liquid was depressurized at a rate slower than 0.5 bar/hour. The group velocity was observed to decrease by more than a factor of 15 in these experiments (they measured a velocity as low as 25 m/s). The same technique was later used to identify the substates of squashing–mode in a magnetic field. The pressure sweeping technique was also used by Northwestern group to study both the $J = 2$ collective modes and the pair–breaking edge, over a wide range of pressures.

3.2. Sound Propagation in Conventional Superconductors

3.2.1. Attenuation in the normal state

Ultrasonic techniques have been commonly used to study normal metals and superconductors. At high temperatures, the dominant attenuation mechanism is the scattering of the sound wave from the dislocations in the lattice. As the temperature is lowered and the electron mean free path becomes longer the interaction of electrons with the lattice contributes to attenuation.

In the hydrodynamic limit $ql << 1$ we have the following expressions[6] for the attenuation coefficient α of longitudinal (L) and transverse (T) sound:

$$\alpha_L = \frac{4}{15}\frac{Nmv_F}{\rho v_l}q^2 l \tag{3.1}$$

and

$$\alpha_T = \frac{1}{5}\frac{Nmv_F}{\rho v_T}q^2 l , \tag{3.2}$$

where q is the sound wavevector and l is the electron mean free path, N, m, v_F are the number density, effective mass and Fermi–velocity of electrons, v_L and v_T are the longitudinal and transverse sound velocities, and ρ is the density. The attenuation is proportional to the mean free path and the square of the frequency ($\omega = qv$). As the temperature falls, the electron mean free path increases, and ultimately saturates because of scattering from imperfections in the crystal.

In the collisionless limit, $ql >> 1$, i.e., for high–impurity samples and/or high frequencies, the attenuation becomes independent of l and is proportional to the frequency, f, the expressions are given by

$$\alpha_L = \frac{\pi}{6}\frac{Nmv_F}{\rho v_T}q = \frac{\pi^2}{3}\frac{Nmv_F}{\rho v_L^2}f \tag{3.3}$$

and

$$\alpha_T = \frac{4}{3\pi}\frac{Nmv_F}{\rho v_T}q = \frac{8}{3}\frac{Nmv_F}{\rho v_T^2}f . \tag{3.4}$$

The limits $ql << 1$ and $ql >> 1$ define regimes where the number of collisions per wavelength is respectively much greater than and much less than unit. These limits are sometimes referred to as the hydrodynamic and collisionless regimes. (Note that the number of collisions per cycle is larger by a factor v_F/v.)

3.2.2. Attenuation in the superconducting state

In most superconducting materials, the electronic contribution to the attenuation drops sharply at T_C, approaching zero as the temperature is lowered. The superconducting state is characterized by the existence of a gap in a single–particle spectrum. The gap increases with decreasing temperature, approaching $\Delta = 1.764 k_B T_C$ at zero temperature. The existence of the energy gap implies that there are no unbound electrons to scatter the sound, and therefore the attenuation at $T = 0$ should go to zero. At finite *T*, thermally excited quasiparticles are present. The distribution of these quasiparticles is governed by the Fermi–distribution, leading to an exponential temperature dependence for the ultrasonic attenuation in the superconducting state. A formal expression for the temperature– dependent attenuation is given by

$$\frac{\alpha_S}{\alpha_N} = \frac{2}{e^{\Delta(T)/k_B T} + 1}, \tag{3.5}$$

where α_S and α_N are the attenuation coefficients in the superconducting and normal states, respectively, and $\Delta(T)$ is the temperature–dependent energy gap.

The above expression is valid only when the attenuation arises solely from thermally excited quasiparticles, i.e. when the phonon energy is smaller than the pair–breaking energy, 2Δ. If $\hbar\omega > 2\Delta$, acoustic phonons can break Cooper pairs, resulting in an additional contribution to the attenuation. Typically, for a transition temperature of the order of 1 K, the pair–breaking energy, $2\Delta(0)$, is of the order of 77 GHz. Thus, pair breaking can be neglected, unless one is working at microwave frequencies, or very close to T_C. In the case of superfluid ^{3}He, however, because of the low T_C, pair breaking plays a significant role.

The experimental dependence of the ultrasonic attenuation and the other properties (e.g. heat capacity and NMR relaxation rate) occur when the gap is isotropic in ***k***–space. If there are line or point zeroes (nodes) in the gap, these properties would display power–low temperature dependence.

3.2.3. Attenuation in the superconducting state in a magnetic field

Let us restrict ourselves by type II superconductors. In them above H_{C1} there is an array of vortex lines. The normal cores of these vortices contribute to the ultrasonic attenuation, and thus, in the superconducting state the attenuation will increase with the field. Initially, as the field is increased, the attenuation remains constant (and very low) as long as the superconductor is in the Meissner state and no magnetic flux enters the sample. Once the magnetic field exceeds H_{C1} the attenuation increases, at first slowly, and than more rapidly. Near the upper critical field the attenuation rises sharply, and at H_{C2} becomes equal to the normal state value (at the prevailing field).

Extensive calculations for the attenuation in the mixed state have been performed[7-9]. The calculations were made for two different limits: the clean limit, in which $l >> \xi$ (i.e. the electrons experience an order parameter averaged over many vortices) and the dirty limit, $l << \xi$ (in which the electrons experience a "local" order parameter. In the clean limit ($l >> \xi$) the attenuation close to H_{C2} is proportional to the square root of the difference in field and is given by

$$\frac{\alpha_n - \alpha_S}{\alpha_n} \approx \left(H_{C2} - H\right)^{1/2}. \tag{3.6}$$

In the dirty limit ($l << \xi$) the attenuation is linearly dependent on the applied magnetic field close to H_{C2}, and one has

$$\frac{\alpha_n - \alpha_S}{\alpha_n} \approx H_{C2} - H \,. \tag{3.7}$$

3.2.4. Velocity at the superconducting transition

At a mean field second–order phase transition, the heat capacity shows a discontinuity; other second derivatives of the free energy (elastic moduli, thermal expansion coefficient) also show an anomaly. The heat capacity, thermal expansion coefficient and compressibility (the three second derivatives of the free energy) are related by the so–called Pippard–Buckingham–Fairbank equation, which has been extensively applied to the study of the properties of superfluid ^4He at the λ–transition.

In the superconducting state is lowered by

$$F_N - F_S = V_0 \frac{H_C^2(T,\varepsilon)}{8\pi} = \frac{\alpha^2(\varepsilon)}{8\pi}[T_C(\varepsilon) - T]^2 . \tag{3.8}$$

Thus, the discontinuity in the heat capacity is

$$C_N - C_S = -T \frac{d^2}{dT^2}(F_N - F_S)\Big|_{T_C} = -\frac{\alpha^2}{4\pi} T_C \,. \tag{3.9}$$

The second derivative of the Helmholtz free energy with respect to strain gives the isothermal elastic moduli

$$c_{ij} = \left(\frac{1}{V_0} \frac{d^2 F}{d\varepsilon_i d\varepsilon_j} \right)\Bigg|_T , \tag{3.10}$$

(here, V_0 is the unstrained volume). Substituting the expression for the free energy, one obtains the velocity jump at the superconducting transition

$$c_{ij}^{S} - c_{ij}^{N} = -\frac{\alpha^2}{4\pi}\frac{\partial T_C}{\partial \varepsilon_i}\frac{\partial T_C}{\partial \varepsilon_j}. \tag{3.11}$$

Chapter IV

Superfluid Phases in ^{3}He

4.1. Introduction: Fermi–systems with Nontrivial Pairing

Among physical systems which low temperature physics studies, the most interesting ones are as follows: the superfluid ^{3}He, heavy fermion systems (HFSC) (UBe_{13}, UPt_3, UPd_2Al_3, $CePd_2Si_2$, $CeIn_3$, $CeNi_2Ge_2$, $CeCu_2Si_2$, U_6Fe etc.), organic superconductors (Behgard's solts M_2X) and the high $-T_C$ superconductors (HTSC). Superfluid ^{3}He was the first Fermi–system with nontrivial pairing (*p*–pairing) to cause great interest. The HFSC are candidates for *p*– or *d*–wave superconductors since the specific heat temperatures dependence in HFSC has power behavior – $C \propto \gamma T(T/T_C)$.

P–pairing is as well realized in the organic superconductors. A strong suppression of superconductivity by the lattice defects and abnormally large values of the second critical magnetic field H_{C2} along one of the critical axis testifies to this. The interest in HTSC is connected first of all with the hopes of their practical uses and the possibility of new mechanisms of superconductivity. But here we would like to note that HTSC are also candidates for nontrivial pairing. This assumption is supported by experirnents which showed a two–fold splitting of the specific heat jump as a function of temperature. So all the above systerns have one common feature – they give us examples of superfluid and/or superconducting Fermi–systems with nontrivial pairing. Such systems have usual a few superfluid (superconducting) phases and a rich spectrum of collective excitations. The investigation of the collective excitation spectrum of these systems is a very important task because it gives us a lot of information about their equilibrium and excited states. However, it is very difficult to construct a theory of collective behavior in such systems because of the complication – associated with the order parameter. For example, in the case of p–pairing the order parameter has 18 degrees of freedom and the same number of collective modes in each ordering phase. Below we consider superfluid ^{3}He an example of superfluid Fermi–system with nontrivial pairing, and construct a microscopic theory of its collective excitations. In conclusion we show how our theory could be applied to determine the type of pairing and the order parameter in HFSC and HTSC.

There are two different methods which are used to investigate the collective excitations in superfluid ^{3}He – the kinetic equation (KE) and the path integral (PI) methods. The main advantage of the path integral method over the kinetic equation, ties in the increased accuracy in calculating the collective mode frequencies. For example, in ^{3}He–B the collective mode (CM) dispersion laws for whole spectrum have been calculated first by Brusov and Popov[1]. The investigation of the stability of the goldstone–modes which require a calculation of the corrections of order k^4 in the general case have been made by the same authors[2]. As well the whole collective mode spectrum has been calculated by Brusov and Popov[3] at zero $\boldsymbol{k}$ by taking the damping of the collective modes into account in ^{3}He–A and a decade later experiments by Dobbs et al[4] show the excellent agreement with these results in opposite to ones obtained by the kinetic equation method. Recently Peter Brusov and Pavel Brusov[5,6] have calculated the whole collective mode spectrum in axial A–phase and planar 2D–phase with dispertion corrections.

The main advantages of the kinetic equation method are connected with the calculation of the coupling strength between zero–sound and the collective mode. A fine example of this is the calculation by Koch and Wolfle[7] of the coupling strength between real squashing–mode and zero–sound, which exists only via very small particle–hole asymmetry.

The cause of such situation is as follow. The application of the path integral method to superfluid were developed by Brusov and Popov to investigate the Bose–spectrum especially. In this way they integrated over all Fermi degrees of freedom and derived the Bose–fields, describing the Cooper pairs near the Fermi–surface only. This made the formalism simpler and gave the possibility to move further in the solution of the problem of the collective mode eigenfrequencies. But such simplification doesn't allow to investigate the interaction between

Fermi– and Bose–degrees of freedom. Of course, it is possible to modify Brusov – Popov procedure to include some Fermi–fields into the model and to calculate the coupling between zero–sound and the collective mode. Inclusion of the Fermi–liquid corrections will lead to complication of path integral scheme. However until now kinetic equation method which considers both the Fermi and Bose–fields was more complicated and have not been very successful in calculating the collective mode spectrum. In our opinion, both of these methods, kinetic equation and path integral, are equivalent. A good example of this is due to Combescot[8], who two years latter then Brusov and Popov[1], obtained the same set of equations for the Bose–spectrum of ^{3}He–B by using KE method instead of the path integral one.

4.2. Properties of Superfluid Phases in ^{3}He

In ^{3}He as well as in other strongly degenerate Fermi–systems the Cooper pairing can occur which leads to superfluidity. If the temperature of the system lowers the percent of the potential energy in the full energy increases as well as the dependence of the system properties on what a kind of interaction takes place. Thus, electrons in metals unite in pairs with zero orbital momentum. In ^{3}He such pairing is impossible because of the strong repulsion of atoms at small distances. This is why the pairing of atoms in ^{3}He may occur only in states with a nonzero orbital momentum ($l \neq 0$).

Pitaevski[9] firstly showed the possibility of a superfluid transition in ^{3}He analysed why there is no superfluidity in ^{3}He at the temperatures of the order of 0.2 K and he developed the Pomeranchuk's idea about the attraction of quasiparticles at the sufficiently large values of the orbital momentum l. The point is that large values of l correspond to large distances, where the Van der Vaals attractive forces act. Such an interaction leads to formation of Cooper pairs with $l \neq 0$. But the phase transition temperature T_C should be very low because of the weakness of the interaction at large distances. Then properties of superfluid phases with $l \neq 0$ were investigated theoretically by Anderson and Morel[10] and by Balian and Werthamer[11]. Nevertheless neither the superfluid phase transition temperature nor the value of l was not predicted.

Experimentally a superfluid transition in ^3He in the temperature region of the order of $10^{-3}K$ was discovered only in 1972 by the – Cornell group in the USA (Osheroff, Richardson and Lee[12]). The utilization of the Pomeranchuk effect played the important role in moving forward into this temperature region. This effect means that the transition of ^3He from the liquid state to the solid one at $T < 0.3K$ is accompanied by the absorption of heat instead of the heat excretion as for ordinary materials. The pressure p in this process must exceed 29 atmospheres because ^3He remains liquid if $p < 29\ bar$. This anomaly in the abovementioned temperature region is connected with the fact that the entropy of the solid ^3He indefined mainly by the chaotic orientation of nuclear spins in ^3He. The liquid ^3He entropy is proportional to the absolute temperature T and this entropy becomes smaller than that of the valid phase at $T < 0.3K$. That gin to squeeze it adiabatically up to the pressure of 29 bar, the ^3He temperature would fall down along with the increase of the percent of the solid phase in the system. This method allows to obtain temperatures up to 1 mK but it gives a possibility to investigate systems only in the neighbourhood of the melting curve[13].

Now the superlow temperatures are obtained not only by using the Pomeranchuk effect but also in the solvate refrigerators of ^3He in ^4He and with the help of the adiabatic nuclear demagnetization. The effeciency of the last method depends on the value of the starting temperature which is obtained using the solvate refrigerators and also on the magnetic field. This method allows to obtain temperatures $T \leq 0.5mK$.[14] Several superfluid phases are discovered in liquid ^3He: the phases A and B and also the phase A_1 which arises in the presence of magnetic field. We give a description of the main properties of these phases according to Khalatnikov[13] and reviews by Maris and Massay[15] and by Wheatley[14,16].

Now the phase diagram of ^3He is studied in details. This diagram can be obtained by different methods, including thermodynamical methods, the nuclear magnetic resonance (NMR) the investigation of the superfluid properties[13], the investigation of the ultrasound absorption[16].

The phase diagram of ^{3}He in the variables T, p, H is presented at the Fig.4.1. In the absence of the magnetic field the superfluid phases are separated from the normal state by the curve of the second kind phase transition $T_C(P)$. The temperature T_C of this transition is equal to 2.6 mK at p = 34 bar, and it decreases to 0.9 mK at p = 0. The A–phase is separated from the B–phase by the lihe of the first kind phase transition T_{AB} which intersects the line $T_C(P)$ in the polycritical point (p = 20 bar, T=2.4 mK).

In magnetic field the normal phase N is separated from the superfluid phases A and B by the layer of a new superfluid phase A_1. The temperature region in which this phase exists rises linearily along with the magnetic field.

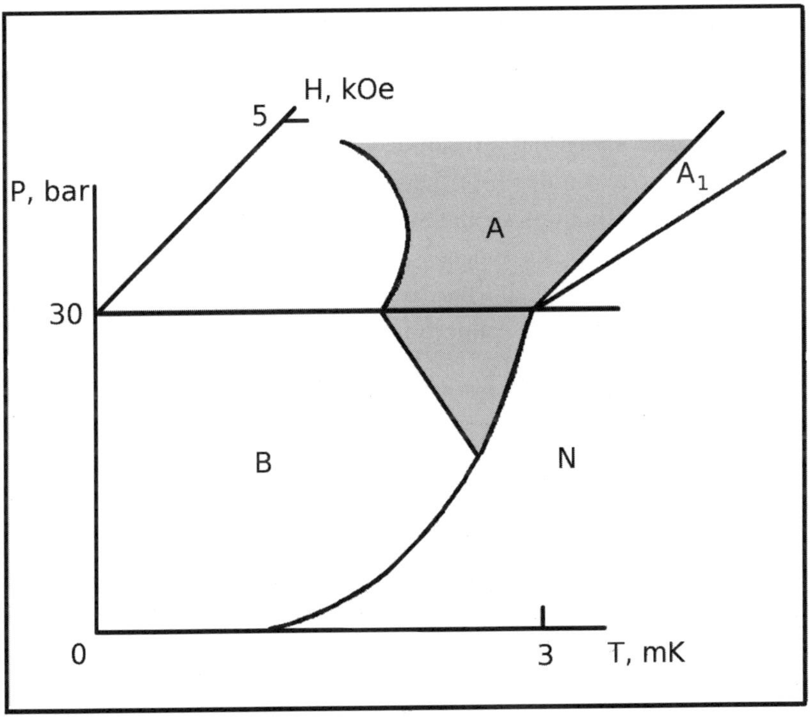

FIG. 4.1. The low temperature phase diagram of ^{3}He.

This region is of the order of 65 mK for the field 10 KE. The line T_{AB} does not intesect the T_C line in the magnetic field. Thus, the A–phase exists for all pressures. The latent heat and hysteresis are absent at the transitions $N \to A_1$, $A_1 \to A$.

A little latent heat (of the order of 20 erg/mole) and a rather large hysteresis (of the order of several tens of milliKelvins) are connected with the B–transition. The temperature of the B–transition depends quadratically on the magnetic field, and the critical field is of the order of 5.5 KE.

^{3}He turns in the solid phase at pressures of the order of 29 bar. At $T = 1$ mK a spin orientation in the solid ^{3}He occurs.

The proof of superfluidity of A, A_1, and B–phases in ^{3}He is obtained in experiments of the viscosity measurement. The viscosity decreases rapidly at $T < T_C$. Another evidence of superfluidity of these discovered phases is an observation of the so–called fourth sound. The fourth sound is the acoustic wave propagation in capillars or in vessels filled by a powder in such a way that a mean distance between powder particles is less than a penetration depth of the normal (viscous) component. In this case only the oscillations of the superfluid component are possible (if such a component exists indeed). The existence of the fourth sound proves the superfluidity, and the measurement of its velocity allows to find the mean density of the superfluid component from the equation

$$\rho_S / \rho = n^2 (c_4^2 / c_1^2),$$

(see Ref. 14), where c_1, c_4 are velocities of the first and of the fourth sound correspondingly, n is the effective reflection index (the empirical value of this index is rather large). The superfluid density becomes nonzero at $T < T_C$, and it increases along with the temperature decreasing.

The behavior of many physical quantities such as the heat capacity, the magnetization, the ultrasound absorption coefficient, the NMR frequancy and also of some others confirms the existence of the phase transition. For example, the heat capacity increases sharply and becomes at 2.9 times greater than that in the normal Fermi liquid immediately below the phase transition. Then this quantity decreases according to the law $\propto T^4$ (see Ref. 13). The investigation of the magnetic properties of the superfluid phases played a very important role in the study of the nature of superfluidity. Nuclear of ^{3}He have nonzero spins. As a result ^{3}He–atoms have magnetic moments and hence interact with the magnetic field. Experiments show that the nuclear magnetic susceptibility in the A–phase practically does not differ from that in the normal Fermi liquid χ_n, but in the B–phase the susceptibility decreases to $0.3\chi_n$, at $T \to 0$. It means that the pairing with odd l occurs in both phases (the full spin s of the pair is equal to unity in this case, which correspond to the triplet pairing). In the case of pairing with an even l (s =0 for the singlet pairing) Cooper pairs have no magnetic momentum and can not react on the magnetic field. In this case the susceptibility would tend to zero when the number of pairs increases along with the decreasing of temperature, but really it does not takes place. The absence of a break in the curve $T_C(P)$ at $H=0$ shows that pairing in both phases occurs in the state with the same value of *l*. Unfortunately, there is no straight experiment which allows to measure a value of *l*. Nevertheless the all totality of the experimental data indicates that the pairing in ^{3}He occurs in the state with l=1 (*p*–pairing).

One more confirmation in the favour of the *p*–pairing is due to the NMR experiments yielding an important information about properties of the superfluid phases. The transverse NMR in the B–phase occurs with the frequency of the Larmor precession $\omega_{\perp 0} = \mu H_0$ (in units with $\hbar = k_B = 1$), where H_0 is a constant magnetic field, μ is a magnetic moment of the nucleas of ^{3}He. In the A –phase the transverse NMR frequency is shifted as compared with the Larmor frequency and it is done by the formula

$$\omega_{\perp 0} = \left(\mu^2 H_0^2 + \Omega_A^2(T)\right)^{1/2} .$$

The value of $\Omega_A(T)$ increases from zero at $T = T_C$ to the quantity of the order of several tens of Gausses at $T = 0$. The longitudinal NMR takes place in both phases and its frequencies are $\omega_{//B} = \Omega_B(T)$, $\omega_{//A} = \Omega_A(T)$ of the same order of magnitude[13,14]. These experiments play an important role in the identification of the superfluid phases. Leggett[17] showed that these experiments can be explained by assuming the *p*–pairing in both phases.

The NMR method is the most exact one but its application has a limited character. Thus, the NMR method can not be always used when investigating orbital motion because yet very weak fields (of the order of the magnetic field of the Earth) can change qualitatively the ***l***–vector in the *A*–phase [16].

In this situation the sound propagation experiments and their interpretation become very important. Let us dwell on these experiments in more details[16] because their interpretation is founded completely on the theory of collective excitations, presented in this book.

The sound in the Fermi liquid can propagate in two different regimes namely hydrodynamical ($\omega\tau << 1$) and collisioneless ($\omega\tau >> 1$) in dependence on values of the sound frequency ω and of the collision time of quasiparticles τ. The collision time increases as T^{-2} at low temperatures, and the collisionless (zero–sound) regime plays the main role at low temperatures. The sound frequency for this regime must obey condition $\omega >> 4MHz$ near the phase transition temperature $T \propto 3mK$.

Up to now experiments were done on the frequences 5, 15, 25 and 63 *MHz*. The main peculiarities of the zero–sound propagation in the *B*–phase are presented on the Fig.4.2 (absorption) and on the Fig.4.3 (velocity)[16]. In contrast to that in the *A*–phase, the sound propagation in the *B*–phase does not depend on the direction of the sound propagation with respect to the direction of the external magnetic field *H*.

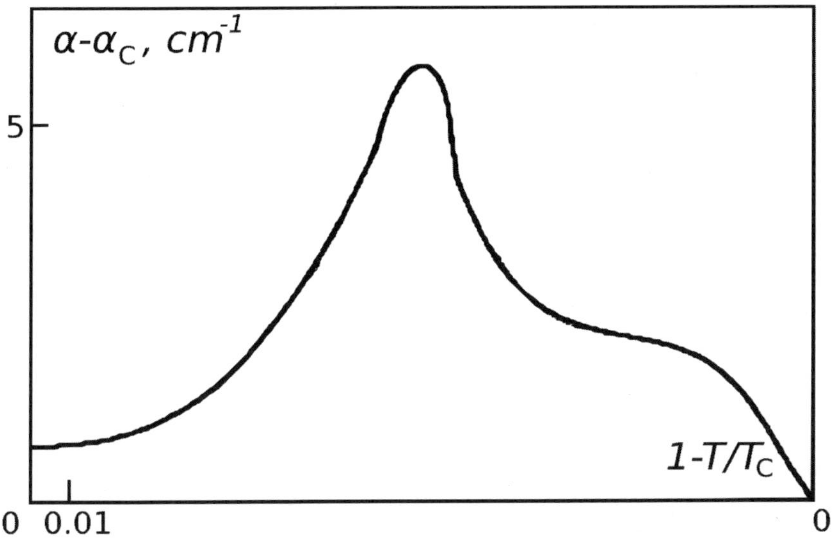

FIG. 4.2. The absorption of the zero–sound in the *B*–phase of ^3He[16].

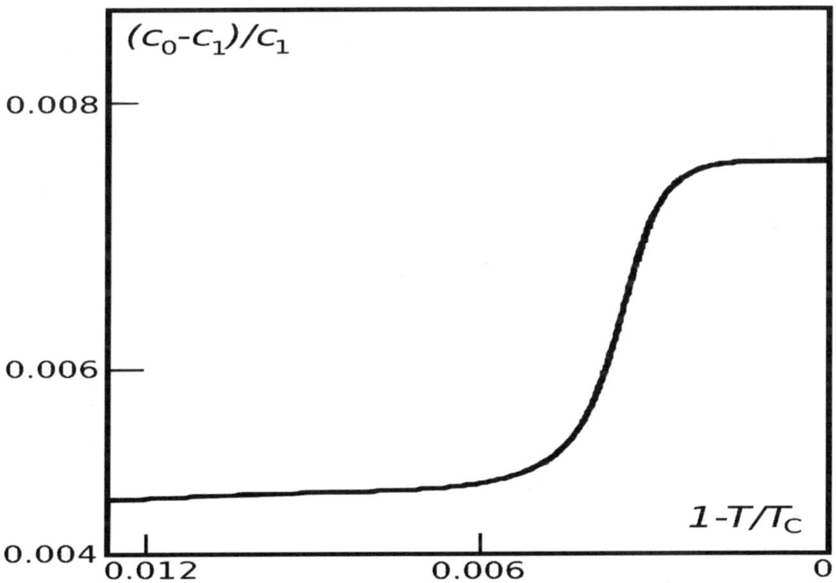

FIG. 4.3. The temperature dependence of the relative velocity of the zero–sound in the B–phase of ^3He[16].

The absorption has a large wide peak arising as a result of excitation of collective modes of the order parameter by zero–sound[18]. It was predicted that the position of the center of the peak corresponds to the temperature defined by the equality

$$\omega = \sqrt{12/5}\Delta(T) \tag{4.1}$$

and the so–called squashing mode (*sq*) is excited. Here

$$\Delta(T) = \left(\frac{2}{3}\frac{\Delta C}{C_N}\right)^{1/2} \pi T_C \left(1 - T/T_C\right)^{1/2} \tag{4.2}$$

is a value of the gap in the Fermi–spectrum. The width of the peak is defined by the quasiparticle collisions, and a small convexity near T_C can be explained by a decay of Cooper pairs[18]. The magnitude of the absorption peak increases along with the increasing of frequency and decreasing of pressure[14].

The zero–sound velocity at $T \approx T_C$ exceeds the velocity of the first sound approximately on 1%. At temperatures corresponding to the center of the absorption peak the sharp decrease of the zero–sound velocity occurs in a narrow temperature interval (see Fig. 4.3). For the further temperature decrease the velocity continues to decrease and it tends to the first sound velocity at $T \to 0$.

Measurements of the absorption and of the velocity of the ultrasound allow to define some important parameters[16], for example, the Landau parameters

$$F_1 = \left(3m^2c^2/k_F^2\right) m^*/m - 1 \tag{4.3}$$

and

$$F_2 = 5\left[\frac{5}{2}(1+F_0)(c_0 - c_1)/c_1 - 1\right] \tag{4.4}$$

where c_1 is the first sound velocity, $(c_0 - c_1)/c_1$ is the relative excess of the zero–sound velocity c_0 with respect to the first sound velocity c_1, k_F is the Fermi–momentum, m, m^* are the atomic mass and the effective mass correspondingly.

Another important quantity is a collision time of quasiparticle, τ_α. It can be defined from the zero–sound absorption according to the formula

$$\tau_\alpha \approx \frac{x}{\omega}\left(1 - \frac{1}{x^2} + \ldots\right), \qquad x = \frac{1}{\alpha_0}\frac{\omega}{c_1}\frac{c_0 - c_1}{c_1}, \tag{4.5}$$

where α_0 is the attenuation coefficient for the zero–sound. The value of - α_0 is connected with a more fundamental quantity, i.e. a lifetime of quasiparticles near the Fermi–surface $\tau(0)$ and also with the parameter of the collision integral α_η, according to the equation

$$\tau_\alpha = \frac{(3/4)\tau(0)}{(1 - \alpha_\eta/2)(1 + F_2/5)}. \tag{4.6}$$

Quantities $\tau(0)$ and α_η are expressed via the probability of the scattering of quasiparticles as follows

$$\tau(0) = \frac{16\pi^2}{m^{*\,3}T^2}\left[\int \frac{d\Omega}{4\pi}\frac{W(\theta,\varphi)}{\cos(\theta/2)}\right]^{-1} \tag{4.7}$$

and

$$1-\frac{\alpha_\eta}{2}=\left[\int\frac{d\Omega}{4\pi}\frac{W(\theta,\varphi)}{\cos(\theta/2)}3\sin^2\varphi\sin^4(\theta/2)\right]\left[\int\frac{d\Omega}{4\pi}\frac{W(\theta,\varphi)}{\cos(\theta/2)}\right]^{-1} \tag{4.8}$$

Here, θ is an angle betwccn initial momenta p_1 and p_2 of the scattering quasiparticles, φ is an angle between the planes formed p_1 and p_2 and by the momenta p_3 and p_4 of the quasiparticle after scattering. Wolfle[18] have shown that the parameter of the collision integral can be found from the measurement of the peak of the collective mode.

The important parameter is also the relaxation time, connected with the viscosity coefficient η

$$\tau_\eta=\frac{5\eta}{(N/V)m^*c_F^2}, \tag{4.9}$$

where c_F is the velocity on the Fermi–surface, N/V is a particle density, the value of τ_η can be defined from experiments on the attenuation of the hydrodynamical sound in the normal Fermi–liquid and also from the viscosity measurements.

Values τ_η and τ_α are connected by the formula

$$\tau_\eta=\tau_\alpha\left(1+F_2/5\right). \tag{4.10}$$

One can see from (4.2)–(4.10) that it is possible to define the Fermi–liquid parameters, and the values of $\alpha_\eta, \tau(0)$, τ_α, τ_η and also the value of the gap in the Fermi–spectrum by combining the measurements of the velocities c_0 and c_1 the effective mass m^*, the zero–sound attenuation in the normal phase above T_C and also the measurements of the form and of the position of the peak due to the collective mode. According to the Wolfle data[18] ($\tau_\alpha T_C^2 = 0.75\ \mu s \cdot mK^2$, $\tau_\eta T_C^2 = 0.81\ \mu s \cdot mK^2$, $(c_0 - c_1)/c_1 = 7.4 \cdot 10^{-3}$, $c_1 = 3.6 \cdot 10^4 cm/s$, $F_0 = 57.1$, $F_2 = 0.4$, $\alpha_0(T_C) = 1.55 cm^{-1}$. These data are in good argument with experiments. Later the second peak of attenuation was discovered at the temperatures defined by the equality[19]

$$\omega = \sqrt{8/5}\Delta(T) \tag{4.11}$$

This peak is due to the ultrasound absorption in the real squashing mode (*rsq*) (Fig.4.4).

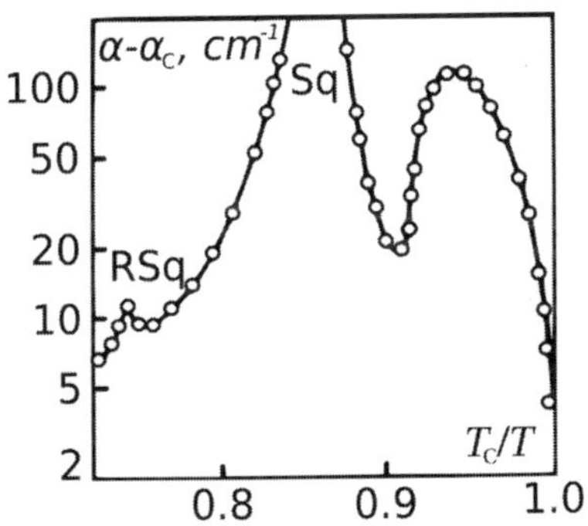

Fig. 4.4.

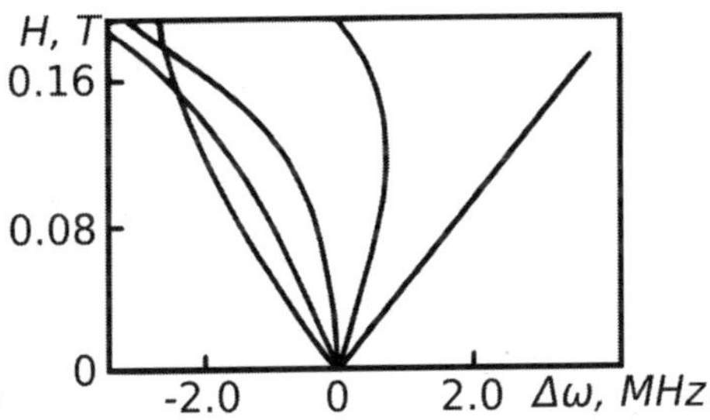

Fig. 4.5.

FIG. 4.4. The zero–sound absorption into the squashing (*sq*) and the real squashing (*rsq*) modes (Mast et al.[20]).

FIG. 4.5. The nonlinear Zeeman effect for the real squashing mode (Shivaram et al.[21]).

Then the splitting of this peak was found[21], which can be explained by the splitting of the *rsq*–branches at nonzero momenta (see Chapter VI). The attenuation peak has a fivefold splitting in magnetic field (the linear Zeeman effect)[22] due to the additional term $aJ_Z H$ $(J_Z = 0, \pm 1, \pm 2)$ in the energy of the mode. Textures lead to the additional splitting of the central peak of the *rsq*–mode with $J_Z = 0$[21]. Intersection of the *rsq*–mode branches (the nonlinear Zeeman effect) occurs in the strong magnetic field (Fig. 4.5). Some peculiarities in the absorption spectrum near the absorption threshold are found by Daniels *et al.* [23], which can be interpreted as a resonance absorption of the zero–sound by the mode with $\omega = 2\Delta(T)$ (the pair–breaking mode), see Chapter X.

Thus, now all the collective modes predicted by the theory were observed in the B–phase, namely a zero–sound, spin waves, *sq*–, *rsq*– and *p* –modes.

In the contrast with the B–phase, the zero–sound propagation in the *A*–phase is characterized by the strong anisotropy. The theory predicts that the attenuation of the zero–sound depends strongly on the angle between the vector of the orbital momentum $\boldsymbol{l}$ and the momentum $\boldsymbol{q}$. The $\boldsymbol{l}$ vector is linked with the spin vector $\boldsymbol{d}$ via the dipole interaction which forces $\boldsymbol{d}$ to be parallel to $\boldsymbol{l}$ ($\boldsymbol{l} \parallel \boldsymbol{d}$). On the other hand the $\boldsymbol{d}$ vector is oriented to be orthogonal to the magnetic field $\boldsymbol{H}$ due to the anisotropy of succeptibility. The point is that Cooper pairs have a zero magnetic succeptibility along $\boldsymbol{d}$ and have a nonzero χ_N $(\chi_N \neq 0)$ in the plane orthogonal to $\boldsymbol{d}$. Besides magnetic field there exist some other factors orienting $\boldsymbol{l}$ such as the flow and the boundary effects. The magnetic field itself does not define the direction of $\boldsymbol{l}$ uniquely.

Velocity and attenuation of the ultrasound depends on $(\boldsymbol{l}, \boldsymbol{q})$, and they are defined uniquely if $(\boldsymbol{H} \parallel \boldsymbol{q})$. For other orientations of the value of $(\boldsymbol{l}, \boldsymbol{q})$ does not defined by $\boldsymbol{H}$ uniquely and the above mentioned factors can influence the ultrasound velocity and attenuation. Thus, if $\boldsymbol{H}$ is not parallel to $\boldsymbol{q}$ the zero–sound propagation is accompanied by the rather interesting time fluctuations[16] connected with the succeptibility fluctuations. For the ultrasound with frequency 10 MHz the fluctuation of the attenuation coefficient is of the order of $2\cdot 10^{-3} H^2 cm^{-1}\cdot (kgs)^{-1}$. Such fluctuations are not observed in zero magnetic field.

The sound attenuation data in the *A*–phase can be presented in the form

$$\alpha - \alpha_C = A\cos^4\theta + B\sin^4\theta + 2C\sin^2\theta\cdot\cos^2\theta ,$$

where θ is the angle between $\boldsymbol{H}$ and $\boldsymbol{q}$. Results for three frequences (5, 15, 25 MHz) are presented in Ref.16. The coefficient *A* corresponds to the case $(\boldsymbol{l},\ \boldsymbol{q}) = 0$, i.e. the case of the state which is well defined from the point of view of the ultrasound propagation. As for the coefficient *C*, one can see the nonuniquence and nonreproducibility of the results for the ultrasound attenuation at the frequency 25 MHz.

The *A*–peak (which exists up to the frequencies of the order of 140 MHz) is a result of the ultrasound absorption[18] caused due to excitation of the so–called "clapping" mode. The maximal absorption exists at the temperature defined by the equality

$$\omega = \frac{2}{5}\left(2\sqrt{6-3}\right)^{1/2}\Delta_0(T) = 1.23\Delta_0(T) , \qquad (4.12)$$

where

$$\Delta_0(T) = \pi T_C\left(\Delta C / C_N\right)^{1/2}(1 - T/T_C)^{1/2} , \qquad (4.13)$$

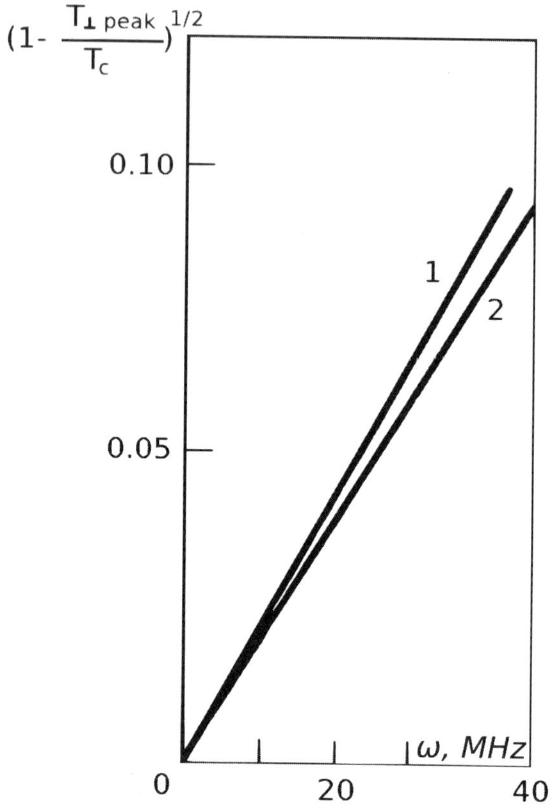

FIG. 4.6. The dependence of the attenuation peak in the *A*–phase of ^{3}He on the frequency of the zero–sound ($\boldsymbol{l} \perp \boldsymbol{q}$) (Wheatley[16]) 1– p = 24.1 bar, 2 – p = 33.5 bar.

The dependence of the reduced temperature of the maximal attenuation $\left(1 - T_{\max} / T_C\right)$ on the sound frequency (at $\boldsymbol{l} \perp \boldsymbol{q}$)[16] (Fig. 4.6) is described by the equality

$$\left(1 - T_{\max} / T_C\right)^{1/2} = a\omega, \tag{4.14}$$

where $a = 4.0 \cdot 10^{-3} MHz$ at p = 2.41 bar and $a = 3.6 \cdot 10^{-3} MHz$ at p = 33.5 bar. These results can be obtained from (4.12) and (4.13) if one knows T_C and $\Delta C / C_N$.

The C–peak is a result of the zero–sound attenuation in one of the "flapping" modes: in the "normal–flapping" mode if the ultrasound frequency $\omega \leq 60 MHz$, and in the "super–flapping" mode at $\omega \geq 80 MHz$, (at frequences $\omega \leq 60 MHz$ the coupling of the "super–flapping" mode with the zero–sound is proportional to $(1 - T / T_C)$ and is very small).

For $\boldsymbol{q} \| \boldsymbol{l}$ ($\boldsymbol{q} \perp \boldsymbol{l}$) there is no coupling of the collective modes with the zero–sound. At $\theta = 0$ ($\boldsymbol{q} \| \boldsymbol{H}$) only the "clapping" mode is coupled with the zero–sound.

Frequencies 5–25 MHz which were used in the early experiments yield the attenuation peak at temperatures near T_C. The extrapolation of these results obtained for temperatures from 0.8 T_C to T_C and formulated in (4.13) in the region $T_C - T \propto T_C$ shows that the attenuation peak near T=0 must be for the ultrasound frequencies of the order of 200–270 MHz depending on the pressure.

Let us note that there exists a strong dependence of the accepted sound signal on the direction of the vector $\boldsymbol{l}$ (respectively to $\boldsymbol{q}$) so this signal can be used for the definition of the orbital momentum orientation.

We have presented the main experimental data for the superfluid phases of ^{3}He. In the next Chapter the theoretical description of these phases is done in the framework of the model which was build up in the functional integration formalism. In the subsequent Chapters of this book we shall give a theoretical interpretation of these experimental data. The main attention is devoted to the theoretical description of two principle experimental methods of investigation of the superfluid ^{3}He namely the NMR method and the ultrasound experiments.

Chapter V

The Model of ^{3}He

5.1. The Path Integral Approach

Method describing below has been suggested by Popov et al.[1] and developed by Brusov and Popov.[2-5]

In the method of functional integration, the initial Fermi–system (^{3}He) is described by anticommuting functions $\chi_s(\mathbf{x},\tau), \overline{\chi}_s(\mathbf{x},\tau)$ defined in the volume $V = L^3$, which are antiperiodic in time, τ, with a period $\beta = T^{-1}$. Here s is the spin index. These functions can be expanded into a Fourier series

$$\chi_s(x) = (\beta V)^{-1/2} \sum_p a_s(p) \exp(i(\omega\tau + \mathbf{k}\cdot\mathbf{x})), \tag{5.1}$$

where $p = (\mathbf{k}, \omega)$; $\omega = (2n+1)\pi T$ are Fermi–frequencies and

$x = (\mathbf{x}, \tau)$.

Let us consider the functional of action for an interacting Fermi–system

$$S = \int_0^\beta d\tau d^3x \sum_s \overline{\chi}_s(\mathbf{x},\tau)\partial_\tau \chi_s(\mathbf{x},\tau) - \int_0^\beta H'(\tau) d\tau, \tag{5.2}$$

which corresponds to the Hamiltonian

$$H'(\tau)=\int d^3x\sum_s(2m)^{-1}\nabla\overline{\chi}_s(\mathbf{x},\tau)\nabla\chi_s(\mathbf{x},\tau)-(\lambda+s\mu_0H)\overline{\chi}_s(\mathbf{x},\tau)\chi_s(\mathbf{x},\tau)$$

$$+\frac{1}{2}\int d^3xd^3yU(\mathbf{x}-\mathbf{y})\sum_{ss'}\overline{\chi}_s(\mathbf{x},\tau)\overline{\chi}_{s'}(\mathbf{y},\tau)\chi_{s'}(\mathbf{y},\tau)\chi_s(\mathbf{x},\tau). \tag{5.3}$$

In order to obtain the effective functional of action, we shall use the method of division of Fermi–fields into "fast" and "slow" fields with subsequent successive integration over these fields. Fast fields χ_{1s} and $\overline{\chi}_{1s}$ are determined by components of expansion (1) either with frequencies $|\omega|>\omega_0$, or with momenta $|k-k_F|>k_0$. The remaining component $\chi_{0s}=\chi_s-\chi_{1s}$ of the Fourier expansion define slow fields χ_{0s}.

Integrating over fast fields, we obtain

$$\int\exp SD\overline{\chi}_{1s}D\chi_{1s}=\exp\widetilde{S}(\chi_{0s},\widetilde{\chi}_{0s}). \tag{5.4}$$

The most general form of $\widetilde{S}$ is the sum of even–order forms of the fields $\chi_{0s},\overline{\chi}_{0s}$:

$$\widetilde{S}=\sum_{n=0}^{\infty}\widetilde{S}_{2n}. \tag{5.5}$$

Neglecting the insignificant constant $\widetilde{S}_0$ and sixth and higher–order terms (this can be done when the layer $|k-k_F|<k_0$ is narrow), we retain only the second– and fourth–order terms $\widetilde{S}_2$ and $\widetilde{S}_4$ describing noninteracting quasiparticles near the Fermi–surface and their paired interaction respectively:

$$\widetilde{S}_2\approx\sum_{s,p}Z^{-1}\left[i\omega-c_F(k-k_F)+s\mu H\right]a_s^+(p)a_s(p). \tag{5.6}$$

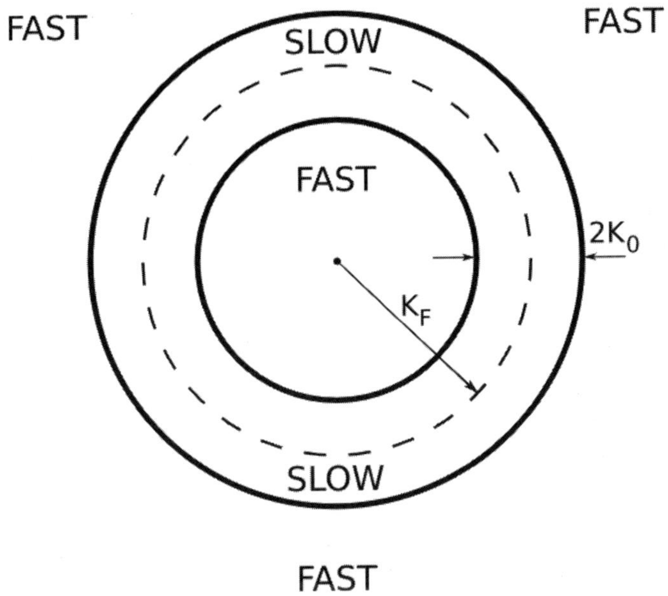

FIG. 5.1. The division of Fermi–fields into "fast" and "slow" fields.

$\tilde{S}_4$ is different for different types of pairing, so starting from here we should split our consideration.

The method of obtaining this functional in the functional integral formalism constitutes an alternative approach to that developed in the Landau theory of Fermi–liquid[6,7].

In Chapter II we have considered a model with g=0, f=const<0 as a simplified model of a superconductor. Hereafter we shall consider a model with

$$f=0,\ g=\text{const}<0. \tag{5.7}$$

as a simplified model of ^{3}He with pairing in the p–state. The use of the Fermi–fields (5.7) (is the most convenient way to describe ^{3}He system. However, significant difficulties appear if we try to use them for description low energy (infrared) phenomena in the superfluid ^{3}He. This is due to the absence of singularities in the single–particle

Green's function $\langle \chi_S(x,\tau)\chi_S(x',\tau')\rangle$ below $E=\Delta$ (Kleinert[8]), where Δ is a gap in the single–particle spectrum $E(k)=\left(\xi^2(k)+\Delta^2\right)^{1/2}$, $\xi(k)=c_F\left(k-k_F\right)$. Thus the description of infrared phenomena (with $E<<\Delta$) such as zero sound, spin waves as so on, is complicated in terms of Fermi–fields. We need to sum up an infinite set of the Feynman diagrams to gain a simple understanding of these phenomena. But there exist Green's functions which describe such kind of excitations directly. They are

$$\langle \overline{\chi}_S(x,\tau)\chi_S(x,\tau)\overline{\chi}_S(x',\tau')\chi_S(x',\tau')\rangle ,$$

$$\langle \overline{\chi}_S(x,\tau)\frac{\sigma_a}{2}\chi_S(x,\tau)\overline{\chi}_S(x',\tau')\frac{\sigma_a}{2}\chi_S(x',\tau')\rangle .$$

Singularities which appear in such complicated Green's functions and are absent in the single particle ones are called collective excitations (CE). The most logical way for their description is the passage from Fermi–fields to Bose–fields which correspond to Cooper pairs of quasi–fermions.

To realize such passage we insert a Gaussian integral of $\exp(\overline{c}Ac)$ with respect to the Bose–field *c*, where $\overline{c}Ac$ is a quadratic form with a certain operator *A* into the integral over the Fermi–fields $\chi_{S0},\overline{\chi}_{S0}$. We then shift the Bose–field by a quadratic form of the Fermi–fields so as to annihilate the quartic form $\widetilde{S}_4$ in the Fermi–fields. The integral over the Fermi–fields is then transformed into a Gaussian integral and is equal to the determinant of the operator $M(c,\overline{c})$, that depends on the Bose–field *c* and $\overline{c}$. We arrive at the functional

$$S_{eff}(c,\overline{c})=\overline{c}Ac+\ln\det\frac{\widehat{M}(c,\overline{c})}{\widehat{M}(0,0)}, \tag{5.8}$$

in which the $\ln\det$ has been regularized by dividing $M(c,\bar{c})$ by the operator $M(0,0) = M(c,\bar{c})_{C=\bar{C}=0}$.

The functional S_{eff} maybe called the effective action (or hydrodynamical action) functional. It defined the point of the phase transition of the initial Fermi–system as a point of the Bose–condensation of the fields c and $\bar{c}$, and S_{eff} determines also the density of the condensate at $T < T_C$ and the spectrum of the collective excitations.

In Chapter II it was demonstrated how this idea works in the case g=0, f=const<0 (a simplified model of a superconductor).

In the case of p –pairing, we need to insert into the integral over the Fermi fields, a Gaussian integral over the complex function $c_{ia}(x,\tau)$ and $\bar{c}_{ia}(x,\tau)$ with the vector index i and the isotopic index a. (i, a = 1, 2, 3) . The Gaussian integral is of the form

$$\int D\bar{c}_{ia}(x,\tau) Dc_{ia}(x,\tau) \exp\left(g_0^{-1} \sum_{p,i,a} \bar{c}_{ia}(p) c_{ia}(p) \right) \tag{5.9}$$

where g_0 is the constant g from Eq. (5.7). It is easily verified that the shift

$$c_{i1}(p) \to c_{i1}(p) + \frac{g_0}{2(\beta V)^{1/2}} \sum_{p1+p2=p} (n_{1i} - n_{2i})[a_+(p_2)a_+(p_1) - a_-(p_2)a_-(p_1)]$$

$$c_{i2}(p) \to c_{i2}(p) + \frac{ig_0}{2(\beta V)^{1/2}} \sum_{p1+p2=p} (n_{1i} - n_{2i})[a_+(p_2)a_+(p_1) + a_-(p_2)a_-(p_1)]$$

$$c_{i3}(p) \to c_{i3}(p) + \frac{g_0}{(\beta V)^{1/2}} \sum_{p1+p2=p} (n_{1i} - n_{2i})\, a_-(p_2)a_+(p_1) \tag{5.10}$$

does indeed eliminate the form $\tilde{S}_4$.

To calculate the Gaussian integral over the Fermi–fields, we introduce a column vector $\psi_a(p)$ with elements

$$\psi_1(p) = a_+(p),\ \psi_2(p) = -a_-(p),\ \psi_3(p) = a_-^+(p),\ \psi_4(p) = a_-^+(p) \tag{5.11}$$

and write down a quadratic form in the Fermi–fields

$$K = \frac{1}{2} \sum_{p1,p2,a,b} \psi_a^+(p_1) M_{ab}(p_1, p_2) \psi_b(p_2) \tag{5.12}$$

The fourth–order matrix $\hat{M}(p_1, p_2)$ has the following elements $M_{ab}(p_1, p_2)$

$$M_{11} = Z^{-1}\left[i\omega - \xi + \mu(\mathbf{H\sigma})\right]\delta_{p_1 p_2},$$

$$M_{22} = Z^{-1}\left[-i\omega + \xi + \mu(\mathbf{H\sigma})\right]\delta_{p_1 p_2},$$

$$M_{12} = (\beta V)^{-1/2}(n_{1i} - n_{2i})c_{ia}(p_1 + p_2)\sigma_a,$$

$$M_{21} = -(\beta V)^{-1/2}(n_{1i} - n_{2i})c_{ia}^+(p_1 + p_2)\sigma_a, \tag{5.13}$$

where $\sigma_a (a = 1,2,3)$ are 2 x 2 Pauli matrices. Integrating over the Fermi–fields

$$\int D\bar{\chi}_0 D\chi_0 \exp K = \left(Det\hat{M}\right)^{1/2}, \tag{5.14}$$

we arrive at the "effective" or "hydrodynamic action" functional

$$S_{\text{eff}} = g_0^{-1}\sum_{p,i,a} c_{ia}^{+}(p)c_{ia}(p) + \frac{1}{2}\ln\det\frac{\hat{M}\left(c_{ia}, c_{ia}^{+}\right)}{\hat{M}(0,0)}. \tag{5.15}$$

This functional contains all the information on the physical properties of the system. In particularly it determines the transition temperature into superfluid state, the order parameter (OP) of the superfluid states, gap equation, collective mode spectrum and many others.

Let us investigate this functional in the Ginzburg–Landau region $|\Delta T| << T_C$, where $\ln\det$ can be expanded in powers of the fields c and $\bar{c}$. Putting

$$\hat{M}(0,0) = G^{-1},\ \hat{M}(c_{ia}, c_{ia}^{+}) = \hat{G} + \hat{u}$$

we retain the first two terms (n=1,2) of the expansion

$$\frac{1}{2}\ln\det\frac{\hat{M}\left(c_{ia}, c_{ia}^{+}\right)}{\hat{M}(0,0)} - \frac{1}{2}Tr\ln(I + \hat{G}\hat{u}) = -\sum_{n=1}^{\infty}\frac{1}{4n}Tr\left(\hat{G}\hat{u}\right)^{2n}.$$

Consider the second–order form. If H=0, the form is diagonal in the isotopic index a and is given by

$$-\sum_{p,a,i,j} A_{ij}(p)c_{ia}^{+}(p)c_{ja}(p) = -\sum_{p,a,i,j}\left(A_{ij}(p) - A_{ij}(0)\right)c_{ia}^{+}(p)c_{ja}(p) -$$
$$-\sum_{p,a,i,j}\left(A_{ij}(0)\right)c_{ia}^{+}(p)c_{ja}(p),$$

where

$$A_{ij}(0) = -\frac{4Z^2}{\beta V}\sum_{p1} n_{1i}n_{1j}(\omega_1^2+\xi_1^2)^{-1} = -\delta_{ij}\frac{2Z^2k_F^2}{3\pi^2 c_F}\int_0^{c_F k_0}\frac{d\xi}{\xi}th\frac{\beta\xi}{2} = -\delta_{ij}A(0),$$

$$A_{ij}(p) - A_{ij}(0) = \frac{Z^2}{2V}\sum_{|k1-kF|<k0}(n_{1i}-n_{2i})(n_{1j}-n_{2j})\times$$

$$\times\left[\frac{1}{i\omega-\xi_1-\xi_2}\left(th\frac{\beta\xi_1}{2}+th\frac{\beta\xi_2}{2}\right)+\frac{1}{\xi_1}th\frac{\beta\xi_1}{2}\right].$$

At small H ($\mu H << T$) it suffices to take into account the increment to $A_{ij}(0)$ and neglect the H dependence of the function $A_{ij}(p) - A_{ij}(0)$ at small p. After summing over the frequencies we get the integrals

$$\int_{-c_F k_0}^{c_F k_0}\frac{d\xi}{\xi\pm\mu H}th\frac{\beta}{2}(\xi\pm\mu H), \quad a=1,\ 2\ ,$$

$$\int_{-c_F k_0}^{c_F k_0}\frac{d\xi}{\xi\pm\mu H}\left(th\frac{\beta}{2}(\xi+\mu H)+th\frac{\beta}{2}(\xi-\mu H)\right), \quad a=1,\ 2\ .$$

Only the last integral which receives an increment $\propto H^2$ depends on H. The H–dependence increments to S_{eff} is of the form

$$-\frac{7\zeta(3)Z^2k_F^2\mu^2H^2}{6\pi^4T^2c_F}\sum_{p,\ i}c_{i3}^+(p)c_{i3}(p).$$

In the fourth–order form we shall hereafter be interested in terms with small 3–momenta of all fields $c_{ia}(p)$ and $c_{ia}^+(p)$ ($c_F|k_i| << T$), which can be written in the form

$$-\frac{7\zeta(3)Z^4k_F^2}{30\pi^4T^2c_F\beta V}\sum_{p1+p2=p3+p4}[2c_{ia}^+(p_1)c_{jb}^+(p_2)c_{ia}(p_3)c_{jb}(p_4)+$$
$$+2c_{ia}^+(p_1)c_{ib}^+(p_2)c_{ja}(p_3)c_{jb}(p_4)+2c_{ia}^+(p_1)c_{jb}^+(p_2)c_{ja}(p_3)c_{ib}(p_4)-$$
$$-2c_{ia}^+(p_1)c_{ja}^+(p_2)c_{ib}(p_3)c_{jb}(p_4)-c_{ia}^+(p_1)c_{ia}^+(p_2)c_{jb}(p_3)c_{jb}(p_4)]$$

We find the Bose–condensation temperature T_C from the condition that the coefficient in the quadratic form at $c_{ia}^+(p)$, $c_{ia}(p)$ vanish for p=0. At H=0 we have the equality

$$0=g_0^{-1}+A(0)=g_0^{-1}+\frac{2Z^2k_F^2}{3\pi^2c_F}\int_0^{c_Fk_0}\frac{d\xi}{\xi\pm\mu H}th\frac{\beta\xi}{2}.$$

The integral with respect to ξ depends logarithmically on k_0. In order for T_C not to depend on k_0 it is necessary that g_0^{-1} depends on k_0 in accordance with the formula

$$g_0^{-1} = g^{-1} - \frac{2Z^2 k_F^2}{3\pi^2 c_F} \ln\frac{k_0}{k_F} \ , \tag{5.16}$$

where g no longer depends on k_0 (we must have here $g<0$).

Substitution of (5.16) in (5.15) yields

$$T_C = \frac{2\gamma}{\pi} c_F k_F \exp\left(-\frac{3\pi^2 c_F}{2|g|Z^2 k_F^2}\right),$$

where $\ln\gamma = C$ is the Euler constant. In the region $\Delta T << T_C$ we have

$$g_0^{-1} + A(0) = \frac{2Z^2 k_F^2}{3\pi^2 c_F} \ln\frac{T_C}{T} \approx \frac{2Z^2 k_F^2}{3\pi^2 c_F} \frac{T_C - T}{T_C} .$$

To find the density of the condensate at $T<T_C$ we make the substitution

$$c_{ia}(p) \to (\beta V)^{1/2} \delta_{p0} a_{ia} \left(\frac{10 T_C |\Delta T|}{7\zeta(3)}\right)^{1/2} \frac{\pi}{Z},$$

$$c_{ia}^{+}(p) \to (\beta V)^{1/2} \delta_{p0} \bar{a}_{ia} \left(\frac{10 T_C |\Delta T|}{7\zeta(3)}\right)^{1/2} \frac{\pi}{Z},$$

which transforms S_{eff} into

$$-\frac{20k_F^2(\Delta T)^2\beta V}{21\zeta(3)c_F}F.$$

and an expression that depends on the matrix A with elements a_{ia} as well as on its hermitian conjugate A^+, its transpose A^T and its complex conjugate A^*:

$$F=-trAA^+ + \nu trA^+AP + (trA^+A)^2 + trAA^+AA^+ + trAA^+A^*A^T -$$
$$-trAA^TA^*A^+ - (1/2)trAA^TtrA^*A^+,$$

where

$$\nu = 7\zeta(3)\mu^2H^2/4\pi^2T_C\Delta T,$$

P is the projector on the third axis which is directed along the magnetic field.

We note that F is invariant to the transformation $A \to e^{i\alpha}UA$, where α is a real parameter and U is a real orthogonal matrix. Minimizing F, we obtain the matrix A that determines the density of the Bose–condensate. The equation $\delta F = 0$, i.e.

$$-A + \nu AP + 2(trA^+A)A + 2AA^+A + 2A^*A^TA - 2AA^TA^* - A^*trAA^T = 0$$

has several nontrivial solutions corresponding to different superfluid phases. We consider the following possibilities

$$A_1 = c_1P,\ \ A_2 = c_2P_2,\ \ \ A_3 = c_3'\,P + c_3''\,P_2,\ \ A_4 = c_4\hat{C}_4,\ \ A_5 = c_5\hat{C}_5,$$
$$A_6 = c_6'\,C_4 + c_6''\,C_5,\ A_7 = c_7\hat{C}_7.$$

Here P_2 is the projector on the two–dimensional subspace and is orthogonal to $P, \hat{C}_4, \hat{C}_5$ and $\hat{C}_7$ are the third–order matrices:

$$\hat{C}_4 = \begin{pmatrix} 1 & 0 & 0 \\ i & 0 & 0 \\ 0 & 0 & 0 \end{pmatrix}, \quad \hat{C}_5 = \begin{pmatrix} 0 & 0 & 0 \\ 0 & 0 & 0 \\ 1 & i & 0 \end{pmatrix}, \quad \hat{C}_6 = \begin{pmatrix} 0 & 0 & 1 \\ 0 & 0 & i \\ 0 & 0 & 0 \end{pmatrix},$$

$$\hat{C}_7 = \begin{pmatrix} 1 & i & 0 \\ i & -1 & 0 \\ 0 & 0 & 0 \end{pmatrix}.$$

The squares of the moduli of the coefficients $|c_i|^2$ as well as the corresponding values of F_i are given by

$$|c_1|^2 = \frac{1}{3}(1-\nu), \; |c_2|^2 = \frac{1}{4}, \; |c_3'|^2 = \frac{1}{5}(1-2\nu),$$

$$|c_3''|^2 = \frac{1}{10}(2+\nu), |c_4|^2 = \frac{1}{4},$$

$$|c_5|^2 = \frac{1}{12}, \; |c_6'|^2 = \frac{1}{8}(2-3\nu), \; |c_6''|^2 = \frac{\nu}{8}, \; |c_7|^2 = \frac{1}{16},$$

$$F_1 = -\frac{1}{6}(1-\nu)^2, \; F_2 = -\frac{1}{4}, F_3 = -\frac{1}{10}(3-2\nu+\nu^2), \; F_4 = -\frac{1}{4},$$

$$F_5 = -\frac{1}{12}, \; F_6 = -\frac{1}{8}(2-4\nu+4\nu^2), \; F_7 = -\frac{1}{8}.$$

At H=0 (ν=0) the minimal of all the variants $F=-0.3$ is obtained for $A_3=c_3I$. This is the *B*–phase. The next values of equal magnitude $F=-0.25$ yields c_2P_2 – planar 2D–phase and $c_4\hat{C}_4$ – the *A*–phase. For A_1–phase ($c_7\hat{C}_7$) $F=-0.125$. The remaining phases are energy wise unprofitable compared with the B, A, 2D and A_1–phases.

When the magnetic field is increased, the value of F for the 2D, *A*– and A_1–phases remain unchanged, while F for the *B* phase increases, the B–phase becoming "deformed". At $\nu=1/2$ we have $A_2=A_3$ and $F_2=F_3$. Moreover, at $\nu>1/2$ the solution A_2 is meaningless, since $|c_3'|^2=(1-2\nu)/5$ becomes negative. Thus, at $\nu=1/2$,

$$H_C^2=2\pi^2T_C\Delta T/7\zeta(3)\mu^2$$

a transition takes place from the *B*–phase into the planar *2D*–phase, whose energy, in the first–order approximation, does not depend on H at all. The continuity of the transition of A_3 into A_2 favors the assumption that this is a second–order phase transition.

The conclusion concerning the phase transition is confirmed by a calculation of the second variation of the F. In order for the phase to be stable to small perturbations, δ^2F must be non–negative. We present the expressions for δ^2F_i for the most interesting cases $i=2,3,4$

$$\delta^2F_2=(\nu-1/2)u_{33}^2+(\nu+1/2)v_{33}^2+\nu(u_{13}^2+u_{23}^2)+(\nu+2)(v_{13}^2+v_{23}^2)+$$
$$+(1/2)[3u_{11}^2+3u_{22}^2+2u_{11}u_{22}+(u_{12}+u_{12})^2]+$$
$$+(1/2)[3v_{12}^2+3v_{21}^2-3v_{12}v_{21}+(v_{11}-v_{22})^2],$$

$$\delta^2 F_3 = \frac{1}{5}(\nu+2)[3u_{11}^2 + 3u_{22}^2 + 2u_{11}u_{22} + (u_{12}+u_{12})^2 + u_{13}^2 + u_{13}^2] +$$
$$+\frac{2}{5}(1-2\nu)[u_{31}^2 + u_{32}^2 + 3u_{33}^2] + \frac{4}{5}\left[\frac{(1-2\nu)(\nu+2)}{2}\right]^{1/2} \times$$
$$\times\left(u_{11}u_{33} + u_{22}u_{33} + u_{13}u_{31} + u_{23}u_{32}\right) + \frac{1}{5}(4-3\nu)(v_{11}^2 + v_{22}^2) +$$
$$+\frac{1}{5}(8-\nu)(v_{12}^2 + v_{21}^2) + \frac{2}{5}(2+\nu)[v_{33}^2 - v_{11}v_{22} - v_{12}v_{21}] + \frac{1}{5}(8+9\nu)(v_{31}^2 + v_{32}^2)$$
$$-\frac{4}{5}\left[\frac{(1-2\nu)(\nu+2)}{2}\right]^{1/2}\left(v_{11}v_{33} + v_{22}v_{33} + v_{13}v_{31} + v_{23}v_{32}\right),$$

$$\delta^2 F_4 = 4\nu\left(u_{13}^2 + u_{23}^2 + u_{33}^2 + v_{13}^2 + v_{23}^2 + v_{33}^2\right) + 2[2(u_{11}+v_{21})^2 +$$
$$+(u_{11}-v_{21})^2 + (u_{12}-v_{22})^2 + (u_{13}-v_{23})^2 + (u_{21}+v_{11})^2 + 2(u_{22}-v_{12})^2 +$$
$$+(u_{22}+v_{12})^2 + 2(u_{23}-v_{13})^2 + (u_{23}+v_{13})^2].$$

Here, $u_{ia} = \mathrm{Re}\,\delta a_{ia}$, $v_{ia} = \mathrm{Im}\,\delta a_{ia}$.

At $\nu < 1/2$ the variation $\delta^2 F_3$ is non–negative and $\delta^2 F_2$ is of alternating sign, while at $\nu > 1/2$ the variation $\delta^2 F_2$ is non–negative and the solution A_3 has no meaning. The variation $\delta^2 F_4$ is non–negative at arbitrary ν.

The second variations $\delta^2 F_1$, $\delta^2 F_5$, $\delta^2 F_7$ turn out to be alternating sign at all ν. The corresponding phases (the one–dimensional $c_1 P$, the "conjugate with the *A*–phase" $c_5 \hat{C}_5$, and the A_1–phase $c_7 \hat{C}_7$,are destroyed in this model by small perturbations, and are therefore not realized. The expression for $\delta^2 F_6$ (as well as for $\delta^2 F$) turns out to be of alternating sign for all $\nu > 0$.

Thus, in the considered model system, the condition for stability against small perturbations is satisfied only for the *A*–phase (for any values of ν), for the *B*–phase (at $H \le H_C$), and for the 2*D*–phase (at $H \le H_C$).

The quadratic form $\delta^2 F_3$ with 18 variables u_{ia} and v_{ia} has at $\nu < 1/2$, four null eigenvectors; the form $\delta^2 F_2$ has, at $\nu > 1/2$, six null vectors; and the form $\delta^2 F_4$ has, at any ν, nine null vectors. This is due to the existence of four Goldstone (gd) modes in the *B*–phase, to the existence of six such modes in the *2D*–phase and nine in the *A*–phase. We note that the presence of at least four $\delta^2 F$ null vectors in the *B*–phase is the consequence of the symmetry of *F*, relative to the transformations with one parameter α, and with three parameters that define the orthogonal matrix *U*.

5.2. Kinetic–equation Method

From a historical point of view, the first method which was used to describe the collective excitations in the superfluid phases of ^{3}He was the kinetic–equation method (KE). Using in this monograph the functional integration method as the main one, we will nevertheless demonstrate briefly the principal futures of the kinctic–equation method, using for this the paper by Wolfle[9]. In normal Fermi–liquids the quasiparticle energy $\varepsilon_k(\boldsymbol{r},\ t)$ may be related to the quasiparticle occupation number $n_k(\boldsymbol{r},\ t)$ via Landau's interaction function $f_{kk'}$. Linearized in the deviation from homogeneity and Fourier transformed, this relation is given by

$$\delta\varepsilon_k(\boldsymbol{q},\ \omega) = \sum_{k'} f_{kk'} \delta n_k(\boldsymbol{q},\ \omega) + \delta\varepsilon^{ext} \tag{5.17}$$

Here, $\delta\varepsilon^{ext}$ is the energy gain of a bare particle in the applied external field.

The quasiparticle distribution function $\delta f_k(\boldsymbol{r},\ t)$ can in turn be related back to $\delta\varepsilon_k$ by making use of the Landau's transport equation

$$\omega n_k - \varepsilon_+ \delta n_k + \delta n_k \varepsilon_- + n_+ \delta\varepsilon_k - \delta\varepsilon_k n_- = I\{\delta n\}, \tag{5.18}$$

where $I\{\delta n\}$ is the collision integral, $n(\varepsilon)$ is the Fermi function, and subscripts $\pm$ denote $k \pm q/2$.

In the collisionless limit ($I\{\delta n\}$=0), solving eq. (5.18) for δn_k and substituting into eq. (5.17), one finds the following selfconsistency equation:

$$\delta\varepsilon_p - \delta\varepsilon_p^{ext} = \sum_k f_{pk} \frac{n_+ - n_-}{\omega - \varepsilon_+ + \varepsilon_-} \delta\varepsilon_k \tag{5.19}$$

The collective modes of the system are obtained as eigenmodes of eq. (5.19). They are eigenoscillations of the mean field acting on a quasiparticle as the result of its interaction with the surrounding quasiparticles.

In the superfluid state the kinetic–equation method has to be modified in two ways. First, the quasiparticle spectrum is altered and, secondly, there appear additional collective degrees of freedom associated with the quantum–mechanical coherence property of the pair correlated state, described by the gap parameter, or off–diagonal mean field $\Delta_k(\boldsymbol{r},\ t)$. The self–consistency equation for the deviation of the gap parameter from equilibrium

$$\delta\Delta_k(\boldsymbol{q},\omega) = \sum_{k'} g_{kk'} \delta g_{k'}(\boldsymbol{q},\omega) \tag{5.20}$$

is coupled to the corresponding self–consistency condition for the diagonal field, which still has the form of Eq. (5.17). Here $g_{kk'}$ is the pair interaction, and

$$\delta g_k(\boldsymbol{q},\omega) = \int dt \exp(i\omega t)\delta\langle c_{k-}(t)c_{k+}(t)\rangle$$

is the off–diagonal distribution function.

The unknown distribution functions δf_k and δg_k are related back to the mean fields by the matrix kinetic equation

$$\delta n - \varepsilon_+ \delta n + \delta n \varepsilon_- + n_+ \delta\varepsilon - \delta\varepsilon\, n_- = I\{\delta n\} \tag{5.21}$$

with

$$\varepsilon_k = \begin{pmatrix} \xi_k & \Delta_k \\ \Delta_k^+ & -\xi_k \end{pmatrix}, \qquad \xi_k = \frac{k^2}{2m^*} - \mu$$

and correspondingly for $\delta\varepsilon$, δn and n.

In the collisionless limit the matrix kinetic equation may be solved for δn in terms of $\delta\varepsilon$ and the result inserted into Eqs. (5.17) and (5.20), yielding

$$\begin{aligned}\delta\varepsilon_p - \delta\varepsilon_p^{ext} = \sum_k f_{pk}\Big\{ &\frac{\eta}{\omega-\eta}(1-\lambda_k)\delta\varepsilon_k - \frac{1}{2}\lambda_k\left(\delta\varepsilon_k - \delta\varepsilon_{k-}\right) + \\ &+\frac{1}{4}(\omega+\eta)\bar{\lambda}_k\left(\delta\Delta_k\Delta_k^+ - \Delta_k\delta\Delta_k^+\right)\Big\} + \\ &+\frac{1}{\omega}\left[\eta\theta_k' + \xi_k(\omega+\eta)\bar{\lambda}_k\right]\left(\delta\Delta_k\Delta_k^+ + \Delta_k\delta\Delta_k^+\right)\end{aligned} \tag{5.22}$$

$$\delta\Delta_p + \sum_k g_{pk}\theta_k\delta\Delta_k =$$

$$\frac{1}{2}\sum_k g_{pk}\{\Delta_k[(\omega+\eta)(1+2\xi_k/\omega)\bar{\lambda}_k + 2\eta\theta'_k/\omega]\delta\varepsilon_k +$$

$$-\lambda_k[\Delta_k\Delta_k^+\delta\Delta_k + \Delta_k\delta\Delta_k^+\Delta_k] - \frac{1}{2}(\omega^2-\eta^2)[1+2\xi_k/\omega\delta\Delta_k] + \frac{\eta^2}{\omega}\theta'_k\,\delta\Delta_k\},$$

(5.23)

where

$$E_k^2 = \xi_k^2 + |\Delta_k|^2,\ \eta = (v_k, q),\ \theta_k = \frac{1}{2E}\tanh\frac{E}{2T},\ v_k = k/m^*,$$

$$\lambda_k = -4|\Delta_k|^2(\omega^2\theta_k + \eta^2\xi_k\theta'_k)/D_k,\quad \bar{\lambda}_k = \lambda_k/|\Delta_k|^2,$$

with $D_k = \omega^2(\omega^2 - 4E_k^2) - \eta^2(\omega^2 - 4\xi_k^2)$.

For later use we define the energy averaged quantity $\lambda(\hat{k}) = \int d\xi\lambda_k$. In Eqs. (5.22) and (5.23) the small particle–hole asymmetric terms have been retained.

The collective modes of the order parameters are given by the eigensolutions of eq. (5.23), regarding the diagonal mean field $\delta\varepsilon_k$ as a fixed external field. Employing the vector notation

$$\Delta(\hat{k}) = iD_j(k)\sigma_j\sigma_2,\ \delta\Delta(\hat{k}) = id_j(k)\sigma_j\sigma_2,$$

and

$$\delta\varepsilon_k = \delta\varepsilon_k^0 + \delta\varepsilon_k^j\sigma_j$$

(bold face character denote 2 x 2 matrices in spin space), replacing $\delta\Delta_p$ on the l.h. s. of eq. (5.23) by $-\delta\Delta_p \sum_k g_{pk}\theta_k$ and assuming a pair interaction,

$$g_{kk'} = g_1 \frac{4\pi}{2l+1} \sum_{\mu} Y^*_{1\mu}(k) Y^*_{1\mu}(k')$$

with l odd, the following equation for general triplet order parameter d_j is obtained (particle–hole symmetry is assumed):

$$\begin{aligned} &\langle Y^*_{1\mu}(\hat{k})\{\overline{\lambda}(\hat{k})\left(\omega^2-\eta^2\right)d_j(\hat{k}) - 2\overline{\lambda}(\hat{k})[D_i(\hat{k})D^*_i(\hat{k})d_j(\hat{k}) - \\ &- D_i(\hat{k})D_i(\hat{k})d^*_j(\hat{k}) + 2D_j(\hat{k})D_i(\hat{k})d^*_j(\hat{k})] + 4\Xi(\hat{k})d_j(\hat{k})\}\rangle = \\ &= 2\langle Y^*_{1\mu}(\hat{k})\overline{\lambda}(\hat{k})\left(\omega+\eta\right)[D_j(\hat{k})\delta\varepsilon^0_k - i\varepsilon_{jlm}D_l(\hat{k})\delta\varepsilon^m_k]\rangle. \end{aligned} \qquad (5.24)$$

Here,

$$\Xi(\hat{k}) = \int d\varepsilon\theta_k - \int d\varepsilon\theta_k \left|D(\hat{k})\right|^2 / \langle \left|D(\hat{k})\right|^2 \rangle$$

and angular brackets $\langle ... \rangle$ denote the average over all directions in momentum space. In Eq. (5.24) d^*_j is the complex conjugate of $d_j(-q,-\omega)$.

We will use the Eq. (5.24) to obtain the collective excitation spectrum in the B– and A–phases of ^{3}He by the kinetic–equation method, following Wolfle[9] (see sections 6.5, 6.6 in the Chapter VI and the section 7.5 in the Chapter VII).

Chapter VI

Collective Excitations in the *B*-phase of ^{3}He

6.1. The Quadratic Form of Action Functional

We will now begin the investigation of the spectrum of collective excitations using the path integral technique[1]. The system in the region $T_C - T \propto T_C$ is described by the effective functional of action

$$S_{\text{eff}} = g^{-1} \sum_{p,i,a} c_{ia}^{+}(p) c_{ia}(p) + \frac{1}{2} \ln \det \frac{\hat{M}\left(c_{ia}, c_{ia}^{+}\right)}{\hat{M}\left(c_{ia}^{(0)}, c_{ia}^{(0)+}\right)}, \tag{6.1}$$

where $c_{ia}^{(0)}$ is the condensate value of Bose–fields c_{ia} and $\hat{M}\left(c_{ia}, c_{ia}^{+}\right)$ is the 4×4 matrix depending on Bose–fields and parameters of quasi–fermions.

$$M_{11} = Z^{-1}\left[i\omega + \xi - \mu(\mathbf{H\sigma})\right]\delta_{p_1 p_2}, \; M_{22} = Z^{-1}\left[-i\omega + \xi + \mu(\mathbf{H\sigma})\right]\delta_{p_1 p_2},$$

$$M_{12} = (\beta V)^{-1/2}(n_{1i} - n_{2i}) c_{ia}(p_1 + p_2)\sigma_a,$$

$$M_{21} = -(\beta V)^{-1/2}(n_{1i} - n_{2i}) c_{ia}^{+}(p_1 + p_2)\sigma_a.$$

We will start from calculation of the collective mode frequencies. To make this in the region $T_C - T \propto T_C$ one should expand the functional $\ln\det$ in powers of the deviations $c_{ia}(p)$ from the condensate value

$c_{ia}^{(0)}(p)$, which is different for different phases and in B–phase has a form

$$c_{ia}^{(0)}(p) = (\beta V)^{1/2} c\delta_{p0}\delta_{ia}\,. \tag{6.2}$$

So, we apply the shift

$$c_{ia}(p) \to c_{ia}^{(0)}(p) + c_{ia}(p)$$

and separate the quadratic form

$$\begin{aligned} &\sum_p A_{ijab}(p)c_{ia}^*(p)c_{jb}(p) + \\ &+\frac{1}{2}\sum_p B_{ijab}(p)\left[c_{ia}(p)c_{jb}(-p) + c_{ia}^*(p)c_{jb}^*(-p)\right] \end{aligned} \tag{6.3}$$

from S_{eff}. The quadratic form in new variables is equal to

$$\frac{1}{g}\sum_{p,i,a} c_{ia}^+(p)c_{ia}(p) - \frac{1}{4}Sp(\widehat{G}\widehat{u})^2\ , \tag{6.4}$$

where

$$u_{p1p2} = \begin{pmatrix} 0 & (n_{1i} - n_{2i})\sigma_a c_{ia}(p_1 + p_2) \\ -(n_{1i} - n_{2i})\sigma_a c_{ia}^+(p_1 + p_2) & 0 \end{pmatrix},$$

$$G^{-1}=\begin{pmatrix} Z^{-1}(i\omega-\xi)\delta_{p1p2} & 2c(\vec{n}\vec{\sigma})\delta_{p1+p2,0} \\ -2c(\vec{n}\vec{\sigma})\delta_{p1+p2,0} & Z^{-1}(-i\omega+\xi)\delta_{p1p2} \end{pmatrix}. \qquad (6.5)$$

Revised G^{-1}, one gets

$$G=\frac{Z}{M}\begin{pmatrix} -(i\omega+\xi)\delta_{p1p2} & \Delta(\vec{n}\vec{\sigma})\delta_{p1+p2,0} \\ -\Delta(\vec{n}\vec{\sigma})\delta_{p1+p2,0} & (i\omega+\xi)\delta_{p1p2} \end{pmatrix}, \qquad (6.6)$$

where

$$M=\omega^2+\xi^2+\Delta^2, \quad \Delta=2cZ.$$

From (6.5) and (6.6) it follows

$$\begin{aligned}&\frac{1}{4}Sp(\widehat{G}\widehat{u})^2=\frac{1}{4}\sum_{p1,p2,p3,p4} tr\,(G_{p1p2}u_{p2p3}G_{p3p4}u_{p4p1})=\\&\frac{Z^2}{4\beta V}\sum_p\{c^+_{ia}(p)c_{jb}(p)2\sum_{p1+p3=p}(M_1M_3)^{-1}(i\omega_1+\xi_1)(i\omega_3+\xi_3)\times\\&\times tr(\sigma_a\sigma_b)(n_1-n_3)_i(-n_1+n_3)_j+4\Delta^2[c_{ia}(p)c_{jb}(-p)+c^+_{ia}(p)c^+_{jb}(-p)]\times\\&\times\sum_{p3-p1=p}(M_1M_3)^{-1}(n_1+n_3)_i(n_1+n_3)_j tr(\vec{n}_1\vec{\sigma})\sigma_a tr(\vec{n}_3\vec{\sigma})\sigma_b\}.\end{aligned} \qquad (6.7)$$

Here, tr means trace with respect to matrix elements. Considering small p, let us make replacement $n_3\to -n_1$, if $p_1+p_3\to p$ and $n_3\to n_1$, if $p_3-p_1\to p$. Making also the replacement $p_1\to -p_1$ in sums with $p_3 - p_1 \to p$ and taking the trace, one gets the quadratic part of S_{eff} in the following form

$$S_{eff} \approx \sum_p c^+_{ia}(p) c_{ja}(p) \left[\frac{\delta_{ij}}{g} + \frac{4}{\beta V} \sum_{p1+p2=p} n_{1i} n_{1j} \varepsilon(-p_1) \varepsilon(-p_2) G(p_1) G(p_2) \right] -$$
$$-\frac{1}{2} \sum_p [c_{ia}(p) c_{jb}(-p) + c^+_{ia}(p) c^+_{jb}(-p)] \times$$
$$\times \frac{4\Delta^2}{\beta V} \sum_{p1+p2=p} n_{1i} n_{1j} (2 n_{1a} n_{1b} - \delta_{ab}) G(p_1) G(p_2) \ , \tag{6.8}$$

where

$$\varepsilon(p) = i\omega - \xi, \quad G(p) = Z(\omega^2 + \xi^2 + \Delta^2)^{-1} .$$

After substitutions

$$c_{ia}(p) = u_{ia}(p) + i v_{ia}(p), \ c^+_{ia}(p) = u_{ia}(p) - i v_{ia}(p), \tag{6.9}$$

the quadratic form decays into two independent forms, one of which depends on $u_{ia}(p)$ and other one depends on $v_{ia}(p)$:

$$-\sum_p \left[A_{ia}(p) u_{ia}(p) u_{ja}(p) + B_{ijab}(p) u_{ia}(p) u_{jb}(p) \right] -$$
$$-\sum_p \left[A_{ia}(p) v_{ia}(p) v_{ja}(p) - B_{ijab}(p) v_{ia}(p) v_{jb}(p) \right] . \tag{6.10}$$

The coefficient tensors A_{ijab}, B_{ijab} are proportional to the integrals of the products of the fermions Green's functions. Most effective in the calculation of these integrals is the Feynman procedure customarily used in relativistic quantum theory. In the present case the procedure is based on the identity

$$(ab)^{-1} = \int_0^1 d\alpha [\alpha a + (1-\alpha)b]^{-2}, \tag{6.11}$$

where $a = \omega_1^2 + \xi_1^2 + \Delta^2$, $b = \omega_2^2 + \xi_2^2 + \Delta^2$.

By this procedure it is easy to evaluate the integrals with respect to the variables ω_1 and ξ_1 and then integrate with respect to the angle variables and the parameter α . For the coefficient tensor A_{ijab} one gets the following expression

$$\begin{aligned} A_{ij}(p) = & -\frac{4Z^2}{\beta V} \sum_{p1+p2=p} n_{1i} n_{1j} \times \\ & \times \int_0^1 d\alpha \left[\frac{(\xi_1 + i\omega_1)(\xi_2 + i\omega_2)}{[\alpha(\omega_1^2 + \xi_1^2) + (1-\alpha)(\omega_2^2 + \xi_2^2) + \Delta^2]^2} - \frac{1}{\omega_1^2 + \xi_1^2 + \Delta^2} \right] \end{aligned} \tag{6.12}$$

eliminating the term $\delta_{ij} g_0^{-1}$ with the aid of the identity

$$\delta_{ij} g_0^{-1} + \frac{4Z^2}{\beta V} \sum_{p1} n_{1i} n_{1j} \frac{1}{\omega_1^2 + \xi_1^2 + \Delta^2} = 0,$$

which determines the gap Δ.

Considering the limit $T \to 0$, we replace the summation in Eq. (6.12) by the integration near the Fermi sphere, in accordance with the rule

$$(\beta V)^{-1}\sum_{p1} \rightarrow k_F^2(2\pi)^{-4}c_F^{-1}\int d\omega_1 d\xi_1 d\Omega_1 , \tag{6.13}$$

where $\int d\Omega_1$ is an integral with respect to the angle variables. We then calculate directly the integrals with respect to ω_1 and ξ_1. We arrive at the formula

$$A_{ij}(p) = \frac{Z^2 k_F^2}{4\pi^3 c_F}\int_0^1 d\alpha \int d\Omega_1 n_{1i} n_{1j}\left[\ln\left(1+\frac{\alpha(1-\alpha)q^2}{\Delta^2}\right)+\frac{\Delta^2+2\alpha(1-\alpha)q^2}{\Delta^2+\alpha(1-\alpha)q^2}\right], \tag{6.14}$$

$$q^2 = \omega^2 + c_F^2(\vec{n}_1\,\vec{k})^2 .$$

A similar procedure yields for $B_{ijab}(p)$

$$B_{ijab}(p) = \frac{Z^2 k_F^2}{4\pi^3 c_F}\int_0^1 d\alpha \int d\Omega_1 n_{1i} n_{1j}(2n_{1i}n_{1b}-\delta_{ab})\frac{\Delta^2}{\Delta^2+\alpha(1-\alpha)q^2}, \tag{6.15}$$

Our aim being an investigation of the nonphonon spectrum, we consider $A_{ij}(p)$, $B_{ijab}(p)$ at small p, but for nonzero ω. Let us consider first the case of $\boldsymbol{k}=0$.

6.2. The Collective Mode Frequencies

To calculate the collective mode frequencies we set $\boldsymbol{k} = 0$ in (6.14) and (6.15). Then the integrals with respect to the angle variables and with respect to the parameter α separate and can be easily calculated:

$$\int d\Omega_1 n_{1i} n_{1j} = \frac{4\pi}{3}\delta_{ij}, \tag{6.16}$$

$$\int d\Omega_1 n_{1i} n_{1j} (2n_{1i} n_{1b} - \delta_{ab}) = \frac{4\pi}{15}(2\delta_{ai}\delta_{bj} + 2\delta_{aj}\delta_{bi} - 3\delta_{ab}\delta_{ij}),$$

$$\int_0^1 d\alpha \left[\ln\left(1 + \frac{\alpha(1-\alpha)\omega^2}{\Delta^2}\right) + \frac{\Delta^2 + 2\alpha(1-\alpha)\omega^2}{\Delta^2 + \alpha(1-\alpha)\omega^2} \right] = f(\omega), \tag{6.17}$$

$$\int_0^1 d\alpha \frac{\Delta^2}{\Delta^2 + \alpha(1-\alpha)\omega^2} = g(\omega),$$

where

$$f(\omega) = (\omega^2 + 2\Delta^2)h(\omega),\ g(\omega) - 2\Delta^2 h(\omega),$$

$$h(\omega) = \frac{\omega}{\omega\sqrt{\omega^2 + 4\Delta^2}} \ln \frac{\sqrt{\omega^2 + 4\Delta^2} + \omega}{\sqrt{\omega^2 + 4\Delta^2} - \omega}. \tag{6.18}$$

Substituting (6.16)–(6.18) in (6.14) and (6.15) we obtain the first approximation formulas for $A_{ij}(p)$, $B_{ijab}(p)$:

$$A_{ij}(p) = \frac{Z^2 k_F^2}{3\pi^2 c_F} \delta_{ij} f(\omega),$$

$$B_{ijab}(p) = \frac{Z^2 k_F^2}{15\pi^2 c_F} (2\delta_{ai}\delta_{bj} + 2\delta_{aj}\delta_{bi} - 3\delta_{ab}\delta_{ij}) g(\omega).$$

Substituting them into (6.10), we arrive at the quadratic form

$$\begin{aligned} &-\frac{Z^2 k_F^2}{15\pi^2 c_F} \sum_p \{5 f(\omega)(u_{ia}^2 + v_{ia}^2) + \\ &+ g(\omega)[2(u_{ia}u_{ai} - v_{ia}v_{ai}) + 2(u_{aa}u_{ii} - v_{aa}v_{ii}) - 3(u_{ia}^2 - v_{ia}^2)]\}. \end{aligned} \tag{6.19}$$

From this equation we can easily obtain the frequencies of all 14 nonphonon modes.

The expression under the summation with respect to p breaks up into several independent forms of the variables

$$(u_{12}, u_{21}),\ (u_{23}, u_{32}),\ (u_{31}, u_{13}),\ (u_{11}, u_{22}, u_{33}),$$

$$(v_{12}, v_{21}),\ (v_{23}, v_{32}),\ (v_{31}, v_{13}),\ (v_{11}, v_{22}, v_{33}).$$

The form with the variables (u_{12}, u_{21}) is given by

$$[5 f(\omega) - 3 g(\omega)](u_{12}^2 + u_{21}^2) + 4 g(\omega) u_{12}\, u_{21}. \tag{6.20}$$

Equating its determinant to zero, we get

$$\begin{vmatrix} 5f(\omega)-3g(\omega) & 2g(\omega) \\ 2g(\omega) & 5f(\omega)-3g(\omega) \end{vmatrix} = 5(f(\omega)-g(\omega))(5f(\omega)-g(\omega)) = 0 \tag{6.21}$$

If the first factor in (6.21) is equal to zero, we arrive at the equation $\omega^2 h\ (\omega)=0$, which yields the *gd*–mode $E^2=0$. The vanishing of the second factor in (6.21) leads to the equation $(5\omega^2+8\Delta^2)h\ (\omega)=0$ and yields the *rsq*–mode $E^2=8\Delta^2/5$. The forms with the variables (u_{23},u_{32}), (u_{31},u_{13}) have at $\boldsymbol{k}$ = 0 the same coefficients as (6.21). They yield two more *gd*–modes $E^2=0$ and two *rsq*–modes $E^2=8\Delta^2/5$.

We consider now the form with variables (u_{11},u_{22},u_{33}),

$$[5f(\omega)+g(\omega)](u_{11}^2+u_{22}^2+u_{33}^2)+4g(\omega)(u_{11}\ u_{22}+u_{22}\ u_{33}+u_{11}\ u_{33})\ .$$

Equating its determinant to zero, we obtain

$$\begin{vmatrix} 5f(\omega)+g(\omega) & 2g(\omega) & 2g(\omega) \\ 2g(\omega) & 5f(\omega)+g(\omega) & 2g(\omega) \\ 2g(\omega) & 2g(\omega) & 5f(\omega)+g(\omega) \end{vmatrix} = \tag{6.22}$$
$$=[5f(\omega)-g(\omega)]^2 5\ [f(\omega)+g(\omega)]=0.$$

The equation $[5f(\omega)-g(\omega)]^2=0$ gives two more *rsq*–modes, while equation $f(\omega)+g(\omega)=0$ is equivalent to equation $(\omega^2+4\Delta^2)h(\omega)=0$ and gives the *pb*–mode $E^2=4\Delta^2$.

The *u*–variables give thus five *rsq*–modes $E^2=8\Delta^2/5$, three *gd*–modes $E^2=0$ and one *pb*–mode $E^2=4\Delta^2$.

Let us now examine the *v*–branches. The forms with the variables (v_{12},v_{21}), (v_{23},v_{32}) and (v_{31},v_{13}), are the same at $\boldsymbol{k}$ = 0. The coefficients of these forms differ from the corresponding coefficients of the u–forms by the substitution $g(\omega)\to -g(\omega)$. Therefore the counterpart of Eq. (6.21) is of the form

$$5[f(\omega)+g(\omega)]^2[5f(\omega)+g(\omega)]=0. \tag{6.23}$$

The equation $f(\omega)+g(\omega)=0$ yields three *pb*–modes $E^2=4\Delta^2$ (one for each of three forms), while equation $5f(\omega)+g(\omega)=0$ is equivalent to $(5\omega^2+12\Delta^2)h(\omega)=0$ and yields the three squashing modes $E^2=12\Delta^2/5$.

The equation corresponding to the form with variables (v_{11},v_{22},v_{33}) is obtained from equation (6.22) by making the substitution $g(\omega)\to -g(\omega)$ and is given by

$$[5f(\omega)+g(\omega)]^2 5\ [f(\omega)-g(\omega)]=0. \tag{6.24}$$

The equation $[5f(\omega)+g(\omega)]^2=0$ gives two more *sq*–modes $E^2=12\Delta^2/5$, and the equation $[f(\omega)-g(\omega)]=0$ gives one more *gd*–modes $E^2=0$.

The *v*–variables thus yield five *sq*–modes $E^2=12\Delta^2/5$, one *gd*–mode $E^2=0$ and three *pb*–modes $E^2=4\Delta^2$.

We write down all the obtained collective mode frequencies together with their corresponding variables:

four *gd*–modes ($E^2=0$): $(u_{12}-u_{21})$, $(u_{23}-u_{32})$, $(u_{31}-u_{13})$, $(v_{11}+v_{22}+v_{33})$;

four *pb*– modes ($E^2=4\Delta^2$): $(v_{12}-v_{21})$, $(v_{23}-v_{32})$, $(v_{31}-v_{13})$, $(v_{11}+v_{22}+v_{33})$;

five *rsq*–modes ($E^2 = 8\Delta^2/5$): $(u_{12}+u_{21})$, $(u_{23}+u_{32})$, $(u_{31}+u_{13})$, $(u_{11}-u_{22})$, $(u_{11}+u_{22}-2u_{33})$;

five *sq*–modes ($E^2 = 12\Delta^2/5$): $(v_{12}+v_{21})$, $(v_{23}+v_{32})$, $(v_{31}+v_{13})$, $(v_{11}-v_{22})$, $(v_{11}+v_{22}-2v_{33})$.

We note, that the *gd*–modes and the *pb*–modes are "dual" in the sense, that the variables corresponding to them differ by the substitution $u_{ia} \leftrightarrow v_{ia}$. In this sense the *sq*–modes and the *rsq*–modes are "dual" too. We note also that the sum of the squares of the frequencies of "dual" modes is equal to $4\Delta^2$.

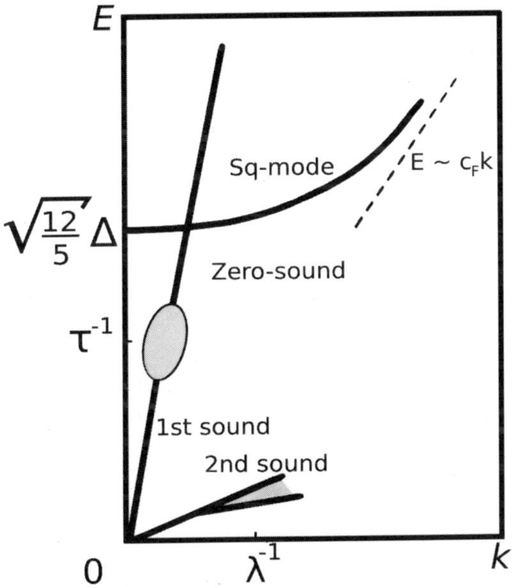

FIG. 6.1. Coupling of the squashing–mode with zero–sound in ^{3}He–*B*.

6.3. Dispersion Corrections to the Collective Mode Spectrum

6.3.1. Dispersion laws for *rsq*– and *sq*–modes

We now calculate the nonphonon branches of the spectrum with corrections of order k^2. The most labor–consuming task is the analysis of the *pb*–modes, in which the coefficients of k^2 are complex. We consider therefore first *sq*– and *rsq*–modes, in which corrections to the spectrum can be obtained by expanding the coefficient tensors $A_{ij}(p)$, $B_{ijab}(p)$ in powers of k^2, confining ourselves to terms $\propto k^2$.

We have

$$\int_0^1 d\alpha \left[\ln\left(1+\frac{\alpha(1-\alpha)q^2}{\Delta^2}\right) + \frac{\Delta^2+2\alpha(1-\alpha)q^2}{\Delta^2+\alpha(1-\alpha)q^2} \right] = ,$$

$$= f(\omega) + \frac{df(\omega)}{d\omega^2} c_F^2 (\vec{n}_1\vec{k})^2 + O(k^4)$$

$$\int_0^1 d\alpha \frac{\Delta^2}{\Delta^2+\alpha(1-\alpha)q^2} = g(\omega) + \frac{dg(\omega)}{d\omega^2} c_F^2 (\vec{n}_1\vec{k})^2 + O(k^4). \tag{6.25}$$

We can thus obtain the corrections to the quadratic form (6.19). Calculating the integrals, corresponding to these corrections, with respect to the angle variables,

$$\int_0^1 d\Omega_1 n_{1i} n_{1j} c_F^2 (\vec{n}_1\vec{k})^2 = \frac{4\pi}{15}\left(k^2\delta_{ij} + 2k_i k_j\right),$$

$$\int_0^1 d\Omega_1 (2n_{1a}n_{1b} - \delta_{ab}) n_{1i} n_{1j} (\vec{n}_1\vec{k})^2 = \frac{4\pi}{105}[k^2\left(2\delta_{ai}\delta_{bj} + 2\delta_{aj}\delta_{bi} - 5\delta_{ab}\delta_{ij}\right) +$$

$$+ 4k_a k_b \delta_{ij} - 10 k_i k_j \delta_{ab} + 2k_a k_i \delta_{bj} + 2k_b k_j \delta_{ai} + 2k_a k_j \delta_{bi} + 2k_b k_i \delta_{aj}]$$

we write the corrections to (6.19) in the form

$$-\frac{Z^2k_F^2c_F}{15\pi^2}\sum_p\{\frac{df(\omega)}{d\omega^2}[k^2(u_{ia}^2+v_{ia}^2)+2k_ik_j(u_{ia}u_{ja}+v_{ia}v_{ja})]+$$

$$+\frac{1}{7}\frac{dg(\omega)}{d\omega^2}[k^2(2(u_{aa}u_{bb}-v_{aa}v_{bb})+2(u_{ia}u_{ai}-v_{ia}v_{ai})-$$

$$-5(u_{ia}^2-v_{ia}^2))+k_ik_j(4(u_{ai}u_{aj}-v_{ai}v_{aj})-10(u_{ia}u_{ja}-v_{ia}v_{ja})+ \tag{6.26}$$

$$+8(u_{ij}u_{aa}-v_{ij}v_{aa})+8(u_{ia}u_{aj}-v_{ia}v_{aj}))]\} \; .$$

Since the B–phase is isotropic, it is sufficient to consider excitations that propagate in an arbitrary direction, say along the third axis. Making the substitutions $k_1=k_2=0$ and $k_3=k$, the corrections (6.26) breaks upinto a sum of forms with the sum of variables as the main form (6.19). Adding (6.19) and (6.26), we obtain (at $k_1=k_2=0$ and $k_3=k$) under $\sum_p$ sign, the sum of the following form

$$(w_{12}^2+w_{21}^2)\left[5f(\omega)\mp 3g(\omega)+c_F^2k^2\left(\frac{df(\omega)}{d\omega^2}\mp\frac{5}{7}\frac{dg(\omega)}{d\omega^2}\right)\right]\pm$$

$$\pm 4w_{12}w_{21}\left[g(\omega)+\frac{1}{7}c_F^2k^2\frac{dg(\omega)}{d\omega^2}\right],$$

$$w_{31}^2\left[5f(\omega)\mp 3g(\omega)+c_F^2k^2\left(3\frac{df(\omega)}{d\omega^2}\mp\frac{15}{7}\frac{dg(\omega)}{d\omega^2}\right)\right]\pm$$
$$\pm 4w_{31}w_{13}\left[g(\omega)+\frac{3}{7}c_F^2k^2\frac{dg(\omega)}{d\omega^2}\right]+$$
$$+w_{13}^2\left[5f(\omega)\mp 3g(\omega)+c_F^2k^2\left(\frac{df(\omega)}{d\omega^2}\mp\frac{1}{7}\frac{dg(\omega)}{d\omega^2}\right)\right],$$

$$(w_{11}^2+w_{22}^2)\left[5f(\omega)\pm g(\omega)+c_F^2k^2\left(\frac{df(\omega)}{d\omega^2}\mp\frac{1}{7}\frac{dg(\omega)}{d\omega^2}\right)\right]\pm$$
$$\pm 4w_{11}w_{22}\left[g(\omega)+\frac{1}{7}c_F^2k^2\frac{dg(\omega)}{d\omega^2}\right],$$
$$w_{33}^2\left[5f(\omega)\pm g(\omega)+c_F^2k^2\left(3\frac{df(\omega)}{d\omega^2}\pm\frac{9}{7}\frac{dg(\omega)}{d\omega^2}\right)\right]\pm$$
$$\pm 4w_{33}(w_{11}+w_{22})\left[g(\omega)+\frac{3}{7}c_F^2k^2\frac{dg(\omega)}{d\omega^2}\right],$$

$$w_{31}^2\left[5f(\omega)\mp 3g(\omega)+c_F^2k^2\left(3\frac{df(\omega)}{d\omega^2}\mp\frac{15}{7}\frac{dg(\omega)}{d\omega^2}\right)\right]\pm 4w_{31}w_{13}\left[g(\omega)+\frac{3}{7}c_F^2k^2\frac{dg(\omega)}{d\omega^2}\right]+$$
$$w_{13}^2\left[5f(\omega)\mp 3g(\omega)+c_F^2k^2\left(\frac{df(\omega)}{d\omega^2}\mp\frac{1}{7}\frac{dg(\omega)}{d\omega^2}\right)\right],$$
$$(w_{11}^2+w_{22}^2)\left[5f(\omega)\pm g(\omega)+c_F^2k^2\left(\frac{df(\omega)}{d\omega^2}\mp\frac{1}{7}\frac{dg(\omega)}{d\omega^2}\right)\right]\pm$$
$$\pm 4w_{11}w_{22}\left[g(\omega)+\frac{1}{7}c_F^2k^2\frac{dg(\omega)}{d\omega^2}\right]+ \tag{6.27}$$
$$+w_{33}^2\left[5f(\omega)\pm g(\omega)+c_F^2k^2\left(3\frac{df(\omega)}{d\omega^2}\pm\frac{9}{7}\frac{dg(\omega)}{d\omega^2}\right)\right]\pm$$
$$4w_{33}(w_{11}+w_{22})\left[g(\omega)+\frac{3}{7}c_F^2k^2\frac{dg(\omega)}{d\omega^2}\right].$$

In the $\pm$ or $\mp$ signs the upper sign corresponds to $w_{ia} = u_{ia}$, and the lower sign corresponds to $w_{ia} = v_{ia}$. The forms with the variables w_{32}, w_{23} are obtained from the second form of (6.27) by making the substitution $(w_{31}, w_{13}) \to (w_{32}, w_{23})$.

It is easy to obtain from (6.27) all the collective modes of spectrum (except *pb*–modes) with corrections of order k^2. Consider, for example, the first of the forms (6.27). Letting $w_{12} = u_{12}, w_{21} = u_{21}$ (taking the upper signs in $\pm$, $\mp$), we equate to zero the determinant of the form

$$5\left[f(\omega) - g(\omega) + \frac{c_F^2 k^2}{5}\frac{d}{d\omega^2}\big(f(\omega) - g(\omega)\big)\right]\times$$
$$\times\left[5f(\omega) - g(\omega) + c_F^2 k^2 \frac{d}{d\omega^2}\left(f(\omega) - \frac{3}{7}g(\omega)\right)\right] = 0. \tag{6.28}$$

If we equate the first factor to zero, we obtain

$$\omega^2 h(\omega) + \frac{c_F^2 k^2}{5}\frac{d}{d\omega^2}\left[\omega^2 h(\omega)\right] = 0,$$

or

$$\omega^2 + \frac{c_F^2 k^2}{5} + \frac{\omega^2 c_F^2 k^2}{5}\frac{d\ln h(\omega)}{d\omega^2} = 0.$$

The last term here is of higher order, since $d\ln h(\omega)/d\omega^2$ is finite as $\omega\to 0$. As a result we obtain the *gd*–mode (the longitudinal spin waves):

$$E^2=\frac{c_F^2k^2}{5}. \tag{6.29}$$

All the remaining *gd*–modes can also be obtained in the approach described here when considering the forms (u_{13},u_{31}), (u_{23},u_{32}) and (v_{11},v_{22},v_{33}).

Equating the second factor in (6.28) to zero, we obtain the equation

$$(5\omega^2+8\Delta^2)h(\omega)+\frac{c_F^2k^2}{7}\frac{d}{d\omega^2}[(5\omega^2+8\Delta^2)h(\omega)]=0,$$

which yields the spectrum branch

$$E^2=\frac{8\Delta^2}{5}+\frac{c_F^2k^2}{5}\left(1-\frac{16\Delta^2}{35}\frac{d\ln h(\omega)}{d\omega^2}\right)\Bigg|_{\omega^2=-\frac{8\Delta^2}{5}}. \tag{6.30}$$

Using the formula

$$\frac{d[\ln h(\omega)]}{d\omega^2}=-\left(2\omega^2\right)^{-1}-\left[2(\omega^2+4\Delta^2)\right]^{-1}-\left[\omega\sqrt{\omega^2+4\Delta^2}\ln\frac{\sqrt{\omega^2+4\Delta^2}+\omega}{\sqrt{\omega^2+4\Delta^2}-\omega}\right]^{-1}$$

and substituting $\omega=i\Delta\sqrt{8/5}$, we obtain

$$d\ln h(\omega)/d\omega^2 = (5/48\,\Delta^2)(1-2\sqrt{6}/arctg\,2\sqrt{6}).$$

This leads to the spectrum branch corresponding to the variable $u_{12}+u_{21}$

$$E^2 = \frac{8\Delta^2}{5} + \frac{c_F^2 k^2}{105}\left(20 + \frac{2\sqrt{6}}{arctg\,2\sqrt{6}}\right). \tag{6.31}$$

Similar calculations are easily made for the remaining modes, with the exception of the *pb*–ones. The simplest way of obtaining the sought formulas is to substitute in the quadratic forms (6.27) value of w_{ia} such that only the variables of interest remains, and then equate the forms to zero. For example, to obtain the *gd*–mode (6.29) we substitute in the first of the forms of (6.27) $w_{12}=u_{12}=u_{21}$ while in the investigation of the nonphonon mode we substitute $w_{12}=u_{12}=-u_{21}$.

We write down the collective mode spectrum together with their corresponding variables. These are the four *gd*–modes:

$$E^2 = \frac{c_F^2 k^2}{5},\ u_{12}-u_{21};$$

$$E^2 = \frac{2c_F^2 k^2}{5},\ u_{13}-u_{31}, u_{23}-u_{32}; \tag{6.32}$$

$$E^2 = \frac{c_F^2 k^2}{3},\ v_{11}+v_{22}+v_{33}.$$

the five *rsq*–modes

$$E^2 = \frac{8\Delta^2}{5} + \frac{c_F^2 k^2}{105}\left(20 + \frac{2\sqrt{6}}{arctg\,2\sqrt{6}}\right), \quad u_{12} + u_{21},\ u_{11} - u_{22},$$

$$E^2 = \frac{8\Delta^2}{5} + \frac{c_F^2 k^2}{210}\left(85 - \frac{2\sqrt{6}}{arctg\,2\sqrt{6}}\right), \quad u_{13} + u_{31},\ u_{23} + u_{32},$$

$$E^2 = \frac{8\Delta^2}{5} + \frac{c_F^2 k^2}{105}\left(50 - \frac{2\sqrt{6}}{arctg\,2\sqrt{6}}\right), \quad u_{11} + u_{22} - 2u_{33}. \tag{6.33}$$

the five *sq*–modes

$$E^2 = \frac{12\Delta^2}{5} + \frac{c_F^2 k^2}{105}\left(20 - \frac{2\sqrt{6}}{\pi - arctg\,2\sqrt{6}}\right), \quad v_{12} + v_{21},\ v_{11} - v_{22},$$

$$E^2 = \frac{12\Delta^2}{5} + \frac{c_F^2 k^2}{210}\left(85 + \frac{2\sqrt{6}}{\pi - arctg\,2\sqrt{6}}\right), \quad v_{13} + v_{31},\ v_{23} + v_{32},$$

$$E^2 = \frac{8\Delta^2}{5} + \frac{c_F^2 k^2}{105}\left(50 + \frac{2\sqrt{6}}{\pi - arctg\,2\sqrt{6}}\right), \quad v_{11} + v_{22} - 2v_{33}. \tag{6.34}$$

6.3.2. Dispersion induced splitting of the *rsq*– and *sq*–mode

The above results show a three–fold dispersion induced splitting of the *sq*– and *rsq*–modes. In 1982, Shivaram et al.[2] (see also Brusov et al.[3]) observed the splitting of the *rsq*–mode experimentally.

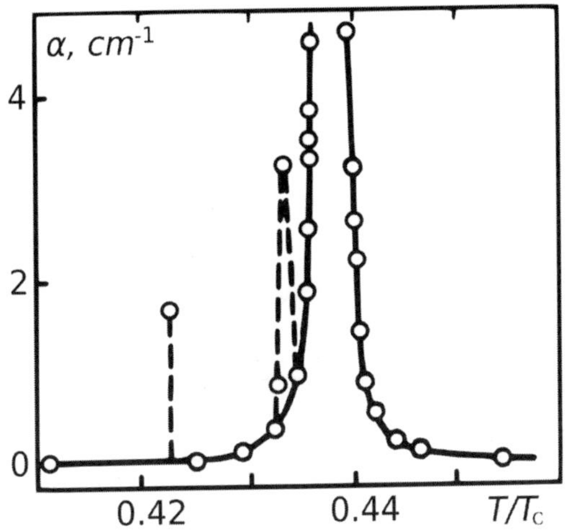

FIG. 6.2. Dispersion induced splitting of the real squashing (*rsq*)–mode.

The ratio of the distances between the branches of *rsq*–modes with $|J_Z|=1,2$ and $J_Z=0$ in theory (1:4) (Brusov and Popov[1]) and in experiments (1:3.8) are in good agreement. It would seem that one could also observe the same effect for the *sq*–mode using the same technique, but the sound absorption into the *sq*–mode is much larger than into the *rsq*–mode and it is difficult to observe the splitting. One of the ways this could be done, would be to decrease the volume of the superfluid ^{3}He (and to decrease the sound absorption into the *sq*–mode). Unfortunately, it is impossible to use for this purpose the method of measuring the group velocity of sound, which has been used by Movshovich et al.[4] for observation of the 4–fold splitting of the *sq*–mode in magnetic field (and later the 5–fold splitting), because it determines the collective mode frequencies at zero momentum. It is however possible to use this method to observe the *rsq*– and *sq*–modes splitting in the presence of electric fields or superflow (see below).

Zhao et al.[5] at Northwestern University could observe two–fold (but not three–fold) splitting of *sq*–mode, using the acoustic impedance technique. This technique turned out to be especially suitable in the case of high absorption.

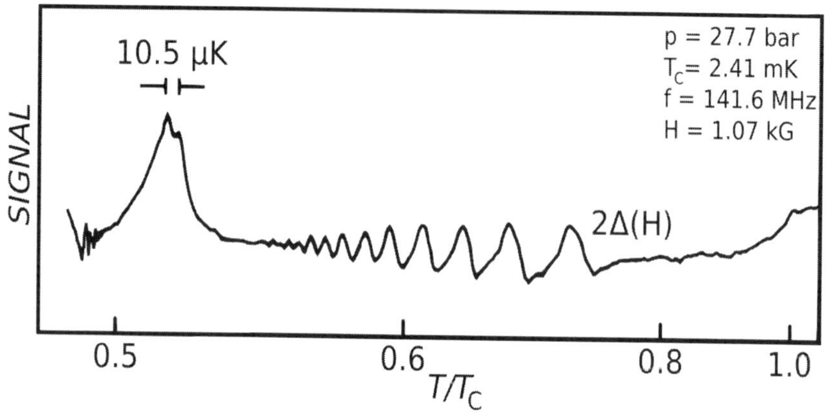

FIG. 6.3. Typical demagnetization traces of the acoustic impedance signal: P=27.7 bars, f=141.6 MHz, H=1.07 kG. The traces are taken as a function of time with the approximate temperatures as shown.

There are three possible explanations of this effect: 1) dispersion induced splitting of *sq*–mode; 2) texture induced splitting of the $J_Z = 0$ branch of *sq*–mode and 3) the existence of other phase near the boundary. In the first case, a three–fold splitting should be observed as predicted by Brusov and Popov[1]. However, the splitting between the $|J_Z| = 1$ and $|J_Z| = 2$ branches are predicted to be only one–quarter of the splitting between the $J_Z = 0$ and $|J_Z| = 2$ branches, and they could not be resolved in this experiment.

The experimentally observed splitting is about four times as large as that predicted by BCS theory (Brusov and Popov[1]). The same situation happened for *rsq*–mode[2,3], and the discrepancy was greatly reduced when the Fermi–liquid corrections (FLC) were taken into account.

In the case of *sq*–mode, the problem is more complicated and the detailed temperature dependence of *sq*–mode frequency should be studied with higher Fermi–liquid correction parameters besides the first non–zero F_2^S.

In the case of two–fold splitting, the texture induced splitting of the $J_Z = 0$ branch of the *sq*–mode or the existence of other phase near the boundary are more favorable explanations. Because of the textures created by the restricted geometry in zero field, a phenomena similar to the texture induced doublet splitting for the central peak of *sq*–mode in a magnetic field observed[2] could occur. But the theory of texture induced splitting of *sq*–mode still needs to be developed. Another possibility is that the additional peak appeared from the collective mode in the 2*D*–phase near the boundary[6]. One peak would appear from the *sq*–mode in *B*–phase and another from the superflapping (*sfl*) mode in the 2*D*–phase. The spectrum of the collective modes in 2*D*–phase have been calculated by Brusov et al.[7] The results showed that part of the spectrum in the 2*D*–phase was the same as in the *A*–phase (e.g., we have *cl*–mode and *pb*–mode). It is also known that super–flapping–mode[8] (see section 5.2) only appear in the our model[1], while the strong coupling effect was taken into account. Estimations showed that the difference between the *sq*–mode in the 2*D*–phase and the superflapping–mode in the 2D phase is of the order of a few tens of μK at $T/T_C = 0.7$. This is in good agreement with the experimental data.

6.4. The Pair–breaking–mode Dispersion Low

We have obtained all the Bose–spectrum branches (except for the *pb*–modes) by expanding the coefficients of the tensors $A_{ij}(p)$, $B_{ijab}(p)$ at small k (6.25). This procedure, however, cannot be used for branches with $\Omega = 2\Delta$, since the function $h(\omega)$ (6.18), and also the functions $f(\omega)$ and $g(\omega)$, have a singularity $\propto \left(\omega^2 + 4\Delta^2\right)^{-1/2}$ at $\omega^2 \to -4\Delta^2$. Therefore the *pb*–modes call for a special investigation.

We start with the branch corresponding to the variable $u_{11} + u_{22} + u_{33}$. We separate the terms corresponding to this variable from the quadratic form (6.8), putting $u_{ia}(p) = c(p)\delta_{ia}, v_{ia} = 0$.

In this case

$$\delta_{ia}\delta_{ja}n_{1i}n_{1j} = \delta_{ia}\delta_{jb}\left(2n_{1a}n_{1b} - \delta_{ab}\right) n_{1i}n_{1j} = 1$$

and (6.8) is transformed into

$$\sum_{p} c^2(p)A(p,\ u_{11} + u_{22} + u_{33}),$$

Where

$$A(p, u_{11} + u_{22} + u_{33}) = \\ = \frac{4Z^2}{\beta V}\sum_{p1+p2=p}\left[\frac{(\xi_1 + i\omega_1)(\xi_2 + i\omega_2) - \Delta^2}{(\omega_1^2 + \xi_1^2 + \Delta^2)(\omega_2^2 + \xi_2^2 + \Delta^2)} - \frac{1}{\omega_1^2 + \xi_1^2 + \Delta^2}\right] \tag{6.35}$$

Using the Feynman procedure to calculate $A(p,\ u_{11} + u_{22} + u_{33})$ and integrating with respect to ω_1 and ξ_1, we obtain

$$A(p, u_{11} + u_{22} + u_{33}) = \frac{Z^2 k_F^2}{4\pi^3 c_F}\int d\Omega_1 \int_0^1 d\alpha\left[\ln\frac{\Delta^2}{\Delta^2 + \alpha(1-\alpha)q^2} - 2\right] \tag{6.36}$$

We obtain the coefficient functions corresponding to the variables $v_{12} - v_{21}$, $v_{13} - v_{31}$, $v_{23} - v_{32}$ in a similar fashion. They can be written in the form

$$A(p, v_{ik} - v_{ki}) = \frac{Z^2 k_F^2}{4\pi^3 c_F}\int d\Omega_1 \left(n_{1i}^2 + n_{1k}^2\right)\int_0^1 d\alpha\left[\ln\frac{\Delta^2}{\Delta^2 + \alpha(1-\alpha)q^2} - 2\right] \tag{6.37}$$

Evaluating integrals with respect to parameter α and angle variables, one has

$$\int_0^1 d\alpha\left[\ln\frac{\Delta^2}{\Delta^2+\alpha(1-\alpha)q^2}-2\right]=-\frac{2a}{b}arctg\frac{b}{a}=-\frac{a}{b}\left(\pi-2arctg\frac{a}{b}\right),$$

(6.38)

where

$$a^2=\Delta^2+q^2/4,\quad b^2=-q^2/4,\quad q^2=\omega^2+c_F^2\left(\vec{n}_1\vec{k}\right)^2,$$

As $\omega^2\to-4\Delta^2$ the quantity b^2 is positive and close to Δ^2. As will be shown below $a^2=O(k^2)$, so that $2arctg\dfrac{a}{b}<<\pi$ and in first–order approximation

$$\int_0^1 d\alpha\left[\ln\frac{\Delta^2}{\Delta^2+\alpha(1-\alpha)q^2}-2\right]=-\frac{a}{b}\left(\pi-2arctg\frac{a}{b}\right)\approx-\frac{\pi a}{\Delta}. \qquad (6.39)$$

This leads to a first order equation for the dispersion of the branch

$$\int a\,d\Omega_1=0. \qquad (6.40)$$

We direct $\boldsymbol{k}$ along the third axis and denote $\cos\theta_1=x$. Then $(\boldsymbol{n}_1,\boldsymbol{k})=kx$ and (6.40) becomes

$$\int_{-1}^{1} dx\ \left(4\Delta^2 - E^2 + c_F^2 k^2 x^2\right)^{1/2} = 0\,,$$

Substituting

$$z^2 = (4\Delta^2 - E^2)/c_F^2 k^2\,, \tag{6.41}$$

we obtain the equation

$$\int_{-1}^{1} dx\left(x^2 + z^2\right)^{1/2} = 0\,,$$

or

$$\left(1+z^2\right)^{1/2} + z^2 \ln\frac{1+\left(1+z^2\right)^{1/2}}{z} = 0\,. \tag{6.42}$$

Also substituting

$$t = 2\ln\frac{1+\left(1+z^2\right)^{1/2}}{z}\,, \tag{6.43}$$

we can rewrite (6.42) in the simple form

$$t + sht = 0\,. \tag{6.44}$$

If t is a nontrivial root of this equation, we obtain, substituting

$$z = [sh(t/2)]^{-1}$$

in

$$E^2 = 4\Delta^2 - c_F^2 k^2 sh^{-2}(t/2). \tag{6.45}$$

Equation (6.44) and the dispersion law (6.45) turn out to be identical to the single nonphonon branch of the spectrum in the Fermi–gas model with scalar point interaction, investigated by Andrianov and Popov[9]. It is indicated there that a physical meaning can be possessed by the branch (6.25), the first to appear upon analytic continuation with respect to the variable E from the upper half plane to the unphysical sheet. This branch is obtained if t is replaced by the smallest (in absolute value) nontrivial $(\neq 0)$ root of (6.44), which is equal to

$$t_1 \approx 2.251 + i \cdot 4.212. \tag{6.46}$$

The obtained branch was called "resonant excitation". It corresponds to the pole of the Bose–fields $c_{ia}(p)$ Green's function which is located near the point $E^2 = 4\Delta^2$.

We can threat similarly the remaining three *pb*–modes. They correspond to the equations

$$\int_{-1}^{1} dx \left(x^2 + z^2\right)^{1/2} \left(1 \mp x^2\right) = 0, \tag{6.47}$$

or

$$\mp 2\left(1+z^2\right)^{3/2} + \left(4 \mp z^2\right)^{1/2} \left[\left(1+z^2\right)^{1/2} + z^2 \ln \frac{z + \left(1+z^2\right)^{1/2}}{z}\right] = 0, \tag{6.48}$$

in which it is necessary to take the minus sign for the variable $v_{12} - v_{21}$ and the plus sign for $v_{13} - v_{31}$ and $v_{23} - v_{32}$. Changing the variable t (6.43), we obtain in place of (6.48)

$$\frac{cht - 2}{2cht - 1} sht + t = 0, v_{12} - v_{21}, \tag{6.49}$$

$$\frac{3cht - 2}{2cht - 3} sht + t = 0, v_{13} - v_{31}, v_{23} - v_{32}. \tag{6.50}$$

As a result, the dispersion laws for all *pb*–modes are given by formula (6.45), where t is one of the nontrivial roots of Eqs. (6.44) for the branch $u_{11} + u_{22} + u_{33}$, (6.49) for $v_{12} - v_{21}$ and (6.50) for $v_{13} - v_{31}, v_{23} - v_{32}$. The branches with direct physical meaning are those appearing first in the analytic continuation from the physical sheet. For Eq. (6.49), the sought nontrivial solution with minimum modulus is

$$t_2 \approx 2.93 + i \cdot 4.22 \tag{6.51}$$

and for Eq. (6.50) the solution is

$$t_2 \approx 1.94 + i \cdot 4.12 \tag{6.52}$$

The coefficients of $c_F^2 k^2$ are complex for all the *pb*–modes as already noted above. This is due to the possibility of the Bose–excitation, to decay into two fermions. The functional integration method makes it possible to calculate all collective modes at small momenta k, accurate to terms $\propto k^2$ (see Table 6.1). The results at $k = 0$ (the frequencies of the collective modes) coincide with those obtained by Vdovin[10], Nagai[11], and Wolfle[8] (see below) by essentially different methods. All 18 collective modes with dispersion correction were first obtained by Brusov and Popov[1]. Vdovin[10] and Nagai[11] obtained dispersion laws for 14 modes and Wolfle[8] – for 9 modes.

Table 6.1

J, J_Z, i or r	variables	Ω^2	α / c_F^2	Coupling to zero sound	Coupling to electromagnetic field
$[0;0;i]$	$v_{11}+v_{22}+v_{33}$	0	1/3	1	0
$[1;0;i]$	$v_{12}-v_{21}$	$4\Delta^2$	$0.111-i0.169$	strong*	weak
$[1;\pm1;i]$	$v_{13}-v_{31}$ $v_{23}-v_{32}$	$4\Delta^2$	$0.353-i0.335$	via gap distortion	weak
$[2;0;i]$	$v_{11}+v_{22}-2v_{33}$	$12\Delta^2/5$	0.502	strong	0
$[2;\pm1;i]$	$v_{13}+v_{31}$, $v_{23}+v_{32}$	$12\Delta^2/5$	0.418	strong	0
$[2;\pm2;i]$	$v_{12}+v_{21}$, $v_{11}-v_{22}$	$12\Delta^2/5$	0.164	strong	0
$[0;0;r]$	$u_{11}+u_{22}+u_{33}$	$4\Delta^2$	$0.237-i0.295$	0	0
$[1;0;r]$	$u_{12}-u_{21}$	0	1/5	0	strong
$[1;\pm1;r]$	$u_{13}-u_{31}, u_{23}-u_{32}$	0	2/5	weak*	strong
$[2;0;r]$	$u_{11}+u_{22}-2u_{33}$	$8\Delta^2/5$	0.442	weak	0
$[2;\pm1;r]$	$u_{13}+u_{31}$ $u_{23}+u_{32}$	$8\Delta^2/5$	0.388	weak	0
$[2;\pm2;r]$	$u_{12}+u_{21}$, $u_{11}-u_{22}$	$8\Delta^2/5$	0.224	weak	0

*Strong" means the coupling exists without particle–hole asymmetry

*"Weak" means the coupling exists via particle–hole asymmetry

6.5. Collective Mode Spectrum Calculated by the Kinetic Equation Method

1) Case of $q=0$

Let us consider the Eq. (5.24) from Chapter V (Wolfle[8]). Introducing the nine complex order parameter components $d_{j\alpha}$ by $d_j = \sum_\alpha d_{j\alpha}\hat{k}_\alpha$ and decoupling d_j and d_j^* by taking the sum and difference of Eq. (5.24) and its complex conjugate, the following equations for the real (u_j) and imaginary (v_j) parts of d_j,

$$\begin{Bmatrix} u_j \\ v_j \end{Bmatrix} = \frac{1}{2}\left[d_j(\boldsymbol{q},\omega) \pm d_j^*(-\boldsymbol{q},-\omega)\right]$$ are found in the limit $q \to 0$:

$$\left(\omega^2 - 4\Delta^2\right)v_{j\alpha} + \frac{4}{5}\Delta^2\left[\delta_{j\alpha}v_{\beta\beta} + v_{j\alpha} + v_{\alpha j}\right] = 2\omega\delta_{j\alpha}\Delta\left\langle \delta\varepsilon_k^0 \right\rangle \tag{6.53}$$

$$\omega^2 u_j - \frac{4}{5}\Delta^2\left[\delta_{j\alpha}u_{\beta\beta} + 2u_{j\alpha}\right] = 2i\omega\varepsilon_{j\alpha m}\Delta\left\langle \delta\varepsilon_k^m \right\rangle \tag{6.54}$$

The quantity Ξ vanishes here due to the isotropy of the quasiparticle energy. λ is isotropic for $q \to 0$ and has dropped out of Eqs. (6.53)–(6.54). Since Eqs. (6.53)–(6.54) are linear in ω^2, there is exactly one eigenmode per variable. Wolfle[8] has obtained from Eqs. (6.53)–(6.54) at $q=0$ the same CE spectrum, which we have obtained above: 4 *gd*–, 4 *pb*–, 5 *sq*– and 5 *rsq*–modes.

2) Case of $q \neq 0$

Wolfle[8] considered the q^2 corrections to CE spectrum but he could calculate only 9 modes in this approximation. He considered the generalization of Eqs. (6.53)–(6.54) to nonzero wave vector. It is given by

$$\sum_{l,\beta} A'_{j\alpha,l\beta} u_{l\beta} = = -2i\omega\Delta\varepsilon_{jlm}\left\langle \hat{k}_\alpha \hat{k}_l (\omega+\eta)\delta\varepsilon_k^m \right\rangle \tag{6.55}$$

This is the dispersion relation of the longitudinal (with respect to $R\hat{q}$, R being the rotation of the orbital against spin variables caused by the dipole interaction) spin wave. In the limit $T \to 0$ and assuming $\omega, c_F q << \Delta$ one has $\lambda(\hat{k},\hat{q}) = 1 + O(q^2)$ and the (unrenormalized) spin wave velocity is $c_{||} = c_F (1/5)^{1/2}$.

The zeros of $\det A'_{13}$ are found as

$$\omega^2 = \frac{8\Delta^2}{5} + c_F^2 q^2 \left(\frac{2}{5} + \frac{4}{25}\Delta^2 \lambda' / \lambda \right) + O\left(q^4\right), \tag{6.56}$$

And

$$\omega^2 = \frac{\lambda_{10} + \lambda_{20}}{\lambda_{10} + \lambda_{00}} c_F^2 q^2 \to \frac{2}{5} c_F^2 q^2, \tag{6.57}$$

with Eq. (6.57) describing the spectrum of transverse spin waves. It is interesting to note that in the limit of large q the *rsq*–modes merge with the respective spin–wave branch to which they couple.

A similar analysis for the imaginary components of $d_{j\alpha}$ yields

$$\omega^2 = \frac{12\Delta^2}{5} + c_F^2 q^2 \left(\frac{1}{5} + \frac{16}{175} \Delta^2 \lambda' / \lambda \right),$$

$$\det A''_{\alpha\alpha} = 0 \quad \to \quad \omega^2 = \frac{12\Delta^2}{5} + c_F^2 q^2 \left(\frac{7}{15} - \frac{16}{175} \Delta^2 \lambda' / \lambda \right),$$

$$\omega^2 = \frac{\lambda_{10}}{\lambda_{00}} c_F^2 q^2 \to \frac{1}{3} c_F^2 q^2,$$

$T \to 0$

$$\det A''_{12} = 0 \quad \to \quad \begin{aligned} \omega^2 &= \frac{12}{5}\Delta^2 + O(q^2) \\ \omega^2 &= 4\Delta^2 + O(q^2) \end{aligned}$$

$$\det A''_{13} = 0 \quad \to \quad \begin{aligned} \omega^2 &= \frac{12}{5}\Delta^2 + O(q^2) \\ \omega^2 &= 4\Delta^2 + O(q^2) \end{aligned} \tag{6.58}$$

6.6. Fermi–Liquid Corrections

The quasiparticle interaction $f_{kk'}$ gives rise to renormalization of the collective frequencies obtained in the last section[8]. This is determined by the coupling of the order parameter equation (5.23) (from Chapter V) to the diagonal mean field equation (5.22). Again, a general analysis of the coupled equations is quite complicated. It is instructive to consider the simpler case of sound propagation at $T = 0$.

In this limit $\lambda = 1$, and Eq. (5.22) takes the form

$$\delta\varepsilon_p = \sum_k f_{pk}\left[-\delta\varepsilon_k + \frac{\omega+\eta}{2\Delta} v_{j\alpha}\hat{k}_j\hat{k}_\alpha\right]. \tag{6.59}$$

Inserting the solution of the order–parameter equation,

$$\sum_\alpha v_{\alpha\alpha} = \frac{2\omega\Delta}{\omega^2 - \frac{1}{3}c_F^2 q^2}\langle(\omega+\eta)\delta\varepsilon_k\rangle.$$

into Eq. (6.59), one derives the renormalized sound dispersion relation

$$(\omega/c_F q)^2 = \frac{1}{3}\left(1+F_0^S\right)\left(1+F_1^S/3\right), \tag{6.60}$$

where F_1^S are dimensionless, spin–symmetric Landau parameters.

Similarly, one finds the renormalized spin wave spectra

$$(\omega/c_F q)^2 = \frac{1}{5}\left(1+\frac{2}{3}F_0^a\right)\times\begin{cases}1 & \text{longitudinal}\\ 2 & \text{transverse}\end{cases}, \tag{6.61}$$

where we have dropped F_1^a.

The frequencies of the nonphonon modes are only affected by F_2 and higher Landau parameters. For example, for *sq*–mode one has

$$\omega^2 = \frac{12\Delta^2}{5}\left(1+\frac{2}{25}\lambda F_2^S\right), \tag{6.62}$$

which, assuming that $F_2^S \approx 0.5$ and noting that $\lambda \leq 1$, cause a frequency shifts of less than 4%.

6.7. Textural Effects on the Squashing Modes

In this section we discuss the influence of textures on the collective mode spectrum in B–phase, following the paper by Halperin and Varoquaux[12] and by Brusov et al.[13] In the B–phase, the texture of the order parameter is specified by a unit vector $\boldsymbol{n} = (n_1, n_2, n_3)$ such that the free energy associated with the presence of walls, magnetic field, superflow, and competition between these perturbations of the order parameter is minimized. The origin of these textures comes from the dipolar interaction which provides a coupling between spin and orbit space through a rotation *R*. The $\boldsymbol{n}$–vector is defined as the axis of this rotation. The angle of rotation, θ, is called the Leggett angle. For small perturbations $\theta = \arccos(-1/4)$. Otherwise, it depends on distortion of the energy gap as indicated by the equation

$$\Delta_1/\Delta_2 = -4\cos\theta. \tag{6.63}$$

Any rotation can be represented in terms of its axis $\boldsymbol{n}$ and angle θ in the form

$$R_{ij} = \delta_{ij}\cos\theta + n_i n_j(1-\cos\theta) + \varepsilon_{ijk} n_k. \tag{6.64}$$

This is particularly convenient for the discussion of B–phase textures. Those that are important for the collective modes generally involve the magnetic field. Considering only the effect of a magnetic field $\boldsymbol{H}$ far away from a wall, the quantization direction with respect to which the azimuthal quantum numbers m_j are defined, and which we call $\boldsymbol{h}$, is along the magnetic field. Close to a wall this quantization direction is rotated from $\boldsymbol{H}$, having components (summing over j)

$$h_i \equiv R_{ij} H_j / H \,. \tag{6.65}$$

There are two effects of the texture on the excitonic order–parameter collective modes that are detected in ultrasonic experiments. First, the coupling strength $\Lambda_{2\pm}(m_j, \varphi)$, between the collective modes and sound is determined by the texture through the angle φ,

$$\cos\varphi = \hat{q}_i R_{ij} \hat{H}_j \equiv \hat{\boldsymbol{q}} \cdot \hat{\boldsymbol{h}} \,. \tag{6.66}$$

The second effect is on the dispersion of the modes as was pointed out in Ref.14 and will be described later on.

Let us consider three regions of the liquid shown in Fig. 6.4: region I, near the wall, where there is competition between the magnetic field and the wall for orientation of the texture; region II in the bulk, a long distance from the wall, where the texture orientation is determined solely by the magnetic field; and region III where the texture accommodates to the constraints imposed on either side in regions I and II. For region I, Smith et al.[15] have shown that the texture is defined by minimizing the surface magnetic free energy,

$$F_H^S \equiv -\xi(\chi_N - \chi_B) \int (s_i R_{ij} H_j)^2 da \tag{6.67}$$

This requires maximizing the integrand with respect to the components of $\boldsymbol{n}$. Here, $\boldsymbol{s}$ is the unit vector normal to the surface. For the specific case in which the magnetic field is transverse to the direction of sound propagation, $\boldsymbol{s} \parallel \boldsymbol{q} \perp \boldsymbol{H}$, we use the notation $\boldsymbol{s}$ = (0,0,1), $\boldsymbol{H}$ = (0,0,0) and find,

$$n_1^2 = n_3^2 = \frac{\Delta_2/\Delta_1}{4+\Delta_2/\Delta_1}, \qquad n_2^2 = \frac{4-\Delta_2/\Delta_1}{4+\Delta_2/\Delta_1}, \quad h = (0,0,1), \tag{6.68}$$

whence $\varphi = 0$ and the quantization direction is clearly independent of Δ_2/Δ_1 and the magnetic field strength. In the bulk, region II, we find the components of $\boldsymbol{n}$ by minimizing the bulk magnetic free energy[15]

$$F_H^B \equiv -\int (n_j H_j)^2 dV \tag{6.69}$$

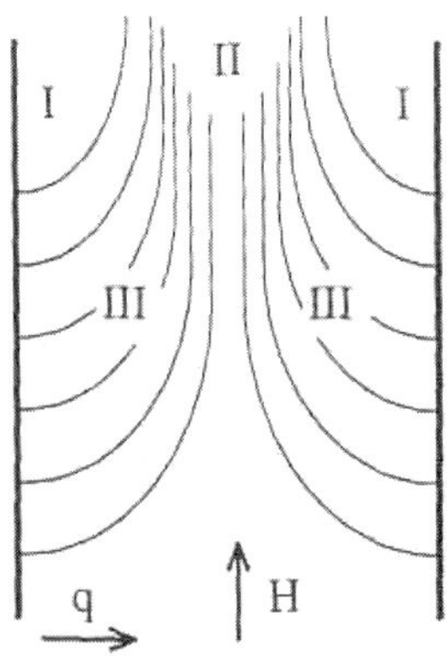

FIG. 6.4. Schematic of one possible textural distribution of the quantization axis $\boldsymbol{h}$ in the B–phase for transverse magnetic fields, $\boldsymbol{q} \perp \boldsymbol{H}$, showing three regions, I, II, and III, where the collective modes couple differently to sound as described by Eqs. (6.70)–(6.73).

The result is $\boldsymbol{n}$ = (1,0,0) and from Eqs. (6.64) and (6.65) we find $\boldsymbol{h}$ = (1,0,0) and then $\varphi = \pi/2$. These results are shown in Fig. 6.4. We can summarize their effect on the coupling strength (from Eq. (4.4) of Ref. 12) as follows

$$\begin{aligned}\Lambda_{2\pm}(m_j, \varphi = 0) &= \Lambda_{2\pm} \qquad m_j = 0 \\ &= 0 \qquad\qquad m_j = \pm 1 \qquad \text{for } \boldsymbol{s} \parallel \boldsymbol{q} \perp \boldsymbol{H}\text{, wall region I,} \\ &= 0 \qquad\qquad m_j = \pm 2\end{aligned} \tag{6.70}$$

$$\begin{aligned}\Lambda_{2\pm}\left(m_j, \varphi = \frac{\pi}{2}\right) &= \frac{1}{4}\Lambda_{2\pm} \qquad m_j = 0 \\ &= 0 \qquad\qquad m_j = \pm 1 \qquad \text{for } \boldsymbol{q} \perp \boldsymbol{H}\text{, bulk region II,} \\ &= \frac{3}{8}\Lambda_{2\pm} \qquad m_j = \pm 2\end{aligned} \tag{6.71}$$

$$\Lambda_{2\pm}(m_i, \varphi = 0) \neq 0 \qquad \text{all } \; m_i \qquad \text{for } \boldsymbol{q} \perp \boldsymbol{H}\text{, region III,} \tag{6.72}$$

For the case in which the magnetic field is aligned with the sound propagation direction, $\boldsymbol{q} \parallel \boldsymbol{H}$, there is a uniform texture with $\boldsymbol{n} = \boldsymbol{h} = \boldsymbol{s}$ throughout the liquid, where $\varphi = \arccos(\hat{\boldsymbol{q}} \cdot \hat{\boldsymbol{h}}) = 0$. The results for the coupling of collective modes to sound (from Eq. (4.4) of Ref. 12) are then

$$\begin{aligned}\Lambda_{2\pm}(m_j, \varphi=0) &= \frac{1}{4}\Lambda_{2\pm} \quad && m_j = 0 \\ &= 0 && m_j = \pm 1 \qquad \text{for, } \boldsymbol{s} \parallel \boldsymbol{q} \parallel \boldsymbol{H} \text{ all regions.} \\ &= 0 && m_j = \pm 2\end{aligned} \tag{6.73}$$

Finally, if the magnetic field is at some intermediate angle with respect to $\boldsymbol{q}$, between 0 and $\pi/2$ there will, in general, be coupling between sound and all of the collective modes, m_j = 0, + 1, ±2. The results for coupling strength, summarized in Eq. (6.70) and (6.73), take the effect of gap distortion into account through the magnetic field dependence of the Leggett angle. However, they do not allow for the possibility that a new superfluid state might be stable near a wall, owing to the competing depairing effect from a transverse magnetic field.

Some of the results listed above are consistent with experimental observations of both real and imaginary squashing modes. In fig. 4, all five modes in the real squashing mode multiplet were observed for the field oriented transverse to the sound propagation direction. Coupling to the m_j = ± 1 components in this experiment where the field is transverse to the sound propagation direction, reflects the existence of an inhomogeneous texture that we can assign to region III, Eq. (6.72). The squashing mode has been extensively studied in a transverse magnetic field by the acoustic impedance method which is sensitive only to the liquid in region I where the attenuation length is much shorter than the $\boldsymbol{n}$–texture bending length[16] $R_C \approx (10\ Oe/H)$ cm. For magnetic fields both parallel and transverse, there is only non–zero coupling to the $m_j = 0$ component in this region, consistent with observations[17].

The second effect of the texture on the excitation of collective modes is through the dispersion of the collective modes. This point has been discussed in Ref. 14, 18, 19. Recall that the dispersion, up to order q^2, is given in Eq. (4.9) relative to the measurement frequency ω as

$$\frac{\Delta\omega(m_j,q,0)}{\omega}=\frac{\omega(m_j,q,0)-\omega(0)}{\omega}=\left(a+bm_j^2\right)\frac{(v_F/c)^2}{(\Delta_0/\hbar\omega)}$$

where the coefficients a and b both depend on the texture,

$$a=a_0+a_1\left(\hat{q}\cdot\hat{h}\right)^2, \qquad b=b_0+b_1\left(\hat{q}\cdot\hat{h}\right)^2 \tag{6.74}$$

The texture dependence of the dispersion was originally presented in Ref. 14, 19 in a different notation,

$$\omega^2\left(m_j=0,q,0\right)=\omega^2(0)+\left[c_1^2+\frac{1}{3}c_2^2+c_2^2\left(\hat{q}\cdot\hat{h}\right)^2\right]q^2,$$

$$\omega^2\left(m_j=\pm1,q,0\right)=\omega^2(0)+\left[c_1^2+\frac{1}{2}c_2^2+\frac{1}{2}c_2^2\left(\hat{q}\cdot\hat{h}\right)^2\right]q^2,$$

$$\omega^2\left(m_j=\pm2,q,0\right)=\omega^2(0)+\left[c_1^2+c_2^2-c_2^2\left(\hat{q}\cdot\hat{h}\right)^2\right]q^2, \tag{6.75}$$

where the connection between the different notations is

$$a_0=\left[\frac{1}{2}(\Delta_0/\hbar\omega)\right]\left[c_1^2+\frac{1}{2}c_2^2\right];\ a_0=\left[\frac{1}{2}(\Delta_0/\hbar\omega)\right]c_2^2;$$

$$b_0=\left[\frac{1}{2}(\Delta_0/\hbar\omega)\right]\frac{1}{6}c_2^2;\ b_1=\left[\frac{1}{2}(\Delta_0/\hbar\omega)\right]\left(-\frac{1}{2}c_2^2\right). \tag{6.76}$$

According to the proposal of Volovik[14], the doublet splitting corresponds to the different frequencies of the $m_j=0$ real squashing mode in different textural environments. More specifically, consider the case of a magnetic field transverse to the sound propagation direction where the coupling strengths are given in Eqs. (6.70) and (6.71). In region I of Fig. 6.4, the $m_j = 0$ mode is the only mode that couples to sound Eq. (6.70). Since $\boldsymbol{h} \parallel \boldsymbol{q}$, its frequency is shifted from $\omega(0)$ by an amount proportional to $a_0 + a_1$. In the bulk, region II, the $m_j = 0$ and $m_j = \pm 2$ modes couple to sound. Since $\boldsymbol{h} \perp \boldsymbol{q}$, the $m_j = 0$ mode is frequency shifted from its value at $\omega(0)$ proportional to just a_0. The difference between these is the doublet splitting,

$$\frac{\Delta\omega_{doublet}}{\omega} = \frac{\omega(m_j = 0, q, 0)_{wall} - \omega(m_j = 0, q, 0)_{bulk}}{\omega} = a_1 \frac{(v_F/c)^2}{(\Delta_0/\hbar\omega)} \tag{6.77}$$

Furthermore, on the basis of symmetry arguments[14,18,19] it has been argued that this splitting is equal to the dispersion splitting in zero magnetic field,

$$\frac{\omega(m_j = \pm 1, q, 0) - \omega(m_j = \pm 2, q, 0)}{\omega} = -3b \frac{(v_F/c)^2}{(\Delta_0/\hbar\omega)}, \tag{6.78}$$

requiring the relationship $a_1/b = -3$. This has been verified in Ref. 20, even in the presence of quasi–particle interactions not included in the earlier work[19]. Direct measurements of the dispersion splitting at P=4.8 bar and $T/T_C = 0.45$ give $b = 0.025 \pm 0.006$. From the above relationship, $a_1/b = -3$, we calculate the doublet splitting to be 0.20 MHz at this temperature and pressure, in excellent agreement with the direct measurements of the doublet splitting given in fig. 22 for the same experimental conditions. This consistency is a dramatic demonstration of the success of the theory.

6.8. Coupling of Order–Parameter Collective Modes to Ultrasound

In experiments[21,22] on the propagation of ultrasound in ^{3}He–B in the temperature regime $0.35 < T/T_C < 0.95$, for frequencies ranging from 12 to 133 MHz in addition to the main attenuation peak, occurring approximately at $\omega = \sqrt{\frac{12}{5}}\Delta(T)$ (squashing mode) as predicted by weak–coupling theory, a second sharp peak is observed at lower temperatures. The area under this second peak is orders of magnitude smaller than the area of the main peak at a given frequency. Koch and Wolfle[23] pointed out that the new collective peak corresponds to a theoretically predicted mode at $\omega = \sqrt{\frac{8}{5}}\Delta(T)$ (real squashing mode), which couples only weakly to sound via particle–hole–symmetry–violating terms. Here we describe the coupling of the real squashing mode to ultrasound following to Ref. 23. It seems that derivation of this coupling strength is one of the most important achievement of the kinetic equation method.

The dynamics of superfluid ^{3}He appears to be reasonably well described by a time–dependent mean–field theory for diagonal and off–diagonal fields in particle–hole space,

$$\delta\varepsilon_k(\boldsymbol{q},t) = \sum_{k'} f_{kk'}\delta n_{k'}(\boldsymbol{q},t) \text{ and } \delta\Delta_k(\boldsymbol{q},t) = \sum_{k'} g_{kk'}\delta g_{k'}(\boldsymbol{q},t)$$

respectively.

Here, $\delta n_k(\boldsymbol{q},t) = \delta\langle c^+_{k_+}(t)c_{k_-}(t)\rangle$ and $\delta g_k(\boldsymbol{q},t) = \delta\langle c_{k_-}(t)c_{k_+}(t)\rangle$ are the diagonal and off–diagonal distribution functions $\boldsymbol{k}_\pm = \boldsymbol{k} \pm \frac{\boldsymbol{q}}{2}$ and $f_{kk'}$ and $g_{kk'}$ are the Fermi–liquid and pair interactions, respectively. Solving the equations of motion for δn_k and δg_k one finds the sound dispersion relation[24]

$$\omega^2 = c_1^2 q^2 \{1 - 2[(c_0 - c_1)/c_1]\xi\}. \tag{6.79}$$

Here, $c_1^2 = \frac{1}{3}(1+F_0^S)\left(1+\frac{1}{3}F_1^S\right)$, $(c_0 - c_1)/c_1 = \frac{2}{5}\left(1+\frac{1}{5}F_2^S\right)\Big/(1+F_0^S)$ and F_l^S are the spin–symmetric Landau parameters. The function ξ is given by $\xi = \frac{5}{2}\sum_{\hat{k}} P_2(\hat{\boldsymbol{k}}\cdot\hat{\boldsymbol{q}})\delta n_k/\delta n$, where P_2 is the l = 2 Legendre polynomial.

Explicitly,

$$\begin{aligned}\xi = \frac{5}{2}\sum_k P_2(\hat{\boldsymbol{k}}\cdot\hat{\boldsymbol{q}})\Bigg\{&\frac{\boldsymbol{v}\cdot\boldsymbol{q}}{\omega - \boldsymbol{v}\cdot\boldsymbol{q}}(1-\lambda_k)\delta\varepsilon_k + \\ &+\frac{\omega}{4|\Delta_k|^2}\lambda_k \frac{1}{2}Tr_\sigma\left[(\delta\Delta_k)\Delta_k^+ - \Delta_k(\delta\Delta_k^+)\right] + \\ &+\frac{\xi_k}{|\Delta_k|^2}\lambda_k \frac{1}{2}Tr_\sigma\left[(\delta\Delta_k)\Delta_k^+ + \Delta_k(\delta\Delta_k^+)\right]\Bigg\}\frac{1}{\delta n},\end{aligned} \tag{6.80}$$

where

$$\lambda_k = \frac{|\Delta_k|^2}{E_k^2 - \omega^2/4}\frac{1}{2E_k}\tanh\frac{E_k}{2T}, \tag{6.81}$$

$E_k = \left(\xi_k^2 + |\Delta_k|^2\right)^{1/2}$, $\xi_k = (k^2/2m^*) - \mu$, and $\boldsymbol{v} = \boldsymbol{k}/m^*$. For further references, we define $\lambda(\omega, T) = \Delta^2\bar{\lambda} = \int_{-\infty}^{\infty}\lambda_k d\varepsilon$.

We parametrize the nonequilibrium gap parameter by

$$\Delta_k = \Delta \sum_{j=1}^{3} i\sigma_j \sigma_2 \boldsymbol{d}_j(\hat{\boldsymbol{k}}),$$

where the σ_j are the Pauli matrices. For l=1 pairing, $d_j(\hat{\boldsymbol{k}}) = d_{ja}\hat{\boldsymbol{k}}_a$. The equilibrium gap parameter for ^{3}He–B may be taken as $d^0_{ja} = \delta_{ja}$ (without accounting of relative spin–orbit rotation). Performing the spin traces and angular integrations in (6.80), one finds that only the combination of d_{ja} given by $d'' = \frac{1}{2}\left(2d''_{zz} - d''_{xx} - d''_{yy}\right)$, where $d''_{ja} = \frac{1}{2}\left[d_{ja}(\boldsymbol{q},\omega) - d^*_{ja}(-\boldsymbol{q},-\omega)\right]$, couples strongly to the sound $\boldsymbol{q} \| \hat{z}$.

The last term in (6.80) couples to the order–parameter variable $d' = \frac{1}{2}\left(2d'_{zz} - d'_{xx} - d'_{yy}\right)$, where $d'_{ja} = \frac{1}{2}\left[d_{ja}(\boldsymbol{q},\omega) + d^*_{ja}(-\boldsymbol{q},-\omega)\right]$, but because of the extra factor ξ_k the energy integral would vanish in the case of exact particle–hole symmetry. This term has been neglected in previous theories. The relative importance of the last term in (6.80) may be expressed by the following dimensionless parameter characterizing the degree of particle–hole–symmetry violation,

$$\eta = \frac{1}{\Delta}\frac{N'(\varepsilon_F)}{N(\varepsilon_F)}\frac{1}{\lambda}\int_0^{\varepsilon_C} d\xi \xi^2 \lambda(\xi) \cong \frac{\Delta}{2\varepsilon_F}\ln\frac{\varepsilon_C}{\Delta} \quad \text{as } T \to 0. \tag{6.82}$$

Here, $N(\varepsilon)$ is the density of states and $\varepsilon_C \cong 0.1\varepsilon_F$ is a cutoff energy simulating the energy dependence of the interaction function $g_{kk'}$. Since very little is known about energy dependence of $N(\varepsilon)$ and $g_{kk'}$, let us restrict ourselves to the order of magnitude estimate (6.81), with the result $\eta \propto 3\times10^{-3}$ to $\eta \propto 5\times10^{-3}$ for pressures from 0 to 12 bar.

The order parameter components d_{ja} may be calculated from the time–dependent gap equation

$$d_j(\hat{p})+(4\pi)^{-1}\int d\Omega_k G(\hat{p}\cdot\hat{k})\left\{\frac{1}{2}\lambda\left[\omega^2+2\eta\omega\Delta-(\boldsymbol{v}\cdot\boldsymbol{q})^2(1+2\eta'\Delta/\omega)-\right.\right.$$
$$\left.-2\Delta^2\right]d_j(\hat{k})-G_1^{-1}d_j(\hat{k})+2\lambda\left[d_j^*(\hat{k})-2\hat{k}_i d_i^*(\hat{k})\hat{k}_j\right]\Big\}=$$
$$=(4\pi)^{-1}\int d\Omega_k G(\hat{p}\cdot\hat{k})\hat{k}_j[\omega+2\eta\Delta+(\boldsymbol{v}\cdot\boldsymbol{q})(1+2\eta'\Delta/\omega)]\delta\varepsilon(\hat{k})$$

(6.83)

Here, $G(\hat{p}\cdot\hat{k})=N(\varepsilon_F)g_{\boldsymbol{p}\cdot\boldsymbol{k}}=-\sum_l(2l+1)G_l P_l(\boldsymbol{p}\cdot\boldsymbol{k})$

is the dimensionless pair interaction and η' is another particle–hole–asymmetry parameter, defined by (6.83) with $\xi^2\lambda_k$ in the integrand replaced by $\{\xi^2\lambda_k+\Delta^2\xi(d/d\xi)[(1/2E)\tanh(E/2T)]\}$. For $T<<T_C$ one finds $\eta'\cong(\omega^2/4\Delta^2)(\Delta/2\varepsilon_F)<<\eta$. We neglect η' in the following. Solving (6.83) for d'' and d', keeping only the $l=1$ component of the pair interaction, expanding powers of $v_F q/\omega$ and η, and substituting the result into (6.80), one finds the quantity ξ in the sound dispersion relation (6.81) as

$$\xi=1-\lambda+\frac{2}{5}\frac{c_1^2q^2}{\omega^2-\frac{12}{5}\Delta^2-\frac{7}{15}(v_Fq)^2+2i\omega\gamma}+$$
$$+8\eta^2\lambda\frac{c_1^2q^2}{\omega^2-\frac{8}{5}\Delta^2-\frac{7}{15}(v_Fq)^2+2i\omega\gamma}. \quad (6.84)$$

Equations (6.84) and (6.79) imply that the sound attenuation $\alpha = -q[(c_0 - c_1)/c_1]\mathrm{Im}(\xi)$ (when $\alpha << g$) has two resonance peaks as a function of temperature, corresponding to oscillations of $d''_{ja}\left(\omega^2 = \frac{12}{5}\Delta^2\right)$ and $d'_{ja}\left(\omega^2 = \frac{8}{5}\Delta^2\right)$ of quadrupolar symmetry, $d_{ja} \propto (\delta_{ja} - 3\hat{q}_j\hat{q}_\alpha)$.

The motion may be visualized as a squashing of the order–parameter structure in the direction $\hat{q}$. The mode involving d'' has been called "squashing" mode (sq). It has been proposed the term "real squashing" mode (rsq) for the new mode with frequency $\left(\omega^2 = \frac{8}{5}\Delta^2\right)$, because it involves the real part of *d*.

In (6.84) the principal effect of quasiparticle collision processes has been included by adding an imaginary damping term $2i\omega\gamma$, to the denominators in (6.84). γ, is related to the energy–dependent quasiparticle lifetime $\tau(E)$ by[24]

$$\gamma = \frac{1}{\lambda}\int_{-\infty}^{\infty} d\xi \frac{1}{\tau(E)}\left(-\frac{df}{dE}\right)\frac{\Delta^2}{E^2 - \omega^2/4},$$

for $\omega\tau >> 1$. Here $f(E)$ is the Fermi–function, γ is essentially the quasiparticle relaxation rate times the relative density of thermal excitations.

It turns out that the damping of the collective modes decreases rather rapidly for $T \to 0$, thus giving rise to enormously sharp and high acoustic attenuation peaks. The structure described by (6.84) is observed experimentally.[21,22] However, the collective–mode frequencies at $T << T_C$ as determined from the peak positions, are found to be 10% (sq) and 20% (rsq) lower than the ideal values $\sqrt{\frac{12}{5}}\Delta(T)$ and

$\sqrt{\frac{8}{5}}\Delta(T)$, respectively, even when gap renormalization is taken into account. In the following, we discuss three effects which can cause such a frequency shift.

(i) *Fermi–liquid corrections.* — It has been shown previously[8] that the squashing–mode frequency is shifted by the Landau parameter F_2^S. in particular at lower *T*: $\omega_{sq}^2 = \frac{12}{5}\Delta^2\left[1+\frac{2}{25}\lambda F_2^S\right]$. A corresponding shift of the rsq–mode is negligible because of the smallness of the coupling parameter η^2. F_2^S may be inferred from the measured values of $(c_0 - c_1)/c_1$. Unfortunately, the evaluation depends crucially on the effective mass parameter m^*/m. Taking m^*/m from Wheatley's tables[25] one finds a small positive $F_2^S \cong 0.5$ whereas employing the recent m^*/m values of Alvesalo *et al.*,[26] which are considerably lower, one finds $F_2^S \cong -1$ at pressures from 0 to 10 bars.

(ii) *Strong–coupling effects.* — Both the squashing and real squashing modes are associated with minima in the static free energy with respect to order–parameter distortions of the appropriate symmetry. One may estimate[27] strong–coupling corrections of the collective–mode frequencies from the strong–coupling corrections of the parameters β_i in the Ginzburg–Landau free energy as

$$(\delta\omega_{sq}/\omega_{sq}) \propto \frac{1}{2}\delta\beta_1/\beta_1 \propto -2\% \text{ and } (\delta\omega_{rsq}/\omega_{rsq}) \propto$$

$$\propto \frac{1}{2}(\delta\beta_3 + \delta\beta_4 + \delta\beta_5)/(\beta_3 + \beta_4 + \beta_5) \propto -5\% \text{ with}$$

insertion of values for β_i appropriate for low pressure from the literature.[28]

(iii) *f–wave pair fluctuations.* — The collective–mode frequencies are also influenced by higher–angular–momentum fluctuations of the order parameter.[29] Solving the gap equation (6.83), and keeping only G_1 and G_3, one finds

$$\omega_1^2 = \Delta^2 \left\{ (2 - b_i) - \left[2 a_i b_i + (2 - b_i)^2 \right]^{1/2} \right\}$$

for i = sq, rsq where $b_i = \left(G_1^{-1} - G_3^{-1} \right) / \lambda(\omega_i), a_{sq} = \frac{12}{5}$ and $a_{rsq} = \frac{8}{5}$. The ω_i approach the weak–coupling values $a_i^{1/2}\Delta$ for $b_i \to -\infty$ as $\left(- b_i \right)^{-1}$, such that even relatively small G_3 can cause sizable shifts of ω_i. With an estimated value of $G_1 \cong 0.25$ at low (high) pressure (when a cutoff energy $\varepsilon_c \propto 50\ mK$ is assumed in the formula for T_C) a value of $G_3 \cong 0.1$ would give rise to a decrease of ω_i by 20%. Sauls and Serene[30] have discussed the possibility of new collective modes with frequencies less than 2Δ by coupling to l = 3 and l = 5 fluctuations. They have found a total–angular–momentum J=4 mode for values of G_3 and G_5 of the order of G_1. There exist many more over–damped modes in the pair–breaking continuum $\omega > 2\Delta$. Unfortunately we have little information on the value of G_3 at present.

The sharpness of the collective resonances at low temperatures gives rise to enormous sound attenuation peaks, which are difficult to measure experimentally. It may be more feasible, then, to determine the group velocity of sound, which varies dramatically near resonance.[21,22] From the dispersion relation (1) and (6), one finds

$$v_g = \frac{d\omega}{dq} = c_1^2 \frac{q}{\omega} \left[1 + 2\left(\frac{c_0 - c_1}{c_1} \right)(1 - \lambda) + 4 \sum_i B_i R_i \right].$$

$$\cdot\left[1+2\sum_i B_i R_i^2\left(1-\frac{4\omega_i^2\gamma^2}{\left(\omega^2-\omega_i^2\right)^2}\right)\right]^{-1}.$$

Here, $i = rs$, rsq and $B_{sq} = \frac{2}{5}\lambda\left(\frac{c_0 - c_1}{c_1}\right)$; $B_{rsq} = 20\eta^2 B_{sq}$,

and $R_i = c_1^2 q^2 \frac{\left(\omega^2-\omega_i^2\right)}{\left(\omega^2-\omega_i^2\right)^2+4\omega_i^2\gamma^2}$. For simplicity we have assumed $\mathrm{Im}(q)/\mathrm{Re}(q) << 1$.

We now briefly compare the theoretical strength and width of the new mode with experiment. From the area under the sound attenuation peak measured in Refs. 32, 21, and 22 at pressures of 3, 9, and 13 bars and sound frequencies of 45, 60, and 84 MHz, respectively, one determines coupling strengths $10^6 B_{rsq}^{\exp} = 2.7,\ 4.5,$ and 1.25, from which the particle–hole–asymmetry parameter η is inferred as $10^3\eta = 4.1,\ 4.7,$ and 3.5, in fair agreement with the estimate (4). The width of the attenuation peaks $\Delta T/T_C$ is found from the first two experiments as $3\cdot 10^{-3}$ and $4\cdot 10^{-3}$, which compare reasonably well with the collision–induced theoretical widths $3\cdot 10^{-3}$ and $6\cdot 10^{-3}$. Here we have used a normal–state relaxation time $\tau_N(0) = (0.6 - 0.01P)\mu\sec(mK)^2$, with P in bars, and the BCS gap function $\Delta(T)$.

At low temperatures, quasiparticle collision damping becomes very small. Another resonance broadening mechanism is introduced by dispersion. Determining this broadening self–consistently with the sound attenuation, one finds a line–width $\Delta\omega_i \propto \omega_i(v_F/c_1)B_i^{1/2}$, which should dominate the width of the rsq peak below $0.5T_C$.

In conclusion, we believe to have shown that the new collective mode may be identified as a $J = 2, J_Z = 0$ mode of the real part of the order parameter, which couples weakly to sound via particle–hole–

asymmetric terms. Tewordt and Schopohl[31] have predicted a linear splitting of this $J = 2$ multiplet in a magnetic field $\boldsymbol{H}$. For nonparallel orientations of $\boldsymbol{H}$ and $\boldsymbol{q}$, the $J_Z \neq 0$ modes couple to the sound as well, giving rise to a fivefold splitting of the sound attenuation peak. This has been observed recently by Avenel, Varoquaux, and Ebisawa.[32] The strong direct evidence in favor of a $J = 2$ mode makes any alternative explanation[30] employing $J = 4$ or $J = 6$ modes unlikely.

In ^{3}He–A the order–parameter collective modes are damped considerably by pair breaking, even at $T << T_C$. The two well–defined p–wave collective modes[2] (the clapping and flapping modes) couple to sound waves already in the case of exact particle–hole symmetry. There is indeed a new mode with frequency $\omega \cong 1.74\,\Delta_0(T)$ where Δ_0 is the maximum of the anisotropic gap (essentially oscillations of the magnitude of the equilibrium gap), which couples to sound only via particle–hole–asymmetric terms. However, this contributes to the sound attenuation $\alpha(T)$ only a broad peak of magnitude about 10^{-3} of the pair–breaking contribution and angular symmetry $P_2(\hat{q}\cdot\hat{l})$ ($\hat{l}$ is the axis of the gap).

Chapter VII

Collective Excitations in the *A*–Phase of ^{3}He

7.1. *A*–Phase of ^{3}He

The *A*–phase is perhaps the most interesting object in superfluid ^{3}He. It gives us an example of an anisotropic superfluid quantum liquid. The main features of the *A*–phase of ^{3}He are connected to the existence of two nodes in the gap of a single–particle spectrum on the Fermi–surface. This leads to existence of chiral fermions, gauge fields, analogs of *W*– and *Z*–bosons, zero–charge phenomenon, the damping of collective excitations (CE) at zero momentum, and to many other consequences for the system.

In this Chapter we construct a microscopic theory of collective excitations in the *A* phase of ^{3}He, investigate the influence of a magnetic field on collective excitations and calculate the dispersion correction to collective mode spectrum using the path–integration (PI) method[1,2].

The most popular method which is used to investigate the collective excitations in superfluid ^{3}He is the kinetic equation (KE) method. The main advantages of the alternative path–integral method over the kinetic equation is the increased accuracy in calculating the collective mode frequencies. For example, in the *B* phase of ^{3}He, the first collective–mode (CM) dispersion laws for the whole spectrum have been calculated by Brusov and Popov. The investigation of the stability of the Goldstone modes, which requires a calculation of the corrections of order k^4 in the general case, have also been made[3]. The whole collective–mode spectrum has been calculated by taking collective–mode damping into account in the *A* phase of ^{3}He[2] and experiments[5,6] show excellent agreement with these results in opposition to those obtained by kinetic–equation method.

The main advantages of the kinetic–equation method are connected with the calculation of the coupling strength between zero sound and the collective–mode. A fine example of this is the calculation by Koch and Wolfle[7,8] of the coupling strength between the real squashing mode and zero sound, which exists only via very small particle–hole asymmetry.

The cause of such a situation is as follows. The application of the path integral method to superfluid ^{3}He was developed by Brusov and Popov[9] especially to investigate the Bose spectrum. In this way they integrated over all Fermi degrees of freedom and derived the Bose fields, describing the Cooper pairs near the Fermi surface only. This made the formalism simpler and raised the possibility of moving closer to the solution of the problem of the collective–mode eigen–frequencies. But such simplification does not allow one to investigate the interaction between Fermi and Bose degrees of freedom. It is possible to modify our procedure to include some Fermi–fields. We show the possibility of including the coupling between zero sound and the collective mode. Inclusion of the Fermi–liquid correction leads to complications in our scheme. However, until now, the kinetic equation method which considered both the Fermi– and Bose–fields was more complicated and has not been very successful in calculating the collective–mode spectrum. In our opinion, both of these methods, kinetic equation and path integral, are equivalent. A good example of this is due to Combescot[10], who subsequent to Brusov and Popov obtained the same set of equations for the Bose–spectrum of ^{3}He–B by using the kinetic–equation method instead of the path–integral method.

7.2. The Collective Mode Spectrum in the Absence of Magnetic Fields

The collective modes (CM) in ^{3}He–A describe the oscillations δA_{ia} of the order parameter (complex 3x3 matrix) around its equilibrium state[2]

$$A_{ia}^{(0)} = \Delta_0 (e_1^i + e_2^i) d_\alpha \qquad (7.1)$$

The unit vectors $\boldsymbol{e}_1$ and $\boldsymbol{e}_2$ describe the orbital part of the order parameter; their vector product determines the orbital anisotropy vector

$\vec{l} = [\vec{e}_1, \vec{e}_2]$ and the unit vector $\boldsymbol{d}$ specifies the spin axis, i.e., the axis of the magnetic anisotropy. The number of collective–mode in ^{3}He–A as well as in any other phase is equal to 18 (3x3x2). In principle all these modes could be observed either in NMR or in sound absorption experiments.

The classification of the collective modes in the *A*–phase of ^{3}He has been done[12] in terms of the irreducible representations of the symmetry group *H* of the equilibrium state Eq. (7.1). In contrast to the modes in the *B* phase of ^{3}He which are characterized by one quantum number *J* and a single parity (with respect to complex conjugation), the modes in ^{3}He–*A* are characterized by two quantum numbers: *Q* and S_z, and two parities P^1 and P^2. The charges *Q* and S_z assume the values $0,\pm1,\pm2$ and $0,\pm1$ respectively. Owing to the parities P^1 and P^2 those modes, which differ in the sign of either S_z or *Q*, turn out to be degenerate. Consequently, if the wave vector $\boldsymbol{k}$ is parallel to the orbital anisotropy axis the spectrum of modes will consist of two fourfold degenerate branches, and twofold–degenerate branches, and two nondegenerate branches (see Table 7.1).

An additional degeneracy of the spectrum of modes is exhibited in the weak–coupling limit, on account of an enlargement of the group *H* owing to hidden symmetry. This leads in particular to the appearance of four additional Goldstone modes, which were first obtained by Brusov and Popov[9].

The calculation of the collective excitation spectrum has been made in numerous papers (see, for example, Ref. 13), where the energies of clapping $E_{cl} = 1.22\Delta_0$, flapping $E_{fl} = 1.56\Delta_0$, and pair–breaking $E_{pb} = 2\Delta_0$ modes were obtained (here Δ_0 is the maximum value of gap $\Delta = \Delta_0 \sin\theta$). These energy values were obtained without taking any collective excitation damping into account. But it is clear that the vanishing of the gap along the orbital anisotropy axis leads to collective excitation damping due to decay into two fermions, because collective excitation with nonzero energy and small momentum can always decay kinematically into two fermions whose momenta are almost opposite and close to $\boldsymbol{l}$. The whole collective excitation spectrum taking into account this damping was obtained first by Brusov and Popov[2].

Table 7.1. Various modes and their respective quantum numbers[12].

Quantum numbers												
Modes	Variables	In the absence of dipole interaction and magnetic field				Taking into account dipole interaction		In magnetic field			In weak coupling	
		Q,	P^2	S_Z	P^1	$\tilde{Q}$	$\tilde{P}^2$	Q	P^2	P^1	$\tilde{S}$	Q
Sound	$u_{23}-v_{12}$	0	−1	0	+1	0	−1	0 0	−1 +1, $u_{11}+v_{21}$	+1 +1		
Spin waves	$u_{11}+v_{21}$ $u_{12}+v_{23}$	0	+1	±1	–	±1	–	0	+, $u_{11}+v_{22}$	−1	1	0
Orbital modes	u_{33}, v_{33}	±1	–	0	–	±1 0	– −1, $u_{31}+v_{32}$	±1	–	−1		
Spin–orbital modes	u_{31}, v_{31} u_{32}, v_{32}	±1	–	±1	–	0 ±2	+1, $v_{31}-u_{32}$ –, $u_{32}+v_{31}$, $u_{31}-v_{32}$,	±1 ±1	–, $u_{32}+v_{32}$ –, $u_{31}+v_{31}$	+1	1	±1
Pseudo sound	$u_{13}+v_{23}$	0	+1	0	+1	0	+1	0 0	+1 −1, $v_{11}-u_{21}$	−1 +1 +1		
Pseudo spin modes	$u_{23}-v_{12}$ $v_{11}-u_{21}$ $u_{23}+v_{13}$ $u_{13}-v_{23}$	0 ±2	−1 –	±1 0	– +1	±1 ±2	– –	0 ±2	−1, $u_{22}-v_{22}$ –	−1 +1	1	0

Clap– ping modes	$u_{11}-v_{21}$ $u_{21}+v_{11}$ $u_{12}-v_{22}$ $u_{22}+v_{12}$	±2	–	±1	–	±1 ±3	– $u_{11}-v_{21}+$ $+u_{22}+v_{12}$, $u_{21}+v_{11}-$ $-u_{12}+v_{22}$ – $u_{11}-v_{21}-$ $-u_{22}-v_{12}$, $u_{21}+v_{11}+$ $+u_{12}-v_{22}$	±2 ±2	– $u_{11}-v_{21}$, $u_{21}+v_{11}$ – $u_{12}-v_{22}$, $u_{22}+v_{12}$	+1 –1	1	±2

Following their paper we derived below the whole collective excitation spectrum in ^{3}He–A without magnetic fields. The spectrum in A–phase as well as in B–phase is described by the "effective" or "hydrodynamic action" functional, obtained by us in previous Chapters:

$$S_{\text{eff}} = g_0^{-1} \sum_{p,i,a} c_{ia}^+(p) c_{ia}(p) + \frac{1}{2} \ln \det \frac{\hat{M}\left(c_{ia}, c_{ia}^+\right)}{\hat{M}\left(c_{ia}^{(0)}, c_{ia}^{(0)+}\right)}. \tag{7.2}$$

For calculation of the collective–mode spectrum in the region $T_C - T \propto T_C$, we expand the functional (½)lndet[M(c,c)/M(0,0)] in Eq. (7.2) in powers of the deviation $c_{ia}(p)$ from the condensate value $c_{ia}^{(0)}(p)$, which is different for different phases. We apply the shift $c_{ia}(p) \to c_{ia}(p) + c_{ia}^{(0)}(p)$ and separate from S_h the quadratic form

$$\sum_p A_{ijab}(p)c^*_{ia}(p)c_{jb}(p)+$$
$$+\frac{1}{2}\sum_p B_{ijab}(p)\left[c_{ia}(p)c_{jb}(-p)+c^*_{ia}(p)c^*_{jb}(-p)\right]. \tag{7.3}$$

This form determines, in first approximation, the Bose–spectrum obtained from the equation

$$\det Q = 0. \tag{7.4}$$

Here, Q is a matrix of quadratic form, determined by the tensor coefficients $A_{ijab}(p), B_{ijab}(p)$ in (7.3). These quantities are proportional to the integrals of the products of the Green's functions of the fermions. Most effective in the calculation of these integrals is the Feynman procedure customarily used in relativistic quantum theory. In the present case the procedure is based on the identity

$$\left[\left(\omega_1^2+\xi_1^2+\Delta^2\right)\left(\omega_2^2+\xi_2^2+\Delta^2\right)\right]^{-1}=$$

$$\int d\alpha\left[\alpha\left(\omega_1^2+\xi_1^2+\Delta^2\right)+(1-\alpha)\left(\omega_2^2+\xi_2^2+\Delta^2\right)\right]^{-2}. \tag{7.5}$$

It is easy to evaluate by this procedure the integrals with respect to the variables ω and ξ, and then with respect to the angle variables and the parameter α. The quadratic part of S_h for the *A* phase of the model is a sum of three quadratic forms, the first of which depends on the variables c_{i1}, the second one – on c_{i2}, and the third one – on c_{i3}. The second and third forms are transformed into the first one by the substitutions $c_{i2}\to c_{i1}$ and $c_{i3}\to ic_{i1}$. The quadratic form of the variables $c_{i1}, c_{j1}, c^+_{i1}, c^+_{j1}$ (with $i,j=1,2$) is

$$S(c_{i1}, c_{j1}, c_{i1}^{+}, c_{j1}^{+}) =$$
$$= c_{i1}^{+} c_{j1} \left[\frac{\delta_{ij}}{g} + \frac{4Z^2}{\beta V} \sum_{p1+p2=p} n_{i1} n_{j1} (\xi_1 + i\omega_1)(\xi_2 + i\omega_2) G_1 G_2 \right] +$$
$$+ c_{i1} c_{j1} \left[\frac{2\Delta_0^2 Z^2}{\beta V} \sum_{p1+p2=p} (n_1 - in_2)^2 n_{1i} n_{1j} G_1 G_2 \right] +$$
$$+ c_{i1}^{+} c_{j1}^{+} \left[\frac{2\Delta_0^2 Z^2}{\beta V} \sum_{p1+p2=p} (n_1 + in_2)^2 n_{1i} n_{1j} G_1 G_2 \right]. \tag{7.6}$$

The functions G_1, G_2, ξ_1, ξ_2 are independent on the azimuthal angle ϕ, thus we can averaged (at $\boldsymbol{k}$ =0) over this angle ϕ the right sides of (7.6). Using as well the equality

$$\frac{\delta_{ij}}{g} + \frac{Z^2}{\beta V} \sum_{p1} n_{i1} n_{j1} G_1 = 0, \tag{7.7}$$

which determines the value of the gap Δ_0, that enters into $G_1 = (\omega_1^2 + \xi_1^2 + \Delta_0^2 \sin\theta_1)$, we arrive to the following quadratic form

$$S(c_{i1}, c_{j1}, c_{i1}^{+}, c_{j1}^{+}) =$$
$$= c_{i1}^{+} c_{j1} \left[\frac{2\delta_{ij} Z^2}{\beta V} \sum_{p1+p2=p} (n_1^2 + n_2^2)(\xi_1 + i\omega_1)(\xi_2 + i\omega_2) G_1 G_2 - G_1 \right] +$$
$$+ c_{i1} c_{j1} \left[b_{ij} \frac{\Delta_0^2 Z^2}{2\beta V} \sum_{p1+p2=p} (n_1^2 + n_2^2) G_1 G_2 \right] +$$
$$+ c_{i1}^{+} c_{j1}^{+} \left[b_{ij}^{*} \frac{\Delta_0^2 Z^2}{2\beta V} \sum_{p1+p2=p} (n_1^2 + n_2^2) G_1 G_2 \right], \tag{7.8}$$

where

$$b_{ij} = \begin{pmatrix} 1 & -i \\ -i & 1 \end{pmatrix}, \quad b^*_{ij} = \begin{pmatrix} 1 & i \\ i & 1 \end{pmatrix}. \tag{7.9}$$

We now investigate the Bose–spectrum branches defined by (7.8) at zero momentum $\boldsymbol{k}$.

We denote the coefficient of δ_{ij} in (7.8) by $f(\omega)$, and the coefficient of b_{ij} by g (ω). We also put $u_1 = \operatorname{Re} c_{11}$, $v_1 = \operatorname{Im} c_{11}, u_2 = \operatorname{Re} c_{21}$, $v_2 = \operatorname{Im} c_{21}$.

The quadratic form of the variables u_1, u_2, v_1, v_2 (at $\boldsymbol{k}$ = O) can then be taken as a sum of two forms:

$$\begin{gathered}\left[\left(f(\omega)+g(\omega)\right)\left(u_1^2+v_2^2\right)-2g(\omega)u_1v_2\right]+ \\ +\left[\left(f(\omega)-g(\omega)\right)\left(v_1^2+u_2^2\right)-2g(\omega)v_1u_2\right],\end{gathered} \tag{7.10}$$

to which the following matrices are corresponding

$$\begin{pmatrix} f(\omega)+g(\omega) & -g(\omega) \\ -g(\omega) & f(\omega)+g(\omega) \end{pmatrix},$$

$$\begin{pmatrix} f(\omega)-g(\omega) & -g(\omega) \\ -g(\omega) & f(\omega)-g(\omega) \end{pmatrix}. \tag{7.11}$$

Equating their determinants to zero, we obtain the equations

$$f(\omega)(f(\omega)+2g(\omega))=0,$$

$$f(\omega)(f(\omega)-2g(\omega))=0. \tag{7.12}$$

We add to (7.12) the equation obtained from an examination of the terms with c_{31} and c_{31}^{+}:

$$h(\omega)=\left[\frac{1}{g}+\frac{4Z^2}{\beta V}\sum_{p1+p2=p} n_3^2(\xi_1+i\omega_1)(\xi_2+i\omega_2)G_1G_2\right]=$$

$$=\frac{2Z^2}{\beta V}\sum_{p1+p2=p} 2n_3^2(\xi_1+i\omega_1)(\xi_2+i\omega_2)G_1G_2-(n_1^2+n_2^2)G_1=0. \qquad (7.13)$$

Three equations (7.12) can be combined into one:

$$\frac{2Z^2}{\beta V}\sum_{p1+p2=p}(n_1^2+n_2^2)\{[(\xi_1+i\omega_1)(\xi_2+i\omega_2)G_1G_2\pm(1,0)\Delta^2]G_1G_2-G_1\}=0 \qquad (7.14)$$

in which $\pm(1,0)\Delta^2=0$ denotes either Δ^2, or $-\Delta^2$, or 0.

Changing from summations to integrals in (7.13) and (7.14) (at T=0) by

$$\frac{1}{\beta V}\sum_{p}=\frac{k_F^2}{(2\pi)^4 c_F}\int d\Omega\; d\omega\; d\xi \qquad (7.15)$$

and substituting the expressions for G_1 and G_2, we can write

$$\frac{2Z^2k_F^2}{(2\pi)^4c_F}\int d\Omega_1 d\omega_1 d\xi_1\left[\frac{2\cos^2\theta(\xi_1+i\omega_1)(\xi_2+i\omega_2)}{(\omega_1^2+\xi_1^2+\Delta^2)(\omega_2^2+\xi_2^2+\Delta^2)}-\frac{\sin^2\theta}{\omega_1^2+\xi_1^2+\Delta^2}\right]=0$$

$$\frac{2Z^2k_F^2}{(2\pi)^4c_F}\int\sin^2\theta d\Omega_1 d\omega_1 d\xi_1\left[\frac{(\xi_1+i\omega_1)(\xi_2+i\omega_2)\pm(1,0)\Delta^2}{(\omega_1^2+\xi_1^2+\Delta^2)(\omega_2^2+\xi_2^2+\Delta^2)}-\right.$$
$$\left.-\frac{1}{\omega_1^2+\xi_1^2+\Delta^2}\right]=0. \tag{7.16}$$

Integrating with respect to ω_1,ξ_1 with the help of the Feynman procedure, we get

$$\frac{Z^2k_F^2}{4\pi^3c_F}\int_0^1 d\alpha\int\cos^2\theta d\Omega_1\left[\ln\frac{\Delta^2}{\Delta^2+\alpha(1-\alpha)\omega^2}-\frac{\alpha(1-\alpha)\omega^2}{\Delta^2+\alpha(1-\alpha)\omega^2}\right]=0,$$

$$\frac{Z^2k_F^2}{4\pi^3c_F}\int_0^1 d\alpha\int\sin^2\theta d\Omega_1\left[\ln\frac{\Delta^2}{\Delta^2+\alpha(1-\alpha)\omega^2}-\right.$$
$$\left.-\frac{2\alpha(1-\alpha)\omega^2+\Delta^2\mp(1,0)\Delta^2}{\Delta^2+\alpha(1-\alpha)\omega^2}\right]=0. \tag{7.17}$$

The integrals with respect to α can be calculated by the following equations

$$\int_0^1 d\alpha\left[\ln\frac{\Delta^2}{\Delta^2+\alpha(1-\alpha)\omega^2}-\frac{\alpha(1-\alpha)\omega^2}{\Delta^2+\alpha(1-\alpha)\omega^2}\right]=$$
$$=1-\frac{\omega^2+2\Delta^2}{\omega\sqrt{\omega^2+4\Delta^2}}\ln\frac{\sqrt{\omega^2+4\Delta^2}+\omega}{\sqrt{\omega^2+4\Delta^2}-\omega},$$

$$\int_0^1 d\alpha\frac{\alpha(1-\alpha)\omega^2}{\Delta^2+\alpha(1-\alpha)\omega^2}=1-\frac{2\Delta^2}{\omega\sqrt{\omega^2+4\Delta^2}}\ln\frac{\sqrt{\omega^2+4\Delta^2}+\omega}{\sqrt{\omega^2+4\Delta^2}-\omega}. \tag{7.18}$$

substituting (7.18) into (7.17), and putting $\omega \to \omega\Delta_0, \cos\theta = x$, we arrive at the equations

$$\int_0^1 dx(1-x)^2 \frac{\omega^2 + 4(1-x)^2}{\omega\sqrt{\omega^2+4(1-x)^2}} \ln\frac{\sqrt{\omega^2+4(1-x)^2}+\omega}{\sqrt{\omega^2+4(1-x)^2}-\omega} = 0,$$

$$\int_0^1 dx(1-x)^2 \frac{\omega^2 + 2(1-x)^2}{\omega\sqrt{\omega^2+4(1-x)^2}} \ln\frac{\sqrt{\omega^2+4(1-x)^2}+\omega}{\sqrt{\omega^2+4(1-x)^2}-\omega} = 0,$$

$$\int_0^1 dx(1-x)^2 \frac{\omega}{\omega\sqrt{\omega^2+4(1-x)^2}} \ln\frac{\sqrt{\omega^2+4(1-x)^2}+\omega}{\sqrt{\omega^2+4(1-x)^2}-\omega} = 0,$$

$$\int_0^1 dx x^2 \left[\frac{\omega^2 + 2(1-x)^2}{\omega\sqrt{\omega^2+4(1-x)^2}} \ln\frac{\sqrt{\omega^2+4(1-x)^2}+\omega}{\sqrt{\omega^2+4(1-x)^2}-\omega} - 1\right] = 0.$$

(7.19)

The first of these equations is the equation $f(\omega) - 2g(\omega) = 0$, the second is $f(\omega) = 0$, and the third is $f(\omega) + 2g(\omega) = 0$, and the fourth is $h(\omega) = 0$. They determine the Bose–spectrum at ***k***=0 following the analytic continuation $i\omega \to E$. The spectrum branches corresponding to the second and fourth equations are doubly degenerated. Taking into account the forms of variables c_{i2} and c_{i3}, leading to the same equations for spectrum, it is necessary to multiply by 3, the multiplicity of each branch in the considered model.

The third and fourth equations in (7.19) have roots ω=0 and correspond to the Goldstone modes. From the first and second equations we obtain the complex energies of nonphonon branches $E_1(k=0)$, $E_2(k=0)$.

Under calculating of nontrivial roots of equations in (7.19) the problem of analytical continuation $i\omega \to E$ of integrated function

$$\begin{aligned}
&\frac{1}{\sqrt{\omega^2+4(1-x)^2}}\ln\frac{\sqrt{\omega^2+4(1-x)^2}+\omega}{\sqrt{\omega^2+4(1-x)^2}-\omega}= \\
&=\frac{1}{\sqrt{4(1-x)^2-E^2}}\ln\frac{\sqrt{4(1-x)^2-E^2}-iE}{\sqrt{4(1-x)^2-E^2}+iE}= \\
&=\frac{1}{\sqrt{4(1-x)^2-E^2}}\ln\frac{E+i\sqrt{4(1-x)^2-E^2}}{E-i\sqrt{4(1-x)^2-E^2}}-\frac{i\pi}{\sqrt{4(1-x)^2-E^2}}
\end{aligned} \tag{7.20}$$

is appearing. One can continue the right size of (7.20) first from the positive imaginary semi–axis $E=i\omega$ ($\omega>0$) into upper (physical) semi–plane, and then into below one through the interval [0,2] of real axis. First term in right hand part of (7.20) is analytic in the vicinity of the interval [0,2]. Contributions into integrals (7.19) from the second term can be evaluated in terms of elementary functions, which can be easy continued into nonphysical leaf through the interval [0,2].

Using as well the formula

$$\ln\frac{E+i\sqrt{4(1-x)^2-E^2}}{E-i\sqrt{4(1-x)^2-E^2}}=-2\ln\frac{E-i\sqrt{4(1-x)^2-E^2}}{2\sqrt{(1-x^2)}},$$

one can write the first and second equations of (7.19) in the following form

$$F_1(E) = -2\int_0^1 dx\left[(1-x^2)\sqrt{4(1-x)^2 - E^2}\ln\frac{E - i\sqrt{4(1-x)^2 - E^2}}{2\sqrt{(1-x^2)}}\right] -$$

$$-\frac{\pi}{128}\{(E^2-4)(E^2-12)\ln[(E+2)/(E-2)] + 4E(12-E^2)\} = 0$$

$$F_2(E) = -2\int_0^1 dx\left[\frac{(1-x^2)(2(1-x)^2 - E^2)}{\sqrt{4(1-x)^2 - E^2}}\ln\frac{E - i\sqrt{4(1-x)^2 - E^2}}{2\sqrt{(1-x^2)}}\right] -$$

$$-\frac{\pi}{256}\{(5E^4 + 24E^2 - 48)\ln[(E+2)/(E-2)] + 4E(12-5E^2)\} = 0$$

(7.21)

The logarithm standing out of integral in (7.21) under analytical continuation into below semiplane through the interval [0,2] is equal to

$$\ln[(E+2)/(E-2)] = -i\pi + \ln(E+2) - \ln(E-2)\,. \tag{7.22}$$

Formulae (7.21) and (7.22) have been used under numerical calculation of the roots $E_1(0)$, $E_2(0)$.

The resulting collective excitation energies are

$$E_1(0) = \Delta_0(1.96 - i\ 0.31)\,,$$

$$E_2(0) = \Delta_0(1.17 - i\ 0.13)\,, \tag{7.23}$$

the second of the modes being doubly degenerate.

The attempts have been made to find the nontrivial roots of the third and forth equations in (7.19) (which have trivial roots at $\omega = 0$). The third equation does not have nontrivial roots. For forth equation the root $E_4(0) = \Delta_0(0.7 - i\ 0.5)$ have been found.

The mode with energy $E_2(0) = \Delta_0(1.17 - i\ 0.13)$ is called "clapping" mode. In Fig. 7.1 we illustrate the motion of vectors of the order parameter corresponding to clapping–mode and so–called flapping–mode, obtained by other authors, which absent in weak coupling approximation, used by us.

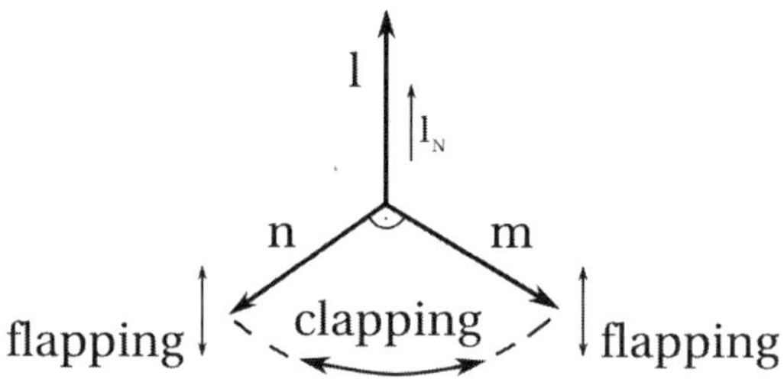

FIG. 7.1. The illustration of the motion of vectors of the order parameter corresponding to clapping–mode and flapping –mode in ^{3}He–A.

The difference between *ReE* here and in Ref. 13 is due to the fact that taking collective–mode damping into account ($\mathrm{Im}\,E \neq 0$) leads via dispersion relations to renormalization of *ReE.*

Dobbs et al[5,6]. have made precise measurements of the clapping–mode frequency. They obtained the value of $E_{cl} = (1.15 \pm 0.01)\Delta_0(T)$, which is in excellent agreement with the results of Brusov and Popov[2,14] $E_{cl} = (1.17 \pm 0.01)\Delta_0(T)$ (Fig. 7.2). This shows mainly that collective–mode damping is significant in obtaining the right value for the clapping–mode frequency.

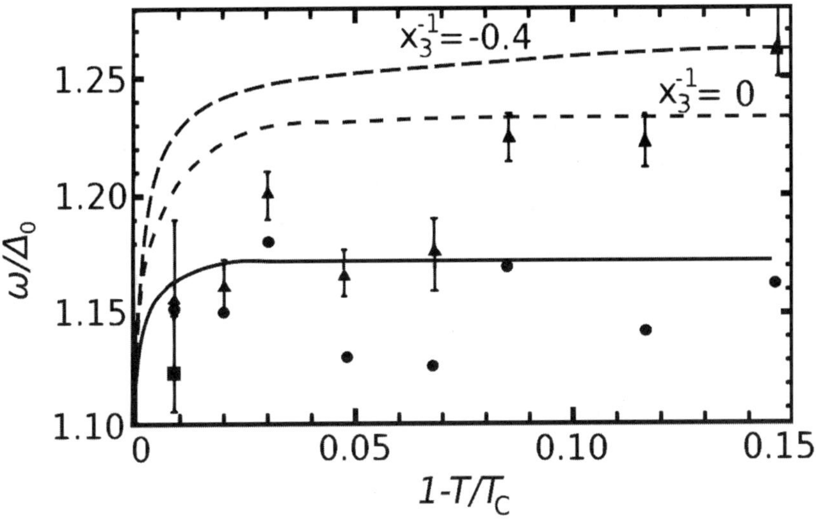

FIG. 7.2. The normalized clapping–mode resonances from Ref. 4 for two choices of A=2.03 (triangles) and 2.64 (circles) [$A=\Delta_0(0)/k_BT_C$]. Two upper curves follow from kinetic equation theory by using the formula $E_{cl}=1.23\Delta_0(T)\{1-(0.005-0.106x_3^{-1}-0.052F_2^2)[\Delta_0(T)/k_BT_C]\}$ at $x_3^{-1}=0$ and $x_3^{-1}=-0.4$ respectively. Solid curve is the result of Ref. 14.

The Brusov and Popov theory[2,14] does not require taking high pairing corrections into account, used in kinetic equation theory to explain the discrepancy between the kinetic equation value of the clapping–mode frequency $E_{cl}=1.23\Delta_0(T)$ and the experimental data. Note that a 6% difference remains between experimental data of Refs. 5, 6 and kinetic–equation theory[13] (Fig. 7.2) in spite of taking higher pairing and Fermi–liquid corrections into account.

The other interesting fact, obtained first by Brusov and Popov[2], is that the number of Goldstone modes in the weak coupling approximation is equal to 9 rather than 5, which takes place in real ^{3}He–A. The existence of four additional quasi–Goldstone spin–orbit modes is a consequence of the latent symmetry of the system and we investigate this equation below in more detail.

7.3. The Latent Symmetry, Additional Goldstone Modes, W–Bosons

We shall show that taking into account strong coupling effects decreases the number of phonon modes from 9 to 5, and that turning on a magnetic field decreases the number of phonon modes from 9 to 6 for weak coupling and from 5 to 4 when strong–coupling effects are taken into account.

We consider, in the Ginzburg–Landau region $|T-T_C| << T_C$ that part F of the action which is independent of the gradients. In the weak–coupling model we have

$$F=-Tr(AA^+)+\nu Tr(AA^+P)+Tr(AA^+)^2+Tr(AA^+AA^+)+ \\ +Tr(AA^+A^*A^T)-Tr(AA^TA^*A^+)-(1/2)Tr(AA^T)Tr(A^+A^*) \tag{7.24}$$

where A (the order parameter) is a complex matrix with elements A_{ij}.

The A–phase in the weak coupling model is described by the order parameter

$$C_0=\frac{1}{2}\begin{pmatrix}1 & 0 & 0\\ i & 0 & 0\\ 0 & 0 & 0\end{pmatrix}, \tag{7.25}$$

and the phonon variables are

$$u_{21}-v_{11}, u_{12}+v_{22}, u_{13}+v_{23}, u_{31}, v_{31}, u_{32}, v_{32}, u_{33}, v_{33} \tag{7.26}$$

where $u_{ia}=\mathrm{Re}\, c_{ia}$ and $v_{ia}=\mathrm{Im}\, c_{ia}$. These variables correspond to the Goldstone modes of the spectrum not only in the Ginzburg–Landau region, but also at all temperatures $T<T_C$. In the limit $T\to 0$, the first three of the variables in (7.26) correspond to sound waves

$E = c_F k / \sqrt{3}$, and the six remaining ones to orbital waves $E = c_F k_{||}$. The phonon spectrum is thus degenerate in the spin index.

To take into account the strong–coupling effects, we consider F with arbitrary coefficients of the fourth–order terms:

$$F = -Tr(AA^+) + vTr(AA^+P) + a(TrAA^+)^2 + bTr(AA^+AA^+) + \\ + cTr(AA^+A^*A^T) + dTr(AA^TA^*A^+) + eTr(AA^T)Tr(A^+A^*) \tag{7.27}$$

The condition $\delta F = 0$, yields in the A phase an order parameter in the form

$$C = \frac{1}{2\sqrt{a+b+d}} \begin{pmatrix} 1 & 0 & 0 \\ i & 0 & 0 \\ 0 & 0 & 0 \end{pmatrix}, \tag{7.28}$$

To find the phonon variables we calculate the second variation $\delta^2 F = 0$,

$$\delta^2 F = -Tr(AA^+) + vTr(AA^+P) + aTr\left[(A^+C)^2 + (C^+A)^2 + 2A^+AC^+C + \right. \\ \left. + 2A^+CC^+A\right] + bTr\left[2AA^+CC^+ + 2AA^+C^+C + A^+CAC^+ + A^+CA^+C\right] + \\ + cTr\left[AA^+C^*C^T + A^+A^*C^TC + A^*A^TCC^+ + A^TAC^+C^* + AC^+A^*C^T + \right. \\ \left. + A^TCA^+C^*\right] + dTr\left[AA^TC^*C^+ + A^TA^*C^+C + A^*A^+CC^T + A^+AC^TC^*\right] + \\ + 4e\left|TrC^TA\right|,$$

(7.29)

where C is the matrix (7.28) and A is a variable matrix $\|a_{ia}\|$. Substituting the values of C, C^*, C^T and C^+ we get

$$\delta^2 F = \nu(a+b+d)(u_{13}^2 + v_{13}^2 + u_{23}^2 + v_{23}^2 + u_{33}^2 + v_{33}^2) + 4a(u_{11} + v_{21})^2 +$$
$$+ 2b[2(u_{11} + v_{21})^2 - (u_{13} - v_{23})^2 - (u_{12} - v_{22})^2 - (u_{22} + v_{12})^2 - (u_{23} + v_{13})^2 -$$
$$- 2(u_{32}^2 + v_{32}^2 + u_{33}^2 + v_{33}^2)] + 2c[2(u_{11} - v_{21})^2 + 2(u_{21} + v_{11})^2 + (u_{21} - v_{22})^2 +$$
$$+ (u_{22} + v_{12})^2 + (u_{13} - v_{23})^2 + (u_{23} + v_{13})^2] + 2d[2(u_{11} + v_{21})^2 - (u_{12} - v_{22})^2$$
$$- (u_{13} - v_{23})^2 - (u_{22} + v_{12})^2 - 2(u_{22} - v_{12})^2 - 2(u_{23} - v_{13})^2 - (u_{23} + v_{13})^2 -$$
$$- 2(u_{23}^2 + v_{23}^2 + u_{33}^2 + v_{33}^2)] + 4e[(u_{11} - v_{21})^2 + (u_{21} + v_{11})^2] \quad (7.30)$$

We consider first the system in a zero magnetic field ($\nu = 0$). Equation (7.30) is the sum of five quadratic forms multiplied by the independent coefficients *a, b, c, d,* and *e.* The variables

$$u_{12} + v_{22}, u_{13} + v_{23}, u_{21} - v_{11}, u_{31}, v_{31}$$

do not enter in any of these forms, which therefore correspond to Goldstone modes. Thus, allowance for the strong–coupling effects decreases the number of phonon branches from 9 to 5. The modes $u_{32}, v_{32}, u_{33}, v_{33}$, which correspond in the weak–coupling approximation to orbital waves, become nonphonon modes when the strong–coupling effects are taken into account.

Expression (7.30) at $\nu \neq 0$ describes the system in a magnetic field. In the weak–coupling approximation the number of phonon modes decreases from 9 to 6, and the variables $u_{13} + v_{23}, u_{33}, v_{33}$, become nonphonon because of the appearance of the gap $\propto \mu H$ in the spectrum. In a system with a strong coupling a mode that becomes nonphonon upon application of a magnetic field is $u_{13} + v_{33}$, (the modes u_{33} and v_{33} in the case of strong coupling remain Goldstone modes at $\nu = 0$ and the number of Goldstone modes decreases from 5 to 4.

To gain an idea of the total Bose–spectrum (including the Goldstone branches) when strong–coupling effects are taken into account, we write (7.30) at $H = 0$ ($\nu = 0$) in the form

$$\delta^2 F = 4(a+b+d)(u_{11}+v_{21})^2 + 4(c+e)[(u_{11}-v_{21})^2 + (u_{21}+v_{11})^2] + $$
$$+ 2(c-b-d)[(u_{21}-v_{22})^2 + (u_{22}+v_{12})^2 + (u_{13}-v_{23})^2 + (u_{23}+v_{13})^2] -$$
$$- 4d[(u_{22}-v_{12})^2 + (u_{23}-v_{13})^2] - 4(b+d)(u_{32}^2 + v_{32}^2 + u_{33}^2 + v_{33}^2) \tag{7.31}$$

For comparison we write down $\delta^2 F$ in the weak–coupling approximation, using $a = b = c = -d = -2e = 1$ in (7.31):

$$\delta^2 F = 4[(u_{11}+v_{21})^2 + (u_{22}-v_{12})^2 + (u_{23}-v_{13})^2] +$$
$$+ 0\cdot(u_{32}^2 + v_{32}^2 + u_{33}^2 + v_{33}^2) + 2[(u_{11}-v_{21})^2 + (u_{21}+v_{11})^2 + \tag{7.32}$$
$$+ (u_{13}-v_{23})^2 + (u_{12}-v_{22})^2 + (u_{22}+v_{12})^2 + (u_{23}+v_{13})^2]$$

The form (7.32) has three eigenvalues equal to 4, corresponding to the variables $u_{11}+v_{21}, u_{22}-v_{12}, u_{23}-v_{13}$. The branches $E_1(0)$ correspond, as $T \to 0$, to these variables. The other nonzero eigenvalue equal to 2 corresponds to six variables: $u_{21}+v_{11}, u_{12}-v_{22}, u_{13}-v_{23}, u_{11}-v_{21}, u_{22}+v_{12}, u_{23}+v_{13}$ and six $E_2(0)$ branches as $T \to 0$.

The calculation of the Bose–spectrum in Ref. 10 yields 6 clapping modes and three $2\Delta_0$, modes, i.e., as many as in the weak–coupling case considered here. Formula (7.31) shows that in the general case allowance for the strong coupling effects leads to splitting. The clapping modes break up into two groups: two branches correspond to the eigenvalue $4(c+e)$ and four correspond to the number $2(c-b-d)$. The three $2\Delta_0$ branches also break up into one branch with eigenvalue $4(a+b+d)$ and two branches with eigenvalue $-4d$. We note that no conclusion can be drawn from the data of Ref. 13 concerning the splitting of the branches.

Branches $u_{32}, v_{32}, u_{33}, v_{33}$, which in a weak–coupling approximation are orbital waves, become the normal flapping modes and the superflapping modes when the strong–coupljng effects are taken into account, as shown by comparison with the data of Ref. 13.

Volovik[15] first showed that the fermions in the ^{3}He–*A* are chiral and a field theory in superfluid ^{3}He–A, which describes the dynamics of chiral fermion excitations interacting with the order–parameter collective boson modes is similar to the theory of the electroweak interaction. The roles of photons and *W* bosons are played by orbital waves and four quasi–Goldstone spin–orbit modes, which we obtained above, respectively.

An equation of the Dirac type for fermions in ^{3}He–*A* near the poles of the Fermi–sphere can be derived from the Bogoliubov equation, in which it is neccssary to take into account the fluctuations δA_{ia} of the order parameter A_{ia} around its equilibrium value

$$A_{ia}^{(0)} = \Delta_0 d_\alpha \left(e_1^i + ie_2^i\right). \tag{7.33}$$

Only the certain combinations of fluctuations act on the fermions. These combinations form a "photon" field and W–field[15]

$$A_1 + iA_2 = -\frac{k_F}{\Delta_0} d_\alpha l^i \delta A_{ia},$$

$$W_1^\alpha + iW_2^\alpha = \frac{k_F}{i\Delta_0} e^{\alpha\beta\gamma} d_\beta l^i \delta A_{ia},$$

$$A_3 = \delta\, k_F, W_3^\alpha = \frac{1}{2k_F} e^{\alpha\beta\gamma} d^\beta \frac{\partial d^\gamma}{\partial t},$$

$$A_0 = k_F \left(\boldsymbol{l}, \boldsymbol{v}_S\right), W_0^\alpha = \frac{1}{2} k_F e^{\alpha\beta\gamma} d_\beta \left(\boldsymbol{l}, \nabla\right) d_\gamma. \tag{7.34}$$

Equations (7.34) also incorporate the effect of the fluctuational spin $\boldsymbol{S} \propto [\boldsymbol{d}, \partial\boldsymbol{d}/\partial t]$, which accounts for a third component of the *W*–field, in precisely the same way as density fluctuations account for a third component, along $\boldsymbol{l}$, of the "photon" field *A*.

In terms of the fields in (7.34), the Bogoliubov equation for the Bogoliubov spinor ψ takes the following form near the poles of the Fermi–sphere:

$$\left[\left(i\frac{\partial}{\partial t} - eA_0 - e\sigma^{\alpha}W_0^{\alpha}\right) + \right.$$
$$\left. + e\left(c_{\parallel}\, l^i\tau_3 + c_{\perp}\left(e_1^i\tau_1 + e_2^i\tau_2\right)\right)\left(\frac{1}{i}\nabla_i - eA_i - e\sigma^{\alpha}W_i^{\alpha}\right)\right]\psi = 0.$$

Here, τ_i and σ^{α} are the Pauli matrices corresponding to the Bogoliubov isospin and ordinary spin. This equation is reminiscent of the Dirac equation for massless chiral fermions in the Weinberg–Salam theory. The primary distinction is in the anisotropy of ^{3}He–A along the $\boldsymbol{l}$ and $\boldsymbol{d}$ axes. The velocity $c_{\parallel} = v_F$ along $\boldsymbol{l}$ is far greater than the transverse velocity $c_{\perp} = \Delta_0 / k_F$, and we have $W^{\alpha}d^{\alpha} = 0$, i.e., there are no *Z*–bosons. The charge $e = (\boldsymbol{k}, \boldsymbol{l})/k$, takes on the values +1 and −1 for fermions near the upper and lower poles, respectively.

An important point is that in the weak–coupling approximation (in which the Fermi spheres with different spin projections do not interact with each other) there is an additional SO(3) symmetry, which combines the "photons" and the *W*–bosons in a single triplet (more precisely, a sextet, when we take into account the polarization of the collective modes). In this approximation, the *W*–bosons, like the Goldstone orbital waves (or "photons"), have no mass in consequence with results obtained above. But as Brosov and Popov showed first[2] (see above) four additional Goldstone modes become nonphonon if we turn on the strong–coupling corrections. It means that the *W*–bosons acquire a mass via the strong–coupling corrections. Consequently, and in contrast with the

Weinberg–Salam theory, the Higgs phenomenon is not required for the appearance of massive W–bosons in ^{3}He–A.

7.4. The Linear Zeeman Effect for Clapping– and Pair–breaking Modes

Below we consider the influence of magnetic fields on the collective excitation spectrum in ^{3}He–A.[16,17]

In accordance with Nasten'ka and Brusov's idea[18] for the investigation of the collective excitation spectrum in the presence of a magnetic field, we must take into account both the additional term in S_{eff} and the distortion of the order parameter. The latter one, in our case, is equal to[19]

$$c_{ia}(p) = c(\beta V)^{1/2}\delta_{p0}\left(\delta_{a1}\alpha_+ + i\delta_{a2}\alpha_-\right)\left(\delta_{i1} + i\delta_{i21}\right). \tag{7.35}$$

Here,

$$\begin{gathered}\alpha_\pm = \frac{\Delta_\uparrow \pm \Delta_\downarrow}{2\Delta},\quad \Delta^2_{\uparrow\downarrow} = N(0)(\tau \pm \eta h)/2\beta_{345},\\ \eta = \frac{N'(0)}{N(0)}T_C \ln\frac{1.14\varepsilon_0}{T_C},\quad h = \frac{\mu_0 H}{T_C},\quad \Delta = 2cZ,\end{gathered} \tag{7.36}$$

is a single–fermion spectrum gap, determined by the gap equation

$$\begin{aligned}g_0^{-1} = -\frac{Z^2}{\left(\alpha_+^2 + \alpha_-^2\right)\beta V}\sum_p\Bigg[&\frac{\left(\alpha_+ + \alpha_-\right)^2\sin^2\theta}{\omega^2 + (\xi - \mu H)^2 + \Delta^2\sin^2\theta\left(\alpha_+ + \alpha_-\right)^2} +\\ &+\frac{\left(\alpha_+ - \alpha_-\right)^2\sin^2\theta}{\omega^2 + (\xi + \mu H)^2 + \Delta^2\sin^2\theta\left(\alpha_+ - \alpha_-\right)^2}\Bigg].\end{aligned} \tag{7.37}$$

We could calculate the collective excitation spectrum of our system in the presence of a magnetic field by the techniques developed by us, but using the deformed order parameter (7.35). Making these calculations below, we obtained 18 equations, which completely determine the collective excitation spectrum in ^{3}He–A in an arbitrary magnetic field H and with arbitrary collective excitation momenta $\boldsymbol{k}$. Then we consider the case of small H and zero momentum of collective excitation, k=0, and calculate the linear corrections to the collective excitation spectrum.

7.4.1. The equations for the collective mode spectrum in an arbitrary magnetic field and at arbitrary collective mode momenta

In first approximation the collective excitation spectrum is determined by the quadratic part of S_{eff}. To obtain it we need to calculate G. One has for G^{-1}

$$G^{-1}=\begin{pmatrix} Z^{-1}(i\omega-\xi+\mu H\sigma_3)\delta_{p1p2}; & \begin{array}{c}2c(n_1+in_2)\times\\ \times(\alpha_+\sigma_1+i\alpha_-\sigma_2)\delta_{p1+p2,0};\end{array} \\ \begin{array}{c}-2c(n_1-in_2)\times\\ \times(\alpha_+\sigma_1-i\alpha_-\sigma_2)\delta_{p1+p2,0}\end{array} & Z^{-1}(-i\omega+\xi+\mu H\sigma_3)\delta_{p1p2} \end{pmatrix}. \tag{7.38}$$

Inverting G^{-1} one gets

$$G^{-1}=\begin{pmatrix} G_{11} & G_{12} \\ G_{21} & G_{22}\end{pmatrix},\ \text{where } G_{11}=\begin{pmatrix} \dfrac{a^+ +b}{d_1} & 0 \\ 0 & \dfrac{a^+ -b}{d_2}\end{pmatrix}\delta_{p1p2},$$

$$G_{12} = \begin{pmatrix} 0 & \dfrac{q_1 - iq_2}{d_1} \\ \dfrac{q_1 + iq_2}{d_2} & 0 \end{pmatrix} \delta_{p1+p2,0},$$

$$G_{21} = \begin{pmatrix} 0 & \dfrac{-q_1^+ + iq_2^+}{d_2} \\ \dfrac{-q_1^+ - iq_2^+}{d_1} & 0 \end{pmatrix} \delta_{p1p2},$$

$$G_{22} = \begin{pmatrix} \dfrac{-a^+ + b}{d_2} & 0 \\ 0 & \dfrac{-a^+ - b}{d_1} \end{pmatrix} \delta_{p1p2}, \tag{7.39}$$

Here,

$$a = Z^{-1}(i\omega - \xi),\ b = Z^{-1}\mu H,\ q_1 = Z^{-1}\Delta(n_1 + in_2)\alpha_+,$$
$$q_2 = iZ^{-1}\Delta(n_1 + in_2)\alpha_-, d_{1,2} = Z^{-2}\left(\omega^2 + (\xi \pm \mu H)^2 + \Delta^2 \sin^2\theta(\alpha_+ \pm \alpha_-)^2\right).$$

Using the expressions (7.39) for G and the following expression for u

$$u_{p1p2} = (\beta V)^{-1/2} \begin{pmatrix} 0 & (n_{1i} - n_{2i})\sigma_a c_{ia}(p_1 + p_2) \\ -(n_{1i} - n_{2i})\sigma_a c^+_{ia}(p_1 + p_2) & 0 \end{pmatrix}$$

(7.40)

one could obtain the quadratic part of S_{eff}

$$g_0^{-1} \sum_{p,i,a} c^+_{ia} c_{ia} + \frac{Z^2}{\beta V} \sum_{p1+p2=p} n_{1i} n_{1j} \{[(n_1 + in_2)^2 \left(-\partial_3 c^+_{i3} c^+_{j3} + \partial_1 (c^+_{i1} c^+_{j1} - c^+_{i2} c^+_{j2}) + \right.$$
$$\left. + i\partial_2 (c^+_{i1} c^+_{j2} + c^+_{i2} c^+_{j1}) + i\partial_2 (c^+_{i1} c^+_{j2} + c^+_{i2} c^+_{j1})\right) + h.c.] + D_1 \left(c^+_{i1} c_{j1} + c_{i1} c^+_{j1} + c^+_{i2} c_{j2} + \right.$$
$$\left. + c_{i2} c^+_{j2}\right) + D_3 \left(c^+_{i3} c_{j3} + c_{i3} c^+_{j3}\right) + iD_2 \left(c^+_{i2} c_{j1} + c_{i1} c^+_{j2} - c^+_{i2} c_{j1} - c_{i1} c^+_{j2}\right)\}.$$

(7.41)

Here,

$$\partial_{1,2} = \frac{(\alpha_+ + \alpha_-)^2}{d_1(1) d_1(2)} \pm \frac{(\alpha_+ - \alpha_-)^2}{d_2(1) d_2(2)}, \quad D_{1,2} = \frac{(a^+(1) + b)(a^+(2) + b)}{d_1(1) d_1(2)} \pm$$
$$\pm \frac{(a^+(1) - b)(a^+(2) - b)}{d_2(1) d_2(2)}, \partial_3 = (\alpha_+^2 + \alpha_-^2) \left(\frac{1}{d_1(2) d_2(1)} + \frac{1}{d_1(1) d_2(2)} \right),$$
$$D_3 = \frac{(a^+(1) + b)(a^+(2) - b)}{d_1(1) d_2(2)} \pm \frac{(a^+(1) - b)(a^+(2) + b)}{d_2(1) d_1(2)}$$

(7.42)

By diagonalization the quadratic form we get its canonical form (here, $u_{ia} = \mathrm{Re}\, c_{ia}, v_{ia} = \mathrm{Im}\, c_{ia}$)

$$
\begin{aligned}
&\left[g_0^{-1} + \frac{2Z^2}{\beta V}\sum_{p1+p2=p} D_3 \cos^2\theta\right]\left(u_{33}^2 + v_{33}^2\right) + \left[g_0^{-1} + \frac{2Z^2}{\beta V}\sum_{p1+p2=p}(D_1 + D_2)\cos^2\theta\right]\times \\
&\times\left(\left(u_{31} + v_{32}\right)^2 + \left(u_{32} + v_{31}\right)^2\right) + \\
&+\left[g_0^{-1} + \frac{2Z^2}{\beta V}\sum_{p1+p2=p}(D_1 - D_2)\cos^2\theta\right]\left(\left(u_{31} - v_{32}\right)^2 + \left(u_{32} - v_{31}\right)^2\right) + \\
&+\left[g_0^{-1} + \frac{Z^2}{\beta V}\sum_{p1+p2=p} D_3 \sin^2\theta\right]\left(\left(v_{31} + u_{32}\right)^2 + \left(u_{13} - v_{23}\right)^2\right) + \\
&+\left[g_0^{-1} + \frac{Z^2}{\beta V}\sum_{p1+p2=p}(-\Delta^2 \sin^2\theta\partial_3 + D_3)\sin^2\theta\right]\left(u_{13} + v_{23}\right)^2 + \\
&+\left[g_0^{-1} + \frac{Z^2}{\beta V}\sum_{p1+p2=p}(\Delta^2 \sin^2\theta\partial_3 + D_3)\sin^2\theta\right]\left(v_{13} - u_{23}\right)^2 + \\
&+\left[g_0^{-1} + \frac{Z^2}{\beta V}\sum_{p1+p2=p}(D_1 + D_2)\sin^2\theta\right]\left(\left(u_{12} + v_{11} + u_{21} + v_{21}\right)^2 + \left(u_{11} + v_{12} - u_{22} - v_{21}\right)^2\right) + \\
&+\left[g_0^{-1} + \frac{Z^2}{\beta V}\sum_{p1+p2=p}(D_1 - D_2)\sin^2\theta\right]\left(\left(u_{12} - v_{11} - u_{21} + v_{22}\right)^2 + \left(u_{11} - v_{12} + u_{22} - v_{21}\right)^2\right) + \\
&+\left[g_0^{-1} + \frac{Z^2}{\beta V}\sum_{p1+p2=p}\sin^2\theta(D_1 + D_2 - \Delta^2 \sin^2\theta(\partial_1 + \partial_2))\right]\left(u_{12} + v_{11} - u_{21} - v_{21}\right)^2 + \\
&+\left[g_0^{-1} + \frac{Z^2}{\beta V}\sum_{p1+p2=p}\sin^2\theta(D_1 - D_2 - \Delta^2 \sin^2\theta(\partial_1 - \partial_2))\right]\left(u_{11} - v_{11} + u_{21} - u_{22}\right)^2 + \\
&+\left[g_0^{-1} + \frac{Z^2}{\beta V}\sum_{p1+p2=p}\sin^2\theta(D_1 + D_2 + \Delta^2 \sin^2\theta(\partial_1 + \partial_2))\right]\left(u_{11} + v_{12} + u_{22} + v_{21}\right)^2 + \\
&+\left[g_0^{-1} + \frac{Z^2}{\beta V}\sum_{p1+p2=p}\sin^2\theta(D_1 - D_2 + \Delta^2 \sin^2\theta(\partial_1 - \partial_2))\right]\left(u_{11} - v_{12} - u_{22} + v_{21}\right)^2 .
\end{aligned}
\tag{7.43}
$$

The equation det$Q$=O, where Q is the matrix of quadratic form (7.41), gives us 18 equations which completely determine 18 collective modes in ^{3}He–A in an arbitrary magnetic field and at arbitrary collective excitation momenta.

7.4.2. The collective mode spectrum for small magnetic fields and zero collective mode momenta (linear Zeeman effect for clapping– and pair–breaking–modes)

Below we consider the case of small H and k=0, and calculate the linear correction to the collective excitation spectrum. Setting k=0 and retaining the first order terms in field, we obtain:

I) $$\int_0^\pi \left(1-(1+2c)I\right)\cos^2\theta\sin\theta\,d\theta=0;\quad u_{33},v_{33};$$

$$\int_0^\pi \left(-1+(1+2c)I\right)\cos^2\theta\sin\theta\,d\theta \pm \gamma H\int_0^\pi \frac{4c}{1+4c}(1+2cI)\cos^2\theta\sin\theta\,d\theta=0;$$

$u_{31}+u_{32}, u_{32}+v_{31}, u_{31}-v_{32}, u_{32}-v_{31};$

II) $$\int_0^\pi I\sin^3\theta\,d\theta=0;\quad u_{23}-v_{13};$$

$$\int_0^\pi I\sin^3\theta\,d\theta \pm \gamma H\int_0^\pi \frac{4c}{1+4c}(2-I)\cos^2\theta\sin^3\theta\,d\theta=0;$$

$u_{11}+v_{12}+u_{22}+v_{21}, u_{11}-v_{12}+u_{22}+v_{21};$

III) $$\int_0^\pi (1+2c)I\sin^3\theta\,d\theta=0;\quad v_{13}+u_{23}, u_{13}-v_{23};$$

$$\int_0^\pi (1+2c)I\sin^3\theta\, d\theta \pm \gamma H\int_0^\pi \frac{4c}{1+4c}(1+2cI)\sin^3\theta\, d\theta = 0;$$

$$u_{12}+v_{11}+u_{21}+v_{22}, u_{11}+v_{12}-u_{22}-v_{21}, u_{12}-v_{11}-u_{21}+v_{22}, u_{11}-v_{12}+u_{22}-v_{21};$$

$$\text{IV)}\ \int_0^\pi (1+4c)I\sin^3\theta\, d\theta = 0;\ u_{13}+v_{23};$$

$$\int_0^\pi (1+4c)I\sin^3\theta\, d\theta \pm \gamma H\int_0^\pi 4cI\sin^3 0\, d\theta = 0;$$

$$u_{12}+v_{11}-u_{21}-v_{22}, u_{12}-v_{11}+u_{21}-v_{22}; \tag{7.44}$$

where

$$I = \frac{1}{\sqrt{1+4c}}\ln\frac{\sqrt{1+4c}+1}{\sqrt{1+4c}-1},\quad c = \frac{\Delta^2\alpha_+^2\sin^2\theta}{\omega^2},\quad \gamma H = \frac{\alpha_-}{\alpha_+}.$$

We have four groups of three or six equations. The I and II groups describe the Goldstone modes. For these modes we need to take into account the quadratic field corrections. Brusov and Popov[9] (see part III above) concluded that in the presence of a magnetic field three out of nine Goldstone modes become nonphonon because of the appearance of the gap $\propto \mu H$ in their spectrum.

The III and IV groups of equations describe the clapping and pair–breaking modes. If we write these equations as $F_0(E) \pm \gamma HF_1(E) = 0$ and try to express E as $E = E_0 \pm \gamma HE_1$ we obtain

$$E = E_0 \pm \gamma HE_1 = E_0\left(-1 \mp \frac{F_1(E_0)}{1-3F_1(E_0)/4}\right) \tag{7.45}$$

Using the values of E_0, obtained by Brusov and Popov[9] we obtain for the energies of clapping and pairbreaking modes:

clapping:

$$E_1 = (1.17 - i0.13)\Delta_0 ,$$

$$E_{2,3} = (1.17 \pm 1.70\gamma H)\Delta_0 - i(0.13 \mp 1.20\gamma H)\Delta_0$$

pair breaking:

$$E_1 = (1.96 - i0.31)\Delta_0 ,$$

$$E_{2,3} = (1.96 \pm 2.04\gamma H)\Delta_0 - i(0.31 \pm 0.06\gamma H)\Delta_0 . \quad (7.46)$$

So for small H we have three–fold splitting of the clapping– and pair–breaking modes, i.e. we have a linear Zeeman effect for these modes. Magnetic fields lift the degeneracy of pair–breaking modes completely, and lift the degeneracy of clapping–modes particularly, each branch of the latter modes remaining twice degenerate.

Note that the magnetic field alters both the real part of the collective–mode energy and the imaginary, i.e., it alters both the frequencies the collective excitations and their attenuation even the linear approximation. In this case the attenuation of some modes grows, and of others decreases. Note that the frequencies and the attenuation of one *pb*–mode and two *cl*–modes in the linear approximation are not changed.

Let us compare the results obtained here with the results of Refs. 20 and 21. As was already mentioned, in addition to the difference in the number of Goldstone–modes, our results are more accurate (by roughly 5%), since we have taken account of the influence of the attenuation of the collective modes on their frequencies. In addition, we have investigated the influence of the magnetic field on the attenuation of the collective excitations, which, naturally, is absent in Refs. 14 and 15.

In Ref. 20 some of the modes are not changed in a magnetic field, while the frequencies of others change from ω_1 to $\left(\omega_1^2+\Omega^2\right)^{1/2}$, (where Ω is the effective Larmor frequency). Such a quadratic (in the field) frequency shift in small fields (instead of a linear shift) was obtained by the authors of Ref. 20 since they did not take account of the deformation of the gap in the Fermi spectrum. The authors of Ref. 21, taking account of the deformation of the gap, obtained a linear splitting of the frequencies of some of the modes

$$\omega_{i\sigma}\left(\frac{T}{T_{C\sigma}}\right)=\frac{T_{C\sigma}}{T_{C0}}\omega_i\left(\frac{T}{T_{C\sigma}}\right),\quad i=nfl,\ sfl,\ scl,\ \sigma=\uparrow,\downarrow, \tag{7.47}$$

where *nfl*– is the normal flapping mode, *sfl*– is the super–flapping mode, and *scl*– is the super clapping mode, while the frequencies of other modes changed from ω_1 to $\left(\omega_1^2+\Omega^2\right)^{1/2}$, i.e., they remained invariant in the linear approximation.

Thus, the authors of Ref. 21, obtained as we did here a threefold splitting of the collective mode spectrum in small fields, and in this sense the conclusions of both papers are similar. However, there exists an entire list of differences including qualitative ones. Thus, in Ref. 21 the collective mode spectrum in ^{3}He–*A* in a magnetic field is obtained from the spectrum without a magnetic field by the substitution $\Delta\rightarrow\Delta\pm\mu H$, inasmuch as in the weak–coupling approximation, used in both works, the subsystems of fermions with spins aligned with and against the field are independent. Our result does not satisfy such a simple condition, and the reason for this in our opinion is the following. Tewordt and Schopohl[20] did not allow for the attenuation of the collective modes which takes place even in the case of zero momenta of the excitations and is connected with the vanishing of the gap at the poles. But the presence of attenuation, i.e., of an imaginary part of the energy (frequency) of the collective mode, renormalizes the real part of the energy of the collective mode. Since the attenuation is different for different modes, the above–mentioned condition for the spectrum to change in a magnetic field, as follows from our results when attenuation is taken into account, must be looked at more closely, i.e., for each collective mode separately and in the form $\Delta \to \Delta \pm \alpha_i \mu H$, where i stands for either *cl–* or *pb–*. But since attenuation (i.e., the imaginary part) depends on the field, α_i can be different for the real and imaginary parts.

Another conclusion that can be drawn from an analysis of the results presented here is that the linear Zeeman effect for the *cl–* and *pb–*modes takes place due only to the deformation of the order parameter (or the particle–hole asymmetry) in the case in which the momenta of the excitations are zero: $\boldsymbol{k} = 0$. For nonzero momenta of the collective excitations there exists the fundamental possibility that the additional term in the hydrodynamic action functional will, without deformation of the gap, lead to linear corrections to the frequencies of the collective modes.

The threefold splitting of the spectra of the *cl–* and *pb–*modes, which we have obtained, can be observed in ultrasound experiments. Note that in the case of the A–phase (in contrast with the case of the B–phase) the deformation of the gap is linear in the field, which should make it possible to observe splitting of the spectrum of the collective modes equal in order of magnitude to the existence region of the A_1–phase in moderate fields.

7.5. Kinetic Equation Results on Collective Modes in *A*–Phase

7.5.1. Sound and the order parameter collective modes

In this section we will describe a kinetic equation approach following to review[23]. A detailed study of the properties of the collective modes requires solving the eigenvalue equations. This has been done using a variety of theoretical approaches, the Green function theory[21,24-27], kinetic theory[7,8,13,28] or semi–phenomenological theories[2,29,30] as well as pure hydrodynamic theory (see reviews[31,32]).

Sound dispersion in the A–phase in the formulation of Wolfle[13,28] and Wolfle and Koch[7,8] is expressed in terms of the response function $\xi^0(\hat{\boldsymbol{q}},\omega)$,

$$\omega^2 \approx c_0^2 q^2 \left[1 - 2\frac{c_0 - c_1}{c_1}\xi^0(\hat{\boldsymbol{q}},\omega)\right] \tag{7.48}$$

The first–sound velocity c_1, is given in terms of the Landau parameters F_0^S, and F_1^S, and the Fermi velocity v_F by

$$c_1^2 = \frac{1}{3}\left(1 + F_0^S\right)\left(1 + \frac{1}{3}F_1^S\right)v_F^2 \tag{7.49}$$

and the difference between the velocities of zero– and first–sounds by

$$\frac{c_0 - c_1}{c_1} = \frac{2}{5}\frac{1 + \frac{1}{5}F_2^S}{F_0^S} + O\left(\frac{v_F^2}{c_1^2}\right) \tag{7.50}$$

The sound velocity and attenuation coefficient can be obtained from the dispersion law, Eq. (7.48), as

$$c = c_0 - (c_0 - c_1)\operatorname{Re}\xi^0(\hat{\boldsymbol{q}},\omega), \quad \alpha = q\left[(c_0 - c_1)/c_0\right]\operatorname{Im}\xi^0(\hat{\boldsymbol{q}},\omega). \tag{7.51}$$

The response function $\xi^0(\hat{\boldsymbol{q}},\omega)$ is a complicated average over the distortion of the gap parameter due to the sound wave and depends on the angle β, between the sound propagation direction $\boldsymbol{q}$ and the orbital anisotropy axis $\boldsymbol{l}$. These distortions of the L = 1 symmetry involve second–order spherical harmonics, Y_2^m. As for any expectation value $\langle\delta\psi|0|\delta\psi\rangle$, the distortion enters as its square, and the final angular dependence of $\xi^0(\hat{\boldsymbol{q}},\omega)$ on β is the same as that of the $\left|Y_2^m\right|^2$.

$$\xi^0 = \xi_1 \cos^4\beta + 2\xi_0 \sin^2\beta\cos^2\beta + \xi_{-1}\sin^4\beta \,. \tag{7.52}$$

In Wolfle's notation, the subscripts 1, 0 and –1, which correspond to the superscripts m of d_j^m are labelled ||, c and $\perp$. They indicate components parallel to $\boldsymbol{l}$, or intermediate (combined), or perpendicular to $\boldsymbol{l}$. The full expressions for $\xi^0_{\parallel}, \xi^0_C$ and $\xi^0_{\perp}$ containing the effect of quasi–particle collision are lengthy and are not reproduced here. When the quasi–particle relaxation time becomes infinite, they simplify significantly,

$$\xi^0_{\parallel} = \frac{45}{4}\left[\langle\lambda\cos^4\theta\rangle - \langle\lambda\cos^2\theta\rangle^2 / \langle\lambda\rangle\right],$$

$$\xi^0_C = \frac{45}{8}\Bigg[3\langle\lambda\sin^2\theta\cos^2\theta\rangle - \langle\lambda\cos^2\theta\rangle\langle\lambda\sin^2\theta\rangle / \langle\lambda\rangle -$$

$$- \frac{\langle\lambda\cos^2\theta\rangle^2}{\langle\lambda\,\mathrm{cotan}^2\theta - 2(\Delta_0/\omega)^2(\lambda+\Xi)\cos^2\theta\rangle}\Bigg],$$

$$\xi_{\perp}^{0} = \frac{45}{16}\left[\frac{3}{2}\langle\lambda\sin^4\theta\rangle - \langle\lambda\sin^2\theta\rangle/\langle\lambda\rangle - \right.$$

$$\left. -\frac{1}{4}\frac{\langle\lambda\sin^2\theta\rangle^2}{\langle\lambda - 2(\Delta_0/\omega)^2\lambda\sin^2\theta\rangle}\right]. \qquad (7.53)$$

In these expressions, θ is the angle between $\boldsymbol{k}$ and $\boldsymbol{l}$, the bras and kets denote the angular average over the Fermi surface,

$$\langle\ldots\rangle = \frac{1}{2}\int_{-1}^{1} d\cos\theta \int_{0}^{2\pi} \ldots d\varphi, \qquad (7.54)$$

and the functions λ and Ξ of ω, T and θ are the following averages over energy,

$$\lambda(\omega, T, \theta) = \int_{-\infty}^{\infty} d\varepsilon_k \frac{\tanh(E_k/2k_BT)}{E_k^2 - \frac{1}{4}\omega^2}\frac{|\Delta_k^2|}{2E_k},$$

$$\Xi(\omega, T, \theta) = \int_{-\infty}^{\infty} d\varepsilon_k \frac{|\Delta_k^2|}{E_k}\frac{\partial}{\partial E_k}\left[\frac{\tanh(E_k/2k_BT)}{2E_k}\right], \qquad (7.55)$$

where E_k and ε_k are given by

$$E_k = \left[\varepsilon_k^2 + \Delta^2\right]^{1/2}, \varepsilon_k = (\hbar^2/2m)(k^2 - k_F^2). \qquad (7.56)$$

The square of the unitary gap is equal to $\Delta_0^2\sin^2\theta$. The function $\langle\lambda\rangle$ is the generalization of the Tsuneto function,

$$\lambda(\omega,T)=\int_{-\infty}^{\infty}d\xi\frac{\Delta^2}{2E}\frac{\tanh(E/2k_BT)}{E^2-(\hbar^2\omega^2/4)}, \tag{7.57}$$

to the anisotropic case.

Sound propagation in the A–phase is governed by Eqs. (7.51)–(7.55) at low temperature, i.e., when quasi–particle collisions become unimportant $T \leq 0.8T_C$. It is markedly anisotropic, as Eq. (7.53) are quite different. Let us discuss these equations. The first two terms on the right–hand side contain the mean effect of pairing, i.e., the transition of the zero–sound velocity from c_0 close to T_C to c_1 at absolute zero, and also pair breaking which gives a singularity in λ when $\omega = 2E_k$. These singularities can be handled as Cauchy principal values[7] and cause no computational problem. However, they have physical significance:

(1) the weak (logarithmic) divergence at the pair breaking edge gives a cusplike feature in the propagation of sound[24,33];

(2) they contribute an imaginary part to the response function $\xi^0(\hat{q},\omega)$ for all angles smaller than θ_{pb} defined by $\hbar\omega = 2\Delta_0 \sin\theta_{pb}$.

Pair–breaking takes place at all frequencies and temperatures in a small cone about the nodes of the gap: there is dissipation in the A–phase down to absolute zero. This prevalent feature of the anisotropic gap strongly affects sound propagation as well as the modes which couple to sound and to which we now turn.

These modes appear as poles in the last terms on the right–hand sides of second and third of Eq. (7.53). They are absent in the first of Eq. (7.53): sound propagating with $\boldsymbol{q}$ in the direction of $\boldsymbol{l}$ ($\beta = 0$) does not couple to collective modes. In the perpendicular direction, only $\xi_\perp^0$ contributes. The corresponding normal variable is $\delta d_z''^{-1}$ which represents a clapping motion. The poles of $\xi_\perp^0$ are given by

$$\left\langle \lambda - 2(\Delta_0/\omega)^2 \lambda \sin^2\theta \right\rangle = 0. \tag{7.58}$$

At intermediate angles β, ξ_C^0 also contributes, with $\delta d_z''^0$ as the normal variable representing a flapping motion. The poles of ξ_C^0 are obtained by solving

$$\left\langle \lambda \cot\mathrm{an}^2\theta - 2(\Delta_0/\omega)^2(\lambda + \Xi)\cos^2\theta \right\rangle = 0 . \tag{7.59}$$

Eigenvalue Eqs. (7.58) and (7.59) have to be solved numerically. There is one root for the so–called clapping mode, Eq. (7.58), given approximately by Wolfie and Koch[7] who included the effect of a finite quasi–particle relaxation time τ,

$$\omega_{cl}^2 \cong [1.23\Delta_0(T)]^2 - i\omega_{cl}\left(0.12\omega_{cl} + \frac{1}{\tau}\right). \tag{7.60}$$

The clapping frequency is proportional to the gap within 1% over the whole temperature range from T_C to absolute zero.

The numerical analysis of Eq. (7.59) (Wolfle[34]) for the flapping motion yields, somewhat surprisingly, three zeros. One, $\omega = 0$, corresponds to gapless orbital waves. The two others are the so–called normal flapping and the superflapping modes. The frequencies and linewidths of these modes, as well as those of the clapping have been calculated by Tewordt and Schopohl[27]. These authors have reanalyzed the problem in detail, taking the dipole–dipole coupling into account and confirm in essence Wolfle's findings.

Some analytical results are available in the limits $T \to T_C$ and $T \to 0$. Near T_C, the normal flapping frequency behaves as[7] $\Delta_0(T)$,

$$\omega_{nfl}^2 \cong \frac{4}{5}\Delta_0^2(T)\left[1 - \frac{56}{3\pi^4}\xi(3)\frac{\Delta_0(T)}{k_B T}\right] - 2i\omega_{nfl}^2\left(0.34 + \frac{\omega_{nfl}}{\tau}\right), \tag{7.61}$$

while, close to absolute zero, ω_{nfl}^2 decreases as T. The exact variation depends on the dipolar coupling and the details of the merging of the normal flapping and the orbital modes[27].

This reentrance of the normal flapping at low temperature is a curious phenomenon which has been verified experimentally by Ling et al.[35]. It gives a clue as to the physical nature of this mode which owes its existence to the presence of normal quasi–particles in the vicinity of the nodes of the gap as will be explained in more detail in section 7.5. The role of the normal fluid in this mode explains its name.

The last zero of Eq. (7.59) leads to a high–frequency mode, which varies as $1.56\Delta_0(T)$ close to T_C and tends toward $2\Delta_0(T)$ at zero temperature. This mode involves motions of $\boldsymbol{l}$ connected with the structure of the superfluid order parameter and has been called the superflapping mode. Being a high–frequency mode, it experiences very severe damping from pair–breaking and its width is very large.

We have attributed three modes to a single normal variable, $\delta d_z''^0$, which[36] gives rise to the orbital Goldstone mode. This situation, which would be ruled out for the small motions of a well–behaved mechanical system, is not implausible here, because Eq. (7.59) is highly non–linear: a perturbation expansion about a given mode frequency will involve frequency–dependent coefficients and the resulting seemingly linear equation may well have more than one root per variable.

Dombre and Combescot[37] have paid attention to this problem and have demonstrated that both Eqs. (7.58) and (7.59) have no genuine root in the lower half complex plane. The restriction to $\mathrm{Im}\,\omega < 0$ comes from causality. This result implies that there is no pole in the response function, and hence, no genuine mode of the order parameter in addition to the orbital Goldstone mode. The clapping and superflapping features are resonances in ξ_0 which give rise to sound absorption peaks. Their frequencies are given by the zeros of the real part of Eqs. (7.58) and (7.59). As the imaginary parts are not overwhelmingly large, in spite of the existence of the pair–breaking cones, the denominators in the last terms of the second and third of Eq. (7.53) for ξ_C^0 and $\xi_\perp^0$ become small and contribute to resonant–like features in the sound propagation. These resonances are clearly evident in an experiment, especially the clapping mode. For want of a better terminology, we shall stick to common, if loose, usage and go on calling such resonances modes. Thus, a satisfactory way out of the somewhat paradoxical situation of several modes pertaining to the same normal variable has been proposed, but no additional light has been cast on the physical origin of these high–frequency quasi–modes.

7.5.2. Orbital waves and sound

The normal flapping mode has not yet been included in the classification of the OPCM. This mode is closely related to propagating fluctuations of $\boldsymbol{l}$. At low frequency, these orbital modes are the Goldstone excitations of the broken rotational symmetry in real space. Most of what is known about orbital dynamics is contained in the paper by Leggett and Takagi[30] and in the reviews by Brinkman and Cross[38] and Volovik[31]. The question of the intrinsic, or spontaneous, orbital angular momentum is a central issue here. It turned out to be quite a subtle affair which has been the object of debate in the literature[29-31]. This question will not be covered in any detail here. We shall simply make use of the fact that this intrinsic angular momentum is very small. Below we describe spin dynamics of the system, following the papers by Halperin and Varoquaux[23].

As recognized by Leggett and Takagi[29,30] based on their description of spin dynamics[39] and with input from microscopic theory, orbital dynamics is governed by three factors:

(1) Oscillations of $\boldsymbol{l}$ are motions of the nodes of the gap. At finite temperature, normal quasi–particles are thermally excited and the distribution function is denser in the vicinity of the nodes than along the equator where the gap is maximum; the normal fluid density is anisotropic. When $\boldsymbol{l}$ changes direction, the normal quasi–particles have to relax back to equilibrium with its new direction. This relaxation is accompanied by dissipation which strongly damps the motion of $\boldsymbol{l}$.

(2) When $\boldsymbol{l}$ is out of equilibrium with the quasi–particle system, the total energy is higher. This creates a potential well in direction $\boldsymbol{l}_n$ defined by the quasi–particle distribution (for as long as the initial distribution lasts) and along which $\boldsymbol{l}$ tends to fall. The role of the normal fluid as providing a restoring force for $\boldsymbol{l}$, aside from being very effective in damping its motion, was recognized in Refs. 40, 41. It has been termed the *normal locking* effect. This typical visco–elastic effect can be evaluated by assuming that the quasi–particle distribution is frozen–in at high frequency, and that the quasi–particle energy is enhanced according to the increase in the non–equilibrium gap (the superfluid component adjusts itself to prevailing local conditions in a time of the order of $\hbar/\Delta$).

Although the angular momentum of a pair of bare atoms in an $L = 1$ state is $\hbar$, the spontaneous angular momentum in the ABM–state is very small. This is due to the fact that Cooper pairs are made of Landau quasi–particles and quasi–holes in nearly equal proportion. The angular momentum arising from the spinning of the quasi–particle part about $\boldsymbol{l}$ is almost exactly balanced by that coming from the quasi–hole part which behaves like a negative mass also spinning about $\boldsymbol{l}$ (or a real mass spinning in the opposite direction). The cancellation which results, decreases the orbital angular momentum by a factor $(\rho_S/\rho)(\Delta/E_F)^2 \cong 10^{-6}$, ρ_S/ρ being the superfluid fraction. This factor would be zero in the case of perfect particle–hole symmetry. The spontaneous orbital angular momentum in the A–phase at rest is aligned along the $\boldsymbol{l}$ vector, and its magnitude for a sample of N atoms is given by

$$\boldsymbol{K}_0 = L_0\boldsymbol{l},\ L_0 = \left(\frac{1}{2}N\hbar\right)(\rho_S/\rho)(\Delta/E_F)^2. \tag{7.62}$$

In off–equilibrium situations, this macroscopic angular momentum, which is associated with the existence of pairs, takes on a dynamics of its own. This additional dynamical variable $\boldsymbol{K}$ is an actual angular momentum on which external torques can act.

The equations of motion of the orbit quantization axis $\boldsymbol{l}$, the intrinsic angular momentum of the Cooper pairs $\boldsymbol{K}$ and the axis of symmetry of the frozen–in normal quasi–particle distribution have been written by Leggett and Takagi[29,30,39] as (see also[36])

$$\dot{\boldsymbol{l}} = -\boldsymbol{l} \times \boldsymbol{\omega}_{K}; \ \dot{\boldsymbol{K}} = -\boldsymbol{l} \times \frac{\partial E}{\partial \boldsymbol{l}} - \frac{\boldsymbol{K} - \boldsymbol{K}_0}{\tau_K}; \ \dot{\boldsymbol{l}}_n = \frac{\boldsymbol{l} - \boldsymbol{l}_n}{\tau_n} \tag{7.63}$$

Last equation of (7.63) simply describes the relaxation of $\boldsymbol{l}_{\mathrm{n}}$ on $\boldsymbol{l}$ with a characteristic time τ which is of the order of the quasi–particle collision time. The first equation of (7.63) describes the precession of $\boldsymbol{l}$ about the driving field $\boldsymbol{\omega}_K$ which must be related to $\boldsymbol{K}$ to account for the coupling between $\boldsymbol{l}$ and $\boldsymbol{K}$,

$$\boldsymbol{\omega}_K = \boldsymbol{K}/\chi_{orb}\,. \tag{7.64}$$

The quantity χ_{orb} is a moment of inertia termed by Leggett and Takagi the “orbital susceptibility at constant normal component”. By analogy with the spin dynamics case, it is defined as the inverse second derivative of the energy with respect to orbital angular momentum, $\chi_{orb}^{-1} = \partial^2 E/\partial K^2$, taken at constant (frozen) normal component; these equations are valid in the collisionless regime. The quantity $\boldsymbol{K}$ has been referred to above as the intrinsic angular momentum of the Cooper pairs in the sense that it does not contain any contributions from macroscopic supercurrents. Its phenomenological definition is not precise and it is best expressed in terms of the microscopic theory (Leggett and Takagi[30]). A visual representation which might be given is the following. The spontaneous orbital angular momentum $\boldsymbol{K}_0$, collinear with $\boldsymbol{l}$, is very small because of particle–hole symmetry. Any off–equilibrium angular momentum collinear with $\boldsymbol{l}$ is small for the same reason. Besides, such an angular–momentum operator will generate rotations about $\boldsymbol{l}$ and interfere with gauge transformations *I*. But rotation of orbits about an axis perpendicular to $\boldsymbol{l}$ are well–behaved transformations which can be generated by $\boldsymbol{K}$ (when perpendicular to $\boldsymbol{l}$) and to which correspond a finite moment of inertia χ_{orb}. A mechanical analog would be an axially symmetric top spinning about $\boldsymbol{l}$ at a low rotation velocity but having a high moment of inertia with respect to transverse rotations.

The energy $E(\boldsymbol{l})$ in the second equation of (7.63) contains several contributions: the normal locking effect, the dipolar coupling to the spin axis $\boldsymbol{d}$, gradient energies in inhomogeneous situations, and the work of externally applied torques, if any. However, orientational energies, like that induced by normal–superfluid counterflows, act on the normal fluid and their effect is felt via $\boldsymbol{l}_n$. The energy density corresponding to these various factors is given by

$$E(\boldsymbol{l})=\frac{g_n(T)}{2}\left[1-(\boldsymbol{l},\boldsymbol{l}_n)^2\right]+\frac{\tilde{g}_d(T)}{2}\left[1-(\boldsymbol{l},\boldsymbol{d})^2\right]+\frac{\hbar^2}{12}(\rho\gamma_b/m^*)[\boldsymbol{l}\times(\nabla\times\boldsymbol{l})]^2. \tag{7.65}$$

The quantities $g_n(T)$ and $\tilde{g}_d(T)$ are the normal locking and the dipolar energies as given by Leggett and Takagi[30] and Leggett[42], respectively, and γ_b is Cross's[41] coefficient for elastic bending energies. To simplify the coefficient of the last term of Eq. (7.65), we introduce the orbital velocity,

$$v_{orb}^2=\hbar^2\rho\gamma_b/6m^*\chi_{orb}. \tag{7.66}$$

The velocity v_{orb} turns out to be of the order of the Fermi–velocity v_F.

For small deviations $\delta\boldsymbol{l}$ of $\boldsymbol{l}$ with respect to its equilibrium position $\hat{z}$, the set of Eqs. (7.63)–(7.66) can be linearized, assuming $\boldsymbol{d}$ to be pinned,

$$\dot{\boldsymbol{l}}=-\hat{z}\times(\boldsymbol{K}/\chi_{orb}), \tag{7.67}$$

$$\dot{\boldsymbol{K}}=-\hat{z}\times\left[g_n(\delta\boldsymbol{l}-\delta\boldsymbol{l}_n)+\tilde{g}_d\delta\boldsymbol{l}-\chi_{orb}v_{orb}^2\frac{\partial^2\delta\boldsymbol{l}}{\partial z^2}\right]-(\boldsymbol{K}-L_0\hat{z})/\tau_K, \tag{7.68}$$

$$\delta \dot{l}_n = (\delta l - \delta l_n)/\tau_n \,. \tag{7.69}$$

Finally, looking for plane wave solutions varying as $\exp[-i(\omega t - \boldsymbol{q}\cdot\boldsymbol{r})]$ and eliminating $\boldsymbol{K}$ and $\boldsymbol{l}_n$, we find the dispersion relation for small oscillations of $\boldsymbol{l}$,

$$\omega_{orb}^2 = -i\omega_{orb}\frac{\chi_{orb}}{\tau_{\boldsymbol{K}}} + \frac{1}{1 + i/\omega\tau_{\boldsymbol{n}}}\frac{g_n}{\chi_{orb}} + \frac{\tilde{g}_d}{\chi_{orb}} + v_{orb}^2 q_z^2 \,. \tag{7.70}$$

Relation (7.70) describes the behavior of the orbit–wave mode at all frequencies in the range of validity of Leggett and Takagi's semi–phenomenological theory. It is both a remarkable and satisfactory result that, starting with a precession equation for $\boldsymbol{l}$ [first of Eq. (7.63)], we finally obtain a linearly polarized oscillation to which there corresponds only one mode, in contrast with the case of spin waves, but in agreement with the group–theoretical classification. This mode is the flapping motion of $\boldsymbol{n}, \delta\boldsymbol{n}_z$ giving an oscillation δl_x of $\boldsymbol{l}$. It is schematically represented in Fig. 7.2.

The normal locking energy $g_{\boldsymbol{n}}(T)$ is given approximately by

$$g_{\boldsymbol{n}}(T) \approx \frac{\rho_n}{\rho}\frac{dn}{dE_F}\frac{\Delta^3(T)}{k_B T_C} \,. \tag{7.71}$$

The quantity $dn/dE_F = 3n/2E_F$ is the density of states (of both spins) per unit energy and unit volume at the Fermi–surface. Equation (7.71) is valid at all temperatures except very close to T_C and near absolute zero[29]. In the same temperature range, we also have $g_n >> \tilde{g}_d$ the normal locking effect dominates the dynamics of $\boldsymbol{l}$. The orbital susceptibility is given by

$$\chi_{orb} \approx \hbar^2 \frac{dn}{dE_F} \frac{\Delta(T)}{k_B T_C} \tag{7.72}$$

for $T \propto T_C$, and, in the $T \to 0$ limit, by

$$\chi_{orb} \approx \hbar^2 \frac{dn}{dE_F} \ln \frac{T}{T_C}. \tag{7.73}$$

Care has to be exercised in this last limit about the proper order with which the small quantities, T or ω go to zero.

In the collisionless limit, setting $1/\tau_n = 0$ in Eq. (7.70), the orbit–mode frequency simply becomes

$$\omega_{orb}^2 = -i\omega_{orb} \frac{\chi_{orb}}{\tau_K} + \frac{g_n}{\chi_{orb}} + \frac{\tilde{g}_d}{\chi_{orb}} + v_{orb}^2 q_z^2 \ . \tag{7.74}$$

Equation (7.74) is nothing more than the dispersion relation of the normal flapping mode, given in Wolfle's approach by Eq. (7.61). From Eqs. (7.71) and (7.73), we can give an approximate expression for $\omega_{orb} \approx (g_n/\chi_{orb})^{1/2} \approx (\rho_n/\rho)^{1/2}\Delta(T)$. The relaxation time τ_K for the pair intrinsic angular momentum contains pair breaking and is always significant. The connection between Leggett and Takagi's approach[30] and microscopic theory for the normal flapping mode (the orbital mode with a gap generated by normal locking) has been investigated in detail by Tewordt and Schopohl[27].

By various cross–checks with microscopic theory, Leggett and Takagi[30] have produced an ad hoc expression for τ_K which includes pair breaking,

$$\frac{1}{\tau_K} = \frac{\pi}{48}\hbar\frac{dn}{dE_F}\frac{(\hbar\omega)^2}{k_BT_C}. \tag{7.75}$$

The normal locking energy term in Eq. (7.70) exhibits the typical visco–elastic behavior that we anticipated in our discussion of the progressive unlocking of $\boldsymbol{l}$ from the anisotropic normal particle distribution as the collision time increases. At low frequency, the orbital mode dispersion is given by

$$\omega_{orb}^2 = -i\omega_{orb}\left[\frac{\chi_{orb}}{\tau_K} + \frac{\tau_n g_n}{\chi_{orb}}\right] + \frac{\tilde{g}_d}{\chi_{orb}} + v_{orb}^2 q_z^2 . \tag{7.76}$$

Damping is strong and orbit waves do not propagate[40], except very close to $T = 0$[43] where the limits leading to Eq. (7.76) require special care. In this $T = 0$ limit, and for $\hbar\omega << (k_BT_C)^2/E_F$ undamped orbital waves are found to propagate with a dispersion law entirely different from Eq. (7.76),

$$\omega_{orb} \cong \left[E_F/\hbar(k_BT_C)^2\right](q_z v_F)^2 \ln(k_BT_C/q_z v_F). \tag{7.77}$$

This interesting result shows that, at very low temperature where the normal locking effect vanishes, the intrinsic orbital angular momentum, given by Eq. (7.62), again plays an important role in the dynamics of $\boldsymbol{l}$.

Strictly speaking, Eq. (7.77) is the dispersion relation of the true Goldstone mode arising from orbital degeneracy. Equation (7.76) refers to a situation where this orbital degeneracy is partially lifted by the presence of the normal fluid which lags behind and defines a reference axis $\boldsymbol{l}_n$. We shall refer to the high–frequency orbital mode, i.e., the normal flapping mode, as the orbital Goldstone mode, in the same manner as for spin waves when the dipolar interaction is taken into account.

Finally, since the orbital degrees of freedom are coupled to the spin modes via the dipolar interaction, hybridization, or mode coupling effects occur. An orbital contribution to the transverse NMR linewidth can be expected in the mode–crossing region[20,29,30].

7.6. Textural Effects in A–Phase

In this section we discuss the textural effects in A–phase, following the papers by Halperin and Varoquaux[23] and by Brusov et al.[44]. Anisotropy is the hallmark of the A–phase. Experimental evidence for the existence of the orbital directrix $\boldsymbol{l}$, and if at all possible, for its motion, was actively sought in the early days of superfluid ^{3}He research. These pioneer's efforts (Lawson et al.[45], Roach et al.[46], Paulson et al.[47,48]) were very successful in demonstrating the existence of anisotropy as well as establishing the conditions under which the A–phase can be oriented and form a quasi–monodomain.

The factors governing the formation of textures can be summarized as follows (Brinkman and Cross[38]). Forces acting on $\boldsymbol{l}$ are due to:

(1) wall depairing effects, which orient $\boldsymbol{l}$ perpendicular to the boundary with a typical energy of the order of the condensation energy;

(2) the effect of the superfluid current;

$$\boldsymbol{J}_s = \rho_s \boldsymbol{v}_s + C_s\left(\vec{\nabla} \times \boldsymbol{l}\right), \tag{7.78}$$

which orients $\boldsymbol{l}$ parallel to $\boldsymbol{v}_s$ when the orbital part of the current $C_s\left(\vec{\nabla} \times \boldsymbol{l}\right)$ is small, i.e., in the direction in which the superflow kinetic energy $\frac{1}{2}\rho_s v_s^2$ is minimized;

(3) the elastic bending energy of the **l**–vector field, contained in Eq. (7.65), as well as additional splay and twist terms, involving $(\vec{\nabla}\times \boldsymbol{l})^2$ and $(\boldsymbol{l}\cdot\vec{\nabla}\times \boldsymbol{l})^2$ respectively;

(4) the dipolar coupling to **d**, favoring alignment.

Forces acting on **d** are due to:

(1) the dipolar coupling to **l**,

(2) the effect of a magnetic field caused by the anisotropy of the susceptibility and which orients **d** in a plane perpendicular to **H** in order to minimize the magnetic energy $\frac{1}{2}\chi(\boldsymbol{d}\cdot\boldsymbol{H})^2$.

In a real–life experimental cell, these various effects compete with one another to produce a variety of textures. Orders of magnitude can be fixed by comparing the orientational energies to the energy of the dipolar coupling between **l** and **d** which is a well–known quantity, given in terms of the longitudinal NMR frequency $\Omega_{\parallel}$ by $\frac{1}{2}\chi(\Omega_{\parallel}/\gamma)^2$, γ being the normal–phase susceptibility tensor. The dipole field $\Omega_{\parallel}/\gamma$ is of the order of 25 G.

For applied fields much larger than $\sim\Omega_{\parallel}/\gamma$, the **d**–vector is locked in a plane perpendicular to the field by the magnetic anisotropy. The **l**–vector tends to align with **d**. The situation in which **l** and **d** are tied to one another is called the dipole–locked case.

If a different orientation is forced locally on **l**, e.g., by the presence of the walls of the experimental cell, gradient energies fix, by comparison with the dipole energy, a characteristic length ξ_d over which the alignment of **l** and *d* is restored. A typical value for the dipole length is $\xi_d \cong 10\mu m$.

Finally, a flow velocity of 1 mm/s produces a torque aligning **l** with $\boldsymbol{v}_s$ of the same order of magnitude as the dipolar torque aligning **l** with **d**. It should be noted that such a flow velocity is not far from reaching the critical velocity in the A–phase. More precise values for these quantities can be found in a study of thermally driven flow by Kleinberg[49].

Competition between these various forces leads to a host of interesting phenomena. This vast field of the physics of anisotropic superfluids has been reviewed by Hall and Hook[32] in general, and by Salomaa and Volovik[50] for the more restrictive, but particularly interesting, case of superfluid ^{3}He under rotation. Although sonic experiments are underway in rotating cryostats at Cornell and at the Technical University of Helsinki, we shall not discuss the interaction of sound with vortices, but shall concentrate here on nearly uniform textures and on their effect on sound propagation. Even this simpler situation is far from being perfectly understood, as we shall see now.

According to Eq. (7.52), sound attenuation and velocity depend on the angle β between the sound propagation wave–vector q and the gap–axis $\boldsymbol{l}$. By application of a magnetic field, $\boldsymbol{d}$, hence $\boldsymbol{l}$ in the dipole–locked situation, can be confined to the plane perpendicular to $\boldsymbol{H}$. However, their direction in this plane will be fixed by other factors, probably related to supercurrents created by residual heat leaks. Following Roach et al.[46] and Ketterson et al.[51], it may be assumed, unless very special care is taken, that $\boldsymbol{l}$ takes all possible orientations in the plane perpendicular to $\boldsymbol{H}$ with more or less equal weight. Letting ψ be the angle between $\boldsymbol{q}$ and $\boldsymbol{H}$ and φ be the angle between $\boldsymbol{l}$ and the intersection of the plane defined by $\boldsymbol{H}$ and $\boldsymbol{q}$ and the plane perpendicular to $\boldsymbol{H}$, we can express β from the inner product between $\boldsymbol{l}$ and $\boldsymbol{q}$: $(\boldsymbol{l}\cdot\boldsymbol{q}) = \cos\beta = \cos\varphi\sin\beta$. Inserting this relation into Eq. (7.52) and averaging over all possible values of φ, we obtain the following mean response function,

$$\left\langle \xi^0 \right\rangle = \frac{1}{8}\left(3\xi^0_{\parallel} + 2\xi^0_C + 3\xi^0_{\perp}\right)\sin^4\psi + \left(\xi^0_{\perp} + \xi^0_C\right)\sin^2\psi\cos^2\psi + \xi^0_{\perp}\cos^4\psi \tag{7.79}$$

Equation (7.79) for $\left\langle \xi^0 \right\rangle$ has the same functional dependence with respect to ψ as Eq. (7.52) for the actual response function with respect to β, but the coefficients are admixtures of the $\parallel$, $\perp$, and c components.

Hence, there will be a contribution from the collective modes, and in particular from the clapping mode which comes from $\xi_\perp^0$, for all field directions.

An anisotropic sound propagation as described by Eq. (7.79), can be taken for example from the work of Roach et al.[46]. The angular functional dependence of Eq. (7.79) is very well verified in these experiments. From fits to the measured curves, the actual response function ξ^0 can be reconstructed.

An extensive study of the temperature dependence of the parallel (||), perpendicular ($\perp$) and combined attenuation coefficients has been performed by Paulson et al.[47,48]. The comparison of their data with theory, including the effect of quasi–particle collisions which cannot be neglected in the vicinity of T_C, has been carried out by Wolfle and Koch[7] and Wolfle[28]. The agreement between theory and experiment, although not perfect, especially for $\alpha_\parallel$, must nonetheless be considered as quite satisfactory in view of the fact that experimenters have no direct handle on the orientation of $\boldsymbol{l}$.

The La Jolla group also studied the relaxation of $\boldsymbol{l}$ toward a new equilibrium direction after a sudden rotation of the magnetic field which orients $\boldsymbol{d}$ (Paulson et al.[47,48], Wheatley[69]). These experiments probe the low–frequency dynamics of the orbital axis and provide values for the Cross–Anderson[52] orbital viscosity, arising from the normal locking visco–elasticity, and for $\tau_K \cong \tau$, the quasi–particle relaxation time (Leggett and Takagi[29,30]). They also revealed a host of strange time–dependent behaviors of the $\boldsymbol{l}$–vector field (Paulson et al.[47,48], Krusius et al.[70] and the review by Hall and Hook[32]) coming from a complex interplay of the $\boldsymbol{l}$–vector field with thermally driven flow, walls and other textures. Orbital dynamics studies in a true zero–flow state have been reported only recently by Reppy and Gammel[53].

The observation of various types of irreproducibilities in the propagation of sound which were already noted in the early work by Lawson et al.[54], as well as the time–dependent behavior mentioned above, called into question the stability of the A–phase textures in the presence of a superflow (Bhattacharyya et al.[55], Cross and Liu[56]). A transition from a uniform to a helical texture below a certain temperature ($T < 0.82T_C$ in the weak–coupling model) was predicted by Fetter[57,58]. Lin–Liu et al.[59] and Saslow and Hu[60] studied the transition toward helical textures induced by a magnetic field at given v_s (see Fetter and Williams[61] for a full account).

Helical textures stem from the term proportional to $(\vec{\nabla} \times \boldsymbol{l})$ in the expression of the supercurrent, Eq. (7.78), and are formed by the winding of $\boldsymbol{l}$ along the flow direction at an apex angle and with a pitch which depend on temperature and superflow velocity.

The experimental search for the formation of helical textures driven by a heat current was first carried out in a systematic fashion by Kleinberg[62]. This author studied in detail the problems associated with residual heat leaks, lack of uniformity in space, and hysteresis using sound anisotropy as a probe for the direction of $\boldsymbol{l}$. He failed to reach firm conclusions as to the existence of helical textures, but did find out that the A–phase was in fact more resistant to distortion by flow than anticipated theoretically.

An extensive experimental study of the helical textures was pursued at the University of Southern California (USC) by Bates et al.[63,64] and Bozler and Gould[65]. These authors used a bellows assembly and a superleak to create a well–controlled superflow velocity and also monitored textural changes by sound propagation. They claim to have established that most of the experimentally probed space was a quasi–uniform texture and that local inhomogeneous structures such as vortex lines occupied only a small fraction of the total volume. This result is, of course, a prerequisite for any study of transitions toward complex structures. They also observed drastic changes in the sound attenuation upon increasing magnetic field at constant v_s at well–defined field thresholds. Although these observations provide conclusive evidence for a textural transition, the observed phase diagram departs sufficiently from the calculated one so as to prevent an unambiguous identification.

Thus, for the time being, there is no direct experimental check of the existence of helical textures. It again seems in the work of Bates et al.[63,64], as it already did in Kleinberg's[62] that the A–phase uniform texture appears a great deal more stable against flow perturbation than theory predicts. Possible explanations for such a fact, discussed by Hall and Hook[32] and based in part on the work of Dow[71], imply the presence of solitons which can apparently appear at very low flow velocity. In the course of the work at USC in which the uniformity and the orientation of the textures was more carefully controlled than in previous studies, Bozler and Gould[65] have noted a discrepancy between the observed sound anisotropy at 30 MHz and under 28.8 bar, and that computed from Eqs. (7.52)–(7.55). This remark is disconcerting in view of the successes of the Serene–Wolfle (Serene[66], Wolfle and Koch[7]) theory. A conclusion that could possibly be drawn and which would follow the line suggested by Hall and Hook[32], is that the pre–existing textures were indeed not so uniform.

Textural changes in a different experimental set–up have been investigated by Hook et al.[67,68]. In this work, the anisotropic superfluid is confined in a slab between two parallel plates. The initial texture is very likely uniform, with $\boldsymbol{l}$ and $\boldsymbol{d}$ lying perpendicular to the plates. The application of a magnetic field in the same direction, i.e., normal to the plate, tends to tilt $\boldsymbol{d}$ at right angle with H. The spin axis $\boldsymbol{d}$ gives in at a certain threshold field, causing a textural transition known as the Freedericksz transition in work on liquid crystals. The field H_F at which the transition occurs can be computed (Hook et al.[68]) from knowledge of the plate spacing $\boldsymbol{d}$, of the bending energy, Eq. (7.65), and of the anisotropy of the magnetic energy.

Two methods were used to detect the change in texture, the motion of a torsional oscillator, and sound propagation anisotropy at 15 MHz. Both methods consistently give Freedericksz fields some 15% lower than estimated. This discrepancy may not be so significant, given the combined uncertainties of the experiment (~ 8%) and of the Landau and the superfluid gap parameters used in the theoretical estimate. In view of the difficulties met by experimenters in the field, we should not understate the success of these preliminary experiments which will undoubtedly provide better controlled textures and more accurately known bending energies and sound anisotropy coefficients.

Sonic studies have proved an effective tool in the study of the quasi–statics and low–frequency dynamics of the orbital degrees of freedom of the A–phase. We now turn to high–frequency properties and to the mode–mode interaction between sound and the collective modes.

Chapter VIII

Identification of ^{3}He–A by Ultrasound Experiments

8.1. Introduction

On the basis of available experimental data Gould[1] has suggested that ^{3}He–A, conventionally identify as the axial phase, may actually be an axi–planar phase. Some experiments have been done to clarify the situation[2], but the problem is still open. While we take no position on the interpretation of the data, we note that the collective mode spectrum structure of the two phases differ and that appropriate measurement could resolve the issue. We investigate this problem within a simple, time–dependent Ginzburg–Landau model as well as by studying the second variation of the free energy functional[3]. Both methods show that the spectrum in the axial phase is degenerate, while it is split in the axi–planar phase. This fact may serve as a sensitive test of the existence of the latter.

8.2. Mermin–Star's Phase Diagram Analysis

If superfluid ^{3}He had an s–wave state, there would be no problem. In this case l=0, s=0 and there are no degrees of freedom for the order parameter, which is a single complex number, and thus there is no alternative superfluid phase. Order parameter has an amplitude, corresponding to the value of energy gap in a single particle spectrum, and a phase, which for a bulk uniform superfluid would be constant and, thus, observable.

For s–wave superfluid in the Ginzburg–Landau region (asymptotically close to T_C) the free energy difference between normal and superfluid phases is, to lowest order,

$$F = -\lambda(1 - T/T_c)|\chi|^2 + \beta|\chi|^4 \tag{8.1}$$

where α and β are material dependent constants. For $T < T_C$ this gives the free energy difference correct to order $(1 - T/T_C)^2$, which by differentiating twice yields the specific heat discontinuity at T_C.

In superfluid ^3He, since $l = 1, s = 1$, the order parameter Δ has both spin structure and angular dependence in momentum space. Using the fact that the angular dependence must be of Y_{1m} symmetry, the order parameter can be recast as a 3x3 matrix of complex constants denoted A_{ij}. Mermin and Star[6] showed that the order parameter enters the lowest order Ginzburg–Landau free energy expansion with only fife unknown coefficients $\beta_1, \ldots, \beta_5$ as

$$\begin{aligned} F = &-\lambda(1 - T/T_c)|\chi|^2 Tr(AA^+) + \frac{1}{2}\beta_0|\chi|^4(\beta_1|Tr(A\tilde{A})|^2 + \\ &+ \beta_2|Tr(AA^+)|^2 + \beta_3 Tr|(A\tilde{A})(A\tilde{A})^*| + \beta_4 Tr|(AA^+)^2| + \\ &+ \beta_5 Tr|(AA^+)(AA^+)^*|), \end{aligned} \tag{8.2}$$

where $\alpha = N(0)$, $\beta_0/\alpha = (21\zeta(3)/40\pi^2)/(k_B T_C)^2$ and $N(0)$ is the density of states of one spin species. Separating out an overall amplitude Δ by requiring that $Tr(AA^+) = 1$ we get for the weak coupling values of the coefficients $\{\beta_1, \ldots, \beta_5\}$ $\{-1,2,2,2,-2\}$.

All we need to do is to take real values of β–parameters and find the order parameter which minimizing the above free energy (8.2). A problem, however, is that we do not know *a priori* what values of β–parameters we have. The β–parameters are the contact points between experiment and theory. At any given pressure they will have some particular set of values. As the pressure is reduced to zero, they will approach their weak coupling values.

In the weak coupling approximation the *B*–phase has the lowest energy, while other possible phases have lower energies than the *B*–phase over much of the five dimensional β–parameter space. The general problem of finding the absolute minimum of the Ginzburg–Landau free energy at an arbitrary location in the five dimensional β–parameter space is unsolved. Mermin and Star[6] showed that there are only six possible states with the static magnetic properties of the *A*–phase. There results in the physically relevant case where $\beta_1 < 0$ are shown in Fig. 8.1.

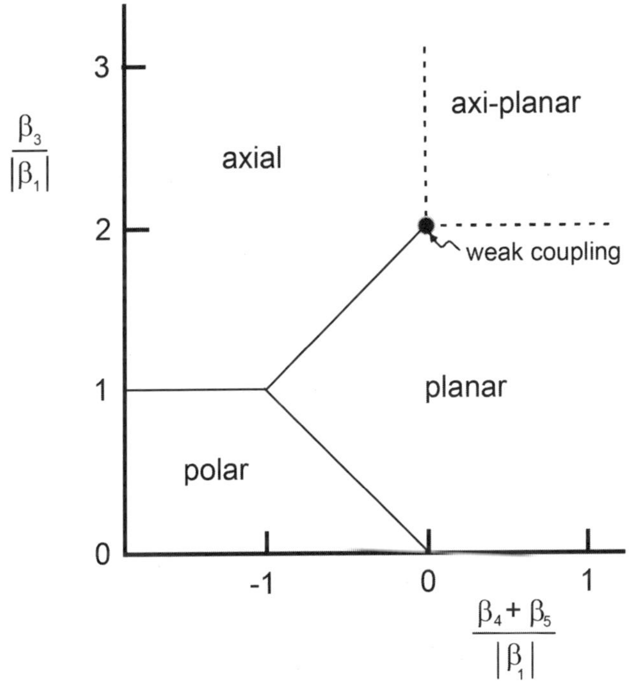

FIG. 8.1. Mermin–Star phase diagram[6] of the possible states explaining the existence of the A–phase. The figure assumes $\beta_1 < 0$, as is required for pressures below the PCP. Solid lines denote first order phase transitions, while dashed lines denote second order phase transitions. Note the location of the weak coupling point at the critical point joining three states of different symmetries.

Only four states remain as possibilities (polar, planar, axial and axi–planar phases). If as a function of pressure, we were able to move through β–parameter five dimensional space from the axial to the planar region, we would undergo a first order phase transition. On the other hand, the order parameters of the axial and the axi–planar phases are continuously connected, so that a transition between them would be of second order.

Note also, that the weak coupling point is at a critical point making the junction of three states – the axial–, the planar– and the axi–planar. The degeneracy of phases at this well–studied point is responsible for remarkable effects which do not survive under accounting the strong coupling corrections: the vanishing of the *A–B* surface energy and the vanishing of the *B*–phase's longitudinal component in a magnetic field.

There are two approaches one can take in determining the order parameter, the first with a theoretical orientation and the second with the experimental one. The first relies on the fact that the β_i's can be calculated from the microscopic scattering amplitude in the normal phase of liquid ^{3}He. Of course, the precise form of the amplitude needs then to be known, but for this one takes one's favorite interaction to generate the form, and uses normal state properties (Fermi–liquid factors, etc.) to fix the coefficients in the scattering amplitude. The superfluid state is then the one that minimizes the Ginzburg–Landau free energy at the predicted position in β–space.

The second approach is to assume nothing about β–parameters, but simply guess that a particular state is the correct one, calculate experimental properties as a function of where within the phase diagram the state is allowed, and compare to real situation. We can exclude a candidate state if the values inferred for the β–parameters from experiment do not lie within the allowed region of the phase diagram. (Alternatively, a non–thermodynamic measurement could find a result incompatible with the candidate state's properties.) This process needs to be repeated for each candidate state until all but one is excluded.

Of course, in practice one uses both approaches. It turns out that the second approach cannot be carried out to its natural conclusion so that the first approach has been allowed to acquire unwarranted prominence.

8.3. Axial Phase

Let us now return to the problem of determining the β–parameters to locate ourselves within one unique state's domain. We consider the situation for pressures below the polycritical point, since there is more experimental data available in this domain. This constraint eliminates the polar state as a possibility because in the region of β–space where it is stable, it always has a lower energy than the *B*–state, contrary to our assumption. Furthermore, since the *A*–*B* transition is a first order transition at all pressures and fields, the planar state is also eliminated as a possibility as it would always manifest a second order *A*–*B* transition.

This leaves only two possibilities: the axial state and axi–planar state. Distinguishing between these states by non–thermodynamic means will undoubtedly be difficult because they are continuously related to each other. Near their boundary in Fig. 8.1 the axi–planar state is simply the axial state with a small distortion toward the planar state.

If we had five distinct thermodynamic measurements, then we could locate ourselves in β–space, answering the question concerning realized phase. Unfortunately, we have only four measurements.

The first one is the specific heat discontinuity at T_C going into the *B*–phase, which yields the following combination of β–parameters:

$$\frac{\Delta C_B}{C_N} = \frac{12}{7\zeta(3)} \frac{5}{3\beta_{12} + \beta_{345}} . \tag{8.3}$$

All of the following expressions depend upon the state we choose as the model for the *A*–phase. For convenience we'll present expressions valid only for the axial state, although analogous forms can be found for the axi–planar state.) A second measurement is the suppression of the *B*–phase in a field which, assuming the *A*–phase is the axial state, yields

$$g(\beta)=\frac{\beta_{245}}{2(-3\beta_{13}+2\beta_{345})}\left[1+\sqrt{\frac{(3\beta_{12}+\beta_{345})(2\beta_{13}-\beta_{345})}{\beta_{245}\beta_{345}}}\right]. \tag{8.4}$$

With only these two measurements we can constrain the A–phase specific heat discontinuity as

$$\left(1-\frac{1}{6g(\beta)}\right)^2 \le \frac{\Delta C_A}{\Delta C_B} \le \left(1-\frac{1}{6g(\beta)}\right), \tag{8.5}$$

which already constrains the range of possible experimental results to a few percents. However, one can do better. The non–linear magnetization versus field characteristic of the B–phase depends upon a specific β–parameter combination. We characterize this by the amount of non–linearity which is exhibited in the largest possible magnetic field, that is the maximum suppression of the reduction of B–phase "susceptibility"

$$S=\frac{\lim\limits_{H\to H_C}\Delta\left(\dfrac{M}{H}\right)_{AB}}{\lim\limits_{H\to 0}\Delta\left(\dfrac{M}{H}\right)_{AB}}=1-\frac{1}{g(\beta)}\frac{(2\beta_{12}+\beta_{345})}{2\beta_{345}}, \tag{8.6}$$

which can be shown to also yield, in combination with Eqs. (8.3) and (8.4), the A–phase specific heat discontinuity directly. Finally, the up/down ratio of the A_1 and A_2 transitions yields

$$U=-\frac{(dT/dH)_1}{(dT/dH)_2}=\frac{\beta_5}{\beta_{245}} \tag{8.7}$$

The Ginzburg–Landau energy

$$F = -\alpha\left(1-\frac{T}{T_C}\right)\Delta^2 Tr\left(AA^+\right)+\frac{1}{2}\beta_0\Delta^4\times\left\{\beta_1\left|Tr\left(A\tilde{A}\right)\right|^2+\beta_2\left[Tr\left(AA^+\right)\right]^2+\right.$$
$$\left.+\beta_3 Tr\left[\left(A\tilde{A}\right)\left(A\tilde{A}\right)^*\right]+\beta_4 Tr\left[\left(AA^+\right)^2\right]+\beta_5 Tr\left[\left(AA^+\right)\left(AA^+\right)^*\right]\right.$$

(here $\alpha = N(0)$, $\beta_0/\alpha = \left(21\zeta(3)/40\pi^2\right)/(k_B T_C)^2$, and $N(0)$ is the density of states of one spin species) applies only to bulk, uniform superfluid without external field. This energy by itself, therefore can only give us one experimental number, the specific heat jump at T_C, yielding one combination of the β–parameters.

More experiments are permitted if we stress the superfluid with perturbing fields. Applying a magnetic field has, as one effect, the suppression of the *B*–phase relative to the *A*–phase. At pressures below polycritical point, applying this field uncovers the *A*–phase near T_C so that a second specific heat measurement can be made, yielding a second combination of the β–parameters.

Further, it is only because the prefactor g_Z to the magnetic field energy

$$F_H = g_Z HAA^+H \tag{8.8}$$

is well known a priori, as well as being directly measurable, that one can measure the suppression of the *B*–phase in a field, and by dividing out the known g_Z, extract yet a third combination of β–parameters.

Finally, in strong magnetic fields, the normal to superfluid transition splits into two second order transitions, named A_1 and A_2. The origin of this splitting is the finite slope to the density of states at the Fermi–energy, usually referred to as "particle–hole asymmetry". Unfortunately, this finite slope is not well known, has never been measured in the normal state, and even if measured, would not enter the Ginzburg–Landau free energy with a known prefactor. What saves us in this case is that the same particle–hole asymmetry prefactor, usually denoted η appears in the expressions for both the A_1 and A_2 transitions, so that

η does not appear in the ratio of the two transitions temperature dependencies. This yields the fourth and, to date, final combination of β–parameters.

This effectively exhausts the possibilities for extracting information about the β–parameters from magnetic field stresses. Experiments looking at apparently unrelated magnetic properties of the *B*–phase (non–linear magnetization versus field, suppression of the longitudinal component of the order parameter, enhancement of the transverse component, shift in the order parameter's rotation angle) all can be shown to be thermodynamically equivalent in information content to the four above experiments.

What about other fields? Including the well–known dipole–dipole energy effects the phase diagram, but only on a temperature scale of $10^{-7} K$. To extract useful information requires measuring fine detail on a smaller scale, that should be difficult. In principle, an electric field could stress the superfluid to yield more information, but in this case prefactor g_E is poorly known (uncertainty in the theoretical estimation of g_E is of order 300%, it is experimentally measured with an uncertainty of 25%, and g_E is very weak that require fields of order $10^5 V/cm$ to reach the dipole–dipole energy scale).

It may be possible to extract useful information regarding β–parameters from a measurement of *A*–*B* surface tension. It is not known how important small errors in non–β–parameters will be in such a measurement. Even if such uncertainties are ignored, there will not be a simple analytic formula connecting an experimental quantity with β–combinations as given in Eqs. (8.3)–(8.7), but rather numerical calculations will be needed. Whether these will show that a measurement will be sensitive to β–parameters has yet to be seen.

So, presently, having four combination of β–parameters at any given pressure, we can confine the β–parameters to a line in five dimensional space. Gould believes that strong coupling effects do not lead to the case, when this line will be exceedingly far from the weak coupling's point in five dimensional space. Finding the point on each pressure's line which minimizes the distance to the weak coupling point yields a set of points which appears to hug the boundary between the axial and axi–planar states in Fig. 8.1 to within experimental resolution. The direction in which β–parameters move in five dimensional space is at an angle of something greater than 90° away from theoretical predictions.

To resolve the issue with two alternative phases one of which describes the *A*–phase Brusov et al.[3] suggested two methods. Below we investigate this problem within a simple, time–dependent Ginzburg–Landau model as well as by studying the second variation of the free energy functional. Both methods show that the spectrum in the axial phase is degenerate, while it is split in the axi–planar phase. This fact may serve as a sensitive test of the existence of the latter.

8.3.1. Ginzburg–Landau model

We write the order parameter of the axi–planar phase of ^{3}He in the following form

$$C=\tilde{\Delta}=\Delta\begin{pmatrix} 0 & n_2 & 0 \\ -n_3 & 0 & in_1 \\ 0 & 0 & 0 \end{pmatrix}+i\begin{pmatrix} A_{11} & A_{12} & A_{13} \\ A_{21} & A_{22} & A_{23} \\ A_{31} & A_{32} & A_{33} \end{pmatrix}, \tag{8.9}$$

where the unit vector $\boldsymbol{n}=\sin\theta\,\cos\varphi\cdot\hat{x}+\sin\theta\,\sin\varphi\cdot\hat{y}+\cos\theta\cdot\hat{z}$ defines the equilibrium axi–planar state and the quantities $A_{\alpha i}$ measure the distortion associated with the excited–state collective modes. We use a simple, oscillatory, time–dependant Ginzburg–Landau model to describe the time dependence of the order parameter involving the Lagrangian

$$\begin{aligned} L=\Lambda\dot{\Delta}_{\alpha i}\dot{\Delta}^*_{\alpha i}-[\alpha\Delta_{\alpha i}\Delta^*_{\alpha i}+\beta_1\Delta_{\alpha i}\Delta_{\alpha i}\Delta^*_{\beta j}\Delta^*_{\beta j}+\beta_2\Delta_{\alpha i}\Delta^*_{\alpha i}\Delta_{\beta j}\Delta^*_{\beta j}+ \\ +\beta_3\Delta_{\alpha i}\Delta_{\beta i}\Delta^*_{\beta j}\Delta^*_{\alpha j}+\beta_4\Delta_{\alpha i}\Delta^*_{\beta i}\Delta_{\beta j}\Delta^*_{\alpha j}+\beta_5\Delta_{\alpha i}\Delta^*_{\beta i}\Delta^*_{\beta j}\Delta_{\alpha j}]. \end{aligned} \tag{8.10}$$

Minimizing the equilibrium potential energy $(D_{\alpha i} = 0)$ with respect to n_i yields

$$n_1^2 = \beta_{113}\beta_{345}/2\left(2\beta_{13}\beta_{345} - \beta_3^2\right) \text{ and } n_2^2 = \beta_{13}\beta_{45}/\left(2\beta_{13}\beta_{345} - \beta_3^2\right), \tag{8.11}$$

where $n_1^2 + n_2^2 + n_3^2 = 1$, $\beta_{ijk...} = \beta_i + \beta_j + \beta_k + ...$; we not the limits $n_2 = 0$ (pure axial), $n_1 = 0$ (pure planar) and $n_1 = n_2 = 0$ (polar). Inseting Eq. (8.9) into Eq. (8.10), retaining only the quadratic terms in $A_{\alpha i}$ and using the Lagrange's equation, we obtain the equations of motion for $A_{\alpha i}$ and $A_{\alpha i}^*$ which determine the collective mode frequencies ω. For the case of arbitrary n_i, these equations factorize into four classes, where C are coupled within each class. We write these equations in matrix form as

$$\sum_j M_{ij}^{(C)} X_j^{(C)} = \Omega^2 X_i^{(C)}, \tag{8.12}$$

where $\Omega^2 = \left(\Lambda\omega^2 - \alpha\right)/\Delta^2$, the superscript C numbers the classes and i (j) denotes the various linear combinations of $A_{\alpha i}$ and $A_{\alpha i}^*$ which couple within a given class. The solutions of Eq. (8.12) define the collective mode response of the axi–planar phase ($n_2 \neq 0$) and the axial phase ($n_2 = 0,\ n_1 = n_3 = 1$). For the axial phase one gets the following results for ω^2:

0 (5 modes); $\alpha\beta_{45}/\Lambda\beta_{245}$ (4 modes); $\alpha(\beta_{45} - \beta_3)/\Lambda\beta_{245}$ (4 modes);

$2\alpha\beta_5/\Lambda\beta_{245}$ (2 modes); $-2\alpha\beta_{13}/\Lambda\beta_{245}$ (2 modes);

$-2\alpha/\Lambda$ (1 mode).

Here, $\Delta^2 = -\alpha/\beta_{245}$.

We see that collective mode spectrum in the axial phase turns out to be degenerate.

For the axi–planar phase the solutions would have to be obtained numerically in general with the exception of the $X^{(1)}$ mode whose degeneracy is lifted for $n_2 \neq 0$. We would expect all the remaining doubly degenerate modes to be also split. Hence, a splitting of degenerate axial phase modes in the axi–planar phase might serve as a sensitive test of the existence of the letter.

8.3.2. The second variation of free energy

The second method for investigating this problem is the study of the second variation of the action S of the system

$$\delta^2 S = -$$

$$\mathrm{tr}AA^+ + \beta_2 \mathrm{tr}[(A^+C)^2 + (C^+A)^2 + 2A^+AC^+C + 2A^+CC^+A] + \beta_4 \mathrm{tr}[2AA^+CC^+$$

$$+2A^+AC^+C + AC^+AC^+ +$$

$$A^+CA^+C] + \beta_5 \mathrm{tr}[AA^+C^*C^T + A^+A^*C^TC + A^*A^TCC^+ + A^TAC^+C^*$$

$$+AC^+A^*C^T + A^TCA^+C^*] +$$

$$\beta_3 \mathrm{tr}[AA^TC^*C^+ + A^TA^*C^+C + A^*A^+CC^T + A^+AC^TC^*] + 4\beta_1 \mathrm{tr}|C^TA|^2, \tag{8.13}$$

which allows to analyze the structure of the collective mode spectrum. Here $D_{\alpha i} = U_{\alpha i} + iV_{\alpha i}$ and C is the equilibrium order parameter.

After all calculations one gets that for the axi–planar phase in the strong coupling limit $\delta^2 S$ is equal to

$$\delta^2S/\Delta^2 = U_{11}^2(\Delta^{-2}+2n_3^2\beta_{2345}+2n_2^2\beta_{2345})+U_{12}^2(\Delta^{-2}+2n_1^2\beta_{2345}+n_2^2(6\beta_{245}+4\beta_{13}))+$$
$$+U_{13}^2(\Delta^{-2}+2n_1^2(\beta_{234}-\beta_5)+2n_2^2\beta_{2345})+U_{21}^2(\Delta^{-2}+2n_1^2(\beta_{245}-\beta_3)+2n_3^2(\beta_{245}+2\beta_{1345}))+U_{22}^2(\Delta^{-2}+2n_1^2(\beta_{245}-\beta_3)+2n_3^2\beta_{2345}+2n_2^2\beta_{2345})+U_{23}^2(\Delta^{-2}+2n_1^2(\beta_{245}+2\beta_1)+$$
$$+2n_3^2\beta_{2345})+U_{31}^2(\Delta^{-2}+2n_3^2\beta_{2345})+U_{32}^2(\Delta^{-2}+2n_2^2\beta_{2345})+U_{33}^2(\Delta^{-2}+2n_1^2(\beta_{234}-\beta_5)+$$
$$V_{11}^2(\Delta^{-2}+2n_3^2(\beta_{234}-\beta_5)+2n_2^2(\beta_{245}-\beta_3))+V_{12}^2(\Delta^{-2}+2n_2^2(\beta_{2345}+2\beta_{15}))+$$
$$+V_{13}^2(\Delta^{-2}+2n_1^2\beta_{2345}+2n_2^2(\beta_{24}-\beta_3))+V_{21}^2(\Delta^{-2}+2n_1^2\beta_{2345}+2n_3^2(\beta_{12345}+\beta_1))+$$
$$V_{22}^2(\Delta^{-2}+2n_1^2\beta_{2345}+2n_3^2(\beta_{245}-\beta_3)+2n_2^2(\beta_{234}-\beta_5))+$$
$$V_{23}^2(\Delta^{-2}+2n_1^2(3\beta_{245}+2\beta_{13})+2n_3^2(\beta_{245}-\beta_3)+2n_2^2\beta_5)+$$
$$V_{31}^2(\Delta^{-2}+2n_3^2(\beta_{234}-\beta_5))+V_{32}^2(\Delta^{-2}+2n_2^2(\beta_{234}-\beta_5))+V_{33}^2(\Delta^{-2}+2n_1^2\beta_{2345})+$$
$$U_{21}V_{32}(-4n_1n_3\beta_2)+U_{32}V_{21}(-4n_1n_3\beta_2)+$$
$$U_{13}V_{22}(4n_1n_2(\beta_{24}+\beta_4-\beta_5))+U_{22}V_{13}(4n_1n_2\beta_{245})+$$
$$U_{11}U_{22}(4n_2n_3\beta_{245})+V_{11}V_{22}(4n_2n_3(\beta_{24}-\beta_5)+U_{11}V_{13}(-4n_1n_3\beta_{2345})+$$
$$U_{21}V_{23}(-4n_1n_3(\beta_2+2\beta_{45}-\beta_3-2\beta_1))+U_{31}V_{33}(-4n_1n_3(\beta_{245}-\beta_3))+$$
$$U_{13}V_{11}(4n_1n_3(\beta_{24}-\beta_{35}))+U_{23}V_{21}(4n_1n_3(\beta_2-\beta_3-2\beta_1))+$$
$$U_{33}V_{31}(4n_1n_3(\beta_{24}-\beta_{35}))+U_{12}V_{21}(-8n_2n_3\beta_1)+$$
$$V_{12}V_{21}(-8n_2n_3\beta_1)+U_{23}V_{21}(8n_1n_2\beta_1)+U_{12}V_{23}(-8n_1n_2\beta_1) \tag{8.14}$$

One can deduce from this expression that the spectrum in the axi–planar phase is nondegenerate.

To get the second variation $\delta^2 S$ of the action S for the axial phase one should put $n_2 = 0,\ n_1 = n_3 = 1$ in Eq. (8.14)

$$\delta^2 S/\Delta^2 = U_{11}^2(\Delta^{-2}+2\beta_{2345})+U_{12}^2(\Delta^{-2}+4\beta_{12345})+U_{13}^2(\Delta^{-2}+2(\beta_{234}-\beta_5))+$$
$$+U_{21}^2(\Delta^{-2}+2(\beta_{12345}+2\beta_{45}-\beta_3))+U_{22}^2(\Delta^{-2}+4\beta_{245})+U_{23}^2(\Delta^{-2}+2(\beta_{12345}-2\beta_3))+$$
$$U_{31}^2(\Delta^{-2}+2\beta_{2345})+U_{32}^2(\Delta^{-2})+U_{33}^2(\Delta^{-2}+2(\beta_{234}-\beta_5))+V_{11}^2(\Delta^{-2}+2(\beta_{234}-\beta_5))+$$
$$+V_{12}^2(\Delta^{-2})+V_{13}^2(\Delta^{-2}+2\beta_{2345})+V_{21}^2(\Delta^{-2}+4\beta_{12345})+V_{22}^2(\Delta^{-2}+4\beta_{245})+$$
$$V_{23}^2(\Delta^{-2}+2(3\beta_{245}+2\beta_{13})+2(\beta_{245}-\beta_3))+V_{31}^2(\Delta^{-2}+2(\beta_{234}-\beta_5))+V_{32}^2(\Delta^{-2})$$
$$+V_{33}^2(\Delta^{-2}+2\beta_{2345})+U_{21}V_{32}(-4\beta_2)+U_{32}V_{21}(-4\beta_2)+U_{11}V_{13}(-4(\beta_{245}-\beta_3))+$$
$$U_{21}V_{23}(-4(\beta_2+2\beta_{45}-\beta_3-2\beta_1))+U_{31}V_{33}(-4(\beta_{245}-\beta_3))+$$
$$U_{13}V_{11}(4(\beta_{24}-\beta_{35}))+$$
$$U_{23}V_{21}(4(\beta_2-\beta_3-2\beta_1))+U_{33}V_{31}(4(\beta_{24}-\beta_{35})). \tag{8.15}$$

From this expression it follows that the spectrum in the axial phase becomes degenerate: in particular, doubly degenerate modes correspond to the following variables (U_{12} and V_{12}), (U_{22} and V_{22}), (U_{11}, V_{13}

and U_{31}, V_{33}), (U_{13}, V_{11} and U_{33}, V_{31}). So at least six modes are degenerate in the axial phase. These results are similar to the ones obtained above by Ginsburg–Landau equations.

8.4. Conclusion

Hence, both methods show that the splitting of degenerate axial phase modes in the axi–planar phase might serve as a sensitive test of the existence of the axi–planar phase. Studies of ^{3}He in aerogel could help with the problem of identification of thc *A*–phase. Aerogel makes it possible that the *A*–phase exists in a weak–coupling regime (at low pressures)[4,5]. Under these conditions the collective mode spectrum can be calculated for the axi–planar phase. The comparison of these results with ultrasound experiments in ^{3}He in aerogel could solve the problem of identification of the *A*–phase.

Why is this issue so important? There will be experimental consequences if the *A*–phase is not the axial–state. The axi–planar state has an extra degree of freedom which would likely show up in flow properties. It is known that flow in the *A*–phase shows unexpected discrepancies from theory. Theoretically, there needs to be reexamination of the connection between textures and superfluidity, and of the kind of structural defects allowed in ordered textures of the axi–planar state.

Chapter IX

Stability of Goldstone–Modes

9.1. Stability of Goldstone–Modes and Their Dispersion Laws

It is known that in ^{4}He the phonon spectrum has a decay character near k = 0. The situation is quite different in the superfluid ^{3}He. Below we investigate the stability of the *gd*–modes in the B–phase and also the stability of orbital waves in A– and 2D–phases against the decay of an excitation into several others (Brusov and Popov[1,2]) .

The stability of the *cl*–modes can be considered with respect to various processes: the pair decay into initial fermions, the decay of a collective Bose–excitation into several Bose–excitations of the same type of into several Bose–excitations of different types, corresponding to different energy–spectrum dispersion laws.

In the isotropic B–phase, the decay of a phonon into individual fermions is forbidden, since the excitation energy is much lower than the binding energy 2Δ of the Cooper pair. The decay of an excitation into two or several excitations of the same type is kinematically forbidden if $d^2E/dk^2 < 0$, and the $E(k)$ curve bends downward away from the tangent uk. This is equivalent to a positive dispersion coefficient γ in the dispersion law $E(k) \cong uk(1-\gamma k^2)$ at small k. Therefore the question of the stabilty of *gd*–modes can be solved by calculating the corrections to the linear dispersion law. The calculation shows the stability of all *gd*–modes in the B–phase.

In the anisotropic phase (A, $2D$) the energy gap of the Fermi–spectrum depends on the direction of the momenturn and vanishes in the selected direction. The decay of the phonon into individual fermions is therefore energetically possible here.

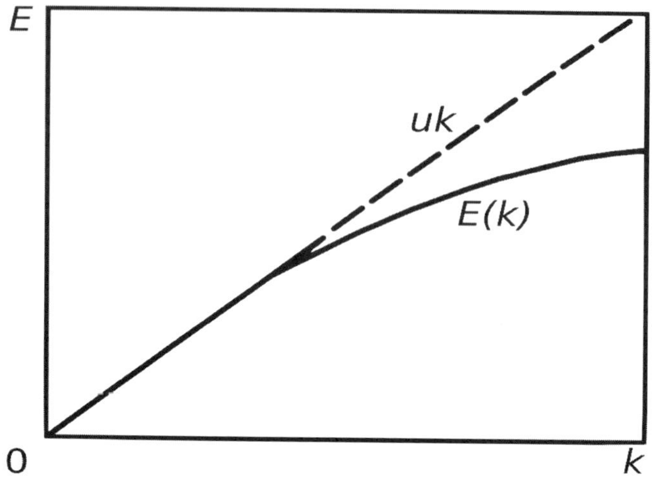

FIG. 9.1. The stable phonon spectrum of collective excitation.

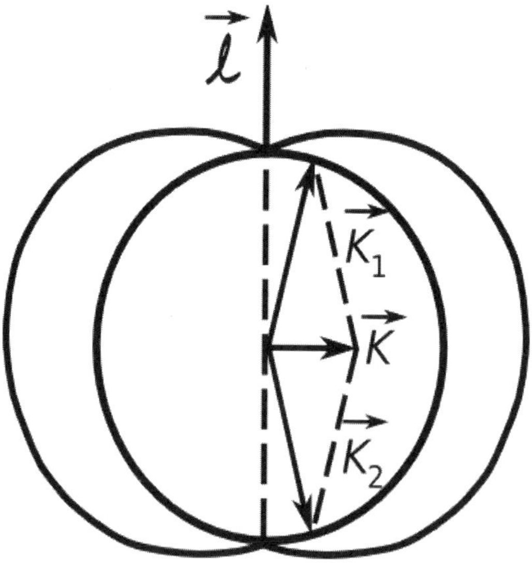

FIG. 9.2. The decay of the phonon (collective excitation) with momentum $\boldsymbol{k}$ into initial fermions with momenta $\boldsymbol{k}_1$ and $\boldsymbol{k}_2$ which directions are close to axis of anisotropy $\boldsymbol{l}$ ($\boldsymbol{H}$) in anisotropic A (2D)–phase.

On the other hand, the problem of stability against the decay into Bose excitations, just as in the *B*–phase, reduces to finding the corrections to the linear dispersion law. A calculation shows that the excitation is stable if its momenturn lies within a certain cone described around a preffered direction and is unstable otherwise.

9.2. Stability of Goldstone–Modes in the B–Phase

The condensate function $c_{ia}^{(0)}(p)$ in the *B*–phase is given by

$$c_{ia}^{(0)}(p) = c(\beta V)^{1/2}\delta_{p0}\delta_{ia},$$

where c is obtained from the equation

$$3g_0^{-1} + \frac{4Z^2}{\beta V}\sum_p \left(\omega^2 + \xi^2 + 4c^2 Z^2\right)^{-1} = 0\,.$$

Making the substitution $c_{ia}(p) \to c_{ia}^{(0)}(p) + c_{ia}(p)$, we separate a quadratic form in the new variables

$$g_0^{-1}\sum_{p,i,a} c_{ia}^{+}(p)c_{ia}(p) - \frac{1}{4}Tr\left(\hat{G}\hat{u}\right)^2$$

in S_{eff}, where

$$u_{p1p2} = (\beta V)^{-1/2} \begin{pmatrix} 0 & (n_{1i} - n_{2i})\sigma_a c_{ia}(p_1 + p_2) \\ -(n_{1i} - n_{2i})\sigma_a c^+_{ia}(p_1 + p_2) & 0 \end{pmatrix} \tag{9.1}$$

$$G^{-1} = \begin{pmatrix} Z^{-1}(i\omega - \xi)\delta_{p1p2} & 2c(\vec{n}\vec{\sigma})\delta_{p1+p2,0} \\ -2c(\vec{n}\vec{\sigma})\delta_{p1+p2,0} & Z^{-1}(-i\omega + \xi)\delta_{p1p2} \end{pmatrix}.$$

Revised G^{-1}, one gets

$$G = \frac{Z}{M} \begin{pmatrix} -(i\omega + \xi)\delta_{p1p2} & \Delta(\vec{n}\vec{\sigma})\delta_{p1+p2,0} \\ -\Delta(\vec{n}\vec{\sigma})\delta_{p1+p2,0} & (i\omega + \xi)\delta_{p1p2} \end{pmatrix}, \tag{9.2}$$

where

$$M = \omega^2 + \xi^2 + \Delta^2, \ \Delta = 2cZ.$$

It follows from (9.1) and (9.2) that

$$\frac{1}{4} Tr(\widehat{G}\widehat{u})^2 = \frac{1}{4} \sum_{p1,p2,p3,p4} tr\,(G_{p1p2} u_{p2p3} G_{p3p4} u_{p4p1}) =$$

$$= \frac{Z^2}{4\beta V} \sum_p \{c^+_{ia}(p) c_{jb}(p) 2 \sum_{p1+p3=p} (M_1 M_3)^{-1} (i\omega_1 + \xi_1)(i\omega_3 + \xi_3) \times$$

$$\times tr(\sigma_a \sigma_b)(n_1 - n_3)_i (-n_1 + n_3)_j + 4\Delta^2 [c_{ia}(p) c_{jb}(-p) + c^+_{ia}(p) c^+_{jb}(-p)] \times$$

$$\times \sum_{p3-p1=p} (M_1 M_3)^{-1} (n_1 + n_3)_i (n_1 + n_3)_j tr(\vec{n}_1 \vec{\sigma})\sigma_a tr(\vec{n}_3 \vec{\sigma})\sigma_b \}$$

Here, tr denotes the trace over matrix indices.

Considering small p, let us make replacement $n_3 \to -n_1$, if $p_1 + p_3 \to p$ and $n_3 \to n_1$, if $p_3 - p_1 \to p$. Making also the replacement $p_1 \to -p_1$ in sums with $p_3 - p_1 \to p$ and taking the trace, we obtain the quadratic part of S_{eff} in the form

$$S_{eff} \approx \sum_p c^+_{ia}(p) c_{ja}(p) \left[\frac{\delta_{ij}}{g} + \frac{4}{\beta V} \sum_{p1+p2=p} n_{1i} n_{1j} \varepsilon(-p_1)\varepsilon(-p_2) G(p_1) G(p_2) \right] -$$
$$-\frac{1}{2}\sum_p [c_{ia}(p)c_{jb}(-p) + c^+_{ia}(p)c^+_{jb}(-p)] \times$$
$$\times \frac{4\Delta^2}{\beta V} \sum_{p1+p2=p} n_{1i} n_{1j} (2n_{1a}n_{1b} - \delta_{ab}) G(p_1) G(p_2), \tag{9.3}$$

where

$$\varepsilon(p) = i\omega - \xi, \quad G(p) = Z(\omega^2 + \xi^2 + \Delta^2)^{-1}.$$

After substitutions

$$c_{ia}(p) = u_{ia}(p) + iv_{ia}(p), \; c^+_{ia}(p) = u_{ia}(p) - iv_{ia}(p),$$

the quadratic form (9.3) breaks into two independent forms, one of which depends on $u_{ia}(p)$ and other one depends on $v_{ia}(p)$:

$$-\sum_p \left[A_{ia}(p) u_{ia}(p) u_{ja}(p) + B_{ijab}(p) u_{ia}(p) u_{jb}(p) \right] -$$
$$-\sum_p \left[A_{ia}(p) v_{ia}(p) v_{ja}(p) - B_{ijab}(p) v_{ia}(p) v_{jb}(p) \right] \tag{9.4}$$

In (9.4), the term corresponding to $p = 0$ is equal to

$$- A_{ij}(0)\left(u_{ia}u_{ja} + v_{ia}v_{ja}\right) - B_{ijab}(0)\left(u_{ia}u_{ja} - v_{ia}v_{jb}\right) =$$
$$= -\frac{2Z^2 k_F^2}{15\pi^2 c_F}\left(u_{ia}u_{ja} + u_{aa}u_{bb} + u_{ia}u_{ai} + 4v_{ia}v_{ia} - v_{aa}v_{bb} - v_{ia}v_{ai}\right). \tag{9.5}$$

The u–form has three zero eigenvectors corresponding to the variables $u_{12} - u_{21}, u_{23} - u_{32}, u_{31} - u_{13}$, while the v–form has single zero vector corresponding to the variable $v_{11} + v_{22} + v_{33}$.

It is precisely these which are the "phonon" variables, and the corresponding branches of collective excitations start out from zero ($E(k) \to 0$, if $k \to 0$).

We now consider the difference $A_{ij}(p) - A_{ij}(0)$. Applying Feynman procedure to the denominators of the Green's functions $G(p_1)$ and $G(p_2)$ in (9.3) we obtain

$$A_{ij}(p) - A_{ij}(0) = -\frac{4Z^2}{\beta V}\sum_{p1+p2=p} n_{1i}n_{1j}\int_0^1 d\alpha \times$$
$$\times\left[\frac{(\xi_1 + i\omega_1)(\xi_2 + i\omega_2)}{\left[\alpha(\omega_1^2 + \xi_1^2) + (1-\alpha)(\omega_2^2 + \xi_2^2) + \Delta^2\right]^2} - \frac{\omega_1^2 + \xi_1^2}{\left(\omega_1^2 + \xi_1^2 + \Delta^2\right)^2}\right]. \tag{9.6}$$

Considering the limit as $T \to 0$, we change the summation in (9.6) to an integration near the Fermi sphere in accordance with the rule

$$(\beta V)^{-1}\sum_{p1} \to k_F^2 (2\pi)^{-4} c_F^{-1}\int d\omega_1 d\xi_1 d\Omega_1,$$

where $\int d\Omega_1$ is an integral with respect to the angle variables. Furthermore we replace $\xi_2 \to -\xi_2$ in (9.6) and make the shifts $\omega_1 \to \omega_1 + (1-\alpha)\omega$ and $\omega_2 \to -\omega_1 + \alpha\omega$, so that $\omega_1 + \omega_2 = \omega$ and also $\xi_1 \to \xi_1 + (1-\alpha) c_F (n_1 k)$ and $\xi_2 \to \xi_1 + \alpha c_F (n_1 k)$.

Expression (9.6) then becomes

$$A_{ij}(p) - A_{ij}(0) = -\frac{4Z^2 k_F^2}{(2\pi)^4 c_F} \int_0^1 d\alpha \int d\omega_1 d\xi_1 d\Omega_1 n_{1i} n_{1j} \times$$

$$\times \left[\frac{\omega_1^2 + \xi_1^2 - \alpha(1-\alpha) q^2}{\left[\alpha(\omega_1^2 + \xi_1^2) + (1-\alpha)(\omega_2^2 + \xi_2^2) + \Delta^2\right]^2} - \frac{\omega_1^2 + \xi_1^2}{\left(\omega_1^2 + \xi_1^2 + \Delta^2\right)^2} \right],$$

where $q^2 = \omega^2 \xi_1 + c_F^2 (n_1 k)^2$. Integrating with respect to the variable $r_1^2 = \omega_1^2 + \xi_1^2$, we obtain

$$A_{ij}(p) - A_{ij}(0) =$$

$$= \frac{Z^2 k_F^2}{4\pi^3 c_F} \int_0^1 d\alpha \int d\Omega_1 n_{1i} n_{1j} \left[\ln\left(1 + \frac{\alpha(1-\alpha) q^2}{\Delta^2}\right) + \frac{\alpha(1-\alpha) q^2}{\Delta^2 + \alpha(1-\alpha) q^2} \right].$$

Expanding the integrand in powers of the small parameter $\alpha(1-\alpha) q^2 / \Delta^2$ and integrating with respect to α, we arrive at the formula

$$A_{ij}(p) - A_{ij}(0) \cong \frac{Z^2 k_F^2}{4\pi^3 c_F} \int d\Omega_1 \left(\frac{q^2}{3\Delta^2} - \frac{q^4}{20\Delta^4} \right) n_{1i} n_{1j}. \tag{9.7}$$

Similar calculations yields for

$$B_{ijab}(p) - B_{ijab}(0) \cong \frac{Z^2 k_F^2}{4\pi^3 c_F} \int d\Omega_1 \left(\frac{q^2}{6\Delta^2} - \frac{q^4}{20\Delta^4} \right) (\delta_{ab} - 2 n_{1a} n_{1b}) n_{1i} n_{1j}. \tag{9.8}$$

The integrals with respect to the angle variable in (9.7) and (9.8) can be easily calculated. Using (9.5), (9.7) and (9.8), we write down the explicit expression for (9.4) in the form

$$
\begin{aligned}
&-\frac{Z^2 k_F^2}{\pi^2 c_F}\sum_p\{\left(\frac{1}{3}\mp\frac{1}{5}\right)w_{ia}w_{ia}\pm\frac{2}{15}\left(w_{aa}w_{bb}+w_{ia}w_{ai}\right)+\\
&+w_{ia}w_{ja}[\frac{1}{3}\delta_{ij}\left(\frac{\omega^2}{3\Delta^2}-\frac{\omega^4}{20\Delta^4}\right)+\frac{c_F^2}{15}\left(k^2\delta_{ij}+2k_ik_j\right)\left(\frac{1}{3\Delta^2}-\frac{\omega^2}{10\Delta^4}\right)-\\
&-\frac{c_F^4}{700\Delta^4}\left(k^4\delta_{ij}+4k^2k_ik_j\right)]\pm w_{ia}w_{jb}[\delta_{ab}[\frac{1}{3}\delta_{ij}\left(\frac{\omega^2}{6\Delta^2}-\frac{\omega^4}{30\Delta^4}\right)+\\
&+\frac{c_F^2}{15}\left(k^2\delta_{ij}+2k_ik_j\right)\left(\frac{1}{6\Delta^2}-\frac{\omega^2}{15\Delta^4}\right)-\frac{c_F^4}{1050\Delta^4}\left(k^4\delta_{ij}+4k^2k_ik_j\right)]+\\
&+\left(\delta_{ab}\delta_{ij}+\delta_{ai}\delta_{bj}+\delta_{aj}\delta_{bi}\right)[\frac{2}{15}\left(-\frac{\omega^2}{6\Delta^2}+\frac{\omega^4}{30\Delta^4}\right)+\\
&+\frac{2k^2c_F^2}{105}\left(-\frac{1}{6\Delta^2}+\frac{\omega^2}{15\Delta^4}+\frac{c_F^2k^4}{4725\Delta^4}\right)]+\\
&+\frac{4c_F^2}{315}\left(\delta_{ij}k_ak_b+\delta_{ab}k_ik_j+\delta_{ai}k_bk_j+\delta_{bj}k_ak_i+\delta_{aj}k_bk_i+\delta_{bi}k_ak_j\right)\times\\
&\times\left(-\frac{1}{2\Delta^2}+\frac{\omega^2}{5\Delta^4}+\frac{c_F^2k^4}{15\Delta^4}\right)+\frac{8c_F^4}{4725\Delta^4}k_ak_bk_ik_j]\}
\end{aligned}
\tag{9.9}
$$

It is understood here, that we first must substitute $w_{ia}=u_{ia}$ and take the upper sign in place of the symbols $\pm$ and $\mp$ and then substitute $w_{ia}=v_{ia}$ and take the lower sign.

Expression (9.9) determines the phonon branches of the spectrum of the system. Since the *B*–phase is isotropic and has no preffered direction, it suffices to consider excitations propagating in any direction, say along the third axis. The quadratic form of the variables w_{ia} in (9.9) breaks up after the substitutions $k_1 = k_2 = 0$ and $k_3 = k$ into a sum of four variables, of which the first depends on w_{12} and w_{21}, the second on w_{13} and w_{31}, the third on w_{23} and w_{32}, and the fourth on w_{11}, w_{22} and w_{33}.

For $w_{ia} = v_{ia}$ the *gd*–mode is determined by the form of the variables v_{11}, v_{22} and v_{33}. This is proportional to the expression

$$a\left(v_{11}^2 + v_{22}^2\right) + bv_{33}^2 + 2cv_{11}v_{22} + 2d\left(v_{11} + v_{22}\right)v_{33}. \qquad (9.10)$$

Here,

$$a = \frac{4}{15} + a_1, \; b = \frac{4}{15} + b_1, \; c = -\frac{2}{15} + c_1, \; d = \frac{2}{15} + d_1,$$

$$a_1 = \frac{11}{90}x^2 + \frac{13}{630}y^2 - \frac{17}{900}x^4 - \frac{19}{1350}x^2y^2 - \frac{1}{900}y^4,$$

$$b_1 = \frac{11}{90}x^2 + \frac{17}{210}y^2 - \frac{17}{900}x^4 - \frac{9}{350}x^2y^2 - \frac{1}{900}y^4,$$

$$c_1 = \frac{1}{45}x^2 + \frac{1}{315}y^2 - \frac{1}{225}x^4 - \frac{2}{1575}x^2y^2 - \frac{1}{4725}y^4,$$

$$d_1 = \frac{1}{45}x^2 + \frac{1}{105}y^2 - \frac{1}{225}x^4 - \frac{2}{525}x^2y^2 - \frac{1}{945}y^4,$$

where $x = \omega/\Delta$, $y = c_F k/\Delta$.

The third order determinant of the form (9.10) is equal to

$(a-c)[b(a+c)-2d^2]=0$. Since the difference $(a-c)$ does not vanish as $\omega\to 0$ and $k\to 0$, we obtain

$$b(a+c)-2d^2=0, \tag{9.11}$$

or

$$\frac{2}{15}(b_1+2a_1+2c_1+4d_1)+b_1(a_1+c_1)-2d_1^2=0.$$

We have

$$b_1+2a_1+2c_1+4d_1=\frac{1}{2}x^2+\frac{1}{6}y^2-\frac{1}{12}x^4-\frac{1}{18}x^2y^2-\frac{1}{60}y^4.$$

$$b_1(a_1+c_1)-2d_1^2=\frac{1}{60}x^4+\frac{13}{945}x^2y^2+\frac{11}{630}y^4.$$

Substitution in (9.11) leads to the equation

$$x^2+\frac{1}{3}y^2+\frac{1}{12}x^4+\frac{2}{21}x^2y^2+\frac{1}{140}y^4=0.$$

Its solution after returning to the variables ω and $\boldsymbol{k}$ and making the substitution $i\omega\to E$ yields the *gd*–mode of the spectrum

$$E=\frac{c_F k}{\sqrt{3}}\left(1-\frac{2c_F^2k^2}{45\Delta^2}\right).$$

It is stable with respect to decay of excitation into two or several excitations of the same type. It is interesting that the dispersion coefficient $\gamma = 2c_F^2/45\Delta^2$ turns out to be two times as large as for the Fermi–gas model with pointlike scalar interaction (Andrianov and Popov[3]).

We consider now the form of the variables u_{ia}, which decay at $k_1 = k_2 = 0$ into four independent forms. Of importance to us are the forms of the variables (u_{12}, u_{21}) and (u_{13}, u_{31}), which are proportional to the expressions

$$\frac{2}{15}(u_{12}+u_{21})^2 + (u_{12}^2+u_{21}^2)\left[\frac{13}{90}x^2 + \frac{19}{630}y^2 - \frac{7}{300}x^4 - \frac{31}{3150}x^2y^2 - \right.$$
$$\left. -\frac{41}{18900}y^4\right] -$$
$$-2u_{12}\,u_{21}\left[\frac{1}{45}x^2 + \frac{1}{315}y^2 - \frac{1}{225}x^4 - \frac{2}{1575}x^2y^2 - \frac{1}{4725}y^4\right],$$

$$\frac{2}{15}(u_{13}+u_{31})^2 + u_{13}^2\left[\frac{13}{90}x^2 + \frac{1}{42}y^2 - \frac{7}{300}x^4 - \frac{23}{3150}x^2y^2 - \frac{1}{750}y^4\right] +$$
$$+u_{31}^2\left[\frac{13}{90}x^2 + \frac{19}{210}y^2 - \frac{7}{300}x^4 - \frac{31}{1050}x^2y^2 - \frac{41}{3780}y^4\right] -$$
$$-2u_{13}\,u_{31}\left[\frac{1}{45}x^2 + \frac{1}{105}y^2 - \frac{1}{225}x^4 - \frac{2}{525}x^2y^2 - \frac{1}{945}y^4\right]. \tag{9.12}$$

The form of the variables (u_{23}, u_{32}) is obtained from the second forms in (9.12) by making the substitution $(u_{13}, u_{31}) \to (u_{23}, u_{32})$. Equating to zero the determinants of the forms (9.12), we obtain the branch of the longitudinal spin waves (the variables (u_{12}, u_{21}))

$$E = \frac{c_F k}{\sqrt{5}}\left(1 - \frac{2c_F^2 k^2}{105\Delta^2}\right)$$

and the double degenerate branch of the transverse spin waves (the variables (u_{13}, u_{31}) and (u_{23}, u_{32}))

$$E = \sqrt{\frac{2}{5}} c_F k \left(1 - \frac{173 c_F^2 k^2}{3360 \Delta^2}\right)$$

Both moqes of spin waves are stable as well as the sound mode.

9.3. Stability of Goldstone–Modes in the Axial A–Phase

The A–phase is anisotropic and has a preffered direction along the common orbital angular momentum of the Cooper pair. The gap is given by

$$\Delta = \Delta_0 \sin\theta\,,\ \Delta_0 = 2cZ\,.$$

The quantity c enters in the condensate wave function

$$c_{ia}^{(0)}(p) = c(\beta V)^{1/2} \delta_{p0} (\delta_{i1} + i\delta_{i2}) \delta_{a3}$$

and satisfies the equation

$$g_0^{-1} + \frac{2Z^2}{\beta V} \sum_p \frac{\sin^2\theta}{\omega^2 + \xi^2 + 4c^2 Z^2 \sin^2\theta} = 0\,. \tag{9.13}$$

Performing a shift by an amount $c_{ia}^{(0)}$ in the functional S_{eff} and separating the quadratic form with respect to the new variables we get

$$
\begin{aligned}
&\sum_{p} c_{ia}^{+}(p)c_{ja}(p)\times \\
&\times\left[g_0^{-1}\delta_{ij} + \frac{4}{\beta V} \sum_{p1+p2=p} n_{1i}n_{1j}(\xi_1 + i\omega_1)(\xi_2 + i\omega_2)G(p_1)G(p_2) \right] + \\
&+\frac{1}{2}\sum_{p,a=1,2} \left(c_{ia}^{+}(p)c_{ja}^{+}(-p) + c_{ia}(p)c_{ja}(-p)\right)\times \\
&\times\frac{4\Delta_0^2}{\beta V}\sum_{p1+p2=p}(n_1 \pm n_2)^2 n_{1i}n_{1j}G(p_1)G(p_2) - \\
&-\frac{1}{2}\sum_{p,a=1,2} \left(c_{i3}^{+}(p)c_{j3}^{+}(-p) + c_{i3}(p)c_{j3}(-p)\right)\times \\
&\times\frac{4\Delta_0^2}{\beta V}\sum_{p1+p2=p}(n_1 \pm n_2)^2 n_{1i}n_{1j}G(p_1)G(p_2),
\end{aligned}
\tag{9.14}
$$

where $G(p) = Z\left(\omega^2 + \xi^2 + 4c^2 Z^2 \sin^2\theta\right)^{-1}$. The plus or minus sign in (9.14) means that it is necessary to take the plus sign when multiplying by $c_{ia}^{+}c_{ja}^{+}$ and minus sign when multiplying by $c_{ia}c_{ja}$.

The form (9.14) is a sum of three independent forms that differ in the value of the isotopic index and go over into one another when the variables are interchanged. Therefore the spectrum in the *A*–phase turns out to be triply degenerate and it suffices to consider one of the three forms (e.g., with $a = 1$).

We take the form with $a = 1$ and separate in it the terms corresponding to $p = 0$:

$$
-\frac{Z^2 k_F^2}{2\pi^3 c_F}\left(3v_{11}^2 + 3u_{21}^2 - 2u_{21}v_{11} + \left(u_{11} - v_{21}\right)^2\right).
$$

It is clear therefore that one can choose as the phonon variables

$$
u = \operatorname{Re} c_{31},\ \ v = \operatorname{Im} c_{31},\ \ w = \frac{1}{2}\operatorname{Re}\left(c_{11} - ic_{21}\right).
$$

The Bose–spectrum is determined by a form that depends on the phonon variables. It is possible to make the substitutions $A_{ij}(p) \to A_{ij}(p) - A_{ij}(0)$ and $B_{ij}(p) \to B_{ij}(p) - B_{ij}(0)$ in this form, since $A_{ij}(0)$ and $B_{ij}(0)$ are equal to zero for the phonon variables. Calculation of $A_{ij}(p) - A_{ij}(0)$ and $B_{ij}(p) - B_{ij}(0)$ is analogous to that carried out above for the B – phase. Using the Feynman procedure and integrating with respect to ω_1 and ξ_1 (as $T \to 0$) we obtain

$$A_{ij}(p) - A_{ij}(p) = -\frac{Z^2 k_F^2}{4\pi^3 c_F}\int_0^1 d\alpha \int n_{1i} n_{1j} d\Omega_1 \times$$

$$\times\left[\ln\frac{\Delta_0^2 \sin^2\theta_1}{\Delta_0^2 \sin^2\theta_1 + \alpha(1-\alpha)q^2} + \frac{\alpha(1-\alpha)q^2}{\Delta_0^2 \sin^2\theta_1 + \alpha(1-\alpha)q^2}\right],$$

$$B_{ij}^{\pm}(p) - B_{ij}^{\pm}(0) =$$

$$= \frac{Z^2 k_F^2}{4\pi^3 c_F}\int_0^1 d\alpha \int n_{1i} n_{1j} d\Omega_1 e^{\pm 2i\varphi_1} \frac{\alpha(1-\alpha)q^2}{\Delta_0^2 \sin^2\theta_1 + \alpha(1-\alpha)q^2}. \tag{9.15}$$

Here, $q^2 = \omega^2 + c_F^2 (n_1 k)^2$, B_{ij}^+ is the coefficient of $c_i^+ c_j^+$ and B_{ij}^- is the coefficient of $c_i c_j$.

If we expand under the integral signs in (9.15) in powers of $\alpha(1-\alpha)q^2 / \Delta_0^2 \sin^2\theta_1$, at small ω and k, and confine ourselves to the first term of the expansion, then we obtain in the calculation of $A_{33}(p) - A_{33}(0)$ a logarithmically diverging integral proportional to

$$\int d\Omega_1 \frac{q^2 n_3^2}{\Delta_0^2 \sin^2\theta_1} = \frac{2\pi\left(\omega^2 + c_F^2 k_3^2\right)}{\Delta_0^2}\int_0^{\pi} \frac{d\theta_1}{\sin\theta_1} + \text{finite part.}$$

The cause of the divergence is that near the poles of the Fermi–sphere

($\theta_1 = 0, \pi$) the parameter $\alpha(1-\alpha)q^2 / \Delta_0^2 \sin^2\theta_1$ is no longer small and.we cannot expand in its terms. What is needed here is an accurate calculation of the first integral of (9.15) at $i = j = 3$, the results of which is

$$A_{33}(p) - A_{33}(0) = \frac{Z^2 k_F^2}{48\pi^2 c_F \Delta_0^2}\left[2p^2 \ln\frac{4\Delta_0^2}{p^2} + \frac{1}{3}p^2 + \frac{2}{3}c_F^2 k^2 - 2c_F^2 k_3^2\right]$$

at small $p^2 = \omega^2 + c_F^2 k_3^2$.

This difficulty does not arise in the calculation of the remaining elements $A_{ij}(p) - A_{ij}(0)$ as well as $B_{ij}(p) - B_{ij}(0)$. However, it is precisely the appearance of logarithms which makes it possible to calculate the corrections to the linear dispersion law in the chosen approximation, where we confine ourselves to phonon variables. For comparison we recall that in the *B*–phase the calculation of the dispersion coefficient called for allowance for the coupling between *gd*– and nonphonon–modes.

As a result we arrive at the following matrix of the phonon variables

$$Q = \frac{Z^2 k_F^2}{48\pi^2 c_F \Delta_0^2} \times$$

$$\times \begin{pmatrix} a(p) + \frac{1}{12}c_F^2\left(k_1^2 - k_2^2\right) & \frac{1}{6}c_F^2 k_1 k_2 & c_F^2 k_1 k_3 \\ \frac{1}{6}c_F^2 k_1 k_2 & a(p) + \frac{1}{12}c_F^2\left(k_2^2 - k_1^2\right) & c_F^2 k_2 k_3 \\ c_F^2 k_1 k_3 & c_F^2 k_2 k_3 & 3\omega^2 + c_F^2 k^2 \end{pmatrix}$$

where

$$a(p) = p^2\left(\ln\frac{4\Delta_0^2}{p^2} + \frac{1}{6}\right) + \frac{1}{3}c_F^2\left(k^2 - 3k_3^2\right).$$

The equation $\det Q = 0$ can be written in the form

$$\left(a(p) - \frac{1}{12}c_F^2 k_\perp^2\right)\left[\left(3\omega^2 + c_F^2 k^2\right)\left(a(p) + \frac{1}{12}c_F^2 k_\perp^2\right) - c_F^4 k_\perp^2 k_{\text{II}}^2\right] = 0. \tag{9.16}$$

where $k_{\text{II}}^2 = k_3^2$ and $k_\perp^2 = k_1^2 + k_2^2$. The equation gives three branches of the spectrum: one with $E^2 \cong c_F^2 k^2/3$ and two with $E^2 \cong c_F^2 k_{\text{II}}^2$. From (9.16) we can obtain the corrections to the linear dispersion law and determine the region of stability of the Bose–spectrum. The result of the solution of (9.16) is

$$E_1(k) = \frac{c_F k}{\sqrt{3}}\left(1 - \frac{\sin^2\theta\cos^2\theta}{2\left(\cos^2\theta - \frac{1}{3}\right)\ln\frac{4\Delta_0^2}{f_1(\theta,k)}}\right),$$

$$E_2(k) = c_F k_{II}\left(1 - \frac{11\cos^2\theta - 3}{24\cos^2\theta\ln\frac{4\Delta_0^2}{f_2(\theta,k)}}\right),$$

$$E_3(k) = c_F k_{II} \left(1 - \frac{51\cos^4\theta - 40\cos^2\theta + 5}{72\cos^2\theta\left(\cos^2\theta - \frac{1}{3}\right)\ln\frac{4\Delta_0^2}{f_3(\theta,k)}} \right), \tag{9.17}$$

where

$$f_1(\theta,k) = c_F^2 k^2 \left(\cos^2\theta - \frac{1}{3}\right),$$

$$f_2(\theta,k) = \frac{1}{12} c_F^2 k^2 \left(11\cos^2\theta - 3\right),$$

$$f_3(\theta,k) = \frac{1}{36} c_F^2 k^2 \frac{51\cos^4\theta - 40\cos^2\theta + 5}{\cos^2\theta\left(\cos^2\theta - \frac{1}{3}\right)}.$$

The obtained equations show that the stability of the spectrum in the *A*–phase depends on the angle θ between the excitation momentum and the preffered direction. The first (acoustic) mode is stable inside the cones $\cos^2\theta > 1/3$, and the second inside the cones $\cos^2\theta > 3/11$. The third mode is stable in the regions

$$\cos^2\theta > \frac{20+\sqrt{145}}{51},\ \frac{1}{3} > \cos^2\theta > \frac{20-\sqrt{145}}{51}.$$

Outside the stability regions, the energy of the excitation becomes complex because of the imaginary parts of the logarithms in (9.17). Physically this is connected with the possibility of the decay of the excitation into constituted fermions whose momenta are close to the preffered direction.

Equations (9.17) for the orbital waves $E_2(k)$ and $E_3(k)$ can be compared with the results of Combsecot[4], where the following dispersion laws were obtained for them

$$\left(\omega^2 - \hat{k}_Z^2 v_F^2\right)\left(\ln\frac{4\delta^2}{\hat{k}_Z^2 v_F^2 - \omega^2} + \frac{1}{5}\right) = \frac{1}{4}\hat{k}_Y^2 v_F^2 + \frac{1}{3}\hat{k}_Y^2 v_F^2\left(\frac{1}{4}F_1^S - 2\right) \tag{9.18}$$

$$\left(\omega^2 - \hat{k}_Z^2 v_F^2\right)\left(\ln\frac{4\delta^2}{\hat{k}_Z^2 v_F^2 - \omega^2} + \frac{1}{6}\right) = = \hat{k}^2 v_F^2\left(\frac{5}{12}\hat{k}_Y^2 - \frac{2}{3}\hat{k}_Z^2 - \hat{k}_Y^2\hat{k}_Z^2 + \frac{1}{2}F_1^S\left(\hat{k}_Y^2 - \hat{k}_Z^2\right)\right), \tag{9.19}$$

where $\hat{k}_Z = \cos\theta$, $\hat{k}_Y = \sin\theta$, $\delta = \Delta_0$ and $v_F = c_F$.

Although the excitation stability investigated in (Combescot[4]) was with respect to decay into fermions, we can decay of orbital excitations into two or several excitations of the same type (with respect to sign of $\partial^2 E / \partial k^2$).

If we neglect in (9.18) the Fermi–liquid correction F_1^S , we obtain an equation that differs from $a(p) - c_F^2 k_\perp^2 / 12 = 0$ only in the addition to the logarithm (1/6 in $a\ (p)$ and 1/5 in (9.18)). The stability region at $F_1^S = 0$ turns out to be the same ($\cos^2\theta > 1/3$) as for the $E_2(k)$ branch in the model system considered here.

For the branch (9.19), the stability takes place inside the cones $\cos^2\theta > \dfrac{25 - \sqrt{385}}{24} \cong 0.22$, and the situation here is greatly different from that of $E_3(k)$, where we have two stability regions.

Thus, one of the modes of the orbital waves, obtained here by a method based on functional integration, coincides with the one obtained by using the kinetic–equation method[4]. As to the second orbital mode, the corrections to the linear spectrum $E = c_F k_{\parallel}$ obtained by Combescot[4] and here (Brusov and Popov[1]) are substantially different and lead to different stability regions. The possible reason is that in our case the mode is coupled with the *gd*–mode $E_1(k)$, whereas no such coupling is considered by Combescot[4].

9.4. Stability of Goldstone Modes in the Planar 2D–Phase

The planar 2*D*–phase is unstable in a zero magnetic field. However, as shown by Alonso and Popov[5] when the external magnetic fields is increased in the considered model to $H = H_C$, the *B*–phase goes over into the 2*D*–phase. Popov et al.[5] have obtained an explicit formula for the critical magnetic field in the Ginzburg–Landau region

$$H_C^2 = 2\pi^2 T_C \Delta T / 7\zeta(3)\mu^2$$

and advanced argument favoring the possibility of a transition of real ^{3}He from the *B*– to the 2*D*–phase in a magnetic field .

We investigate here the Bose spectrum of the 2*D*–phase and show that the *gd*–modes are determined by the same equation and are given by the same formulas as in the *A*–phase, and differ only in the degeneracy multiplicity (2 in the 2*D*–phase and 3 in the *A*–phase).

The preffered direction in the 2*D*–phase is the direction of the external magnetic field. The condensate wave function is of the form

$$c_{ia}^{(0)}(p) = c(\beta V)^{1/2}\delta_{p0}(\delta_{i1}\delta_{a1} + \delta_{i2}\delta_{a2}),$$

where c satisfies an equation that coincides with (9.13) for the *A*–phase. The form of the gap $\Delta = \Delta_0 \sin\theta$ also coincides with the one existing in the *A*–phase.

Following a shift $c_{ia}(p) \to c_{ia}^{(0)}(p) + c_{ia}(p)$ and separation of the quadratic form with respect to the new variables, we obtain the expression

$$\sum_p c_{ia}^+(p) c_{ja}(p) \left[\frac{\delta_{ij}\delta_{ab}}{g} + \frac{2Z^2}{\beta V} \sum_{p1+p2=p} n_{1i} n_{1j} tr_2 (A_1 - B_1\sigma_3)\sigma_a (A_2 + B_2\sigma_3)\sigma_b \right] -$$
$$-\sum_p c_{ia}^+(p) c_{jb}^+(-p) \frac{Z^2}{\beta V} \sum_{p1+p2=p} n_{1i} n_{1j} tr_2 (C_1\sigma_1 + D_{-1}\sigma_2)\sigma_a (C_2\sigma_1 + D_2\sigma_2)\sigma_b$$
$$-\sum_p c_{ia}(p) c_{jb}(-p) \frac{Z^2}{\beta V} \sum_{p1+p2=p} n_{1i} n_{1j} tr_2 (\overline{C}_1\sigma_1 + \overline{D}_{-1}\sigma_2)\sigma_a (\overline{C}_2\sigma_1 + \overline{D}_2\sigma_2)\sigma_b . \tag{9.20}$$

Here, tr_2 denotes the trace of a second–order matrix, and the functions A, B, C and D are given by

$$A = M^{-1}\left[-(i\omega + \xi)\left(\omega^2 + \xi^2 + \mu^2 H^2 + \Delta_0^2 \sin^2\theta\right) + 2\xi\mu^2 H^2 \right],$$

$$B = M^{-1}\mu H\left[\omega^2 + \xi^2 + \mu^2 H^2 + \Delta_0^2 \sin^2\theta - 2\xi(i\omega + \xi) \right],$$

$$C = M^{-1}\Delta_0\left[n_1\left(\omega^2 + \xi^2 + \mu^2 H^2 + \Delta_0^2 \sin^2\theta\right) - 2i\xi\mu H n_2 \right],$$

$$D = M^{-1}\Delta_0\left[n_2\left(\omega^2 + \xi^2 + \mu^2 H^2 + \Delta_0^2 \sin^2\theta\right) + 2i\xi\mu H n_1 \right],$$

$$M = \left(\omega^2 + \xi^2 + \mu^2 H^2 + \Delta_0^2 \sin^2\theta\right) - 4\xi^2\mu^2 H^2,$$

$$n_1 = \sin\theta\cos\varphi, \quad n_2 = \sin\theta\sin\varphi .$$

Examination of those terms of (9.20), which correspond to $p = 0$ shows that the phonon variables of the system are

$$u=(u_{12}-u_{21})/2\,,\ v=(v_{11}+v_{22})/2\,,\ u_{31},u_{32},v_{31},v_{32}\,.$$

We set in (9.20) all the nonphonon variables equal to zero, making the substitution

$$u_{i3}=v_{i3}=0,u_{11}=u_{22}=0,v_{12}=v_{21}=0,u_{12}=-u_{21}=u,v_{11}=v_{22}=v,$$
$$u_{31}=u_{31},u_{32}=u_{32},v_{31}=v_{31},v_{32}=v_{32}. \tag{9.21}$$

Subtracting from the coefficient tensors their values at $p = 0$ and calculating the trace, tr_2, we obtain in place of (9.20)

$$\sum_p\{\sum_{a=1,2}c^+_{ia}(p)c_{ja}(p)\frac{4Z^2}{\beta V}\sum_{p1+p2=p}n_{1i}n_{1j}(A_1A_2+B_1B_2-A_1A_{-1}-B_1B_{-1})-$$
$$-(c_{i1}(p)c_{j1}(-p)-c_{i2}(p)c_{j2}(-p))\times$$
$$\times\frac{2Z^2}{\beta V}\sum_{p1+p2=p}n_{1i}n_{1j}(\overline{C}_{-1}\overline{C}_2-\overline{D}_{-1}\overline{D}_2-\overline{C}_{-1}\overline{C}_{-1}-\overline{D}_{-1}\overline{D}_{-1})-$$
$$-(c^+_{i1}(p)c^+_{j1}(-p)-c^+_{i2}(p)c^+_{j2}(-p))\times$$
$$\times\frac{2Z^2}{\beta V}\sum_{p1+p2=p}n_{1i}n_{1j}(C_{-1}C_2-D_{-1}D_2-C_{-1}C_{-1}-D_{-1}D_{-1})-$$
$$-c_{i1}(p)c_{j2}(-p)\frac{4Z^2}{\beta V}\sum_{p1+p2=p}n_{1i}n_{1j}(\overline{C}_{-1}\overline{D}_2-\overline{D}_{-1}\overline{C}_2-2\overline{C}_{-1}\overline{D}_{-1})-$$
$$-c^+_{i1}(p)c^+_{j2}(-p)\frac{4Z^2}{\beta V}\sum_{p1+p2=p}n_{1i}n_{1j}(C_{-1}D_2-D_{-1}C_2-2C_{-1}D_{-1})\} \tag{9.22}$$

It is understood that the substitutions (9.21) have been made in (9.22). The expression $A_1A_2+B_1B_2-A_1A_{-1}-B_1B_{-1}$ can be represented in the form of a sum of two terms, one of which depends on $\xi_1+\mu H$ and the other on $\xi_1-\mu H$. Then, after replacing the integration variable $\xi_1\to\xi_1-\mu H$ in the first term and $\xi_1\to\xi_1+\mu H$ in the second term, we arrive at expressions that do not depend on *H*. We can therefore replace

$$Z^2\left(A_1A_2+B_1B_2-A_1A_{-1}-B_1B_{-1}\right)\to\left(i\omega_1+\xi_1\right)\left(i\omega_2+\xi_2\right)G_1G_2-$$
$$-\left(\omega_1^2+\xi_1^2\right)G_1^2,$$

where $G_i=Z\left(\omega_1^2+\xi_1^2+\Delta_0^2\sin^2\theta\right)^{-1}$.

We similarly have

$$Z^2\left(C_{-1}C_2-D_{-1}D_2-C_{-1}C_{-1}-D_{-1}D_{-1}\right)\to\Delta_0^2\left(n_1^2-n_2^2\right)\left(G_1G_2-G_1^2\right),$$

$$Z^2\left(\overline{C}_{-1}\overline{C}_2-\overline{D}_{-1}\overline{D}_2-\overline{C}_{-1}\overline{C}_{-1}-\overline{D}_{-1}\overline{D}_{-1}\right)\to\Delta_0^2\left(n_1^2-n_2^2\right)\left(G_1G_2-G_1^2\right),$$

$$Z^2\left(C_{-1}D_2+D_{-1}C_2-2C_{-1}D_1\right)\to2\Delta_0^2n_1n_2\left(G_1G_2-G_1^2\right),$$

$$Z^2\left(\overline{C}_{-1}\overline{D}_2+\overline{D}_{-1}\overline{C}_2-2\overline{C}_{-1}\overline{D}_1\right)\to2\Delta_0^2n_1n_2\left(G_1G_2-G_1^2\right).$$

We can now rewrite (9.22) in the form

$$\sum_p \{ \sum_{a=1,2} \left(u_{ia} u_{ja} + v_{ia} v_{ja} \right) \times$$
$$\times \frac{4}{\beta V} \sum_{p1+p2=p} n_{1i} n_{1j} \left[(i\omega_1 + \xi_1)(i\omega_2 + \xi_2) G_1 G_2 - \left(\omega_1^2 + \xi_1^2 \right) G_1^2 \right]$$
$$- \frac{4\Delta^2}{\beta V} \sum_{p1+p2=p} n_{1i} n_{1j} \left(G_1 G_2 - G_1^2 \right) \times$$
$$\times \left[\left(n_1^2 - n_2^2 \right) \left(u_{i1} u_{j1} - u_{i2} u_{j2} - v_{i1} v_{j1} + v_{i2} v_{j2} \right) + 4 n_1 n_2 \left(u_{i1} u_{j2} - v_{i1} v_{j2} \right) \right] \}. \tag{9.23}$$

We have obtained the sum of two forms, one of which depends only on u and the other one – only on v, while v–form is obtained from the u–form by the substitutions $u \to v$, $u_{31} \to -v_{32}$ and $u_{32} \to v_{31}$. This shows that the spectrum is doubly degenerate.

Comparing the v–form in (9.23) with the form of the variables in the A–phase (9.14), we see that they coincide if we replace $w = \left(u_{11} + v_{21} \right)/2$ in (9.14) by $v = \left(v_{11} + v_{22} \right)/2$, $u = u_{31}$ by v_{31} and $v = v_{31}$ by v_{32}.

Thus, in the considered approximation (neglecting the coupling between the *gd*– and honphonon–modes) the *gd*–modes of the spectrum in the *A*– and 2*D*–phases coincide and differ only in the degeneracy multiplicity (3 in the *A*–phase and 2 in the 2*D*–phase). This neglect does not influence the stability of the spectrum. We note, that no such agreement is obtained for the nonphonon modes, which depend on *H* in the 2*D*–phases.

In conclusion note, that it is possible to check the stability of *gd*–modes in ^{3}He–A and ^{3}He–B by measuring the angular width of a narrow sound beam.

Chapter X

Influence of Dipole Interaction and Magnetic Field on Collective Excitations

10.1. The Influence of the Dipole Interaction on the Collective Excitations

The spin–spin interaction in ^{3}He has a dipole character. In spite of its smallness ($E_D \propto 10^{-7} K$) it leads to some interesting effects. Such phenomena, as the frequency shift of the transverse nuclear magnetic resonance (NMR) in the superfluid A–phase and the longitudinal NMR in the A– and B–phases, are caused by the dipole interaction. A gap of the order of Ω_B) arises in the spectrum of one of the Goldstone modes in the B–phase (a branch of longitudinal spin waves $E = c_F k / \sqrt{5}$). This gap is connected with the dipole interaction as well as a nonzero oscillator force for the sq–modes (Brusov and Popov[1]), which vanishes if we neglect the dipole interaction. These examples show the important role of the dipole interaction.

In this Chapter we generalize the ^{3}He model in order to take the dipole interaction into account (Brusov[2]). This model is described by the effective action functional in which the dipole interaction is included. We are interested in the following problems: arising new branches of collective excitations corresponding to the ***H***–field introduced below, the ***H***–field Bose–condensation (i.e. a possibility of the ^{3}He transition into the ferromagnetic state due to the dipole interaction) and also an influence of the dipole interaction on collective excitations, their velocities and their stability. Such problems can be considered using the approximation (5.7) for the Hamiltonian (5.3) independently on the relation between a size of the dipole interactions and the potential corrections corresponding to the approximation (5.7).

A part of action responsible for the dipole interaction has a form

$$S_{CC}=-\frac{\mu^2}{2}\int d\tau\, d^3xd^3y\left[\left(\delta_{ij}-3\left(\boldsymbol{r}_i,\boldsymbol{r}_j\right)r^{-2}\right)r^{-3}-4\pi\delta_{SS'}\delta_{ij}\delta(\boldsymbol{x}-\boldsymbol{y})\right]\times$$

$$\times\overline{\chi}_{0S}(\boldsymbol{x},\tau)\frac{\sigma_i}{2}\chi_{0S}(\boldsymbol{x},\tau)\overline{\chi}_{0S'}(\boldsymbol{y},\tau)\frac{\sigma_j}{2}\chi_{0S'}(\boldsymbol{y},\tau)=\frac{\mu^2}{2}\int d\tau\, d^3xd^3y\times$$

$$\times\left[\frac{\partial^2 r^{-1}}{\partial x_i\partial x_j}+4\pi\delta_{SS'}\delta_{ij}\delta(\boldsymbol{x}-\boldsymbol{y})\right]\times$$

$$\times\overline{\chi}_{0S}(\boldsymbol{x},\tau)\frac{\sigma_i}{2}\chi_{0S}(\boldsymbol{x},\tau)\overline{\chi}_{0S'}(\boldsymbol{y},\tau)\frac{\sigma_j}{2}\chi_{0S'}(\boldsymbol{y},\tau),$$

where $r=|x-y|$, and the second term in the square brackets is the so–called "contact term".

Now we are going to describe the dipole interaction by means of the magnetic field generated by the quasiparticle magnetic moment μ. In order to do it we insert the following functional integral with respect to the variable $\boldsymbol{H}(x,\tau)$ obeying the equation $div\boldsymbol{H}=0$

$$\int DH_1DH_2DH_3\delta(div\boldsymbol{H})\exp\left(-(8\pi)^{-1}\int d\tau d^3xH^2(\boldsymbol{x},\tau)\right)$$

into the integral (5.4) with respect to slow Fermi–fields $\chi_{0S},\overline{\chi}_{0S}$.

After such insertion we get

$$\int D\overline{\chi}_{0S}D\chi_{0S}D\boldsymbol{H}\delta(div\boldsymbol{H})\exp\left(\widetilde{S}+S_{CC}-(8\pi)^{-1}\int d\tau d^3xH^2(\boldsymbol{x},\tau)\right)\tag{10.1}$$

Then we perform the shift transformation

$$\boldsymbol{H}(\boldsymbol{x},\tau)\to\boldsymbol{H}(\boldsymbol{x},\tau)-$$
$$-\mu\int d^3y\overline{\chi}_{0S}(\boldsymbol{y},\tau)\left[\left(\boldsymbol{\sigma}-3\boldsymbol{r}(\boldsymbol{\sigma},\boldsymbol{r})\right)r^{-3}-2\pi\boldsymbol{\sigma}\delta(\boldsymbol{x}-\boldsymbol{y})\right]\chi_{0S}(\boldsymbol{y},\tau),$$

which cancels the dipole part of the action S_{CC}. We then rewrite this shift in terms of components of the *H*–field:

$$H_k(\boldsymbol{x},\tau) \to H_k(\boldsymbol{x},\tau) +$$
$$+\mu\int d^3y \bar{\chi}_{0S}(\boldsymbol{y},\tau)\sum_{ii}\frac{\sigma_i}{2}\frac{\partial^2 r^{-1}}{\partial x_k \partial x_i}\chi_{0S}(\boldsymbol{y},\tau) + 2\pi\bar{\chi}_{0S}(\boldsymbol{x},\tau)\sigma_k\bar{\chi}_{0S}(\boldsymbol{x},\tau).$$

Squaring each component and then summing over *k*, we get

$$-(8\pi)^{-1}H^2(\boldsymbol{x},\tau) \to -(8\pi)^{-1}H^2(\boldsymbol{x},\tau) -$$
$$-\frac{\mu^2}{2}\int d^3y \frac{\partial^2 r^{-1}}{\partial x_k \partial x_i}\bar{\chi}_{0S}(\boldsymbol{y},\tau)\frac{\sigma_i}{2}\chi_{0S}(\boldsymbol{y},\tau)\bar{\chi}_{0S'}(\boldsymbol{x},\tau)\frac{\sigma_j}{2}\chi_{0S'}(\boldsymbol{x},\tau)$$
$$-2\pi\mu^2\sum_k\left(\bar{\chi}_{0S}(\boldsymbol{x},\tau)\frac{\sigma_k}{2}\bar{\chi}_{0S}(\boldsymbol{x},\tau)\right)^2 - \frac{\mu}{2}H(\boldsymbol{x},\tau)\bar{\chi}_{0S}(\boldsymbol{x},\tau)\sigma\bar{\chi}_{0S}(\boldsymbol{x},\tau)$$

(10.2)

Substituting (10.2) into (10.1), we obtain

$$\int D\bar{\chi}_{0S}D\chi_{0S}D\boldsymbol{H}\,\delta(div\boldsymbol{H})\times$$
$$\times\exp\left(\tilde{S}\quad\int d\tau d^3x\left[(8\pi)^{-1}H^2(\boldsymbol{x},\tau) + \frac{\mu}{2}\boldsymbol{H}(\boldsymbol{x},\tau)\bar{\chi}_{0S}(\boldsymbol{x},\tau)\boldsymbol{\sigma}\chi_{0S}(\boldsymbol{x},\tau)\right]\right).$$

(10.3)

Then if we use the momentum representation according to the equation

$$H(\boldsymbol{x},\tau) = (\beta V)^{-1/2}\sum_p H(p)e^{ipx}\,,\ \chi_{0S}(x) = (\beta V)^{-1/2}\sum_p a_S(p)e^{ipx}\,,$$

where

$$p = (\boldsymbol{k}, \omega), \quad x = (\boldsymbol{x}, \tau), \quad px = \omega\tau + \boldsymbol{kx} ,$$

we get

$$\int D\overline{\chi}_{0S} D\chi_{0S} D\boldsymbol{H}\, \delta(div\boldsymbol{H}) \times$$
$$\times \exp\left(\tilde{S} - \int d\tau d^3x \left[(8\pi)^{-1} H^2(\boldsymbol{x},\tau) + \frac{\mu}{2} \boldsymbol{H}(\boldsymbol{x},\tau) \overline{\chi}_{0S}(\boldsymbol{x},\tau) \boldsymbol{\sigma} \chi_{0S}(\boldsymbol{x},\tau) \right] \right) \tag{10.4}$$

instead of (10.3). Then let us go from Fermi–fields to Bose–fields by inserting the Gaussian functional integral with respect to auxiliary Bose–fields $c_{ia}(p)$, corresponding to Cooper pairs of fermions

$$\int Dc_{ia}^{+} Dc_{ia} \exp\left(g_0^{-1} \sum_{p,i,a} c_{ia}^{+}(p) c_{ia}(p) \right)$$

into the integral (10.4). Then let us perform the shift of Bose–fields, annihilating the quartic form $\tilde{S}_4$ in $\tilde{S}$

$$c_{i1}(p) \to c_{i1}(p) + \frac{g_0}{2(\beta V)^{1/2}} \sum_{p1+p2=p} (n_{1i} - n_{2i}) [a_{+}(p_2) a_{+}(p_1) - a_{-}(p_2) a_{-}(p_1)]$$

$$c_{i2}(p) \to c_{i2}(p) + \frac{ig_0}{2(\beta V)^{1/2}} \sum_{p1+p2=p} (n_{1i} - n_{2i}) [a_{+}(p_2) a_{+}(p_1) + a_{-}(p_2) a_{-}(p_1)]$$

$$c_{i3}(p) \to c_{i3}(p) + \frac{g_0}{(\beta V)^{1/2}} \sum_{p1+p2=p} (n_{1i} - n_{2i})\, a_{-}(p_2) a_{+}(p_1) .$$

After this shift the integral over the slow fields $a_S(p), a_S^+(p)$ is Gaussian. Evaluating this integral we obtain the integral with respect to Bose–fields $c_{ia}(p), c_{ia}^+(p), H(p)$. The integrand is equal to

$$\exp S_{eff}\left[c_{ia}(p), c_{ia}^+(p), H(p)\right],$$

where

$$S_{eff}(c,\bar{c}) = g_0^{-1}\sum_{p,i,a} c_{ia}^+(p)c_{ia}(p) -$$
$$-(8\pi)^{-1}\sum_p \left(\boldsymbol{H}(p), \boldsymbol{H}(-p)\right) + \frac{1}{2}\ln\det\frac{\hat{M}(c_{ia}^+, c_{ia}, H)}{\hat{M}(0,0,0)} \tag{10.5}$$

is an effective action functional and the fourth–order matrix $\hat{M}(p_1, p_2)$ has the following elements $M_{ab}(p_1, p_2)$

$$M_{11} = Z^{-1}\left[i\omega - \xi + \mu(\mathbf{H}\boldsymbol{\sigma})\right]\delta_{p_1 p_2} + Z^{-1}(\beta V)^{-1/2} H(p_1 - p_2)\mu\frac{\sigma}{2},$$

$$M_{22} = Z^{-1}\left[-i\omega + \xi + \mu(\mathbf{H}\boldsymbol{\sigma})\right]\delta_{p_1 p_2} - Z^{-1}(\beta V)^{-1/2} H(p_2 - p_1)\mu\frac{\sigma}{2},$$

$$M_{12} = (\beta V)^{-1/2}(n_{1i} - n_{2i})c_{ia}(p_1 + p_2)\sigma_a,$$

$$M_{21} = -(\beta V)^{-1/2}(n_{1i} - n_{2i})c_{ia}^+(p_1 + p_2)\sigma_a,$$

where $\sigma_a (a = 1,2,3)$ are 2 x 2 Pauli matrices.

The functional (10.5) describes all physical properties of a model system. Particularly, it describes all phenomena connected with NMR. Moreover it is not founded on the phenomenological assumptions as the Leggett–Takagi equation[3] which are usually applied to the NMR description.

The Ginzburg–Landau region

In the region $T \approx T_C$ the $\ln\det$ functional can be expanded in power series on fields, which play a role of the order parameters, and H (which is small due to weakness of the dynamical interaction generated this field). Thus we have

$$\frac{1}{2}\ln\det\frac{\widehat{M}(c_{ia}^{+},c_{ia},H)}{\widehat{M}(0,0,0)} = -\frac{1}{4}Tr\left(\hat{G}\hat{u}\right)^2 - \frac{1}{8}Tr\left(\hat{G}\hat{u}\right)^4,$$

where

$$\hat{G}^{-1} = \hat{M}(0,0,0) = Z^{-1}\delta_{p1p2}\begin{pmatrix} i\omega-\xi & 0 \\ 0 & -i\omega+\xi \end{pmatrix},$$

$$u_{p1p2} = Z^{-1}(\beta V)^{-1/2}\times$$

$$\times\begin{pmatrix} H(p_1-p_2)\mu\dfrac{\sigma}{2} & Z(n_{1i}-n_{2i})\sigma_a c_{ia}(p_1+p_2) \\ -Z(n_{1i}-n_{2i})\sigma_a c_{ia}^{+}(p_1+p_2) & -H(p_2-p_1)\mu\dfrac{\sigma}{2} \end{pmatrix}.$$

We put the external field to be equal to zero. Then we obtain

$$Tr\left(\hat{G}\hat{u}\right)^2 = \sum_{p1,p2,p3,p4} trG_{p1p2}u_{p2p3}G_{p3p4}u_{p4p1} = Z^2\mu^2(\beta V)^{-1}\times$$

$$\times \sum_{p1-p2=p} G(p_1)G(p_2)(H(p),H(-p))-$$

$$-16(\beta V)^{-1} \sum_{p1+p2=p} G(p_1)G(p_2)n_{1i}n_{1j}c^+_{ia}(p)c_{ia}(p).$$

Here,

$$G = Z(i\omega-\xi)^{-1}$$

and momenta p are small. At $T > T_C$ we have

$$S_{eff} = g_0^{-1}\sum_{p,i,a} c^*_{ia}(p)c_{ia}(p)-(8\pi)^{-1}\sum_p (H(p)H(-p))-$$

$$-\mu^2(4\beta V)^{-1} \sum_{p1-p2=p} (H(p)H(-p))(i\omega_1-\xi_1)^{-1}(i\omega_2-\xi_2)^{-1}+$$

$$+4Z^2(\beta V)^{-1} \sum_{p1+p2=p} (i\omega_1-\xi_1)^{-1}(i\omega_2-\xi_2)^{-1}n_{1i}n_{1j}c^*_{ia}(p)c_{ja}(p)=$$

$$=\sum_p\left[g_0^{-1}\delta_{ij}+\frac{4Z^2}{\beta V}\sum_{p1+p2=p}(i\omega_1-\xi_1)^{-1}(i\omega_2-\xi_2)^{-1}n_{1i}n_{1j}\right]c^*_{ia}(p)c_{ja}(p)-$$

$$-\sum_p\left[(8\pi)^{-1}+\frac{\mu^2}{4\beta V}\sum_{p1+p2=p}(-i\omega_1+\xi_1)^{-1}(i\omega_2+\xi_2)^{-1}\right]H(p)H(-p).$$

Equating to zero a coefficient in front of $H^2(0)$, we get an equation for the critical temperature T_H for the H–field:

$$1+\frac{2\pi\mu^2}{\beta V}\sum_{p1}(i\omega_1-\xi_1)^{-2}=0.$$

Summing over ω_1 according to the formula

$$\beta^{-1}\sum_{\omega 1}\left(i\omega_1-\xi_1\right)^{-2}=-\beta e^{\beta\xi_1}(e^{\beta\xi_1}+1)^{-2}$$

and then integrating with respect to the momentum k_1, we get

$$1-\frac{\mu^2k_F^2}{\pi c_F}=0.$$

The expression $1-\mu^2k_F^2/\pi c_F$ does not vanish because $\mu^2k_F^2/\pi c_F<<1$. This is why the Bose–condensation of the H–field is impossible and the system can not go into the ferromagnetic state due to the dipole interaction. Let us note that the dipole interaction can lead to the weak ferromagnetism in the *A*–phase[3] of ^{3}He, but for obtaining this effect it is necessary to take the gap deformation in the Fermi–spectrum into account (see Chapter XII). The spectrum of collective excitations of the ***H***–field can be defined by equating to zero a coefficient in front of $(\boldsymbol{H}(p),\boldsymbol{H}(-p))$.

The equation for the spectrum has the following form

$$1+\frac{2\pi\mu^2}{\beta V}\sum_{p_1+p_2=p}\left(-i\omega_1+\xi_1\right)^{-1}\left(-i\omega_2+\xi_2\right)^{-1}=0.$$

Evaluating the sum over ω_1 and ξ_1 we come to the equation

$$1-\frac{\mu^2 k_F^2}{(2\pi)^2 c_F}\int\frac{c_F^2(\boldsymbol{nk})^2 d\Omega}{\omega^2+c_F^2(\boldsymbol{nk})^2}=0.$$

One can show that this equation has no solutions. Thus there are no magnetic excitations in the system at $T>T_C$.

At $T<T_C$ we have to separate the condensate $c_{ia}^{(0)}(p)$ which has a different form for different phases. We consider the isotropic B–phase in the low temperature limit $T\to 0$.

Collective excitations in the B–phase at $T_C-T\propto T_C$

The condensate function $c_{ia}^{(0)}(p)$ in the B–phase has a following form

$$c_{ia}^{(0)}(p)=(\beta V)^{1/2}c\delta_{p0}\delta_{ia}$$

where c can be defined from the equation

$$3g_0^{-1}+4Z^2(\beta V)^{-1}\sum_p\left(\omega^2+\xi^2+4c^2\boldsymbol{Z}^2\right)^{-1}=0.$$

After the shift $c_{ia}(p)\to c_{ia}(p)+c_{ia}^{(0)}(p)$ we have to consider the quadratic part of new variables in S_{eff}

$$S_{eff}(c,\bar{c})=g_0^{-1}\sum_{p,i,a}c_{ia}^{+}(p)c_{ia}(p)-(8\pi)^{-1}\sum_p(\boldsymbol{H}(p),\boldsymbol{H}(-p))-\frac{1}{4}Tr\left(\hat{G}\hat{u}\right)^2$$

where

$$u_{p1p2} = (\beta V)^{-1/2}\begin{pmatrix} Z(H(p_1-p_2),\sigma)\frac{\mu}{2} & (n_{1i}-n_{2i})\sigma_a c_{ia}(p_1+p_2) \\ -(n_{1i}-n_{2i})\sigma_a c^{+}_{ia}(p_1+p_2) & -Z(H(p_2-p_1),\sigma)\frac{\mu}{2} \end{pmatrix}$$

$$\hat{G}_{p1p2} = \frac{Z}{M}Z^{-1}\begin{pmatrix} -(i\omega_1+\xi_1)\delta_{p1p2} & 2cZ(n,\sigma)\delta_{p1+p2=0} \\ -2\bar{c}Z(n,\sigma)\delta_{p1+p2=0} & (i\omega_1+\xi_1)\delta_{p1p2} \end{pmatrix},$$

$$M = \omega^2+\xi^2+4c^2Z^2 = \omega^2+\xi^2+\Delta^2 = \omega^2+\varepsilon^2 .$$

Evaluating

$$\frac{1}{4}Tr\left(\hat{G}\hat{u}\right)^2 = \frac{1}{4}\sum_{p1,p2,p3,p4} tr\left(G_{p1p2}u_{p2p3}G_{p3p4}u_{p4p1}\right),$$

we obtain the following expression for $(S_{eff})_2$

$$\begin{aligned}(S_{eff})_2 = &-\frac{Z^2}{4\beta V}\sum_p\{\sum_{p1+p2=p}\frac{\mu^2(-i\omega_1+\xi_1)(i\omega_2+\xi_2)}{Z^2(\omega_1^2+\varepsilon_1^2)(\omega_2^2+\varepsilon_2^2)}(\boldsymbol{H}(p),\boldsymbol{H}(-p))+\\ &+\sum_{p1+p2=p}\Delta^2(\omega_1^2+\varepsilon_1^2)^{-1}(\omega_2^2+\varepsilon_2^2)^{-1}\{(\boldsymbol{H}(p),\boldsymbol{H}(-p))-\\ &-2(\boldsymbol{n}_1,\boldsymbol{H}(p))(\boldsymbol{n}_2,\boldsymbol{H}(-p))]-8ic\mu\sum_{p1+p2=p}\frac{i\omega+(\xi_1+\xi_2)}{(\omega_1^2+\varepsilon_1^2)(\omega_2^2+\varepsilon_2^2)}n_{1i}\times\\ &\times[((\boldsymbol{n}_1,\boldsymbol{H}(p))_a c^*_{ia}(p)+(\boldsymbol{n}_1,\boldsymbol{H}(-p))_a c_{ia}(p))]-(8\pi)^{-1}\sum_p(\boldsymbol{H}(p),\boldsymbol{H}(-p))\}-\\ &-\sum_p\{c^*_{ia}(p)c_{ja}(p)\left[-g_0^{-1}\delta_{ij}-\frac{4Z^2}{\beta V}\sum_{p1+p2=p}n_{1i}n_{1j}\frac{(i\omega_1+\xi_1)(i\omega_2+\xi_2)}{(\omega_1^2+\varepsilon_1^2)(\omega_2^2+\varepsilon_2^2)}\right]-\\ &-\frac{1}{2}(c_{ia}(p)c_{jb}(-p)+c^*_{ia}(p)c^*_{jb}(-p))\frac{4Z^2\Delta^2}{\beta V}\sum_{p1+p2=p}n_{1i}n_{1j}\frac{(2n_{1a}n_{1b}-\delta_{ab})}{(\omega_1^2+\varepsilon_1^2)(\omega_2^2+\varepsilon_2^2)}\}\end{aligned} \tag{10.6}$$

This quadratic form defines the Bose–spectrum in the B–phase of ^{3}He in which the dipole interaction is taken into account. Let us rewrite it as follows

$$
\begin{aligned}
(S_{eff})_2 = \sum_p \{ -A_{ij}(p)c^*_{ia}(p)c_{ja}(p) - \\
-\frac{1}{2}B_{ijab}(p)\left(c_{ia}(p)c_{jb}(-p) + c^*_{ia}(p)c^*_{jb}(-p)\right) + \\
+C(p)H_i(p)H_i(-p) + D_{iak}(p)\left(c^*_{ia}(p)H_k(p) + c_{ia}(p)H_k(-p)\right)\} .
\end{aligned}
\tag{10.7}
$$

After substitution $c_{ia}(p) = u_{ia}(p) + iv_{ia}(p)$, $c^*_{ia}(p) = u_{ia}(p) - iv_{ia}(p)$ the quadratic form (10.7) splits in two independent forms of variables $u_{ia}(p)$ and $v_{ia}(p)$ respectively

$$
\begin{aligned}
&\sum_p \{ -A_{ij}(p)u_{ia}(p)u_{ja}(-p) - B_{ijab}(p)u_{ia}(p)u_{jb}(-p) + \\
&+\frac{1}{2}C(p)H_i(p)H_i(-p) + D_{iak}(p)\left(u_{ia}(-p)H_k(p) + u_{ia}(p)H_k(-p)\right)\} , \\
&\sum_p \{ -A_{ij}(p)v_{ia}(p)v_{ja}(-p) + B_{ijab}(p)v_{ia}(p)v_{jb}(-p) + \\
&+\frac{1}{2}C(p)H_i(p)H_i(-p) - iD_{iak}(p)\left(v_{ia}(-p)H_k(p) - v_{ia}(p)H_k(-p)\right)\}
\end{aligned}
\tag{10.8}
$$

Let us consider the first form. It is appropriate to go to phonon variables according to the formula

$$u_{12} = (u + u_1)/2, \quad u_{32} = (v_1 - v)/2, \; u_{11} = (w' + w_1')/3,$$

$$u_{21} = (u_1 - u)/2, \quad u_{32} = (t + t_1)/2, \; u_{22} = (3w_2' - w_1' + 2w')/6,$$

$$u_{23} = (v + v_1)/2, \quad u_{13} = (t_1 - t)/2, \; u_{33} = -(3w_2' + w_1' - 2w')/6. \tag{10.9}$$

The Bose–spectrum is defined by the equation

$$\det Q = 0, \tag{10.10}$$

where Q is a matrix of the quadratic form (10.7) in the phonon variables (10.9).

Evaluating the coefficients of the quadratic form and then expanding them in power series in small quantities ω^2, k^2, we can then solve the equation (10.10) and obtain a dispersion law for the transverse spin waves in which the dipole interaction is taken into account.

When calculating a dispersion law for the longitudinal spin waves which has a gap at $k = 0$ and also corrections to frequencies of phonon branches we have to take small only k and it is necessary to expand the coefficient functions in power series on k^2 (ω remains finite at $k \to 0$).

The equation for the spectrum of the transverse spin waves is as follows

$$\omega^2 + \frac{2}{5}c_F^2 k^2 - \frac{1}{2\Delta^2}\left(\frac{\omega^4}{3} + \frac{4}{15}\omega^2 c_F^4 k^4 + \frac{19}{40}c_F^4 k^4\right) - \\ -12x_1\omega^2\left(\frac{1}{3} - \frac{\omega^2}{18\Delta^2} - \frac{c_F^2 k^2}{45\Delta^2}\right) = 0, \tag{10.11}$$

where $x_1 = \mu^2 k_F^2 / \pi c_F$.

Taking only the terms $\propto \omega^2,\ k^2$ into account, we get

$$\omega^2 + \frac{1}{5}c_F^2 k^2 - \frac{4}{3}x_1\omega^2 = 0$$

After the analytic continuation $i\omega \to E$ we obtain the following dispersion law

$$E = \sqrt{\frac{2}{5}}c_F k\left(1 - \frac{2}{3}\frac{\mu^2 k_F^2}{\pi c_F}\right)^{-1}.$$

The velocity of transverse spin waves increases if we take a dipole interaction into account.

If we take also the fourth order terms in (10.11) in consideration we get the dispersion law of the form

$$E = \sqrt{\frac{2}{5}}\frac{c_F k}{\left(1 - \frac{2}{3}\frac{\mu^2 k_F^2}{\pi c_F}\right)}\left(1 - \frac{173 c_F^2 k^2}{3360\Delta^2}\right).$$

Thus, in the first approximation the dipole interaction does not influence the stability of the transverse spin waves (Brusov and Popov[5]), but only renormalizes the velocity of excitations.

Now let us consider the second quadratic form

$$\sum_p \{-A_{ij}(p)v_{ia}(p)v_{ja}(-p)+B_{ijab}(p)v_{ia}(p)v_{jb}(-p)+$$

$$+\frac{1}{2}C(p)H_i(p)H_i(-p)-iD_{iak}(p)\big(v_{ia}(-p)H_k(p)-v_{ia}(p)H_k(-p)\big)\}$$

and then go to new variables as follows

$$v_{12}=(u'+u_1')/2,\quad v_{32}=(v_1'-v')/2,\ v_{11}=(w+w_1)/3,$$

$$v_{21}=(u_1'-u')/2,\quad v_{31}=(t'+t_1')/2,\ v_{22}=(3w_2-w_1+2w)/6,$$

$$v_{23}=(v+v_1)/2,\quad v_{13}=(t_1'-t')/2,\ v_{33}=-(3w_2+w_1-2w)/6.$$

Here, w is a phonon variable.

The part of the quadratic form, connected with the ***H***–field, has a form

$$\frac{Z^2c\mu\omega^2k_F^2}{2\pi^2c_F\Delta^2}\left\{-i\left[\frac{1}{3}-\frac{1}{18\Delta^2}\left(\omega^2+\frac{1}{5}c_F^2k^2\right)\right]\big(v_{12}(-p)H_3(p)-\right.$$

$$-v_{13}(-p)H_2(p)-v_{21}(-p)H_3(p)+v_{23}(-p)H_1(p)\big)-$$

$$-i\left[\frac{1}{3}-\frac{1}{18\Delta^2}\left(\omega^2+\frac{1}{5}c_F^2k^2\right)-\frac{c_F^2k^2}{45\Delta^2}\right]\big(v_{13}(-p)H_2(p)-v_{32}(-p)H_1(p)\big)+$$

$$+i\left[\frac{1}{3}-\frac{1}{18\Delta^2}\left(\omega^2+\frac{1}{5}c_F^2k^2\right)-\frac{c_F^2k^2}{45\Delta^2}\right]\big(v_{12}(p)H_3(-p)-$$

$$-v_{13}(p)H_2(-p)-v_{21}(p)H_3(-p)+v_{23}(p)H_1(-p)\big)+$$

$$\left.+i\left[\frac{1}{3}-\frac{1}{18\Delta^2}\left(\omega^2+\frac{1}{5}c_F^2k^2\right)-\frac{c_F^2k^2}{45\Delta^2}\right]\big(v_{13}(p)H_2(-p)-v_{32}(p)H_1(-p)\big)\right\}$$

Calculating the addend to the matrix of ν–variables due to this form one can show that the addend to the matrix element corresponding to the sound branch is equal to zero. Thus, the dipole interaction does not influence the velocity and stability of sound waves $E = c_F k / \sqrt{3}$.

Using obtained dispersion laws one can see that the dipole interaction does not lead to gaps in the spectrum of spin waves. The reason of this fact is that we do not take the deformation of the order parameter due to dipole interaction into account. The investigation of the order parameter deformation is done in the Chapter XII.

10.2. The Influence of the Magnetic Field on the Collective Excitations

In this section we investigate the influence of the magnetic field on the structure of the Bose–spectrum in the *A*– and *B*–phases of ^{3}He particularly on the number of Goldstone–modes in them. This problem can be solved by investigating the symmetry of the Lagrangian and of the states corresponding to these phases in the magnetic field. However we shall consider this question by investigating properties of the $\delta^2 F$ form. It was shown in Chapter VII that the number of Goldstone–modes diminishes from nine to six in the weak coupling approximation and from five to four if we take strong coupling effects into account after switching the magnetic field on.

Let us begin with the *B*–phase. Here the form $\delta^2 F$ is as follows

$$\delta^2 F = \frac{1}{5}(2+\nu)\left[3u_{11}^2 + 3u_{22}^2 + 2u_{11}u_{22} + (u_{12}+u_{21})^2 + u_{13}^2 + u_{23}^2\right] +$$
$$+\frac{2}{5}(1-2\nu)(u_{31}^2 + u_{32}^2 + 3u_{33}^2) + \frac{4}{5}\left[\frac{1}{2}(1-2\nu)(2+\nu)\right]^{1/2} \times$$
$$\times (u_{11}u_{33} + u_{22}u_{33} + u_{13}u_{31} + u_{23}u_{32}) + \frac{1}{5}(4-3\nu)(v_{11}^2 + v_{22}^2) +$$
$$+\frac{1}{5}(8-\nu)(v_{12}^2 + v_{21}^2) + \frac{2}{5}(2+\nu)(v_{33}^2 - v_{11}v_{22} - v_{12}v_{21}) +$$
$$+\frac{1}{5}(8+9\nu)(v_{13}^2 + v_{23}^2) + \frac{8}{5}(1-2\nu)(v_{31}^2 + v_{32}^2) -$$
$$-\frac{4}{5}\left[\frac{1}{2}(1-2\nu)(2+\nu)\right]^{1/2}(v_{11}v_{33} + v_{22}v_{33} + v_{13}v_{31} + v_{23}v_{32}) \tag{10.12}$$

This quadratic form splits in two independent ones depending on u_{ia} and v_{ia} respectively. After reducing them to canonical forms we get the expression for the *u*–form:

$$\frac{1}{5}(2+\nu)(u_{12}+u_{21})^2 + \left(\sqrt{\frac{1}{5}(2+\nu)}\ u_{13} + 2\sqrt{\frac{1}{10}(1-2\nu)}\ u_{31}\right)^2 +$$
$$+\left(\sqrt{\frac{1}{5}(2+\nu)}\ u_{23} + 2\sqrt{\frac{1}{10}(1-2\nu)}\ u_{32}\right)^2 + \frac{1}{16}(1-2\nu)u_{33}^2 +$$
$$+\frac{5}{3(2+\nu)}\left[\frac{3}{5}(2+\nu)u_{11} + \frac{1}{5}(2+\nu)u_{22} + \frac{2}{5}\sqrt{\frac{1}{2}(1-2\nu)(2+\nu)}\ u_{33}\right]^2 +$$
$$+\frac{15}{8(2+\nu)}\left[\frac{8}{15}(2+\nu)u_{22} + \frac{4}{15}\sqrt{\frac{1}{2}(1-2\nu)(2+\nu)}\ u_{33}\right]^2 \tag{10.13}$$

and the expression for the v–form

$$
\begin{aligned}
&\frac{1}{5}(2+\nu)(u_{12}+u_{21})^2+\left(\sqrt{\frac{1}{5}(2+\nu)}\;u_{13}+2\sqrt{\frac{1}{10}(1-2\nu)}\;u_{31}\right)^2+\\
&+\left(\sqrt{\frac{1}{5}(2+\nu)}\;u_{23}+2\sqrt{\frac{1}{10}(1-2\nu)}\;u_{32}\right)^2+\\
&+\frac{5}{4-3\nu}\left[\frac{4-3\nu}{5}v_{11}-\frac{1}{5}(2+\nu)v_{22}-\frac{2}{5}\sqrt{\frac{1}{2}(1-2\nu)(2+\nu)}\;v_{33}\right]^2+\\
&+\frac{4}{(3-\nu)8(1-2\nu)}\left[\frac{(3-\nu)(1-2\nu)}{4-3\nu}v_{22}-\frac{3-\nu}{4-3\nu}\sqrt{\frac{1}{2}(1-2\nu)(2+\nu)}\;v_{33}\right]^2\\
&+\frac{1}{5}(8-\nu)\left(v_{12}-\frac{2+\nu}{8-\nu}v_{21}\right)^2+\frac{4(3-\nu)}{8-\nu}v_{21}^2+\\
&+\frac{1}{5}(8+9\nu)\left[v_{13}-\frac{4}{8+9\nu}\sqrt{\frac{1}{2}(1-2\nu)(2+\nu)}\;v_{31}\right]^2+\frac{8}{5}(1-2\nu)(1+\nu)v_{31}^2+\\
&+\frac{1}{5}(8+9\nu)\left[v_{23}-\frac{4}{8+9\nu}\sqrt{\frac{1}{2}(1-2\nu)(2+\nu)}\;v_{32}\right]^2+\frac{8}{5}(1-2\nu)(1+\nu)v_{32}^2.
\end{aligned}
\tag{10.14}
$$

There exist three phonon variables $u_{12}-u_{21}$, $u_{23}-u_{32}$, $u_{13}-u_{31}$ in the u –form at $\nu=0$, corresponding to three Goldstone (phonon) modes of the Bose–spectrum (spin waves).

The v–form gives at $\nu=0$ only one Goldstone–mode (the sound), corresponding to the phonon variable $v_{11}+v_{22}+v_{33}$. One can see from (10.14) that this variable remains a phonon variable also after the magnetic field switches on.

Thus, the investigation of the second variation form $\delta^2 F$ for the B–phase shows that a number of Goldstone–modes does not change (in the contrast with situation in the *A*–phase).

Now let us consider the magnetic field influence on the nonphonon modes in the *A*–phase and the *B*–phase. The form $\delta^2 F$ in the *A*–phase has a following form in the magnetic field (after reducing to the canonical form):

$$\delta^2 F = 4\nu(u_{33}^2 + v_{33}^2) + 2(1+2\nu)\left(u_{13} - \frac{1}{1+2\nu} v_{23}\right)^2 + \\ + \frac{1}{2}\frac{(2\nu-1)(3+2\nu)}{1+2\nu} v_{23}^2 + 4(u_{11}+v_{21})^2 + 2(u_{11}-v_{21})^2 + \\ + 2(u_{12}-v_{22})^2 + 2(u_{21}+v_{11})^2 + 4(u_{22}-v_{12})^2 + 2(u_{22}+v_{12})^2. \tag{10.15}$$

The analysis of the forms (10.13), (10.14), (10.15) shows that in the *B*–phase all the nonphonon branches in the magnetic field obtain addends of the order of μH. In the *A*–phase only three modes having gaps at *H*=0, obtain analogous addends. These modes correspond to the variables $u_{13} - v_{23}$, $u_{23} + v_{23}$, $u_{23} - v_{13}$. Magnetic field does not influence six other nonphonon modes in the *A*–phase corresponding to the variables $u_{11} + v_{21}$, $u_{11} - v_{21}$, $u_{21} - v_{22}$, $u_{21} + v_{11}$, $u_{22} - v_{12}$, $u_{22} + v_{12}$.

10.3. Conclusion

In this Chapter the ^{3}He model is generalized in order to take the dipole interaction into account. The effective (hydrodynamical) action functional which determines physical properties of a model system and particularly completely describes the NMR effects. It is shown that the magnetic field connected with a dipole interaction does not condensed, and there are no magnetic excitations. A renormalized velocity of transverse spin waves in the isotropic *B*–phase was calculated at

$T_C - T \propto T_C$. It is shown that a dipole interaction does not influence the velocity and stability of sound waves.

The influence of the external magnetic field on the Bose–spectrum structure in the *A*– and *B*–phases is investigated. It is shown that the number of Goldstone–modes is not changed after switching on the magnetic field, and the energy of each nonphonon mode gets an addend $\propto \mu H$.

The number of Goldstone–modes in the *A*–phase diminishes in the presence of the magnetic field due to arising of gaps $\propto \mu H$, and an energy of each of three nonphonon modes obtains an addend of the same order. The main results of this Chapter are firstly published in the paper by Brusov[2].

Chapter XI

The Influence of the Electric Field on the Collective Excitations in ^3He and ^4He

11.1. The Energy Spectrum and Hydrodynamics of ^4He in a Strong Electric Field (Macroscopic Approach)

In this Chapter we consider some properties of superfluid ^4He and ^3He in strong electric fields. The outer electric field polarizes the helium atoms, which begin to interact each other. Long range dipole interaction forces change the sound velocity and make it anlsotropic as well as ρ_S / ρ_n ratio (the ratio of the superfluid density to the normal one). Such effects are considered below in sections 11.1, 11.2 for ^4He and in sections 11.3–11.5 for ^3He. A in strong electric field must also make the second sound velocity anisotropic as well as the vortex excitation spectrum. Similar effects must also arise in ^3He in strong magnetic fields.

One can describe the phonon part of the energy spectrum of ^4He in electric fields by using the Landau hydrodynamical Hamiltonian[1]

$$\hat{H} = \int \left(\frac{1}{2} \hat{v} \rho \hat{v} + \varepsilon_E(\rho) \right) dV \tag{11.1}$$

where

$$\varepsilon_E(\rho) = \varepsilon_0(\rho) - \varepsilon E^2 / 8\pi$$

is the energy of a unit of volume of ^4He in the field[2] E, $\varepsilon_0(\rho)$ is the same energy with zero field E, ε is the dielectric constant of ^{4}He. In order to describe elementary excitations in such a system we write ρ as a sum $\rho_0 + \delta\rho$, where ρ_0, is a constant, $\delta\rho$ is a fluctuation[3]. The value of ρ_0 can be find from the equation, $\partial\varepsilon_E(\rho)/\partial\rho = 0$. Expanding g $\varepsilon_E(\rho)$ in powers of $\delta\rho$ up to $(\delta\rho)^2$, one can write down

$$\varepsilon_E(\rho) = \varepsilon_E(\rho_0) + \frac{1}{2}\frac{\partial\varepsilon_E(\rho)}{\partial\rho^2}(\delta\rho)^2 \tag{11.2}$$

Where

$$\frac{\partial^2\varepsilon_E(\rho)}{\partial\rho^2} = \frac{\partial\mu_E}{\partial\rho} = \frac{1}{\rho}\frac{dp}{d\rho} = \frac{m}{\rho}C_E^2 \tag{11.3}$$

$$C_E^2 = C^2 - \frac{E^2}{8\pi}\frac{\rho_0}{m}\frac{\partial^2\varepsilon}{\partial\rho^2} \tag{11.4}$$

Using Fourier expansions

$$\delta\rho = V^{-1/2}\sum_k \rho_k \exp(i\boldsymbol{kr}), \quad \boldsymbol{v} = V^{-1/2}\sum_{\boldsymbol{k}} v_k \exp(i\boldsymbol{kr}), \tag{11.5}$$

we find from the continuity equation $\dot{\rho} + div\boldsymbol{v} = 0$ in the k – representation

$$\boldsymbol{v}_k = i\frac{\dot{\rho}_k}{\rho_0}\frac{\boldsymbol{k}}{k^2}\ . \tag{11.6}$$

If we substitute (11.2) into (11.1) and take Eqs. (11.3)–(11.6) into account we may rewrite $\hat{H}$ in the form

$$\hat{H} = \int \varepsilon_E(\rho_0)dV + \sum_k \left(\frac{\dot{\rho}_k \dot{\rho}_{-k}}{2\rho_0 k^2} + \frac{C_E^2}{2\rho_0} \rho_k \rho_{-k} \right). \tag{11.7}$$

The Hamiltonian describes sound excitations and C_E^2 (Eq. (11.4)) is the square of the sound velocity in the electric field. Equation (11.4) contains $\partial^2 \varepsilon / \partial \rho^2 = 0$. One can find this second derivative from the well known Clausius–Massotti formula

$$\frac{\varepsilon - 1}{\varepsilon + 2} = \frac{4\pi}{3} \chi ,$$

where $\chi = \alpha\rho$, α is the susceptibility of the ^{4}He atom.

Using this formula, we obtain the following result

$$C_E^2 = C_0^2 - \frac{4\pi}{3} \rho_m \left(\frac{\alpha E}{m} \right)^2 \left(1 - \frac{4\pi}{3} \chi \right)^{-3} .$$

Note that C_E vanishes for a critical value of E_C

$$E_C = \frac{mC_0}{2\alpha} \left(\frac{\pi\rho}{3} \right)^{-1/2} \approx 3.1 \cdot 10^8 \ V/cm .$$

Here, we use $\alpha = 0.196 \cdot 10^{-24} \ cm^3$ (Ref. 4), $m = 6.648 \cdot 10^{-24} \ g$, $\rho_m = 0.146 \ cm^{-3}$, $C_0 = 2.8 \cdot 10^4 \ cm/s$ (Ref. 5).

Now let us try to take into account the interaction between polarized ^{4}He atoms, by adding to the Hamiltonian (11.7) the interaction Hamiltonian

$$\hat{H}_D = \frac{1}{2}\iint d^3 r_1 d^3 r_2 \delta\rho(\boldsymbol{r}_1)\delta\rho(\boldsymbol{r}_2)\big((\boldsymbol{d}(\boldsymbol{r}_1),\boldsymbol{d}(\boldsymbol{r}_2)) - 3(\boldsymbol{d}(\boldsymbol{r}_1),\boldsymbol{n}_{12})(\boldsymbol{d}(\boldsymbol{r}_2),\boldsymbol{n}_{12})\big)\times$$
$$\times|\boldsymbol{r}_1 - \boldsymbol{r}_2|^{-3} \tag{11.8}$$

Due to small susceptibility we can take $\boldsymbol{d} = \alpha\boldsymbol{E}$ in (11.8). Using Eq. (11.5) for $\delta\rho$ in eq.(11.8) we obtain

$$\hat{H} = \int \varepsilon(\rho_0)dV +$$
$$+\sum_k \left(\frac{\dot{\rho}_k \dot{\rho}_{-k}}{2\rho_0 k^2} + \frac{1}{2\rho_0}\left[C_0^2 + 4\pi\left(\frac{\alpha E}{m}\right)^2 \rho_0 \left(\cos^2\theta - \frac{1}{3}\right)\right]\rho_k\rho_{-k}\right).$$

Here, θ is the angle between $\boldsymbol{E}$ and $\boldsymbol{k}$. This Hamiltonian implies the following formula

$$C_E^2 = C_0^2 + 4\pi\left(\frac{\alpha E}{m}\right)^2 \rho_0\left(\cos^2\theta - \frac{1}{3}\right) \tag{11.9}$$

for the sound velocity. According to (11.9) the sound velocity has its maximum in a "polar" region and minimurn in an "equatorial" region. In the following section we obtain an analogous formula frorn the pure microscopic approach for the Bose–gas model.

We now discuss some hydrodynamical consequences of the results connected with the anisotropy of the energy spectrum in the electric field.

Let us consider the momentum, transferred by phonon excitations

$$\boldsymbol{K} = \int \boldsymbol{p} n\big(\varepsilon(\boldsymbol{p}) - (p, \boldsymbol{v}_n - \boldsymbol{v}_S)\big)\frac{d^3 p}{(2\pi)^3},$$

where $n(\varepsilon) = \left(\exp(\beta\varepsilon - 1\right)$. For small $\boldsymbol{v}_n - \boldsymbol{v}_S$ one obtains

$$\boldsymbol{K} = \rho_{n\perp}(\boldsymbol{v}_n - \boldsymbol{v}_S)_\perp + \rho_{n||}(\boldsymbol{v}_n - \boldsymbol{v}_S)_{||}, \tag{11.10}$$

where $\left(\boldsymbol{v}_n - \boldsymbol{v}_S\right)_{||}$ is the projection of $\boldsymbol{v}_n - \boldsymbol{v}_S$ on the $\boldsymbol{E}$ – direction,

$$(\boldsymbol{v}_n - \boldsymbol{v}_S)_\perp = \boldsymbol{v}_n - \boldsymbol{v}_S - (\boldsymbol{v}_n - \boldsymbol{v}_S)_{||}.$$

The coefficients $\rho_{n||}, \rho_{n\perp}$ in (11.10) denote the normal component densities for movements of the liquid along $\boldsymbol{E}$ and across $\boldsymbol{E}$ respectively. After simple calculation we obtain

$$\rho_{n||} = \frac{c_0^5}{c_\perp^2 c_{||}^3}\rho_{n0}, \quad \rho_{n\perp} = \frac{c_0^5}{c_\perp^4 c_{||}}\rho_{n0}, \tag{11.11}$$

where $\rho_{n0} = 2\pi^2 T^4 / 45c^5$ is the normal density in the case of zero field, $c_{||}$ and $c_\perp$ are phonon velocities along and across $\boldsymbol{E}$:

$$c_{||}^2 = c_0^2 + \frac{8\pi}{3}\left(\frac{\alpha E}{m}\right)^2 \rho_0, \quad c_\perp^2 = c_0^2 - \frac{4\pi}{3}\left(\frac{\alpha E}{m}\right)^2 \rho_0. \tag{11.12}$$

The most interest consequence of (11.10)–(11.12) is an increase of the normal density for liquid helium moving perpendicular to the electric field for $\boldsymbol{E}$ is of the order of E_C. This phenomenon may decrease or stop the superfluid flow in thin capillars in which only the superfluid component can move. As a result the electric field may be used as a regulator of the superfluid flow.

11.2. Superfluid Bose–Systems in the Electric Field Microscopic Approach)

We have investigated the influence of the electric field on the properties of the superfluid Bose–liquid in the framework of the semiphenomenological theory. For the microscopic consideration we shall use the method of successive integration over fast and slow variables. This method which was described in the section 1.4, turns out to be useful in application to the superfluid Bose–systems. For the case of the Bose–system in the electric field the action can be written down as

$$S = \int d\tau\, d^3x \{\overline{\psi}(\mathbf{x},\tau)\partial_\tau \psi(\mathbf{x},\tau) - (2m)^{-1}\nabla\overline{\psi}(\mathbf{x},\tau)\nabla\psi(\mathbf{x},\tau) +$$
$$+ (\lambda + \alpha_0 E^2)\overline{\psi}(\mathbf{x},\tau)\partial_\tau \psi(\mathbf{x},\tau)\} -$$
$$-\frac{1}{2}\int d\tau\, d^3x d^3y U_E(\mathbf{x}-\mathbf{y})\overline{\psi}(\mathbf{x},\tau)\overline{\psi}(\mathbf{y},\tau)\psi(\mathbf{y},\tau)\psi(\mathbf{x},\tau), \qquad (11.13)$$

where

$$U_E(\mathbf{x}-\mathbf{y}) = U(\mathbf{x}-\mathbf{y}) + (\boldsymbol{d}_1, \boldsymbol{d}_2)r^{-3} - 3(\boldsymbol{d}_1, \boldsymbol{r})(\boldsymbol{d}_2, \boldsymbol{r})r^{-5}$$

is the pair interaction potential plus the addend describing the interaction of induced dipole momenta of Bose–particles in the electric field $\boldsymbol{d}_1 = \boldsymbol{d}_2 = \alpha \boldsymbol{E}$, where α is a bare (nonrenormalized) susceptibility of a particles in the electric field, $\boldsymbol{r} = \boldsymbol{x} - \boldsymbol{y}$.

After integration with respect to the "fast" fields which are defined by the formula

$$\psi_1(\mathbf{x},\tau) = (\beta V)^{-1/2} \sum_{|k|>k_0,\omega} a(\boldsymbol{k},\omega)\exp i(\omega\tau + \boldsymbol{kx}), \qquad (11.14)$$

where $\omega = \omega_n = 2\pi n/\beta$, we get

$$\int \exp S D\overline{\psi}_1 D\psi_1 = \exp \widetilde{S}[\overline{\psi}_0, \psi_0].$$

Here, the functional $\widetilde{S}$ depends on slow fields $\chi_0(\boldsymbol{x},\tau)$, $\overline{\chi}_0(\boldsymbol{x},\tau)$. Its definition differs from that of $\chi_1(\boldsymbol{x},\tau)$ (11.14) by changing the inequality $|k| > k_0$ by $|k| < k_0$.

For the low density system (a nonideal Bose–gas) the action in the first approximation has a following form

$$\begin{aligned} S = \beta V p_0 &+ \int d\tau\, d^3x \{\overline{\psi}_0(\mathbf{x},\tau)\partial_\tau \psi_0(\mathbf{x},\tau) - \\ &- (2m)^{-1}\nabla\overline{\psi}_0(\mathbf{x},\tau)\nabla\psi_0(\mathbf{x},\tau) + (\mu + \alpha E^2)\overline{\psi}_0(\mathbf{x},\tau)\partial_\tau\psi_0(\mathbf{x},\tau)\} \\ &- \frac{1}{2}\int d\tau\, d^3x d^3y t_E(\mathbf{x}-\mathbf{y})\overline{\psi}_0(\mathbf{x},\tau)\overline{\psi}_0(\mathbf{y},\tau)\psi_0(\mathbf{y},\tau)\psi_0(\mathbf{x},\tau) \end{aligned} \qquad (11.15)$$

Here, p_0 is a pressure of the interactioning Bose–system of particles with momenta $k > k_0$. The terms of the second order and of the forth order in (11.15) differs from those in the starting functional (11.13) by changing fields $\chi(\boldsymbol{x},\tau)$, $\overline{\chi}(\boldsymbol{x},\tau)$ by $\chi_0(\boldsymbol{x},\tau)$, $\overline{\chi}_0(\boldsymbol{x},\tau)$ and also by changing the nonrenormalized susceptibility α_0 and the chemical potential λ by their renormalized values α and μ and the potential $U_E(\boldsymbol{x}-\boldsymbol{y})$ must be changed by

$$t_E(\mathbf{x}-\mathbf{y}) = t_0\delta(\mathbf{x}-\mathbf{y}) + \alpha^2\left(E^2 r^{-3} - 3(\boldsymbol{E},\boldsymbol{r})^2 r^{-5}\right),$$

Where t_0 a *t*–matrix (a scattering amplitude at zero energy and momentum).

Using the momentum representation and omitting the constant which is not important in what follows, we get

$$\sum_p \left(i\omega - \frac{k^2}{2m} + \mu + \alpha E^2 \right) a^*(p)\, a(p)$$

$$-(2\beta V)^{-1} \sum_{p_1+p_2=p_3+p_4} t\,(p_1 - p_2) a^*(p_1) a^*(p_2) a\,(p_4) a\,(p_3), \tag{11.16}$$

where $a(p) = a(\boldsymbol{k},\omega)$, $p = (\boldsymbol{k},\omega)$, $t(p) = t_0 + 4\pi\alpha^2 E^2(\cos^2\theta - 1/3)$, θ is the angle between the vectors $\boldsymbol{k}$ and $\boldsymbol{E}$.

The phase transition means the existence of Bose–condensate in the system. One can find its density, putting $a(p) = a^*(p) = (\rho_0 \beta V)^{1/2} \delta_{p0}$ in (11.16) and demanding the obtained expression

$$\beta V \left(\rho_0 (\mu + \alpha E^2) - \frac{t_0}{2} \rho_0^2 \right)$$

be maximal. So we get

$$\rho_0 = (\mu + \alpha E^2)/t_0. \tag{11.17}$$

Let us make the shift

$$a(p) \to a(p) + (\rho_0 \beta V)^{1/2} \delta_{p0}. \tag{11.18}$$

The Bose–excitation spectrum may be obtained from the quadratic part $\tilde{S}$ after the shift (11.18). This quadratic part of the action is as follows

$$\sum_p \left(i\omega - \frac{k^2}{2m} + \mu + \alpha E^2 \right) a^*(p)\, a(p) -$$

$$-(2\beta V)^{-1} \sum_{p_1+p_2=p_3+p_4} t\,(p_1 - p_2) a^*(p_1) a^*(p_2) a\,(p_4) a\,(p_3).$$

Using (11.17) one can rewrite this quadratic form as

$$\sum_p \left(i\omega - \frac{k^2}{2m} + \mu + \alpha E^2 \right) a^*(p)\, a(p)$$

$$-\frac{1}{2} \sum_p \rho_0 t(p) \left[a\,(p) a\,(-p) + a^*(p) a^*(-p) \right].$$

Equating the determinant of this form to zero, we come to the equation

$$\det \begin{pmatrix} i\omega - \frac{k^2}{2m} - \rho_0 t(p) & -\rho_0 t(p) \\ -\rho_0 t(p) & i\omega - \frac{k^2}{2m} - \rho_0 t(p) \end{pmatrix} = 0$$

and after the analytic continuation $i\omega \to E$ we get

$$E^2(k) = \left(\frac{k^2}{2m} \right)^2 + \frac{k^2}{2m} \rho_0 t(p) =$$
$$= \left(\frac{k^2}{2m} \right)^2 + \frac{k^2}{2m} \rho_0 \left[t_0 + 4\pi\alpha^2 E^2 \left(\cos^2\theta - 1/3 \right) \right] \qquad (11.19)$$

For the case $E = 0$ we get the well known Bogoliubov formula of energy spectrum for the Bose–gas

$$E^2(k) = \left(\frac{k^2}{2m}\right)^2 + c_0^2 k^2 .$$

This spectrum has a phonon character, i.e. it is linear in k at small k. The value c_0 has a meaning of the sound velocity ($c_0^2 = \rho_0 t_0 / m$). In the presence of the electric field we also have a phonon–type spectrum, but the sound velocity depends on a direction according to the formula

$$c^2 = c_0^2 + 4\pi\alpha^2 E^2 \rho_0 m^{-1}\left(\cos^2\theta - 1/3\right)$$

coinciding with the result (11.9) of the semiphenomenological theory[10]. The sound velocity has its maximum in a "polar" region and minimum in an "equatorial" region. The sound velocity in the "equatorial" directions ($\theta = \pi/2$) vanishes at a critical field

$$E_C = 3mc_0^2 / 4\pi\alpha^2 \rho_0 = 3t_0 / 4\pi\alpha^2 .$$

It leads to disappearance of superfluidity at $E > E_C$.

11.3. The Effective Action Functional for the Superfluid ^{3}He in the Electric Field

Atoms of ^{3}He have no electric dipole momenta in the absence of the electric field, and the electric dipole interaction is equal to zero. An extemal electric field polarizes ^{3}He atoms and the interaction arises between their induced electric momenta. This interaction as well as the interaction of magnetic dipoles, leads to some interesting consequences. Thus, Delrieu[6] and Maki[7] showed that the electric field of the order of 10^4 V/cm in the anisotropic *A*–phase caused the orientationa1 effect, and orbital momenta of Cooper pairs become to be perpendicular to the field. Fomin et al. [8] showed that Fermi–liquid corrections diminish orientation effect which begins to take place only in fields of the one or two orders times larger (it depends on pressure) than it was predicted earlier[6,7].

Here we shall consider the influence of the dipole interaction on the spectrum of collective modes in ^{3}He. In order to investigate this problem we consider at first the effective (hydrodynamical) action functional for the ^{3}He–type system in the electric field.

Then in the subsquent sections we shall obtain the Bose–spectrum of the *B*– and *A*–phases ^{3}He in the electric field[9-14].

^{3}He in the electric field $\boldsymbol{E}$ can be described by the action functional

$$S = \int d\tau\, d^3x \sum_S \{\bar{\chi}_S(\mathbf{x},\tau)\partial_\tau \chi_S(\mathbf{x},\tau) -$$
$$-(2m)^{-1}\nabla\bar{\chi}_S(\mathbf{x},\tau)\nabla\chi_S(\mathbf{x},\tau) + (\mu_0 + \alpha_0 E^2)\bar{\chi}_S(\mathbf{x},\tau)\partial_\tau\chi_S(\mathbf{x},\tau)\} -$$
$$-\frac{1}{2}\int d\tau\, d^3x d^3y U_E(\mathbf{x}-\mathbf{y}) \sum_{S,S'} \bar{\chi}_S(\mathbf{x},\tau)\bar{\chi}_{S'}(\mathbf{y},\tau)\chi_{S'}(\mathbf{y},\tau)\chi_S(\mathbf{x},\tau),$$

where

$$U_E(\mathbf{x}-\mathbf{y}) = U(\mathbf{x}-\mathbf{y}) + \alpha_0^2\left(E^2 r^{-3} - 3(\boldsymbol{E},\boldsymbol{r})^2 r^{-5}\right),$$

is a potential with the addend corresponding to the interaction of induced electric momenta of ^{3}He–atoms in the electric field, α_0, μ_0 are nonrenormalized (bare) values of susceptibility and of the chemical potential, $\chi_S(\mathbf{x},\tau), \bar{\chi}_S(\mathbf{x},\tau)$, are anticommuting Fermi–fields

$$\chi_S(\boldsymbol{x},\tau) = (\beta V)^{-1/2} \sum_p \exp(ipx) a_S(p), \tag{11.20}$$

where $px = \boldsymbol{kx} + \omega\tau$. Let us average $\exp S$ with respect to "fast" Fermi–fields $\chi_1, \bar{\chi}_1$ with $|k - k_F| > k_0$ or $|\omega| > \omega_0$ in (11.20). We obtain

$$\int \exp S D\bar{\chi}_{1s} D\chi_{1s} = \exp \tilde{S}(\chi_{0s}, \tilde{\chi}_{0s}).$$

For the above described ^{3}He model which takes the induced dipole–dipole interaction into account we get

$$\begin{aligned}
\tilde{S} &= \sum_{p,S} Z^{-1}\left(i\omega - c_F(k-k_F) + \alpha E^2\right) a_S^*(p)\ a_S(p) - \\
&- \frac{g_0}{2\beta V} \sum_{p_1+p_2=p_3+p_4} (\boldsymbol{n}_1 - \boldsymbol{n}_2, \boldsymbol{n}_3 - \boldsymbol{n}_4)\ [2a_+^*(p_1)a_-^*(p_2)a_-(p_4)a_+(p_3) + \\
&+ a_+^*(p_1)a_+^*(p_2)a_+(p_4)a_+(p_3) + a_-^*(p_1)a_-^*(p_2)a_-(p_4)a_-(p_3)] - \\
&- \frac{\alpha^2}{2}\sum_{S,S'} \int d\tau d^3x d^3y [\left(E^2(\boldsymbol{x}-\boldsymbol{y})^2 - 3(\boldsymbol{E},(\boldsymbol{x}-\boldsymbol{y})^2\right)|\boldsymbol{x}-\boldsymbol{y}|^{-5} - \\
&- 4\pi\delta_{SS'}E^2\delta(\boldsymbol{x}-\boldsymbol{y})]\bar{\chi}_{0S}(\boldsymbol{x},\tau)\bar{\chi}_{0S'}(\boldsymbol{y},\tau)\chi_{0S'}(\boldsymbol{y},\tau)\chi_{0S}(\boldsymbol{x},\tau).
\end{aligned}$$

(11.21)

The last addend in (11.21) describes the dipole–dipole interaction in which $-4\pi\delta_{SS'}E^2\delta(\boldsymbol{x}-\boldsymbol{y})$ is the so–called contact term.

It is appropriate to go to the description of the dipole interaction between induced dipole momenta of quasiparticles $\alpha\boldsymbol{E}$ via the electric field $\boldsymbol{E}(\boldsymbol{x},\tau)$. In order to do it we insert the Gaussian integral with respect to the transverse vector Bose–field $\boldsymbol{E}(\boldsymbol{x},\tau)$

$$\int \prod_{x,\tau} \delta(div\boldsymbol{E}) \prod_i dE_i(\boldsymbol{x},\tau) \exp\left[-\frac{1}{8\pi}\int d\tau d^3x E^2(\boldsymbol{x},\tau)\right]$$

into the functional integral of $\exp\tilde{S}$ with respect to the "slow" Fermi–fields $\chi_{0S}(\mathbf{x},\tau), \bar{\chi}_{0S}(\mathbf{x},\tau)$. Then after the shift

$$\boldsymbol{E}(\boldsymbol{x},\tau) \to \boldsymbol{E}(\boldsymbol{x},\tau) + \alpha\nabla(\boldsymbol{E},\nabla)\int d^3y \sum_S \nabla\overline{\chi}_{0S}(\boldsymbol{y},\tau)\nabla\chi_{0S}(\boldsymbol{y},\tau)|\boldsymbol{x} - \boldsymbol{y}|^{-1}$$
$$+ 4\pi\alpha\boldsymbol{E}\sum_S \overline{\chi}_{0S}(\boldsymbol{x},\tau)\chi_{0S}(\boldsymbol{x},\tau) ,$$

which cancels the last addend in (11.21), we obtain a functional integral of $\exp\tilde{\tilde{S}}$ with respect to $\chi_{0S}(\mathbf{x},\tau), \overline{\chi}_{0S}(\mathbf{x},\tau)$ and also with respect to the transverse field $\boldsymbol{E}(\boldsymbol{x},\tau)$, where

$$\tilde{\tilde{S}} = \tilde{S} - \int d\tau d^3x \left\{ E^2(\boldsymbol{x},\tau)/8\pi - \alpha(\boldsymbol{E},\boldsymbol{E}(\boldsymbol{x},\tau))\sum_S \nabla\overline{\chi}_{0S}(\boldsymbol{x},\tau)\nabla\chi_{0S}(\boldsymbol{x},\tau) \right\}.$$

Now, let us go from the Fermi–fields $\chi_{0S}(\mathbf{x},\tau), \overline{\chi}_{0S}(\mathbf{x},\tau)$ to the Bose–fields, describing Cooper pairs of fermions near the Fermi–sphere. In order to do it we insert the Gaussian functional integral

$$\int \exp\left[g_0^{-1} \sum_{p,i,a} c_{ia}^*(p) c_{ia}(p) \right] D\overline{c}_{ia} Dc_{ia}$$

with respect to complex fields $c_{ia}(p), \overline{c}_{ia}(p)$ (with Fourer coefficients $c_{ia}(p), c_{ia}^*(p)$) into the functional integral over the fields $\chi_{0S}(\boldsymbol{x},\tau), \overline{\chi}_{0S}(\boldsymbol{x},\tau), \boldsymbol{E}$.

Now let us perform the shift of Bose–fields $c_{ia}(p), c_{ia}^*(p)$ on the quadratic forms of Fermi–fields which cancels the quartic form of Fermi–fields. Then after taking the Gaussian integral over Fermi–fields $\chi_{0S}(\mathbf{x},\tau), \overline{\chi}_{0S}(\mathbf{x},\tau)$ we obtain the integral

$$\int \exp S_{eff}\left(c_{ia}(p), c_{ia}^*(p), \boldsymbol{E}\right) D\overline{c}_{ia} Dc_{ia} D\mu(E) ,$$

where

$$S_{eff} = g_0^{-1} \sum_{p,i,a} c_{ia}^*(p)\, c_{ia}(p) - (8\pi)^{-1} \sum_{p} (\boldsymbol{E}(p), \boldsymbol{E}(\text{-}p)) +$$
$$+ \frac{1}{2} \ln \det \hat{M}\left[c_{ia}(p), c_{ia}^*(p), E\right] / \hat{M}[0,0,0],$$

$$D\bar{c}_{ia} D c_{ia} D\mu(E) = \prod_{p,i,a} dc_{ia}^*(p) c_{ia}(p) \prod_{p} \delta((\boldsymbol{k}, \boldsymbol{E})) \prod_{i} dE_i(p) \tag{11.22}$$

and $\hat{M}$ in (11.22) is an operator defined by the formula

$$\hat{M} =$$
$$= \begin{pmatrix} \begin{array}{l} Z^{-1}(i\omega_1 - \xi_1 + \alpha E^2)\delta_{p1p2} + \\ + \alpha(\beta V)^{-1/2}(\boldsymbol{E}, \boldsymbol{E}(p_1 - p_2)); \end{array} & (\beta V)^{-1/2}(n_{1i} - n_{2i})\sigma_a c_{ia}(p_1 + p_2) \\ -(\beta V)^{-1/2}(n_{1i} - n_{2i})\sigma_a c^*(p_1 + p_2) & \begin{array}{l} Z^{-1}(-i\omega_1 + \xi_1 - \alpha E^2)\delta_{p1p2} - \\ - \alpha(\beta V)^{-1/2}(\boldsymbol{E}, \boldsymbol{E}(p_2 - p_1)) \end{array} \end{pmatrix}$$

The S_{eff} functional (the effective or hydrodynamical action) determines all the physical properties of a model system, and, particular, the spectrum of Bose–excitations of the system in the electric fields.

11.4. The Influence of the Electric Field on the Bose–Spectrum in the B–Phase

In order to calculate the Bose–spectrum we have to do the shift of the field

$$c_{ia}(p) \to c_{ia}(p) + c_{ia}^{(0)}(p)$$

according to the method developed above. Here $c_{ia}^{(0)}(p)$ is a condensate value of the field $c_{ia}(p)$. Then we have to take the quadratic form on the field $E(p)$ and on fields $c_{ia}(p), c^*_{ia}(p)$, which are fluctuations of the initial fields around its condensate values $c^{(0)}_{ia}(p), c^{(0)*}_{ia}(p)$.

This quadratic form is as follows

$$S_{eff} = g_0^{-1} \sum_{p,i,a} c^*_{ia}(p)\, c_{ia}(p) - (8\pi)^{-1} \sum_p (\boldsymbol{E}(p), \boldsymbol{E}(-p)) - \frac{1}{4} Tr\left(\hat{G}\hat{u}\right)^2, \tag{11.23}$$

Green's function of the corresponding superfluid phase of ^{3}He

$$u_{p1p2} = \begin{pmatrix} \alpha(\boldsymbol{E}, \boldsymbol{E}(p_1 - p_2)) & (n_{1i} - n_{2i})\sigma_a c_{ia}(p_1 + p_2) \\ -(n_{1i} - n_{2i})\sigma_a c^*(p_1 + p_2) & -\alpha(\boldsymbol{E}, \boldsymbol{E}(p_1 - p_2)) \end{pmatrix}. \tag{11.24}$$

Let us note that this equation for u_{p1p2} resembles the corresponding equation (20.70) for the model ^{3}He (see Chapter XX) and it can be obtained from (20.70) by changing

$$-\alpha t_{BF}\left[b(p_1 - p_2) + b^*(p_2 - p_1)\right] \to \alpha(\boldsymbol{E}, \boldsymbol{E}(p_1 - p_2)).$$

The most labour–consuming calculation in (11.23) is that of $-\frac{1}{4} Tr\left(\hat{G}\hat{u}\right)^2$ Here, $\hat{G}$ is a fermion Green's function for the *B*–phase, which has a form

$$\hat{G} = \frac{Z}{M_1} \begin{pmatrix} -(i\omega_1 + \xi_1)\delta_{p1p2} & \Delta(\boldsymbol{n}, \boldsymbol{\sigma})\delta_{p1+p2,0} \\ -\Delta(\boldsymbol{n}, \boldsymbol{\sigma})\delta_{p1+p2,0} & (i\omega_1 + \xi_1)\delta_{p1p2} \end{pmatrix},$$

where $M_1 = \omega_1^2 + \xi_1^2 + \Delta^2$, and $\hat{u}$ is defined by (11.24).

After calculation we get

$$-(8\pi)^{-1}\sum_p (\boldsymbol{E}(p), \boldsymbol{E}(-p)) - \sum_p (\boldsymbol{E}, \boldsymbol{E}(p))(\boldsymbol{E}, \boldsymbol{E}(-p))C(p) -$$
$$-\sum_p \left\{ c_{ia}^*(p)c_{ja}(p)A_{ij}(p) + \frac{1}{2}\left[c_{ia}(p)c_{jb}(-p) + c_{ia}^*(p)c_{jb}^*(-p)\right]B_{ijab}(p)\right\} -$$
$$-\sum_p (\boldsymbol{E}, \boldsymbol{E}(p))\left[c_{ia}^*(-p) - c_{ia}(p)\right]D_{ia}(p), \tag{11.25}$$

where

$$A_{ij}(p) = -g_0^{-1}\delta_{ij} - \frac{4Z^2}{\beta V}\sum_{p1+p2=p} M_1^{-1}M_2^{-1}n_{1i}n_{1j}(\xi_1 + i\omega_1)(\xi_2 + i\omega_2),$$

$$B_{ijab}(p) = \frac{4Z^2}{\beta V}\sum_{p1+p2=p} M_1^{-1}M_2^{-1}n_{1i}n_{1j}(2n_{1a}n_{1b} - \delta_{ab}),$$

$$C(p) = \frac{2Z^2\alpha^2}{\beta V}\sum_{p1+p2=p} M_1^{-1}M_2^{-1}\left[(\xi_1 + i\omega_1)(\xi_2 - i\omega_2) - \Delta^2\right],$$

$$D_{ia}(p) = \frac{2i\omega Z^2\alpha\Delta}{\beta V}\sum_{p1+p2=p} M_1^{-1}M_2^{-1}n_{1i}n_{1j}. \tag{11.26}$$

The functions $C(p), D_{ia}(p)$ in (11.26) differ only by constant multiplier from corresponding functions in the system ^{3}He–^{4}He (see

Chapter XX). Thus at small k we have $D_{ia}(p) \propto \omega\delta_{ia}$ as well as for the case of ^{3}He–^{4}He. It leads to the fact, that only the variable

$$c_{ii}(p) - c_{ii}^{*}(p) = 2iv(p)$$

exists in the last addend in (11.25) describing the interaction of collective modes with the electric field. This variable corresponds to the sound mode. Thus, in the first approximation the electric field influences only the sound (acoustic) mode of collective excitations and it does not influence other collective modes. As it will be shown in the Chapter XII, this result is connected with the fact that here we do not take the deformation of the order parameter in the electric field into account. This deformation leads to more important consequences and, particularly, to the splitting of modes with J=2 due to corrections $\propto E^2$ in the spectrum. In order to calculate the spectrum of sound mode in the electric field, let us do the substitution in (11.25):

$$c_{ia}(p) = c_{ia}(-p) = i\delta_{ia}v(p) = -c_{ia}^{*}(-p) = -c_{ia}^{*}(p)\,, \qquad (11.27)$$

which leads the form (11.25) into the form

$$-(8\pi)^{-1}\sum_p (\boldsymbol{E}(p), \boldsymbol{E}(-p)) - \sum_p (\boldsymbol{E}, \boldsymbol{E}(p))(\boldsymbol{E}, \boldsymbol{E}(-p))C(p) -$$
$$-\sum_p v^2(p)\left[A_{ii}(p) - B_{ijij}(p)\right] - \sum_p (\boldsymbol{E}, \boldsymbol{E}(p) - \boldsymbol{E}(-p))v(p)D_{ii}(p)\,.$$

This expression shows that the dynamical variables are

$$v(p), (\boldsymbol{E}(p), \boldsymbol{E}_\perp) = a(p)\boldsymbol{E}_\perp, (\boldsymbol{E}(-p), \boldsymbol{E}_\perp) = a(-p)\boldsymbol{E}_\perp,$$

where $\boldsymbol{E}_\perp = E\sin\theta$ is a transverse component of the electric field $\boldsymbol{E}$ (which are orthogonal to the excitation momentum $\boldsymbol{k}$).

Now, we can use the expression for

$$
\begin{aligned}
&A_{ii}(p)-B_{ijij}(p)= \\
&=-3g_0^{-1}-\frac{4Z^2}{\beta V}\sum_{p1+p2=p} M_1^{-1}M_2^{-1}\left[(\xi_1+i\omega_1)(\xi_2+i\omega_2)+\Delta^2\right]= \\
&=\frac{4Z^2}{\beta V}\sum_{p1+p2=p}\left[\frac{1}{\omega_1^2+\xi_1^2+\Delta^2}-\frac{(\xi_1+i\omega_1)(\xi_2+i\omega_2)+\Delta^2}{\left(\omega_1^2+\xi_1^2+\Delta^2\right)\left(\omega_2^2+\xi_2^2+\Delta^2\right)}\right]
\end{aligned}
$$

at small $\omega,\ \boldsymbol{k}$. Then we can go to the integral over the neighborhood of the Fermi–sphere and use the Feynman trick. Thus, we get for small $\omega,\ k$

$$
\begin{aligned}
A_{ii}(p)-B_{ijij}(p)=\frac{4Z^2k_F^2}{(2\pi)^4c_F}\int_0^1 d\alpha\int d\Omega d\omega_1 d\xi_1\Bigg[\frac{1}{\omega_1^2+\xi_1^2+\Delta^2}- \\
-\frac{\omega_1^2+\xi_1^2+\Delta^2-\alpha(1-\alpha)(\omega^2+c_F^2(\boldsymbol{nk})^2}{\left[\omega_1^2+\xi_1^2+\Delta^2+\alpha(1-\alpha)(\omega^2+c_F^2(\boldsymbol{nk})^2\right]^2}\Bigg]=
\end{aligned}
$$

$$
\begin{aligned}
&=-\frac{Z^2k_F^2}{4\pi^3c_F}\int_0^1 d\alpha\int d\Omega\left[\ln\left(1+\frac{\alpha(1-\alpha)(\omega^2+c_F^2(\boldsymbol{nk})^2)}{\Delta^2}\right)-\right. \\
&\left.-\frac{2\alpha(1-\alpha)(\omega^2+c_F^2(\boldsymbol{nk})^2)}{\Delta^2+\alpha(1-\alpha)(\omega^2+c_F^2(\boldsymbol{nk})^2)}\right]\cong\frac{Z^2k_F^2}{2\pi^3c_F\Delta^2}\left(\omega^2+\frac{1}{3}c_F^2k^2\right).
\end{aligned}
\tag{11.28}
$$

We have also

$$C(p) \cong C(0) = \frac{2Z^2\alpha^2 k_F^2}{(2\pi)^4 c_F} \int d\Omega d\omega_1 d\xi_1 \frac{\left(-2\Delta^2\right)}{\left(\omega^2+\xi_1^2+\Delta^2\right)^2} = -\frac{Z^2\alpha^2 k_F^2}{\pi^2},$$

$$D_{ii}(p) = \frac{2i\omega Z^2 \alpha k_F^2 \Delta}{(2\pi)^4 c_F} \int \frac{d\Omega d\omega_1 d\xi_1}{\left(\omega^2+\xi_1^2+\Delta^2\right)^2} = \frac{i\omega Z^2 \alpha k_F^2}{2\pi^2 c_F \Delta}. \tag{11.29}$$

Substituting (11.28), (11.29) into the quadratic form (11.27), we get a quadratic form of the variables $a(p), a(-p), v(p)$ with the matrix

$$\begin{pmatrix} (4\pi)^{-1} - \dfrac{Z^2\alpha^2 k_F^2 E_\perp}{\pi^2 c_F} & 0 & \dfrac{\omega Z^2 \alpha\, k_F^2 E_\perp}{2\pi^2 c_F \Delta} \\ 0 & (4\pi)^{-1} - \dfrac{Z^2\alpha^2 k_F^2 E_\perp}{\pi^2 c_F} & \dfrac{\omega Z^2 \alpha\, k_F^2 E_\perp}{2\pi^2 c_F \Delta} \\ -\dfrac{\omega Z^2 \alpha\, k_F^2 E_\perp}{2\pi^2 c_F \Delta} & -\dfrac{\omega Z^2 \alpha\, k_F^2 E_\perp}{2\pi^2 c_F \Delta} & \dfrac{Z^2 k_F^2}{2\pi^2 c_F \Delta}\left(\omega^2 + \dfrac{1}{3}c_F^2 k^2\right) \end{pmatrix} \tag{11.30}$$

Equating a determinant of this matrix to zero, we obtain the equation

$$\left[(4\pi)^{-1} - \frac{Z^2\alpha^2 k_F^2 E_\perp}{\pi^2 c_F}\right] \frac{Z^2 k_F^2}{\pi^2 c_F \Delta^2} \times$$
$$\times \left\{\left[(4\pi)^{-1} - \frac{Z^2\alpha^2 k_F^2 E_\perp}{\pi^2 c_F}\right]\left(\omega^2 + \frac{1}{3}c_F^2 k^2\right) + \frac{\omega^2 Z^2 \alpha^2 k_F^2 E_\perp^2}{\pi^2 c_F}\right\} = 0.$$

We can here cancel $\dfrac{Z^2 k_F^2}{\pi^2 c_F \Delta^2}\left[(4\pi)^{-1} - \dfrac{Z^2\alpha^2 k_F^2 E_\perp}{\pi^2 c_F}\right]$ and get the following equation

$$\frac{\omega^2}{4\pi}+\frac{1}{3}c_F^2k^2\left[(4\pi)^{-1}-\frac{Z^2\alpha^2k_F^2E_\perp}{\pi^2c_F}\right]=0\,. \tag{11.31}$$

Changing $i\omega \to E$ we obtain the spectrum in the electric field

$$E^2(k)=\frac{1}{3}c_F^2k^2\left(1-\frac{4Z^2\alpha^2k_F^2E^2\sin^2\theta}{\pi\ c_F}\right). \tag{11.32}$$

from (11.31). This formula shows that the sound velocity in the direction orthogonal to the electric field diminishes, and the velocity along the field is not changed.

11.5. The Influence of the Electric Field on the Bose–Spectrum in the *A*–Phase

The Bose–spectrum in the *A*–phase is defined by the quadratic part of the functional S_{eff} as it is the case for the *B*–phase. It is given by the same formula (11.28) as in the *B*–phase but with different $\hat{G}$:

$$\hat{G}=\frac{Z}{M_1}\begin{pmatrix} -(i\omega_1+\xi_1)\delta_{p1p2} & \Delta_0\sigma_1(n_1+in_2)\delta_{p1+p2,0} \\ -\Delta_0\sigma_1(n_1-in_2)\delta_{p1+p2,0} & (i\omega_1+\xi_1)\delta_{p1p2} \end{pmatrix},$$

where $M_1=\omega_1^2+\xi_1^2+\Delta^2(n_1^2+n_2^2)=\omega_1^2+\xi_1^2+\Delta^2\sin^2\theta$.

After calculation we get the following expression for the quadratic part of S_{eff}

$$-(8\pi)^{-1}\sum_p (\boldsymbol{E}(p),\boldsymbol{E}(\text{-}p))-\sum_p (\boldsymbol{E},\boldsymbol{E}(p))(\boldsymbol{E},\boldsymbol{E}(\text{-}p))C(p)-$$

$$-\sum_p c^*_{ia}(p)c_{ja}(p)A_{ij}(p)-\frac{1}{2}\left[c_{ia}(p)c_{jb}(-p)B^-_{ijab}(p)+\right.$$

$$\left.+c^*_{ia}(p)c^*_{jb}(-p)B^+_{ijab}(p)\right]-\sum_p (\boldsymbol{E},\boldsymbol{E}(p))\left[c^*_{ia}(-p)D^+_{ia}(p)+c_{ia}(p)D^-_{ia}(p)\right], \tag{11.33}$$

where

$$A_{ij}(p)=-g_0^{-1}\delta_{ij}-\frac{4Z^2}{\beta V}\sum_{p1+p2=p}M_1^{-1}M_2^{-1}n_{1i}n_{1j}(\xi_1+i\omega_1)(\xi_2+i\omega_2),$$

$$B^{\pm}_{ijab}(p)=\frac{4Z^2\Delta_0^2}{\beta V}\sum_{p1+p2=p}M_1^{-1}M_2^{-1}(2\delta_{1a}\delta_{1b}-\delta_{ab})(n_1\pm in_2)^2 n_{1i}n_{1j},$$

$$C(p)=\frac{2Z^2\alpha^2}{\beta V}\sum_{p1+p2=p}M_1^{-1}M_2^{-1}\left[(\xi_1+i\omega_1)(\xi_2-i\omega_2)-\Delta^2\right],$$

$$D^{\pm}_{ia}(p)=\delta_{a1}\frac{4Z^2\alpha\Delta_0}{\beta V}\sum_{p1+p2=p}M_1^{-1}M_2^{-1}n_i(n_1\pm in_2)(i\omega_1+\xi_1)=$$

$$=\delta_{a1}\frac{2i\omega Z^2\alpha\Delta_0}{\beta V}\sum_{p1+p2=p}M_1^{-1}M_2^{-1}n_i(n_1\pm in_2).$$

Here, the functions $A_{ij}(p), B^{\pm}_{ijab}(p), C(p)$ are even on p and $D^{\pm}_{ia}(p)$ is odd on p.

The quadratic form (11.33) for the *A*–phase has analogous that for the *B*–phase. Particularly, the last form describes an interaction between the field $\boldsymbol{E}(p)$ and fields $c_{ia}(p), c^*_{ia}(p)$. At small $p=(\boldsymbol{k},\omega)$ we have

$$D^{\pm}_{ia}(p) \propto \omega\delta_{a1}(\delta_{i1} \pm i\delta_{i2}).$$

It means that only the mode, corresponding to the variable $[c_{11}(p)-ic_{21}(p)]-[c^*_{11}(-p)+ic^*_{11}(p)]$ interacts with the electric field $\boldsymbol{E}(p)$. This variable corresponds to the sound mode of collective excitations. We arrive to conclusion, as that for the *B*–phase, namely: in the first approximation the electric field influences only the sound mode and and does not influence the other collective modes.

In order to define the sound spectrum in the *A*–phase, we do like in the case of the *B*–phase in the preceeding section. Let us put

$$c_{21}(p)=c^*_{21}(p)=u(p)=-ic_{11}(-p)=c^*_{11}(-p)$$

in the quadratic form (11.33). Thus we obtain the form

$$\begin{aligned}
&-(8\pi)^{-1}\sum_p(\boldsymbol{E}(p),\boldsymbol{E}(\text{-}p))-\sum_p(\boldsymbol{E},\boldsymbol{E}(p))(\boldsymbol{E},\boldsymbol{E}(\text{-}p))C(p)-\\
&-\sum_p u^2(p)\Bigg\{-2g_0^{-1}-\frac{4Z^2}{\beta V}\sum_{p1+p2=p}M_1^{-1}M_2^{-1}\times\\
&\times\left[(\xi_1+i\omega_1)(\xi_2+i\omega_2)+\Delta_0^2\left(n_1^2+n_2^2\right)\right]\left(n_1^2+n_2^2\right)\Bigg\}-\\
&-\sum_p(\boldsymbol{E},\boldsymbol{E}(p))\,u(p)\frac{4\omega Z^2\alpha\Delta_0}{\beta V}\sum_{p1+p2=p}M_1^{-1}M_2^{-1}\sin^2\theta
\end{aligned} \tag{11.34}$$

instead of (11.33). At small $p=(\boldsymbol{k},\omega)$ we have

$$-2g_0^{-1}-\frac{4Z^2}{\beta V}\sum_{p1+p2=p}M_1^{-1}M_2^{-1}\left[(\xi_1+i\omega_1)(\xi_2+i\omega_2)+\Delta_0^2\left(n_1^2+n_2^2\right)\right]\left(n_1^2+n_2^2\right)=$$

$$=\frac{4Z^2}{\beta V}\sum_{p1+p2=p}\left(n_1^2+n_2^2\right)\left\{M_1^{-1}-M_1^{-1}M_2^{-1}\left[(\xi_1+i\omega_1)(\xi_2+i\omega_2)+\Delta_0^2\left(n_1^2+n_2^2\right)\right]\right\}=$$

$$=\frac{4Z^2k_F^2}{(2\pi)^4c_F}\int_0^1 d\alpha\int d\Omega\left(n_1^2+n_2^2\right)d\omega_1 d\xi_1\times$$

$$\times\left[\frac{1}{\omega_1^2+\xi_1^2+\Delta^2}-\frac{\omega_1^2+\xi_1^2+\Delta^2-\alpha(1-\alpha)(\omega^2+c_F^2(\boldsymbol{nk})^2}{\left[\omega_1^2+\xi_1^2+\Delta^2+\alpha(1-\alpha)(\omega^2+c_F^2(\boldsymbol{nk})^2)\right]^2}\right]=$$

$$=\frac{Z^2k_F^2}{4\pi^3c_F}\int_0^1 d\alpha\int d\Omega\left(n_1^2+n_2^2\right)\left[\frac{2\alpha(1-\alpha)(\omega^2+c_F^2(\boldsymbol{nk})^2}{\Delta^2+\alpha(1-\alpha)(\omega^2+c_F^2(\boldsymbol{nk})^2)}+\right.$$

$$\left.+\ln\left(1+\frac{\alpha(1-\alpha)(\omega^2+c_F^2(\boldsymbol{nk})^2}{\Delta^2}\right)\right]\approx\frac{Z^2k_F^2}{2\pi^2c_F\Delta_0^2}\left(\omega^2+\frac{c_F^2k^2}{3}\right),$$

$$C(p)\approx C(0)=\frac{2Z^2\alpha^2k_F^2}{(2\pi)^4c_F}\int\frac{\left(-2\Delta^2\right)d\Omega d\omega_1 d\xi_1}{\left(\omega_1^2+\xi_1^2+\Delta^2\right)^2}=-\frac{Z^2\alpha^2k_F^2}{\pi^2c_F},$$

$$\frac{4\omega Z^2\alpha\Delta_0}{\beta V}\sum_{p1+p2=p}\frac{\left(n_1^2+n_2^2\right)}{M_1M_2}\approx\frac{4\omega Z^2\alpha\Delta_0k_F^2}{(2\pi)^4c_F}\int\frac{d\Omega d\omega_1 d\xi_1\left(n_1^2+n_2^2\right)}{\left(\omega_1^2+\xi_1^2+\Delta_0^2\right)}=\frac{\omega Z^2\alpha k_F^2}{\pi^2\Delta_0c_F}$$

(11.35)

A comparison of (11.34), (11.35) and the corresponding equations (11.28), (11.29) for the *B*–phase shows that the quadratic form of the variables $a(p),a(-p)$, $u(p)$ has the same matrix (11.30) as the form of the variables $a(p),a(-p)$, $v(p)$ in the *B*–phase. It leads to the same formula (11.22) for the spectrum of the sound mode in the electric field. In the first approximation the electric field does not influence other modes of collective excitation.

Brusov[11] has investigated the influence of the electric field on the spectrum of collective modes in the superfluid ^{3}He films, where analogous effects of anisotropy of the sound propagation take place which diminish the sound velocity in the direction orthogonal to that of the electric field.

Chapter XII

The Order Parameter Distortion and Collective Modes in ^{3}He–B

We investigate the influence of gap distortion, caused by the dipole interaction or by magnetic or electric fields on the order–parameter collective modes in ^{3}He–B by the path integral technique. The dipole–interaction–induced gap distortion splits the pair breaking, squashing, and real squashing modes at zero momentum $\boldsymbol{q}$. Furthermore, a branch crossing of these modes with different l_z appears at nonzero q. Electric and magnetic fields also produce gap distortion with similar consequences.

12.1. The External Perturbations and the Order Parameter Distortions

The condensate wave function of superfluid ^{3}He, or the order parameter (OP), determines all the main properties of the superfluid state. External fields and the dipole interaction cause a deformation of the order parameter, which leads to a number of interesting effects. Such phenomena as the longitudinal NMR in the *A*– and *B*–phases and the shift of the transverse NMR frequency in the *A*–phase are connected with the dipole–interaction–induced distortion of the order parameter, while the nonlinear field splitting of the real squashing mode in the *B*–phase–which has been observed–is connected with the magnetic–field–induced distortion.

In this Chapter, we use the path integral method to investigate the influence of the order parameter distortion on collective modes (CM) in superfluid ^{3}He–B as caused by either dipole interaction (DI) or by magnetic field (MF) or electric field (EP) in the weak–coupling approximation[1-3].

We shall show, that the gap distortion induced by dipole interaction splits the pair–breaking $E = 2\Delta$ mode at zero momentum q and makes possible the existence of pair breaking modes as resonances and the observation of them. A similar splitting takes place for the squashing (*sq*) and real squashing (*rsq*) modes. Furthermore, a branch of these modes with different l_Z appears at nonzero q. In addition, this distortion generates gaps in the spectrum of the Goldstone modes associated with NMR phenomena. Electric and magnetic fields also produce gap distortion with similar consequences. In the case of the magnetic field we derive, in particular, the well–known nonlinear field splitting of the real squashing modes as well as of the squashing modes.

Dipole interaction, magnetic field and electric field

The order parameter in ^{3}He–B is proportional to the matrix, R_{ij}, which rotates the spin space with respect to the orbital space

$$A_{ij} = \Delta(T) R_{ij}(\hat{n}, \theta) e^{i\phi}. \tag{12.1}$$

The energy gap $\Delta(T)$ is isotropic, whereas the rotation axis $\hat{n}$ and the rotation angle θ are arbitrary in the absence of dipole interaction and external fields.

The dipole interaction

$$F_D = g_D \left(A_{ii} A_{jj}^* + A_{ij} A_{ji}^* - \frac{2}{3} A_{ij} A_{ij}^* \right)$$

introduces anisotropy in the energy gap, so that its values parallel (Δ_2) and perpendicular (Δ_1) to the direction of $\hat{n}$ (which remains arbitrary) are different:

$$\Delta_1^2 - \Delta_2^2 = \frac{5}{2} \Omega_B^2. \tag{12.2}$$

The angle θ is also fixes by dipole interaction $\theta = \arccos(-\Delta_2/\Delta_1) \cong \arccos(-1/4)$. Here g_D is the dipolar constant and Ω_B is the longitudinal NMR frequency.

A moderately strong magnetic field, $\boldsymbol{H}$ with the energy F_Z

$$F_{cond} >> F_Z = g_Z H_i \left(AA^+\right)_{ij} H_j >> F_D$$

causes a similar anisotropy of the energy gap

$$\begin{aligned}\Delta_1^2 &= \Delta^2 + \frac{1}{2}\left(3\beta_{12} + \beta_{345}\right)^{-1} \beta_{345}^{-1} \beta_{12} g_Z H^2 \\ \Delta_2^2 &= \Delta^2 - \frac{1}{2}\left(3\beta_{12} + \beta_{345}\right)^{-1} \beta_{345}^{-1} \left(2\beta_{12} + \beta_{345}\right) g_Z H^2,\end{aligned} \tag{12.3}$$

where $\Delta^2 = \left(6\beta_{12} + 2\beta_{345}\right)^{-1}\alpha$ is the gap in the absence of the magnetic field and $\beta_{ijk} = \beta_i + \beta_j + \beta_k$ (the β_i are the coefficients in the expansion of F_{cond}). In addition to this, the magnetic field fixes the direction of $\hat{\boldsymbol{n}} \| \boldsymbol{H}$, because of the same corrections to the order parameter, and the dipole interaction fixes $\theta = \arccos(-\Delta_2/\Delta_1) \cong \arccos(-1/4)$.

An electric field $\boldsymbol{E}$ with the energy

$$F_E = -g_E \left(E_i A_{ki} A_{kj}^* E_j - \frac{1}{3} E^2 A_{ki} A_{ki} \right) \tag{12.4}$$

has a similar effect[5] (we take $\boldsymbol{E}$ to be parallel to the $\hat{z}$–axis). However, unlike the case of dipole interaction or magnetic field, the electric field causes the gap to increase in the direction of the field $\boldsymbol{E}$. In the case of strong electric field ($F_E >> F_D$) one has

$$\Delta_1^2 = \Delta^2 - (6\beta_{345})^{-1} g_E E^2, \ \Delta_2^2 = \Delta^2 + (6\beta_{345})^{-1} g_E E^2. \tag{12.5}$$

The electric field $\boldsymbol{E}$ requires that $\hat{n}$ be perpendicular to $\boldsymbol{E}$ and the dipole interaction specifies the angle

$$\theta = \arccos(-\Delta_1/2(\Delta_1 + \Delta_2)) \cong \arccos(-1/4).$$

Superfluid flow

Volhardt and Maki[4] observed that even small temperature gradients can inducc appreciable flows in ^{3}He, and this is responsible for the practical interest in the study of homogeneous superfluid flows. In addition to causing a hydrodynamic shift of the longitudinal NMR frequency in the *A*– and *B*–phases[5,6] these flows couple the zero sound mode to at least one of the four branches of the real squashing modes with $J_Z = \pm 1, \pm 2$ (Sauls[7]). This results in a threefold splitting of the spectrum for absorption of zero sound by the real squashing mode, when the excitation momentum $\boldsymbol{k}$ is nonzero (this splitting has been observed experimentally[1-3,8]. A superfluid flow of energy[9]

$$F_{flow} = \frac{\rho_S}{5|\Delta|^2} v_{Si} A_{jl}^* v_{Sl} A_{lj} \tag{12.6}$$

depends only on orbital coordinates, and (12.6) is independent of the direction $\hat{n}$. However the flow slightly distorts the order parameter because of the pair–breaking effect; the magnitude of the distortion is of the order of $\rho_S v_S^2 / 2F_{cond}$. Due to this distortion, a flow of energy[9,10]

$$F_{flow} = -a(\boldsymbol{n}, \boldsymbol{v}_S)^2 \tag{12.7}$$

will cause $\hat{n}$ to be parallel to $\boldsymbol{v}_S$. (This situation is similar to the orienting effect of magnetic fields[9,11]. In addition to orienting $\hat{n}$, the flow also causes the gap width Δ_2 and Δ_1 parallel and normal to $\boldsymbol{v}_S$ to be unequal. In what follows we will examine how the order parameter is altered by gap distortion $(\Delta_1 \neq \Delta_2)$ with $\hat{n} \| \boldsymbol{v}_S$.

Kleinert[12,13] has shown that for $T \approx T_C$ gap distortion by superfluid flow splits each of the real squashing– and squashing–modes into three groups of branches. We will analyze how a homogeneous flow alters the collective modes at temperatures $0 << T << T_C/2$.

Rotation effects

Nuclear magnetic resonance (NMR) experiments have revealed vortices[14] and the gyromagnetic effect[15] in rotating superfluid ^{3}He. The rotation generates a vortex lattice of density

$$n_\Omega = 4m_3\Omega/h,$$

which alters the order parameter. In addition to orienting the vector $\hat{n}$, vortices also distort the gap in a manner similar to that described above[10,14], and the same is true of the gyromagnetic effect. We will consider these two effects together when an external magnetic field is present.

The gap is distorted both by the local flow around the vortices[10] and by the vortex cores themselves[14] (in the latter case because the susceptibility is anisotropic inside the cores). If the distance between vortices is less than the characteristic magnetic length ξ_H (as was the case in the Helsinki experiments, where the angular velocity of rotation was $\Omega \approx 1\, rad/s$), we can find the contribution of the local flow to the free energy by averaging the local $r^{-1}\, \boldsymbol{v}_S$–field near vortices[10,14]

$$F_{vx}^{(1)} = \frac{2}{5} a\lambda^{(1)} (\Omega_i R_{ik} H_k)^2, \tag{12.8}$$

where

$$a \approx (\xi/\xi_D)^2 (\chi_\perp - \chi_\parallel), \; \lambda^{(1)} = \pi\xi_D^2 A_{cell}^{-1} \ln(\xi^{-1} A_{cell}^{1/2}).$$

Since the area of an elementary cell is

$$A_{cell} = h/4m_3\,\Omega,$$

$\lambda^{(1)}$ is proportional to Ω. Here ξ is a coherence length, ξ_D is a dipole length, $\chi_\perp$ and $\chi_\parallel$ are the susceptibilities normal and parallel to $\hat{n}$.

If the vortices are symmetric, their cores contribute an amount[14]

$$F_{vx}^{(2)} = \frac{2}{5} a\lambda^{(2)} (\Omega_i R_{ik} H_k)^2 \tag{12.9}$$

to the energy. Here

$$a\lambda^{(2)} \cong \Delta_\chi (A_{core}/A_{cell})$$

(like $\lambda^{(1)}$) is proportional to the angular velocity Ω.

The gyromagnetic energy is given by[16]

$$F_{gm} = \frac{4}{5} a\tilde{\lambda} (\Omega_i R_{ik} H_k), \tag{12.10}$$

where $\tilde{\lambda}$ is proportional to the spontaneous magnetization and to the number of vortices (or to Ω).

All three of the contributions (12.8)–(12.10) have been measured as functions of temperature and pressure and were found to be proportional to Ω [14,15,17].

Once the gyromagnetic and vortex contributions to the free energy are known, we can calculate also the ensuring gap distortion. If we also include the contribution from the magnetic field $\boldsymbol{H}$, we find the resulting total free energy

$$F_Z + F_{vx}^{(1)} + F_{vx}^{(2)} + F_{gm} \tag{12.11}$$

causes the longitudinal and transverse width Δ_2 and Δ_1 to differ

$$\Delta_1^2 = \Delta^2 + \Omega_0^2, \Delta_2^2 = \Delta^2 + \alpha_0 \Omega_0^2,$$

$$\Omega_0^2 = (10\beta_0)^{-1} \left\{ H^2 \left[g_Z + \frac{2}{5} \frac{a}{\Delta^2} \left(\lambda^{(1)} + \lambda^{(2)} \right) \right] \pm \frac{4}{5} \frac{a}{\Delta^2} \tilde{\lambda} H \right\}, \alpha_0 = -4. \tag{12.12}$$

One can describe the order parameter distortion for all cases, considered above, by using a unified approach with the order parameter matrix in the form (for the cases of magnetic field and electric field only, see[1-3])

$$A_{ia} = \left[\Lambda^{1/2} R(\hat{n}, \theta) \right]_{ia} e^{i\phi}. \tag{12.13}$$

Here, Λ is a diagonal matrix with the elements $\lambda_1, \lambda_2, \lambda_3$ and

$$\lambda_1 = \Delta_1^2 + \Omega_0^2, \lambda_2 = \Delta_2^2 = \Delta^2 + \alpha_0 \Omega_0^2. \tag{12.14}$$

For dipole interaction,

$$\Omega_0^2 = \frac{5}{6} \Omega_B^2, \alpha_0 = -2, \quad \hat{n} \| \hat{z}; \tag{12.15}$$

for magnetic field,

$$\Omega_0^2 = (g_Z / 10\beta_0) H^2, \alpha_0 = -4, \quad \hat{n} \| \hat{z}; \tag{12.16}$$

for electric field,

$$\Omega_0^2 = -(g_Z/16\beta_0)H^2, \alpha_0 = -2, \ \hat{n} \perp \hat{z}; \tag{12.17}$$

for superfluid flows Ω_0^2 and α_0 depends on T and v_S^2, while $\hat{n}$ and v_S are both parallel to the $\hat{z}$ –axis; for rotations

$$\Omega_0^2 = (10\beta_0)^{-1}\left\{H^2\left[g_Z + \frac{2}{5}\frac{a}{\Delta^2}\left(\lambda^{(1)} + \lambda^{(2)}\right)\right] \pm \frac{4}{5}\frac{a}{\Delta^2}\tilde{\lambda}H\right\}, \tag{12.18}$$
$$\alpha_0 - -4, \ \hat{n}\|\boldsymbol{H}\|\hat{z}.$$

We set $\theta = \arccos(-1/4)$ in all of the above cases, except for superfluid flows, where we take $\theta = 0$ for simplicity.

12.2. The Collective Mode Spectrum Under the Order Parameter Distortion

The collective mode spectrum for all these cases of gap distortion has been calculated by Brusov[1-3,40]. In the presence of external perturbations to obtain the quadratic part of S_{eff}, which determines the collective–mode spectrum, we must take a shift in Bose–fields $c_{ia}(p) \to c_{ia}(p) + c_{ia}^{(0)}(p)$, where the condensate wave function $c_{ia}^{(0)}(p)$ for dipole interaction, magnetic field, superfluid flows and rotations is given by

$$c_{ia}^{(0)}(p) = \begin{pmatrix} \Delta_1\cos\theta & -\Delta_1\sin\theta & 0 \\ \Delta_1\sin\theta & \Delta_1\cos\theta & 0 \\ 0 & 0 & \Delta_2 \end{pmatrix}, \tag{12.19}$$

while for electric field

$$c_{ia}^{(0)}(p) = \begin{pmatrix} \Delta_1 & 0 & 0 \\ 0 & \Delta_1 \cos\theta & -\Delta_1 \sin\theta \\ 0 & \Delta_2 \sin\theta & \Delta_2 \cos\theta \end{pmatrix}. \tag{12.20}$$

For calculating the collective modes in the *B*–phase with a deformed order parameter, we have used the path integral technique, in which we describe the initial fermions by Fermi–fields and transform to Bose–fields $c_{ia}(\boldsymbol{x},\tau)$, which correspond to Cooper pairs. In terms of these Bose–fields we construct the functional of "hydrodynamical action"

$$S_h = g_0^{-1} \sum_{p,i,a} c_{ia}^{+}(p) c_{ia}(p) + \frac{1}{2} \ln \det \frac{\hat{M}\left(c_{ia}, c_{ia}^{+}\right)}{\hat{M}\left(c_{ia}^{(0)}, c_{ia}^{(0)+}\right)}.$$

Here, $c_{ia}(p)$ is the Fourier transform of $c_{ia}(\boldsymbol{x},\tau)$, a negative constant g is proportional to the pair scattering amplitude of the quasiparticles, and $\hat{M}$ is an operator dependent on the Bose–fields and quasifermion parameters.

This functional contains all the information on the physical properties of the system. In particularly it determines the transition temperature into superfluid state, the order parameter (OP) of the superfluid states, gap equation, the collective mode spectrum and many others. To calculate the collective mode spectrum at low temperatures (in the region $T_C - T \approx T_C$) we should expand *lndet* in S_h in powers of the deviation of $c_{ia}(p)$ from the condensate wave function $c_{ia}^{(0)}(p)$. In the first approximation the Bose–spectrum is determined by the quadratic form of S_h, obtained after the shift $c_{ia}(p) \to c_{ia}(p) + c_{ia}^{(0)}(p)$

$$\sum_p A_{ijab}(p)c^*_{ia}(p)c_{jb}(p)+\frac{1}{2}\sum_p B_{ijab}(p)\left[c_{ia}(p)c_{jb}(-p)+c^*_{ia}(p)c^*_{jb}(-p)\right] \tag{12.21}$$

Equation for the Bose–spectrum is det$Q$=0, where Q is the matrix of the quadratic form (12.21).

The tensor coefficients $A_{ijab}(p)$ and $B_{ijab}(p)$ are proportional to the integrals (sums) of the products of the Green's functions of the fermions and are given by

$$A_{ijab}=\delta_{iab}\left\{\frac{\delta_{ij}}{g}+\frac{4Z^2}{\beta V}\sum_{p1+p2=p} n_{1i}n_{1j}\frac{(i\omega_1+\xi_1)(i\omega_2+\xi_2)}{\left[\omega_1^2+\xi_1^2+\Delta^2(\theta')\right]\cdot\left[\omega_2^2+\xi_2^2+\Delta^2(\theta')\right]}\right\},$$

$$B_{ijab}=\frac{4Z^2}{\beta V}\sum_{p1+p2=p} n_{1i}n_{1j}\frac{(2f_af_b-f^2\delta_{ab})}{\left[\omega_1^2+\xi_1^2+\Delta^2(\theta')\right]\cdot\left[\omega_2^2+\xi_2^2+\Delta^2(\theta')\right]}.$$

Here,

$$\Delta^2(\theta')=\Delta_1^2\left(n_1^2+n_2^2\right)+\Delta_2^2n_3^2=\Delta^2+\Omega_0^2\left[\alpha_0+\left(n_1^2+n_2^2\right)(1-\alpha_0)\right],$$

$$n_1^2+n_2^2=\sin^2\theta',$$

$$f=\{n_1\Delta_1, n_2\Delta_1\cos\theta'+n_3\Delta_2\sin\theta', -n_2\Delta_1\sin\theta'+n_3\Delta_2\cos\theta'\}$$

for the electric field $\boldsymbol{E}$ and

$$f=\{\Delta_1(n_1\cos\theta'+n_2\sin\theta'), \Delta_1(-n_1\sin\theta'+n_2\cos\theta'), n_3\Delta_2\}$$

in the remaining cases.

Here, V is the volume of the system, $\beta=T^{-1}$, $\xi=c_F(k-k_F)$, $n_i=k_i/k_F$, Z is normalization constant, and $\omega=(2n+1)\pi T$.

The results are as follows (gd–Goldstone, pb–pair breaking, sq–squashing, rsq–real squashing modes)

12.2.1. Dipole interaction

In this case the squared energies E^2 for the collective modes are given by the following equations:

pair breaking–modes

$$E^2 = 4\Delta_0^2(T) - \Omega_B^2 \quad \text{for } (1,\pm 1, i) \text{ branches},$$

$$E^2 = 4\Delta_0^2(T) + \frac{2}{3}\Omega_B^2 \quad \text{for } (1,0,i) \text{ branches},$$

$$E^2 = 4\Delta_0^2(T) \quad \text{for } (0,0,r) \text{ branches},$$

real squashing modes

$$E^2 = (8/5)\Delta_0^2(T) - \frac{1}{2}\Omega_B^2 \quad \text{for } (2,\pm 1, r) \text{ branches},$$

$$E^2 = (8/5)\Delta_0^2(T) + \Omega_B^2 \quad \text{for } (2,\pm 2, r) \text{ branches},$$

$$E^2 = (8/5)\Delta_0^2(T) - \Omega_B^2 \quad \text{for } (2,0,r) \text{ branches},$$

squashing modes

$$E^2 = (12/5)\Delta_0^2(T) - \frac{5}{22}\Omega_B^2/10 \quad \text{for } (2,\pm 1, i) \text{ branches},$$

$$E^2 = (12/5)\Delta_0^2(T) + 5\Omega_B^2/10 \quad \text{for } (2,\pm 2, i) \text{ branches},$$

$$E^2 = (12/5)\Delta_0^2(T) - \frac{5}{11}\Omega_B^2/10 \quad \text{for } (2,0,i) \text{ branch}.$$

(12.22)

In parenthesis we have given $(J, J_Z, r$ or $i)$, where J is the total moment of a Cooper pair and J_Z is its projection on the z–axis, and , r or i means the real or imaginary parts of order parameter, respectively.

A gap of order Ω_B appears in the spectrum of the longitudinal spin wave (1, 0, r). This mode can be excited by longitudinal NMR, that means that we could describe the longitudinal NMR phenomena microscopically in the language of collective mode. (Note that the spin wave velocities will merely be rescaled if the dipole interaction is treated solely in terms of its contribution to free energy (i.e., if the gap distortion is neglected), in this case no gap are present in the Goldstone mode spectrum and there is no change in the frequency of nonphonon modes.)

For $T/T_C \approx 0.27$, Ω_B^2 varies from $9.69\cdot 10^{10}\,Hz^2$ at $P = 32bar$ to $6.51\cdot 10^{10}\,Hz^2$ at $P = 19bar$.[19] There is no gap in the spectrum of the transverse spin waves $(1,\pm 1, r)$ and in the Bogoliubov–Anderson sound $(0,0,i)$, that is a consequence of the broken gauge symmetry. As it was shown by Brusov[44], the velocity of Bogoliubov–Anderson sound is not changed by the dipole interaction.

The pair breaking–modes with $E = 2\Delta$ split into a set of three modes with energies lying between

$$E_{\max} = 2\Delta_{\max} = 2\left(\Delta^2 + \frac{5}{6}\Omega_B^2\right)^{1/2}$$

and $$E_{\min} = 2\Delta_{\min} = 2\left(\Delta^2 - \frac{5}{3}\Omega_B^2\right)^{1/2}. \qquad (12.23)$$

As in A–phase[20], where the gap $\Delta = \Delta_{\max}\sin\theta$ is anisotropic too, the collective excitations with energies less than $2\Delta_{\max}$ attenuate moderately and can be regarded as resonances. By the analogy with the *A*–phase we can say that pair breaking–modes can exist as resonances. As Schopohl and Tewordt have shown[21], among pair breaking–modes only one mode, $(0,0,r)$, is coupled with density (a sound wave), and an additional peak could be experimentally observed in the ultrasound absorption spectrum at $E \approx 2\Delta$.

It has been shown[40], that all imaginary $J = 1$ modes (pair breaking–modes) couple to spin density (an electromagnetic waves) owing to the broken of particle–hole symmetry (weak coupling). This implies that two peaks should be observed in NMR absorption experiments near the continuum edge $E \approx 2\Delta$ (at pressure $P = 29$ bar and $T \approx 0.3T_C$, the energies of these modes $E \approx 1.994\Delta$). Note that the observability of these two modes splitting is limited by two factors: the collision–induced line broadening and smallness of the relative distance between peaks $\Omega_B^2/6\omega_0^2 \approx 3 \cdot 10^{-6}$, which is of the order of resolution of the NMR measurements. We recall, that if the dipole interaction is not taken into account, the energies of all pairbreaking–modes are exactly equal 2Δ and the dispersion coefficients γ are complex[22]. Physically, this is a consequence of the fact that the Bose–excitations can decay into their constituent fermions.

The dipole interaction leads to a splitting of both the squashing and the real squashing modes. Modes of each type are obtained in three sets: two modes with $J_Z = \pm 1$, one mode with $J_Z = 0$, and two modes with $J_Z = \pm 2$. The relative distance between the mode energies of any two groups is $\propto \Omega_B^2/2\omega_0^2 \approx 1.3 \cdot 10^{-5}$, which is comparable with the current sensitivity of ultrasound experiments. Thus, it is possible to observe the dipole interaction induced splitting of the squashing and the real squashing modes.

The splitting of the squashing and the real squashing modes also gives rise to their branches crossing for nonvanishing momenta $\boldsymbol{k}$. The dispersion law for these modes is of the form $E^2 = \Omega_1^2 + \gamma k^2$, for nonzero momentum $\boldsymbol{k}$. Because[22] $\gamma(J_{Z1}) > \gamma(J_{Z2})$, for $J_{Z1} < J_{Z2}$, the branch $J_Z = 0$ crosses the $J_Z = \pm 1$ and $J_Z = \pm 2$ branches and branches $J_Z = \pm 1$ cross the branches $J_Z = \pm 2$ at nonzero $\boldsymbol{k}$ (for the both squashing and real squashing modes).

12.2.2. Magnetic fields

The magnetic field induced gap distortion leads to corrections of order $\propto H^2$ in the collective mode spectrum. For the $J = 2$ modes $E^2(0) \equiv E^2$ takes the following values:

$$\text{rsq:}\quad E^2 = \frac{8}{5}\Delta^2 - 2\tilde{H}^2,\ (2,\pm 1, r);\qquad E^2 = \frac{8}{5}\Delta^2 + \tilde{H}^2,\ (2,\pm 2, r);$$

$$E^2 = \frac{8}{5}\Delta^2 - 3\tilde{H}^2,\ (2,0,r);$$

$$\text{sq:}\quad E^2 = \frac{12}{5}\Delta^2 - \frac{21}{10}\tilde{H}^2,\ (2,\pm 1, i);\quad E^2 = \frac{12}{5}\Delta^2 - \frac{2}{3}\tilde{H}^2,\ (2,\pm 2, i);$$

$$E^2 = \frac{12}{5}\Delta^2 - \frac{19}{10}\tilde{H}^2,\ (2,0,i). \tag{12.24}$$

Here, $\tilde{H}^2 = (g_Z/10\beta_0)H^2$ and the coefficient $g_Z/10\beta_0$ is of the order[23] of a few GHz/T^2.

Thus the magnetic field splits the squashing and the real squashing modes into sets containing three groups of modes each. This field–induced splitting leads to two kinds of branch crossings for the modes. If we also include the linear splitting of the squashing and the real squashing–modes in strong enough magnetic field, we find crossing in the $J_Z = 2$ and $J_Z = 1$ branches of the real squashing–mode as well as in the $J_Z = 0,1$ branches and in the $J_Z = -1,-2$ branches of the squashing mode. The crossing of real squashing–mode branches with $J_Z = 1,2$ was predicted earlier[29] and was observed[23] in the fields $\propto 0.15T$.

Another kind of branch crossing of the squashing and the real squashing–modes occurs at nonzero momenta $\boldsymbol{k}$. For the real squashing–mode branches $J_Z = 0$ crosses the branches $J_Z = \pm 1, \pm 2$, branches $J_Z = \pm 1$ crosses the branches $J_Z = \pm 2$. For the squashing–mode, the branches $J_Z = 0, \pm 1$ crosses the branches $J_Z = \pm 2$.

The magnetic field also splits the pair breaking–mode into a set of three groups of modes with energies

$$E^2 = 4\Delta^2 - \frac{4}{3}\widetilde{H}^2, \ (0,0,r); \ E^2 = 4\Delta^2, \ (1,0,i); \\ E^2 = 4\Delta^2 - 2\widetilde{H}^2, \ (0,\pm 1, i), \tag{12.25}$$

which lie between

$$E_{\max} = 2\Delta_{\max} = 2\left(\Delta^2 + \widetilde{H}^2\right)^{1/2} \text{ and } E_{\min} = 2\Delta_{\min} = 2\left(\Delta^2 - 4\widetilde{H}^2\right)^{1/2}.$$

This leads to the possibility of existence of pair breaking–modes as resonances.

The magnetic field creates also a gap $\Delta^2 = 2\widetilde{H}^2$ in the spectrum of the transverse spin waves, but does not change the spectrum of the longitudinal spin waves.

12.2.3. Electric fields

Let us discuss a little more detailed the case of electric field. As it was shown in Chapter XI without accounting the gap distortion the frequencies of all nonphonon modes remain unperturbed and only the sound velocity is changed[44]. Below we show that gap distortion leads to much more significant consequences: to splitting of all nonphonon modes (the squashing–, the real squashing– and the pair breaking–modes) and estimate the electric field value, which is necessary for observation of this splitting.

Working from the equation $\det Q = 0$, evaluating the integrals over the frequencies and momenta of the quasifermions, and restricting the analysis to terms $\propto E^2$, since the corrections for the field are small in comparison with the frequencies of collective modes, we find the following equations for the spectrum (the corresponding variables are listed at the right; $u_{ia} = \mathrm{Re}\, c_{ia}$, $v_{ia} = \mathrm{Im}\, c_{ia}$):

$$\int_0^1 (1-x^2) I (1+2c_1) dx = 0, \quad u_{21}+u_{12}, u_{11}-u_{22},$$

$$\int_0^1 I\left\{1+3x^2+4\left[\left(c_1(1-x^2)\right)^{1/2}-2\left(c_2x^2\right)^{1/2}\right]^2\right\}dx = 0, \quad u_{11}+u_{22}-2u_{33}$$

$$\int_0^1 I\left\{1+x^2+4\left[\left(c_2(1-x^2)\right)^{1/2}+\left(c_1x^2\right)^{1/2}\right]^2\right\}dx = 0, \quad u_{13}+u_{31}, u_{23}+u_{32}$$

$$\int_0^1 (1-x^2) I (1+4c-2c_1) dx = 0, \quad v_{21}+v_{12}, v_{11}-v_{22},$$

$$\int_0^1 I\left\{(1+3x^2)(1+4c)-4\left[\left(c_1(1-x^2)\right)^{1/2}-2\left(c_2x^2\right)^{1/2}\right]^2\right\}dx = 0,$$

$v_{11}+v_{22}-2v_{33}$

$$\int_0^1 I\left\{(1+x^2)(1+4c)-4\left[\left(c_2(1-x^2)\right)^{1/2}+\left(c_1x^2\right)^{1/2}\right]^2\right\}dx = 0,$$

$v_{13}+v_{31}, v_{23}+v_{32}$

$$\int_0^1 (1-x^2) I dx = 0, \quad u_{21}-u_{12},$$

$$\int_0^1 I\left\{1+x^2+4\left[\left(c_2(1-x^2)\right)^{1/2}-\left(c_1x^2\right)^{1/2}\right]^2\right\}dx = 0, \quad u_{13}-u_{31}, u_{23}-u_{32},$$

$$\int_0^1 I\left\{1+4c-4\left[\left(c_1(1-x^2)\right)^{1/2}+\left(c_2x^2\right)^{1/2}\right]^2\right\}dx = 0, \quad v_{11}+v_{22}+v_{33},$$

$$\int_0^1 (1-x^2) I (1+4c_1) dx = 0, \;\; v_{21} - v_{12},$$

$$\int_0^1 I\left\{(1+x^2)(1+4c) - 4\left[(c_2(1-x^2))^{1/2} - (c_1 x^2)^{1/2}\right]^2\right\} dx = 0,$$

$$v_{13} - v_{31}, v_{23} - v_{32}$$

$$\int_0^1 I\left\{1 + 4\left[(c_1(1-x^2))^{1/2} + (c_2 x^2)^{1/2}\right]^2\right\} dx = 0, \;\; u_{11} + u_{22} + u_{33}. \tag{12.26}$$

Here,

$$I = \frac{1}{(1+4c)^{1/2}} \ln \frac{1+(1+4c)^{1/2}}{1-(1+4c)^{1/2}},$$

$$c_n = \frac{\Delta_0^2 + \alpha_n E^2}{\omega^2 + (c_F(k,n))^2}\left[2 - n + (2n-3)x^2\right], \;\; n = 1,2, \;\; c = c_1 + c_2.$$

Results for k=0

1) All four Goldstone modes (sound, a longitudinal spin wave and two transverse spin waves) remain unperturbed.
2) There is a threefold splitting of the real squashing–modes:

$$E_0^2 = (8/5)\Delta^2(T) + \Gamma_+ E^2, \;\; u_{11} + u_{22} - 2u_{33},$$

$$E_1^2 = (8/5)\Delta^2(T) + \frac{1}{2}\Gamma_+ E^2, \;\; u_{13} + u_{31}, u_{23} + u_{32},$$

$$E_2^2 = (8/5)\Delta^2(T) - \Gamma_+ E^2, \;\; u_{12} + u_{21}, u_{11} - u_{22}. \tag{12.27}$$

The subscript on the energy of a mode is equal to $|J_Z|$;

$$\Gamma_+ = \frac{8}{105}\left(3+\frac{\sqrt{6}}{5\left(\pi - arctg\sqrt{6}\right)}\right)\frac{g_E}{\beta_0} = \frac{1}{4}\frac{g_E}{\beta_0}, \tag{12.28}$$

where $\beta_0 = \beta_{345}$ in the weak coupling approximation and $u_{ia} = \mathrm{Re}\, c_{ia}$. The modes with projections $\pm J_Z$ of the total angular momentum remain degenerate.

3) There is a threefold splitting of the squashing–modes:

$$E_0^2 = (12/5)\Delta^2(T) + \Gamma_- E^2, \quad v_{11} + v_{22} - 2v_{33},$$

$$E_1^2 = (12/5)\Delta^2(T) + \frac{1}{2}\Gamma_- E^2, \quad v_{13} + v_{31}, v_{23} + v_{32},$$

$$E_2^2 = (12/5)\Delta^2(T) - \Gamma_- E^2, \quad v_{12} + v_{21}, v_{11} - v_{22}, \tag{12.29}$$

where

$$\Gamma_- = \frac{2}{35}\left(1-\frac{2\sqrt{6}}{5 arctg\sqrt{6}}\right)\frac{g_E}{\beta_0} = 0.016\frac{g_E}{\beta_0}, \quad v_{ia} = \mathrm{Im}\, c_{ia}. \tag{12.30}$$

The threefold splitting of the real squashing– and squashing–modes is very reminiscent of the dispersion–induced splitting of these modes, which has been predicted independently by Vdovin[25] and Brusov and Popov[22]. Splitting of these modes in a magnetic field has been observed experimentally by Daniels et al.[26] The frequencies fall in the order $E_0 > E_1 > E_2$. The ratio (1:4) of the differences between the frequencies of the branches with $J_Z = 0$ and $|J_Z| = 1,2$ is the same for the dispersion–induced splitting of the $J = 2$ modes and for the splitting of these modes in an electric field.

4) There is a threefold splitting of the pair breaking–modes:

$$E_{0+}^2 = 4\Delta^2(T), \;\; u_{11} + u_{22} + u_{33},$$

$$E_{\pm 1}^2 = 4\Delta^2(T) + \Gamma_0 E^2, \;\; u_{13} - u_{31}, v_{23} - v_{32},$$

$$E_2^2 = 4\Delta_0^2(T) - 2\Gamma_0 E^2, \;\; v_{12} - v_{21}, \tag{12.31}$$

where

$$\Gamma_0 = \frac{2}{15}\frac{g_E}{\beta_0} = 0.133\frac{g_E}{\beta_0}. \tag{12.32}$$

The ratio of the frequency differences between the $E_{\pm 1}$ mode and the E_0 and E_{0+} modes is 1:3. The energies of all the pairbreaking–modes lie between

$$E_{\max} = 2\Delta_{\max} = 2\left(\Delta^2 + \frac{5}{2}\Gamma_0 E^2\right)^{1/2}$$

$$\text{and} \;\; E_{\min} = 2\Delta_{\min} = 2\left(\Delta^2 - \frac{5}{4}\Gamma_0 E^2\right)^{1/2}. \tag{12.33}$$

This result tells us that all three branches of the pair breaking–mode are moderately damped and could be observed in ultrasound experiments, as resonances at the absorption edge, in addition to the absorption of zero–sound as a result of pair–decay processes. This conclusion follows from the analogy with the case of the *A*–phase[20], (mentioned above), where the gap $\Delta = \Delta_{\max}\sin\theta$ is anisotropic too and the collective excitations with energies less than $2\Delta_{\max}$ attenuate moderately and can be observed as resonances.

The pair breaking–modes can thus be observed in an electric field, as resonances at the absorption edge, in the manner in which they have been observed by Danniels et al.[26] in a magnetic field.

We can compare the maximum splitting $\delta\omega_{\max}$ for nonphonon modes. Using the formula

$$\omega = \omega_0\left(1+\frac{E^2\Gamma}{2\omega_0}\right), \tag{12.34}$$

where Γ is equal to any of $\Gamma_0, \Gamma_+, \Gamma_-$, we find the results:

$$pb: \quad \delta\omega_{\max} = \frac{3}{2}\frac{\Gamma_0 E^2}{\omega_{pb}} = \frac{1}{5}\frac{g_E E^2}{\beta_0\omega_{pb}} = 0.10\frac{g_E E^2}{\beta_0\Delta},$$

$$sq: \quad \delta\omega_{\max} = \frac{\Gamma_- E^2}{\omega_{sq}} = 0.016\frac{g_E E^2}{\beta_0\omega_{sq}} = 0.01\frac{g_E E^2}{\beta_0\Delta},$$

$$rsq: \delta\omega_{\max} = \frac{\Gamma_+ E^2}{\omega_{rsq}} = \frac{1}{4}\frac{g_E E^2}{\beta_0\omega_{rsq}} = 0.20\frac{g_E E^2}{\beta_0\Delta}. \tag{12.35}$$

We see that that the maximum splitting ratios are

$$\delta\omega^{rsq}_{\max} : \delta\omega^{pb}_{\max} : \delta\omega^{sq}_{\max} = 20:10:1. \tag{12.36}$$

It would thus be easiest to observe the spitting of the spectrum in an electric field in the case of the real squashing– and the pair breaking–modes.

The estimation of the maximum splitting of the spectrum

An electric field causes a threefold splitting of the spectrum of all nonphonon modes (the real squashing–, squashing– and the pair–breaking–modes), leaving the Goldstone–modes without a gap. The electric field strength which would be required for an observation of this splitting can be found from the following general arguments. Equating the dipole energy and the electric dipole energy, $g_D \approx g_E E^2$, we find $E \approx 1.5 \cdot 10^4 V/cm$. In order to observe effects of an electric field in acoustic experiments, we would need fields stronger by a factor of $\sqrt{10}$, i.e. $E \approx 5 \cdot 10^4 V/cm$. The Fermi–liquid corrections increases these fields to $E \approx 5 \cdot 10^5 - 5 \cdot 10^6 V/cm$ (depending on pressure). Since the critical field E_C in ^{3}He is[48] $E_C \approx 2.7 \cdot 10^6 V/cm$ a threefold splitting of the spectrum of nonphonon modes could be seen in part of the phase diagram (at not too high pressures).

We can make more precisely estimation of the maximum splitting for the real squashing–mode, using (12.35).

For the real squashing–mode we find from (12.35)

$$\delta\omega_{\max} = 0.20 \frac{g_E E^2}{\beta_0 \Delta}, \tag{12.37}$$

$$\Delta^2 = (6\beta_{12} + 2\beta_{345})^{-1} \alpha, \tag{12.38}$$

$$\alpha = \frac{1}{3} N(0)(1 - T/T_C). \tag{12.39}$$

Taking account of the temperature dependence of the gap

$$\Delta = \Delta_0 (1 - T/T_C)^{1/2}, \tag{12.40}$$

we find

$$\beta_0 = \frac{1}{15}\frac{N(0)}{\Delta_0^2} = 0.20\frac{g_E E^2}{\beta_0 \Delta}. \tag{12.41}$$

From[47] we have

$$g_E = 4\alpha^2 (\gamma\hbar)^{-2} g_D, \tag{12.42}$$

$$g_D = \frac{\pi}{10}\left[N(0)\gamma\hbar \ln\frac{1.13\hbar\omega}{k_B T_C}\right]^2. \tag{12.43}$$

Here, g_D is the dipole interaction constant. From (12.41) and (12.42) we have

$$\frac{g_E}{\beta_0 \Delta} = 6\pi\alpha^2 N(0)\left[\ln\frac{1.13\hbar\omega}{k_B T_C}\right]^2 \Delta_0. \tag{12.44}$$

Here, we have[47,49]

$$\alpha = 2\cdot 10^{-25}\,cm^3, \quad N(0) = (0.54 - 1.26)\cdot 10^{38}\,erg^{-1}\cdot cm^{-3}$$

at a zero pressure and at the melting surface,

$$\hbar\omega/k_B \approx 0.7K, \quad \Delta_0 = \Delta_{BCS} = 1.76T_C. \tag{12.45}$$

The Fermi–liquid corrections have been ignored in (12.44); they are taken into account below. Using (12.44), we can estimate the maximum splitting of the real squashing–mode in an electric field.

1) If we ignore Fermi–liquid corrections we have

$$\delta\omega_{\max} = 0.20\frac{g_E E^2}{\beta_0 \Delta} = 1.2\pi\alpha^2 N(0)\left[\ln\frac{1.13\hbar\omega}{k_B T_C}\right]^2 \Delta_0 E^2. \tag{12.46}$$

We choose

$$E = 10^5 V/cm = \frac{1}{3} \cdot 10^3\ sgs/cm, P = 10 bar. \tag{12.47}$$

We then have

$$\Delta_0 = 1.76 k_B T_C \approx 4.44 \cdot 10^{-19} erg,\ \ N(0) = 0.8 \cdot 10^{38} erg^{-1} \cdot cm^{-3},$$

$$\ln \frac{1.13 \hbar \omega}{k_B T_C} = 6.08. \tag{12.48}$$

Substituting (12.47) and (12.48) into (12.46), we find $\delta\omega_{\max} \approx 3.3 kHz$.

This estimate of the splitting is close to the sensitivity of ultrasound experiments (5–10 kHz). Consequently, fields of the order of $E = 10^5 V/cm$ would be required to observe the splitting of the spectrum of collective modes in ultrasound experiments, if the Fermi–liquid corrections are ignored.

2) We now take the Fermi–liquid corrections into account. Fomin et al.[43] have shown, that the Fermi–liquid corrections reduce the energy associated with the electric field by a factor $< R^2 >= 4 \cdot 10^{-3} - 4.8 \cdot 10^{-5}$ at a zero pressure and at the melting surface, respectively.

For $\delta\omega_{\max}$ we thus have the formula

$$\delta\omega_{\max} = 1.2 \pi \alpha^2 N(0) \left[\ln \frac{1.13 \hbar \omega}{k_B T_C} \right]^2 \Delta_0. \tag{12.49}$$

At critical field $E_C \approx 2.7 \cdot 10^6 V/cm$ we find the following estimates for the maximum splitting of the real squashing–mode in an electric field:

$$\delta\omega_{\max} \approx 4.1 \ kHz, \ \ P = 0 \ bar,$$

$$\delta\omega_{\max} \approx 1.1 \ kHz, \ \ P = 10 \ bar. \tag{12.50}$$

Taking into account the sensitivity of ultrasound experiments (5–10 kHz for ultrasound frequency of order 100 MHz), we conclude that it would be possible in principle to observe splitting of the real squashing–mode in fields close to the breakdown field and at low pressures.

We have found a fairly crude estimate of the splitting here, assigning several of the parameters values which are approximations in Ginzburg–Landau region. If more realistic values of the parameters were used in expression (12.49), there might be an increase in $\delta\omega_{\max}$, and it might be simpler to actually observe the predicted splitting (and to observe it over a broader pressure range and in weaker fields).

An experimental study of the effect on an electric field on the spectrum of collective excitations in the superfluid ^{3}He–B should be targeted on the observation of the following features[1-3]:

a) splitting of the real squashing– and squashing–mode in fields close to the breakdown field $E = 5\cdot 10^5 \div 2.7\cdot 10^6 V/cm$ and at low pressures (near P=0);
b) the change of distance between three dispersion–induced peaks in the sound absorption spectrum for the real squashing– and squashing–modes under the switching of electric fields;
c) the resonant sound absorption at the absorption edge (see below);
d) a shift of the longitudinal NMR frequency in fields of order $E = 1.5\cdot 10^5 \div 1.5\cdot 10^6 V/cm$ under experimental accuracy of order 10^{-5};
e) crossing of the branches of the real squashing– and squashing–modes with different values of $|J_Z|$ at nonzero momentum $\boldsymbol{k}$: under sweeping of the field $\boldsymbol{E}$ value it should be possible to observe the disappearance of one peak at some value of E.

Note that in ultrasound experiments in the presence of electric fields one could use the method of measuring of the group velocity of sound[42], because the collective mode frequencies are splitting now at zero momentum $\boldsymbol{k}$.

12.2.4. Superfluid flow

The influence of superflow on the collective excitation spectrum is dependent on temperature. At $T/T_C \le 0.1$, the spectrum remains almost unchanged, but for higher temperatures several important changes appear. For the collective mode spectrum in the presence of superflow one has[18]

rsq:

$$E^2 = \frac{8}{5}\Delta^2 + \frac{1}{10}(7\alpha+9)\Omega_0^2,\ (2,\pm1,r);$$
$$E^2 = \frac{8}{5}\Delta^2 + \frac{3}{20}(\alpha+10)\Omega_0^2,\ (2,\pm2,r);$$
$$E^2 = \frac{8}{5}\Delta^2 + \frac{3}{10}(3\alpha+2)\Omega_0^2,\ (2,0,r); \tag{12.51}$$

sq:

$$E^2 = \frac{12}{5}\Delta^2 + \frac{1}{10}(6\alpha+11)\Omega_0^2,\ (2,\pm1,i);$$
$$E^2 = \frac{12}{5}\Delta^2 + \frac{1}{2}(2\alpha+3)\Omega_0^2,\ (2,0,i);$$
$$E^2 = \frac{12}{5}\Delta^2 + \frac{3}{5}(\alpha+3)\Omega_0^2,\ (2,\pm2,i); \tag{12.52}$$

pb:

$$E^2 = 4\Delta^2 + \frac{2}{5}(2\alpha+3)\Omega_0^2,\ (1,\pm1,i);$$
$$E^2 = 4\Delta^2 + \frac{2}{5}(\alpha+4)\Omega_0^2,\ (1,0,i);$$
$$E^2 = 4\Delta^2 + \frac{2}{3}(\alpha+2)\Omega_0^2,\ (0,0,r). \tag{12.53}$$

Here,

$$\alpha = -4.2,\ \Omega_0^2 = 0.004\Delta_0^2,\ \text{at } T/T_C = 0.03;$$
$$\alpha = -8.4,\ \Omega_0^2 = 0.004\Delta_0^2,\ \text{at } T/T_C = 0.05;$$
$$(\Delta_0 = \Delta_{BCS} = 1.76T_C). \tag{12.54}$$

a) We have a three–fold splitting of the real squashing– and squashing–modes. One could observe this splitting at $v < v_C$. At velocities close to the critical value (1 mm/s), such splitting is of the order of $10^{-2}\Delta_0^2$ for squares of frequencies.

b) This three–fold splitting leads to one additional new physical effect (see above for other cases). If we take into account the nonzero momentum of excitation we can see that one could observe the crossing of the branches of the real squashing– and squashing–modes with different value of J_Z. So the branch with $|J_Z| = 0$ could be observed to cross the branches with $|J_Z| = 1$ and with $|J_Z| = 2$, and the branches with $|J_Z| = 1$ could cross the branches with $|J_Z| = 2$.

c) The four pairbreaking–modes with $E \cong 2\Delta$ are split into three groups (the modes with $J_Z = \pm 1$ remain degenerate). Because the mode energies lie between $E_{\max} = 2\Delta_{\max} = 2\Delta_1$ and $E_{\min} = 2\Delta_{\min} = 2\Delta_2$, they can be excited as resonances in ultrasound and NMR experiments (the $(0,0,r)$ mode and the modes with $J = 1$ respectively).

d) It is possible to observe the hydrodynamical shifts in NMR experiments

Note that all this superflow induced effects increase with temperature.

12.2.5. Rotational effects (vortices and gyromagnetism)

We can express the results for the pairbreaking–, real squashing– and squashing–modes in the following form[1-3]

$$E^2 = A\Delta^2 + Bg(\Omega)H^2 + Ca'\lambda H , \tag{12.55}$$

where

$$g(\Omega) = (25\beta_0)^{-1}\left[g_Z + \frac{2}{5}\frac{a^2}{\Delta^2}\left(\lambda^{(1)} + \lambda^{(2)}\right)\right], \quad a' = \frac{2a}{25\beta_0\Delta^2}, \tag{12.56}$$

and $A = 4,\ 12/5,\ 8/5$ for the pairbreaking–, real squashing– and squashing–modes, respectively. The magnitude and sign of the coefficients B and C differ for the different modes.

We can use the data from[14,15,17] to estimate the magnitude of the gap distortion by vortices and gyromagnetic effects at the frequencies of the nonphonon modes excited in ultrasound experiments.

This data imply that $\lambda/\Omega \propto (1 \div 3)s/rad$ (depending on the pressure) for $T = 0.5T_C; \lambda/\Omega$ increases rapidly with T and P, and $\lambda/H\Omega \propto (0.04 \div 0.07)s/rad$. If we use the estimate

$$g_Z = \frac{7\xi(3)\chi}{24\pi^2\Delta^2} \approx 35\cdot 10^{-3}\frac{\chi}{\Delta^2}, \tag{12.57}$$

we find that the vortex and gyromagnetic contributions are equal to 10^{-4} and $2\cdot 10^{-6}$ times of the contribution from the term $g_Z H^2$, respectively. For the fields $\propto 1600G$ employed in the nonlinear Zeeman effect experiments[23], the term $g_Z H^2$ shifts the real squashing–mode frequency by $2\cdot 10^{-2}$ for $J_Z = 0$ branch, so that the vortex and gyromagnetic–induced shifts are $\propto 2 \cdot 10^{-6}$ and $2 \cdot 10^{-8}$, respectively. Because the resolution in ultrasound frequency[1-3], these shifts (associated with the gap distortion) are completely negligible. However, they may become appreciable at low pressures and $T \approx 0$.

Rotation does not alter the spectrum of the longitudinal spin waves. For the transverse spin wave spectrum we have

$$E^2 = 5g(\Omega)H^2 \pm a'\tilde{\lambda}H \tag{12.58}$$

for $\boldsymbol{k}$=0.

Estimates show that for $H \propto 300G$, the relative gyromagnetic, vortex and $g_Z H^2$ contributions are of the order $2 \cdot 10^{-7}$, 10^{-5} and 10^{-1} at the precession frequency $\propto 1MHz$ since the resolution of NMR experiments[14,15] is $\propto 10^{-5}$, the gyromagnetic gap distortion is clearly completely undetectable, although in principle the gap distortion caused by vortices may be observable in NMR experiments.

12.3. Sound Experiments at the Absorption Edge

Soon after the discovery of superfluidity of ^{3}He, Paulson et al.[45] have observed a wide sound absorption spectrum near the absorption edge via the pair breaking process.

In 1980 Brusov and Popov[22,40] have calculated in the absence of external perturbations, the energy of the $E = 2\Delta$ modes at nonzero momentum and showed that the energies of all these modes were complex that was connected to the pair breaking by phonon absorption. In 1983 Daniels et al.[26] experimentally observed the resonant absorption of zero–sound at the absorption edge in the presence of magnetic fields. At first their results were interpreted as indicating the excitation of a new mode with $J = 4$ and $E \approx 1.82\Delta$ (one introduced the *f*–pairing into system). But in 1984 two papers were presented at LT–17, which showed that in these experiments the pairbreaking–mode – a mode, which energy is equal to $E = 2\Delta$ in the absence of perturbations – is excited by zero sound and not the $J = 4$ mode. Shopohl and Tewordt[46] reached such conclusions by comparing the coupling between zero sound and the modes with the absorption actually observed. They showed that the magnetic field induced gap distortion leads to coupling of zero sound with the $(1;-1;i)$ pairbreaking–mode (such a coupling is otherwise not present even if particle–hole symmetry is violated). The absorption coefficient is proportional to gap distortion $\alpha \propto (\Delta_1 - \Delta_2)/\Delta$. Brusov and Popov[3] showed at LT–17 that the frequencies of the pairbreaking–mode lie between $2\Delta_{\max}$ and $2\Delta_{\min}$ in the presence of a magnetic field or an electric field or taking the dipole interaction into account (when we have a gap distortion $\Delta \rightarrow \Delta_1, \Delta_2$). This means that one could observe these modes as resonances (as in case of the *A*–phase, where gap is also anisotropic and excitations with energy $E < 2\Delta_{\max}$ are only moderately damped and therefore can be treated as resonances). Brusov[1-3,44] has shown that not only magnetic fields, but electric fields, superfluid flow as well as rotational effects can also give rise to pairbreaking–mode resonances and lead to resonant absorption of zero sound near the absorption edge in addition to the absorption associated with the decay of Cooper pairs. Until now, only resonant absorption of the pairbreaking–mode in a magnetic field has been observed, but it could be possible to observe it in the presence of superflow (at velocities less critical one $v_S \propto 1\ mm/s$). If one will increase the electric field strength or the superflow velocity the absorption peak will increase and move to low temperatures.

12.4. Subdominant f–Wave Pairing Interactions in Superfluid ^{3}He

While the dominant type of pairing in superfluid ^{3}He is the p–pairing the possibility of higher order subdominant pairing is discussed during decades. The existence of higher order pairing can lead to a couple effects on the collective mode spectrum: it alters the collective mode frequencies as well as leads to appearance of new modes near the absorption edge. The observation of the both effects requires very precise sound experiments, and, in particular, near the absorption edge. Such precision measurements of collective mode frequencies in superfluid ^{3}He–B have been made recently at Northwestern University[27] (by Bill Halperin group) and they turned out to be sensitive to quasiparticle and f–wave pairing interactions. Measurements were performed at various pressures using interference of transverse sound in an acoustic cavity. The authors[27] fit the measured collective mode frequencies, which depend on the strength of f–wave pairing and the Fermi–liquid parameter F_2^S to theoretical predictions and discuss what implications these values have for observing new order parameter collective modes. In current section we will discuss the higher order pairing, following to these papers.[27]

The frequency of a collective mode in superfluid ^{3}He–B can depend, in part, on the strength of subdominant f–wave pairing interactions. In addition to the known p–wave pairing that defines the equilibrium state of superfluid ^{3}He there may be nontrivial contributions from f–wave interactions[28] that appear in the dynamics of the order parameter. Depending on the strength of these interactions, Sauls and Serene[28] have predicted the existence of modes with total angular momentum $J=4$. The *f*–wave pairing is parameterized by $x_3^{-1}=\ln^{-1}\left(T_{Cf}/T_C\right)$, where T_C is the transition temperature due to p–wave interactions and T_{Cf} is the hypothetical transition temperature due to f–wave interactions that would occur in the absence of p–wave pairing. This form for the f–wave interaction parameter, x_3^{-1}, is chosen such that it is zero if they are negligible, and large but negative if they are significant and the f–wave pairing interaction is attractive. Sauls and Serene[28] have shown that this f–wave pairing influences the frequency spectrum of collective modes, namely, the real (+) and imaginary (–) squashing modes labeled as, $J=2^{\pm}$, each having a total angular momentum *J*=2; the plus and minus signs refer to the fact that different components of the order parameter are involved in each case, either real or imaginary. These squashing modes correspond to time–dependent, anisotropic, momentum–space deformations (or squashings) of the energy gap amplitude, $\delta\Delta^{+}\left(k,T\right)$. In the absence of quasiparticle interactions and f–wave pairing these modes have frequencies proportional to the gap[29] which we take from the weak–coupling plus model of Rainer and Serene[30]:

$$\Omega_{2^\pm} = a_\pm \Delta^+(T, P), \tag{12.59}$$

where $a_+ = \sqrt{8/5}$ and $a_- = \sqrt{12/5}$. In general, however, $a_\pm$ will have weak temperature and pressure dependences. To date, measurements of the squashing mode frequency have not been as precise as those of the real squashing mode and hence determination of x_3^{-1} from this mode has been inexact.

Authors[27] used transverse acoustic cavity technique that takes advantage of the existence of transverse acoustic standing waves near the collective mode in order to measure precisely the squashing mode frequency. They obtained a frequency resolution during experiments of $\Delta\nu/\nu \approx 2\times10^{-9}$ from 85 to 125 MHz using a cw, frequency–modulated, acoustic–impedance spectrometer[31].

Transverse sound measurements using acoustic–impedance techniques in ^{3}He are reviewed by Halperin and Varoquaux[29]. The acoustic impedance is $Z = \rho/\omega q = \rho C$, where C is the complex phase velocity and $q = \omega/c + i\alpha$. Hence the acoustic impedance is simultaneously sensitive to changes in the (normal fluid) density ρ, phase velocity c, and attenuation α. Transverse sound has been observed as a propagating mode in superfluid ^{3}He[33,34] following the prediction of Moores and Sauls[35], the only liquid for which this has been demonstrated. As a result, the quartz transducer, which forms one side of a cavity, detects an impedance modulated by the sound wave reflected from the opposite cavity wall. The interference between outgoing and reflected waves gives rise to an oscillatory response in impedance as the velocity changes near the mode frequency.

The acoustic cavity response as a function of temperature was measured for a sequence of pressures at fixed acoustic frequency 100 MHz at various pressures near 9 bar. As the temperature is lowered the squashing mode frequency approaches, and then crosses, the fixed transverse sound frequency. Correspondingly, according to the theoretical dispersion represented in Eq. (12.60), the phase velocity increases and transverse sound propagates with lower attenuation. In fact, it appears from this equation as though the velocity diverges at the crossing if viewed sufficiently far from the crossing itself. As the velocity increases so does the wavelength and for each half–wavelength that leaves the cavity there is one oscillation. As the squashing mode frequency crosses the measurement frequency the order parameter mode resonantly absorbs acoustic energy and the interference pattern is extinguished. This corresponds to the sharp upward bend with decreasing temperature in the traces. As the temperature is lowered further, transverse sound propagates again and the interference pattern reappears. Finally, as the temperature is lowered still further the transverse sound becomes highly attenuated. The details of the oscillations on the low temperature side of the mode and the acoustic impedance near the crossing point have not been observed previously and are not predicted by Eq. (12.60).

In order to determine the exact crossing temperature for the imaginary squashing mode with transverse sound, one need to plot the temperature difference of sequential extrema in the interference pattern, δT, and extrapolate independently from both the high and low temperature sides of the squashing mode. The temperature at which δT goes to zero, where the velocity diverges, marks this crossing point. The slopes of the lines in the figure are determined by the temperature dependence of the phase velocity of transverse sound. It was found that they are different on the high and low temperature sides by about a factor of 2. The determinations of $a_{-}(T,P)/\sqrt{12/5}$ from the mode crossings, which have been made with a resolution of 5 to 15 μK, resulting in an uncertainty in a_{-} between 0.1% and 0.25%.

These data can be compared with the theory for transverse sound propagation in ^{3}He as given by Moores and Sauls[35]. They have shown that the dispersion of transverse sound is

$$\left(\frac{\omega}{qv_f}\right)^2 = \frac{F_1^S}{15}(1-\lambda) + \frac{2F_1^S}{75}\lambda \frac{\omega^2}{(\omega+i\Gamma)^2 - \Omega_{2^-}^2 - \frac{2}{5}q^2v_f^2}, \qquad (12.60)$$

where v_f is the Fermi velocity, ω is the measurement frequency, $\Omega_{2^-} = a_-(T,P)\Delta^+(T,P)$ is the squashing mode frequency, and $\lambda(\omega,T)$ is the Tsuneto function. $\Gamma(T)$ is the width of the mode, with an approximate form of $\Gamma(T) \cong \Gamma_C\sqrt{T/T_C}\exp\left[-\frac{\Delta(T)}{T}\right]$ and $\Gamma_C \propto 10^6 - 10^7 Hz$ .At temperatures low compared with T_C, the first term goes to zero like the normal fluid density, and the second term dominates. From Eq. (12.60), it can be seen that transverse sound couples off–resonantly to the squashing mode at frequencies above $\Omega_{2^-}(T)$. Below the squashing mode, owing to nonzero $\Gamma(T)$ transverse sound continues to propagate but highly attenuated. This allows to identify the position of the squashing mode as a specific point along the trace of acoustic impedance. From Eq. (12.60) we see that there are no significant dispersion corrections to the mode frequency, a consequence of the fact that transverse sound propagates only through off–resonant coupling to the squashing mode.

The influence of quasiparticle interactions, F_2^S, and f–wave pairing, x_3^{-1}, on the imaginary squashing mode frequency, was calculated by Sauls and Serene[28],

$$\Omega_{2^-}^2 - \frac{12}{5}\Delta^{+2} + \frac{3}{5}F_2^S\left(\Omega_{2^-}^2 - 4\Delta^{+2}\right)\lambda + \frac{1}{4}x_3^{-1}\Omega_{2^-}^2\left(\Omega_{2^-}^2 - 4\Delta^{+2}\right)\lambda/\Delta^{+2} = 0 \qquad (12.61)$$

Authors[28] have found that at zero pressure $a_-(T,P)/\sqrt{12/5} = 1$. This is expected if quasiparticle interactions become insignificant. There is reasonable evidence that both f–wave pairing interactions and F_2^S, become small at low pressure[29]. However, any discrepancy between the Greywall temperature scale[32] and the absolute temperature scale will shift the data. Possible inaccuracy in the temperature scale was estimated by Greywall[32] to be better than 1%. The level of spectroscopic precision is very high. Nonetheless, one can at most state that the combined effect of quasiparticle interactions, f–wave pairing, and inaccuracy in the temperature scale is negligible at $p = 0$.

There is a reasonable consensus in published data at intermediate pressures near 10 bar and above; although less so at low pressure: i.e., $F_2^S = 0.17$, 0.34, and 0.5 at P=0, 15, and 20 bar and $F_2^S = 0$ at P=0. Taking previous work[29] into account one can estimate that F_2^S could be larger but likely not more than +0.25 for any pressure affecting present determination of x_3^{-1}. In spite of these inaccuracies it is nonetheless clear that the *f*–wave pairing interaction becomes negative (attractive) with higher pressure.

The values of x_3^{-1} from our measurements can be compared with those from other techniques. Acoustic real squashing mode measurements[36] yield values of x_3^{-1} that roughly start at zero at zero pressure and increase to $\propto -0.25$ at 20 bar. These results depend on the Fermi liquid parameter F_2^a, which is not well established. Analysis of the acoustic Faraday effect[34] yields a value of $x_3^{-1} = -0.375$ at 4.3 bar[37]. An analysis of longitudinal acoustic squashing mode data[38] was performed to extract x_3^{-1}. It was concluded that x_3^{-1} was $\propto 0.2$ at 0 bar, but decreased as the pressure increased becoming negative near 5 bar, then leveling off at $\propto -0.2$. The absolute values are generally much larger than what was inferred from the data of Ref.27 but let us note that longitudinal sound measurements are inherently less precise since this sound mode couples so strongly to the squashing mode. Authors[27] have found the pressure dependence of x_3^{-1} to be about a factor of 6 smaller than in these previous works independent of our estimated inaccuracy in the temperature scale or in F_2^S. They concluded that the predicted[1]

$J=4$ modes may exist but, if so, only very near or slightly below $2\Delta^{+}(T,P)^{2}$. Some of the modes in the nine–fold multiplet will couple to transverse sound and our transverse acoustic cavity technique should allow resolution of these collective modes at high frequencies very close to the particle–hole continuum. In addition, application of a magnetic field will be helpful for observation of $J=4$ modes whose frequencies decrease by the Zeeman effect.

Fourier transform longitudinal acoustics experiments near the gap edge were performed at low pressure by Masuhara et al.[39]. They observed the onset of anomalously high attenuation at an energy (frequency) 4% lower than was expected from weak–coupling BCS theory, $2\Delta_{BCS}$. Since it is unlikely that the superfluid ^{3}He–B order parameter is smaller than $2\Delta_{BCS}$, their results mean that either the temperature scale is imprecisely known or there is a new collective mode near the gap edge. Davis et al.[27] measurements of the squashing mode rule out that inaccuracy in temperature of this magnitude is a possible explanation and suggest that these authors have observed attenuation from higher order $J=4$ order parameter collective modes.

In summary, Davis et al.[27] have made high precision measurements of an order parameter collective mode using interference of transverse sound in an acoustic cavity. Taking into account strong coupling effects we interpret our data to determine values of the *f*–wave pairing interaction strength that are much smaller than those from previous work. Despite inaccuracy in the parameters required for the analysis that reduce our resolution, they have set limits on the pressure dependence of the strength of *f*–wave interactions in superfluid ^{3}He. Increased accuracy in measurements of F_2^S will help to refine this conclusion. Lastly, the high resolution of the transverse acoustic cavity technique should make it possible to observe the predicted $J=4$ order parameter collective modes.

Chapter XIII

Splitting of the Squashing Mode and the Method of Superfluid Velocity Measurement in ^{3}He–B

13.1. A Doublet Splitting of the Squashing Mode in Superfluid ^{3}He–B

The squashing–mode in ^{3}He–B[1,2] is a dramatic manifestation of its unconventional L = 1, S = I superfluidity. This $J = 2$ $(J = L + S)$ mode can undergo a five–fold Zeeman splitting in the presence of a magnetic field or a three–fold splitting due to finite wave vector, electric fields, or superflow. The Zeeman splitting was observed by the Cornell group[3]. The study of the squashing–mode is complicated by the enormously high attenuation that occurs near the mode. Two ways to approach this experimental problem are to use a shortpath–length cell (for propagation experiments) or the acoustic impedance technique, which is especially sensitive to the high–attenuation region.

Below we report a study[13,14] of the squashing–mode with the cw acoustic impedance technique in a sound cell containing a pair of *X*–cut quartz transducers with a fundamental frequency of 12.8 MHz. The data to be reported here were taken at frequencies of 115.8 and 141.6 MHz, corresponding to the ninth and tenth harmonics. The transducers were separated by a pair of gold–plated tungsten wires; the resulting separation was 190.5 μm but the single–ended technique employed resulted in a round–trip path length of 381 μm. Other experimental details see in Ref. 4.

A typical temperature trace is shown in Fig. 13.1.

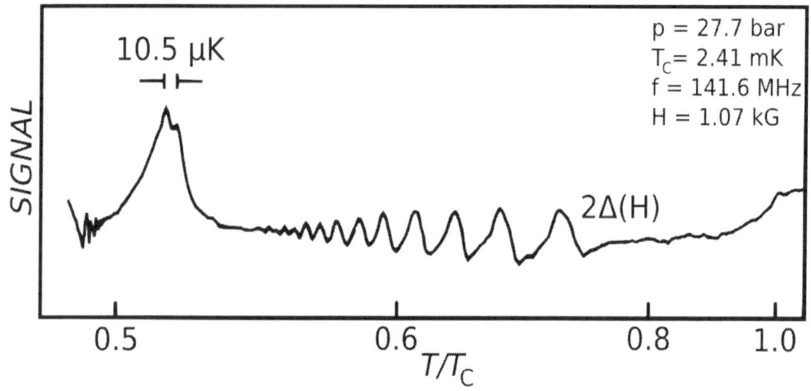

FIG. 13.1. Typical demagnetization traces of the acoustic impedance signal[13,14]: P=27.7 bars, f=141.6 MHz, H=1.07 kG. The traces are taken as a function of time with the approximate temperatures as shown.

The steplike feature in the trace corresponds to the normal–to–superfluid transition. As the temperature is further decreased, a clear onset of oscillations is observed, conesponding to the disappearance of acoustic pair breaking. The oscillations themselves arise from a continuous change in the standing–wave pattern in the cell caused by a shift in the phase velocity of zero sound associated with the approach of the collective mode (to be discussed next); as we approach the squashing–mode, the oscillations gradually die away because of the increased attenuation, and become closer together due to a more rapid change of the phase velocity.

The new phenomenon is the doublet splitting observed at the squashing–mode peak. The behavior of this splitting has been studied for pressures in the range from 19.2 to 27.7 bars in zero magnetic field. Measurements with the magnetic field perpendicular to $\boldsymbol{q}$ (the sound propagation direction) were performed up to 1.36 kG at a single pressure of 27.3 bars. The doublet splitting of the squashing–mode was observed in both zero and finite magnetic field; thus, there was no threshold value of the field required to produce the splitting and, furthermore, no substantial magnetic–field dependence of the splitting (at the fixed pressure of 27.7 bars) was observed below 1.36 kG.

A Lorentzian fit of the form

$$y = \frac{\beta \cos\varphi}{(T - T_0)^2 + (w)^2} + \frac{(\beta/w)(T - T_0)\sin\varphi}{(T - T_0)^2 + (w)^2} \tag{13.1}$$

yields a good representation of the data; here T_0, w and φ are, respectively, the peak position, the half width at half maximum (HWHM), and the phase of the detected signal. The area under a Lorentzian is given by $\pi\beta/w$. Figure 13.2 shows an example of a conventional nonlinear least–squares fit to a temperature sweep performed at a pressure of 19.2 bars with a sound frequency of 115.8 MHz in zero field. In order to best represent the splitting feature, the fit was performed in the range 1.51 to 1.55 mK, although the resulting curve is plotted over a wider temperature range. Two Lorentzian lines were assumed which are depicted by the dot–dashed line (*A*) and the dashed line (*B*). The resultant trace, shown as the solid line, is in reasonably good agreement with the experimental data, which are depicted by the solid circles. The fitting yields the following parameters:

$$\varphi = 25.5^\circ;$$
$$\beta_A = 1.3\cdot 10^{-4}, T_{0A} = 1.533 mK, w_A^{-1} = 1.73\cdot 10^2 K^{-1};$$
$$\beta_B = 1.37\cdot 10^{-4}, T_{0B} = 1.517 mK, w_B^{-1} = 0.91\cdot 10^2 K^{-1}.$$

The coupling strength λ of the collective mode components will be assumed to be proportional to the area under the corresponding Lorentzian; by this criteria the ratio of the coupling strength of these two components is $\lambda_A/\lambda_B \propto 0.18$. We emphasize that $T_{0A} > T_{0B}$, so that $\omega_{A(T)} > \omega_{B(T)}$, (for a given temperature), where ω_A and ω_B are respective frequencies of peaks *A* and *B*. Also, since peak *A* couples much more weakly, than peak B with sound, peak A cannot be the $J_Z = 0$ component of the squashing–mode. Avenel et al.[5] have generated a model to interpret the acoustic impedance signal when only one of the squashing–mode components couples to zero sound; this model could be generalized. We adopted a Lorentzian form here for simplicity.

A pressure dependence of the zero–field splitting (using a sound frequency of 141.6 MHz) was observed, showing clearly that the splitting increases as the pressure is increased. The T/T_C dependence of the splitting (at the frequency studied) is plotted in Fig. 13.2.

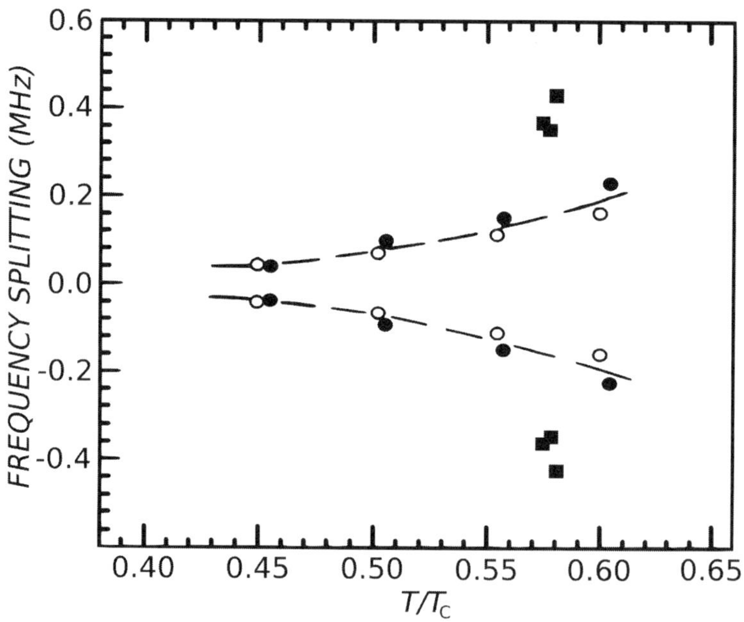

FIG. 13.2. The measured temperature splitting is converted to a frequency splitting and plotted here against T/T_C (circles)[13,14]. The squares represent data taken from a different demagnetization of faster rate. The dashed lines are guides to the eye.

Furthermore, the splitting near 27 bars at f = 141.6 M Hz was studied with two different demagnetization rates when the data were taken: 14 G/min (open circles) and 20 G/min (squares). The corresponding cooling rates are respectively $\propto 6$ and $\propto 11\mu K/\min$. Evidently, cooling with different rates causes different thermal gradients, and hence different heat flows inside the cell, which is basically a 7–in.–long cylindrical silver tube placed on the nuclear stage along the field direction. Therefore, Fig. 13.2 unambiguously tells us that the observed splitting increases with increasing thermal gradient inside the acoustic cell. This new feature of the squashing–mode has not been resolved for sound frequencies of 90.1 and 64.3 MHz.

In short, the observed doublet splitting of the squashing–mode is strongly pressure and thermal gradient dependent but independent of magnetic field on the range studied).

It was unexpected that a doublet splitting rather than a threefold splitting was observed (as was the case for the dispersion–induced splitting for the real squashing–mode[6]); for a total angular momentum of J =2 a threefold splitting would be induced by dispersion, superflow, or an electric field (which is not the case here).

There arise two possibilities: either a twofold splitting has been observed or only two components of a threefold splitting have been resolved. Two arguments that would support the existence of a twofold splitting are (a) a texture effect induced by the restricted geometry, or (b) the possible existence of some other phase near the transducer interface. Fujita et al.[7] have shown that the B–phase may evolve into a 2D–phase, with an order parameter $A_{ij}(p) \propto \Delta(T)(\delta_{i1}\delta_{a1} + \delta_{i2}\delta_{a2})$ close to a boundary. Calculations on the spectrum of the collective modes in such a 2D–phase[8] show that part of the spectrum in the 2D–phase is the same as in the A–phase (e.g., the clapping–mode, pair breaking–, or the superflapping–mode). It was estimated that the difference between the squashing–mode in ^{3}He–B and the superflapping–mode of the 2D–phase is of the order of a few tens of mK at T/T_C = 0.7. This is quite close to the experimental results. (Note, however, that our recent study[x] has shown that not $2D$–phase, but deformed B–phase is realized in the vicinity of the boundary, see Chapter XII).

A three–fold splitting could arise from either dispersion–induced splitting (DIS) or superflow–induced splitting (SIS)[9,10]. However, note the following two features of the dispersion–induced splitting: the splitting decreases with increasing T/T_C and the mode spectrum has the ordering $\omega_0 > \omega_1 > \omega_2$, where the subscript is equal to $|J_Z|$. Apparently these two features are contradictory to observed results.

In the case of superflow–induced splitting, the superflow has a twofold effect on the order parameter: It aligns the direction of the vector $\boldsymbol{n}$ along the superflow velocity $\boldsymbol{v}_S$; it also leads to a gap distortion transverse, $\Delta_\perp^2 = \Delta^2 + \Omega^2$, and parallel, $\Delta_{||}^2 = \Delta^2 + \alpha\Omega^2$, to the direction of superflow $\boldsymbol{v}_S$, where α and Ω^2 are functions of the superflow velocity and reduced temperature (T/T_C). The calculations performed by Brusov[10] and Nasten'ka and Brusov[11] give the following frequencies for the superflow–induced splitting of the squashing–mode:

$$E^2 = \frac{12}{5}\Delta^2 + \frac{1}{2}(2\alpha + 3)\Omega_0^2, \ (J = 2, J_Z = 0);$$

$$E^2 = \frac{12}{5}\Delta^2 + \frac{1}{10}(6\alpha + 11)\Omega_0^2, \ (J = 2, J_Z = \pm 1);$$

$$E^2 = \frac{12}{5}\Delta^2 + \frac{3}{5}(\alpha + 3)\Omega_0^2, \ (J = 2, J_Z = \pm 2), \tag{13.2}$$

where the branches of the squashing–mode with $|J_Z| = 1,2$ couple to the zero sound via the texture which in this case is created by the simultaneous effect of superflow and restricted geometry.

In order to show the semiquantitative relation between the superflow–induced splitting values of the squashing–mode and the reduced temperature T/T_C an estimate was made to span the temperature range $0.3 \le T/T_C \le 0.6$. Estimates have been made for two different $\boldsymbol{v}_S$ values for each fixed value of T/T_C. The results are listed in Table13.1 and display the following features: (a) The frequency spectrum has the ordering $\omega_0 < \omega_1 < \omega_2$; (b) the splitting increases with increasing T/T_C for the same $\boldsymbol{v}_S$; and (c) at fixed T/T_C the splitting increases with $\boldsymbol{v}_S$, which, from the theory, should increase with increasing thermal gradient. Features (a) and (b) are probably unique to superflow. The agreement of all three features with the experimental results strongly suggests the superflow interpretation. In the case of superflow–induced splitting, there are two possible reasons for the observation of a doublet (rather than a threefold) splitting of the squashing–mode: (a) The coupling between the sound and the $J_Z = \pm 2$ component is too weak to observe; or (b) the $J_Z = 0$ and $J_Z = \pm 1$ components are too close to resolve (which is not inconsistent with the very strong coupling observed for the *B*–peak). The coupling strength of the $|J_Z| = 2$ modes with sound is not known, so we cannot address the first possibility. But from Table I we can conclude that the second possibility seems very likely, because ω_1 and ω_2 are quite close each other.

In conclusion, observed a doublet splitting of the squashing–mode[13,14] which, most probably, should be ascribed to superflow. From the measured values of the splitting, it is possible to estimate the superflow velocity $\boldsymbol{v}_S$. Since the superfluid velocity is a difficult quantity to measure directly, one might use the doublet splitting as a probe of the superfluid velocity in future experiments designed to study superflow–induced phenomena.[12] The discussed above results strongly suggest that such experiments are now possible.

Table 13.1. The calculated values of the superflow–induced splitting of the squashing–mode $\left[\Delta_{BCS}(0) = 1.764 k_B T_C\right]$.

T/T_C	v_S (mm/s)	α	$\dfrac{\Omega^2}{\Delta^2_{BCS}(0)}$	$\dfrac{\omega_2 - \omega_0}{\Delta_{BCS}(0)}$	$\dfrac{\omega_1 - \omega_0}{\Delta_{BCS}(0)}$
0.3	14.8	-21	0.003	0.008(5)	0.007(8)
0.3	21.0	-6.7	0.024	0.023	0.017(7)
0.4	13.7	-12	0.006	0.010	0.008(7)
0.4	19.4	-6.5	0.036	0.034	0.026
0.5	12.6	-12.5	0.010	0.020	0.015(9)
0.5	17.7	-16.25	0.020	0.044	0.040(6)
0.6	12.2	-33.5	0.004	0.019	0.018(4)
0.6	15.9	-19.25	0.016	0.038	0.036

13.2. The Method of Superfluid Velocity Measurement in ^{3}He–*B*

In a previous section we discussed the structure of the ultrasound absorption spectrum of the squashing–mode and suggested that this phenomenon is induced by superflow. In this section we describe a method of determining the superfluid velocity using ultrasound[14]: the order parameter collective mode splitting may be a measure of the superfluid velocity, v_S, which is not easy to obtain. Using broad–band, piezoelectric, plastic–film transducers, one is not restricted to odd harmonics of the fundamental transducer resonance frequency, i.e. one may monitor the superfluid velocity continuously as a function of thermal gradient, pressure, and temperature. The frequency would be swept through the (split) resonance and the response monitored via a nonresonant, acoustic–impedance technique. Figure 13.3 can be used to determine the dependence of the maximum splitting, $\Delta\omega_{2-0}$, on T/T_C and on the superfluid velocity, v_S (obtained from the Brusov–Nasten'ka–Kleinert theory[9-11]). Measuring the splitting and knowing the temperature, it is easy to find superfluid velocity. For example, for the splitting obtained in Northwestern experiments[13,14] the value of superfluid velocity turns out to be of order 1.7 mm/s (at T/T_C =0.45) and

3 mm/s (at T/T_C =0.6). Note that these experiments are performed at finite $\boldsymbol{q}$ (the sound propagates perpendicular to the thermal gradient), while the theory is presently limited to $\boldsymbol{q}$=0. If the theory will be extended to finite $\boldsymbol{q}$, this will allow a more accurate comparison with experiment.

Note that there is some uncertainty in the values of $\boldsymbol{v}_S$ obtained, because we do not know which branch of the squashing–mode ($|J_Z| = 1$ or $|J_Z| = 2$) contributes the second attenuation peak. This modes are very close to each other (as can easily seen from Table 13.1), and thus uncertainty is not high.

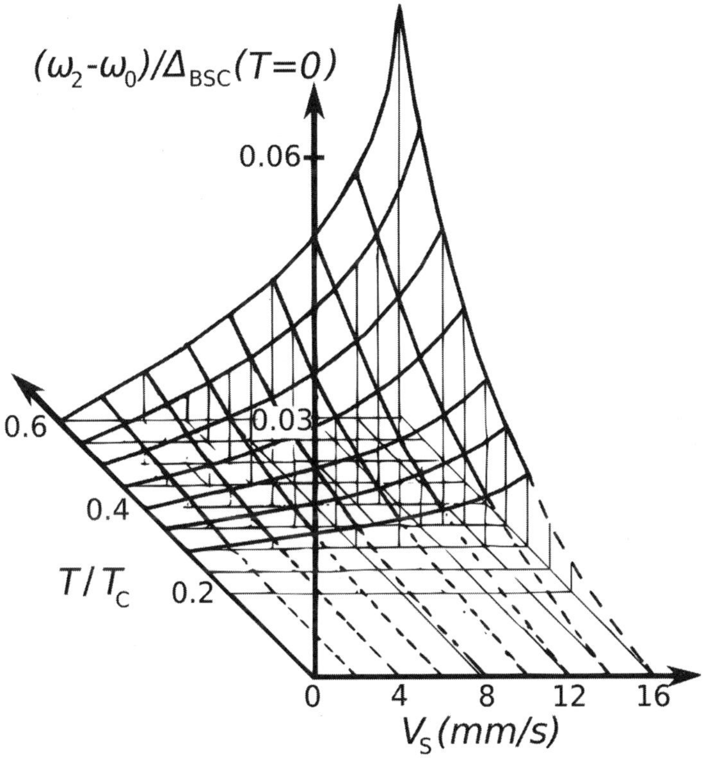

FIG. 13.3. Dependence of the relative maximum splitting , $\Delta\omega_{2-0}$, on T/T_C and on the superfluid velocity, $\boldsymbol{v}_S$ [13,14].

Chapter XIV

Superfluid Phase of ^{3}He–B Near the Boundary

In this Chapter we analyze the old and recent transverse sound experiments in superfluid ^{3}He–B and solve the old problem of superfluid quantum liquids in confined geometry: what is the boundary state of ^{3}He–B. We pay special attention to difference of transverse sound experiments data from ones of longitudinal sound experiments. We consider a few possible explanations of above experimental data: existence of a new superfluid phase in the vicinity of a boundary, excitation of different branches of squashing mode by longitudinal and transverse sounds and deformation of B–phase near the boundary. We come to conclusion that last possibility seems the most likely and boundary state of ^{3}He–B is deformed B–phase, supposed by Brusov and Popov[3] twenty years ago for case of presence of external perturbations like magnetic and electric fields. Our result[14,15] means, that influence of wall or, generally speaking, of confined geometry does not lead to existence of a new phase near the boundary, as it was supposed many yeas ago and seemed up to now, but like other external perturbations (magnetic and electric fields etc.) wall deforms the order parameter of B–phase and this deformation leads to very important consequences. In particular, frequencies of collective modes in the vicinity of boundary are changed up to 20%.

14.1. Introduction

The problem of boundary state of superfluid quantum liquid with complex order parameter like superfluid ^{3}He is quite interesting and has a long history[10–13]. One of the problems here is a particular superfluid phase realizing in the vicinity of a boundary.

Ultrasound experiments used longitudinal sound have played an important role in investigation of superfluid ^{3}He. By these experiments all collective modes in both A– and B–phases of ^{3}He (eighteen modes in each phase) have been observed[1,2]. Longitudinal sound experiments have also helped to identify the superfluid phases in bulk ^{3}He. While these experiments give information about whole volume of liquid, transverse sound experiments via strong damping of transverse sound can be a probe of the state in the vicinity of the boundary. Thus the problem of boundary state could be solved by transverse sound experiments. In past, a few possibilities for a "boundary" state have been considered: A–phase (with l–vector perpendicular to surface) and 2D–phase (existing in magnetic field). It has been shown that former phase is unlikely.

We consider here a few additional superfluid phases as well as alternative picture exploiting Brusov et al.[3] idea concerning deformation of a gap in B–phase. Our analysis of existing experimental data leads to conclusion that deformed B–phase is realized near the boundary[14,15].

14.2. Transverse Sound Experiments

More than ten year ago Ketterson et al.[4,5] have developed the transverse acoustic impedance technique and used it to investigate the superfluid ^{3}He–B. That time the interest in transverse sound in superfluid ^{3}He has been caused by theoretical prediction by Moorse and Sauls[6] that the collective modes with $J=2^-$ (squashing–modes) can provide additional mechanism for supporting a transverse response. Ketterson et al.[4,5] used a fixed sound frequency and swiped temperature. They were able to provide simultaneously measurements of longitudinal and transverse responses, that have allowed them to compare data from two these measurements. It turned out that while there is a correlation between the longitudinal and transverse responses they did not coincide exactly: there are peaks in the imaginary transverse acoustic impedance at temperatures above the squashing (sq)–mode, as well as below. At temperatures above the sq–mode these features might be interpreted as due to a standing wave pattern[4]. Based on this assumption, authors[4] turned out to be able to obtain the change in the phase velocity associated with these oscillations. The nature of the contribution at temperatures below the sq–mode was not clear that time as well as now. In this paper we will concentrate on interpretation of last features, accounting that they have been observed in recent transverse sound experiments as well[7]. These recent Japanese experiments are very important, because they provide us with some numerical data allowing us to distinguish different possibilities. The main results of these experiments are as follows. The data have been obtained for two transverse sound frequencies: 28 MHz and 47 MHz. While temperature dependant squashing mode frequency is equal to $\omega_{sq} = \sqrt{12/5}\Delta_0(T) \approx 1.55\Delta_0(T)$ (here $\Delta_0(T)$ is a gap in a single particle spectrum, which is isotropic in B–phase; in 2D–phase as well as other phases, considering below, $\Delta_0(T)$ is a maximum gap in a single particle spectrum) at both frequencies, one has peak at $\omega \approx 1.25\Delta_0(T)$ for transverse sound frequency 28 MHz and at $\omega \approx 1.35\Delta_0(T)$ for transverse sound frequency 47 MHz. We consider a few possible interpretation of above experimental data: existence of new superfluid phases in the vicinity of a boundary, excitation of different branches of squashing mode by longitudinal and transverse sounds and deformation of B–phase by the wall. We come to conclusion that last possibility seems the most likely.

14.3. Possible New Phases Near the Boundary

For theoretical description of superfluid phases of ^{3}He one uses the order parameter – value, which is nonzero below T_c and is equal to zero above T_c. We can use as the order parameter the anomalous Green's function – wave function of Cooper pairs – $F_{\alpha\beta}(\boldsymbol{k})$, where α, β are spin indexes, $\boldsymbol{k}$ is momentum of Cooper pairs. Because wave function of Cooper pairs is symmetric spinor of rank 2, it can be decomposed by the basis of symmetric unitary second order matrixes $i(\boldsymbol{\sigma}\sigma_y)_{\alpha\beta}$

$$F_{\alpha\beta}(\boldsymbol{k}) = id(\boldsymbol{k})(\boldsymbol{\sigma}, \sigma_y)_{\alpha\beta}, \tag{14.1}$$

where $\boldsymbol{\sigma} = (\sigma_1, \sigma_2, \sigma_3)$ are Pauli matrices.

Vector $\boldsymbol{d}(\boldsymbol{k})$ depends on direction in momentum space $\boldsymbol{n} = \boldsymbol{k} / k_F$ only. Under pairing with orbital moment $l = 1$ this dependence is described by the combinations of spherical harmonics with $l = 1$, which can be considered as components of the unit vector $\boldsymbol{n}$. Thus,

$$d_i = A_{ij} n_j . \tag{14.2}$$

Here, *j* is spin (isotope) index, *i* is vector index. Complex matrix (3x3) A_{ij} is the order parameter in superfluid phases of ^{3}He.

As possible "boundary" states we consider 2D–phase and phases with order parameters proportional to

$$\begin{pmatrix} 100 \\ 0-10 \\ 000 \end{pmatrix}, \begin{pmatrix} 010 \\ 100 \\ 000 \end{pmatrix}, \begin{pmatrix} 010 \\ -100 \\ 000 \end{pmatrix} \text{ and } \begin{pmatrix} 110 \\ ii0 \\ 000 \end{pmatrix}. \tag{14.3}$$

2D–phase

2D–phase with order parameter $\begin{pmatrix} 100 \\ 010 \\ 000 \end{pmatrix}$ exists at magnetic fields H>H$_C$ [8].

Without magnetic fields in the vicinity of the wall the gradient terms in free energy can stabilized this phase, thus it could be consider as a candidate for "boundary" state.

The collective mode spectrum in 2D–phase has been calculated by Brusov et al., who have used path integral technique[8]. In magnetic fields it consists of 18 collective modes, among them six Goldstone modes, four clapping–modes $E=(1.17-i0.13)\Delta_0$, two pairbreaking–modes $E=(1.96-i0.31)\Delta_0$ and six modes, depending on magnetic fields: two $E=2\mu H$, two $E^2=(1.96-i0.31)\Delta_0^2+4\mu^2H^2$, $E^2=(0.518)\Delta_0^2+4\mu^2H^2$ and $E^2=(0.495)\Delta_0^2+4\mu^2H^2$.

This spectrum contains a clapping mode with frequency $E=(1.17-i0.13)\Delta_0$. If we have B–phase in the most part of sample and 2D–phase in the vicinity of a boundary, thus our signal of transverse (longitudinal) sound is averaged signal from both regions. In bulk we have squashing mode with frequency $\omega_{sq}=\sqrt{12/5}\Delta_0\approx1.55\Delta_0$ and we have clapping mode with frequency $E=1.17\Delta_0$ in the vicinity of the boundary. It could be easy shown that detecting signal will have frequency depending on both squashing and clapping mode frequencies and the amplitudes of two these modes. If one has two signals of sound absorption $y_1=-a(x-x_0)^2+b$ and $y_2=-a'(x-x'_0)^2+b'$ with peaks at frequencies x_0 and x'_0 than resulting signal $y=-a(x-x_0)^2+b-a'(x-x'_0)^2+b'$ will has a maximum at frequency $x=(ax_0+a'x'_0)/(a+a')$ depending on both frequencies and the amplitudes of both signals. For example, at a = 0 (or $a<<a'$) $x=x'_0$, at $a'=0$ (or $a'<<a$) $x=x_0$ and at $a=2a'$ $x=(2x_0+x'_0)/3$.

If this picture is valid, it should be two resonances in sound absorption: one from squashing–mode in bulk B–phase and other one from clapping–mode in boundary 2D–phase. Via large uncertainty in determination of the temperature of the peaks these two peaks can be seems as one peak with features, described above. For longitudinal sound experiments main signal absorption takes place in a bulk liquid (region of "boundary" phase is much smaller than bulk phase one), thus attenuation amplitude peaked at squashing–mode frequency $\omega_{sq}=\sqrt{12/5}\Delta_0\approx1.55\Delta_0$. In case of transverse sound experiments region of "boundary" phase becomes comparable to bulk phase region, thus one should expect attenuation amplitude peaked at frequency intermediate between squashing mode frequency $\omega_{sq} = \sqrt{12/5}\Delta_0 \approx 1.55\Delta_0$ and clapping mode frequency, $\omega_{cl} = 1.17\Delta_0$. From this point of view observed peak frequencies $\omega \approx 1.25\Delta_0(T)$ and $\omega \approx 1.35\Delta_0(T)$ are understandable. But let us consider dependence of peak frequency on sound frequency. The particular value of peak frequency depends on attenuation amplitudes of sound into these two modes, which finally depend on the relative regions occupied by bulk and "boundary" phases. We have from Japanese experiment[7] that peak frequency increases with sound frequency: this means that part of bulk liquid involve into considering process increases with frequency of transverse sound. This fact contradicts to the hydrodynamic attenuation mechanism of transverse sound, which predicts increased attenuation with frequency as ω^2. Thus we should state that supposed picture when we have B–phase in bulk and 2D–phase near the boundary is inconsistent with Japanese data[7] and we should rule out a such possibility.

Phases $\begin{pmatrix} 100 \\ 0-10 \\ 000 \end{pmatrix}$, $\begin{pmatrix} 010 \\ 100 \\ 000 \end{pmatrix}$ and $\begin{pmatrix} 010 \\ -100 \\ 000 \end{pmatrix}$.

Let us consider these three phases. Spectra for them are identical and for them we get the following set of equations (at zero momentum of excitations)

$$\int_0^1 dx(1-x^2)(1+\frac{4\Delta^2}{\omega^2})J = 0 \text{ (2 modes)}$$

$$\int_0^1 dx(1-x^2)(1+\frac{6\Delta^2}{\omega^2})J = 0 \text{ (3 modes)}$$

$$\int_0^1 dx(1-x^2)(1+\frac{8\Delta^2}{\omega^2})J=0 \text{ (1 mode)} \tag{14.4}$$

$$\int_0^1 dx(2-x^2)(1+\frac{4\Delta^2}{\omega^2})(J-1)=0 \text{ (1 mode)}$$

$$\int_0^1 dx(2-x^2)(1+\frac{6\Delta^2}{\omega^2})(J-1)=0 \text{ (2 modes)}$$

$$\int_0^1 dx(1-x^2)(1-\frac{2\Delta^2}{\omega^2})J=0 \text{ (2 modes)}$$

$$\int_0^1 dx(1-x^2)J=0 \text{ (2 modes)}$$

$$\int_0^1 dx(1-x^2)(1-\frac{4\Delta^2}{\omega^2})J=0 \text{ (2 modes)}$$

$$\int_0^1 dxx^2\left[(1-\frac{2\Delta^2}{\omega^2})J-1\right]=0 \text{ (2 modes)}$$

$$\int_0^1 dxx^2(J-1)=0 \text{ (1 mode).}$$

Here, $J = \frac{1}{\sqrt{1+4\Delta^2/\omega^2}} \ln \frac{1-\sqrt{1+4\Delta^2/\omega^2}}{1+\sqrt{1+4\Delta^2/\omega^2}} = 0$ and number in brackets shows the degeneracy of branch.

Solving these equations numerically, one gets for high frequency modes

$$E \;= \Delta_0(T)(1.83 - i\,0.06);\, E \;= \Delta_0(T)(1.58 - i\,0.04);$$

$$E \;= \Delta_0(T)(1.33 - i\,0.10);$$

$$E \;= \Delta_0(T)(1.33 - i\,0.08);\, E \;= \Delta_0(T)(1.28 - i\,0.04);$$

$$E \;= \Delta_0(T)(1.09 - i\,0.22); \tag{14.5}$$

$$E \;= \Delta_0(T)(0.71 - i\,0.05);\, E \;= \Delta_0(T)(0.33 - i\,0.34);$$

$$E \;= \Delta_0(T)(0.23 - i\,0.71).$$

The last two modes have imaginary parts of the same order as real ones. This means that they are damped very strongly and could not be considered as resonances. Among the obtained spectra there are modes, which frequencies are closed to observed ones $\omega \approx 1.25\Delta_0(T)$ and $\omega \approx 1.35\Delta_0(T)$. These are well determined modes (imaginary part of frequency is much less than real part)

$$E \;= \Delta_0(T)(1.33 - i\,0.10);\, E \;= \Delta_0(T)(1.33 - i\,0.08);$$

$$E \;= \Delta_0(T)(1.28 - i\,0.04). \tag{14.6}$$

But there is no reason, why at one sound frequency one branch is excited while at other frequency another branch is excited. Thus, at this step we should rule out these phases as a candidate for "boundary" state.

Phase $\begin{pmatrix} 110 \\ ii0 \\ 000 \end{pmatrix}$.

For spectrum of collective modes in this phase we get the following set of equations (at zero momentum of excitations)

$$\int_0^1 dx x^2 (1+\frac{2\Delta^2}{\omega^2})(J-1)=0 \text{ (6 modes)}$$

$$\int_0^1 dx(1-x^2)(1+\frac{2\Delta^2}{\omega^2})J=0 \text{ (4 modes)} \tag{14.7}$$

$$\int_0^1 dx(1-x^2)(1+\frac{\Delta^2}{\omega^2})J=0 \text{ (4 modes)}$$

$$\int_0^1 dx(1-x^2)(1+\frac{3\Delta^2}{\omega^2})J=0 \text{ (4 modes)}.$$

From these equations the following high frequency modes have been found

$$E = \Delta_0(T)(1.55-i\,0.32);\, E = \Delta_0(T)(1.2-i\,0.06);$$

$$E = \Delta_0(T)(0.62 - i\,0.05); \tag{14.8}$$

$$E = \Delta_0(T)(0.4 - i\,0.55);\; E = \Delta_0(T)(0.3 - i\,1.0).$$

The last two modes have imaginary parts of the same order as real ones. They are damped very strongly and could not be considered as resonances.

Among the collective modes of this state there is only one $E = \Delta_0(T)(1.2 - i\,0.06)$, which frequency is closed to one of observed peaks $\omega \approx 1.25\Delta_0(T)$. We can not explain the existence of peak at $\omega \approx 1.35\Delta_0(T)$. Thus, similar to previous case, we should rule out these phases as a candidate for "boundary" state.

14.4. Different Branches of Squashing Mode

One more possible reason for difference in longitudinal and transverse sound experiments data has been suggested by Ketterson et al.[4] They supposed, that in accordance to[6] longitudinal and transverse sounds can be coupled to different branches of squashing–mode: namely, in the absence of magnetic field transverse sound couples to $J_Z = \pm 1$ modes while longitudinal sound couples to the J_Z=0 mode. While frequencies of all branches of squashing–mode are the same and equal to $\omega_{sq} = \sqrt{12/5}\Delta_0 \approx 1.55\Delta_0$, dispersion corrections for branches with different projections of total moment of Cooper pairs J turns out to be different. These dispersion corrections have been calculated by numerous authors, but complete calculations for all 18 collective modes in B–phase have been done by Brusov and Popov[8], who have obtained the following results for considering branches of squashing–mode

$$\omega_{sq}^2 = 12/5\Delta_0^2(T) + 0.418c_F^2k^2(2,\pm 1, i)$$

and

$$\omega_{sq}^2 = 12/5\Delta_0^2(T) + 0.502c_F^2 k^2 (2,0,i).$$

From these results it follows that longitudinal resonance should takes place at higher frequencies than transverse one and this explains attenuation peak at temperatures below the sq–mode in experiments on transverse sound attenuation.

However if we will compare the real splitting of the squashing mode branches via dispersion corrections (via nonzero momentum of Cooper pairs) and difference between squashing–mode peak (from longitudinal experiments) and transverse sound data it will turn out that former one is much less than later one. To estimate the dispersion induced splitting of the squashing mode branches we can refer to similar splitting of real squashing mode observed by Shivaram et al.[9]

This splitting turned out of order 0.01 T_C. While the splitting of sq–mode will be a little bit larger it should be the same order of magnitude. From the other side the difference between sq–mode peak (from longitudinal experiments) and transverse sound data is $0.2\Delta_0(T)$–$0.3\Delta_0(T)$ depending on transverse sound frequency. Accounting that T_C is the same order of magnitude as $\Delta_0(0)$ we come to conclusion that coupling longitudinal and transverse sounds to different branches of squashing–mode can not explain the observed features.

14.5. Deformed *B*–Phase

Presence of boundary leads to deformation of the order parameter. Component of vector $\boldsymbol{d}$, which is perpendicular to the boundary becomes zero at the boundary in case of mirror, as well as diffusive reflection of atoms. By other words, Cooper pairs tend to move in the plane parallel to the boundary. We will consider *x*–*y* plane as a boundary. While *A*–phase in a slab geometry should have the same order parameter as in bulk case (only $\vec{l}$ should be parallel to $\hat{z}$), in case of *B*–phase order parameter of bulk *B*–phase can not satisfy to boundary condition, if it remains nondeformed. This deformation is coupled to appearance of additional gradient energy and decreasing of condensation energy.

More than twenty years ago Brusov and Popov[8] have investigated the influence of gap distortion, caused by dipole interaction or by different type of external perturbations such as magnetic or electric fields, on the order parameter collective modes in B–phase. They have shown, that consequences of such influence are quite significant: it changes the frequencies of all collective modes and (that is especially important for us) splits the pairbreaking, squashing and real squashing modes at zero momentum $\boldsymbol{q}$ of collective excitations. At nonzero $\boldsymbol{q}$ they predicted a branch crossing of these modes with different J_Z.

Let us summarize their results and see what we have for sq–mode.
In presence of external perturbations such as magnetic or electric fields or boundary order parameter has the following form

$$A_{ij} = \left[\Lambda^{1/2} R(\hat{n},\theta)\right]_{ij} e^{i\varphi}, \tag{14.9}$$

where Λ is diagonal matrix with elements

$$\lambda_1, \lambda_1, \lambda_2 \text{ and } \lambda_1 = \Delta_1^2 = \Delta^2 + \Omega^2, \lambda_2 = \Delta_2^2 = \Delta^2 + \alpha\Omega^2. \tag{14.10}$$

We can rewrite order parameter as

$$A_{ij} \approx \begin{pmatrix} \Delta_1 \cos\theta & -\Delta_1 \sin\theta & 0 \\ \Delta_1 \sin\theta & \Delta_1 \cos\theta & 0 \\ 0 & 0 & \Delta_2 \end{pmatrix}. \tag{14.11}$$

Here, $\theta = \arccos(-1/4)$. In case of dipole interaction and electric fields $\alpha = -2$, in case of magnetic fields $\alpha = -4$. In case of confined geometry boundary will suppress the gap in a single particle spectrum in perpendicular direction, thus we suppose α to be negative in the vicinity of a boundary. Thus, gap Δ_1 along the boundary will be bigger than gap

Δ in a bulk liquid, while gap Δ_2 perpendicular boundary will be less, than Δ.

Brusov and Popov results[8] for the sq–mode are as follows:

$$E^2 = (12/5)\Delta_0^2(T) + (6\alpha + 11)\Omega^2/10 \quad \text{for } (2,\pm1,i) \text{ branches,}$$

$$E^2 = (12/5)\Delta_0^2(T) + (2\alpha + 3)\Omega^2/2 \quad \text{for } (2,0,i) \text{ branch,} \qquad (14.12)$$

$$E^2 = (12/5)\Delta_0^2(T) + 3(\alpha + 3)\Omega^2/5 \quad \text{for } (2,\pm2,i) \text{ branches.}$$

The results for $(2,0,i)$ branch is insufficient because in longitudinal sound experiments signal comes from whole volume of liquid and influence of region near the boundary is small or even negligible in this case.

The essential for us is result for $(2,\pm1,i)$ branches, which are excited in transverse sound experiments. This result can explain observed features of transverse sound experiments: resonant absorption of transverse sound below squashing mode frequency, dependence of absorption frequency on transverse sound one. Thus, from results for $(2,\pm1,i)$ branches it follows, that $(6\alpha+11)\Omega^2/10 < 0$, thus $\alpha < -11/6 \approx -2$. We can estimate the value of extra term in sq–mode spectrum appearing via boundary influence (through the gap distortion). From experiments[7] it follows that $\left|(6\alpha+11)\Omega^2/8\sqrt{15}\Delta\right|$ should be of order $0.3\Delta(T)$ at sound frequency 28MHz and of order $0.2\Delta(T)$ at sound frequency 47MHz.

14.6. Conclusion

We have analyzed the old and recent transverse sound experiments in superfluid ^{3}He–B, where some peaks in transverse sound absorption have been observed. These peaks are in disagreement with squashing–mode frequency, obtained from longitudinal sound experiments. We consider a few possible explanations of above experimental data: existence of a new superfluid phase in the vicinity of a boundary, excitation of different branches of squashing mode by longitudinal and transverse sounds and deformation of B–phase by the wall. Our analysis of existing experimental data leads to conclusion that deformed B–phase is realized near the boundary. This important conclusion solves considering issue. Our result means, that influence of wall or generally speaking of confined geometry does not lead to existence of a new phase near the boundary, as it was supposed many yeas ago and seemed up to now, but like other external perturbations (magnetic and electric fields etc.) wall deforms the order parameter of B–phase and this deformation leads to very important consequences. In particular, frequencies of collective modes in the vicinity of boundary are changed up to 20%. Their branches could cross at nonzero momentum of collective excitations. Knowledge of the collective mode spectrum together with experimental data on transverse sound allows estimate parameters of gap deformation, caused by the boundary.

Farther transverse sound experiments at different sound frequencies are desirable to make picture clearer. These experiments turned out to be very important for farther investigation of superfluid ^{3}He in confined geometry.

Discussed problem is very important and has very closed connection with study of superfluid ^{3}He in aerogel[13], where aerogel could be treated as some effective surface.

Chapter XV

Collective Excitations in the Planar $2D$–Phase of Superfluid ^{3}He

15.1. The Planar 2D–Phase of Superfluid ^{3}He

The superfluid phases of ^{3}He in addition to the isotropic B–phase, the anisotropic A–phase and the A_1–phase also include a $2D$–phase[1], known as planar and having the order parameter

$$c_{ia}^{(0)}(p) = c(\beta V)^{1/2}\,\delta_{p0}\left(\delta_{i1}\delta_{a1} + \delta_{i2}\delta_{a2}\right),$$

or in a matrix form

$$c_{ia}^{(0)}(p) = c(\beta V)^{1/2}\,\delta_{p0}\begin{pmatrix} 1 & 0 & 0 \\ 0 & 1 & 0 \\ 0 & 0 & 0 \end{pmatrix}.$$

In opposite to the A–, B– and A_1–phases $2D$–phase has not yet been observed, but its existence under certain conditions was deduced by many researchers. In particular, Popov et al.[2] predicted a phase transition from the B– to the $2D$–phase at $H = H_C$ and proved the stability of the $2D$–phase to small perturbations for $H > H_C$. Fujita et al.[3] by considering the B–phase in semi–bounded space, have shown that a $2D$–phase is realized on the boundary: in this situation it is energetically more favored than the A–phase. One of the possible explanations of the double splitting of the squashing mode in the B–phase, observed experimentally,[4] was an assumed existence on the sell boundary of a

2D–phase, one of the collective modes of which leads to the appearance of a second peak in ultrasound absorption (our recent study[5,6], however, has shown that not 2D–phase, but deformed B–phase is realized in the vicinity of the boundary, see Chapter XIV).

These examples suffice to understand the importance of investigation of the planar 2D–phase and particularly the spectrum of its collective excitations. Below we calculate this spectrum by path integral technique.

All properties of superfluid ^{3}He are determined by the functional of the hydrodynamical action, S_h, given by

$$S_h = g^{-1} \sum_{p,i,a} c^{+}_{ia}(p) c_{ia}(p) + \frac{1}{2} \ln \det \frac{\hat{M}\left(c_{ia}, c^{+}_{ia}\right)}{\hat{M}\left(c^{(0)}_{ia}, c^{(0)+}_{ia}\right)}. \tag{15.1}$$

Here, $c_{ia}(p)$ is the Fourier transform of the Bose–field $c_{ia}(\boldsymbol{x},\tau)$ describing the Cooper pairs of the quasifermions on the Fermi–surface, the operator $\hat{M}$ is given by

$$\hat{M} = \begin{pmatrix} Z^{-1}(i\omega - \xi + \mu H \sigma_3)\delta_{p1p2}; & \begin{matrix} (\beta V)^{-1/2}(n_{1i} - n_{2i})\sigma_a \times \\ \times c_{ia}(p_1 + p_2)\delta_{p1+p2,0}; \end{matrix} \\ \begin{matrix} -(\beta V)^{-1/2}(n_{1i} - n_{2i}) \times \\ \times \sigma_a c_{ia}(p_1 + p_2)\delta_{p1+p2,0} \end{matrix} & Z^{-1}(-i\omega + \xi + \mu H \sigma_3)\delta_{p1p2} \end{pmatrix} \tag{15.2}$$

where $\xi = c_F(k - k_F), -n_i = k_i / k_F$, H is the magnetic field and μ is the magnetic moment of the quasiparticle, σ_a (a=1,2,3) are 2x2 Pauli–matrixes, and $\omega = (2n+1)\pi T$ are the Fermi–frequencies.

Expanding the functional (15.1) in the Ginzburg–Landau region $T_C - T \propto T_C$ in powers of the fields c and c^+ we obtain

$$S_h = -\frac{20k_F^2(\Delta T)^2 \beta V}{21\xi(3)c_F}\Pi, \tag{15.3}$$

where

$$\begin{aligned}\Pi = & -Tr(AA^+) + \nu Tr(AA^+P) + (TrAA^+)^2 + Tr(AA^+AA^+) + \\ & + Tr(AA^+A^*A^T) - Tr(AA^TA^*A^+) - \frac{1}{2}Tr(AA^T)Tr(A^+A^*).\end{aligned} \tag{15.4}$$

Here,

$$\nu = 7\xi(3)\mu^2 H^2 / 4\pi^2 T_C\, \Delta T, \tag{15.5}$$

P is the projector on the third axis along which the field is directed. Minimizing Π, we obtain the matrix A that determines the condensate density. The equation $\delta\Pi = 0$ or

$$\begin{aligned}& -A + \nu AP + 2(trA^+A)A + 2AA^+A + 2AA^*A^TA - 2AA^TA^* - \\ & - A^*trAA^T = 0\end{aligned} \tag{15.6}$$

has several nontrivial solutions corresponding to the superfluid phases. One of them has an order parameter

$$\frac{1}{2}\begin{pmatrix} 1 & 0 & 0 \\ 0 & 1 & 0 \\ 0 & 0 & 0 \end{pmatrix}. \tag{15.7}$$

This is, in fact, the planar $2D$–phase. Calculations of the second variation $\delta^2\Pi$ yields

$$\delta^2 F_2 = (\nu - 1/2)u_{33}^2 + (\nu + 1/2)v_{33}^2 + \nu(u_{13}^2 + u_{23}^2) + \\ + (\nu + 2)(v_{13}^2 + v_{23}^2) + (1/2)[3u_{11}^2 + 3u_{22}^2 + 2u_{11}u_{22} + \\ + (u_{12} + u_{12})^2] + (1/2)[3v_{12}^2 + 3v_{21}^2 - 3v_{12}v_{21} + (v_{11} - v_{22})^2]. \quad (15.8)$$

Here, $u_{ia} = \mathrm{Re}\, c_{ia}$, $v_{ia} = \mathrm{Im}\, c_{ia}$.

For $\nu < 1/2$ the second variation $\delta^2 \Pi$ is of alternating sign, while for $\nu > 1/2$ it is non–negative. This means that the $2D$–phase is stable in a magnetic field $H > H_C = \nu = [\pi\mu 2T_C \Delta T / 7\xi(3)]^{1/2}$. As indicated above at $H = H_C$ a phase transition takes place from the B– to the $2D$–phase.

15.2. Collective Modes in ^{3}He–2D at Zero Momenta of Excitations

To calculate the collective mode spectrum in ^{3}He–$2D$ at $T \to 0$ the functional S_h must be expanded in terms of fluctuations of the fields $c_{ia}(p)$. Making in S_h a shift $c_{ia}(p) \to c_{ia}(p) + c_{ia}^{(0)}(p)$ we get the quadratic part of the functional S_h

$$\sum_p c_{ia}^+(p) c_{jb}(p) \times \\ \times \left[\frac{\delta_{ij}\delta_{ab}}{g} + \frac{2Z^2}{\beta V} \sum_{p1+p2=p} n_{i1} n_{j1} tr(A_1 - B_1\sigma_3)\sigma_a (A_2 + B_2\sigma_3)\sigma_b \right] - \\ - \sum_p c_{ia}^+(p) c_{jb}(-p) \frac{Z^2}{\beta V} \sum_{p1+p2=p} n_{i1} n_{j1} tr(C_1\sigma_1 + D_1\sigma_2)\sigma_a (C_2\sigma_1 + D_2\sigma_2)\sigma_b - \\ - \sum_p c_{ia}^+(p) c_{jb}(-p) \times \\ \times \frac{Z^2}{\beta V} \sum_{p1+p2=p} n_{i1} n_{j1} tr(\overline{C}_1\sigma_1 + \overline{D}_1\sigma_2)\sigma_a (\overline{C}_2\sigma_1 + \overline{D}_2\sigma_2)\sigma_b, \quad (15.9)$$

where the coefficients are given by

$$
\begin{aligned}
&A = A_1 = M^{-1}\left[-(i\omega-\xi)(\omega^2+\xi^2+\mu^2H^2+\Delta^2)+2\xi\mu^2H^2\right],\\
&B = -B_1 = M^{-1}\mu H\left[(\omega^2+\xi^2+\mu^2H^2+\Delta^2)-2\xi(i\omega+\xi)\right],\\
&C = \overline{C}_1 = M^{-1}\Delta_0\left[n_1(\omega^2+\xi^2+\mu^2H^2+\Delta^2)-2i\xi\mu Hn_2\right],\\
&D = \overline{D}_1 = M^{-1}\Delta_0^2\left[n_2(\omega^2+\xi^2+\mu^2H^2+\Delta^2)+2i\xi\mu Hn_1\right],\\
&M = (\omega^2+\xi^2+\mu^2H^2+\Delta^2)^2-4\xi^2\mu^2H^2,\\
&\Delta = \Delta_0\sin\theta.
\end{aligned}
\tag{15.10}
$$

The equation for gap in the $2D$–phase is

$$
g^{-1} = \frac{2Z^2}{\beta V}\sum_p \frac{\sin^2\theta}{\omega^2+\xi^2+\Delta^2}\ . \tag{15.11}
$$

After calculating the quadratic form coefficients (15.19) by taking the trace and replacing g with the aid of Eq. (15.11) we obtain from the equation $\det Q=0$ the following equations for the collective mode spectrum

$$
\int_0^1 (1-x^2)I(c)(1+4c)dx = 0,\ \ v_{11}+v_{22}\pm(u_{12}+u_{21}); \tag{15.12}
$$

$$
\begin{aligned}
&\int_0^1 (1-x^2)I(c)(1+2c)dx = 0,\\
&u_{11}+u_{22}\pm(v_{12}+v_{21}), v_{11}-v_{22}\pm(u_{12}-u_{21});
\end{aligned}
\tag{15.13}
$$

$$\int_0^1 \left(1-x^2\right) I(c)dx = 0, \quad v_{11} - v_{22} \pm \left(u_{12} - u_{21}\right); \tag{15.14}$$

$$\int_0^1 \left[(1+2c)I(c) - 1\right] x^2 dx = 0, \quad u_{31} \pm v_{32} \ , \ v_{31} \pm u_{32}; \tag{15.15}$$

$$\int_0^1 x^2 \left[(1+4c_+)I(c_+) + (1+4c_-)I(c_-) - 2\right] dx = 0, \quad u_{33}; \tag{15.16}$$

$$\int_0^1 x^2 \left[I(c_+) + I(c_-) - 2\right] dx = 0, \quad v_{33}; \tag{15.17}$$

$$\int_0^1 \left(1-x^2\right)\left[(1+4c_+)I(c_+) + (1+4c_-)I(c_-)\right] dx = 0, \quad u_{13}, u_{23}; \tag{15.18}$$

$$\int_0^1 \left(1-x^2\right)\left[I(c_+) + I(c_-)\right] dx = 0, \quad v_{13}, v_{23}. \tag{15.19}$$

Here,

$$I(c) = \frac{1}{(1+4c)^{1/2}} \ln \frac{(1+4c)^{1/2}+1}{(1+4c)^{1/2}-1},$$

$$c_{\pm} = \frac{\Delta_0^2\left(1-x^2\right)}{\omega^2 + \left[c_F(\boldsymbol{n},\boldsymbol{k}) \pm 2\mu H\right]^2}, c = \frac{\Delta_0^2\left(1-x^2\right)}{\omega^2 + c_F^2(\boldsymbol{n},\boldsymbol{k})^2},$$

$$u_{ia} = \operatorname{Re} c_{ia}, \quad v_{ia} = \operatorname{Im} c_{ia}. \tag{15.20}$$

Let us examine Eqs. (15.12)–(15.19) at zero momenta (***k***=0) of the collective excitations. In this case Eqs. (15.12)–(15.14) coincide with those obtained by Brusov and Popov for the A–phase without a magnetic field, while Eqs. (15.17)–(15.19) go over into aforementioned Brusov–Popov equations for an A–phase without a magnetic field following the substitution $\omega^2 + 4\mu^2 H^2 \to \omega^2$. These equations can thus be solved by using the results of Ref. 7. Finding also the roots of Eqs. (15.15) and (15.16), we obtain the following result of the collective mode spectrum at ***k***=0, as listed in Table 15.1.

Table 15.1. High-frequency mode spectrum in 2D–phase at $\boldsymbol{k} = 0$.

No.	Type	Frequency	Variables
6	Goldstone	$E = 0$	$u_{31} \pm v_{32}$, $v_{31} \pm u_{32}$ $v_{11} + v_{22} \pm (u_{12} + u_{21})$
4	clapping	$E = (1.17 - i \cdot 0.13)\Delta$	$v_{11} - v_{22} \pm (u_{12} - u_{21})$ $u_{11} + u_{22} \pm (v_{12} + v_{21})$
2	pairbreaking	$E = (1.96 - i \cdot 0.31)\Delta$	$u_{11} - u_{22} \pm (v_{12} - v_{21})$
2	quasi–Goldstone	$E = 2\mu H$	u_{13}, u_{23}
2	quasi–pairbreaking	$E^2 = (1.96 - i \cdot 0.31)^2 \Delta^2 + 4\mu^2 H^2$	v_{13}, v_{23}
1		$E^2 = (0.518)^2 \Delta^2 + 4\mu^2 H^2$	u_{33}
1		$E^2 = (0.495)^2 \Delta^2 + 4\mu^2 H^2$	v_{33}

Thus, the spectrum of a planar 2*D*–phase in a magnetic field contains modes similar to those in the *A*–phase without a magnetic field, as well as a number of new modes. The former consist of six Goldstone–modes, four clapping–modes, and two pairbreaking–modes. Two quasi Goldstone–modes and two quasi–pairbreaking–modes are obtained from the Goldstone– and pairbreaking–modes respectively by substituting $E^2 \to E^2 - 4\mu^2 H^2$. The gap in the quasi Goldstone–mode spectrum is $\propto 2\mu H$. Finally, we obtained two new modes having no analogs in the *A*–phase. They correspond to the variables u_{33} and v_{33}, are not degenerate, and the difference between their frequencies is small. Interestingly, whereas for the clapping– and pairbreaking–modes there exists in the *A*–phase a linear Zeeman effect (threefold splitting in a magnetic field), the frequencies of this modes in the 2*D*–phase are independent of the magnetic field, while the energies of the quasi–pairbreaking modes and of the two "new" modes are quadratic in the field. Note also that the energies of all the nonphonon modes, except the two "new" ones, have imaginary parts due, just as in *A*–phase, to the vanishing of a Fermi–spectrum gap in a special direction (that of the magnetic field). The frequencies of all the nonphonon modes of the spectrum turn out to be complex, in view of the possible decay of the collective excitations into the initial fermions (owing to the vanishing of the Fermi–spectrum gap along the field direction). Just as in the *A*– and *B*–phases, collective modes can be excited in the 2*D*–phase in ultrasound and NMR experiments.

Note that notwithstanding some similarity between the spectra of the *A*– and 2*D*–phases, they also have substantial differences that can possibly help identify the 2*D*–phase. Just as in the latter, there exist some nonphonon modes absent from the *A*–phase (and also from the *B*–phase), and the behavior of the spectrum (and even of the analog modes) in the 2*D*–phase and in the *A*–phase is quite difference: in the *A*–phase we have a linear splitting of the pairbreaking– and clapping–modes, while in the 2*D*–phase one part of the spectrum is independent of the field, whereas the other part has a quadratic field dependence.

Collective mode spectrum in the 2*D*–phase was studied as well by Hirashima et al.[8], who however, considered a 2*D*–phase without a magnetic field. Since the 2*D*–phase is stable only for $H > H_C$, the meaning of their calculations is not clear. Obviously, they could not obtain six collective modes with frequencies dependent on the magnetic field. Comparing nonetheless results of Brusov and Popov with those of Ref. 8 we note the following:

1. The main conclusions of both studies, that the 2*D*–phase spectrum coincides in part with the *A*–phase spectrum, but modes, typical of the 2*D*–phase are present and are close to one another.
2. The correspondence between that fraction of the modes which is the same in both phases as in the *A*–phase spectrum investigation by the kinetic–equation[9] and path–integration–methods[2]. A frequency $\omega_{cl} = 1.23\Delta_0(T)$ was thus obtained in Ref.8 for the clapping–mode, as against $\omega_{cl} = 1.17\Delta_0(T)$ in paper by Brusov et al.[10], in much better agreement with the experiments in the *A*–phase (see Ref. 11 an the citations therein). The reason is that we have taken into account the collective mode damping due to decay of Cooper pairs in view of the vanishing of the Fermi–spectrum gap (see Ref. 11 for details).
3. In Ref. 8 one mode was obtained, typical only of the 2*D*–phase and having an energy somewhat lower than that of the super–flapping mode at all temperatures. This new mode is due to spin waves with a coupling coefficient $O(k^2)$. It is neither resonant no diffuse. Note, that Brusov et al.[7] have obtained not the super–flapping mode, but additional Goldstone–modes whose appearance is due to the presence of latent symmetry. As noted above, Brusov et al.[10] have obtained in a magnetic field two modes, that are indicative only of the 2*D*–phase. Their frequencies are close to one another and depend on the field.

In this Chapter, we have determined the compete collective excitation spectrum in ^{3}He–2*D* at zero momentum of collective excitations. But knowledge of the collective excitation spectrum at zero momentum of excitations is not enough. First of all in sound experiments, which are used to study the collective excitation spectrum the collective modes are created with nonzero momentum $\boldsymbol{k}$, and for more detailed comparison of the theoretical results with sound experiment data we must take the dispersion corrections into account. From the other hand taking of the dispersion corrections into account can lead to the lift of the degeneracy of the collective modes similar to case of ^{3}He–*B*, where the dispersion–induced splitting of collective modes takes place[12] and has been observed experimentally[13]. In the next Chapter we calculate the dispersion corrections to collective mode spectrum in planar 2*D*–phase, as well as in axial *A*–phase.

Chapter XVI

Dispersion Induced Splitting of the Collective–Mode Spectrum in Axial– and Planar–Phases of Superfluid ^{3}He

A renewal of the interest to investigation of superfluid ^{3}He is caused by the intensive study during last two decades of superfluid ^{3}He in porous media like aerogel, Vycor glasses etc[1]. In such complex systems superfluid phase analogs to axial A–phase has been observed. Another cause of such an interest is related to the study of superfluid ^{3}He in confined geometry, where a very old problem of the boundary state in the isotropic *B*–phase has been solved recently[2]. One more cause of the renewal of the interest to superfluid ^{3}He relates to recent study of the possibility of subdominant f–wave pairing in superfluid ^{3}He (in addition to dominant p–wave pairing) [24,25].

The whole collective mode spectrum in axial– and planar–phases of superfluid ^{3}He with dispersion corrections is calculated in this Chapter. In axial *A*–phase the degeneracy of clapping–modes depends on the direction of the collective mode momentum $\boldsymbol{k}$ with respect to the vector $\boldsymbol{l}$ (mutual orbital moment of Cooper pairs), namely: the mode degeneracy remains the same as in case of zero momentum $\boldsymbol{k}$ for $\boldsymbol{k}\|\boldsymbol{l}$ only. For any other directions there is a three–fold splitting of these modes, which reaches maximum for $\boldsymbol{k} \perp \boldsymbol{l}$ [26].

In planar $2D$–phase, which exists in magnetic field (at $H > H_C$) we find that for clapping–modes the degeneracy depends on the direction of the collective mode momentum $\boldsymbol{k}$ with respect to the external magnetic field $\boldsymbol{H}$, namely: the mode degeneracy remains the same as in case of zero momentum $\boldsymbol{k}$ for $\boldsymbol{k}\|\boldsymbol{H}$ only. For any other directions different from this one (for example, for $\boldsymbol{k} \perp \boldsymbol{H}$) there is two–fold splitting of these modes[27].

The obtained results[26,27] means that new interesting features can be observed in ultrasound experiments in axial– and planar–phases: the change of the number of peaks in ultrasound absorption into clapping–mode. One peak, observed for these modes by Ling et al. [8] will split into two peaks in planar–phase and into three peaks in axial–phase under change the ultrasound direction with respect to the external magnetic field $\boldsymbol{H}$ in planar–phase and with respect to the vector $\boldsymbol{l}$ in axial– phase.

In planar–phase some Goldstone–modes in magnetic field become massive (quasi–Goldstone) and have similar two–fold splitting under change the ultrasound direction with respect to the external magnetic field $\boldsymbol{H}$.

The obtained results as well will be useful under interpretation of the ultrasound experiments in axial– and planar–phases of superfluid ^{3}He.

16. 1. Introduction

A study of the collective excitations in superfluid ^{3}He plays a very important role. A lot of very interesting features of the collective mode spectrum still remain experimentally unexplored[3,4]. The collective excitation spectrum in different phases of superfluid ^{3}He at zero momentum of collective excitation has been calculated by many authors. But in sound experiments, where the collective modes are excited, one has the collective excitations with nonzero momentum $\boldsymbol{k}$. So, the knowledge of the collective excitation spectrum at zero momentum $\boldsymbol{k}$ is not sufficient to compare the theoretical results with the sound experiment data and we must take into account the dispersion corrections.

In present Chapter we calculate the whole collective mode spectrum in axial– and planar–phases of superfluid ^{3}He with dispersion corrections. We show that in axial *A*–phase the degeneracy of clapping–modes depends on the direction of the collective mode momentum $\boldsymbol{k}$ with respect to the vector $\boldsymbol{l}$ (mutual orbital moment of Cooper pairs): the mode degeneracy remains the same as in the case of zero momentum $\boldsymbol{k}$ for $\boldsymbol{k} \| \boldsymbol{l}$ only. For any other direction, there is a threefold splitting of these modes, which reaches maximum for $\boldsymbol{k} \perp \boldsymbol{l}$.

In a planar $2D$–phase, which exists in the magnetic field (at $H > H_C$), we find that for clapping– and quasi–Goldstone–modes the degeneracy depends on the direction of the collective mode momentum $\boldsymbol{k}$ with respect to the external magnetic field $\boldsymbol{H}$: the mode degeneracy remains the same as in case of zero momentum $\boldsymbol{k}$ for $\boldsymbol{k} \| \boldsymbol{H}$ only. For any other direction different from this one (for example, for $\boldsymbol{k} \perp \boldsymbol{H}$) there is a twofold splitting of these modes.

16.2. Axial Phase

In the bulk superfluid ^{3}He the *A*–phase gives us an example of an anisotropic superfluid quantum liquid and thus is perhaps the most interesting object in superfluid ^{3}He. The main features of the *A*–phase of ^{3}He are connected to the existence of two nodes in the gap of a single–particle spectrum on the Fermi–surface. This leads to the existence of chiral fermions, gauge fields, analogs of *W* and Z bosons, zero–charge phenomenon, the damping of collective excitations (CE) at zero momentum, and to many other consequences for the system[5].

The collective excitation spectrum in ^{3}He–A at zero momentum of collective excitation has been calculated by many authors, in particular by Wolfle[6] using kinetic equation method and by Brusov and Popov[7] using path integration technique. The latter authors have taken the damping of collective excitations into account that led to some differences in results of Ref. 7 and 6. The precise experiments on measurement of the clapping–mode frequency[8] are in excellent agreement with the Brusov and Popov theory[9].

The case of nonzero momentum $\boldsymbol{k}$ has been considered by Wolfle[6], Combescot[10], Brusov and Popov[11]. Wolfle has considered for simplicity the case of $\boldsymbol{k} \| \boldsymbol{l}$, where 3x3 matrices are diagonal, and obtained the dispersion lows for Goldstone modes: sound mode $E = c_F k / \sqrt{3}$, orbital waves $E = c_F k_{||}$, and diffusive mode, associated with the breakdown of the rotation symmetry in orbital spaces and quadratic dispersion corrections to the spectrum of nonphonon modes, for example, to normal–flapping (nfl)–mode

$$E_{nfl}^2 = E_{nfl}^2(k=0) + \alpha(c_F k)^2 .$$

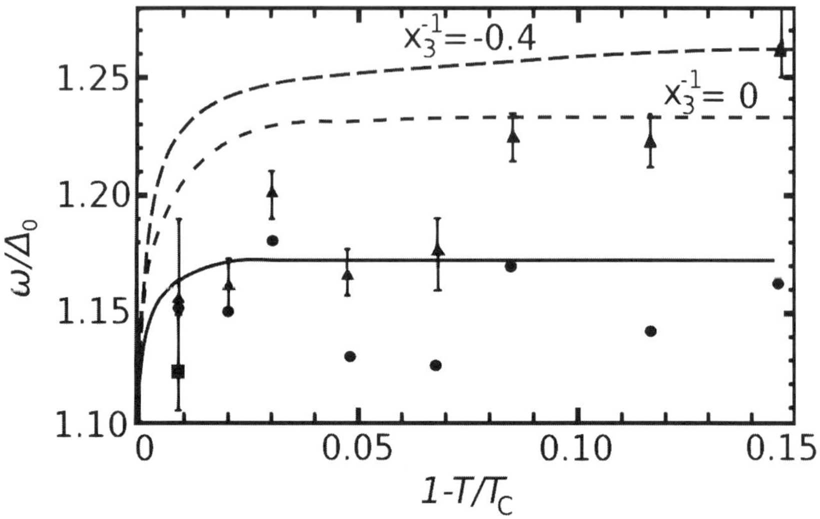

FIG. 16.1. The normalized clapping–mode resonances[8] for two choices of A=2.03 (triangles) and 2.64 (circles) [$A = \Delta_0(0)/k_B T_C$]. Two upper curves follow from kinetic equation theory[6] by using the formula $E_{cl} = 1.23\Delta_0(T)\{1-(0.005-0.106x_3^{-1}-0.052F_2^2)[\Delta_0(T)/k_B T_C]\}$ at $x_3^{-1}=0$ and $x_3^{-1}=-0.4$ respectively. Solid curve is the result of Ref. 9.

Combescot[7] pointed out that for T=0 the small particle – hole asymmetric terms change the dispersion low of orbital mode from linear to quadratic in the low frequency regime $E << T_C(T_C/T_F)$.

Brusov and Popov[11] have investigated the stability of Goldstone–modes with respect to decay into several Bose–excitations. They have shown that stability of Goldstone–modes depends on the angle θ between the excitation momentum $\boldsymbol{k}$ and $\boldsymbol{l}$: the mode is stable when its momentum $\boldsymbol{k}$ lies inside some cones and unstable when outside them.

16.2.1. The model of superfluid ^{3}He

We remind shortly the model of superfluid ^{3}He, which will be used for calculation of the dispersion corrections to the collective mode spectrum in axial– and planar–phases of superfluid ^{3}He. In the method of functional integration, the initial Fermi–system (^{3}He) is described by anticommuting functions $\chi_s(\mathbf{x},\tau), \overline{\chi}_s(\mathbf{x},\tau)$ defined in the volume $V = L^3$, which are antiperiodic in time τ with a period $\beta = T^{-1}$. Here s is the spin index. These functions can be expanded into a Fourier series

$$\chi_s(x) = (\beta V)^{-1/2} \sum_p a_s(p) \exp(i(\omega\tau + \mathbf{k}\cdot\mathbf{x})), \tag{16.1}$$

where $p = (\mathbf{k},\omega)$; $\omega = (2n+1)\pi T$ are Fermi–frequencies and $x = (\mathbf{x},\tau)$. Let us consider the functional of action for an interacting Fermi–system

$$S = \int_0^\beta d\tau d^3x \sum_s \overline{\chi}_s(\mathbf{x},\tau)\partial_\tau \chi_s(\mathbf{x},\tau) - \int_0^\beta H'(\tau)d\tau, \tag{16.2}$$

which corresponds to the Hamiltonian

$$H'(\tau) = \int d^3x \sum_s (2m)^{-1} \nabla\overline{\chi}_s(\mathbf{x},\tau)\nabla\chi_s(\mathbf{x},\tau) - (\lambda + s\mu_0 H)\overline{\chi}_s(\mathbf{x},\tau)\chi_s(\mathbf{x},\tau) +$$

$$+\frac{1}{2}\int d^3x d^3y U(\mathbf{x}-\mathbf{y}) \sum_{ss'} \overline{\chi}_s(\mathbf{x},\tau)\overline{\chi}_{s'}(\mathbf{y},\tau)\chi_{s'}(\mathbf{y},\tau)\chi_s(\mathbf{x},\tau). \tag{16.3}$$

In order to obtain the effective functional of action, we shall use the method of division of Fermi–fields into "fast" and "slow" fields with subsequent successive integration over these fields. Fast fields χ_{1s} and

$\bar{\chi}_{1s}$ are determined by components of expansion (16.1) either with frequencies $|\omega| > \omega_0$, or with momenta $|k - k_F| > k_0$. The remaining components $\chi_{0s} = \chi_s - \chi_{1s}$ of the Fourier expansion define slow fields χ_{0s}. Integrating over fast fields, we insert into the integral over the Fermi–fields, a Gaussian integral over the complex function $c_{ia}(x,\tau)$ and $\bar{c}_{ia}(x,\tau)$ with the vector index i and the isotopic index a (i, a = 1, 2, 3)

$$\int D\bar{c}_{ia}(x,\tau) Dc_{ia}(x,\tau) \exp\left(g_0^{-1} \sum_{p,i,a} \bar{c}_{ia}(p) c_{ia}(p) \right) \tag{16.4}$$

We then shift the Bose–fields by a quadratic form of the Fermi–fields so as to annihilate the quartic form in the Fermi–fields. After integrating over slow fields we arrive at the "effective" or "hydrodynamic action" functional

$$S_{\text{eff}} = g_0^{-1} \sum_{p,i,a} c_{ia}^{+}(p) c_{ia}(p) + \frac{1}{2} \ln \det \frac{\hat{M}\left(c_{ia}, c_{ia}^{+}\right)}{\hat{M}\left(c_{ia}^{(0)}, c_{ia}^{(0)+}\right)}. \tag{16.5}$$

It defines the point of the phase transition of the initial Fermi–system as a point of the Bose–condensation of the fields c and $\bar{c}$, and S_{eff} determines also the density of the condensate at $T < T_C$ and the spectrum of the collective excitations.

The fourth–order matrix $\hat{M}(p_1, p_2)$ has the following elements $M_{ab}(p_1, p_2)$:

$$M_{11} = Z^{-1}\left[i\omega - \xi + \mu(\mathbf{H}\boldsymbol{\sigma})\right]\delta_{p_1 p_2},$$

$$M_{22} = Z^{-1}\left[-i\omega + \xi + \mu(\mathbf{H}\boldsymbol{\sigma})\right]\delta_{p_1 p_2},$$

$$M_{12} = (\beta V)^{-1/2}(n_{1i} - n_{2i})c_{ia}(p_1 + p_2)\sigma_a,$$

$$M_{21} = -(\beta V)^{-1/2}(n_{1i} - n_{2i})c^+_{ia}(p_1 + p_2)\sigma_a, \tag{16.6}$$

where

$$\xi = c_F(k - k_F), n_i = k_i/k_F, \beta = T^{-1} \tag{16.7}$$

H is the magnetic field and μ is the magnetic moment of the quasi–particle, V–volume, σ_a (a=1,2,3) are 2 x 2 Pauli–matrixes, and $\omega = (2n+1)\pi T$ are the Fermi frequencies.

In the first approximation the quadratic form of action functional describes the collective excitations in A–phase in case of nonzero momenta as well. We use it to calculate the dispersion corrections.

16.2.2. The equations for the collective mode spectrum in an arbitrary magnetic field and at arbitrary collective mode momenta

We will now begin the investigation of the spectrum of collective excitations. To make this in the region $T_C - T \propto T_C$ one should expand the functional $\ln\det$ in powers of the deviations $c_{ia}(p)$ from the condensate value $c^{(0)}_{ia}(p)$, which is different for different phases and in A–phase has a form

$$c^{(0)}_{ia}(p) = (\beta V)^{1/2} c\delta_{p0}(\delta_{i1} + i\delta_{i2})\delta_{a1} \tag{16.8}$$

So, we apply the shift

$$c_{ia}(p) \to c^{(0)}_{ia}(p) + c_{ia}(p) \tag{16.9}$$

and separate the quadratic form

$$\sum_p A_{ijab}(p)c_{ia}^*(p)c_{jb}(p) + \frac{1}{2}\sum_p B_{ijab}(p)\left[c_{ia}(p)c_{jb}(-p) + c_{ia}^*(p)c_{jb}^*(-p)\right] \tag{16.10}$$

from S_{eff}. After calculation of this quadratic form we obtain the following result[2] (here, $u_{ia} = \mathrm{Re}\, c_{ia}, v_{ia} = \mathrm{Im}\, c_{ia}$)

$$\left[g_0^{-1} + \frac{2Z^2}{\beta V}\sum_{p1+p2=p} D_3 \cos^2\theta\right]\left(u_{33}^2 + v_{33}^2\right) +$$

$$+\left[g_0^{-1} + \frac{2Z^2}{\beta V}\sum_{p1+p2=p}(D_1 + D_2)\cos^2\theta\right]\left(\left(u_{31} + v_{32}\right)^2 + \left(u_{32} + v_{31}\right)^2\right) +$$

$$+\left[g_0^{-1} + \frac{2Z^2}{\beta V}\sum_{p1+p2=p}(D_1 - D_2)\cos^2\theta\right]\left(\left(u_{31} - v_{32}\right)^2 + \left(u_{32} - v_{31}\right)^2\right) +$$

$$+\left[g_0^{-1} + \frac{Z^2}{\beta V}\sum_{p1+p2=p} D_3 \sin^2\theta\right]\left(\left(v_{31} + u_{32}\right)^2 + \left(u_{13} - v_{23}\right)^2\right) +$$

$$\left[g_0^{-1} + \frac{Z^2}{\beta V}\sum_{p1+p2=p}(-\Delta^2 \sin^2\theta\partial_3 + D_3)\sin^2\theta\right]\left(u_{13} + v_{23}\right)^2 +$$

$$+\left[g_0^{-1} + \frac{Z^2}{\beta V}\sum_{p1+p2=p}(\Delta^2 \sin^2\theta\partial_3 + D_3)\sin^2\theta\right]\left(v_{13} - u_{23}\right)^2 +$$

$$+\left[g_0^{-1} + \frac{Z^2}{\beta V}\sum_{p1+p2=p}(D_1 + D_2)\sin^2\theta\right]\times$$

$$\times\left(\left(u_{12} + v_{11} + u_{21} + v_{21}\right)^2 + \left(u_{11} + v_{12} - u_{22} - v_{21}\right)^2\right) +$$

$$+\left[g_0^{-1} + \frac{Z^2}{\beta V}\sum_{p1+p2=p}(D_1 - D_2)\sin^2\theta\right]\times$$

$$\times\left(\left(u_{12} - v_{11} - u_{21} + v_{21}\right)^2 + \left(u_{11} - v_{12} + u_{22} - v_{21}\right)^2\right) +$$

$$+\left[g_0^{-1} + \frac{Z^2}{\beta V}\sum_{p1+p2=p}\sin^2\theta(D_1 + D_2 - \Delta^2\sin^2\theta(\partial_1 + \partial_2))\right]\times$$

$$\times\left(u_{12} + v_{11} - u_{21} - v_{21}\right)^2 +$$

$$
\begin{aligned}
&+\left[g_0^{-1}+\frac{Z^2}{\beta V}\sum_{p1+p2=p}\sin^2\theta(D_1-D_2-\Delta^2\sin^2\theta(\partial_1-\partial_2))\right]\times \\
&\times\left(u_{11}-v_{11}+u_{21}-u_{22}\right)^2+ \\
&+\left[g_0^{-1}+\frac{Z^2}{\beta V}\sum_{p1+p2=p}\sin^2\theta(D_1+D_2+\Delta^2\sin^2\theta(\partial_1+\partial_2))\right]\times \qquad (16.11) \\
&\times\left(u_{11}+v_{12}+u_{22}+v_{21}\right)^2+ \\
&+\left[g_0^{-1}+\frac{Z^2}{\beta V}\sum_{p1+p2=p}\sin^2\theta(D_1-D_2+\Delta^2\sin^2\theta(\partial_1-\partial_2))\right]\times \\
&\left(u_{11}-v_{12}-u_{22}+v_{21}\right)^2.
\end{aligned}
$$

Here,

$$
\begin{aligned}
&a=Z^{-1}(i\omega-\xi),\ b=Z^{-1}\mu H,\ q_1=Z^{-1}\Delta(n_1+in_2)\alpha_+, \\
&q_2=iZ^{-1}\Delta(n_1+in_2)\alpha_-, d_{1,2}=Z^{-2}\left(\omega^2+(\xi\pm\mu H)^2+\Delta^2\sin^2\theta(\alpha_+\pm\alpha_-)^2\right),
\end{aligned}
$$

$$
\begin{aligned}
&\partial_{1,2}=\frac{(\alpha_++\alpha_-)^2}{d_1(1)d_1(2)}\pm\frac{(\alpha_+-\alpha_-)^2}{d_2(1)d_2(2)}, \\
&D_{1,2}=\frac{(a^+(1)+b)(a^+(2)+b)}{d_1(1)d_1(2)}\pm\frac{(a^+(1)-b)(a^+(2)-b)}{d_2(1)d_2(2)}, \\
&\partial_3=(\alpha_+^2+\alpha_-^2)\left(\frac{1}{d_1(2)d_2(1)}+\frac{1}{d_1(1)d_2(2)}\right), \\
&D_3=\frac{(a^+(1)+b)(a^+(2)-b)}{d_1(1)d_2(2)}\pm\frac{(a^+(1)-b)(a^+(2)+b)}{d_2(1)d_1(2)}.
\end{aligned}
$$

The equation det$Q$=0, where Q is the matrix of quadratic form (16.11), gives us 18 equations which completely determine the 18 collective modes in ^{3}He–A in an arbitrary magnetic field and at arbitrary collective excitation momenta.

16.2.3. The dispersion corrections to collective mode spectrum in ^{3}He–A

Considering the limit $T \to 0$, we replace the summation in equation det$Q$=0, by the integration near the Fermi–sphere, in accordance with the rule

$$(\beta V)^{-1} \sum_{p1} \to k_F^2 (2\pi)^{-4} c_F^{-1} \int d\omega_1 d\xi_1 d\Omega_1 , \qquad (16.12)$$

where $\int d\Omega_1$ is an integral with respect to the angle variables. We then calculate directly the integrals with respect to ω_1 and ξ_1. Putting H=0 and considering small momenta of collective modes, we obtain the whole collective mode spectrum in A–phase of superfluid ^{3}He with dispersion corrections.

It consists of nine ***Goldstone modes***, among them – three sound modes

$$E = c_F k / \sqrt{3} , \qquad v_{13} - u_{23}, u_{11} - v_{21}, u_{22} - v_{12} ,$$

and six orbital waves

$$E = c_F k_{||} , \qquad u_{33}, v_{33}, u_{32}, v_{32}, u_{31}, v_{31} ; \qquad (16.13)$$

six ***clapping modes***

$$u_{11} + v_{21}, u_{22} + v_{12}, u_{23} + v_{13} ,$$

$$E_{cl}^2 = \Delta_0^2 (1.17 - i0.13)^2 + c_F^2 \left[k_{||}^2 (0.159 - i0.090) + k_\perp^2 (0.284 - i0.061) \right]$$

$v_{11}+u_{21}, v_{22}-u_{12}$,

$$E_{cl}^2=\Delta_0^2(1.17-i0.13)^2+c_F^2\left[k_{||}^2(0.159-i0.090)+k_\perp^2(0.557-i0.029)\right]$$

$u_{13}-v_{23}$

$$E_{cl}^2=\Delta_0^2(1.17-i0.13)^2+c_F^2\left[k_{||}^2(0.159-i0.090)+k_\perp^2(0.694-i0.013)\right] \tag{16.14}$$

Three ***pairbreaking modes***

$u_{13}+v_{23}, u_{12}-v_{11}, u_{21}+v_{22}$,

$$E_{pb}^2=\Delta_0^2(1.96-i0.31)^2+c_F^2\left[k_{||}^2(0.096-i0.0004)+k_\perp^2(0.898-i0.509)\right] \tag{16.15}$$

Here, $k_\perp^2=k_1^2+k_2^2$, $k_{||}^2=k_3^2=(\boldsymbol{k},\boldsymbol{l})^2$.

We have obtained the dispersion corrections for whole collective mode spectrum for arbitrary direction of collective excitation momentum. As we expected, the dispersion corrections (except the ones for Goldstone–modes) turn out to be complex as well as the frequencies (energies) of collective excitations. This is related to the fact that the excitation with nonzero energy and small momentum $\boldsymbol{k}$ can decay into initial fermions (see Fig.16.2).

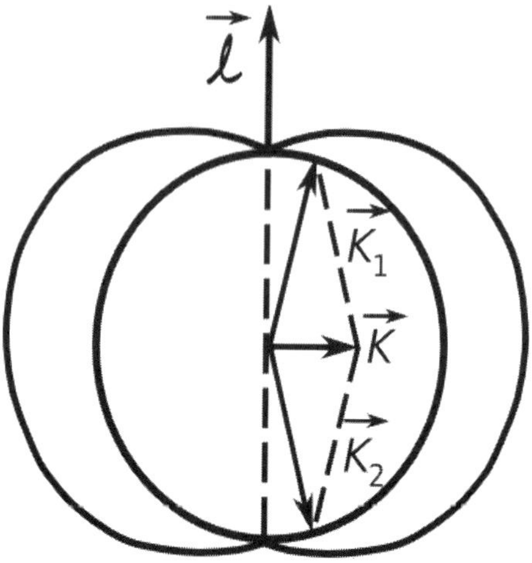

FIG. 16.2. The decay of the phonon (collective excitation) with momentum $\boldsymbol{k}$ into initial fermions with momenta $\boldsymbol{k}_1$ and $\boldsymbol{k}_2$, the directions of which are close to axis of anisotropy $\boldsymbol{l}$ ($\boldsymbol{H}$) in anisotropic A– (2D–) phase.

It is interesting to note that for case of $\boldsymbol{k} \| \boldsymbol{l}$ (which has been considered by Wolfle[6]) the clapping–mode remains fully degenerated while for other directions of collective excitation momentum, a threefold splitting of clapping–mode takes place. Note, that the value of splitting increases with $k_\perp^2$ and reaches maximum at $\boldsymbol{k} = \boldsymbol{k}_\perp$ or at $\boldsymbol{k} \perp \boldsymbol{l}$. The maximal ratio of the splitting of the clapping mode is equal to $\Delta\alpha_1 c_F^2 k^2 / \Delta\alpha_2 c_F^2 k^2 = 0.117 : 0.059 \approx 2:1$.

16.3. Planar Phase

The superfluid phases of ^{3}He, in addition to the isotropic B–phase, the anisotropic A–phase and the A_1–phase, also include a $2D$–phase[12,13], known as planar and having the order parameter

$$c_{ia}^{(0)}(p) = c(\beta V)^{1/2} \delta_{p0} \left(\delta_{i1}\delta_{a1} + \delta_{i2}\delta_{a2} \right) \tag{16.16}$$

or in a matrix form

$$c_{ia}^{(0)}(p) = c(\beta V)^{1/2}\delta_{p0}\begin{pmatrix} 1 & 0 & 0 \\ 0 & 1 & 0 \\ 0 & 0 & 0 \end{pmatrix}. \qquad (16.17)$$

Contrary to the *A*–, *B*– and A_1–phases 2*D*–phase has not yet been observed, but its existence under certain conditions was deduced by many researchers. In particular, Popov et al.[12] predicted a phase transition from the *B*– to the 2*D*–phase at $H = H_C$ and proved the stability of the 2*D*–phase to small perturbations for $H > H_C$. Fujita et al.[14] by considering the *B*–phase in semibounded space have shown that a 2*D*–phase is realized on the boundary: in this situation it is energetically more favored than the *A*–phase. One of the possible explanations of the double splitting of the squashing mode in the *B*–phase, observed experimentally,[15] was an assumed existence on the sell boundary of a 2*D*–phase, one of the collective modes of which leads to the appearance of a second peak in ultrasound absorption (Brusov's et al. recent study[2], however, has shown that not 2*D*–phase, but a deformed *B*–phase is realized in the vicinity of the boundary, see Ref. 2 for details).

These examples suffice to understand the importance of investigation of the planar 2*D*–phase and particularly the spectrum of its collective excitations. Below we calculate this spectrum by path integral technique.

16.3.1. Stability of 2*D*–phase

All properties of superfluid ^{3}He are determined by the functional of the hydrodynamical action, S_h, given by

$$S_h = g^{-1}\sum_{p,i,a} c_{ia}^{+}(p)c_{ia}(p) + \frac{1}{2}\ln\det\frac{\hat{M}\left(c_{ia}, c_{ia}^{+}\right)}{\hat{M}\left(c_{ia}^{(0)}, c_{ia}^{(0)+}\right)} \qquad (16.18)$$

Here, $c_{ia}(p)$ is the Fourier transform of the Bose–field $c_{ia}(\boldsymbol{x},\tau)$ describing the Cooper pairs of the quasi–fermions on the Fermi–surface, the operator $\hat{M}$ is given by

$$\hat{M}=\begin{pmatrix} Z^{-1}(i\omega-\xi+\mu H\sigma_3)\delta_{p1p2}; & (\beta V)^{-1/2}(n_{1i}-n_{2i})\times \\ & \times\sigma_a c_{ia}(p_1+p_2)\delta_{p1+p2,0}; \\ -(\beta V)^{-1/2}(n_{1i}-n_{2i})\times & Z^{-1}(-i\omega+\xi+\mu H\sigma_3)\delta_{p1p2} \\ \times\sigma_a c_{ia}(p_1+p_2)\delta_{p1+p2,0} & \end{pmatrix}, \tag{16.19}$$

where $\xi=c_F(k-k_F), n_i=k_i/k_F$, H is the magnetic field and μ is the magnetic moment of the quasi–particle, σ_a (a=1,2,3) are 2x2 Pauli–matrixes, and $\omega=(2n+1)\pi T$ are the Fermi–frequencies. Expanding the functional (16.18) in the Ginsburg–Landau region $T_C-T\propto T_C$ into powers of the fields c and c^+ we obtain

$$S_h=-\frac{20k_F^2(\Delta T)^2\beta V}{21\xi(3)c_F}\Pi, \tag{16.20}$$

where

$$\Pi=-Tr(AA^+)+\nu Tr(AA^+P)+(TrAA^+)^2+Tr(AA^+AA^+)+$$
$$+Tr(AA^+A^*A^T)-Tr(AA^TA^*A^+)-\frac{1}{2}Tr(AA^T)Tr(A^+A^*). \tag{16.21}$$

Here,

$$\nu=7\xi(3)\mu^2H^2/4\pi^2T_C\,\Delta T,$$

P is the projector on the third axis along which the field is directed. Minimizing Π, we obtain the matrix A that determines the condensate density. The equation $\delta\Pi = 0$ or

$$-A + \nu AP + 2(trA^{+}A)A + 2AA^{+}A + 2A^{*}A^{T}A - 2AA^{T}A^{*} - A^{*}trAA^{T} = 0 \tag{16.22}$$

has several nontrivial solutions corresponding to the superfluid phases. One of them has an order parameter

$$\frac{1}{2}\begin{pmatrix} 1 & 0 & 0 \\ 0 & 1 & 0 \\ 0 & 0 & 0 \end{pmatrix}. \tag{16.23}$$

This is in fact the planar $2D$–phase. Calculations of the second variation $\delta^2\Pi$ yield

$$\delta^2 F_2 = (\nu - 1/2)u_{33}^2 + (\nu + 1/2)v_{33}^2 + \nu(u_{13}^2 + u_{23}^2) + (\nu + 2)(v_{13}^2 + v_{23}^2) +$$
$$(1/2)[3u_{11}^2 + 3u_{22}^2 + 2u_{11}u_{22} + (u_{12} + u_{12})^2] +$$
$$+ (1/2)[3v_{12}^2 + 3v_{21}^2 - 2v_{12}v_{21} + (v_{11} - v_{22})^2]. \tag{16.24}$$

Here, $u_{ia} = \mathrm{Re}\, c_{ia}$, $v_{ia} = \mathrm{Im}\, c_{iu}$.

For $\nu < 1/2$, the second variation $\delta^2\Pi$ is of an alternating sign, while for $\nu > 1/2$ it is non–negative. This means that $2D$–phase is stable in a magnetic field $H > H_C = [\pi\mu 2T_C \Delta T / 7\xi(3)]^{1/2}$. As indicated above, at $H = H_C$ a phase transition takes place from the B– to the $2D$–phase.

16.3.2. The equations for collective mode spectrum in ^{3}He–2*D*

To calculate the collective mode spectrum in ^{3}He–2*D* at $T \to 0$, the functional S_h must be expanded in terms of fluctuations of the fields $c_{ia}(p)$. Making in S_h a shift $c_{ia}(p) \to c_{ia}(p) + c_{ia}^{(0)}(p)$ and providing the calculations which are similar to the ones for the axial phase, we obtain from the equation det *Q*=0 the following equations for the collective mode spectrum:[13]

$$\text{pb:} \quad \int_0^1 (1-x^2) I(c)(1+4c)dx = 0, \quad u_{11}+u_{22}, \quad v_{12}-v_{21}; \tag{16.25}$$

$$\text{cl:} \quad \int_0^1 (1-x^2) I(c)(1+2c)dx = 0, \; v_{11}-v_{22}, u_{12}+u_{21},$$

$$u_{11}-u_{22}, v_{12}+v_{21}; \tag{16.26}$$

$$\text{Gd:} \quad \int_0^1 (1-x^2) I(c)dx = 0, \quad u_{12}-u_{21}, v_{11}+v_{22} \tag{16.27}$$

$$\text{Gd:} \quad \int_0^1 \left[(1+2c)I(c)-1\right]x^2 dx = 0, \; u_{31}, u_{32}, v_{31}, v_{32} \tag{16.28}$$

$$\int_0^1 x^2 \left[(1+4c_+)I(c_+) + (1+4c_-)I(c_-) - 2\right]dx = 0, \quad u_{33}; \tag{16.29}$$

$$\int_0^1 x^2 \left[I(c_+) + I(c_-) - 2\right]dx = 0, \quad v_{33}; \tag{16.30}$$

qGd:

$$\int_0^1 (1-x^2)[(1+4c_+)I(c_+)+(1+4c_-)I(c_-)]dx = 0, \quad u_{13}, u_{23}; \qquad (16.31)$$

qpb: $$\int_0^1 (1-x^2)[I(c_+)+I(c_-)]dx = 0, \quad v_{13}, v_{23}. \qquad (16.32)$$

Here,

$$I(c) = \frac{1}{(1+4c)^{1/2}} \ln \frac{(1+4c)^{1/2}+1}{(1+4c)^{1/2}-1},$$
$$c_{\pm} = \frac{\Delta_0^2(1-x^2)}{\omega^2+[c_F(\boldsymbol{n},\boldsymbol{k})\pm 2\mu H]^2}, c = \frac{\Delta_0^2(1-x^2)}{\omega^2+c_F^2(\boldsymbol{n},\boldsymbol{k})^2}, \qquad (16.33)$$

$u_{ia} = \operatorname{Re} c_{ia}, \; v_{ia} = \operatorname{Im} c_{ia}$.

Let us examine (16.25)–(16.32) at zero momenta of the collective excitations ($\boldsymbol{k}$=0). In this case (16.25)–(16.27) coincide with those obtained by Brusov and Popov[16] for the A–phase without the magnetic field, while (16.30)–(16.33) go over into aforementioned Brusov–Popov equations for an A–phase without the magnetic field, following the substitution $\omega^2+4\mu^2H^2 \to \omega^2$. These equations can thus be solved by using the results of Ref. 13. Finding also the roots of (16.28) and (16.29), we obtain the following result of the collective mode spectrum at $\boldsymbol{k}$=0, which are listed in Table 16.1.

Table 16.1[13]. Collective mode spectrum in planar 2D–phase.

No.	Type	Frequency	Variables
6	Goldstone	$E=0$	$u_{31}, u_{32}, v_{31}, v_{32}$, $u_{12}-u_{21}, v_{11}+v_{22}$
4	clapping	$E=(1.17-i\cdot 0.13)\Delta$	$v_{11}-v_{22}, u_{12}+u_{21}$, $u_{11}-u_{22}, v_{12}+v_{21}$
2	pairbreaking	$E=(1.96-i\cdot 0.31)\Delta$	$u_{11}+u_{22}, v_{12}-v_{21}$
2	quasi–Goldstone	$E=2\mu H$	u_{13}, u_{23}
2	quasi–pairbreaking	$E^2=(1.96-i\cdot 0.31)^2\Delta^2+$ $+4\mu^2H^2$	v_{13}, v_{23}
1		$E^2=(0.518)^2\Delta^2+4\mu^2H^2$	u_{33}
1		$E^2=(0.495)^2\Delta^2+4\mu^2H^2$	v_{33}

Thus, the spectrum of a planar $2D$–phase in the magnetic field contains modes similar to those in the A–phase without the magnetic field, as well as a number of new modes. The former consist of six Goldstone–modes, four clapping–modes, and two pairbreaking–modes. Two quasi–Goldstone–modes and two quasi–pairbreaking–modes are obtained from the Goldstone–and pairbreaking–modes, respectively, by substituting $E^2 \to E^2-4\mu^2H^2$. The gap in the quasi–Goldstone–mode spectrum is $\propto 2\mu H$. Finally, we obtained two new modes having no analogs in the *A*–phase. They correspond to the variables u_{33} and v_{33}, are not degenerate, and the difference between their frequencies is small. Interestingly, whereas for the clapping– and pairbreaking–modes there exists in the *A*–phase a linear Zeeman effect (threefold splitting in a magnetic field), the frequencies of this modes in the 2*D*–phase are independent of the magnetic field, while the energies of the quasi–pairbreaking–modes and of the two "new" modes are quadratic in the field. Note also that the energies of all the nonphonon modes, except the two "new" ones, have imaginary parts due, just as in *A*–phase, to the vanishing of a Fermi–spectrum gap in a special direction (that of the magnetic field). The frequencies of all the nonphonon modes of the spectrum turn out to be complex, in view of the possible decay of the collective excitations into the initial fermions (owing to the vanishing of the Fermi–spectrum gap along the field direction). Just as in the *A*– and *B*–phases, collective modes can be excited in the 2*D*–phase in ultrasound and NMR experiments.

Note that notwithstanding some similarity between the spectra of the *A*– and 2*D*–phases, they also have substantial differences, that can possibly help identify the 2*D*–phase. Just as in the latter, there exist some nonphonon modes absent in the *A*–phase (and also in the *B*–phase), and the behavior of the spectrum (and even of the analog modes) in the 2*D*–phase and in the *A*–phase is quite difference: in the *A*–phase we have a linear splitting of the pairbreaking–and clapping–modes, while in the 2*D*–phase one part of the spectrum is independent of the field, whereas the other part has a quadratic field dependence.

Collective mode spectrum in the 2*D*–phase was studied as well by Hirashima et al.[17], who however, considered a 2*D*–phase without the magnetic field. Since the 2*D*–phase is stable only for $H > H_C$, the meaning of their calculations is not clear. Obviously, they could not obtain six collective modes with frequencies dependent on the magnetic field. Comparing nonetheless results of Brusov et al. [13] with those of Ref. 17, we note the following:

1. The main conclusions of both studies, that the $2D$–phase spectrum coincides in part with the A–phase spectrum, but modes, typical of the $2D$–phase, are present and are close to each another.
2. The correspondence between that fraction of the modes which is the same in both phases as in the A–phase spectrum investigation by the kinetic–equation[18] and path–integration–methods. A frequency $\omega_{cl} = 1.23\Delta_0(T)$ was thus obtained in Ref. 18 for the clapping–mode, as against $\omega_{cl} = 1.17\Delta_0(T)$ in Brusov et al.[16] paper, in much better agreement with the experiments in the A–phase (see Ref. 19 an the citations therein). The reason is that Brusov et al.[16] have taken into account the collective mode damping due to decay of Cooper pairs in view of the vanishing of the Fermi–spectrum gap (see Ref. 20 for details).
3. In Ref. 13 one mode was obtained, typical only of the $2D$–phase and having an energy somewhat lower than that of the super–flapping mode at all temperatures. This new mode is due to spin waves with a coupling coefficient $O(k^2)$. It is neither resonant no diffuse. Note, that Brusov et al.[13,16] have obtained not the super–flapping mode, but additional Goldstone–modes whose appearance is due to the presence of latent symmetry. As noted above, Brusov et al.[13] have obtained in a magnetic field two modes, that are indicative only of the $2D$–phase. Their frequencies are close to each other and depend on the field.

16.3.3. The equations for collective mode spectrum in ^{3}He–$2D$ with dispersion corrections

In the previous section we determined the complete collective excitation spectrum in ^{3}He–$2D$ at zero momentum of collective excitations. But knowledge of the collective excitation spectrum at zero momentum of excitations is not enough. First of all in sound experiments, which are used to study the collective excitation spectrum the collective modes are created with nonzero momentum $\boldsymbol{k}$, and for more detailed comparison of the theoretical results with sound experiment data we must take the dispersion corrections into account. On the other hand taking of the dispersion corrections into account can lead to the lift of the degeneracy of the collective modes similarly to the case of ^{3}He–*B*, where the dispersion–induced splitting of collective modes takes place[21] and has been observed experimentally[22]. Below we calculate the dispersion corrections for collective mode spectrum in planar 2*D*–phase.

The quadratic form (1.2) from Ref. 13 describes the collective excitations in 2*D*–phase in case of nonzero momenta as well, we can use it to calculate the dispersion corrections. For this we need to take traces and than to make replacement

$$(\beta V)^{-1}\sum_{p1} \rightarrow k_F^2(2\pi)^{-4}c_F^{-1}\int d\omega_1 d\xi_1 d\Omega_1 = \\ = k_F^2(2\pi)^{-4}c_F^{-1}\int d\omega_1 d\xi_1 \sin\theta d\theta d\varphi. \tag{16.34}$$

After integrating in (1.2) from Ref. 13 with respect to ω_1, ξ_1, we must expend the coefficients of quadratic form in powers of k up to k^2. Then equating det$Q$ to zero, where Q is the matrix of the coefficients of the quadratic form, we get the following equations for the collective mode spectrum

Goldstone modes

$$\int_0^1 (1-x^2)\left[I + \left((B-4A)(1-x^2)\frac{k_\perp^2}{2} + (B-2A)x^2k_\parallel^2\right)\frac{c_F^2}{\omega^2}\right]dx = 0,$$

$v_{11}+v_{22},\ u_{12}-u_{21}$

$$\int_0^1 x^2\left[(1+2c)I-1+\left((B-A)(1-x^2)\frac{k_\perp^2}{2}+Bx^2k_\parallel^2\right)\frac{c_F^2}{\omega^2}\right]dx=0,\quad v_{32},u_{31};$$

$$\int_0^1 x^2\left[(1+2c)I-1+\left((B+A)(1-x^2)\frac{k_\perp^2}{2}+Bx^2k_\parallel^2\right)\frac{c_F^2}{\omega^2}\right]dx=0,\quad u_{32},v_{31}$$

(16.35)

Clapping modes

$$\int_0^1 (1-x^2)\left[(1+2c)I+\left((B-2A)(1-x^2)\frac{k_\perp^2}{2}+Bx^2k_{||}^2\right)\frac{c_F^2}{\omega^2}\right]dx=0,$$

$u_{11}-u_{22}, v_{12}+v_{21}$

$$\int_0^1 (1-x^2)\left[(1+2c)I+\left((B+2A)(1-x^2)\frac{k_\perp^2}{2}+Bx^2k_{||}^2\right)\frac{c_F^2}{\omega^2}\right]dx=0,$$

$v_{11}-v_{22}, u_{12}+u_{21}$

(16.36)

Pairbreaking modes

$$\int_0^1 (1-x^2)\left[(1+4c)I+\left((B+4A)(1-x^2)\frac{k_\perp^2}{2}+(B+2A)x^2k_{||}^2\right)\frac{c_F^2}{\omega^2}\right]dx=0,$$

$u_{11}+u_{22}, v_{12}-v_{21}$

(16.37)

Quasi-Goldstone modes

$$\int_0^1 \left(1-x^2\right)\left[I_H + F_2\left(\left(1-x^2\right)\frac{3k_\perp^2}{4} + x^2k_\parallel^2\right)\frac{c_F^2}{\omega^2}\right]dx = 0, \quad u_{13};$$

$$\int_0^1 \left(1-x^2\right)\left[I_H + F_2\left(\left(1-x^2\right)\frac{k_\perp^2}{4} + x^2k_{||}^2\right)\frac{c_F^2}{\omega^2}\right]dx = 0, \quad u_{23}. \qquad (16.38)$$

Quasi-Pairbreaking modes

$$\int_0^1 \left(1-x^2\right)\left[(1+4c_H)I_H + F_1\left(\left(1-x^2\right)\frac{3k_\perp^2}{4} + x^2k_{||}^2\right)\frac{c_F^2}{\omega^2}\right]dx = 0, \quad v_{13};$$

$$\int_0^1 \left(1-x^2\right)\left[(1+4c_H)I_H + F_1\left(\left(1-x^2\right)\frac{k_\perp^2}{4} + x^2k_{||}^2\right)\frac{c_F^2}{\omega^2}\right]dx = 0, \quad v_{23}.$$

(16.39)

Mode $E^2 = (0.518)^2\Delta^2 + 4\mu^2H^2$

$$\int_0^1 x^2\left[I_H - 1 + F_2\left(\left(1-x^2\right)\frac{k_\perp^2}{2} + x^2k_{||}^2\right)\frac{c_F^2}{\omega^2}\right]dx = 0, \quad u_{33}.$$

(16.40)

Mode $E^2 = (0.495)^2\Delta^2 + 4\mu^2H^2$

$$\int_0^1 x^2\left[(1+4c_H)I_H - 1 + F_1\left(\left(1-x^2\right)\frac{k_\perp^2}{2} + x^2k_{||}^2\right)\frac{c_F^2}{\omega^2}\right]dx = 0, \quad v_{33}.$$

(16.41)

Here,

$$k_{\perp}^{2}=k_{1}^{2}+k_{2}^{2},\ k_{||}^{2}=k_{3}^{2},\ \omega_{H}^{2}=\omega^{2}+(2\mu H)^{2},$$
$$c=\Delta_{0}\left(1-x^{2}\right)/\omega^{2},\ c_{H}=\Delta_{0}\left(1-x^{2}\right)/\omega_{H}^{2}, \tag{16.42}$$
$$I=I(c)=\frac{1}{(1+4c)^{1/2}}\ln\frac{(1+4c)^{1/2}+1}{(1+4c)^{1/2}-1},\ I_{H}=I(c_{H}),$$
$$u_{ia}=\operatorname{Re}c_{ia},\ v_{ia}=\operatorname{Im}c_{ia},$$

$$A=c(1-I(1+2c))/(1+4c),\ A=c\left(1+2c-4c^{2}I\right)/(1+4c),$$
$$F_{1}=\frac{(2\mu H)^{2}}{\omega_{H}^{2}}\left(-2-20c_{H}-48c_{H}^{2}+I_{H}8c_{H}\left(1+7c_{H}+7c_{H}^{2}\right)\right)/\left(1+4c_{H}\right)^{2}+$$
$$+B+2A,$$
$$F_{2}=\frac{(2\mu H)^{2}}{\omega_{H}^{2}}\left(-2-4c_{H}-I_{H}8c_{H}\left(1+c_{H}\right)\right)/\left(1+4c_{H}\right)^{2}+B-2A. \tag{16.43}$$

For each equation we have pointed out the corresponding variables.

16.3.4. The collective mode spectrum in ^{3}He–2*D* with dispersion corrections

Solving (16.35)–(16.41) we obtain the 18 collective modes in planar 2*D*–phase with dispersion corrections

1. *Goldstone–modes*

The spectrum of Goldstone–modes consists from two sound modes with $E(k)=\frac{c_F k}{\sqrt{3}}$ and from four orbital waves with $E(k)=c_F k_{\|}$. For latter ones there is a twofold splitting under taking into account the next terms of the expansion[20]

$$E_1(k)=\frac{c_F k}{\sqrt{3}}\left(1-\frac{\sin^2\theta\cos^2\theta}{2\left(\cos^2\theta-\frac{1}{3}\right)\ln\frac{4\Delta_0^2}{f_1(\theta,k)}}\right),$$

$$E_2(k)=c_F k_\parallel\left(1-\frac{11\cos^2\theta-3}{24\cos^2\theta\ln\frac{4\Delta_0^2}{f_2(\theta,k)}}\right), \tag{16.44}$$

$$E_3(k)=c_F k_\parallel\left(1-\frac{51\cos^4\theta-40\cos^2\theta+5}{72\cos^2\theta\left(\cos^2\theta-\frac{1}{3}\right)\ln\frac{4\Delta_0^2}{f_3(\theta,k)}}\right),$$

where

$$f_1(\theta,k)=c_F^2k^2\left(\cos^2\theta-\frac{1}{3}\right),\qquad f_2(\theta,k)=\frac{1}{12}c_F^2k^2\left(11\cos^2\theta-3\right),$$

$$f_3(\theta,k)=\frac{1}{36}c_F^2k^2\frac{51\cos^4\theta-40\cos^2\theta+5}{\cos^2\theta\left(\cos^2\theta-\frac{1}{3}\right)}. \tag{16.45}$$

The obtained equations show that the stability of the spectrum in the *2D*–phase depends on the angle θ between the excitation momentum and the preferred direction. The first (acoustic) mode is stable inside the cones $\cos^2\theta>1/3$, and the second inside the cones $\cos^2\theta>3/11$. The third mode is stable in the regions

$$\cos^2\theta > \frac{20+\sqrt{145}}{51}, \quad \frac{1}{3} > \cos^2\theta > \frac{20-\sqrt{145}}{51}. \tag{16.46}$$

Outside the stability regions, the energy of the excitation become complex because of the imaginary parts of the logarithms in (16.44). Physically this is related to the possibility of the decay of the excitation into constituted fermions whose momenta are close to the preferred direction (see Fig. 16.1).

Note, that in the considered approximation (neglecting the coupling between the Goldstone– and honphonon–modes) the Goldstone–modes of the spectrum in the A– and $2D$–phases coincide and differ only in the degeneracy multiplicity (3 in the A–phase and 2 in the $2D$–phase).

2. *Clapping modes*

$$E_{cl} = \Delta_0\left\{(1.17 - i0.13) + \frac{c_F^2}{\Delta_0^2}\left[k_{||}^2(0.14 - i0.06) + k_\perp^2(0.26 + i0.05)\right]\right\}$$

$$u_{11} - u_{22}, v_{12} + v_{21},$$

$$E_{cl} = \Delta_0\left\{(1.17 - i0.13) + \frac{c_F^2}{\Delta_0^2}\left[k_{||}^2(0.14 - i0.06) + k_\perp^2(0.23 + i0.05)\right]\right\}$$

$$v_{11} - v_{22}, u_{12} + u_{21}. \tag{16.47}$$

3. *Pairbreaking–modes*

$$E_{pb} = \Delta_0\left\{(1.96 - i0.31) + \frac{c_F^2}{\Delta_0^2}\left[k_{||}^2(0.15 + i0.02) + k_\perp^2(0.93 + i0.90)\right]\right\}$$

$$u_{11} + u_{22}, v_{12} - v_{21}. \tag{16.48}$$

4. *Quasi–Goldstone–modes*

$$E_{qgd} = 2\mu H + \frac{c_F^2}{\Delta_0}\left[\frac{1}{3}k_{||}^2 + \frac{1}{2}k_{\perp}^2\right], \qquad u_{13},$$

$$E_{qgd} = 2\mu H + \frac{c_F^2}{\Delta_0}\left[\frac{1}{3}k_{||}^2 + \frac{1}{6}k_{\perp}^2\right], \qquad u_{23}.$$

(16.49)

5. *Quasi–pairbreaking–modes*

$$E_{qpb} = \Delta_0\left[(1.96 - i0.31)^2 + \frac{(2\mu H)^2}{\Delta_0^2}\right]^{-1/2}[(1.96 - i0.31)^2 + \frac{(2\mu H)^2}{\Delta_0^2} +$$
$$+ \frac{c_F^2}{\Delta_0^2}\left[k_{||}^2(0.093 + i0.0004) + k_{\perp}^2(0.1009 - i0.0001)\right] +$$
$$+ \frac{(2\mu H)^2 c_F^2}{\Delta_0^4}\left[k_{||}^2(0.2573 - i0.4796) + k_{\perp}^2(0.1884 - i0.6175)\right]]$$

v_{13}, v_{23}.

(16.50)

6. *Mode* $E^2 = (0.518)^2\Delta^2 + 4\mu^2 H^2$

$$E = \Delta_0\left[(0.518)^2 + \frac{(2\mu H)^2}{\Delta_0^2}\right]^{-1/2}[(0.518)^2 + \frac{(2\mu H)^2}{\Delta_0^2} +$$
$$+ \frac{c_F^2}{\Delta_0^2}\left[0.468k_{||}^2 + 0.016k_{\perp}^2\right] + \frac{(2\mu H)^2 c_F^2}{\Delta_0^4}\left[\alpha_{||}k_{||}^2 + \alpha_{\perp}k_{\perp}^2\right]], \quad u_{33}.$$

(16.51)

7. *Mode* $E^2 = (0.495)^2 \Delta^2 + 4\mu^2 H^2$

$$E = \Delta_0 \left[(0.495)^2 + \frac{(2\mu H)^2}{\Delta_0^2} \right]^{-1/2} \left[(0.495)^2 + \frac{(2\mu H)^2}{\Delta_0^2} + \right. \\ \left. + \frac{c_F^2}{\Delta_0^2} \left[0.467 k_{||}^2 + 0.016 k_\perp^2 \right] + \frac{(2\mu H)^2 c_F^2}{\Delta_0^4} \left[\beta_{||} k_{||}^2 + \beta_\perp k_\perp^2 \right] \right], \quad \nu_{33} .$$

(16.52)

Note, that terms with coefficients $\alpha_{||}, \alpha_\perp$ and $\beta_{||}, \beta_\perp$ are of the higher order of smallness, because we suppose that both momentum of the collective mode and magnetic field are small.

In this Chapter we have obtained the whole collective–mode spectrum in ^{3}He–2*D* with dispersion corrections. It is interesting to note, that for clapping–modes the degeneracy depends on the direction of the collective mode momentum ***k*** with respect to the external magnetic field ***H***, namely: the mode degeneracy remains the same as in the case of zero momentum ***k*** for $\boldsymbol{k} \| \boldsymbol{H}$ only. For any other direction different from this one (for example, $\boldsymbol{k} \perp \boldsymbol{H}$) there is a twofold splitting of these modes.

16.4. Conclusion

We have calculated the whole collective mode spectrum in axial– and planar–phases of superfluid ^{3}He with dispersion corrections. The obtained results imply that new interesting features can be observed in ultrasound experiments in axial– and planar–phases: the change of the number of peaks in ultrasound absorption into clapping–mode. One peak, observed for these modes by Ling et al.[8] in axial–phase, will split into two peaks in planar–phase and into three peaks in axial–phase under change of ultrasound direction with respect to the external magnetic field ***H*** in planar–phase and with respect to the vector ***l*** in axial–phase.

In planar–phase, some Goldstone–modes in the magnetic field become massive (quasi–Goldstone) and have similar twofold splitting under the change of ultrasound direction with respect to the external magnetic field ***H***.

The obtained results[26,27] as well will be useful under interpretation of the ultrasound experiments in axial– and planar–phases of superfluid ^{3}He, because knowledge of the dispersion corrections allows to make a more careful comparison between experimental data and theoretical results.

Chapter XVII

Collective Excitations in Polar–Phase of Superfluid ^{3}He

Among the superfluid phases, appearing in phase classification of the superfluid ^{3}He, there is so called polar one[1], having the order parameter

$$c_{ia}^{(0)}(p) = c(\beta V)^{1/2}\,\delta_{p0}\delta_{i3}\delta_{a3}, \tag{17.1}$$

or in a matrix form

$$c_{ia}^{(0)}(p) = c(\beta V)^{1/2}\,\delta_{p0}\begin{pmatrix} 0 & 0 & 0 \\ 0 & 0 & 0 \\ 0 & 0 & 1 \end{pmatrix}. \tag{17.2}$$

The gap Δ in the single–particle spectrum is described by the expression

$$\Delta^2 = \Delta_0^2 \cos^2\theta. \tag{17.3}$$

Via this form of the gap the polar–phase has a line of gap nodes. Note, that this phase is the only phase in superfluid ^{3}He that has a line of gap nodes. This fact leads to very significant consequences. In this Chapter we investigate the influence of such form of the gap on collective mode spectrum. It will be interesting to compare this spectrum with spectrum of isotropic *B*–phase as well as with one of the axial *A*–phase and with planar 2*D*–phase, where gap has points of gap nodes.

It is clear that because of the disappearance of the gap along the equator and the possibility of decay of the collective excitations into initial fermions, the energies of all high–frequency modes (with $E \neq 0$ at $\boldsymbol{k} = 0$) should be complex. The imaginary parts of these energies determine the damping of these collective modes.

17.1. Calculation of the Collective Mode Spectrum

The method of calculation of the collective mode spectrum in the polar–phase is similar to one, used by us in previous Chapters for other phases. In this method all properties of superfluid ^{3}He are determined by the functional S_h of the hydrodynamical action, given by

$$S_h = g^{-1} \sum_{p,i,a} c_{ia}^{+}(p) c_{ia}(p) + \frac{1}{2} \ln \det \frac{\hat{M}\left(c_{ia}, c_{ia}^{+}\right)}{\hat{M}\left(c_{ia}^{(0)}, c_{ia}^{(0)+}\right)} \tag{17.4}$$

Here, $c_{ia}(p)$ is the Fourier transform of the Bose–field $c_{ia}(\boldsymbol{x}, \tau)$ describing the Cooper pairs of the quasifermions on the Fermi–surface, the operator $\hat{M}$ is given by

$$\hat{M} = \begin{pmatrix} Z^{-1}(i\omega - \xi + \mu H \sigma_3)\delta_{p1p2}; & \begin{array}{c} (\beta V)^{-1/2}(n_{1i} - n_{2i}) \times \\ \times \sigma_a c_{ia}(p_1 + p_2)\delta_{p1+p2,0}; \end{array} \\ \begin{array}{c} -(\beta V)^{-1/2}(n_{1i} - n_{2i}) \times \\ \times \sigma_a c_{ia}(p_1 + p_2)\delta_{p1+p2,0} \end{array} & Z^{-1}(-i\omega + \xi + \mu H \sigma_3)\delta_{p1p2} \end{pmatrix} \tag{17.5}$$

where $\xi = c_F(k - k_F), n_i = k_i / k_F$, H is the magnetic field and μ is the magnetic moment of the quasiparticle, σ_a (a=1,2,3) are 2x2 Pauli–matrixes, and $\omega = (2n+1)\pi T$ are the Fermi–frequencies.

At low temperatures $T \to 0$ the functional (17.4) must be expanded in terms of fluctuations of the fields $c_{ia}(p)$ above the condensate values $c_{ia}^{(0)}(p)$. In this temperature regions the Bose–spectrum of the system is determined in the first approximation by the quadratic part of the functional S_h obtained by the shift $c_{ia}(p) \to c_{ia}(p) + c_{ia}^{(0)}(p)$, where for the polar–phase $c_{ia}^{(0)}(p) = c(\beta V)^{1/2}\delta_{p0}\delta_{i3}\delta_{a3}$ and is obtained from the equation $\det Q = 0$, where Q is the matrix of the quadratic form.

To calculate the quadratic part of the functional S_h one needs to put

$$\hat{M}(c_{ia}^{+(0)}, c_{ia}^{(0)}) = \hat{G}^{-1},\ \hat{M}(c_{ia}, c_{ia}^{+}) = \hat{G} + \hat{u} \tag{17.6}$$

and to keep the first two terms (n=1,2) of the expansion

$$\ln\det\frac{\hat{M}\left(c_{ia}, c_{ia}^{+}\right)}{\hat{M}\left(c_{ia}^{+(0)}, c_{ia}^{(0)}\right)} = Sp\ln(I + \hat{G}\hat{u}) = -\sum_{n=1}^{\infty}\frac{1}{2n}Sp\left(\hat{G}\hat{u}\right)^{2n}. \tag{17.7}$$

In case of the polar–phase[2,3]

$$\hat{G}^{-1} = Z^{-1}\begin{pmatrix} (i\omega - \xi)\delta_{p1p2} & \Delta\delta_{p1+p2,0} \\ -\Delta\delta_{p1+p2,0} & (-i\omega + \xi)\delta_{p1p2} \end{pmatrix};$$

$$\hat{u} = u_{p1p2} = \begin{pmatrix} 0 & (n_{1i} - n_{2i})\sigma_a c_{ia}(p_1 + p_2) \\ -(n_{1i} - n_{2i})\sigma_a c_{ia}^{+}(p_1 + p_2) & 0 \end{pmatrix} \tag{17.8}$$

Here, $\Delta = \Delta_0|\cos\theta|, \Delta_0 = 2cZ$.

Finding $\hat{G}$,

$$\hat{G} = \frac{Z}{M}\begin{pmatrix} -(i\omega+\xi)\delta_{p1p2} & \Delta\delta_{p1+p2,0} \\ -\Delta\delta_{p1+p2,0} & (i\omega+\xi)\delta_{p1p2} \end{pmatrix} \tag{17.9}$$

and calculating $-\sum_{n=1}^{\infty}\frac{1}{2n}Sp\left(\hat{G}\hat{u}\right)^2$ we get the quadratic part of the functional S_h

$$\sum_{p,i,a} c_{ia}^{+}(p)c_{ja}(p)\left[\frac{\delta_{ij}}{g}+\frac{4Z^2}{\beta V}\sum_{p1+p2=p} n_{i1}n_{j1}M_1^{-1}M_2^{-1}(i\omega_1+\xi_1)(i\omega_2+\xi_2)\right]-$$
$$-\sum_{p,i,a}\left[c_{ia}^{+}(p)c_{ja}^{+}(-p)+c_{ia}(p)c_{ja}(-p)\right]\frac{2Z^2}{\beta V}\sum_{p1+p2=p}\Delta^2 n_{i1}n_{j1}M_1^{-1}M_2^{-1}, \tag{17.10}$$

Here, $M_i = \omega_i^2 + \xi_i^2 + \Delta^2$.

Taking the limit $T \to 0$, we transform from summing to integrating in the vicinity of the Fermi–surface by the rule

$$(\beta V)^{-1}\sum_{p1} \to k_F^2(2\pi)^{-4}c_F^{-1}\int d\omega_1 d\xi_1 d\Omega_1 . \tag{17.11}$$

After calculation of the coefficients of the quadratic form, one has[2,3]

$$S_h \approx -\frac{Z^2k_F^2}{2\pi^2 c_F}\int_0^1 dx\{\left(1-x^2\right)\left[J\left(1-4\Delta^2/q^2\right)-2\right]\times$$
$$\times\left(u_{13}^2+u_{23}^2+v_{11}^2+v_{21}^2+v_{22}^2+v_{12}^2\right)+ \tag{17.12}$$
$$+\left(1-x^2\right)(J-2)\left(v_{13}^2+v_{23}^2+u_{11}^2+u_{21}^2+u_{22}^2+u_{12}^2\right)+$$
$$+2x^2J\left(1-4\Delta^2/q^2\right)\left(u_{33}^2+v_{31}^2+v_{32}^2\right)+2x^2J\left(v_{33}^2+u_{31}^2+u_{32}^2\right)\}.$$

Here, $J = \frac{1}{b^2}\ln\frac{1+b}{1-b}$,

$$b^2 = 1 + 4\Delta^2/q^2, \ u_{ia} = \mathrm{Re}\,c_{ia}, \ v_{ia} = \mathrm{Im}\,c_{ia}, \ q^2 = \omega^2 + c_F^2(\boldsymbol{k},\boldsymbol{n}_1)^2. \tag{17.13}$$

Equating determinant of the matrix Q of this quadratic form to zero (det Q=0), we obtain the following equations for the collective mode spectrum[2,3]

$$\int_0^1 (1-x^2) J(b)\left[(1-4\Delta^2/q^2)-2\right] dx = 0, \ u_{13}, u_{23}, v_{11}, v_{21}, v_{22}, v_{12};$$

$$\int_0^1 (1-x^2)\left[J(b)-2\right] dx = 0, \ v_{13}, v_{23}, u_{11}, u_{21}, u_{22}, u_{12};$$

$$\int_0^1 x^2 J(b)\left[(1-4\Delta^2/q^2)-2\right] dx = 0, \ u_{33}, v_{31}, v_{32};$$

$$\int_0^1 x^2 J(b) dx = 0, \ v_{33}, u_{31}, u_{32}. \tag{17.14}$$

Putting $\boldsymbol{k}$=0 and solving these equations we get the collective–excitation spectrum[2,3]. We find six modes

$$E = \Delta_0(1.20 - i\cdot 1.75), \ (v_{13}, v_{23}, u_{11}, u_{21}, u_{22}, u_{12}) \tag{17.15}$$

from the second equation and three Goldstone–modes $E = 0, \ (u_{33}, v_{31}, v_{32})$. We did not find the roots for the first and fourth equations.

17.2. Conclusion

In conclusion, let us briefly discuss the obtained results. Six high–frequency–modes $E = \Delta_0(1.20 - i \cdot 1.75)$, corresponding to variables $v_{13}, v_{23}, u_{11}, u_{21}, u_{22}, u_{12}$ have a large enough imaginary part, of the same order as the real one. This is coupled to the fact that in the polar phase, the gap disappears at the equator line which is opposite to the case of the axial– and planar–phases, where the gap has point nodes, and the imaginary parts of the mode energies are small compared with the real parts. In this connection, we recall that the energies of all high–frequency–modes in the isotropic *B*–phase have real energies at zero momenta of the collective excitations. We can compare results by Brusov et al.[1-3], described above to those of Ref. 4. There authors solved the same problems by using the kinetic equation method. They got the same energy $E = 1.20\Delta_0$ for high–frequency–mode but they did not obtain the imaginary part of this mode. This is the usual short coming of the kinetic equation method as compared with the path integration one. The authors of Ref. 4 showed that the energy $E = 1.20\Delta_0(T)$ increases with T reaching $E = 2\Delta_0(T)$ as $T \to T_C$ and this mode becomes high damping at least at $T \approx T_C$. For this temperature region this conclusion is clear from physical reasons without calculations. They also found a few Goldstone–modes. They did not find solutions for some equations and found them only at $T = T_C$, where these modes have energies $E = 2\Delta_0$.

Chapter XVIII

Collective Mode Spectrum in A_1–Phase of Superfluid ^{3}He

18.1. Calculation of the Collective Mode Spectrum

Three superfluid phases have been discovered in ^{3}He: the isotropic B–phase, the anisotropic A–phase and the A_1–phase[1]. The latter phase exists due to the fact that the superconducting transition temperature for fermions with spins oriented along the magnetic field ($T_{C\uparrow}$) and against the field ($T_{C\downarrow}$)[2,3]

$$T_{C\downarrow\uparrow} = \omega_0 \exp\left(-\frac{1}{V_0 N(\varepsilon_F \mp \mu H)} \right) \tag{18.1}$$

are different: $T_{C\downarrow} < T_{C\uparrow}$. For this reason, the particles with spins directed along the field are the first to undergo a superconducting transition in a magnetic field upon decreasing of the temperature at $T = T_{C\uparrow}$, which is followed by a superconducting transition in the particles with spins directed against the field at $T = T_{C\downarrow}$. The A_1–phase exists in the temperature region confined between $T_{C\uparrow}(H)$ and $T_{C\downarrow}(H)$.

The condensate wave function (order parameter) in this phase takes the form[2,3]

$$c_{ia}^{(0)}(p) = c(\beta V)^{1/2}\,\delta_{p0}\left(\delta_{a1} + i\delta_{a2}\right)\left(\delta_{i1} + i\delta_{i2}\right), \tag{18.2}$$

or in a matrix form

$$c_{ia}^{(0)}(p) = c(\beta V)^{1/2}\delta_{p0}\begin{pmatrix}1 & i & 0\\ i & -1 & 0\\ 0 & 0 & 0\end{pmatrix}. \tag{18.3}$$

Below we obtain the complete collective mode spectrum in ^3He–A_1–phase, which is determined in the first approximation by the quadratic part of the functional of effective action, S_h. To obtain the required quadratic form, we expand the functional S_h in powers of the fields $c_{ia}(p)$ after the shift $c_{ia}(p) \to c_{ia}(p) + c_{ia}^{(0)}(p)$:

$$\ln\det\frac{\hat{M}\left(c_{ia}, c_{ia}^{+}\right)}{\hat{M}\left(c_{ia}^{(0)}, c_{ia}^{(0)+}\right)} = \frac{1}{2}Tr\ln(I + \hat{G}\hat{u}) = -\sum_{n=1}^{\infty}\frac{1}{4n}Tr\left(\hat{G}\hat{u}\right)^{2n} \tag{18.4}$$

and omit all the terms except the one with $n = 1$. The expression for $\hat{G}^{-1}$ has the form

$$G^{-1} = $$
$$= \begin{pmatrix} Z^{-1}(i\omega - \xi + \mu H\sigma_3)\delta_{p1p2} & 2c(n_1 + in_2)(\sigma_1 + i\sigma_2)\delta_{p1+p2,0} \\ -2c(n_1 + in_2)(\sigma_1 + i\sigma_2)\delta_{p1+p2,0} & Z^{-1}(-i\omega + \xi + \mu H\sigma_3)\delta_{p1p2} \end{pmatrix} \tag{18.5}$$

After the inversion of $\hat{G}^{-1}$, we obtain[2,3]

$$\hat{G} = \begin{pmatrix} G_{11} & G_{12} \\ G_{21} & G_{22} \end{pmatrix}, \tag{18.6}$$

where

$$G_{11}=\begin{pmatrix}(a^{+}+b)/d_1 & 0\\ 0 & (a^{+}-b)/d_2\end{pmatrix}\delta_{p1p2},$$

$$G_{22}=\begin{pmatrix}(-a^{+}+b)/d_2 & 0\\ 0 & (-a^{+}-b)/d_1\end{pmatrix}\delta_{p1+p2,0},$$

$$G_{12}=\begin{pmatrix}0 & 2g/d_1\\ 0 & 0\end{pmatrix}\delta_{p1+p2,0},\quad G_{12}=\begin{pmatrix}0 & 0\\ -2g/d_1 & 0\end{pmatrix}\delta_{p1p2}. \tag{18.7}$$

Here,

$$a=Z^{-1}(i\omega-\xi)\ ;\ b=Z^{-1}\mu H\ ;\ g=Z^{-1}\Delta(n_1+in_2)\ ;$$

$$d_1=Z^{-2}\left[\omega^2+(\xi-\mu H)^2+4\Delta^2\sin^2\theta\right];\ d_2=Z^{-2}\left[\omega^2+(\xi+\mu H)^2\right] \tag{18.8}$$

and d_1^{-1} and d_2^{-1} are in fact Green's functions of quasi–fermions with spins directed along and against the field, respectively. Using these functions, we can demonstrate once again the presence of the gap $\tilde{\Delta}=2\Delta\sin\theta$ in the one–particle spectrum of quasifermions with spins directed along the field and its absence for the component with opposite spins. Using expression (18.7) and expression

$$u_{p1p2}=\frac{1}{\sqrt{\beta V}}\begin{pmatrix}0 & (n_{1i}-n_{2i})\sigma_a c_{ia}(p_1+p_2)\\ -(n_{1i}-n_{2i})\sigma_a c_{ia}^{+}(p_1+p_2) & 0\end{pmatrix} \tag{18.9}$$

for u_{p1p2}, we obtain the following quadratic form[2,3]

$$S_h = \frac{1}{g}\sum_{p,i,a} c_{ia}^+(p)c_{ia}(p) +$$
$$+\frac{Z}{\beta V}\sum_{p1+p2=p} n_{1i}n_{1j}\{-\Delta^2(n_1+in_2)^2 D_0[c_{i1}^+(p)c_{j1}^+(p) - c_{i2}^+(p)c_{j2}^+(p) +$$
$$+ic_{i1}^+(p)c_{j2}^+(p) + ic_{i2}^+(p)c_{j1}^+(p) + c.c] + D_1[c_{i1}^+(p)c_{j1}(p) + c_{i1}(p)c_{j1}^+(p) +$$
$$+c_{i2}^+(p)c_{j2}(p) + c_{i2}(p)c_{j2}^+(p)] + D_3[c_{i3}^+(p)c_{j3}(p) + c_{i3}(p)c_{j3}^+(p)] +$$
$$+iD_2[c_{i1}^+(p)c_{j2}(p) + c_{i1}(p)c_{j2}^+(p) - c_{i2}^+(p)c_{j1}(p) - c_{i2}(p)c_{j1}^+(p)]\}. \tag{18.10}$$

Here,

$$D_0 = \frac{4}{d_1(p_1)d_1(p_2)},$$
$$D_{1,2} = \frac{(a^+(p_1)+b)(a^+(p_2)+b)}{d_1(p_1)d_1(p_2)} \pm \frac{(a^+(p_1)-b)(a^+(p_2)-b)}{d_2(p_1)d_2(p_2)}, \tag{18.11}$$
$$D_3 = \frac{(a^+(p_1)+b)(a^+(p_2)-b)}{d_1(p_1)d_2(p_2)} + \frac{(a^+(p_1)-b)(a^+(p_2)+b)}{d_2(p_1)d_1(p_2)}.$$

Diagonalizing the quadratic form (18.19), we obtain[2,3]

$$S_h = \left[\frac{1}{g} + \frac{2Z^2}{\beta V}\sum D_3 \cos^2\theta\right]\left(u_{33}^2 + v_{33}^2\right) +$$

$$+\left[\frac{1}{g} + \frac{2Z^2}{\beta V}\sum (D_1 + D_2)\cos^2\theta\right]\left(\left(u_{31} + v_{32}\right)^2 + \left(u_{32} + v_{31}\right)^2\right) +$$

$$+\left[\frac{1}{g} + \frac{2Z^2}{\beta V}\sum (D_1 - D_2)\cos^2\theta\right]\left(\left(u_{31} - v_{32}\right)^2 + \left(u_{32} - v_{31}\right)^2\right) +$$

$$+\left[\frac{1}{g} + \frac{2Z^2}{\beta V}\sum D_3 \sin^2\theta\right]\times$$

$$\times\left(\left(u_{13} - v_{23}\right)^2 + \left(u_{13} + v_{23}\right)^2 + \left(u_{23} - v_{13}\right)^2 + \left(u_{23} + v_{23}\right)^2\right) +$$

$$+\left[\frac{1}{g} + \frac{2Z^2}{\beta V}\sum (D_1 - D_2)\sin^2\theta\right]\left(\left(u_{12} - v_{11} - u_{21} + v_{22}\right)^2 +\right.$$

$$\left.\left(u_{11} - v_{12} + u_{22} - v_{21}\right)^2 + \left(u_{21} - v_{11} + u_{12} - v_{22}\right)^2 + \left(u_{11} - v_{12} - u_{22} + v_{22}\right)^2\right) +$$

$$+\left[\frac{1}{g} + \frac{2Z^2}{\beta V}\sum (D_1 + D_2)\sin^2\theta\right]\times$$

$$\times\left(\left(u_{12} + v_{11} + u_{21} + v_{22}\right)^2 + \left(u_{11} + v_{12} - u_{22} - v_{21}\right)^2\right) +$$

$$+\left[\frac{1}{g} + \frac{2Z^2}{\beta V}\sum \left[(D_1 + D_2) - \Delta^2 D_0 \sin^2\theta\right]\sin^2\theta\right]\left(u_{12} + v_{11} - u_{21} - v_{22}\right)^2 +$$

$$+\left[\frac{1}{g} + \frac{2Z^2}{\beta V}\sum \left[(D_1 + D_2) + \Delta^2 D_0 \sin^2\theta\right]\sin^2\theta\right]\left(u_{11} + v_{12} + u_{22} + v_{21}\right)^2. \tag{18.12}$$

Here, $u_{ia} = \mathrm{Re}\, c_{ia}$, $v_{ia} = \mathrm{Im}\, c_{ia}$.

Equating the coefficients of this quadratic form to zero (because they are not coupled), we obtain 18 equations defining the complete spectrum of collective modes of the system. Eliminating the constant g^{-1} by using the equation

$$g^{-1} + \frac{2Z^2}{\beta V} \sum_p \frac{\sin^2\theta}{\omega^2 + (\xi - \mu H)^2 + 4\Delta^2 \sin^2\theta} = 0 \tag{18.13}$$

for the gap and calculating the sum (integrals) with respect to frequencies and momenta of quasifermions, we can reduce the equations for the spectrum to the following form (for zero momenta $\boldsymbol{k}$ of collective modes) [2,3]:

$$\int_0^1 (1-x^2) I dx = 0, \quad u_{11} + v_{12} + u_{22} + v_{21}; \tag{18.14}$$

$$\int_0^1 (1-x^2) I (1+4c) dx = 0, \quad u_{12} + v_{11} - u_{21} - v_{22}; \tag{18.15}$$

$$\int_0^1 (1-x^2) I (1+2c) dx = 0, \quad u_{21} + v_{11} + u_{21} + v_{22}\ , \quad u_{11} + v_{12} - u_{22} - v_{21}; \tag{18.16}$$

$$\int_0^1 (1-x^2) \ln c\, dx = 0, \quad \pm u_{21} + v_{11} - u_{12} \mp v_{22}, \quad u_{11} - v_{12} \pm u_{22} \pm v_{21}; \tag{18.17}$$

$$\int_0^1 \left[\ln \frac{4\Delta^2(1-x^2)}{\omega^2 + (2\mu H)^2} - \ln\left(1 - \frac{4\Delta^2(1-x^2)}{\omega^2 + (2\mu H)^2}\right) \right] (1-x^2) dx = 0, \tag{18.18}$$

$u_{13} \pm v_{23}$, $u_{23} \pm v_{13}$

$$\int_0^1 [(1+2c)I - 1] 2x^2 dx = 0, \quad u_{31} + v_{32}\ , \quad u_{32} + v_{31}; \tag{18.19}$$

$$\int_0^1 x^2 \ln c\, dx = 0, \quad u_{31} - v_{32}\ , \quad u_{32} - v_{31}\,; \tag{18.20}$$

$$\int_0^1 \left[\ln \frac{4\Delta^2\left(1-x^2\right)}{\omega^2 + (2\mu H)^2} - \ln\left(1 - \frac{4\Delta^2\left(1-x^2\right)}{\omega^2 + (2\mu H)^2} \right) \right] x^2 dx = 0, \quad u_{33}\ , v_{33}\,. \tag{18.21}$$

Here,

$$I = \frac{1}{(1-4c)^{1/2}} \ln \frac{(1+4c)^{1/2}+1}{(1-4c)^{1/2}-1}, \quad c = \frac{4\Delta_0^2}{\omega^2}\left(1-x^2\right). \tag{18.22}$$

Expressions (18.14)–(18.21) contain, in addition to the equations, the variables corresponding to them. These variables can be obtained from the diagonal quadratic form (18.10). They determine the combinations of the component of the order parameter (having 18 degrees of freedom), which oscillate in a given collective mode. The number of variables determines the degeneracy of the given mode. Equations (18.14)–(18.16) have the same form as for the A–phase of ^{3}He.[4] They define one Goldstone (gd) and two clapping–modes existing in the A–phase. Equations (18.18), (18.19) and (18.21) give eight modes with energy $E = 2\mu H$, while equations (18.17) and (18.20) imaginary energies (frequencies).

Let us write down the complete spectrum of collective modes in the A_1–phase[2,3]:

$$E = 0, \quad u_{11} + v_{12} + u_{22} + v_{21}\,;$$

$$E = (1.96 - i \cdot 0.31)\Delta, \quad u_{12} + v_{11} - u_{21} - v_{22}\,;$$

$$E = (1.17 - i \cdot 0.13)\Delta, \quad u_{21} + v_{11} + u_{21} + v_{22}\ , \quad u_{11} + v_{12} - u_{22} - v_{21}\,;$$

$$E = i\Delta 2 \exp(-5/6), \quad \pm u_{21} + v_{11} - u_{12} \mp v_{22}, \quad u_{11} - v_{12} \pm u_{22} \pm v_{21}\,;$$

$$E = 2\mu H, \quad u_{13} \pm v_{23}, \quad u_{23} \pm v_{13}, \quad u_{21} + v_{22}, \quad u_{22} + v_{21}, \quad u_{33}, \quad v_{33}\,;$$

$$E = i\Delta 2 \exp(-4/3), \quad u_{31} - v_{32}\ , \quad u_{32} - v_{31}. \tag{18.23}$$

18.2. Conclusion

We obtained the collective mode spectrum in the A_1–phase of ^{3}He. The spectrum contains the modes determined earlier for the *A*–phase of ^{3}He[4,5]: Goldstone ($E = 0$), clapping ($E = (1.17 - i \cdot 0.13)\Delta$) and pairbreaking ($E = (1.96 - i \cdot 0.31)\Delta$) modes as well as the modes determined by Brusov et al.[6,7] in the planar 2*D*–phase ($E = 2\mu H$). The frequencies of six modes are found to be imaginary, which is apparently due to the fact that the A_1–phase turns out to be unstable with respect to small perturbations in the given model[1]. Nevertheless the results obtained for other modes can be useful for interpreting experimental results on NMR and ultrasound absorption, since the modes with a real spectrum actually exist in the A_1–phase.

Chapter XIX

Superfluidity of Two–Dimensional and One–Dimensional Systems

19.1. Phase Transitions in Two–Dimensional Systems

It is well–known, that the heat motion of the long–wavelength phonons destroys the long–range order in solids, because the average quadratic deviation of atoms from their equilibrium positions increases logarithmically along with the increasing of dimensions of the system.

It is known also, that there is no spontaneous magnetization in two dimensional magnetics with spins with more than one degree of freedom, and also the average order parameter in the Bose–liquid is equal to zero[1].

On the other hand there exist many examples of phase transitions in two–dimensional systems, for instance, the gas–solid phase transition in the system of rigid discs, phase transitions in *X–Y* models and so on.

The absence of the order parameter and on the other hand existence of phase transitions in two–dimensional systems lead to necessity to introduce another definition of the long range order, namely the topological long–range order.

Let us explain this idea for the example of solid. At this case the disappearance of topological long–range order is connected with transition from the elastic response to the viscid one for small external perturbations. Explanation of such a transition can be done in the framework of the dislocation melting theory. This theory supposes that liquid has a local structure like the structure of solid near melting point, but in the equilibrium configuration there is some concentration of dislocations which can make to the surface under the influence of small tensions. As a result a viscous flow arises.

There are no free dislocations in solids in the equilibrium state (they can be only on the surface), and the system is rigid. The disappearance of isolated dislocations in the solid state at low temperatures is connected with the logarithmical increasing of their energy when the dimension of the system increases. Nevertheless the pair of dislocations with equal but opposite directed Burgers vectors has a finite energy and can be excited thermically. If the temperature increases pairs become unstable and their dissociation turns out. Appearance of isolated dislocations leads to viscous response of the system at small perturbations.

The analogous definition of the topological long–range order may be done in other cases, for instance for superfluid systems, the *X–Y* model and so on.

It is not difficult to estimate T_C for the topological ordering. In the solid the dislocation energy is equal to

$$E = \frac{\nu b^2(1+\tau)}{4\pi}\ln\frac{A}{A_0}$$

where $\boldsymbol{b}$ is the Burgers vector, A is an area of the system, ν is a two–dimensional elastisity modulus.

The entropy of the system depends on A in the similar way

$$S = \ln\frac{A}{A_0} + O(1)$$

At low temperatures the energy E dominates in the free energy $F = E - TS$, and the probability of arising a free dislocation in the system becomes infinitesimally small. At higher temperatures dislocations can arise spontaneously in the system, because an entropy term dominates in F.

Now we can define the critical temperature as a temperature for which the free energy changes its sign (and dislocations arise):

$$T_C = \nu\, b^2 (1+\tau) / 4\pi$$

For the *X–Y* model

$$T_C = \pi J \,,$$

where J is the spin–spin interaction constant.
For superfluid system:

$$T_C = \pi \rho / 2m \,.$$

The pioneer works by Berezinsky[2], Popov[3,4] and Kosterlitz and Thouless[1] played the leading part in the explanation of the nature of the phase transitions in two–dimensional systems.

In sections 19.2 and 19.3 we shall consider two–dimensional superfluid Bose–systems, in section 19.4 the superfluidity in the one–dimensional Bose–systems in discussed. Then (in sections 19.5–19.12) we shall investigate the two–dimensional Fermi–systems (^{3}He–films).

19.2. Two–Dimensional Superfluidity

The superfluidity of a three–dimensional Bose–system is connected with the macroscopic filling of the lowest quantum level–the condensate. In the two–and one–dimensional systems at $T \neq 0$ the condensate does not appear. This statement follows from the well–known Bogoliubov k^{-2} – theorem[5]. Here we shall prove this theorem in the framework of the functional integration method, using the idea of successive integration over fast and slow fields (see Chapter I, section 1.4). The particle density can be expressed via the Green's function $G(p)$ in the p–representation. In order to obtain $G(p)$ let us evaluate the functional $S_{eff}[\psi_0, \overline{\psi}_0]$, which is defined as follows

$$\exp S_{eff}\left[\psi_0,\overline{\psi}_0\right] = \int D\overline{\psi}_1 D\psi_1 \exp S$$

The functional integral in the right–hand side has a meaning of the partition function for a system of fast particles, describing by the fast field ψ_1, which is located in the slow field ψ_0.

To carry out $S_{eff}\left[\psi_0,\overline{\psi}_0\right]$one can develop a perturbation theory that has no divergencies at small momenta, as the sums over momenta are cut at the lower limit k_0.

At the second stage that is integration over slow fields $\overline{\psi}_0(\boldsymbol{x},\tau)$, $\psi_0(\boldsymbol{x},\tau)$, it is convenient to introduce new variables namely a density $\rho(\boldsymbol{x},\tau)$ and a phase $\varphi(\boldsymbol{x},\tau)$ according to equations

$$\psi_0(\boldsymbol{x},\tau) = (\rho(\boldsymbol{x},\tau))^{1/2}\exp(i\varphi(\boldsymbol{x},\tau)),$$

$$\overline{\psi}_0(\boldsymbol{x},\tau) = (\rho(\boldsymbol{x},\tau))^{1/2}\exp(-i\varphi(\boldsymbol{x},\tau)),$$

while

$$d\overline{\psi}_0(\boldsymbol{x},\tau)d\psi_0(\boldsymbol{x},\tau) = d\rho(\boldsymbol{x},\tau)d\varphi(\boldsymbol{x},\tau)$$

The presence of a condensate at low temperatures demonstrates itself in the fact that the main contribution to the integral over the density $\rho(\boldsymbol{x},\tau)$ is given by values close to some positive ρ_0, which has a meaning of the condensate density. The quantity ρ_0 can be determined from the extremality condition applied to the functional $S_{eff}\left[\psi_0,\overline{\psi}_0\right]$, which for $\psi_0 = \overline{\psi}_0 = \sqrt{\rho_0}$ becomes a function of ρ_0. For the calculation of $S_{eff}(\rho_0)$ one can use the perturbation theory in which the summation over internal momenta is cut at some lower limit k_0. This is the reason why $S_{eff}[\psi_0, \overline{\psi}_0]$, and, consequently, also ρ_0, depends on k_0. We shall call the quantity $\rho_0(k_0)$ the bare condensate. In the case of the three–dimensional Bose–system, $\rho_0(k_0)$ is only slightly different from the exact ρ_0, defined as the limit of the average $\langle\psi(\boldsymbol{x},\tau)\ \overline{\psi}(\boldsymbol{x}',\tau')\rangle$ as $|\boldsymbol{x} - \boldsymbol{x}'| \to \infty$. It will be shown below that the bare condensate $\rho_0(k_0)$ may exist also in a two–dimensional Bose–system at $T > 0$ and the superfluidity may exist even when the exact $\rho_0 = 0$. Assuming that the (bare) condensate exists it is suitable to introduce the variable $\pi(\boldsymbol{x},\tau)$ by the formula

$$\pi(\boldsymbol{x},\tau) = \rho(\boldsymbol{x},\tau) - \rho_0(\boldsymbol{x},\tau)$$

and to investigate S_{eff} as the functional of variables $\varphi(\boldsymbol{x},\tau)$ and $\pi(\boldsymbol{x},\tau)$.

Let us consider in greater detail the calculation of the functional S_{eff}. The starting functional S after the substitution $\psi = \psi_0 + \psi_1$ takes a form

$$\int_0^\beta d\tau \int d^3x \left[\overline{\psi}_0(\boldsymbol{x},\tau)\partial_\tau \overline{\psi}_0(\boldsymbol{x},\tau) - (2m)^{-1}\nabla\,\overline{\psi}_0(\boldsymbol{x},\tau)\nabla\,\psi_0(\boldsymbol{x},\tau) + \right.$$
$$\left. + \lambda\overline{\psi}_0(\boldsymbol{x},\tau)\psi_0(\boldsymbol{x},\tau)\right] +$$
$$+ \int_0^\beta d\tau \int d^3x \left[\overline{\psi}_1(\boldsymbol{x},\tau)\partial_\tau \psi_1(\boldsymbol{x},\tau) - (2m)^{-1}\nabla\,\overline{\psi}_1(\boldsymbol{x},\tau)\nabla\,\psi_1(\boldsymbol{x},\tau) + \right.$$
$$\left. + \lambda\overline{\psi}_1(\boldsymbol{x},\tau)\psi_1(\boldsymbol{x},\tau)\right] -$$

$$-\frac{1}{2}\int_0^\beta d\tau \int d^3x d^3y U(\boldsymbol{x}-\boldsymbol{y})[\overline{\psi}_0(\boldsymbol{x},\tau)\overline{\psi}_0(\boldsymbol{y},\tau)\psi_0(\boldsymbol{y},\tau)\psi_0(\boldsymbol{x},\tau)+$$
$$+2\overline{\psi}_1(\boldsymbol{x},\tau)\overline{\psi}_0(\boldsymbol{y},\tau)\psi_0(\boldsymbol{y},\tau)\psi_0(\boldsymbol{x},\tau)+$$
$$+2\overline{\psi}_0(\boldsymbol{x},\tau)\overline{\psi}_0(\boldsymbol{y},\tau)\psi_0(\boldsymbol{y},\tau)\psi_1(\boldsymbol{x},\tau)+2\overline{\psi}_1(\boldsymbol{x},\tau)\psi_1(\boldsymbol{x},\tau)\overline{\psi}_0(\boldsymbol{y},\tau)\psi_0(\boldsymbol{y},\tau)+$$
$$+2\overline{\psi}_1(\boldsymbol{x},\tau)\psi_1(\boldsymbol{y},\tau)\overline{\psi}_0(\boldsymbol{y},\tau)\psi_0(\boldsymbol{x},\tau)+\overline{\psi}_1(\boldsymbol{x},\tau)\overline{\psi}_1(\boldsymbol{y},\tau)\psi_0(\boldsymbol{y},\tau)\psi_0(\boldsymbol{x},\tau)+$$
$$+\overline{\psi}_0(\boldsymbol{x},\tau)\overline{\psi}_0(\boldsymbol{y},\tau)\psi_1(\boldsymbol{y},\tau)\psi_1(\boldsymbol{x},\tau)+2\overline{\psi}_0(\boldsymbol{x},\tau)\overline{\psi}_1(\boldsymbol{y},\tau)\psi_1(\boldsymbol{y},\tau)\psi_1(\boldsymbol{x},\tau)+$$
$$+2\overline{\psi}_1(\boldsymbol{x},\tau)\overline{\psi}_1(\boldsymbol{y},\tau)\psi_1(\boldsymbol{y},\tau)\psi_0(\boldsymbol{y},\tau)+\overline{\psi}_1(\boldsymbol{x},\tau)\overline{\psi}_1(\boldsymbol{y},\tau)\psi_1(\boldsymbol{y},\tau)\psi_1(\boldsymbol{x},\tau)].$$

Here we took into account that the crossing terms in the quadratic form $\propto \overline{\psi}_0 \cdot \psi_1, \overline{\psi}_1 \cdot \psi_0$ vanish when integrated over the volume V.

In order to express the explicit dependence of the functional S_{eff} on the phase variable $\varphi(\boldsymbol{x},\tau)$ we perform the transformation of variables in the integral over $\psi_1, \overline{\psi}_1$:

$$\psi_1(\boldsymbol{x},\tau) \to \psi_1(\boldsymbol{x},\tau)\exp(i\varphi(\boldsymbol{x},\tau)),$$

$$\overline{\psi}_1(\boldsymbol{x},\tau) \to \overline{\psi}_1(\boldsymbol{x},\tau)\exp(-i\varphi(\boldsymbol{x},\tau)),$$

where $\varphi(\boldsymbol{x},\tau)$ is the phase of slow field. After such a transformation the part of the action that contains potential actually depends only on $\rho(\boldsymbol{x},\tau)=\rho_0(k_0)+\pi(\boldsymbol{x},\tau)$, the modulus squared of the function $\psi_0(\boldsymbol{x},\tau)$. The quadratic form of the variables $\psi_1, \overline{\psi}_1$ will look as follows after the above–mentioned transformation

$$\int_0^\beta d\tau \int d^3x[\overline{\psi}_1(\boldsymbol{x},\tau)\partial_\tau\psi_1(\boldsymbol{x},\tau)-(2m)^{-1}\nabla\overline{\psi}_1(\boldsymbol{x},\tau)\nabla\psi_1(\boldsymbol{x},\tau)+$$
$$+\frac{i}{2m}(\overline{\psi}_1(\boldsymbol{x},\tau)\nabla\psi_1(\boldsymbol{x},\tau)-\nabla\overline{\psi}_1(\boldsymbol{x},\tau)\psi_1(\boldsymbol{x},\tau))\nabla\varphi(\boldsymbol{x},\tau)+ \tag{19.1}$$
$$+\overline{\psi}_1(\boldsymbol{x},\tau)\psi_1(\boldsymbol{x},\tau)\left(\lambda-(2m)^{-1}(\nabla\varphi(\boldsymbol{x},\tau))^2+i\partial_\tau\varphi(\boldsymbol{x},\tau)\right)]$$

The perturbation theory can be constructed regarding the quadratic form

$$\int_0^\beta d\tau \int d^3x \left[\overline{\psi}_1(\boldsymbol{x},\tau)\partial_\tau \psi_1(\boldsymbol{x},\tau) - (2m)^{-1}\nabla \overline{\psi}_1(\boldsymbol{x},\tau)\nabla \psi_1(\boldsymbol{x},\tau) + \right.$$
$$\left. + \lambda \overline{\psi}_1(\boldsymbol{x},\tau)\psi_1(\boldsymbol{x},\tau)\right],$$

which describes the ideal gas of Bose–particles with momenta $k > k_0$ as the nonperturbed action, while other parts of the action S depending on $\psi_1, \overline{\psi}_1$ as a perturbation. Thus, we get a perturbation theory which differs from the standard one (see Chapter 1) in the following respects:

1) The second–and third–order vertices, which describe the interaction with the slow field now depend on the variable $\rho(\boldsymbol{x},\tau)$.

2) The second–order vertices describing the interaction with the φ field emerge. They are determined by φ–depending terms in the quadratic form in (19.1).

3) All sums (integrals) over momenta are cut at the lower limit k_0.

Let us expand the expression of S_{eff} into the functional series of variables $\varphi(\boldsymbol{x},\tau)$, $\pi(\boldsymbol{x},\tau)$:

$$S_{eff}[\varphi,\pi] = S_{eff}[0,0] + \frac{1}{2}\int d^3x d^3x' d\tau d\tau' [a_{11}(\boldsymbol{x},\tau|\boldsymbol{x}',\tau')\varphi(\boldsymbol{x},\tau)\varphi(\boldsymbol{x}',\tau') +$$
$$+ 2a_{12}(\boldsymbol{x},\tau|\boldsymbol{x}',\tau')\varphi(\boldsymbol{x},\tau)\pi(\boldsymbol{x}',\tau') + a_{22}(\boldsymbol{x},\tau|\boldsymbol{x}',\tau')\pi(\boldsymbol{x},\tau)\pi(\boldsymbol{x}',\tau')]. \tag{19.2}$$

There are no linear terms in this expansion. Indeed, as we have seen, S_{eff} depends on the gradients of the phase. The coefficient functions of $\nabla\varphi(\boldsymbol{x},\tau)$ and $\partial_\tau\varphi(\boldsymbol{x},\tau)$ are translationally invariant and, thus vanish after integration with $\nabla\varphi(\boldsymbol{x},\tau)$ and $\partial_\tau\varphi(\boldsymbol{x},\tau)$. The coefficient function at $\pi(\boldsymbol{x},\tau)$ is proportional to $\partial S_{eff}/\partial\rho_0$ and is equal to zero thanks to the choice of ρ_0.

If the momentum k_0 is infinitesimally small, one can restrict oneself to the terms of second order in $\varphi(\boldsymbol{x},\tau)$ and $\pi(\boldsymbol{x},\tau)$ in the expansion (19.2). In this case, the low frequency asymptotic behavior of the Green's function is determined by coefficient functions a_{11}, a_{12} and a_{22}. If the condensate is present, it is suitable to substract from the Green's function its asymptotic part as $|\boldsymbol{x}-\boldsymbol{x}'|\to\infty$ and to investigate the expression

$$\tilde{G}(\boldsymbol{x},\tau|\boldsymbol{x}',\tau') = -\langle\psi(\boldsymbol{x},\tau)\overline{\psi}(\boldsymbol{x}',\tau')\rangle + \langle\psi(\boldsymbol{x},\tau)\rangle\langle\overline{\psi}(\boldsymbol{x}',\tau')\rangle =$$
$$= \rho_0\left[1 - \frac{\left\langle\sqrt{\rho(\boldsymbol{x},\tau)\rho(\boldsymbol{x}',\tau')}\exp i(\varphi(\boldsymbol{x},\tau)-\varphi(\boldsymbol{x}',\tau'))\right\rangle}{\left\langle\sqrt{\rho(\boldsymbol{x},\tau)}\exp i\varphi(\boldsymbol{x},\tau)\right\rangle\left\langle\sqrt{\rho(\boldsymbol{x}',\tau')}\exp(-i\varphi(\boldsymbol{x}',\tau'))\right\rangle}\right]$$

Let us restrict ourselves to the terms of maximally second order in the limit $k_0\to 0$ in expansion (19.2). For sufficiently large $|\boldsymbol{x}-\boldsymbol{x}'|$ one can neglect the difference between the variables $\rho(\boldsymbol{x},\tau), \rho(\boldsymbol{x}',\tau')$ and the bare condensate density $\rho_0(k_0)$. Indeed, the averages $<\varphi\pi>, <\pi\pi>$ with different arguments $(\boldsymbol{x},\tau), (\boldsymbol{x}',\tau')$ vanish for $|\boldsymbol{x}-\boldsymbol{x}'|\to\infty$ faster than $<\varphi\varphi>$, and the mean values $<\varphi\pi>, <\pi\pi>$ with identical arguments (either $(\boldsymbol{x},\tau)$ or $(\boldsymbol{x}',\tau')$) appear both in the numerator and in the denominator and therefore mutually cancel.

Finally, we obtain the formula

$$\tilde{G}(\boldsymbol{x},\tau|\boldsymbol{x}',\tau') \approx \rho_0\left[1 - \frac{\langle\exp i(\varphi(\boldsymbol{x},\tau)-\varphi(\boldsymbol{x}',\tau'))\rangle}{\langle\exp i\varphi(\boldsymbol{x},\tau)\rangle\langle\exp(-i\varphi(\boldsymbol{x}',\tau'))\rangle}\right]$$

containing the functional integrals of the Gaussian type. The values of these integrals are determined by the coefficient functions a_{11}, a_{12} and a_{22}, which can be regarded as matrix elements of an operator A. They can be expanded in terms of matrix elements of the inverse operator A^{-1}:

$$\langle \exp i(\varphi(\boldsymbol{x},\tau)-\varphi(\boldsymbol{x}',\tau')) \rangle = \exp\left\{-\frac{1}{2}\left(A^{-1}\right)_{11}\left(\boldsymbol{x},\tau|\boldsymbol{x},\tau\right)-\frac{1}{2}\left(A^{-1}\right)_{11}\left(\boldsymbol{x}',\tau'|\boldsymbol{x}',\tau'\right)+\right.$$

$$\left.+\frac{1}{2}\left(A^{-1}\right)_{11}\left(\boldsymbol{x},\tau|\boldsymbol{x}',\tau'\right)+\frac{1}{2}\left(A^{-1}\right)_{11}\left(\boldsymbol{x}',\tau'|\boldsymbol{x},\tau\right)\right\},$$

$$\langle \exp i\varphi(\boldsymbol{x},\tau) \rangle = \exp\left\{-\frac{1}{2}\left(A^{-1}\right)_{11}\left(\boldsymbol{x},\tau|\boldsymbol{x},\tau\right)\right\},$$

$$\langle \exp(-i\varphi(\boldsymbol{x},\tau)) \rangle = \exp\left\{-\frac{1}{2}\left(A^{-1}\right)_{11}\left(\boldsymbol{x}',\tau'|\boldsymbol{x},\tau\right)\right\}.$$

Thus the Green's function $\tilde{G}$ in the limit $|\boldsymbol{x} - \boldsymbol{x}'| \to \infty$ then takes the form

$$\rho_0\left\{1-\exp\left[\frac{1}{2}\left(A^{-1}\right)_{11}\left(\boldsymbol{x},\tau|\boldsymbol{x}',\tau'\right)+\frac{1}{2}\left(A^{-1}\right)_{11}\left(\boldsymbol{x}',\tau'|\boldsymbol{x},\tau\right)\right]\right\} \approx$$

$$\approx -\frac{\rho_0}{2}\left\{\left(A^{-1}\right)_{11}\left(\boldsymbol{x},\tau|\boldsymbol{x}',\tau'\right)+\left(A^{-1}\right)_{11}\left(\boldsymbol{x}',\tau'|\boldsymbol{x},\tau\right)\right\}$$

Here, the inequality $\left|\left(A^{-1}\right)_{11}\left(\boldsymbol{x},\tau|\boldsymbol{x}',\tau'\right)\right| << 1$, valid for large $|\boldsymbol{x} - \boldsymbol{x}'|$ has been used.

So we come to the following formula for the Fourier coefficient $\tilde{G}(p)$ of the function $\tilde{G}\left(\boldsymbol{x},\tau|\boldsymbol{x}',\tau'\right)$:

$$\tilde{G}(p) = -\frac{\rho_0 a_{22}(p)}{a_{11}(p)a_{22}(p) - a_{12}(p)a_{12}(-p)}$$

The Fourier coefficient of functions a_{11}, a_{12} and a_{22}, which appear in this formula, can be obtained by double differentiation of the functional $S_{eff}[\varphi, \pi]$ with respect to φ, π.

The following exact diagrammatic equations determine each of the Fourier coefficients a_{11}, a_{12}, a_{22} as a sum of an infinite number of diagrams of the modified diagram technique:

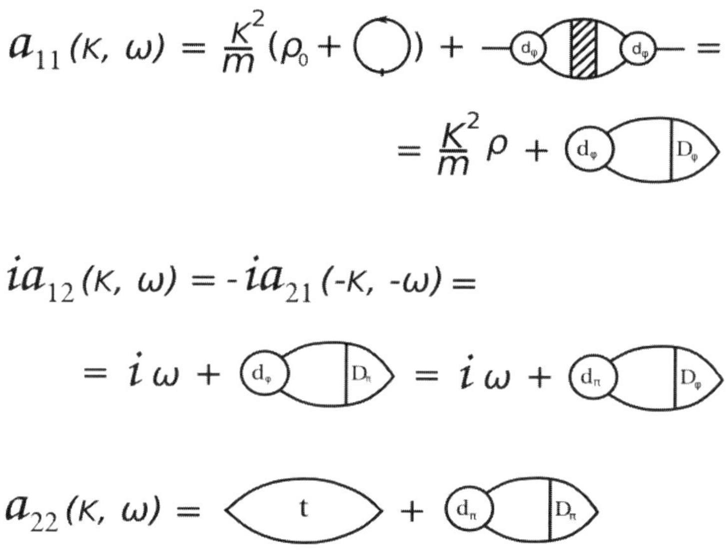

(19.3)

The internal lines of the diagrams correspond to full normal and anomalous Green's functions. They can be distinguished by arrows at the ends of the lines which point in the same direction in the case of normal Green's functions and in opposite directions in the case of anomalous ones. The diagrams (19.3) should be undestood with summation over all possible combinations of arrow directions accomplished.

After the analysis of the arrow directions in diagrams (19.3) different parts of the following form arise

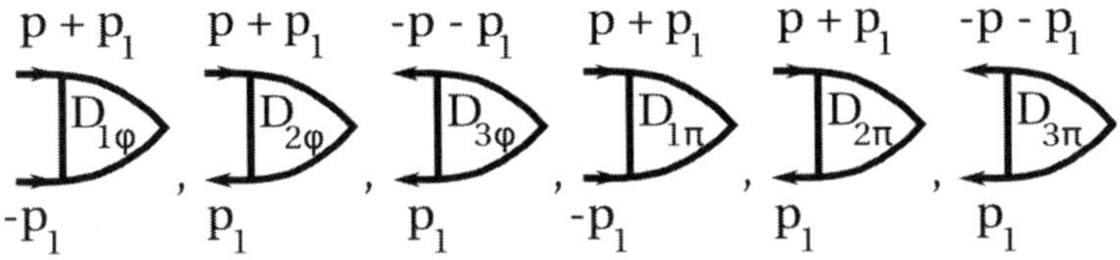

They can be obtained by differentiating normal and anomalous self–energy parts, calculated in the external field ψ_0 with respect to arguments φ, π. The results of differentiation with respect to φ are denoted by $D_{j\varphi}(j=1,2,3)$, and with respect to π – by $D_{j\pi}$.

The contributions of irreducible (in the sense that the interaction vertex with external field cannot be disconnected by cutting one or two internal lines) diagrams to $D_{j\varphi}$ and $D_{j\pi}$, will be denoted by $d_{j\varphi}$ and $d_{j\pi}$ respectively. The contribution of irreducible (in the same sense) diagrams to the expressions a_{22}, which can be obtained by double differentiating of $S_{eff}[\varphi,\pi]$ with respect to π, will be denoted by t. Let us mention that the expressions for irreducible diagrams $d_{j\varphi}$ are known exactly – they are given by the formulae

$$d_{1\varphi} = d_{3\varphi} = 0,\ d_{2\varphi} = i\omega - (\boldsymbol{k}, \boldsymbol{k}_1)/m \tag{19.4}$$

where $(\boldsymbol{k},\omega), (\boldsymbol{k}_1,\omega_1)$ are four momenta of the external field φ and of the internal line of the diagram incoming to the vertex $d_{2\varphi}$ respectively.

Expression (19.4) is determined by the quadratic form of $\psi_1, \overline{\psi}_1$ in (19.1). The diagram  in (19.3), which defined the density of noncondensed particles ρ_1, is determined by the integral of the expression $\overline{\psi}_1\psi_1(\nabla\varphi)^2$. To ρ_1 one must odd the condensate density ρ_0 which has its origin in the expression $-(2m)^{-1}\nabla\overline{\psi}_0\nabla\psi_0$ after separation of the component proportional to $(\nabla\varphi)^2$. The sum $\rho_0 + \rho_1$

is the full density ρ. The term $i\omega$ in the formula (19.3) for ia_{12} corresponds to the term $\overline{\psi}_0 \partial_\tau \psi_0$ in the action integral.

Let us stress that the diagrammatic equations (19.3) do not use the low–density assumption and, thus, they are asymptotically exact in the limit $\boldsymbol{k}, \omega \to 0$. What is actually used is only the smallness of momentum k with respect to k_0.

From the formulae derived so far one can easily obtain the asymptotic behavior of Green's functions at $\omega = 0$. Indeed, the expression $a_{12}(p)$ is proportional to k^2 for $\omega = 0$ and small k while $a_{22}(p)$ goes to a constant $a_{22}(0)$. As $a_{11}(\boldsymbol{k},0) \propto k^2$, $\tilde{G}(\boldsymbol{k},0) \approx -\rho_0 / a_{11}(\boldsymbol{k},0)$ for small k. The second term in formula (11.3) for a_{11} is for $\omega = 0$ equal to $-(mV)^{-1}(\boldsymbol{k}, \boldsymbol{K})$, where $\boldsymbol{K}$ is the average momentum in the coordinate system moving with the velocity $\boldsymbol{v}$. According to the definition of the normal component we have $\boldsymbol{K} = V\rho_n \boldsymbol{k}$. Finally, the formula for a_{11} takes the form:

$$a_{11} \approx \frac{k^2}{m}(\rho - \rho_n) = \frac{\rho_S k^2}{m},$$

where $\rho_S = \rho - \rho_n$ is the superfluid component density. This leads to the well–known Bogoliubov asymptotic formula[5] for the function

$$\tilde{G}(\boldsymbol{k},0) \approx -\rho_0 m / \rho_S k^2$$

From this formula and from formula (1.23) for the average number of particles with momentum $\boldsymbol{k}$ it follows, that for $k \to 0$

$$N_k = -T \sum_{\omega} G(\boldsymbol{k}, \omega) = \frac{T\rho_0 m}{\rho_S k^2} + O(1) \tag{19.5}$$

as the terms with $\omega \neq 0$ in (19.5) give a finite contribution for $k \to 0$. From (19.5) it follows that the condensate cannot emerge in two–and one–dimensional systems for the temperatures $T \neq 0$ as the function k^{-2} is not integrable at small momenta in one– and two–dimensional k–space.

The fact, that there is no condensate, does not exclude the possibility of the existence of superfluidity. Conjectures suggesting such a possibility are based either on variational calculation of the ground–state wave function (the method of collective variables), or on the method of wherent states. Berezinski[2] developed a method which uses the transformation from the continuous system to the discrete set of planar rotators. Let us apply to two–and one–dimensional systems the method of successive functional integration over fast and slow fields. The results confirm the existence of superfluidity in two–dimensional Bose–systems, and indicate that superfluidity exists, even in one–dimensional systems. The basic physical idea is that particles with small momenta behave like a "bare condensate" that is responsible for the creation of a superfluid density ρ_s with an order of magnitude equal to the full density of the system ρ. The functional integration method enables us to turn this idea into reality and to construct the effective action characteristic for superfluid systems.

Let us consider a two–dimensional Bose–system, in which the existence of condensate at zero temperature is possible. The condensate density can be found from the equation $\partial S_{eff} / \partial \rho_0 = 0$, where ∂S_{eff} is the effective action of the system which can be obtained by integration over fast fields. We can expand ∂S_{eff} into the functional series of variables $\varphi(x,\tau), \pi(x,\tau) = \rho(x,\tau) - \rho_0(k_0)$. At low temperatures the coefficient functions of this expression can be put equal to the correspondent functions at zero temperature. The leading components in ∂S_{eff}, which depend on the variables $\varphi(x,\tau), \pi(x,\tau)$ are given as a quadratic form of the variables $\varphi(x,\tau), \pi(x,\tau)$. The coefficient functions of the quadratic form can be expressed in terms of thermodynamical function. This is the case not only for a Bose–gas, but for an arbitrary Bose–system. In order to make it clear let us consider the quadratic form (19.1). We can say that the system of fast particles described by the fields $\psi_1(\boldsymbol{x},\tau),\overline{\psi}_1(\boldsymbol{x},\tau)$ interacts with the following slowly varying fields: the field of nonhomogeneous condensate $\rho(\boldsymbol{x},\tau)$, the field of nonhomogeneous chemical potential

$$\lambda(\boldsymbol{x},\tau)=\lambda+i\partial_\tau\varphi(\boldsymbol{x},\tau)-(2m)^{-1}(\nabla\varphi(\boldsymbol{x},\tau))^2$$

and the velocity field $\boldsymbol{v}(\boldsymbol{x},\tau)=m^{-1}\nabla\varphi(\boldsymbol{x},\tau)$.

In the first (quasihomogeneous) approximation, the inhomogeneities introduced by the slowly varying functions $\varphi(\boldsymbol{x},\tau),\rho(\boldsymbol{x},\tau)$ can be incorporated by the formula

$$S_{eff}\cong\int d\tau d\boldsymbol{x}\boldsymbol{p}(\lambda(\boldsymbol{x},\tau),\rho(\boldsymbol{x},\tau),\boldsymbol{v}(\boldsymbol{x},\tau))$$

where $p(\lambda,\rho_0(k_0),\boldsymbol{v})$ is the pressure in the homogeneous system with chemical potential λ, condensate density $\rho_0(k_0)$ in a coordinate system moving with the velocity $\boldsymbol{v}$.

Let us consider the case of very small temperatures T or large β. It can be shown that in that case $p(\lambda,\rho_0(k_0),\boldsymbol{v})$ does not depend on $\boldsymbol{v}$, and we can use the Taylor expansion of $p(\lambda,\rho_0(k_0),0)$

$$p(\lambda+\delta\lambda,\rho_0(k_0)+\pi,0)=p(\lambda,\rho_0(k_0),0)+p_\lambda\delta\lambda+p_{\rho 0}\pi+$$
$$+\frac{1}{2}p_{\lambda\lambda}(\delta\lambda)^2+p_{\lambda\rho 0}\delta\lambda\pi+\frac{1}{2}p_{\rho 0\rho 0}\pi^2+\ldots$$

Here, $p_{\rho 0}=0$ due to the choice of $\rho_0(k_0)$. Thus we come to the following expression for S_{eff}

$$S_{eff} = p(\lambda, \rho_0(k_0),0)\beta V + \int d\tau dx\{ p_\lambda (i\partial_\tau \varphi - (2m)^{-1}(\nabla\varphi)^2) +$$

$$+\frac{1}{2} p_{\lambda\lambda}(i\partial_\tau \varphi - (2m)^{-1}(\nabla\varphi)^2)^2 +$$

$$+ p_{\lambda\rho 0}(i\partial_\tau \varphi - (2m)^{-1}(\nabla\varphi)^2)\pi + \frac{1}{2} p_{\rho 0\rho 0}\pi^2 + \ldots\}$$

The quadratic form in this expansion is

$$\int d\tau dx \left(-\frac{p_\lambda}{2m}(\nabla\varphi)^2 - \frac{1}{2} p_{\lambda\lambda}(\partial_\tau \varphi)^2 + i p_{\lambda\rho 0}\pi\partial_\tau \varphi + \frac{1}{2} p_{\rho 0\rho 0}\pi^2 \right) \tag{19.6}$$

The coefficients in (19.6) are expressed in terms of the derivatives of $p(\lambda, \rho_0(k_0),0)$, i.e. in terms of thermodynamical functions. Equation (19.6) gives us the exact expression of the quadratic part of S_{eff} for any superfluid Bose–system at low temperatures. It describes noninteracting sound excitations with energy proportional to momentum. Corrective terms to (19.6) lead to deviation of the energy spectrum from linearity and to interaction between excitations.

For a dilute Bose–gas model the main corrections to (19.6) come from the kinetic energy form of the slow field ψ_0

$$\int d\tau dx \frac{|\nabla \psi_0|^2}{2m} = -\int d\tau dx \frac{(\nabla \psi_0)^2}{2m} \left\{ \rho_0 (\nabla\varphi)^2 + \pi(\nabla\varphi)^2 + \frac{(\nabla\pi)^2}{4(\rho_0 + \pi)} \right\}$$

The first term in the integral in the right–hand side was taken into account in the quadratic form (19.6). The second and the third terms can be regarded as corrections, where we may further neglect π in the denominator as compared with ρ_0. Adding the above mentioned corrections to the functional (19.6) we get a functional

$$\int d\tau dx \left(-\frac{p_\lambda}{2m} (\nabla \varphi)^2 - \frac{1}{2} p_{\lambda\lambda} (\partial_\tau \varphi)^2 + i p_{\lambda\rho 0} \pi \partial_\tau \varphi + \right.$$

$$\left. + \frac{1}{2} p_{\rho 0 \rho 0} \pi^2 - \frac{\pi (\nabla \varphi)^2}{2m} - \frac{(\nabla \pi)^2}{8m\rho_0} \right),$$

which has a meaning of an effective (hydrodynamical) action.

Regarding the integral of the expression $-\pi(\nabla\varphi)^2/2m$ as a perturbation with respect to the quadratic part of this functional we can develop a new perturbation theory which has no divergencies at small energies and momenta. The application of this perturbation theory to evaluation of the low frequency spectrum yields at $T \neq 0$ the spectrum with two sound branches (the first and the second sounds) characteristic for a superfluid Bose–system.

Let us now find asymptotic behavior of the one–particle Green's function

$$-\langle \psi_0(\boldsymbol{x},\tau)\overline{\psi}_0(\boldsymbol{x}',\tau')\rangle = \left\langle \sqrt{\rho(\boldsymbol{x},\tau)\rho(\boldsymbol{x}',\tau')} \exp i(\varphi(\boldsymbol{x},\tau) - \varphi(\boldsymbol{x}',\tau')) \right\rangle_0 \tag{19.7}$$

in the limit $r = |\boldsymbol{x} - \boldsymbol{x}'| \to \infty$ for a two–dimensional Bose–system.

Here, $\langle \ldots \rangle_0$ denotes the averaging over slow fields weighted by S_{eff}.

Suppose that the momentum k_0 which distinguishes between "large" and "small" momenta is sufficiently small, so that we could neglect in the first approximation the fact that variables $\rho(\boldsymbol{x},\tau), \rho(\boldsymbol{x}',\tau')$ are different from the zero temperature condensate density ρ_0, and put S_{eff} equal to the quadratic form (19.6). In this case the integral becomes Gaussian and for the Green's function (19.7) we obtain the expression

$$-\rho_0 \exp\left\langle -\frac{1}{2}(\varphi(\boldsymbol{x},\tau)-\varphi(\boldsymbol{x}',\tau'))^2 \right\rangle_0 , \tag{19.8}$$

where

$$\left\langle \frac{1}{2}(\varphi(\boldsymbol{x},\tau)-\varphi(\boldsymbol{x}',\tau'))^2 \right\rangle_0 = \\ = \frac{1}{2\beta V} \sum_{\omega,|k|<k0} \frac{m}{\rho(k^2+\omega^2/c^2)} \left|\exp(i(\omega\tau+\boldsymbol{kx}) - \exp(i(\omega\tau'+\boldsymbol{kx}')\right|^2 \tag{19.9}$$

The formula

$$\langle \varphi(p)\varphi(-p)\rangle_0 = \frac{m}{\rho(k^2+\omega^2/c^2)} \tag{19.10}$$

has been used for the calculation of the average $\langle\varphi\varphi\rangle_0$ in the p–presentation. Here c^2 is the squared sound velocity

$$c^2 = \frac{p_\lambda}{m}\left(p_{\lambda\lambda} - \frac{p_{\lambda\rho 0}^2}{p_{\rho 0\rho 0}}\right)^{-1} = \frac{1}{m}\frac{dp}{d\rho} .$$

The condition $p_{\rho 0} = 0$ implies that $p_{\lambda\lambda} - \dfrac{p_{\lambda\rho 0}^2}{p_{\rho 0\rho 0}}$ equal to full derivative $d^2 p(\lambda,\rho_0(\lambda))/d\lambda^2$). The large distance $r\to\infty$ asymptotic behavior of expression (19.9) is

$$\frac{mc^2}{2\rho(2\pi)^2} \int_{k<k0} \frac{d^2k}{k}(1-\cos(\boldsymbol{k},\boldsymbol{x}-\boldsymbol{x}'))cth\frac{1}{2}\beta ck = \frac{m}{2\pi\beta\rho}\ln\frac{r}{\beta c} + const \tag{19.11}$$

From this expression it follows that for $r \to \infty$ the Green's function vanishes like $r^{-\alpha}$ with the exponent

$$\alpha = \frac{m}{2\pi\beta\rho}. \tag{19.12}$$

Now we shall show that the power–like asymptotic behavior of Green's function remains valid even the temperature increases. Exponent then has the form

$$\alpha = \frac{m}{2\pi\beta\rho_S} \tag{19.13}$$

that differs from (19.12) by replacing the full density with the superfluid density ρ_S. In the limit $T \to 0$ $\rho = \rho_S$ and we come from (19.11) to (19.12).

For $T \neq 0$ the solution of the equation $\partial S_{eff} / \partial \rho_0 = 0$ has a meaning of the bare condensate density $\rho_0(k_0)$. As we shall show $\rho_0(k_0) \propto k_0^{\alpha}$. Let us investigate the expansion of S_{eff} into the functional series of variables $\varphi(\boldsymbol{x},\tau), \pi(\boldsymbol{x},\tau) = \rho(\boldsymbol{x},\tau) - \rho_0(k_0)$. The coefficients of $(\nabla\varphi)^2 / 2m$ in the quadratic form (19.6) is for $T \neq 0$ equal to

$$-\left(\rho_0(k_0) + \rho_1(k_0) - \rho_n(k_0)\right),$$

where $\rho_0(k_0)$ and $\rho_1(k_0)$ are the densities of particles with $k < k_0$ and $k > k_0$, respectively, and ρ_n is the particle number density of the normal component with $k > k_0$. In the limit $k_0 \to 0$ this expression can be put equal to $\rho_n - \rho = -\rho_S$, where ρ_S is the superfluid component density.

Let us return to the Green's function. The consideration applied in order to pass from (19.7) to (19.8) can be used here too. It is only necessary to replace ρ_0 by $\rho_0(k_0)$ in (19.8) and ρ by ρ_S in (19.9), (19.10). The leading contribution to the sum over frequencies (19.9) is given by the term with $\omega = 0$. As a result the formula (6.11) should be replaced by the following one

$$\frac{m}{\beta\rho_S(2\pi)^2}\int\limits_{k<k0}\frac{d^2k}{k^2}\left(1-\cos(\boldsymbol{k},\boldsymbol{x}-\boldsymbol{x}')\right)=\frac{m}{2\pi\beta\rho_S}\ln k_0r+C_1$$

where C_1 is a k_0 – independent constant.

The asymptotic behavior of the Green's function turns out to be

$$G \approx -\rho_0(k_0)(k_0r)^{-\alpha}\exp(-C_1).$$

This function is k_0–independent if one assumes that $\rho_0(k_0) \propto k_0^{\alpha}$. Such an assumption is self–consistent. It leads to the formula $G \approx -a_1 r^{-\alpha}$ with a_1 independent of k_0. This applies the asymptotic formula for the particle number $N(k)$ in the limit of small k of the form $N(k) \approx a_2 k^{\alpha-2}$. This formula leads to the expression for the density $\rho_0(k_0)$:

$$\rho_0(k_0)=(2\pi)^{-2}\int\limits_{k<k0}N(k)d^2k=a_3k_0^{\alpha}$$

that has been supposed to hold.

The above investigation shows that the one–particle Green's function (19.7) of a two–dimensional Bose–system behaves as $r^{-\alpha}$ when $r=|\boldsymbol{x}-\boldsymbol{x}'|\to\infty$ for sufficiently low temperatures.

In a three–dimensional case the one–particle Green's function has a nonzero limit $-\rho_0$ at $r=|\boldsymbol{x}-\boldsymbol{x}'|\to\infty$, where ρ_0 is a condensate density. In two–dimensional systems, one can speak about the long–range correlations – the one–particle Green's function decreases as an inverse power of r, and not exponentially as in the case of a sufficiently high temperature. The temperature of transition from the exponential to the power–like behavior of the Green's function is evidently the temperature of phase transition to the superfluid state.

So for the excitations of the quantum vortex type have not been taken into account. Such excitations play the important role in phase transitions in two–dimensional systems, as it has mentioned in section 1. A description of vortices in the formalism of functional integral is given in the following section, as well as a description of their role in phase transitions of Bose–systems to the superfluid state. At the conclusion of this section we shall find Green's functions, thermodynamical functions and the equation of the phase transition curve for a two–dimensional nonideal Bose–gas of low density.

In the three–dimensional case the corresponding results can be expressed by vurtue of t_0 – the zero energy and zero momentum t–matrix which describes the scattering of two Bose–particles in vacuum.

In the case of two–dimensions the t–matrix defined by the formula

$$t(k_1,k_2,z)+(2\pi)^{-2}\int d^2k_3 u(k_1-k_3)\left(\frac{k_3^2}{m}-z\right)^{-1} t(k_3,k_2,z)=u(k_1-k_2) \tag{19.14}$$

has for small k_1, k_2 and z an asymptotic form

$$t \cong 4\pi\left(m\ln(-\varepsilon_0/z)\right)^{-1} \tag{19.15}$$

and vanishes in the limit $k_1, k_2, z \to 0$. The asymptotic formula (19.15) is valid, if $|k_1|,|k_2| << r_0^{-1}, |z| << m^{-1}r_0^2$, where r_0 is the effective radius of the potential. The quantity ε_0 in (19.15) is of order $m^{-1}r_0^2$, and, consequently $|\ln(-\varepsilon_0/z)| >> 1$.

In order to prove (19.15) we shift the second term in the left–hand side of (19.15) to the right–hand side and then add the same expression to both sides. We obtain the following equation

$$t(k_1,k_2,z)+(2\pi)^{-2}\int d^2k_3 u(k_1-k_3)\left(\frac{k_3^2}{m}-z\right)^{-1}t(k_3,k_2,z)=$$
$$=u(k_1-k_2)+(2\pi)^{-2}\int d^2k_3 u(k_1-k_3)\times \tag{19.16}$$
$$\times\left\{\left(\frac{k_3^2}{m}-z_0\right)^{-1}-\left(\frac{k_3^2}{m}-z\right)^{-1}\right\}t(k_3,k_2,z).$$

Here, z_0 is an arbitrary parameter. The operator acting on the t–matrix in the left–hand side of (19.16) can be inverted. This is equivalent to the replacing the potential on the right–hand side by the t–matrix

$$t(k_1,k_2,z)=t(k_1,k_2,z_0)+$$
$$+(2\pi)^{-2}\int d^2k_3 t(k_1,k_3,z_0)\left\{\left(\frac{k_3^2}{m}-z_0\right)^{-1}-\left(\frac{k_3^2}{m}-z\right)^{-1}\right\}t(k_3,k_2,z)$$
$$\tag{19.17}$$

The passage from (19.14) to (19.17) represents the transformation of the t–matrix equation

$$t_z+uR_zt_z=u \tag{19.18}$$

to the Hilbert identity

$$t_z-t_{z0}=t_{z0}(R_{z0}-R_z)t_z$$

where $R_z = (H_0 - z)^{-1}$ is the resolvent of the unperturbed Schrödinger operator (without a potential), t_z and t_{zo} are the operators of the t–matrix and of the potential energy.

In the asymptotic domain, $|k_1|,|k_2| << r_0^{-1}, |z|,|z_0| << m^{-1} r_0^2$ one can assume that t is momentum independent. This brings formula (19.17) to the form

$$t(z)-t(z_0) \cong (2\pi)^{-2} t(z_0) t(z) \int d^2 k_3 \left\{ \left(\frac{k_3^2}{m} - z_0 \right)^{-1} - \left(\frac{k_3^2}{m} - z \right)^{-1} \right\} =$$

$$= \frac{m}{4\pi} t(z_0) t(z) \ln \frac{z}{z_0}$$

from which (19.15) follows.

For the exclusion of the potential and the transition to the t–matrix we apply the following method. First of all we integrate the functional exp$S$ over variables $a(p), a^*(p)$ with momenta $k > k_0'$, where k_0' satisfies the inequalities

$$\max(|\lambda|, T) << \frac{(k_0')^2}{2m} << \varepsilon_0 .$$

The result of this integration is the functional of variables $a(p), a^*(p)$ with $k < k_0'$, which will be denoted $\exp S'$ in what follows. For a system of low density, the first approximation for S' has the form

$$S' = \sum_{\omega, |k| < k_0'} \left(i\omega - \frac{k^2}{2m} + \lambda \right) a^*(p)\, a(p) -$$

$$-(2\beta V)^{-1} \sum_{\substack{p_1+p_2=p_3+p_4 \\ |k_i| < k_0'}} t' a^*(p_1) a^*(p_2) a(p_3) a(p_4)$$

and differs from formula for S by the fact that the summations over momenta are cut at an upper limit k_0' and that the potential is replaced by the *t*–matrix, in definition of which (19.14) the integral over k_3 is cut at the lower limit k_0'.

Such a *t*–matrix (denoted t') has the form

$$t' = t'(\omega_1 + \omega_2) = 4\pi\left\{ m \ln\left(\varepsilon_0 \left[\frac{(k_0')^2}{m} - i\omega_1 - i\omega_2 \right]^{-1} \right) \right\}^{-1}.$$

Now, let us integrate the functional $\exp S'$ over the variables $a(p), a^*(p)$ with momenta k satisfying the condition $k_0 < k < k_0'$. Here, the variables $a(p), a^*(p)$ with $k < k_0$. have the meaning of "inhomogeneous condensate" with respect to integration variables $a(p), a^*(p)$ with $k \in [k_0, k_0']$. The order of magnitude of the quantity k_0 will be estimated at the end of this section. In any case for all values of temperatures lower than the phase transition temperature we obtain $k_0 << (m\lambda)^{1/2}$.

When calculated the thermodynamical functions, we assume that the condensate is homogeneous in the first approximation; we can to take it into account by changing the variables with $k < k_0$ in S' as follows

$$a(p), a^*(p) \to (\rho_0(k_0)\beta V)^{1/2}\, \delta_{p0} \tag{19.19}$$

The quantity $\rho_0(k_0)$ in (19.18) has the meaning of the density of the particles with $k < k_0$. This density, as well as the density $\rho_1(k_0)$ of the particles with $k > k_0$, depends on the value of k_0–momentum, which distinguishes the slow and fast variables. The total density $\rho = \rho_0(k_0) + \rho_1(k_0)$, however, must be k_0–independent.

The first approximation to $\rho_0(k_0)$ is determined by the condition, that the contribution to S' from terms do not contain $a(p), a^*(p)$ after the change (19.18) should be found from the maximum of the expression

$$\beta V\left(\lambda\rho_0(k_0) - \frac{1}{2}t'(0)\rho_0^2(k_0)\right)$$

This expression is maximal if

$$\rho_0(k_0) = \lambda/t'(0) = (m\lambda/4\pi)\ln(m\varepsilon_0/k_0'\ )^2 .$$

After the transformation (19.18) one can use the pertubation theory with normal and anomalous Green's functions, which is similar to that for three–dimensional systems[21]. Main contribution to the self–energy parts *A*, *B* (see §5.3 in Ref. 21) is given by the second–order vertices. We obtain

$$A \approx 2t'(0)\rho_0(k_0) = 2\lambda, \quad B \approx t'(0) .$$

The normal and anomalous Green's functions are in the first approximation given by the equations

$$G=-\frac{i\omega+\frac{k^2}{2m}+\lambda}{\omega^2+\varepsilon^2(k)},\quad G_1=\frac{\lambda}{\omega^2+\varepsilon^2(k)}, \tag{19.20}$$

where

$$\varepsilon^2(k)=\left(\frac{k^2}{2m}\right)^2+\frac{\lambda}{m}k^2.$$

The density $\rho_1(k_0)$ can be calculated according to the formula

$$\rho_1(k_0)=-(\beta V)^{-1}\sum_{\substack{\omega,|k|>k0\\ \varepsilon\to+0}}\exp(i\omega\varepsilon)G(p)=$$
$$=\frac{1}{2(2\pi)^2}\int_{k>k0}d^2k\left(\frac{\frac{k^2}{2m}+\lambda}{\varepsilon(k)}cth\frac{\beta\varepsilon(k)}{2}-1\right). \tag{19.21}$$

We get the integral that diverges logarithmically as $k_0\to 0$.This divergence must cancel the divergence of the quantity $\rho_0(k_0)$ calculated in the second order of perturbation theory. We can determine $\rho_0(k_0)$ from the equality

$$\lambda=A(0)-B(0) \tag{19.22}$$

which is equivalent to the equation $\partial S_{eff}/\partial\rho_0=0$. This equation has been used for the determination of the second order approximation to ρ_0 also in the case of the three–dimensional Bose–gas[21]. It is sufficient to take into account the following diagrams for the self–energy parts

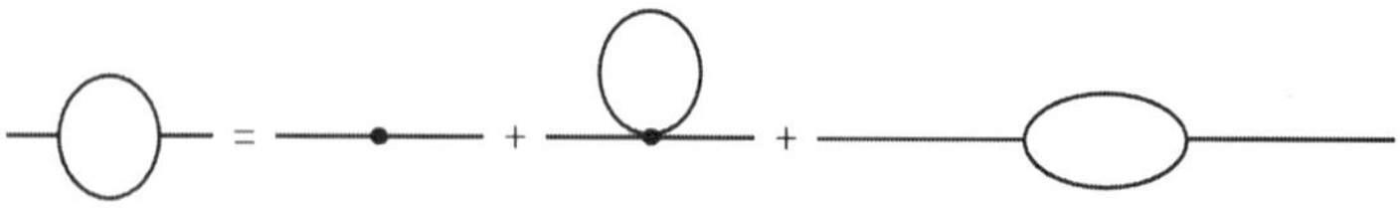

(19.23)

To give the diagram (19.23) any concrete meaning one has to attach arrows to the incoming parts of the diagrams and to sum over all possible combinations of arrow directions at the ends of internal lines. Here, the lines with arrows pointing in the same directions correspond to normal Green's functions, while those with arrows pointing in the opposite directions correspond to anomalous ones.

Equation (19.22) can be written in the form

$$\lambda = t'(0)\left(\rho_0(k_0)+\rho_1(k_0)\right)+t'(0)(\beta V)^{-1}$$
$$\sum_{\omega,k\,0<|k|<k'0} G_1(p) - 2\rho_0(k_0)(t'(0))^2(\beta V)^{-1} \sum_{\omega,k\,0<|k|<k'0}\left(G(p)G(-p)-G_1^2(p)\right)$$

It leads to the expression for $\rho_0(k_0)$

$$\rho_0(k_0)=\frac{m\lambda}{4\pi}\left(\ln\frac{\varepsilon_0}{\lambda}-2\right)-\frac{1}{(2\pi)^2}\int_{k>k0} d^2k\,\frac{\frac{k^2}{2m}+\lambda}{\varepsilon(k)}\frac{1}{\exp(\beta\varepsilon(k))-1}. \tag{19.24}$$

Contrary to the first approximation (19.19), expression (19.24) does not depend on the momentum k_0'. Summing (19.21) and (19.24), we obtain the full density

$$\rho=\frac{m\lambda}{4\pi}\left(\ln\frac{\varepsilon_0}{\lambda}-1\right)-\frac{1}{(2\pi)^2}\int d^2k\,\frac{k^2}{2m\varepsilon(k)}\frac{1}{\exp(\beta\varepsilon(k))-1} \tag{19.25}$$

The integral in (19.25) converges even if the cut–off momentum k_0 approaches zero. Thus, for the density we obtained an expression independent on k_0, k_0'. These quantities are auxiliary and they can be defined only of the order of magnitude.

Let us now write down the expression analogous to (19.25) for the pressure

$$p = \frac{m\lambda^2}{8\pi}\left(\ln\frac{\varepsilon_0}{\lambda} - 1\right) - \frac{1}{\beta(2\pi)^2}\int d^2k \ln(1 - \exp(-\beta\varepsilon(k))) \qquad (19.26)$$

and the formulae for the normal and superfluid component densities

$$\rho_n = \frac{\beta}{2m(2\pi)^2}\int d^2k k^2 \frac{\exp(\beta\varepsilon(k)}{(\exp(\beta\varepsilon(k)) - 1)^2},$$

$$\rho_S = \frac{m\lambda}{4\pi}\left(\ln\frac{\varepsilon_0}{\lambda} - 1\right) - \\ - \frac{1}{(2\pi)^2}\int\frac{d^2k}{2m}\left(\frac{k^2}{\varepsilon(k)}\frac{1}{\exp(\beta\varepsilon(k)) - 1} - \frac{\beta k^2 \exp(\beta\varepsilon(k)}{(\exp(\beta\varepsilon(k)) - 1)^2}\right). \qquad (19.27)$$

The density ρ_n can be easily calculated by evaluating the average value of momentum in the system of coordinates moving with the velocity $\boldsymbol{v}$. Such calculations can be performed using the Green's functions obtained from (19.20) by changing $i\omega \to i\omega + (\boldsymbol{v}, \boldsymbol{k})$.

Formulae (19.25)–(19.27) determine the thermodynamics below the phase transition. The equation of the phase transition curve can be found from the condition

$$\rho = \rho_n \qquad (19.28)$$

indicating that the superfluid density ρ_S vanishes. The solution of equation (19.28) may exist only if $T >> \lambda$ because if $T \leq \lambda$, then $\rho_n \leq m\lambda$, $\rho \propto m\lambda \ln(\varepsilon_0/\lambda)$ and $\rho >> \rho_n$. For $T >> \lambda$ equation (19.28) can be written as

$$\frac{m\lambda}{4\pi} \ln \frac{\varepsilon_0}{\lambda} = \frac{mT}{\pi} \ln \frac{T}{\lambda}. \tag{19.29}$$

From this formula we obtain the following expression for the transition temperature

$$T_C = \frac{\lambda}{4} \frac{\ln(\varepsilon_0/\lambda)}{\ln \ln(\varepsilon_0/\lambda)}.$$

If $T > T_C$, then the system is in the normal state and one can use the standard perturbation theory for calculation of the Green's functions and the thermodynamical functions. The one–particle Green's function is given by the formula

$$G = \left(i\omega - \frac{k^2}{2m} + \lambda' \right)^{-1}$$

where

$$\lambda' = \lambda - \frac{2}{\pi\, m \ln(\varepsilon_0/|\lambda'|)^2} \int \frac{d^2k}{\exp(\beta\left(\frac{k^2}{2m} + |\lambda'|\right) - 1} \tag{19.30}$$

and the addend to (19.30) is determined by the second diagram in the right–hand side of (19.23).

Let us now estimate the order of magnitude of the momentum k_0 for the two dimensional Bose–gas. We require that for $k > k_0$ the second order approximation to the self–energy

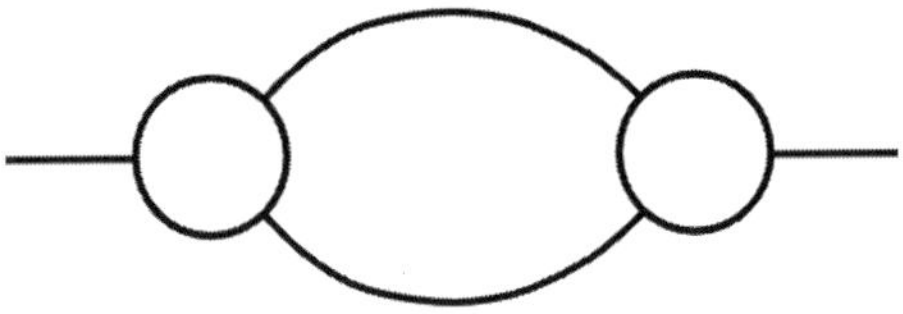

(19.31)

is not larger than the leading term which is of order λ.

Putting for simplicity the external frequency ω of this diagram equal to zero, we consider in our estimation of the order of k_0 only the term with $\omega_1 = 0$ in the sum over internal frequencies.

The expression corresponding to diagram (19.31) can be then estimated by

$$\rho_0(k_0)t'(0)^2 T \int d^2k_1 G(\boldsymbol{k}_1, 0) G(\boldsymbol{k} - \boldsymbol{k}_1, 0) \tag{19.32}$$

Here, $G(\boldsymbol{k}\ ,0) \propto mk^{-2}$ due to (19.20). If we take into account that the integration in (19.32) is carried out over the domain $k_1 > k_0$, $|k - k_1| > k_0$, we can estimate the integral by $m^2 k_0^{-2}$. As $\rho_0(k_0) \le m\lambda \ln(\varepsilon_0/\lambda)$, $t'(0) \propto [m \ln(\varepsilon_0/\lambda)]^{-1}$, we arrive to the conclusion that diagram (19.31) does not exceed in the order of the magnitude the value

$$mT\lambda [k_0^2 \ln(\varepsilon_0/\lambda)]^{-1}.$$

The ratio of diagram (19.31) to the first approximation function does not exceed the value $mT[k_0^2 \ln(\varepsilon_0/\lambda)]^{-1}$. From these relations, the lower estimate of k_0 follows

$$\frac{k_0^2}{m} >> \frac{T}{\ln(\varepsilon_0/\lambda)}.$$

We can also estimate k_0 from above requiring that the "standard" perturbation theory with normal and anomalous Green's functions is still applicable for $k^2/m \propto \lambda$. Finally, we obtain for k_0 the restricting conditions

$$\frac{T}{\ln(\varepsilon_0/\lambda)} << \frac{k_0^2}{m} << \lambda.$$

The inequality $T << \lambda \ln(\varepsilon_0/\lambda)$ following from these conditions holds for all temperatures lower than the phase transition temperature, T_C, as, according to (19.29), $T_C << \lambda \ln(\varepsilon_0/\lambda)$.

The approach to the calculation of Green's functions and thermodynamical functions of a two–dimensional Bose–gas developed above can be applied if the quantity $|1-T/T_C|$ is not very small. For temperatures $T < T_C$ the condition $\alpha << 1$ must hold in order to ensure the applicability of the perturbation theory. This condition means that the probability of creation of quantum vortices is small.

19.3. Quantum Vortices

This section is devoted to the description of quantum vortices. We shall discuss also their role in the superfluid phase transition. This role was understood in papers by Kosterlitz and Thouless[1], Berezinsky[2] and Popov[3,4]. It is well–known the existence of a periodical lattice of quantum vortices in a superconductor located in a magnetic field, quantum vortices in rotating HeII, rotating superfluid ^{3}He. In this section we present the description of quantum vortices in the functional integral formalism. We will concentrate our attention mainly upon the two–dimensional Bose–system, and show that the system of phonons and vortices occuring in such Bose–system at low temperatures is equivalent to two–dimensional relativistic electrodynamics, in which the phonons corresponds to photons and the vertices play the role of charged particles. At sufficiently low temperatures the vortices may exist only in the form of coupled pairs with opposite signs. It will be shown that the power behavior of one–particle Green's functions at large distance remains untouched even if the vortex pairs are taken into account.

Then we will discuss the role of quantum vortices in the phase transition from the superfluid to normal state. In a two–dimensional system, such a phase transition is connected with a dissociation of the coupled vortex pairs. A similar approach to the three–dimensional system leads to the conclusion that phase transition into the normal state is accompanied here by creation of long vortex filaments.

Our starting point is a method of successive integration over fast and slow fields with different perturbation schemes employed at these two stages.

After integration over the fast fields we obtain the functional $S_{eff}\left[\psi_0,\overline{\psi}_0\right]$. In the previous section we have obtained the following expression for S_{eff} :

$$S_{eff} = \int d\tau d^2x \left(ip_{\lambda\rho 0}\pi\partial_\tau\varphi - \frac{p_\lambda}{2m}(\nabla\varphi)^2 - \frac{1}{2}p_{\lambda\lambda}(\partial_\tau\varphi)^2 + \frac{1}{2}p_{\rho 0\rho 0}\pi^2 - \right.$$

$$\left. - \frac{(\nabla\pi)^2}{8m\rho_0} - \frac{\pi(\nabla\varphi)^2}{2m} \right) \tag{19.33}$$

(at the limit of $T \to 0$). Let us note that if $T \neq 0$ the coefficient in front of $-(2m)^{-1}(\nabla\varphi)^2$ in (19.33) has the meaning of superfluid density ρ_S which coincides with the full density $\rho = p_\lambda$ in the limit $T \to 0$. Taking this into account we will replace p_λ by ρ_S, if $T \neq 0$.

When we derived (19.33), we took into account only the field functions $\psi_0(\boldsymbol{x},\tau),\overline{\psi}_0(\boldsymbol{x},\tau)$, which had no zeros in $(\boldsymbol{x},\tau)$ space (more precisely $|\psi_0(\boldsymbol{x},\tau)|^2=\rho(\boldsymbol{x},\tau)$ was close to $\rho_0(k_0)$). Now let us consider the case when $\psi_0(\boldsymbol{x},\tau)$ may vanish at some discrete set of points in the x–plane (at a fixed τ). If we go around such a point, the phase variable $\varphi(\boldsymbol{x},\tau)$ acquires an increment $2\pi n$ where n is an integer. We shall consider only the points with $n=\pm 1$, which will be called the centers of quantum vortices rotating in positive or negative directions. The points with $|n|>1$ can be regarded as bound systems of $|n|$ vortices rotating with the same direction. Such a system are not stable and decay to single vortices with $|n|=1$.

Now we are going to take into account the contribution of vortex–like configurations of $\psi_0(\boldsymbol{x},\tau),\overline{\psi}_0(\boldsymbol{x},\tau)$ fields into the functional integral. From the above discussion it is clear that incorporating of vortices is equivalent to the integration over the functions $\varphi(\boldsymbol{x},\tau)$ that acquire an increment $\pm 2\pi$ after circumventing the "singular points", i.e. zeros of the functions $\psi_0(\boldsymbol{x},\tau),\overline{\psi}_0(\boldsymbol{x},\tau)$.One has to integrate over the variables the density $\rho(\boldsymbol{x},\tau)$ and the phase $\varphi(\boldsymbol{x},\tau)$, determined by the ambiguity conditions, as well as over the paths of the vortex centers in the $(\boldsymbol{x},\tau)$–space.

In the first approximation, we can neglect the terms $(\nabla\pi)^2,\pi(\nabla\varphi)^2$ in (19.33) that lead to the deviation of the phonon spectrum from linearity and to the phonon–phonon interaction. Integrating then $\exp S_{eff}$ over the variable π, we obtain the following action functional of the φ field

$$-\int d\tau d^2x\frac{\rho_S}{2m}\left((\nabla\varphi)^2+\frac{1}{c^2}(\partial_\tau\varphi)^2\right) \qquad (19.34)$$

where c is the sound velocity. The change $p_\lambda = \rho \to \rho_S$ is explained above. It ensures the validity of the formulae even ρ_S is considerably different from ρ.

Expression (19.34) represents the action of a relativistic system in Euclidean variables, where the sound velocity c plays a role of velocity of light. Continuing the analogy with relativity, we shall show that if vortices are taken into account, the action (19.34) coincides in principle with the action of two–dimensional (i.e. (2+1)–dimensional relativistic electrodynamics, where phonons play the role of photons, and vortices play the role of charged particles). Using new coordinates $x_1 = x,\ x_2 = y,\ x_3 = c\tau$ $x_1 = x,\ x_2 = y,\ x_3 = c\tau$ we arrive at the "relativistic action"

$$-\frac{\rho_S}{2mc}\int d^3x(\nabla\varphi)^2 \tag{19.35}$$

where $\nabla\varphi$ is the three–dimensional gradient of the phase variable $\varphi(x)$.

In the presence of vortices $\varphi(x)$ is not a single–valued function. We can reduce it in a single valued function by performing the shift transformation

$$\varphi(x) \to \varphi(x) + \varphi_0(x) \tag{19.36}$$

Here, $\varphi_0(x)$ is the multivalued part of $\varphi(x) + \varphi_0(x)$, obeying the Laplace equation $\Delta\varphi_0(x) = 0$, and the new $\varphi(x)$ is single–valued. To find the function $\varphi_0(x)$ we notice that its three–dimensional gradient $\nabla\varphi_0(\boldsymbol{x}) = \boldsymbol{h}(\boldsymbol{x})$ is the solution of the magnetostatic problem in three–dimensional space defined by the equations

$$curl\boldsymbol{h}(\boldsymbol{x}) = 2\pi\cdot\boldsymbol{j},\ \ div\boldsymbol{h}(\boldsymbol{x}) = 0 \tag{19.37}$$

Here, $\boldsymbol{j}(\boldsymbol{x})$ is the current density corresponding to linear unit currents slowing along the trajectories of the vortex centers. The function $\varphi_0(x)$ is a multi–valued scalar potential of the magnetic field generated by the system of linear currents. The gradient squared $(\nabla\varphi)^2$ in (19.35) transforms into the sum $(\nabla\varphi)^2+(\nabla\varphi_0)^2$ after the shift transformation (19.36). The integral of the first term describes the nonintegrating field and is irrelevant in what follows. The integral of $(\nabla\varphi)^2=h^2$ is proportional to the energy of the magnetic field of the system of linear currents.

The magnetostatic problem (19.37) can be solved by means of the vector potential $\boldsymbol{a}(\boldsymbol{x})(curl\boldsymbol{a}=\boldsymbol{h}, div\boldsymbol{a}=0)$ For a system of linear currents the vector potential is the sum of contribution from currents

$$\boldsymbol{a}(\boldsymbol{x})=\frac{1}{2}\sum_i\int\frac{d\boldsymbol{l}_i(y)}{|\boldsymbol{x}-\boldsymbol{y}|}.$$

The action obtained from (19.35) after the change $\varphi\to\varphi_0$ can be expressed as a double sum of contribution from the currents

$$-\frac{\pi\rho_S}{2mc}\sum_{i,k}\iint\frac{(d\boldsymbol{l}_i(\boldsymbol{x}),d\boldsymbol{l}_k(\boldsymbol{y}))}{|\boldsymbol{x}-\boldsymbol{y}|}. \tag{19.38}$$

The terms with $i=k$ in (19.38) diverge as $\boldsymbol{x}$ close to $\boldsymbol{y}$. This divergence is the result of the approximation in which the vortices are considered as point–like objects and the corresponding currents as linear. In order to remove divergences, it is necessary to take into account the finite size of vortices. To accomplish this, we surround the centers of vortices by circles of radius r_0, which is greater than the vortex core but less than the average vortex distance.

The influence of the finite vortex core size leads to the substitution

$$\frac{\pi\rho_S}{2mc}\sum_i \iint \frac{(d\boldsymbol{l}_i(\boldsymbol{x}), d\boldsymbol{l}_k(\boldsymbol{y}))}{|\boldsymbol{x}-\boldsymbol{y}|} \to \frac{E_v(r_0)}{c}\sum_i \int dS_i \,.$$

Here, $dS = |d\boldsymbol{l}| = \sqrt{dS^2}$, and $E_v(r_0)$ is a part of the vortex energy located in the circle of radius r_0. The expression for $E_v(r_0)$ depends on r_0 logarithmically

$$E_v(r_0) = \frac{\pi\rho_S}{m}\ln\frac{r_0}{a}. \tag{19.39}$$

It is natural to call the quantity a in (19.39) the vortex core radius. It is of the order of $(\lambda m)^{-1/2}$, where λ is the chemical potential and m is the Bose–particle mass. In order to determine a, one can use, for example, the solution of Ginzburg–Landau equations which describes the vortex structure given by Pitayevski[5].

Let us now transform the integrals over $|\boldsymbol{x}-\boldsymbol{y}| > r_0$ in (19.38). We introduce a new vector potential $\boldsymbol{A}(\boldsymbol{x})$ the expansion of which

$$\boldsymbol{A}(\boldsymbol{x}) = \int_{|\boldsymbol{k}|<\tilde{k}0} \exp(i\boldsymbol{kx})\boldsymbol{a}(\boldsymbol{k})d^3k$$

is bounded by momenta less that $\tilde{k}_0 \propto r_0^{-1}$. The action (19.38) may be transformed into the following form

$$S' = -m_v(r_0)c\sum_i \int dS_i - ig\int (\boldsymbol{A}, \boldsymbol{j})d^3x - \frac{1}{2c}\int (curl\boldsymbol{A})^2 d^3x, \tag{19.40}$$

where

$$m_v(r_0) = E_v(r_0)c^{-2}$$

is the vertex "mass" and the coefficient

$$g = 2\pi\sqrt{\frac{\rho_S}{mc^2}}$$

has a meaning of a coupling constant.

The action (19.40) written in Euclidean variables describes a system of charged particles interacting with the electromagnetic field $\boldsymbol{A}(\boldsymbol{x})$ (momenta of this field are bounded by the upper limit $\tilde{k}_0 \propto r_0^{-1}$). The functional $\exp S'$ must be integrated over the field $\boldsymbol{A}(\boldsymbol{x})$ and over the trajectories of charged particles. This is the correct procedure for quantization of a system described by the action (19.40). The integration of $\exp S'$ over the field $\boldsymbol{A}(\boldsymbol{x})$ can be performed exactly using the shift $\boldsymbol{A}(\boldsymbol{x}) \to \boldsymbol{A}(\boldsymbol{x}) + A_0(\boldsymbol{x})$, which makes the linear form of $\boldsymbol{A}(\boldsymbol{x})$ in (19.40) vanish. Then we come back to the action (19.38) and prove the correctness of the expression (19.40).

If we want to describe the motion of vortices with velocities much smaller than c, we have to make a non–relativistic approximation to the action (10.40). In this approximation we get

$$m_v(r_0)c\sum_i \int dS_i = \frac{E_v(r_0)}{c}\int dS_i \cong E_v(r_0)\int_0^\beta \left(1 + \frac{1}{2c^2}\left(\frac{dx_i}{d\tau}\right)^2\right)d\tau =$$

$$= \beta E_v(r_0) + \int_0^\beta \frac{m_v(r_0)}{2} v_i^2(\tau)d\tau.$$

The contribution of the scalar potential $A_0(\boldsymbol{x})$ to the action (19.40) can be transformed to the term describing the direct interaction of charged particles through logarithmic potential. Finally the action (19.40) in nonrelativistic approximation acquires the form

$$-\sum_i \left(\beta E_v(r_0) + \int_0^\beta d\tau \left(\frac{m_v(r_0)}{2} v_i^2(\tau) + i\frac{g_i}{c}(\boldsymbol{v}_i, \boldsymbol{A}_i) \right) \right) -$$

$$-\frac{1}{2}\int d\tau d^2x \left((\partial_1 A_2 - \partial_2 A_1)^2 + c^{-2}(\partial_\tau A)^2 \right) + \tag{19.41}$$

$$+\frac{1}{4\pi}\int d\tau d^2x j_0(\boldsymbol{x},\tau) j_0(\boldsymbol{y},\tau) \ln|\boldsymbol{x}\text{-}\boldsymbol{y}|.$$

Here, $\boldsymbol{A}_i(\boldsymbol{x})$ is the vector potential in the centre of the i–th vortex moving with the velocity $\boldsymbol{v}_i$ and having the charge g_i, where

$$\frac{g^2}{4\pi} = \frac{\pi\rho_S}{m}$$

and the function

$$j_0(\boldsymbol{x},\tau) = \sum_i g_i \delta(\boldsymbol{x}\text{-}\boldsymbol{x}_i(\tau))$$

has a meaning of the charge–density.

Let us now consider some of the consequences of the equivalence between the system of phonons and vortices and the two–dimensional electrodynamics.

If the temperatures are sufficiently low the vortices in the system may exist only in the form of opposite sign pairs coupled by long range logarithmical potential.

We shall show that the including of these coupled pairs does not affect the power character of the asymptotic behavior of one–particle Green's functions at large distances. Let us take the quadratic form in (19.33) (with ρ_λ replaced by ρ_S) as an effective action of $\exp S$. We restrict ourselves to integration over functions $\rho(\boldsymbol{x},\tau)$, $\varphi(\boldsymbol{x},\tau)$

independent of τ. One can show that the contribution of the τ – dependent functions yields in the first approximation a correction to the coefficient in front of $r^{-\alpha}$. We shall not make the explicit calculation of this coefficient here. We shall only calculate the exponent in the case when the bound pairs of vortices are considered. The problem leads to functional integration of the expression

$$\exp\left\{-\frac{\rho_S\beta}{2m}\int d^2z(\nabla\varphi(z))^2+i\varphi(\boldsymbol{x})-i\varphi(\boldsymbol{y})\right\} \tag{19.42}$$

If the vortices occur, then the functions $\varphi(z)$ acquire increments $\pm 2\pi$ corresponding to each circumventing of the vortex and are therefore ambiguous. We can make them unique by means of cuts connecting the vortices of each pair. Besides, it is suitable to suppose that the vortices as well as the "sources" $\boldsymbol{x}$ and $\boldsymbol{y}$, are surrounded by the circles of radius a (where a is the radius of the vortex core), and integrate over the domain outside these circles.

Let us shift the integration variable $\varphi(z)\to\varphi(z)+\varphi_0(z)$, where

$$\varphi_0(z)=\frac{im}{2\pi\beta\rho_S}\ln\frac{|z\text{-}\boldsymbol{y}|}{|z\text{-}\boldsymbol{x}|}\equiv i\alpha\chi_0(z)$$

If the vertices are present, then the expression

$$-\frac{i}{2\pi}\int div(\varphi\nabla\chi_0)d^2z$$

emerging after the transformation of the integrand in (19.42) contains not only the integrals over circumferences of the circles surrounding the "sources" $\boldsymbol{x}$ and $\boldsymbol{y}$, but also a sum of integrals along side the cuts. This leads to the additional multiplier

$$\exp\left\{i\sum_i r_i\left|\nabla\chi_{0i}\right|\sin\theta_i\right\}, \tag{19.43}$$

where $\boldsymbol{r}_i$ is the separation between vortices of the i –th pair, $\left|\nabla\chi_{0i}\right|$ is the modulus of the vector $\nabla\chi_{0i}$ in the point of location of the pair, and θ_i is the angle between the vectors $\boldsymbol{r}_i$ and $\nabla\chi_{0i}$. Taking the average of the multiplier (19.43) with respect to the orientation of vectors $\boldsymbol{r}_i$, we obtain

$$\exp\left\{-\frac{1}{2}\sum_i r_i^2\left|\nabla\chi_{0i}\right|^2\right\}$$

For large $r=\left|\boldsymbol{x}\text{-}\boldsymbol{y}\right|$ we have

$$\frac{1}{2}\sum_i r_i^2(\nabla\chi_{0i})^2=\frac{S_v}{S}\int\frac{(\nabla\chi_{0i})^2}{4\pi}d^2z=\frac{S_v}{S}\ln\frac{r}{a} \tag{19.44}$$

Here, S_v/S is the average relative area occupied by vortex pairs (the area occupied by the *i*–th pair with vortices separated by $\boldsymbol{r}_i$ is assumed to be $2\pi\ r_i^2$).

Formula (19.44) shows that the power character of the asymptotic behavior remains valid, even if the vortices are taken into account, and in this case the exponent α acquires the addend

$$\Delta\alpha=S_v/S\,.$$

This addend is small with respect to α provided α itself is sufficiently small ($\Delta\alpha/\alpha\to 0$ as $T\to 0$), and it has a meaning of a relative area occupied by vortex pairs.

Let us now study the role of quantum vortices in the phase transition from the superfluid to the normal state. When the temperature increases, the number of coupled vortex pairs increases and the mean distance between pairs decreases. Finally at some temperature T_C, the coupled pairs start to dissociate. Above the dissociation temperature, single vortices as well as the coupled pairs appear in the system and we have a plasma–like state. It is natural to identity this dissociation phase transition with the phase transition from the superfluid to the normal state.

For $T > T_C$, the long–range correlations disappear in the plasma–like state due to the characteristic Debye screening. In particular, the correlator $\langle\psi(\boldsymbol{x},\tau)\overline{\psi}(\boldsymbol{x}',\tau')\rangle$ decreases exponentially for $T > T_C$. Notice moreover that the vanishing of the second sound for $T > T_C$ can be interpreted as a transformation of the second sound branch into the plasma oscillations branch.

At the same time the quantity ρ_S defined as the coefficient of $-(\nabla\varphi)^2/2m$ in the effective action does not vanish even for $T > T_C$. This coefficient is analogous to the quantity ρ_S introduced by Berezinski[2] for the model of planar rotators. For $T < T_C$, this coefficient is in fact equivalent to macroscopic superfluid density defined everywhere except a small phase transition domain, in which an intensive creation of quantum vortices occurs. Thus, we arrive to the conclusion that the description of a two–dimensional Bose–system in terms of normal and superfluid components with quantum vortices is possible both below and above the phase transition point and that it becomes meaningless only if the vortex core radius is of the same order as the mean vortex separation.

The method of description of quantum vortices developed above can be extended to three–dimensional Bose–system, too. In this case, the quantum vortices again correspond to the zeros of $\psi(\boldsymbol{x},\tau)$ functions, as it was in the two–dimensional systems. The complex function $\psi(\boldsymbol{x},\tau)$ defined in the three–dimensional space $\boldsymbol{x}\in V$ acquires zero values alongside lines (sets of dimension 1), and in the four–dimensional space–time (x,τ) on two–dimensional surfaces. One has to integrate first over the functions $\psi,\overline{\psi}$ equal to zero on two–dimensional surfaces, and then over the surface configurations.

Here, we restrict ourselves only to a qualitative estimation of the role of the quantum vortices in the phase transition. The conclusions are analogous to those in the papers by Bycling[7] and by Wiegel[8], where the authors use the analogy with the Ising model. At low temperatures specific excitations may occur in a nonrotating Bose–system which have the form of vortex rings. The closed lines where functions $\psi,\overline{\psi}$ vanish correspond to these rings in the formalism of the functional integral. When the temperature rises, the number of vortex rings in a volume unit increases and the mean distance between them decreases. When the mean distance between the rings becomes comparable with the mean ring length, the system exhibits the tendency to create very long rings. It is natural to assume that the phase transition between the superfluid and normal state is connected with the creation of vortex lines of infinite length (in a real system the lines which begin and end on the vessel walls). A long vortex line does not emerge immediately, but rather by subsequent joining and an increasing of vortex rings of finite length. Thus, if the number of vortex rings per volume unit is sufficiently large, the probability of the existence of infinitely long vortex lines may be anything but infinitesimally small.

Notice that the quantity ρ_S defined as the coefficient of $-(\nabla\varphi)^2/2m$ in the effective action is different from zero even above the phase transition point, as well as in two–dimensional system. In other words, the description of a system in terms of normal and superfluid components filled with quantum vertices is possible above the phase transition, too. In connection with this analysis, an idea emerges that the difference between normal and superfluid liquid is due to different characters of vortex excitations. In the normal state there exist not only vortex rings but also (infinitely) long vortex filaments.

Thus, the above qualitative consideration shows that in a three–dimensional system the phase transition is related to quantum vortices, much the same way as in two dimensions. Closed vortex rings of a three dimensional system correspond to coupled vortex pairs of two dimension and long vortex lines are analogous of single vortices.

19.4. One–Dimensional Systems

Can the superfluidity occur in one–dimensional Bose systems? Let us consider a one–dimensional gas of Bose–particles interacting through a repulsive δ–function potential

$$u(x-y)=g\delta(x-y).$$

For such a system, the exact expression for the ground state energy is obtained by Lieb and Liniger[9]. The thermodynamical functions found by C.N.Yang and C.P.Yang[10] have no singularities at finite temperatures, and, consequently, in such a system the phase transition does not occur for nonzero temperature. The possibility of superfluidity of this model at $T=0$ has been studied by Sonin[11].

We shall show that in this system long–range correlations decreasing like a power of r are possible at $T=0$. Particles with small momenta create the bare condensate which give rise to a superfluid density of the same order as full density ρ. One can convince oneself of this by applying the method of derivation of the effective action to the one–dimensional case. In this case, the equation $\partial S_{eff}/\partial\rho_0=0$ determines the bare condensate density $\rho_0(k_0)$, which has the power like asymptotic behavior

$$\rho_0(k_0)\propto k_0^{\gamma},\ \gamma=\frac{mc}{2\pi\rho}.$$

The exponent is expressed in terms of the density and the sound velocity. The asymptotic behavior of Green's function for $r\to\infty$ and low temperatures (in particular, at $T=0$) is given by the formula

$$G \approx -\rho_0(k_0)\exp\left\{-\frac{m}{2\rho\beta V}\sum_{\omega,|k|<k0}\left(k^2+\frac{\omega^2}{c^2}\right)^{-1}\times\right. \tag{19.45}$$
$$\left.\times\left|\exp\left(i(\omega\tau+\boldsymbol{kx}\right)-\exp\left(i(\omega\tau'+\boldsymbol{kx}'\right)\right|^2\right\}$$

analogous to (19.8), (19.9). At $T=0$ formula (19.45) takes the form

$$G \approx -\rho_0(k_0)\exp\left\{-\frac{mc}{4\pi\rho}\ln k_0^2\left(r^2+c^2\delta^2\right)+const\right\},$$

where $\delta=\tau-\tau'$, or, equivalently,

$$G \approx -a\left(r^2+c^2\delta^2\right)^{-\gamma/2} \tag{19.46}$$

The selfconsistence of the assumption $\rho_0(k_0)\propto k_0^\gamma$ is evident. Indeed, if $\rho_0(k_0)\propto k_0^\gamma$, then the coefficient at (19.46) does not depend on k_0 and from (19.46) we obtain that $N(k)\propto k^{\gamma-1}$, $\rho_0(k_0)\propto k_0^\gamma$. If $T\neq 0$, then expression (19.45) yields the exponentially decreasing Green's function

$$G \propto \exp\left(-mr/2\beta\rho\right)$$

in the limit $r\to\infty$.

We can see, that in a one–dimensional Bose–system, the long–range correlations decreasing like a power of r (and not exponentially) may exist only if $T=0$. It is therefore natural to conclude that a one–dimensional system can posses the property of superfluidity only at zero temperature. This conclusion is in agreement with the result of the exact themodynamical calculations[10] which do not indicate the violation of analyticity of thermodynamical functions for nonzero values of temperatures.

19.5. Superfluidity in Fermi–Films. Singlet Pairing

Recently, many experiments were done with few atomic ^{3}He–films (having a different rate of filling, including monoatomic) absorbed on the sublayer. It makes the search of possible phase transitions (especially in the superfluid state) in such films very urgent. Both single and triplet types of pairing may occur in different situations. The s–pairing is more preferable in the incomplete filled layer when the mean distance between ^{3}He atoms are large as compared with the size of hard cores, and the influence of the cores may be neglected. At the case of the completely filled layer or for the several–atomic layers the *p*–pairing is preferable.

Kurihar[12] has considered the superfluidity in ^{3}He–films with *s*–pairing. The ^{3}He–film lies on the layer of the superfluid ^{4}He which is situated on the sublayer (for instance, it may be quartz)

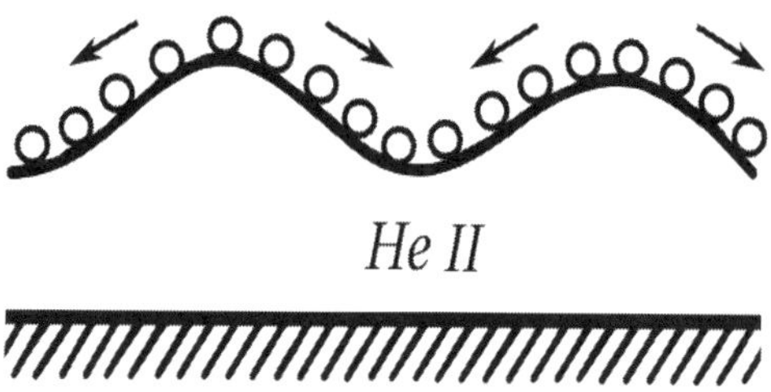

FIG. 19.1.The mechanism of pairing of ^{3}He atoms situated on the ^{4}He films.

The pairing mechanism of ^{3}He atoms is due to the exchange by the quanta of the third sound between ^{3}He–qiasiparticles. Oscillations of the ^{4}He surface (the third sound) lead to the modulation of the Van der Vaals potential of the sublayer which is proportional to D^{-3}, where D is a thickness of the ^{4}He layer. The potential at the hollows of a wave is more intensive than that on the crests. As a result, the density of ^{3}He atoms increases in hollows. This leads to the correlations between the ^{3}He density and the third sound waves.

The system of ^{3}He–film–^{4}He which is analogous qualitatively to the electron–phonon system can be described by the Hamiltonian

$$\hat{H} = \sum_{k,\sigma} \varepsilon_k a_{k\sigma}^+ a_{k\sigma} + \sum_q \omega_q b_q^+ b_q + N^{-1/2} \sum_{q,k,\sigma} g_q \left(b_q + b_{-q}^+\right) a_{k+q,\sigma}^+ a_{k\sigma} ,$$

where $a_{k\sigma}, b_q$ are annihilation operators of the quasifermions of ^{3}He and of the third sound quanta, ε_k, ω_q are their energies, g_q is the coupling function, N is a number of ^{4}He–atoms. The coupling function g_q has a form of

$$g_q = m_4 c^2 \left[q^2 / 2m_4 \omega_q\right]^{1/2}$$

in the case of a monoatomic layer. Here m_4 is the mass of ^{4}He–atom and

$$c = \left[3u_4 h_4 / m_4 n_4 (d + h_4)\right]^{1/2}$$

is the third sound velocity, which is expressed via the Van der Vaals energy u_4, the mean three–dimensional density of ^{4}He, n_4 and the thicknesses of the superfluid layer of ^{4}He, h_4 and that of the solid layer d.

The attraction analogous to that in the electron–phonon system arises between ^{3}He quasiparticles. This attraction is characterized by the potential

$$V_{k,q} = g_q^2 D(\boldsymbol{q}, \omega) = 2 g_q^2 \omega_q / \left[\left(\varepsilon_{k+q} - \varepsilon_k\right)^2 - \omega_q^2\right],$$

$$- 2 g_q^2 / \omega_q = - m_4 c^2 ,$$

where $D(\boldsymbol{q}, \omega) = 2\omega_q / \left(\omega^2 - \omega_q^2\right)$ is the Green's function for the third sound quantum. However this analogy of our system and a superconductor is not complete. This point is that the Debye energy E_D in metals are much smaller than the Fermi–energy E_F. But in our system the situation is inverse. Thus, for a two–atomic layer of ^{4}He we have $E_D \propto 5K$, and $E_F \propto 0.1K$ for the submonoatomic ^{3}He layer. That is why the attraction between ^{3}He atoms does not confined by only the neighborhood of the Fermi circumference, but it takes place for each state. Let us note that the analogous situation was observed also in some superconductors such as $CeCuSi_2$, where $\varepsilon_F \propto 10\,K, \omega_D \propto 200\,K, T_c \propto 0.6K$.

The estimate for a dimensionless coupling constant yields

$$\lambda = \lambda_0 D^{-4}$$

where

$$\lambda_0 = 3m_3 u_4 / \pi n_4 a ,$$

where

a is a mean distance between ^{4}He atoms, $D = h_4 + d$. For the quartz sublayer we have $\varepsilon_D \cong 14.5K$, hence $\lambda_0 = 11$,

Let us note that λ may be sufficiently large, for example, $\lambda \approx 1$ for $D \approx 1.8$ and it can be increased, if we take a sublayer with the greater Van der Vaals energy. In our case, when $\varepsilon_D >> \varepsilon_F$, there is no theory for such a situation and it is hard to calculate the phase transition temperature T_C . We can take only the qualitative estimate for T_C using the BCS formula

$$T_C \propto \varepsilon_F \exp(-1/\lambda)$$

in which the Debye energy ε_D is replaced by the Fermi–energy ε_F . This estimate yields $T_C = 10\,mK$ for the ^{3}He–film on the quartz sublayer with two layers of ^{4}He, but T_C decreases rapidly with the increase of D.

Let us make several remarks on the nature of the superfluid transition. It is natural to suppose that this transition should be of the Kosterlitz–Thouless type (KT) connected with the dissociation of vortex–antivortex pairs at $T = T_C$. It is essential for such a transition that the vortex density is rather small near the transition point. In order to clear is it the case let us calculate the energy of a separate vortex and compare it with the phase transition temperature T_C. The order parameter inside the vortex can be written in the form

$$\tilde{\Delta}(r) = \Delta \cdot th(r / \xi)$$

The core energy can be calculated as a difference of the condensate energies with a core and without a core:

$$E_C = \int_0^{2\pi} 2\pi r dr \frac{N(0)}{2\rho S}\left[\Delta^2 - \tilde{\Delta}^2(r)\right] = \xi^2 \Delta^2 \int_0^{\infty} \frac{x dx}{ch^2 x},$$

where $N(0)/\rho S$ is the density of states per one particle, $N(0) = m_3 S/\pi$. Integrating over x and substituting $\xi = c_F/\pi\Delta$, we obtain

$$E_C = \frac{2\ln 2}{\pi^2} e^{1/\lambda}.$$

If we use the above BCS formula for T_C, we shall have

$$\frac{E_C}{T_C} = \frac{2\ln 2}{\pi^2} e^{1/\lambda}.$$

At the case of a weak coupling $E_C >> T_C$, but for a strong coupling $E_C << T_C$. We have $E_C = T_C$ at $\lambda = \lambda^* = \left[\ln\left(\pi^2/2\ln 2\right)\right]^{-1} = 0.509$. It corresponds to $D^* = 2$ for a quartz sublayer.

It implies that the phase transition into the singlet superfluid state is the transition of the Kosterlitz–Thouless type for large D $\left(D > D^*\right)$ but the character of this transition changes for small D. It would be interesting to check experimentally this change of the phase transition character with changing of thickness of ^{4}He–film.

At the conclusion let us stress once more, that the singlet pairing discussed above should take place in submonoatomic films of ^{3}He. In this case a singlet electroneutral superfluidity is realized. At the case of large ^{3}He density the repulsion of hard cores becomes important. This leads to the triplet pairing of ^{3}He atoms in such films, and a triplet electroneutral superfluidity takes place. The study of such a type of superfluidity will be done at the next section.

19.6. Triplet Pairing. Thick Films

Thus, in the case of the completely filled layer or of the several ^{3}He atomic layer of ^{3}H the phase transition into the superfluid state with the triplet pairing should take place. The method of investigation of superfluidity in the ^{3}He–films essentially depends on the ratio of the film thickness d to the coherence length ξ.

If $d >> \xi$, we can apply the three–dimensional model taking the boundaries into account. If $d < \xi$, we may consider this system as a two–dimensional one and use an approach in the framework of the two–dimensional model.

In this section the first case is considered, i.e. the case of a sufficiently thick film $d > \xi$ according to the paper by Fujita et al[25]. The subsequent sections of this Chapter are devoted to the second case, i.e. to the really two–dimentional ^{3}He.

Thus, in the bulk ^{3}He the *A*– and the *B*–phases are described by the following order parameters:

A–phase: $$A_{i\mu} = \lambda \hat{S}^{(3)}_{\mu}\left(\hat{l}^{(1)}_{i} + i\hat{l}^{(2)}_{i}\right),$$

B–phase: $$A_{i\mu} = \sum_{p=1}^{3} \lambda_p \hat{S}^{(p)}_{\mu}\hat{l}^{(p)}_{i},$$

where $\lambda_1 = \lambda_2 = \lambda_3$, $\left(\hat{l}^{(1)}, \hat{l}^{(2)}, \hat{l}^{(3)}\right)$ and $\left(\hat{S}^{(1)}, \hat{S}^{(2)}, \hat{S}^{(3)}\right)$ are the orthogonal sets of vectors in the orbital and spin space correspondingly. As it was shown above, in the weak–coupling approximation the *B*–phase has a minimum of the free energy for all values of pressures and temperatures. Paramagnetic effects stabilize the *A*–phase at the region of high temperatures and pressures.

The presence of boundaries deforms the order parameter. The component of the A_μ vector, which is perpendicular to the boundary, vanishes on the boundary for specular and diffuse types of the atomic reflection. In other words Cooper pairs have a tendency to move at planes parallel to the boundary (in what follows we choose the (x, y)–plane as the boundary).

The *A*–phase in the ^{3}He – film (if $d >> \xi$) is described by the same order parameter as that in the bulk ^{3}He, if the direction of the vector $\left(\hat{l}^{(3)} = \hat{l}^{(1)} \times \hat{l}^{(2)}\right)$ is chosed to be parallel to z–axis.

In the case of the *B*–phase the order parameter in the bulk ^{3}He does not obey the boundary condition if it is not deformed. The deformation is connected with the appearance of the additional energy and decreasing the condensation energy. Thus, the deformation of the order parameter in the *B*–phase leads to the change of the free energy of the ^{3}He–film. When the film thickness is less than some D_C the *B*–phase is not energetically preferable than the *A*–phase even for the region of pressures and temperatures where it was stable in the bulk ^{3}He, and the phase transition from the *B*– to the *A*–phase occurs at $D = D_C$.

In order to find D_C, let us compare free energies of both states.

The free energy density, in which paramagnon effects are taken into account, look as

$$F = F_2 + F_4 + F_k ,$$

where

$$F_2 = \alpha\Delta_0^2\left(-\left|d_{i\mu}\right|^2\right), F_4 = \left[\alpha\Delta_0^2/(10-2\delta)\right]\left\{\beta_1\left|\left(d_{i\mu}\right)^2\right|^2 + \beta_2\left(d_{i\mu}^* d_{i\nu}^* d_{j\mu} d_{j\mu}\right) +\right.$$

$$\left. + \beta_3\left(d_{i\mu}^* d_{j\mu}^* d_{i\nu} d_{j\nu}\right) + \beta_4\left(\left|d_{i\mu}\right|^2\right)^2 + \beta_5\left(d_{i\mu}^* d_{j\nu}^* d_{i\nu} d_{j\mu}\right)\right\},$$

$$F_k = \alpha\Delta_0^2\xi^2\left\{\left|\nabla_i d_{i\mu}\right|^2 + \left|\nabla_i d_{j\mu}\right|^2 + \nabla_i d_{j\mu}^* \nabla_j d_{i\mu}\right\},$$

$$\Delta_0^2 = \frac{10\left(\pi T_C\right)^2}{7(5-\delta)\zeta(3)}\left(1 - \frac{T}{T_C}\right), \quad \alpha = \frac{N(0)}{3}\left(1 - \frac{T}{T_C}\right), \quad \beta_1 = -(1+0.1\delta),$$

$$\beta_2 = (2 - 0.05\delta), \quad \beta_3 = -(2 + 0.7\delta),$$

$$\beta_4 = (2 + 0.2\delta), \quad \beta_5 = (2 - 0.55\delta),$$

$$\xi = \left(7\zeta(3)/160\right)^{1/2}\left(c_F/\pi T_C\right)\left(1 - T/T_C\right)^{-1/2} \equiv \xi_0\left(1 - T/T_C\right)^{-1/2},$$

δ is the paramagnon parameter.

The free energy density is equal to

$$F^B = -\frac{3}{2}\alpha\Delta_0^2 \qquad \text{in the } B\text{–phase;}$$

$$F^A = -\lambda_0^2\alpha\Delta_0^2 \qquad \text{in the } A\text{–phase;}$$

where

$$\lambda_0^2 = (50 - 10\delta)/(40 - 21\delta).$$

Now, it is necessary to impose the boundary conditions. The one possibility is to take one of vectors $\hat{l}^{(p)}$ to be normal to the boundary. The corresponding coefficient λ_p must vanish on the boundary. In this case we shall obtain the so–called planar or 2*D*–phase at the boundary (if two other coefficients are equal to each other). The another possible boundary state is the *A*–phase.

We shall compare the boundary energies for both cases: $B \to 2D$ and $B \to A$. Let us suppose for the simplicity the film is located in the upper half–space $(z > 0)$. The order parameter in the transition region look as follows

$B \to A$

$$A_{i\mu} = f(z)\begin{pmatrix} 1 & 0 & 0 \\ 0 & 1 & 0 \\ 0 & 0 & 1 \end{pmatrix} + (1 - f(z))\begin{pmatrix} 1 & 0 & 0 \\ i & 0 & 0 \\ 0 & 0 & 0 \end{pmatrix},$$

$B \to 2D$

$$A_{i\mu} = f(z)\begin{pmatrix} 1 & 0 & 0 \\ 0 & 1 & 0 \\ 0 & 0 & 1 \end{pmatrix} + (1 - f(z))\lambda_0'\begin{pmatrix} 1 & 0 & 0 \\ 0 & 1 & 0 \\ 0 & 0 & 0 \end{pmatrix}$$

where $(\lambda_0')^2 = (50 - 10\delta)/(40 - 11\delta)$. The boundary conditions are

$f = 0$ at $z = 0$ and $f = 1$ at $z \to \infty$.

In the first case we obtain the free energy in the form

$$F = \alpha\Delta_0^2\left\{A_1\left(\frac{df}{dz}\right)^2 - f^2\left[B_1(1-f)^2 + C_1\right]\right\},$$

where

$$A_1 = (1-\lambda_0)^2 + \lambda_0^2 + 4,$$
$$B_1 = \beta_1\left[(\lambda_0-1)^2 - \lambda_0^2 + 2\right]^2 + \beta_4\left[(\lambda_0-1)^2 + \lambda_0^2 + 2\right]^2 +$$
$$+\beta_2\left[(\lambda_0-1)^4 + \lambda_0^4 + 2 - 2(\lambda_0-1)^2\lambda_0^2\right] +$$
$$+\beta_{35}\left[(\lambda_0-1)^4 + \lambda_0^2 + 2 + 2(\lambda_0-1)^2\lambda_0^2\right],$$
$$C_1 = 3(3\beta_{14} + \beta_{235}) - 4\beta_{345}\lambda_0^4,\ \ \beta_{ij} = \beta_i + \beta_j,$$
$$\beta_{ijk} = \beta_i + \beta_j + \beta_k.$$

Here, z is normalized to the coherence length ξ. Minimizing F with respect to f we obtain the Euler–Lagrange equation. It is not hard to solve it and to find the surface energy density:

$$\frac{\sigma_A}{|F^B|\xi} =$$
$$= \frac{2}{3}\sqrt{A_1}\left\{\frac{2}{3B_1}\left[C_1^{3/2} - (B_1 + C_1)^{3/2}\right] + (B_1 + C_1)^{1/2} + \frac{C_1}{\sqrt{B_1}}\, sh^{-1}\sqrt{\frac{B_1}{C_1}}\right\}.$$

In the second case ($B \to 2D$) the surface energy σ_{2D} may be obtained from σ_A by changing $A_1, B_1, C_1 \to A_2, B_2, C_2$ where

$$A_2 = 2(\lambda_0'-1)^2 + 3\lambda_0'^2,$$
$$B_2 = \beta_{12}\left[2(\lambda_0'-1)^2 + 1\right]^2 + \beta_{235}\left[2(\lambda_0'-1)^4 + 1\right],$$
$$C_2 = 3(3\beta_{14} + \beta_{235}) - 2(2\beta_{14} + \beta_{235})\lambda_0'^4.$$

Calculations[26] of σ_{2D} and σ_A were done for $0.05 < \delta < 0.25$. According to these results always $\sigma_A = \sigma_{2D}$. It means that the transition $B \to 2D$ on the boundary is preferable.

Let us look for the order parameter in the presence of a surface. We shall consider both cases namely the specular reflection and the diffusion scattering.

Specular reflection

The order parameter in the *B*–phase is not changed and that in the *B*–phase is deformed, and we have

$$\lambda_1 = \lambda_2 = u(z), \;\; \lambda_3 = v(z)\,.$$

We see, that the free energy is expressed in terms of $u(z)$ and $v(z)$. Instead of solution variational equations we can choose for $u(z)$ and $v(z)$ the following test–functions

$$u(z) = \begin{cases} u_0\left(1 + \eta \cos^2\left(\pi z/2 z_0\right)\right), & 0 \le z \le z_0, \\ u_0, & z_0 \le z \le d/2, \end{cases}$$

$$v(z) = \begin{cases} v_0 \sin\left(\pi z/2 z_0\right), & 0 \le z \le z_0, \\ v_0, & z_0 \le z \le d/2. \end{cases}$$

For a given ratio of the film thickness D to the coherence length ξ we have to minimize the following free energy, averaging over the film

$$\overline{F} = \alpha\Delta_0^2\{-\left(2 < u^2 > + < v^2 >\right)^2 + \left(10 - 2\delta\right)^{-1}[\left(8 - 2.2\delta\right) < u^4 >$$
$$+ \left(3 - 1.2\delta\right) < v^4 > + \left(4 - 0.4\delta\right) < u^2 v^2 > + \xi^2\left(2 < u'^2 > + 3 < v'^2 >\right)]\}$$

with respect to u_0, η, v_0, z_0. Here u', v' are derivatives with respect to z, the bar means the average over the film. Then we can find the critical thickness D_C from the condition $\overline{F}_B = \overline{F}_A$ (Table 19.1). At the weak coupling limit $(\delta \to 0)$ D_C does not exist, because at $D, v_0 \to 0, u(z) \to const$ and we obtain the $2D$–phase with the same energy as that of the A–phase. If $\delta \neq 0$, the $2D$–phase energy exceeds that of the A–phase, and we have a finite D_C. This D_C increases along with the increase of the paramagnon effects.

Diffusion scattering

Here both phases (A– and B–) are deformed. We put $\lambda = \lambda_0 w(z)$ for the A–phase and choose the test function $w(z)$ in the following form

$$w(z) = \begin{cases} w_0 \sin\left(\pi z/2 z_0'\right), & 0 \le z \le z_0', \\ w_0, & z_0' \le z \le d/2. \end{cases}$$

with two variational parameters z_0', w_0. For the average free energy we obtain

$$\overline{F} = \lambda_0^2 \alpha \Delta_0^2 \left(-2 < w^2 > + < w^4 > + 2\xi^2 < w'^2 >\right), \quad w' = dw/dz$$

For the B–phase we choose

$$u(z) = \begin{cases} u_0 \sin\left(\pi z/2 z_0\right), & 0 \le z \le z_0, \\ u_0, & z_0 \le z \le d/2, \end{cases}$$

$$v(z) = \begin{cases} v_0 \sin\left(\pi z/2 z_0\right), & 0 \le z \le z_0, \\ v_0, & z_0 \le z \le d/2. \end{cases}$$

The results for z_0', w_0 and u_0, v_0, z_0 are listed in the Table 19.1.

It is easy to find the superfluid density in the *A*– and *B*–phases:

$$\overline{\rho}_{Sx}^{B} = \overline{\rho}_{Sy}^{B} = \frac{1}{5}\left(4 < u^2 > + < v^2 >\right)\rho_{S0},$$

$$\overline{\rho}_{Sx}^{A} = \overline{\rho}_{Sy}^{A} = w^2 \rho_{S\perp},$$

where ρ_{S0} is superfluid density in the the bulk B–phase, $\rho_{S\perp}$ the superfluid density direction orthogonal to the $\boldsymbol{l}$–vector in the bulk A–phase. The values of $\overline{\rho}_{Sx}^{A}$ and $\overline{\rho}_{Sx}^{B}$ at $D = D_C$ are listed in the Table 19.1. The third sound velocity $c^{(3)}$ is proportional to $\sqrt{\overline{\rho}_{Sx}}$ in the longwavelength limit, its changing for the transition is shown in the Table 19.1.

Let us consider the NMR in the film. We have the following expressions for the dipole energy F_D and the magnetic energy F_H:

$$F_D = \frac{g_D}{5}\left\{-\frac{2}{3}\sum_p \lambda_p^2 + \sum_{p,p'} \lambda_p \lambda_{p'}\left(\left(\hat{s}^{p}\hat{l}^{p}\right)\left(\hat{s}^{p'}\hat{l}^{p'}\right) + \left(\hat{s}^{p}\hat{l}^{p'}\right)\left(\hat{s}^{p}\hat{l}^{p'}\right)\right)\right\},$$

$$F_H = const + \sum_p \left(\lambda_p \hat{s}^{p} H\right)^2,$$

where $g_D = \frac{1}{2}\pi\gamma^2\Delta_0^2$.

The rotation matrix $R_{pp'} = \hat{s}^{p}\hat{l}^{p'}$ may be defined by the rotation axis $\hat{n}$ and the rotation angle θ of the orbital space with respect to the spin space. If $\lambda_1 = \lambda_2 = u > v = \lambda_3$, as it is the case for our test functions, F_D has a minimum if $\hat{n} \| \hat{s}^{(B)}$ (and $\hat{n} \| \hat{l}^{(3)}$).

This minimal value is

$$F_D = \frac{2g_D}{5}\left(4 < u^2 > \cos^2\theta + 2 < uv > \cos\theta - < v^2 >\right).$$

Minimizing F_D with respect to the angle θ, we obtain

$$\cos\theta = - < uv > / 4 < u^2 > .$$

This angle θ changes between 104° (this is its bulk or Leggett value of θ and 90° (when δ, d and T change).

Let us consider small oscillations around configurations

$$\boldsymbol{H} \| \hat{n} \| \hat{S}^{(B)} \| \hat{l}^{(3)} \| \hat{z} \text{ and } \theta = \arccos\left(- < uv > / 4 < u^2 >\right).$$

The small rotation of the spin basis, described by the vector $\boldsymbol{\omega}(\omega_x, \omega_y, \omega_z)$ leads to the increase of the dipole energy

$$\delta F_D = \frac{g_D}{5}\left(\omega_x^2 + \omega_y^2\right) \times$$
$$\times \left\{ \frac{1}{< u^2 >} \left[(< u^2 >)^2 + (< uv^2 >)^2 - 2 < u^2 >< v^2 > \right] \right\} +$$
$$+ \frac{g_D}{5} \omega_z^2 \left\{ \frac{1}{2 < u^2 >} \left[16 (< u^2 >)^2 - (< uv >)^2 \right] \right\}.$$

This implies the following values for the longitudinal and transverse NMR:

$$\left(\Omega_z^{(B)}\right)^2 = \frac{1}{15 < u^2 >} \left[16 (< u^2 >)^2 - (< uv >)^2 \right] \Omega_B^2, \; \left(\Omega_x^{(B)}\right)^2 = \left(\Omega_y^{(B)}\right)^2 =$$
$$= \frac{1}{15 < u^2 >} \left[(< u^2 >)^2 + (< uv >)^2 - 2 < u^2 >< v^2 > \right] \Omega_B^2,$$

where Ω_B is the frequency of the longitudinal NMR in the bulk B–phase.

One can show that the NMR frequency in the *A*–phase is proportional to $\sqrt{< w^2 >}$. The values $\Omega_{x,y}^{(A),(B)}$ and $\Omega_{z}^{(A),(B)}$ for the $A \to B$ transition are listed in the Tables 19.1.

Table 19.1.a. The values of variational parameters and observables. The specular reflaction.

δ	$\frac{D_C}{2}$	z_0/ξ	u_0	η	v_0	$\rho_{SX}^{(B)}/\rho_{S0}$	$\frac{C_3(A)}{C_3(B)}$	$\Omega_{XY}^{(B)}/\Omega_B$	$\Omega_Z^{(B)}/\Omega_B$
0.1	5.3	4.9	1.06	0.05	0.75	1.0055	1.014	0.33	1.11
0.2	7.0	4.5	1.03	0.08	0.89	1.0085	1.031	0.28	1.09
0.3	10.5	6.3	1.02	0.11	0.95	1.0088	1.0088	0.22	1.07

Table 19.1.b. The values of variational parameters and observables. The diffusion scattering

δ	$\frac{D_C}{2}$	$\frac{z_0}{\xi}$	u_0	v_0	$\frac{z'_0}{\xi}$	w_0	$\frac{\rho_{SX}^{(B)}}{\rho_{S0}}$	$\frac{\rho_{SX}^{(A)}}{\rho_{S\perp}}$	$\frac{C_3(A)}{C_3(B)}$	$\frac{\Omega_A}{\Omega_B}$	$\frac{\Omega_{X,Y}^{(B)}}{\Omega_B}$	$\frac{\Omega_Z^{(B)}}{\Omega_B}$	$\frac{\Omega_{X,Y,Z}^{(B)}}{\Omega_B}$
0.1	4.9	2.8	1.06	0.62	2.4	1.10	0.700	0.720	1.032	0.59	0.27	0.92	0.48
0.2	6.3	2.9	1.05	0.76	2.5	1.13	0.762	0.766	1.039	0.60	0.23	0.93	0.51
0.3	9.0	3.1	1.03	0.86	2.5	1.17	0.833	0.830	1.054	0.61	0.19	0.95	0.55

19.7. Model of ^{3}He–Film

In the case of ^{4}He the superfluid properties of films are essentially identical to those of a bulk liquid because the superfluid coherence lengths ξ $(10^{-8}\,cm)$ is much smaller than the saturated film thickness d $(3 \cdot 10^{-6}\,cm)$.

In contrast to this, the coherence length ξ in superfluid ^{3}He ($5 \cdot 10^{-6}\, cm$ at p=0) can be comparable to film thickness d. Thus ^{3}He is a more interesting object to study some two dimensional effects such as 2D–superfluidity.

There are two different criterions of two–dimensionalities: weak and strong. Weak one is determined by general relation $d < \xi$: in this case one can speak about 2D–system. Strong criterions follow from specific heat measurements. They show that the behavior of the film becomes "three–dimensional" in the case of thickness of three and more layers, and remains "two–dimensional" for one–atomic and two–atomic layers. In addition to the investigation of the thermo–physical properties and NMR experiments, an important method of study of the films is the study of the spectrum of collective excitations, like it is the case in three–dimensional ^{3}He.

We consider a model of ^{3}He–film – the two–dimensional analog of the three–dimensional ^{3}He–model described above. As it is shown below several superfluid phases turn out to be possible in the model. Two of them, denoted below by a and b, are energetically advantageous and stable relative to small perturbations. The Bose–spectrum is studied by the method developed above for the three–dimensional – case. The spectrum contains both Goldstone modes (gd), the number of which is different for different phases, and also non–phonon modes (nph) which have an energy gap at $k = 0$.

The ^{3}He–film model is described by the effective action functional

$$S_{\text{eff}} = g_0^{-1} \sum_{p,i,a} c_{ia}^{+}(p) c_{ia}(p) + \frac{1}{2} \ln \det \frac{\hat{M}\left(c_{ia}, c_{ia}^{+}\right)}{\hat{M}(0,0)} \tag{19.47}$$

where $\hat{M}\left(c_{ia}, c_{ia}^{+}\right)$ is the 4×4 matrix depending on Bose–fields and parameters of quasi–fermions

$$M_{11} = Z^{-1}\left[i\omega + \xi - \mu(\boldsymbol{H\sigma})\right]\delta_{p_1 p_2},$$

$$M_{22} = Z^{-1}\left[-i\omega + \xi + \mu(\boldsymbol{H\sigma})\right]\delta_{p_1 p_2},$$

$$M_{12} = (\beta V)^{-1/2}(n_{1i} - n_{2i})c_{ia}(p_1 + p_2)\sigma_a,$$

$$M_{21} = -(\beta V)^{-1/2}(n_{1i} - n_{2i})c^{+}_{ia}(p_1 + p_2)\sigma_a \tag{19.48}$$

Here, the index i takes on two values, $i = 1$ and 2 and it is this which distinguishes the two–dimensional and the three–dimensional systems models (remind, that for 3D–systems i = 1, 2 and 3). In all other respects, the notation used here is identical with that used above.

It is known that in two–dimensional systems at finite temperature there can be no Bose–condensate. However, superfluidity is possible even without the Bose–condensate[1,3,4]. This is connected with long–range correlations, which are damped at $T < T_C$ not exponentially but more slowly. Furthermore, a number of results obtained under the naive assumption of the existence of a condensate remain in force even in the more exact analysis, which take into account the fact that the Bose–condensate is actually "smeared out" by the long–wave fluctuations. This applies, in particular, to the temperature of the phase transition T_C, which can be found for the model (19.47) from the condition of the appearance of nontrivial solutions of the equation $\delta S_{eff} = 0$.

We calculate T_C and consider the different nontrivial solution of the equation $\delta S_{eff} = 0$, which correspond to the various superfluid phases. The calculation of the second variation $\delta^2 S_{eff}$ allows us to investigate the stability of the phases relative to small perturbations. These phases differ in the form of the order parameter, which in the two–dimensional model is a 2x3 matrix.

Note that the considerations developed below on the form of the order parameter in the different superfluid phases are certainly valid at $T = 0$, when the Bose–condensate is actually present.

We obtain all the branches of the Bose–spectrum in the a and b phases (12 branches in each phase). We discuss the behavior of the correlation functions $\langle c_{ia}(\boldsymbol{x},\tau)\bar{c}_{jb}(\boldsymbol{x}',\tau')\rangle$ at finite temperatures $T < T_C$.

19.8. Superfluid Phases of a Two–Dimensional Superfluid ^{3}He

We consider a system described by the functional S_{eff} (19.47), first at $|T - T_C| << T_C$ (in the Ginzburg–Landau region). In this region, expanding S_{eff} in powers $c_{ia}(\boldsymbol{x},\tau)$ and $c^*_{ia}(\boldsymbol{x},\tau)$ and limiting ourselves to terms of second and fourth order we get

$$S_{eff} = \sum_p A_{ij}(p)c^*_{ia}(p)c_{ja}(p) - \frac{7\zeta(3)Z^2\mu^2H^2k_F}{4\pi^3T^2}\sum_p c^*_{i3}(p)c_{i3}(p) -$$
$$-\frac{7\zeta(3)Z^4k_F}{16\pi^3c_FT^3\beta V}\sum_{p1+p2=p3+p4}[2c^*_{ia}(p_1)c^*_{jb}(p_2)c_{ia}(p_3)c_{jb}(p_4)+$$
$$+2c^*_{ia}(p_1)c^*_{jb}(p_2)c_{ia}(p_3)c_{jb}(p_4)+2c^*_{ia}(p_1)c^*_{jb}(p_2)c_{ja}(p_3)c_{ib}(p_4)-$$
$$-2c^*_{ia}(p_1)c^*_{ja}(p_2)c_{ib}(p_3)c_{jb}(p_4)-c^*_{ia}(p_1)c^*_{ia}(p_2)c_{jb}(p_3)c_{jb}(p_4)],$$

where

$$A_{ij}(p) = \delta_{ij}g_0^{-1} + \frac{4Z^2}{\beta V}\sum_{p1+p2=p} n_{1i}n_{1j}\frac{1}{(i\omega_1 - \xi_1)(i\omega_2 - \xi_2)}.$$

Here, we take it into account that the magnetic field $\boldsymbol{H}$ is directed perpendicular to the film–along the third axis.

We find the phase transition temperature T_C by equating to zero $A_{ij}(0)$ (the value of the coefficient function $A_{ij}(p)$ at p=0).

We obtain the equation

$$\delta_{ij} g_0^{-1} + \frac{4Z^2}{\beta V} \sum_{p1} n_{1i} n_{1j} \left(\omega_1^2 + \xi_1^2\right)^{-1} = 0 .$$

Calculating the sum over the frequencies, we rewrite the equation in the form

$$g_0^{-1} + \frac{Z^2 k_F}{\pi\, c_F} \int_0^{c_F k_0} \frac{d\xi}{\xi} th \frac{\beta \xi}{2} = g_0^{-1} + \frac{Z^2 k_F}{\pi\, c_F}\left(C + \ln \frac{2 c_F \beta k_0}{\pi} \right) = 0, \tag{19.49}$$

where the integral depends logarithmically on k_0 and C is the Euler constant. Therefore g_0^{-1} should also depend logarithmically on k_0:

$$g_0^{-1} = g^{-1} + \frac{Z^2 k_F}{\pi\, c_F} \ln \frac{k_F}{k_0},$$

where g no longer depends on k_0. This leads to a formula for T_C:

$$T_C = \frac{2 c_F k_F}{\pi} \exp\left(C - \frac{\pi\, c_F}{Z^2 k_F |g|} \right).$$

We now consider the possibilities for a condensate function at $T < T_C$. Substituting

$$c_{ia}(p) = c_{ia}^{(0)}(p) = (\beta V^{1/2})\delta_{p0} b_{ia}$$

in (19.49) and then making the substitution

$$b_{ia} = 4\left(\frac{T_C \Delta T}{7\zeta(3)}\right)^{1/2} a_{ia}$$

we get

$$S_{eff} = -\beta V \frac{16\pi^2 T_C \Delta T}{7\zeta(3)} F,$$

where

$$F = -trAA^+ + \nu\ trA^+AP + (trA^+A)^2 + trAA^+AA^+ +$$
$$+ trAA^+A^*A^T - trAA^TA^*A^+ - (1/2)trAA^T trA^*A^+,$$

$$\nu = 7\zeta(3)\mu^2 H^2 / 4\pi^2 T_C \Delta T \tag{19.50}$$

Equation (19.50) is identical in form with that arising in the three–dimensional system. The difference is that the matrix A with elements a_{ia} for the two–dimensional system is a 2 x 3 matrix. The matrix P is the projector on the third axis:

$$P = \begin{pmatrix} 0 & 0 & 0 \\ 0 & 0 & 0 \\ 0 & 0 & 1 \end{pmatrix}.$$

Minimizing F, we obtain the following equation for the condensate matrix A

$$-A+\nu\ AP+2(trAA^{+})A+2AA^{+}A+2A^{*}A^{T}A-2AA^{T}A^{*}-$$
$$-\left(trAA^{T}\right)A^{*}=0.$$

This equation has several solutions, corresponding to the different superfluid phases. We consider the possibilities

$$A_1=\frac{1}{2}\begin{pmatrix}1 & 0 & 0\\ i & 0 & 0\end{pmatrix},\quad A_2=\frac{1}{2}\begin{pmatrix}1 & 0 & 0\\ 0 & 1 & 0\end{pmatrix},$$

$$A_3=\frac{1}{4}\begin{pmatrix}1 & i & 0\\ i & -1 & 0\end{pmatrix},\quad A_4=\frac{1}{\sqrt{3}}\begin{pmatrix}1 & 0 & 0\\ 0 & 0 & 0\end{pmatrix},$$

$$A_5=\frac{1}{\sqrt{3}}\begin{pmatrix}0 & 0 & 0\\ 0 & 1 & 0\end{pmatrix},\quad A_6=\left(\frac{1-\nu}{3}\right)^{1/2}\begin{pmatrix}0 & 0 & 1\\ 0 & 0 & 0\end{pmatrix},$$

$$A_7=\left(\frac{1-\nu}{4}\right)^{1/2}\begin{pmatrix}0 & 0 & 1\\ 0 & 0 & i\end{pmatrix}.$$

The corresponding values of F are equal to

$$F_1=-\frac{1}{4},\ F_2=-\frac{1}{4},\ F_3=-\frac{1}{8},\ F_4=-\frac{1}{6},\ F_5=-\frac{1}{6},$$
$$F_6=-(1-\nu)^2/6,\ F_7=-(1-\nu)^2/4.$$

For the first five phases, the quantity F does not depend on H. The minimum value of $F=-1/4$ is reached for phases with matrices A_1 and A_2.

We call the phase with the matrix

$$A_1 = \frac{1}{2}\begin{pmatrix} 1 & 0 & 0 \\ i & 0 & 0 \end{pmatrix},$$

the a–phase and the phase with the matrix

$$A_2 = \frac{1}{2}\begin{pmatrix} 1 & 0 & 0 \\ 0 & 1 & 0 \end{pmatrix}$$

the b–phase.

We note that the a–phase is identical with the superfluid phase considered by Stein and Cross[13].

We calculate the second variation $\delta^2 F$. If it is non–negative then the corresponding phase is stable relative to small perturbations. Knowledge of the quadratic form $\delta^2 F$ allows us to determine the phonon variables and determine the change in the number of phonon variables upon switching on the magnetic field.

For the a–phase $\delta^2 F$ has the form

$$\begin{aligned} \delta^2 F &= \nu\left(u_{13}^2 + u_{23}^2 + v_{13}^2 + v_{23}^2\right) + \\ &+ \frac{1}{2}\left[2(u_{11} + v_{21})^2 + (u_{11} - v_{21})^2 + (u_{21} + v_{11})^2\right] + \\ &+ \frac{1}{2}\left[2(u_{22} - v_{12})^2 + (u_{12} - v_{22})^2 + (u_{22} + v_{12})^2\right] + \\ &+ \left[2(u_{23} - v_{13})^2 + (u_{13} - v_{23})^2 + (u_{23} + v_{13})^2\right] \end{aligned} \tag{19.51}$$

where u_{ia} and v_{ia} are the real and imaginary parts of c_{ia}. It follows from (19.51) that the phonon variables in the a–phase are

$$u_{21} - v_{11}, u_{12} + v_{22}, u_{13} + v_{23} \text{ for } H = 0 \ (\nu = 0),$$
$$u_{21} - v_{11}, u_{12} + v_{22} \text{ for } H \neq 0 \ (\nu \neq 0) \tag{19.52}$$

This means that in the *a*–phase at H = 0 there exist three *gd*–mode while the remaining nine branches of the Bose–spectrum are non–phonon. Upon switching on the magnetic field, the *gd*–mode $u_{13} + v_{23}$ acquire a gap. Calculation of $\delta^2 F$ for the *b*–phase yields

$$\delta^2 F = \nu\left(u_{13}^2 + u_{23}^2\right) + (\nu + 2)\left(v_{13}^2 + v_{23}^2\right) + \frac{1}{2}\left[3u_{11}^2 + 3u_{22}^2 + 2u_{11}u_{22} + \left(u_{12} + u_{21}\right)^2 + \right.$$
$$\left. + \left(v_{11} - v_{22}\right)^2 + 3v_{12}^2 + 3v_{21}^2 - 2v_{12}v_{21}\right].$$

This expression shows that in the *b*–phase the phonon variables will be

$$u_{12} - u_{21}, v_{11} + v_{22}, u_{13}, u_{23} \text{ for } H = 0 \ (\nu = 0),$$

$$u_{12} - u_{21}, v_{11} + v_{22} \text{ for } H \neq 0 \ (\nu \neq 0). \tag{19.53}$$

In the *b*–phase at $H = 0$ there exist four *gd*–modes and eight *nph*–modes. Upon switching on of the magnetic field, the branches u_{13} and u_{23} acquire gaps and become non–phonon. Formulae (19.50) and (19.51) also demonstrate the stability of the phases *a*– and *b*– relative to small perturbations.

In the considered model, both phases *a*– and *b*– have equal free energies, which does not permit us to give preference to one of them over the other. As is seen from formulas (19.52) and (19.53), the phonon variables in these two phases are essentially different. Calculation of the Bose–spectrum also gives results that are significantly different for the *a*– and *b*–phases.

Prediction of superfluid phases in ^{3}He–films has stimulated their further theoretical investigation and as well experimental search for the superfluid transition, which has been discovered five years after prediction.

In the weak coupling approximation, *a*– and *b*–phases have equal free energies. To investigate the relative stability of *a*– and *b*–phases it is necessary to come besides the BCS approximation frames. In [40] the difference of free energies of *a*– and *b*–phases ΔF has been calculated in next besides BCS approximation on T_C/ε_F. In two–dimensional case it turns out possible to calculate ΔF not only in Ginzburg–Landau region, but at arbitrary temperature. Sign of ΔF depends on the type of quasiparticle interaction, however, in considering approximation it does not depends on temperature.

Via mentioned above equality of free energies of *a*– and *b*–phases (in the weak coupling approximation) accounting of the next approximation corrections turns out to be significant for the phase diagram form even at the region of the parameter values, where the weak coupling approximation works nice.

Because Fermi–spectrum gap is isotropic in both phases corresponding normal Green's functions

$$G_{\alpha\beta}(k,\omega_n)=\frac{1}{Z}\frac{i\omega_n+\xi}{\omega_n^2+\xi^2+\Delta^2}\delta_{\alpha\beta},\quad \xi=c_F(k-k_F),\quad \omega_n=2\pi nT \tag{19.54}$$

are the same for both phases at equal value of Δ. Anomalous Green's functions

$$F_{\alpha\beta}(k,\omega_n)=-\frac{1}{Z}\frac{(i\sigma^p\sigma^y)_{\alpha\beta}A_{pj}\hat{k}_j}{\omega_n^2+\xi^2+\Delta^2},\quad \hat{k}_j=\frac{k_j}{k} \tag{19.55}$$

are different only by the form of matrix A_{pj} from the numerator.

Rainer and Serene[14] investigated the next (third) on T_C/ε_F order of the smallness corrections to BCS approximation (BCS approximation itself gives the free energy of the second order on T_C/ε_F . As they have shown, to calculate such corrections, it is enough to consider the diagrams which consist from two vertex functions, connecting by the four lines (see Fig. 19.1). It turns out that one can put Green's functions the same as in BCS approximation and the vertex functions the same as in normal Fermi–liquid (their dependence on frequency and momentum modulus (but not on direction) can be neglected).

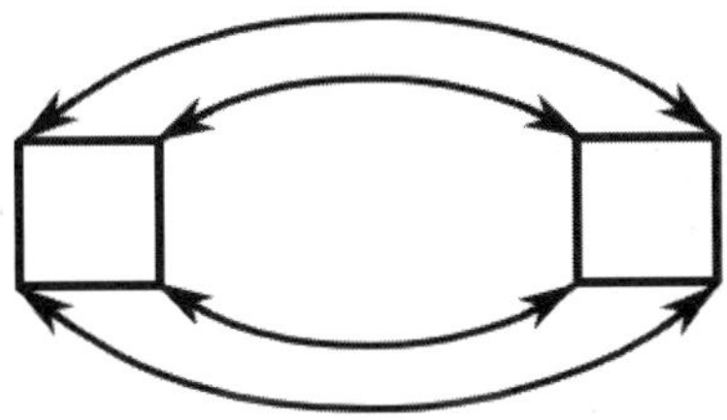

FIG. 19.2. The diagram giving the main contribution to the corrections to the BCS approximation energy[22].

If we would like to compare the free energies of *a*– and *b*–phases, it is obvious, that diagrams, which do not contain the anomalous Green's functions, are the same. The diagrams, which contain two anomalous Green's functions each, give the contribution, proportional to $A_{pj}A^*_{pj}$, which are the same as well.

Thus, in considering approximation the difference of free energies of *a*– and *b*–phases ΔF is determined by the diagram with four anomalous Green's functions, shown on Fig. 19.1. The contribution to the free energy, corresponding to this diagram is as following

$$-\frac{1}{8}T^3 \sum_{\omega n1,\omega n2,\omega n3} \int \frac{d^2k_1}{(2\pi)^2}\frac{d^2k_2}{(2\pi)^2}\frac{d^2k_3}{(2\pi)^2} \cdot \{\Gamma^{(4)}_{\alpha\beta\gamma\delta}(\boldsymbol{k}_1,\boldsymbol{k}_2;\boldsymbol{k}_3,\boldsymbol{k}_1+\boldsymbol{k}_2-\boldsymbol{k}_3)\cdot$$
$$\cdot \overline{F}_{\alpha\lambda}(\boldsymbol{k}_1,\omega_{n1})\overline{F}_{\beta\mu}(\boldsymbol{k}_2,\omega_{n2})\overline{F}_{\gamma\nu}(\boldsymbol{k}_3,\omega_{n3})\overline{F}_{\alpha\lambda}(\boldsymbol{k}_1+\boldsymbol{k}_2-\boldsymbol{k}_3,\omega_{n1}+\omega_{n2}+\omega_{n3})\cdot$$
$$\cdot \Gamma^{(4)}_{\lambda\mu\nu\rho}(\boldsymbol{k}_1,\boldsymbol{k}_2;\boldsymbol{k}_3,\boldsymbol{k}_1+\boldsymbol{k}_2-\boldsymbol{k}_3)$$

(19.56)

Substituting to (19.56) anomalous Green's functions from (19.55) for *a*– and *b*–phases and summing by spin index and integrating on momentum, we find in main order on T_C/ε_F the difference of free energies of *a*– and *b*–phases ΔF_{a-b}

$$\Delta F_{a-b} = -\gamma\tilde{Y}(\Delta,T)\int\frac{d\psi}{2\pi}|\sin\psi|\left[\Gamma_a^2(\hat{k}_1,\hat{k}_2;\hat{k}_1,\hat{k}_2)-\Gamma_a^2(\hat{k}_1,-\hat{k}_1;\hat{k}_2,-\hat{k}_2)\right]$$

(19.57)

where

$$\tilde{Y}(\Delta,T)=$$
$$=T^3\Delta^4\sum_{n1,n2,n3}\left\{(\omega_{n1}^2+\Delta^2)(\omega_{n1}^2+\Delta^2)(\omega_{n1}^2+\Delta^2)\left[(\omega_{n1}+\omega_{n2}+\omega_{n3})^2+\Delta^2\right]\right\}^{-1/2},$$

$\gamma \propto N(0)\varepsilon_F^{-3}$, $\tilde{\Gamma}_a$– antisymmetric on spin index part of dimensionless amplitude of scattering of quasiparticles with momenta, lying on Fermi–surface, ψ– angle between $\hat{k}_1$ and $\hat{k}_2$. Expression (19.57) sign depends on $\tilde{\Gamma}_a$. If it is predominated the scattering in the channel particle–hole, thus *a*–phase is more stable, but if predominated is the scattering in the channel particle–particle, thus *b*–phase becomes more stable. In three–dimensional case the phase diagram form and the thermodynamic parameters shows that the scattering in the channel particle–hole is bigger. But there is no reasons to suppose, that the quasiparticle interaction at sub–monomolecular layer of ^{3}He at whole range of concentrations possesses by the same qualitative characteristics, as in case of three–dimensional ^{3}He. Thus, under the change of concentration (and accordingly of the scattering amplitude) it is possible the change of the integral (19.57) sign, that makes possible the phase transition between *a*– and *b*–phases.

Note, that ΔF_{a-b} sign, in considering approximation does not depend on temperature. This means, that the phase transition line between *a*– and *b*–phases (if such a phase transition takes place) should be almost parallel to the temperature–axis at the phase diagram in variables "temperature–concentration".

In three–dimensional case the similar comparing of the free energies of *A*– and *B*–phases could not be made at arbitrary temperature by a such simple way. The reason is that the gap in *A*–phase is anisotropic and free energy depends significantly on exact form of the gap angle dependence. Thus, in three–dimensional case the comparing of the free energies is possible only in the vicinity of T_C and to do this, one need to calculate the coefficients of Ginzburg–Landau expansion.

In Ref. 22 the phase transitions, which could take place in *a*– and *b*–phases under increasing of the temperature, were studied as well. It was shown, that in *a*–phase under accounting of the spin–orbit interaction the transition from *a*–phase to fully disordered state consists of three different phase transitions: two Ising– and one Kosterlitz–Thouless transition. In case of *b*–phase the transition into disordered state goes over two phase transitions. Similar to case of *a*–phase (under condition $T_C \ll \varepsilon_F$) the temperatures of these transitions are just a little bit less, than T_C in BCS approximation.

19.9. Bose–Spectrum of the *a*–Phase

At $T = 0$ a Bose–condensate can exist in two–dimensional superfluid systems. Separating it out with the help of the shift $c_{ia}(p) \to c_{ia}(p) + c_{ia}^{(0)}(p)$, we consider the quadratic part of the functional S_{eff} in the new variables, c_{ia} and c_{ia}^{*}, the deviations of the old variables from their condensate values $c_{ia}^{(0)}$ and $c_{ia}^{(0)*}$. This quadratic form allows us to find the Bose–spectrum in the first approximation if we set the determinant of the quadratic form equal to zero.

For the *a*–phase, the condensate function $c_{ia}^{(0)}$ has the form

$$c_{ia}^{(0)}(p) = (\beta V)^{1/2} \delta_{p0} c \delta_{a1} (\delta_{i1} + i\delta_{i2}) \tag{19.58}$$

Here, C is a constant, determined from the condition of a maximum $S_{eff}\left[c_{ia}^{(0)}(p), c_{ia}^{(0)*}(p)\right]$. Substituting (19.58) in (19.47) we obtain

$$2\beta V|c|^2 g_0^{-1} + \sum_{p1} \ln\left(\omega_1^2 + \xi_1^2 + 4|c|^2 Z^2\right) \Big/ \left(\omega_1^2 + \xi_1^2\right) = 0. \tag{19.59}$$

The condition for a maximum (19.55) is the equation

$$g_0^{-1} + \frac{2Z^2}{\beta V} \sum_{p1} \left(\omega_1^2 + \xi_1^2 + 4|c|^2 Z^2\right)^{-1} = 0. \tag{19.60}$$

It allows us to express the energy gap at $T = 0$

$$\Delta = 2|c|Z$$

in terms of the transition temperature T_C. It is easy to verify the result (for example, by taking the difference of Eqs. (19.49) at $T = T_C$ and

(19.60) at $T=0$) that Δ and T_C are connected by the "universal formula" $\Delta = \pi\, T_C \gamma$ in the same way as in the BCS model ($\gamma = \exp C$, C is Euler's constant).

We will now construct the quadratic form for the *a*–phase.

This form describes both the *gd*– and the nonphonon–Bose excitations in the *a*–phase at *H*=0 and is the sum of three forms, of which the first depends on c_{i1}, the second on c_{i2} and the third on c_{i3}. The second and the third forms go over into the first after the substitutions $c_{i2} \to ic_{i1}$ and $c_{i3} \to ic_{i1}$. Therefore, the Bose–spectrum of the *a*–phase is triply degenerate, as in the A–phase model of three–dimensional ^{3}He. We consider one of the three independent forms, for example, the form with *a*=1. It has the form

$$\sum_p c_{i1}^* c_{j1} \left[g_0^{-1}\delta_{ij} + \frac{4Z^2}{\beta V} \sum_{p1+p2=p} \frac{n_{1i} n_{1j} (i\omega_1 + \xi_1)(i\omega_2 + \xi_2)}{(\omega_1^2 + \xi_1^2 + \Delta^2)(\omega_2^2 + \xi_2^2 + \Delta^2)} \right] -$$

$$-\frac{1}{2}\sum_p \left[c_{i1}(p) c_{j1}(-p) + c_{i1}^*(p) c_{j1}^*(-p) \right] \times$$
$$\times \frac{4Z^2}{\beta V} \sum_{p1+p2=p} \frac{n_{1i} n_{1j} (n_1 \pm i n_2)^2}{(\omega_1^2 + \xi_1^2 + \Delta^2)(\omega_2^2 + \xi_2^2 + \Delta^2)}. \tag{19.61}$$

Here, $(n_1 \pm in_2)^2$ means that $(n_1 + in_2)^2$ is multiplied by $c_{i1}^*(p) c_{j1}^*(-p)$ and $(n_1 - in_2)^2$ by $c_{i1}(p) c_{j1}(-p)$.

The phonon variable in the form (19.61) is $\nu_{21} - \nu_{11}$. This result which was already obtained above for the Ginzburg–Landau region, is true for all $T < T_C$. Actually, if we separate out the terms in (19.61) with fixed *p* and then set $c_{11} = -i$ and $c_{21} = 1$ in them, we obtain

$$\frac{2}{g_0}+\frac{4Z^2}{\beta V}\sum_{p1+p2=p}\frac{(i\omega_1+\xi_1)(i\omega_2+\xi_2)(n_1^2+n_2^2)}{(\omega_1^2+\xi_1^2+\Delta^2)(\omega_2^2+\xi_2^2+\Delta^2)}-$$

$$-\frac{4Z^2}{\beta V}\times$$

$$\times\sum_{p1+p2=p}\frac{(n_1-in_2)^2(-n_1^2-2in_1n_2+n_2^2)+(n_1+in_2)^2(-n_1^2+2in_1n_2+n_2^2)}{(\omega_1^2+\xi_1^2+\Delta^2)(\omega_2^2+\xi_2^2+\Delta^2)}=$$

$$=\frac{2}{g_0}+\frac{4Z^2}{\beta V}\sum_{p1+p2=p}\frac{(i\omega_1+\xi_1)(i\omega_2+\xi_2)+\Delta^2}{(\omega_1^2+\xi_1^2+\Delta^2)(\omega_2^2+\xi_2^2+\Delta^2)}.$$

Equating this expression to zero, we obtain

$$g_0^{-1}+\frac{4Z^2}{\beta V}\sum_{p1+p2=p}\frac{(i\omega_1+\xi_1)(i\omega_2+\xi_2)+\Delta^2}{(\omega_1^2+\xi_1^2+\Delta^2)(\omega_2^2+\xi_2^2+\Delta^2)}=0\,. \tag{19.62}$$

At $p=0$, this equation goes over into (19.60). Therefore (19.62) has the root $p=0$, while the corresponding branch of the spectrum begins from zero. Its calculation is similar to the calculation for the three–dimensional case (see above) with the replacement of an integral over the Fermi–sphere by an integral over neighborhood of the "Fermi–circle", and gives the result

$$E=c_F k/\sqrt{2}\,. \tag{19.63}$$

The complete phonon spectrum in the a–phase at H=0 consists of three branches of (19.53), corresponding to the variables

$$u_{21}-v_{11},u_{21}+v_{22},u_{13}+v_{23}$$

We proceed to the nonphonon branches of the spectrum. We first find all the branches of the spectrum at $\boldsymbol{k}$ = 0. For this purpose, we consider the terms in (19.61) with $\boldsymbol{k} = 0$ and finite $\omega \neq 0$, which can be written down in the form

$$(u_{11}^2 + u_{21}^2 + v_{11}^2 + v_{21}^2) f(\omega) - \\ - 2\left[u_{11}^2 - u_{21}^2 - v_{11}^2 + v_{21}^2 + 2(u_{11}v_{21} + u_{21}v_{11})\right] g(\omega), \tag{19.64}$$

where

$$f(\omega) = g_0^{-1} + \frac{2Z^2}{\beta V} \sum_{p1} \frac{(i\omega_1 + \xi_1)(i\omega_2 + \xi_2)}{(\omega_1^2 + \xi_1^2 + \Delta^2)(\omega_2^2 + \xi_2^2 + \Delta^2)},$$

$$g(\omega) = \frac{Z^2\Delta^2}{\beta V} \sum_{p1} (\omega_1^2 + \xi_1^2 + \Delta^2)^{-1} (\omega_2^2 + \xi_2^2 + \Delta^2)^{-1}.$$

The quadratic form (19.64) of the variables $u_{11}, u_{21}, v_{11}, v_{21}$ is the sum of two independent forms:

$$(u_{11}^2 + v_{21}^2)(f(\omega) - g(\omega)) - 2u_{11}v_{21}g(\omega),$$

$$(u_{21}^2 + v_{11}^2)(f(\omega) + g(\omega)) - 2u_{21}v_{11}g(\omega).$$

Setting the determinants of these forms equal to zero, we obtain the equations

$$\det\begin{pmatrix} f(\omega)-g(\omega) & -g(\omega) \\ -g(\omega) & f(\omega)-g(\omega) \end{pmatrix} = f(\omega)\big(f(\omega)-2g(\omega)\big)=0,$$

$$\det\begin{pmatrix} f(\omega)+g(\omega) & -g(\omega) \\ -g(\omega) & f(\omega)+g(\omega) \end{pmatrix} = f(\omega)\big(f(\omega)+2g(\omega)\big)=0.$$

Thus, the quantities $E(0)$ (the values of the Bose–spectrum at k=0) are determined from the equations

$$f(\omega)=0,\;\; f(\omega)+2g(\omega)=0,\;\; f(\omega)-2g(\omega)=0.$$

The equation $f(\omega)+2g(\omega)=0$ has the root $\omega=0$ since

$$f(0)+2g(0)=g_0^{-1}+\frac{2Z^2}{\beta V}\sum_{p1}\left(\omega_1^2+\xi_1^2+\Delta^2\right)^{-1}=0$$

by virtue of (19.52). The *gd*–mode of the spectrum (19.59) corresponds to this equation, and the nonphonon–modes to the remaining two.

The equation, $f(\omega)=0$, can be reduced to the form

$$\left(\omega^2+2\Delta^2\right)F(\omega)=0 \tag{19.65}$$

where

$$F(\omega)=\frac{1}{\omega\left(\omega^2+4\Delta^2\right)^{1/2}}\ln\frac{\left(\omega^2+4\Delta^2\right)^{1/2}+\omega}{\left(\omega^2+4\Delta^2\right)^{1/2}-\omega}=0\,. \tag{19.66}$$

Substituting $i\omega\to E$, we find the root

$$E = \sqrt{2}\Delta$$

corresponding to the variables $u_{11} - v_{21}$ and $u_{21} + v_{11}$. The equation $f(\omega) - 2g(\omega) = 0$ reduces to the form

$$\left(\omega^2 + 4\Delta^2\right)F(\omega) = 0$$

and has the root

$$E = 2\Delta \qquad (19.67)$$

after the substituting $i\omega \to E$. The branch (19.67) corresponds to the variable $u_{11} + v_{21}$.

Taking it into account that in the *a*–phase each of the branches of the Bose–spectrum is triply degenerate, we can rewrite the results obtained thus far for the Bose–spectrum in the form

$$E = c_F k/\sqrt{2}\,,\ u_{21} - v_{11}, u_{21} + v_{22}, u_{13} + v_{23}\ ,\ \text{(3 modes)}$$

$$E = \sqrt{2}\Delta\,,\ u_{11} - v_{21}, u_{22} + v_{12}, u_{23} + v_{13}, u_{21} + v_{11}, u_{12} - v_{22}, u_{13} - v_{23}, \qquad \text{(6 modes)}$$

$$E = 2\Delta\,,\ u_{11} + v_{21}, u_{22} - v_{12}, u_{23} - v_{13}.\ \text{(3 modes)} \qquad (19.68)$$

Here, the branches of the spectrum are written down along with the variables to which they correspond.

The next step is to obtain the corrections of relative order $c_F^2 k^2 \Delta^{-2}$ to the modes (19.68). We first consider the non–phonon modes. We begin with the mode $u_{11} - v_{21}$. It is not difficult to see that account of corrections $\propto k^2$ gives the following equation in place of (19.65):

$$\left(\omega^2 + 2\Delta^2\right)F(\omega) + \frac{1}{2}c_F^2 k^2 \frac{d}{d\omega^2}\left[\left(\omega^2 + 2\Delta^2\right)F(\omega)\right] = 0,$$

where $F(\omega)$ is the function (19.66). We then get the formula

$$E^2 = 2\Delta^2 + c_F^2 k^2/2.$$

The same result is obtained for the variable $u_{21} + v_{11}$.

We now consider the branch $E = 2\Delta$, which corresponds to the variable $u_{11} + v_{21}$. The equation for it can be reduced to the form

$$\int_0^{2\pi} d\varphi \int_0^1 d\alpha \left[-\ln\left(1 + \alpha(1-\alpha)\left(\omega^2 + c_F^2(\boldsymbol{n},\boldsymbol{k})^2\right)/\Delta^2\right) - 2\right] = 0 \qquad (19.69)$$

This equation cannot be solved by expansion in $c_F^2(\boldsymbol{n},\boldsymbol{k})^2$ under the integral. The situation is analogous to that encountered for the branch $E = 2\Delta$ in *B*–phase of a three–dimensional system of the ^{3}He type (see above). Equation (19.69) itself differs from the corresponding equation of three–dimensional theory by the fact that in place of the integral $\int d\Omega$ over the solid angle, we have in (19.69) the integral $\int d\varphi$ over the planar angle φ. This leads to the result that the dispersion law for the branch with $u_{11} + v_{21}$, in place of the equation

$$\int_0^1 \left(-z + x^2\right)^{1/2} dx = 0$$

is determined in the two–dimensional case by the equation

$$\int_0^1 \left(\frac{-z+x^2}{1-x^2} \right)^{1/2} dx = 0\,,$$

where

$$z = \left(E^2 - 4\Delta^2\right) / c_F^2 k^2 . \tag{19.70}$$

The root of this equation turns to be complex and its value

$$z = 0.500 - 0.433i$$

we obtained on a computer. Knowing z, we can find E from the formula

$$E^2 = 4\Delta^2 + z c_F^2 k^2 ,$$

which follows from (19.70). The complex root means that the corresponding Bose–excitation is unstable and decay into the fermions of which it is composed. We recall that the branches $E = 2\Delta$ had complex increments $\propto k^2$ also in the three–dimensional theory (see above).

For completion of the study of Bose–spectrum of the a–phase at small κ, it remains to find the dispersion of the gd–mode of the spectrum.

The result has the form

$$E = \frac{c_F k}{\sqrt{2}} \left(1 - \frac{5 c_F^2 k^2}{96\Delta^2} \right).$$

It shows the stability of the gd–modes relative to decay.

We write out once sgain the results obtained in this section for the Bose–spectrum in a–phase:

$$E^2 = \frac{c_F^2 k^2}{2}\left(1 - \frac{5c_F^2 k^2}{96\Delta^2}\right),\ u_{21} - v_{11}, u_{21} + v_{22}, u_{13} + v_{23}\ ,\ \text{(3 modes)}$$

$$E^2 = 2\Delta^2 + c_F^2 k^2/2\,,$$
$$u_{11} - v_{21}, u_{22} + v_{12}, u_{23} + v_{13}, u_{21} + v_{11}, u_{12} - v_{22}, u_{13} - v_{23},\ \text{(6 modes)}$$

$$E^2 = 4\Delta^2 + (0.500 + i0.433)c_F^2 k^2,\ u_{11} + v_{21}, u_{22} - v_{12}, u_{23} - v_{13}\,. \quad \text{(3 modes)}$$

These equations represent a refinement of equations (19.68).

19.10. Bose–Spectrum of the *b*–Phase

We now carry out the investigation of the Bose–spectrum of the *b*–phase, confining ourselves to the scheme developed for the *a*–phase above.

The condensate function of the *b*–phase has the form

$$c_{ia}^{(0)}(p) = c(\beta V)^{1/2}\delta_{p0}\delta_{ia} \tag{19.71}$$

where c is a constant, determined from the condition of a maximum $S_{eff}\left[c_{ia}^{(0)}(p), c_{ia}^{(0)*}(p)\right]$. Substituting (19.71) in (19.47), we obtain the same equation (19.59), which we had for the *a*–phase, while the condition for its maximum is identical with Eq. (19.60). The energy gap $\Delta = 2|c|Z$ at $T = 0$ is given by the same formula $\Delta = \pi T_C/\gamma$ as for the *a* –phase. The calculations, which are similar to those carried out for the *a*–phase, give the following quadratic form for the *a*–phase:

$$\sum_p c_{ia}^* c_{ja}\left[g_0^{-1}\delta_{ij} + \frac{4Z^2}{\beta V}\sum_{p1+p2=p}\frac{n_{1i}n_{1j}(i\omega_1 + \xi_1)(i\omega_2 + \xi_2)}{(\omega_1^2 + \xi_1^2 + \Delta^2)(\omega_2^2 + \xi_2^2 + \Delta^2)}\right] -$$

$$-\frac{1}{2}\sum_{p}\left[c_{ia}(p)c_{jb}(-p)+c_{ia}^{*}(p)c_{jb}^{*}(-p)\right]\times$$
$$\times\frac{4Z^{2}}{\beta V}\sum_{p1+p2=p}\frac{n_{1i}n_{1j}}{\left(\omega_{1}^{2}+\xi_{1}^{2}+\Delta^{2}\right)\left(\omega_{2}^{2}+\xi_{2}^{2}+\Delta^{2}\right)}\times$$
$$\times\left[n_{1}^{2}\left(2\delta_{a1}\delta_{b1}-\delta_{ab}\right)+n_{2}^{2}\left(2\delta_{a2}\delta_{b2}-\delta_{ab}\right)+2n_{1}n_{2}\left(\delta_{a1}\delta_{b2}+\delta_{a2}\delta_{b1}\right)\right].$$
(19.72)

The tensor coefficients A_{ij} and B_{ijab} in (19.72) are real, since the complex expressions $(i\omega_1+\xi_1)(i\omega_2+\xi_2)$ in A_{ij} become real after summation. Therefore the form (19.72) splits into a sum of two independent forms, of which the first depends on $u_{ia}=\operatorname{Re}c_{ia}$, the second on $v_{ia}=\operatorname{Im}c_{ia}$. Furthermore, it is seen that the form of the variables $u_{i3}(v_{i3})$ is independent of the form of the variables $u_{ia}(v_{ia})$ with a=1,2.

We first investigate the Bose–spectrum corresponding to the variables u_{i3}, v_{i3}. Removing from (19.72) the form corresponding to these variables and carrying out the necessary calculations, we obtain two *gd*–modes of the spectrum

$$E=c_F k\sqrt{3}/2,\ u_{13},$$

$$E=c_F k/2,\ u_{23},$$

and two non–phonon branches

$$E=2\Delta,\ v_{13}, v_{23} \tag{19.73}$$

We proceed to the spectrum for the variables $u_{ia}(v_{ia})$ with a=1,2. The terms with p=0 for these variables in (19.73) can be written in the form

$$-\frac{2Z^2\Delta^2}{\beta V}\left[\sum_{p1}\left(\omega_1^2+\xi_1^2+\Delta^2\right)^{-2}\right]\cdot Q_0, \tag{19.74}$$

where

$$\begin{aligned}Q_0 &= 3u_{11}^2+3u_{22}^2+2u_{11}u_{22}+\left(u_{12}+u_{21}\right)^2+\\&+3v_{12}^2+3v_{21}^2-3v_{12}v_{21}+\left(v_{12}-v_{21}\right)^2\end{aligned} \tag{19.75}$$

It follows from (19.74) and (19.75) that the phonon variables will be $u_{12}-u_{21}$ and $v_{11}+v_{22}$, the same as obtained above for the Ginzburg–Landau region. To them correspond the branches

$$E=c_F k/\sqrt{2}\,,\ u_{12}-u_{21}, v_{11}+v_{22}$$

We now consider the form of the variables $u_{ia}(v_{ia})$ with a=1,2 in the case k = 0, in order to find the energies $E(0)$ of the modes. We can write the form of the variables u_{ia} as

$$\begin{aligned}&(u_{11}^2+u_{12}^2+u_{21}^2+u_{22}^2)f(\omega)+\\&+\left(u_{12}^2+u_{21}^2-u_{11}^2-u_{22}^2-2u_{11}u_{22}-2u_{12}u_{21}\right)g(\omega),\end{aligned} \tag{19.76}$$

where $f(\omega)$ and $g(\omega)$ were determined above. The expression (19.76) is the sum of the two independent forms:

$$(u_{11}^2+u_{22}^2)(f(\omega)-g(\omega))-2u_{11}u_{22}g(\omega),$$

$$(u_{12}^2+u_{21}^2)(f(\omega)+g(\omega))-2u_{12}u_{21}g(\omega). \tag{19.77}$$

The matrices of the forms (19.77) are the same as for the forms, obtained in a–phase. Using the results, obtained for a–phase, we can immediately write down the results

$$E = \sqrt{2}\Delta\,,\; u_{11} - u_{22}, u_{12} + u_{21},$$

$$E = 2\Delta\,,\; u_{11} + u_{22}.$$

The form of the variables v_{ia} differs from (19.77) by the substitution $u_{ia} \to v_{ia}$, $g(\omega) \to -g(\omega)$ and has the form

$$\begin{aligned}&(v_{11}^2 + v_{12}^2 + v_{21}^2 + v_{22}^2) f(\omega) + \\ &+ \left(v_{11}^2 + v_{22}^2 - v_{21}^2 - v_{12}^2 + 2v_{11}v_{22} + 2v_{12}v_{21}\right) g(\omega).\end{aligned}$$

This is the sum of two independent forms:

$$(v_{11}^2 + v_{22}^2)(f(\omega) + g(\omega)) + 2v_{11}v_{22}\,g(\omega),$$

$$(v_{12}^2 + v_{21}^2)(f(\omega) - g(\omega)) + 2v_{12}v_{21}\,g(\omega).$$

It then follows that

$$E = \sqrt{2}\Delta\,,\; v_{11} - v_{22}, v_{12} + v_{21},$$

$$E = 2\Delta\,,\; v_{12} - v_{21}.$$

The corrections of relative order k^2 to the branches of the spectrum that have been found can be obtained in a fashion similar to what was done above for the a–phase.

In conclusion, we write down all the branches of the Bose–spectrum for the b–phase, together with the corresponding variables:

$$E^2 = \frac{c_F^2 k^2}{2}\left(1 - \frac{5c_F^2 k^2}{48\Delta^2}\right),\ u_{21} - u_{21}, v_{11} + v_{22};$$

$$E^2 = \frac{3c_F^2 k^2}{4}\left(1 - \frac{c_F^2 k^2}{72\Delta^2}\right),\ u_{13};$$

$$E^2 = \frac{c_F^2 k^2}{4}\left(1 - \frac{c_F^2 k^2}{48\Delta^2}\right),\ u_{23};$$

$$E^2 = 2\Delta^2 + c_F^2 k^2/2,\ u_{11} - u_{22}, u_{12} + u_{21}, v_{11} - v_{22}, v_{12} - v_{21};$$

$$E^2 = 4\Delta^2 + (0.500 - i0.433)c_F^2 k^2,\ u_{11} + u_{22}, v_{12} + v_{21};$$

$$E^2 = 4\Delta^2 + (0.152 - i0.218)c_F^2 k^2,\ v_{13};$$

$$E^2 = 4\Delta^2 + (0.849 - i0.216)c_F^2 k^2,\ v_{23}.$$

Here,

$$z_1 = 0.500 - i0.433,$$

$$z_2 = 0.152 - i0.218,$$

$$z_3 = 0.849 - i0.216$$

are the roots of equations

$$\int_0^1 \left(\frac{-z+x^2}{1-x^2} \right)^{1/2} \begin{pmatrix} 1 \\ x^2 \\ 1-x^2 \end{pmatrix} dx = 0,$$

which arise in consideration of the branches with $E \cong 2\Delta$.

19.11. The Two–Dimensional Superfluidity Must Exist!

The correlation functions

$$\left\langle c_{ia}(\mathbf{x},\tau)\bar{c}_{jb}(\mathbf{x}',\tau')\right\rangle \tag{19.78}$$

of the Bose–fields introduced above tend in the case $T = 0$ and in the limit $r = |\mathbf{x}-\mathbf{x}'| \to \infty$ to a constant $< c_{ia} >< \bar{c}_{jb} >$. At $0 < T < T_C$, the situation should be similar to that existing in theory of a two–dimensional superfluid, where the correlator of the Bose–field $\langle \psi(\mathbf{x},\tau)\bar{\psi}(\mathbf{x}',\tau')\rangle$ falls off as the power $r^{-\alpha}$, where $\alpha = m\beta/2\pi\rho_S$, m is the mass of the Bose–particle, ρ_S is the density of the superfluid component[3,4]. This result for the two–dimensional Bose–system is most simply obtained by going over to polar coordinates[3,4] $\psi(\mathbf{x},\tau) = (\rho(\mathbf{x},\tau))^{1/2}\exp(i\varphi(\mathbf{x},\tau))$, see section 19.2 above. In this case it turns out that the decrease $\propto r^{-\alpha}$ is determined by the phase correlator $\langle \exp i[\varphi(\mathbf{x},\tau)-\varphi(\mathbf{x}',\tau')]\rangle$, while the correlator $\left\langle [\rho(\mathbf{x},\tau)\rho(\mathbf{x}',\tau')]^{1/2}\right\rangle$ tends to some constant value.

For our system of tensor Bose–fields $c_{ia}(\mathbf{x},\tau)$, it is also natural to introduce the analog of polar coordinates, which turn out to be different for the different phases. Here the number of "angular" variables is equal to the number of *gd*–modes of the system.

In order to explain the transition to the new coordinates, we note that the expression (19.50) (at H=0) is invariant relative to the transformations

$$A \to e^{i\varphi} u_2 A u_3 ,$$

where u_2 and u_3 are orthogonal matrices of second and third orders, $e^{i\varphi}$ is the phase factor. For $A = A_1$ (the a –phase) multiplication of A_1 on the left by $u_2 = \exp(i\sigma_2\psi)$ is equivalent to multiplication by the phase factor $e^{i\psi}$, which can be combined with $e^{i\varphi}$. Multiplication of A_1 on the right by $u_3 = U$ gives the matrix

$$A_1 U = \frac{1}{2}\begin{pmatrix} u_{11} & u_{12} & u_{13} \\ iu_{11} & iu_{12} & iu_{13} \end{pmatrix}$$

It is expressed in terms of the elements of the first row (u_{11}, u_{12}, u_{13}) of the orthogonal matrix U, which can be regarded as the component of the unit vector $\boldsymbol{n}$ $(\boldsymbol{n}^2 = 1)$. Thus the role of the angular variables in a–phase is played by the phase φ and by the unit vector $\boldsymbol{n}$ (the element of the sphere S^2).

The action functional can depend only on the gradients of the angular variables, but not on the variables themselves. For slowly changing fields, the part of the action that is dependent on the angular variables has the form

$$S_{\varphi,n} = -\int (a\partial_i\varphi\partial_i\varphi + b\partial_i n_a \partial_i n_a)\, d^2x \tag{19.79}$$

We shall assume that φ and $\boldsymbol{n}$ depend only on $\boldsymbol{x}$ (but not on τ), assuming that we have integrated over the non–zero Fourier components.

We note that the coefficients a and b in (19.79) increase without bound at $T \to 0$.

The consideration that have been given show that the significant parts of the correlators (19.78) are the averages

$$\langle \exp i[\varphi(\boldsymbol{x},\tau) - \varphi(\boldsymbol{x}',\tau')] \rangle, \ \langle n_a(\boldsymbol{x}) n_b(\boldsymbol{x}') \rangle \tag{19.80}$$

The first of these is characteristic for two–dimensional superfluid systems and decay as $r^{-\alpha}$, $\alpha = (4\pi a)^{-1}$. For the second of the averages (19.80) (the correlator of the $\boldsymbol{n}$–field of two–dimensional Euclidean theory) Polyakov[15] has obtained a formula (in the notation used here)

$$\langle \boldsymbol{n}\ (\boldsymbol{x}) \boldsymbol{n}\ (\boldsymbol{x}') \rangle = \left(1 - \frac{1}{4\pi b} \ln \frac{r}{a_1} \right)^2 , \tag{19.81}$$

where a_1 is some constant with the dimensionality of length. The formula (19.81) loses its meaning for large distances, since its right hand side increases without limit as $r = |\boldsymbol{x} - \boldsymbol{x}'| \to \infty$. The reason for the limited applicability of (19.81) is that in its derivation by the method of the renormalization group, the "instantons" existing in the theory, which would correct the behavior of the correlator at large distances, were not taken into account. Up to the present time, the question of the asymptote of the correlator (19.81) as $r \to \infty$ remains open. It is possible that the correlator decays in power–law fashion as $r \to \infty$.

The question of the asymptote of the correlators becomes more simple for a system in a magnetic field. Turning on the magnetic field reduces the symmetry group which, at $H \neq 0$, can be rewritten in the form

$$A \to e^{i\varphi} u_2 A \tilde{u}_3, \ \tilde{u}_3 = \begin{pmatrix} \tilde{u}_2 & 0 \\ 0 & 1 \end{pmatrix}$$

where u_2 and $\tilde{u}_2$ are two orthogonal matrices of second order. In particular, for the a–phase, the replacement of u_3 by $\tilde{u}_3$ leads to an $\boldsymbol{n}$–field with two components: $n_1 = \cos\psi, n_2 = \sin\psi$, while the action takes the form

$$-\int \left(a\partial_i\varphi\partial_i\varphi + b\partial_i\psi\partial_i\psi\right) d^2x$$

The correlator of the $\boldsymbol{n}$–field $(\cos\psi, \sin\psi)$ decays in power–law fashion:

$$\begin{aligned}
&\langle \boldsymbol{n}\,(\boldsymbol{x}), \boldsymbol{n}\,(\boldsymbol{x}')\rangle = \langle \cos\psi(\boldsymbol{x})\cos\psi(\boldsymbol{x}') + \sin\psi(\boldsymbol{x})\sin\psi(\boldsymbol{x}')\rangle = \\
&= \langle \exp i(\psi(\boldsymbol{x}) - \psi(\boldsymbol{x}'))\rangle \propto r^{-\tilde{\alpha}}, \\
&\tilde{\alpha} = (4\pi b)^{-1}
\end{aligned}$$

A similar consideration can also be carried out for the b–phase. Here the multiplication of $A = A_2$ on the left by an orthogonal matrix of second order u_2 is equivalent to multiplication on the right by the third–order matrix

$$\tilde{u}_3 = \begin{pmatrix} \tilde{u}_2 & 0 \\ 0 & 1 \end{pmatrix}$$

which we combine with u_3. Multiplication of A_2 on the right with $u_3 = U$ gives the result

$$A_2 U = \frac{1}{2}\begin{pmatrix} u_{11} & u_{12} & u_{13} \\ u_{21} & u_{22} & u_{23} \end{pmatrix}$$

which is expressed in terms of the first two rows of the matrix U, which we represent as the components of two mutually orthogonal fields $\boldsymbol{n}_1, \boldsymbol{n}_2\left((\boldsymbol{n}_1, \boldsymbol{n}_2) = 0\right)$. The angular coordinates in the b–phase are $\varphi, \boldsymbol{n}_1, \boldsymbol{n}_2\left(\boldsymbol{n}_1^2 = \boldsymbol{n}_1^2 = 1, (\boldsymbol{n}_1, \boldsymbol{n}_2) = 0\right)$ and the part of the action corresponding to them can be written in the form

$$- \int \left(a\partial_i\varphi\partial_i\varphi + b\left(\partial_i n_{1a}\partial_i n_{1a} + \partial_i n_{2a}\partial_i n_{2a}\right)\right) d^2x . \tag{19.82}$$

The nontrivial part of the correlators (19.80) is determined by average

$$\langle n_i(\boldsymbol{x}) n_j(\boldsymbol{x}')\rangle ;\ i, j = 1,2 .$$

The problem of the asymptotic behavior of these averages still remains open. However, it is easily resolved for a system in a field, when, only the first two components of the fields are different from zero, so that $n_{11} = n_{22} = \cos\psi, n_{12} = -n_{21} = \sin\psi$. The action (19.82) takes the form

$$- \int \left(a\partial_i\varphi\partial_i\varphi + 2b\partial_i\psi\partial_i\psi\right) d^2x$$

while the correlator (19.79) decays as $r^{-\gamma}$, $\gamma = (8\pi b)^{-1}$.

Summing up, we can say that the theory of two–dimensional superfluid systems of the ^{3}He– type at $T \neq 0$ is connected with the difficult and interesting problem of the asymptote of the correlators of the $\boldsymbol{n}$–fields of two–dimensional Euclidean theory. The situation is completely clear only for systems in a magnetic field, where the correlators decay in power–law fashion.

19.12. New Possibility for the Search of 2D–Superfluidity in ^{3}He–Films

Now we will discuss some experimental conclusions follow our theoretical investigation of ^{3}He–films.

In 1980, Brusov and Popov have predicted the existence of 2D–superfluidity in the ^{3}He–films[20,24]. They have predicted the existence of two superfluid phases (*a*– and *b*–) in these films, have described the order parameter in each phase and have calculated the whole collective excitation spectrum, which consists of the 12 modes in each phase. In 1985 Sachrajda et al.[19], have discovered the 2D–superfluidity in ^{3}He–films. They have received the superfluid ^{3}He–film with the thickness d less then coherent length ξ. Unfortunately, there was a temperature gradient in their apparatus, thus they actually have a case $d > \xi$, i.e. 3D–superfluidity.

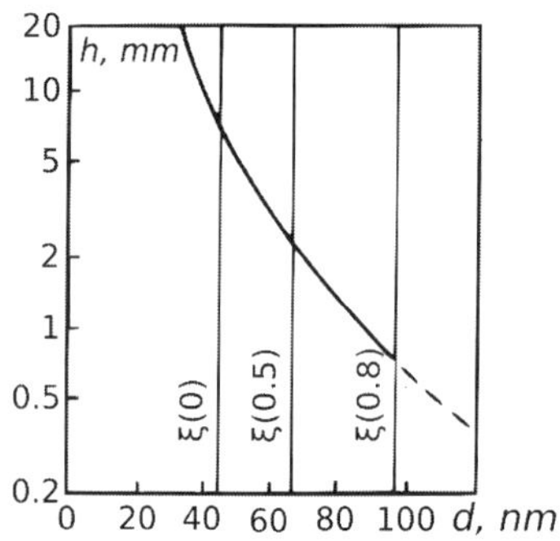

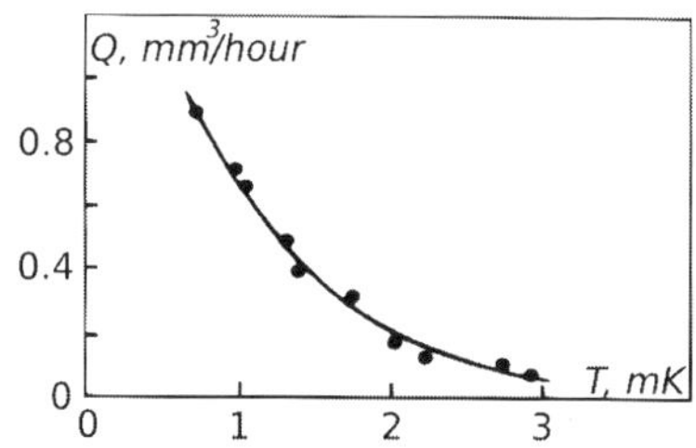

FIG. 19.3. The dependence of film thickness d on its height h above the menisc[19].
FIG. 19.4. The dependence of mass flow Q on the temperature T[19].

New experiments which have been done by the same group in 1987 and by Davis et al.[23] showed that the superflow rate dependences on the film thickness and temperature and superflow stops when d becomes of order $\xi(T)$. More careful experiments by the same groups show the existence of superflow in films with thickness less then less then coherent length ξ. This means the existence of two–dimentional superfluidity in ^3He–films. There are two other possibilities for further search of 2D–superfluidity. The first is connected with studying it at a lower temperature, than has been done before ($T < 0.6mK$). The other possibility is connected with the conclusions have made above. Brusov and Popov have proved[20,24] that 2D–superfluidity does exist in the presence of magnetic field (see previous section 19.11). We suggest to make the experiments for search 2D–superfluidity in ^3He–films in the presence of magnetic field directed along the film.

Chapter XX

Bose–Spectrum of Superfluid Solutions ^3He–^4He

20.1. Superfluidity of ^3He, Dissolved in ^4He

Solutions of ^3He in the superfluid ^4He are investigated theoretocally and experimantally for a long time[1]. Here, the problem of the phase transition of the Fermi–component of the solution into the superfluid state due to Cooper pairs formation arise naturally[2]. Up to now there is no experimental evidence for such a transition, but there exist many theoretical papers, devoted to this problem (see references in Ref. 3). In many works the formation of Cooper pairs with *s*–pairing (i.e. with zero orbital and spin moments) in the three–dimensional system has been considered. Bardeen et al.[2] obtained the estimate $2\cdot 10^{-6}\,K$ for the superfluid phase transition temperature T_C at zero pressure. The estimate by Landau et al.[4] (1970) is $T_C \propto 10^{-4}\,K$ for pressures from 10 to 20 atmospheres. Hoffberg[5] showed that the *s*–pairing is preferable at low ^3He concentrations ($0 < x < 6.6\%$). According to this work the phase transition temperature is maximal (of the order of $10^{-4}\,K$) for p = 20 atmospheres at $x = 1.2\%$. The application of the semi–phenomenological Landau Fermi–liquid theory to the case of *s*–pairing is presented in Ref. 6. The Bose–spectrum of collective excitations for this case was calculated by Andrianov and Popov[7], the *s*–pairing of ^3He in the two–dimensional system was considered by Bashkin[8]. However in many physically interesting cases it is necessary to consider the pairing with a nonzero orbital momentum. For example, the model estimates of the *s*–wave scattering amplitude for the system of the surface levels of He, which was predicted theoretically by Andreyev[9] and discovered experimentally[10], lead to contradictory results for the sign of interaction[8].

As far as there exists the possibility for the pairing of ^{3}He atoms with a nonzero orbital momentum in ^{3}He–^{4}He solutions it should be interesting to consider a model system of Bose–and Fermi–gases with the p–pairing in the Fermi–component of the system. Hoffberg[5] has showed that the the ^{3}He p–pairing in the three–dimensional system is preferable if the ^{3}He–concentration exceeds 6.6% and the pressure is equal to 20 atmospheres. In this case the phase transition temperature increases from $10^{-4}\,K$ for $x \geq 7.6\%$ to $10^{-3}\,K$ for $x \cong 8\%$ and T_C reaches its maximum $1.5\cdot 10^{-2}\,K$ at the phase separation concentration $x = 8.85\%$.

We consider both these cases in this Chapter[18,25,26]. The case of the s–pairing is considered in the sections 20.2 and 20.3, and the p–pairing in three–dimensional and two–dimensional systems in the sections 20.4–20.7.

20.2. The Case of s–Pairing in ^{3}He. Effective Action Functional of the ^{3}He–^{4}He Solutions

In this section, the effective action functional for the case of the s–pairing is obtained. This functional is expressed in terms of natural "physical" density–phase variables. This functional is used for the investigation of the Bose–spectrum of this system. The "unperturbed" spectrum defined by the quadratic form of these density–phase variables, consists of two sound branches. The first one corresponds to the oscillations of the phase of the Fermi–component (Bogoliubov sound), the second one describes the phase–density oscillations of the Bose–component. Besides these sound excitations, the system has a set of the so–called resonant excitations near the decay threshold of the Cooper pair 2Δ (see details in Ref. 11, 12).

We shall derive the effective action functional for the model of a nonideal Fermi–Bose–gas at low temperatures $(T << T_{CB}, T_{CF})$, where T_{CB}, T_{CF} are temperatures of the superfluid phase transitions at the Fermi– and Bose–components correspondingly.

We use the functional integration method. Functional integrals for the Fermi–Bose–systems are integrals over the space $B \otimes F$, where B is the space of complex functions $\psi(\boldsymbol{x},\tau), \overline{\psi}(\boldsymbol{x},\tau)$ $(\boldsymbol{x} \in V, \tau \in [0,\beta])$ with the periodic boundary conditions, F is the space of anticommuting functions $\chi_S(\boldsymbol{x},\tau), \overline{\chi}_S(\boldsymbol{x},\tau)$ ($s = \pm$ is the spin index).

The partition function is the average of $\exp S$ over the space $B \otimes F$, where $S = S_\psi + S_\chi + S_{\psi\chi}$ is the "action" of the system,

$$S_\psi = \int d^3x d\tau \left(\overline{\psi}\partial_\tau \psi - (2m_B)^{-1} \nabla \overline{\psi} \nabla \psi + \lambda_B \overline{\psi}\psi\right)$$

$$-\frac{1}{2}\int d\tau\, d^3x d^3y u_B(\boldsymbol{x}-\boldsymbol{y}) \overline{\psi}(\boldsymbol{x},\tau)\overline{\psi}(\boldsymbol{y},\tau)\psi(\boldsymbol{y},\tau)\psi(\boldsymbol{x},\tau) \tag{20.1}$$

is a part of the action, corresponding to Bose–particles,

$$S_\chi = \int d^3x d\tau \left(\overline{\chi}_S \partial_\tau \chi_S - (2m_F)^{-1} \nabla \overline{\chi}_S \nabla \chi_S + \lambda_F \overline{\chi}_S \chi_S\right)$$

$$-\frac{1}{2}\int d\tau\, d^3x d^3y u_F(\boldsymbol{x}-\boldsymbol{y}) \sum_{S,S'} \overline{\chi}_S(\boldsymbol{x},\tau)\overline{\chi}_{S'}(\boldsymbol{y},\tau)\chi_{S'}(\boldsymbol{y},\tau)\chi_S(\boldsymbol{x},\tau) \tag{20.2}$$

corresponds to Fermi–particles, and

$$S_{\psi\chi} = -\int d\tau\, d^3x d^3y u_{BF}(\boldsymbol{x}-\boldsymbol{y}) \overline{\psi}(\boldsymbol{x},\tau)\psi(\boldsymbol{y},\tau) \sum_S \chi_S(\boldsymbol{y},\tau)\chi_S(\boldsymbol{x},\tau) \tag{20.3}$$

describes the interaction between Bose– and Fermi–particles.

We write down S as $S = S_0 + S_1$, where S_0 is a quadratic form, S_1 is a quartic forms of the fields $\psi(\boldsymbol{x},\tau), \overline{\psi}(\boldsymbol{x},\tau), \chi_S(\boldsymbol{x},\tau), \overline{\chi}_S(\boldsymbol{x},\tau)$ and then expand $\exp S = \exp S_0 \exp S_1$ into the series of powers of S_1, it is not hard to build up the standard temperature diagram techniques (see Ref. 13 and Chapter I) in application to the Fermi–Bose–system.

These standard diagram techniques are not applicable immediately below the phase transition temperature of one of components into the superfluid state. We have to modify the calculation of the functional integral. It can be done according to the scheme developed by Popov for the Bose–system[14,15] (see also Chapter I).

In application to the Fermi–Bose–system at the first stage we have to integrate over the fast parts $\psi_1(\boldsymbol{x},\tau), \chi_{S1}(\boldsymbol{x},\tau)$ of the Bose–field $\psi(\boldsymbol{x},\tau)$ and the Fermi–field $\chi_S(\boldsymbol{x},\tau)$. We define them by the following equations

$$\psi_1(\boldsymbol{x},\tau) = \psi(\boldsymbol{x},\tau) - \psi_0(\boldsymbol{x},\tau), \tag{20.4}$$

$$\psi_0(\boldsymbol{x},\tau) = (\beta V)^{-1/2} \sum_{\omega_B, k<k_{0B}} \exp[i(\omega_B\tau + \boldsymbol{kx})\psi(\boldsymbol{k},\omega_B)], \tag{20.5}$$

$$\chi_{S1}(\boldsymbol{x},\tau) = \chi_S(\boldsymbol{x},\tau) - \chi_{S0}(\boldsymbol{x},\tau), \tag{20.6}$$

$$\chi_{S0}(\boldsymbol{x},\tau) = (\beta V)^{-1/2} \sum_{\omega_F, |k-k_F|<k_{0F}} \exp[i(\omega_F\tau + \boldsymbol{kx})\chi_S(\boldsymbol{k},\omega_F)], \tag{20.7}$$

where $\omega_B = 2\pi n/\beta, \omega_F = (2n+1)\pi/\beta$. The momenta k_{0B}, k_{0F} distinguished "fast" particles from "slow" ones are defined only of the order of magnitude for a concrete system.

We denote the result of the integration over $\psi_1(\boldsymbol{x},\tau), \chi_{S1}(\boldsymbol{x},\tau)$ as

$$\exp\tilde{S}[\psi_0, \bar{\psi}_0, \chi_{S0}, \bar{\chi}_{S0}] = \int D\bar{\chi}_{S1} D\chi_{S0} D\bar{\psi}_1 D\psi_1 \exp S[\psi, \bar{\psi}, \chi_S, \bar{\chi}_S] \tag{20.8}$$

The right hand side of (20.8) has a meaning of the partition function for Fermi particles with momenta $|k - k_F| > k_{0F}$ and Bose–particles with momenta $|k| > k_{0B}$ in the external fields $\psi_0(\boldsymbol{x}, \tau), \chi_{S0}(\boldsymbol{x}, \tau)$. The functional $\tilde{S}$ for the case of a low–density system can be calculated with the help of perturbation theory. In the first approximation we have

$$\tilde{S} = \tilde{S}_B + \tilde{S}_F + \tilde{S}_{BF}, \tag{20.9}$$

$$\tilde{S}_B = p_{0B}\beta V + \int d^3x d\tau \left(\overline{\psi}_0 \partial_\tau \psi_0 - (2m_B)^{-1} \nabla \overline{\psi}_0 \nabla \psi_0 + \mu_B \overline{\psi}_0 \psi_0\right) -$$

$$-\frac{1}{2}\int d\tau\, d^3x d^3y t_B(\boldsymbol{x} - \boldsymbol{y}) \overline{\psi}_0(\boldsymbol{x}, \tau) \overline{\psi}_0(\boldsymbol{y}, \tau) \psi_0(\boldsymbol{y}, \tau) \psi_0(\boldsymbol{x}, \tau),$$

$$\tilde{S}_F = p_{0F}\beta V + \int d^3x d\tau \left(\overline{\chi}_{S0} \partial_\tau \chi_{S0} - (2m_F)^{-1} \nabla \overline{\chi}_{S0} \nabla \chi_{S0} + \mu_F \overline{\chi}_{S0} \chi_{S0}\right)$$

$$-\int d\tau\, d^3x t'_F \sum_{s,s'} \overline{\chi}_{0+}(\boldsymbol{x}, \tau) \overline{\chi}_{0-}(\boldsymbol{y}, \tau) \chi_{0-}(\boldsymbol{y}, \tau) \chi_{0+}(\boldsymbol{x}, \tau),$$

$$\tilde{S}_{BF} = p_{0BF}\beta V - \int d\tau\, d^3x t_{BF} \overline{\psi}_0(\boldsymbol{x}, \tau) \psi_0(\boldsymbol{y}, \tau) \sum_S \overline{\chi}_{S0}(\boldsymbol{y}, \tau) \chi_{S0}(\boldsymbol{x}, \tau)$$

$$\tag{20.10}$$

Here, p_{0B}, p_{0F} are pressures of a nonideal Bose–gas (Fermi–gas) with $|k| > k_{0B}$ ($|k - k_F| > k_{0F}$), p_{0BF} is a part of the pressure arising due to the interaction between Bose– and Fermi–particles, μ_B, μ_F are renormalized chemical potentials. The quantities t_B, t_F, t_{BF} are *t*–matrices for the pair Bose–Bose, Fermi–Fermi and Bose–Fermi interactions which are obtained by the ladder summation of corresponding diagrams. In the gas approximation these *t*–matrices are constants (they do not depend on momenta and energies). Let us note that the interaction of particles with the same spin do not contribute in $\tilde{S}$ in the first approximation, because the change of the antisymmetrized potential by the antisymmetrized constant t–matrix yields a zero result. The quantity t'_F depends logarithmically on the parameter k_{0F} (see Ref. 11, 12 and also Chapter II). However, this k_{0F}–dependence disappears from the final answers.

Now we can transform the integral over the Fermi–field $\chi_{S0}(\boldsymbol{x},\tau),\bar{\chi}_{S0}(\boldsymbol{x},\tau)$ into the integral over the auxiliary Bose–field $c(\boldsymbol{x},\tau),\bar{c}(\boldsymbol{x},\tau)$ describing Cooper pairs. In order to do it, let us introduce the Gaussian integral

$$\int D\bar{c}Dc\exp\left[\int (t'_F)^{-1}\bar{c}(\boldsymbol{x},\tau),c(\boldsymbol{x},\tau)d^3x d\tau\right] \tag{20.11}$$

into the integral over $\chi_{S0}(\boldsymbol{x},\tau),\bar{\chi}_{S0}(\boldsymbol{x},\tau)$. We shall assume $t'_F<0$ (attraction). Then after the transformation

$$\begin{aligned} c(\boldsymbol{x},\tau) &\to c(\boldsymbol{x},\tau)+t'_F\chi_{0+}(\boldsymbol{x},\tau)\chi_{0-}(\boldsymbol{x},\tau), \\ \bar{c}(\boldsymbol{x},\tau) &\to \bar{c}(\boldsymbol{x},\tau)+t'_F\bar{\chi}_{0-}(\boldsymbol{x},\tau)\bar{\chi}_{0+}(\boldsymbol{x},\tau) \end{aligned} \tag{20.12}$$

in the integral over $c(\boldsymbol{x},\tau),\bar{c}(\boldsymbol{x},\tau)$ we obtain the integral of the

$$\exp\tilde{\tilde{S}}\left[\chi_{S0},\bar{\chi}_{S0},\psi_0,\bar{\psi}_0,c,\bar{c}\right] \tag{20.13}$$

over the fields $\chi_{S0},\bar{\chi}_{S0},\psi_0,\bar{\psi}_0,c,\bar{c}$,

where

$$\tilde{\tilde{S}} = \int d^3x d\tau \Big[\sum_S \left(\overline{\chi}_{S0} \partial_\tau \chi_{S0} - (2m_F)^{-1} \nabla \overline{\chi}_{S0} \nabla \chi_{S0} + \lambda_F \overline{\chi}_{S0} \chi_{S0} \right) +$$

$$\overline{c} \chi_{0+} \chi_{0-} + \chi_{0-} \chi_{0+} c + (t'_F)^{-1} \overline{c} c \Big] + \tilde{S}_B [\psi_0, \overline{\psi}_0] + \tilde{S}_{BF} [\psi_0, \overline{\psi}_0, \chi_{S0}, \overline{\chi}_{S0}] . \tag{20.14}$$

In what follows it is appropriate to go to the density–phase variables according to the formulae

$$\psi_0(x,\tau) = \sqrt{\rho_B(x,\tau)} \exp(i\varphi_B(x,\tau)),$$

$$\overline{\psi}_0(x,\tau) = \sqrt{\rho_B(x,\tau)} \exp(-i\varphi_B(x,\tau)),$$

$$c(x,\tau) = \sqrt{\rho_F(x,\tau)} \exp(i\varphi_F(x,\tau)),$$

$$\overline{c}(x,\tau) = \sqrt{\rho_F(x,\tau)} \exp(-i\varphi_F(x,\tau)), \tag{20.15}$$

and then to introduce new variables

$$\pi_{B(F)}(x,\tau) = \rho_{B(F)}(x,\tau) - \rho_{0B(F)}(x,\tau). \tag{20.16}$$

The parameters $\rho_{0B(F)}(x,\tau)$ can be obtained from conditions $\partial S_{eff} / \partial \rho_{0B(F)}(x,\tau) = 0$ (S_{eff} is defined by (20.14)).

Let us make the change of variables

$$\chi_{S0}(x,\tau) = \chi_{S0}(x,\tau) \exp(i\varphi_F(x,\tau)/2),$$

$$\overline{\chi}_{S0}(x,\tau) = \overline{\chi}_{S0}(x,\tau) \exp(-i\varphi_F(x,\tau)/2). \tag{20.17}$$

After change the quadratic form of the variables $\chi_{S0}, \overline{\chi}_{S0}$ looks as follows

$$\sum_S [\overline{\chi}_{S0}\partial_\tau\chi_{S0} - (2m_F)^{-1}\nabla\overline{\chi}_{S0}\nabla\chi_{S0} -$$
$$-\frac{i}{4m_F}\left((\overline{\chi}_{S0}\nabla\chi_{S0} - \nabla\overline{\chi}_{S0}\chi_{S0}), \frac{\nabla\varphi_F}{2}\nabla\right) + \tag{20.18}$$
$$+\left(\tilde{\mu} - (8m_F)^{-1}(\nabla\varphi_F)^2 + \frac{i}{2}\partial_\tau\varphi_F - t_{BF}\pi_B\right)\overline{\chi}_{S0}\chi_{S0}],$$

where $\tilde{\mu} = \lambda_F - t'_F\rho_F(|k - k_F| > k_{0F}) - t_{BF}\rho_{0B}$. Now the integral over $\chi_{S0}, \overline{\chi}_{S0}$ is Gaussian and it formally is equal to the determinant of the matrix-differential operator $\hat{M}$:

$$\hat{M} =$$
$$\begin{pmatrix} \partial_\tau - \frac{\nabla^2}{2m_F} - \left(\tilde{\mu} - \frac{(\nabla\varphi_F)^2}{8m_F} + \frac{i}{2}\partial_\tau\varphi_F\right) & \\ + t_{BF}\pi_B + \frac{i}{4m_F}((\nabla - \nabla'), \nabla\varphi_F); & \sqrt{\rho_F(\boldsymbol{x},\tau)} \\ & \partial_\tau + \frac{\nabla^2}{2m_F} + \left(\tilde{\mu} - \frac{(\nabla\varphi_F)^2}{8m_F} + \frac{i}{2}\partial_\tau\varphi_F\right) \\ \sqrt{\rho_F(\boldsymbol{x},\tau)} & - t_{BF}\pi_B - \frac{i}{4m_F}((\nabla - \nabla'), \nabla\varphi_F) \end{pmatrix}$$
$$\tag{20.19}$$

which acts on the column $\begin{pmatrix} \chi_{0+} \\ \overline{\chi}_{0-} \end{pmatrix}$. Let us rewrite (20.19) as

$$\hat{M} = \begin{pmatrix} G_+^{-1} + F(\boldsymbol{x},\tau) & \sqrt{\rho_{0F}(\boldsymbol{x},\tau)} + Q(\boldsymbol{x},\tau) \\ \sqrt{\rho_{0F}(\boldsymbol{x},\tau)} + Q(\boldsymbol{x},\tau) & G_-^{-1} - F(\boldsymbol{x},\tau) \end{pmatrix}, \tag{20.20}$$

where

$$F(\boldsymbol{x},\tau)=\frac{(\nabla\varphi_F)^2}{8m_F}-\frac{i}{2}\partial_\tau\varphi_F+\frac{i}{4m_F}((\nabla-\nabla'),\nabla\varphi_F)+t_{BF}\pi_B(\boldsymbol{x},\tau),$$

$$Q(\boldsymbol{x},\tau)=\frac{\pi_F(\boldsymbol{x},\tau)}{2\rho_{0F}^{1/2}}-\frac{\pi_F^2(\boldsymbol{x},\tau)}{8\rho_{0F}^{3/2}}+\dots \tag{20.21}$$

and the Green's functions are defined by equations

$$\left(\partial_\tau\mp\frac{\nabla^2}{2m_F}\mp\tilde{\mu}\right)G_\pm(x,x')=\delta(x-x')=\delta(\boldsymbol{x}-\boldsymbol{x}')\delta(\tau-\tau'). \tag{20.22}$$

We can regularize $\det\hat{M}$ dividing it by $\det\hat{M}_0$, where $\hat{M}_0$ is the $\hat{M}$ –operator with zero values of the variables φ_F,π_F. Thus, we come to the functional integral of the functional $\exp S_{eff}$ over the physical variables $\varphi_B,\varphi_F,\pi_B,\pi_F$, where S_{eff} is equal to

$$S_{eff}=\int(t'_F)^{-1}\rho_F(\boldsymbol{x},\tau)d^3xd\tau+\ln\det\frac{\hat{M}(\varphi_F,\pi_F,\pi_B)}{\hat{M}(0,0,0)}+S_{eff}(\varphi_B,\pi_B) \tag{20.23}$$

Here, $S_{effB}=S_{eff}(\varphi_B,\pi_B)$ is the effective action functional for the Bose–gas, obtained by Popov[14,15] and in Chapter II. Its explicit form is written below. The functional (20.4) is the effective action for the nonideal Fermi–Bose–gas in the case when both its components are in the superfluid state.

20.3. Bose–Spectrum of the ^{3}He–^{4}He Solution

The effective action (20.23) is appropriate for the calculation of the longwavelength spectrum of the system. In the first approximation this spectrum is defined by the quadratic part of variables $\varphi_B(x,\tau),\varphi_F(x,\tau),\pi_B(x,\tau),\pi_F(x,\tau)$ in the functional (20.23). We write the functional

$$\ln\det\hat{M}(\varphi_F,\pi_F,\pi_B)/\hat{M}(0,0,0) \tag{20.24}$$

as a sum of two terms

$$\ln\det\hat{M}(\varphi_F,\pi_F,\pi_B)/\hat{M}(0,\Delta,0)+\ln\det\hat{M}(0,\Delta,0)/\hat{M}(0,0,0) \tag{20.25}$$

where $\Delta=\rho_{0F}^{1/2}$. Then let us expand the first term as follows

$$\ln\det\hat{M}(\varphi_F,\pi_F,\pi_B)/\hat{M}(0,\Delta,0)=Tr\ln\left(I+\hat{G}\hat{u}\right)=$$
$$\sum_{n=1}^{\infty}\frac{(-1)^{n-1}}{n}\int\prod_{i=1}^{n}dx_i tr\left(\hat{G}(1,2)\hat{u}(2)\cdot\ldots\cdot\hat{G}(n,1)\hat{u}(1)\right), \tag{20.26}$$

where

$$\hat{G}=\begin{pmatrix}G_+ & \tilde{G}\\ \tilde{G} & G_-\end{pmatrix};\ \hat{u}=\begin{pmatrix}F & Q\\ Q & -F\end{pmatrix} \tag{20.27}$$

and $G_\pm,\tilde{G}$ functions in p–representation are

$$G_\pm=(-i\omega\pm\xi(\boldsymbol{k}))/\left(\omega^2+\varepsilon^2(\boldsymbol{k})\right),\tilde{G}=\Delta/\left(\omega^2+\varepsilon^2(\boldsymbol{k})\right), \tag{20.28}$$

where

$$\varepsilon^2(\boldsymbol{k}) = \xi^2(\boldsymbol{k}) + \Delta^2$$

Let us confine ourselves only by terms with n=1,2 in (20.26) and consider only a quadratic form of variables $\varphi_F(\boldsymbol{x},\tau), \pi_F(\boldsymbol{x},\tau), \pi_B(\boldsymbol{x},\tau)$. So for n=1 we have

$$\int dx tr\left(\hat{G}(1,1)\hat{u}(1)\right) = \int dx\left[(G_+ - G_-)F + 2\tilde{G}Q\right], \tag{20.29}$$

where $x = (\boldsymbol{x},\tau), dx = d^3x d\tau$.

Then we can go in (20.29) to the momentum representation for the Green's functions $G_\pm, \tilde{G}$ and fields $\varphi_F(\boldsymbol{x},\tau), \pi_F(\boldsymbol{x},\tau), \pi_B(\boldsymbol{x},\tau)$ and write (20.29) as

$$\sum_p \left\{(G_+ - G_-)\frac{k^2}{2m_F}\varphi_F(p)\varphi_F(-p) - \frac{\tilde{G}}{4\Delta^3}\pi_F(p)\pi_F(-p)\right\}. \tag{20.30}$$

For the term with n=2 in (20.26)

$$-\frac{1}{2}\int dx_1 dx_2 tr\left(\hat{G}(1,2)\hat{u}(2)\hat{G}(2,1)\hat{u}(1)\right) \tag{20.31}$$

we shall operate in an analogous way, namely after calculating $tr\left(\hat{G}\hat{u}\hat{G}\hat{u}\right)$ we substitute the expressions (20.21) instead of F, Q. Taking only the terms, quadratic in fields φ and π into account, we go to the p–representation. As a result we obtain for the sum of terms with n=1 and n=2:

$$\int dx_1 tr\left(\hat{G}(1,1)\hat{u}(1)\right) - \frac{1}{2}\int dx_1 dx_2 tr\left(\hat{G}(1,2)\hat{u}(2)\hat{G}(2,1)\hat{u}(1)\right) \Rightarrow$$

$$\Rightarrow -\frac{1}{2}\sum_p \{R_{11}(p)\varphi_F(p)\varphi_F(-p) + R_{14}(p)\varphi_F(p)\pi_B(-p) + \qquad (20.32)$$

$$+R_{22}(p)\pi_F(p)\pi_F(-p) + R_{44}^{BF}(p)\pi_B(p)\pi_B(-p)\}.$$

It can be shown that the coefficients R_{12}, R_{21} (in front of the variables $\varphi_F \cdot \pi_F$) and R_{24}, R_{42} corresponding to $\pi_F \cdot \pi_B$ disappear. The coefficients $R_{11}, R_{14}, R_{22}, R_{44}$ are defined by the formulae

$$R_{11}(p) = (\beta V)^{-1}\sum_{p1}\left\{\left(i\omega + \frac{(\mathbf{kk}_1)}{m_F}\right)^2 r_{p1}^{(1)}(p) + Sp_{p1}(G_+ - G_-)\frac{k^2}{m_F}\right\},$$

$$R_{14}(p) = -2t_{BF}(\beta V)^{-1}\sum_{p1}\left(i\omega + \frac{(\mathbf{kk}_1)}{m_F}\right) r_{p1}^{(1)}(p),$$

$$R_{22}(p) = (\beta V)^{-1}\sum_{p1}\left(\tilde{G}/2\Delta^3\right) + \left(4\Delta^2\right)^{-1}(\beta V)^{-1}\sum_{p1} r_{p1}^{(2)}(p), \qquad (20.33)$$

$$R_{44}(p) = t_{BF}^2(\beta V)^{-1}\sum_{p1} r_{p1}^{(1)}(p)\ ,$$

where

$$r_{p1}^{(1)}(p) = G_+(p_1)G_+(p+p_1) + G_-(p_1)G_-(p+p_1) - 2\tilde{G}(p_1)\tilde{G}(p+p_1),$$
$$r_{p1}^{(2)}(p) = G_+(p_1)G_-(p+p_1) + G_-(p_1)G_+(p+p_1) + 2\tilde{G}(p_1)\tilde{G}(p+p_1).$$

$$(20.34)$$

The general form of the quadratic part of S_{eff} can be obtained if we supplement the functional (20.32) by the effective action functional of the Bose–field $S_{eff}(\varphi_B, \pi_B)$ of the form

$$-\frac{1}{2}\sum_p\{R_{33}(p)\varphi_B(p)\varphi_B(-p)+R_{34}(p)\varphi_B(p)\pi_B(-p)+ \\ +R_{44}^B(p)\pi_B(p)\pi_B(-p)\} \tag{20.35}$$

with $R_{33}(p)=\frac{\rho_B k^2}{m_B}, R_{34}(p)=\omega, R_{44}^B=\omega+\frac{k^2}{4m_B\rho_{0B}}.$ (20.36)

The "unperturbed" Bose–spectrum of the system defined by the functional (20.32) + (20.35) can be obtained from the equation

$$\det\hat{R}(p)=0, \tag{20.37}$$

where

$$\hat{R}(p)=\begin{pmatrix} R_{11} & 0 & 0 & R_{14} \\ 0 & R_{22} & 0 & 0 \\ 0 & 0 & R_{33} & R_{34} \\ R_{41} & 0 & R_{43} & R_{44} \end{pmatrix} \tag{20.38}$$

is formed by the coefficients of the full quadratic form which obey the symmetry conditions

$$R_{14}(p)=-R_{41}(-p), R_{34}(p)=-R_{43}(-p), \tag{20.39}$$

and

$$R_{44}(p)=R_{44}^{BF}(p)+R_{44}^B(p). \tag{20.40}$$

The equation (20.37) can be rewritten in the following form

$$R_{22}(p)\left[R_{11}(p)\left(R_{33}(p)R_{44}(p)-R_{34}^2(p)\right)-R_{33}(p)R_{34}^2(p)\right]=0. \tag{20.41}$$

This equation splits in two ones. The equation $R_{22}(p)=0$ coincides with the equation obtained for a Bose–spectrum of the superfluid Fermi–gas[11,12] (see also Chapter II). It was shown that this equation has a set of roots corresponding to the so–called resonant excitations, which tend to 2Δ as $k\to 0$. The second equation defines phonon modes of the Bose–spectrum of the sistem. In order to find them, let us calculate the coefficients of the matrix $\hat{R}$ in the limit of small k,ω:

$$\begin{aligned}
R_{11}(p) &\approx \frac{m_F k_F}{\pi^2}\left(\frac{k_F^2}{3m_F^2}k^2+\omega^2\right),\\
R_{14}(p) &\approx \frac{2t_{BF}m_F k_F}{\pi^2}\omega,\\
R_{22}(p) &\approx \frac{m_F k_F}{(2\pi\Delta)^2}\left[1+O\left(\frac{\omega^2}{\Delta^2},\frac{\alpha k^2}{\Delta^2}\right)\right],\\
R_{44}(p) &\approx -\frac{m_F k_F t_{BF}^2}{(\pi)^2}\left(1-\frac{\omega^2}{6\Delta^2}\right)+\left(t_B+\frac{k^2}{4m_B\rho_{0B}}\right).
\end{aligned} \tag{20.42}$$

Using (20.42), one can write down the equation for phonon modes in the following form:

$$\left(E^2-v_F^2k^2\right)\left(E^2-v_B^2k^2\right)+\delta^2E^2k^2=0, \tag{20.43}$$

where v_F^2, v_B^2, δ^2 are expressed in terms of the parameters of the system. In the low density approximation (and if $\rho_F << \rho_B$)

$$\delta^2 << \left(v_F^2=\frac{c_F^2}{3}, v_B^2=\frac{t_B\rho_B}{m_B}\right). \tag{20.44}$$

Thus, the spectrum consists of two phonon branches

$$E^{(1)} = v_F k, \quad E^{(2)} = v_B k. \tag{20.45}$$

In the general case, as it was seen from (20.44) the variables φ, π of both component are "coupled". If we neglect this "coupling" one can see that the first branch of (20.45) is defined by the coefficient $R_{11}(p)$ of the matrix $\hat{R}$ and it corresponds to the phase oscillations φ_F. The second branch corresponds to the joint oscillations of the phase φ_B and the density π_B.

20.4. The Case of *p*–pairing. The Effective Action Functional of the ^{3}He–^{4}He Solution

In the sections 20.4–20.7 a three–dimensional model of the ^{3}He–^{4}He solution is considered and also a two–dimensional model of ^{3}He–^{4}He film built by Brusov and Popov[18,25,26]. For the three–dimensional system we shall consider the possible formation of superfluid A– and B–phases of the Fermi–component. In the two–dimensional case, we consider superfluid *a*– and *b*–phases of the Fermi–component, such phases were introduced by Brusov and Popov[17,19,20] for a purely two–dimensional Fermi–system – a model of ^{3}He–film. It is these phases that are energetically the most advantageous.

We use the approach of successive functional integration over the fast and slow variables. After the integration over the fast fields and transition to new collective coordinates, we obtain the effective action, which describes collective Bose–excitations of the system. In the first approximation, their spectrum is determined by the quadratic part of this functional, which is different for different superfluid phases.

The Bose–spectrum of a three–dimensional Fermi–Bose–system consist of 19 modes, for two–dimensional system, there are 13. One of the 19 (respectively, 13) modes is the sound mode of the Bose–subsystem, the remaining 18 (respectively, 12) correspond to different collective excitations of the Fermi–subsystem. We shall consider the influence of the interaction between the Fermi– and Bose–subsystems on the collective Bose–excitations. We show that in a first approximation the interaction leads to a "coupling" of the sound modes of the Fermi– and Bose–subsystems, for those velocities a biquadratic equation is obtained.

In this section the effective action functional is constructed for models of three– and two–dimensional systems of ^3He–^4He type. In the subsequent sections 20.5–20.7 we investigate the quadratic part of the functional, which determines the Bose–spectrum for different cases corresponding to different superfluid phases, at $T = 0$.

In order to construct the effective action functional we start from the same form of initial action S as that for the s–pairing in the section 20.2. Integrating $\exp S$ over the fast fields, we obtain $\exp \tilde{S}$, where $\tilde{S}$ is the action of the slow variables. This $\tilde{S}$ is different from the $\tilde{S}$ for the case of the s–pairing because we have to take into account also the scattering of fermions with the same spins in the case of the p–pairing. We take the action $\tilde{S}$ in the form

$$\begin{aligned}
&\tilde{S} = \tilde{S}_{\psi} + \tilde{S}_{\chi} + \tilde{S}_{\psi\chi} = \sum_{p} \varepsilon_B(p) b^*(p) b(p) - (\beta V)^{-1} \times \\
&\times \sum_{p1+p2=p3+p4} t_B(p_1, p_2, p_3, p_4) b^*(p_1) b^*(p_2) b(p_3) b(p_4) + \\
&\sum_{p,s} \varepsilon_{F,s}(p) a_s^*(p) a_s(p) - (\beta V)^{-1} \sum_{p1+p2=p3+p4} t_0(p_1, p_2, p_3, p_4) \times \\
&\times a_+^*(p_1) a_-^*(p_2) a_-(p_4) a_+(p_3) - (2\beta V)^{-1} \sum_{p1+p2=p3+p4} t_1(p_1, p_2, p_3, p_4) \times \\
&\times [2a_+^*(p_1) a_-^*(p_2) a_-(p_4) a_+(p_3) + a_+^*(p_1) a_+^*(p_2) a_+(p_4) a_+(p_3) + \\
&+ a_-^*(p_1) a_-^*(p_2) a_-(p_4) a_-(p_3)] - \\
&- (\beta V)^{-1} \sum_{p1+p2=p3+p4} t_{BF}(p_1, p_2, p_3, p_4) a_s^*(p_1) a_s(p_3) b^*(p_2) b(p_4).
\end{aligned} \tag{20.46}$$

Here, we have taken into account the quadratic forms of slow fields ψ_0, χ_{s0}, describing noninteracting Bose–particles with low momenta and noninteracting Fermi–particles with momenta in the neighborhood of the Fermi–sphere, and also the fourth order terms describing the pair interactions between particles. We have omitted in (20.46) constant terms which do not depend on slow fields and also the high order terms, which are responsible for the triple, quartic and so on interactions.

In (20.46)

$$\varepsilon_B(p) = i\omega - \frac{k^2}{2m_B} + \lambda_B,\ \varepsilon_{FS}(p) = Z^{-1}\left(i\omega - c_F\left(k - k_F\right) + s\mu H\right) \tag{20.47}$$

and t_B is the scattering amplitude of Bose–particles, t_0 and t_1 are those for Fermi–particles, the amplitude t_{BF} describes the scattering of Bose–particle by Fermi–particle. These amplitudes are obtained by the ladder summation of corresponding diagrams. The amplitude $t_0\left(p_1, p_2, p_3, p_4\right)$ is symmetric and $t_1\left(p_1, p_2, p_3, p_4\right)$ antisymmetric for each of the permutations $p_1 \leftrightarrow p_2, p_3 \leftrightarrow p_4$. In the neighbourhood we can put $\omega_i = 0, \boldsymbol{k}_i = k_F \boldsymbol{n}_i$. The amplitudes t_0 and t_1 depend only on two invariants, and we can write down them in the form

$$t_0 = f((\boldsymbol{n}_1, \boldsymbol{n}_2), (\boldsymbol{n}_1\text{-}\boldsymbol{n}_2, \boldsymbol{n}_3\text{-}\boldsymbol{n}_4)),$$

$$t_1 = (\boldsymbol{n}_1\text{-}\boldsymbol{n}_2, \boldsymbol{n}_3\text{-}\boldsymbol{n}_4)\, \mathrm{g}((\boldsymbol{n}_1, \boldsymbol{n}_2), (\boldsymbol{n}_1\text{-}\boldsymbol{n}_2, \boldsymbol{n}_3\text{-}\boldsymbol{n}_4)), \tag{20.48}$$

where f and g are even functions of the second argument.

Using the gaseous approximation, we may consider the values t_B, t_{BF}, f and g to be constants (do not depending on momenta and energies). In what follows we put $f = 0$. Thus we shall have instead of (20.46)

$$\tilde{S} = \sum_{p} \varepsilon_B(p) b^*(p) b(p) - t_B (2\beta V)^{-1} \times$$
$$\times \sum_{p1+p2=p3+p4} b^*(p_1) b^*(p_2) b(p_3) b(p_4) + \sum_{p,s} \varepsilon_{F,s}(p) a_s^*(p) a_s(p) -$$
$$- g_0 (2\beta V)^{-1} \sum_{p1+p2=p3+p4} (\boldsymbol{n}_1 - \boldsymbol{n}_2, \boldsymbol{n}_3 - \boldsymbol{n}_4)[2a_+^*(p_1) a_-^*(p_2) a_-(p_4) a_+(p_3) +$$
$$+ a_+^*(p_1) a_+^*(p_2) a_+(p_4) a_+(p_3) + a_-^*(p_1) a_-^*(p_2) a_-(p_4) a_-(p_3)] -$$
$$- t_{BF} (\beta V)^{-1} \sum_{p1+p2=p3+p4} a_s^*(p_1) a_s(p_3) b^*(p_2) b(p_4). \tag{20.49}$$

Then, we introduce the Bose–fields $c_{ia}(p)$, describing Cooper pairs of fermions[21]. For this, we insert in the integral over the slow Bose– and Fermi– fields a Gaussian integral with respect to the new Bose–field

$$\int Dc_{ia}^* Dc_{ia} \exp\left(g_0^{-1} \sum_{p,i,a} c_{ia}^*(p) c_{ia}(p) \right) \tag{20.50}$$

We then make a shift of the fields:

$$c_{i1}(p) \to c_{i1}(p) +$$
$$+ \frac{g_0}{2(\beta V)^{1/2}} \sum_{p1+p2=p} (n_{1i} - n_{2i})[a_+(p_2) a_+(p_1) - a_-(p_2) a_-(p_1)],$$
$$c_{i2}(p) \to c_{i2}(p) +$$
$$+ \frac{i g_0}{2(\beta V)^{1/2}} \sum_{p1+p2=p} (n_{1i} - n_{2i})[a_+(p_2) a_+(p_1) + a_-(p_2) a_-(p_1)],$$
$$c_{i3}(p) \to c_{i3}(p) + \frac{g_0}{(\beta V)^{1/2}} \sum_{p1+p2=p} (n_{1i} - n_{2i})\, a_-(p_2) a_+(p_1), \tag{20.51}$$

which annihilates the quartic form of Fermi–fields. Let us note that we have to consider the amplitude g_0 to be negative. Namely in this case the Gaussian integral (20.50) has a meaning.

After the shift (20.51) we obtain a new action

$$\tilde{S} = \sum_p \varepsilon_B(p) b^*(p) b(p) - t_B (2\beta V)^{-1} \sum_{p1+p2=p3+p4} b^*(p_1) b^*(p_2) b(p_3) b(p_4) +$$
$$+ \sum_{p,s} \varepsilon_{F,s}(p) a_s^*(p) a_s(p) + g_0^{-1} \sum_{p,i,a} c_{ia}^*(p) c_{ia}(p) + \frac{1}{2\sqrt{\beta V}} \sum_{p1+p2=p} (n_{1i} - n_{2i}) \times$$
$$\times \{ [c_{i1}^*(p)(a_+(p_2)a_+(p_1) - a_-(p_2)a_-(p_1)) + c.c.] + [ic_{i2}^*(p)(a_+(p_2)a_+(p_1) +$$
$$+ a_-(p_2)a_-(p_1)) + c.c.] + 2[c_{i3}^*(p) a_-(p_2) a_+(p_1) + c.c.] \}$$

(20.52)

The action $\tilde{S}$ is quadratic in the Fermi fields $a_S(p), a_S^*(p)$ and we can integrate with respect to these variables (the integral is Gaussian). After the Gaussian integration with respect to the Fermi–fields we arrive to the functional of the effective action

$$S_{eff} = \sum_p \varepsilon_B(p) b^*(p) b(p) - t_B (2\beta V)^{-1} \sum_{p1+p2=p3+p4} b^*(p_1) b^*(p_2) b(p_3) b(p_4)$$
$$+ g_0^{-1} \sum_{p,i,a} c_{ia}^*(p) c_{ia}(p) + \frac{1}{2} \ln\det \frac{\hat{M}[c_{ia}, c_{ia}^*, b, b^*]}{\hat{M}[0,0,0,0]}.$$

(20.53)

Here, $\hat{M}$ is the operator with matrix elements $\hat{M}_{a,b}(p_1, p_2)$, which can be conveniently written in the form of a fourth–order matrix

$$\hat{M} =$$

$$= \begin{pmatrix} Z^{-1}(i\omega - \xi + \mu(\boldsymbol{H},\boldsymbol{\sigma}))\delta_{p1p2} - & \\ -\frac{t_{BF}}{\beta V}\sum_p b^*(p)b(p+p_1-p_2); & \frac{n_{1i}-n_{2i}}{\sqrt{\beta V}}\sigma_a c_{ia}(p_1+p_2); \\ & Z^{-1}(-i\omega + \xi + \mu(\boldsymbol{H},\boldsymbol{\sigma}))\delta_{p1p2} + \\ -\frac{n_{1i}-n_{2i}}{\sqrt{\beta V}}\sigma_a c_{ia}^*(p_1+p_2); & +\frac{t_{BF}}{\beta V}\sum_p b^*(p)b(p+p_2-p_1) \end{pmatrix}$$

(20.54)

The effective action functional (20.53) contains all the information about the physical properties of the model system and, in particular, the Bose–spectrum of the collective excitations.

We assume that both the Bose–and the Fermi–components of the system are superfluid. In the considered formalism, the phase transition to the superfluid state is a Bose–condensation of the fields $b(p)$ and $c_{ia}(p)$.

We separate the condensates by the shift transformations

$$b(p) \to b(p) + (\beta V)^{1/2}\alpha\delta_{p0},$$

$$c_{ia}(p) \to c_{ia}(p) + c_{ia}^{(0)}(p), \tag{20.55}$$

where $(\beta V)^{1/2}\alpha\delta_{p0}$, and $c_{ia}^{(0)}(p)$,are the condensate functions of the fields $b(p)$ and $c_{ia}(p)$. The function $c_{ia}^{(0)}(p)$is different for the different superfluid phases in which the Fermi–component can be.

If we take a pure Bose–system (first two terms in (20.53)) and make the shift $b(p) \to b(p) + (\beta V)^{1/2}\alpha\delta_{p0}$, then from the condition of a maximum of S_{eff} at $b(p) = (\beta V)^{1/2}\alpha\delta_{p0}$, we obtain

$$\alpha^2 = \rho_{0B} = \lambda_B / t_B \tag{20.56}$$

After the shift, the quadratic part of S_{eff} is

$$\begin{aligned}&\sum_p \left(i\omega - \frac{k^2}{2m_B} + \lambda_B - 2\alpha^2 t_B \right) b^*(p) b(p) - \\ &-\left(\alpha^2 t_B/2\right) \sum_p \left(b(p)b(-p) + b^*(p)b^*(-p) \right) .\end{aligned} \tag{20.57}$$

The spectrum determined by the form (20.57) is the Bogoliubov spectrum

$$E^2 = \frac{\lambda_B k^2}{m_B} + \left(\frac{k^2}{2m_B} \right)^2 \tag{20.58}$$

which has a phonon form $E = ck$ at $k \to 0$, where $c^2 = \lambda_B / t_B = t_B \alpha^2 / m_B$.

For the Fermi–Bose system, it is also necessary to separate the condensate $c_{ia}^{(0)}(p)$ of the fields $c_{ia}(p)$ and to take the dependence of the expression $\frac{1}{2} \ln \det\left(\hat{M}/\hat{M}_0\right)$ on the fields $b(p)$, $b^*(p)$.

The forms of the condensate functions $c_{ia}^{(0)}(p)$ are different for different superfluid phases. In the next two sections we consider the *B*– and *A*–phases of the three–dimensional system (Sections 20.5 and 20.6) and then the *a*– and *b*–phases of the two–dimensional systems corresponding to ^{3}He–^{4}He films.

20.5. Bose–Spectrum of a Solution of the Type ^{3}He–*B*–^{4}He

In the *B*–phase for $H = 0$ the condensate function of the field $c_{ia}(p)$ has a form

$$c_{ia}^{(0)}(p) = (\beta V)^{1/2}\,\delta_{p0}\delta_{ia}c \tag{20.59}$$

the condensate function of the field $b(p)$ is $(\beta V)^{1/2}\alpha\delta_{p0}$. The parameters c and α are determined from the condition of a maximum of S_{eff} by the equations

$$\partial_{\alpha}S_{eff} = 0, \tag{20.60}$$

(after replacement in S_{eff} of the fields $b(p)$ and $c_{ia}(p)$ by their condensate values). The sum of the first two terms in the formula (20.53) for S_{eff} after such a substitution is equal to

$$\beta V\left(\lambda_B\alpha^2 - \frac{1}{2}\alpha^4 t_B\right) \tag{20.61}$$

and the derivative with respect to α of this expression is

$$2\beta V\left(\lambda_B\alpha - t_B\alpha^3\right). \tag{20.62}$$

It is somewhat difficult to find the derivative with respect to α of $\frac{1}{2}\ln\det\left(\hat{M}/\hat{M}_0\right)$, since not only the integral over the momenta near the Fermi–sphere contribute to it. We find this derivative by using the following arguments. The substitution $b(p) = b^*(p) = (\beta V)^{1/2}\alpha\delta_{p0}$ reduces the contribution of the b–field to $\frac{1}{2}\ln\det\left(\hat{M}/\hat{M}_0\right)$ to the addition $-Zt_{BF}\alpha^2$ to the chemical potential μ_F of the Fermi–particles.

Since the functional integral of $\exp\tilde{S}$ with respect to the Fermi–fields is proportional to $\exp(-\beta\Omega)$, the contribution to $\partial_\alpha S_{eff}$ in which we are interested can be written in the form

$$-\beta(\partial\Omega/\partial\mu)\partial_\alpha\left(-Zt_{BF}\alpha^2\right)=-2\alpha\beta V\rho_F Zt_{BF} \tag{20.63}$$

in view of the fact that $\partial\Omega/\partial\mu=-N_F=-V\rho_F$, where ρ_F is the density of the Fermi–particles. Adding (20.62) and (20.63) and equating to zero, we obtain an equation for α

$$\alpha\left(\lambda_B-t_B\alpha^2-\rho_F Zt_{BF}\right)=0. \tag{20.64}$$

This equation has the trivial solution $\alpha=0$ and can have a nontrivial one

$$\alpha^2=\left(\lambda_B-\rho_F Zt_{BF}\right)/\lambda_B. \tag{20.65}$$

Now let us consider the second equation in (20.60), namely, $\partial_c S_{eff}=0$. It is not hard to see that it has the same form as for a pure Fermi–system, namely,

$$c\left[3g_0^{-1}+\left(Z^2/\beta V\right)\sum_P\left(\omega^2+\xi^2+\Delta^2\right)^{-1}\right]=0, \tag{20.66}$$

where

$$\Delta=2cZ,\ \ \xi=\frac{k^2-k_F^2}{2m_F}+Zt_{BF}\alpha^2\cong c_F\left(k-k_F\right). \tag{20.67}$$

We now write down the part of S_{eff}, quadratic in the fluctuations of the fields $b(p)$, $b^*(p), c_{ia}(p), c^*_{ia}(p)$ around their condensate values $(\beta V)^{1/2}\alpha\delta_{p0}, c^{(0)}_{ia}(p)$:

$$\sum_p\left(i\omega - \frac{k^2}{2m_B} + \lambda_B - 2\alpha^2 t_B\right) b^*(p)b(p) -$$
$$-\left(\alpha^2 t_B/2\right)\sum_p \left(b(p)b(-p) + b^*(p)b^*(-p)\right) +$$
$$+ g_0^{-1}\sum_{p,i,a} c^*_{ia}(p)c_{ia}(p) - \frac{1}{4}Tr\left(\hat{G}\hat{u}\right)^2 - Zt_{BF}\rho_F\sum_p b^*(p)b(p) \qquad (20.68)$$

Here, the first two terms derive from the first two terms in (20.53), which depend only on $b(p)$, $b^*(p)$. The last term in (20.46) derives from correction quadratic in $b(p)$, $b^*(p)$ to the diagonal elements of the operator $\hat{M}$ and can be obtained by the same arguments as the correction (20.63) to $\partial_\alpha S_{eff}$.

The most laborious part is the calculation of $-\frac{1}{4}Tr\left(\hat{G}\hat{u}\right)^2$, where $\hat{G}$ is the inverse of the operator

$$G^{-1} = \begin{pmatrix} Z^{-1}(i\omega_1 - \xi_1)\delta_{p1p2} & 2c(\boldsymbol{n},\boldsymbol{\sigma})\delta_{p1+p2,0} \\ -2c(\boldsymbol{n},\boldsymbol{\sigma})\delta_{p1+p2,0} & Z^{-1}(-i\omega_1 + \xi_1)\delta_{p1p2} \end{pmatrix} \qquad (20.69)$$

$$\hat{u}_{p1p2} = (\beta V)^{-1/2} \times$$
$$\times \begin{pmatrix} -\alpha t_{BF}\left(b(p_1 - p_2) + b^*(p_2 - p_1)\right) & (n_{1i} - n_{2i})\sigma_a c_{ia}(p_1 + p_2); \\ -(n_{1i} - n_{2i})\sigma_a c^*_{ia}(p_1 + p_2); & \alpha t_{BF}\left(b(p_1 - p_2) + b^*(p_2 - p_1)\right) \end{pmatrix}. \qquad (20.70)$$

Inverting $\hat{G}^{-1}$, we obtain

$$\hat{G} = \frac{Z}{M_1}\begin{pmatrix} -(i\omega_1 + \xi_1)\delta_{p1p2} & \Delta(\boldsymbol{n},\boldsymbol{\sigma})\delta_{p1+p2,0} \\ -\Delta(\boldsymbol{n},\boldsymbol{\sigma})\delta_{p1+p2,0} & (i\omega_1 + \xi_1)\delta_{p1p2} \end{pmatrix} \tag{20.71}$$

where

$$M_1 = \omega_1^2 + \xi_1^2 + \Delta^2 \tag{20.72}$$

Calculation $-\frac{1}{4}Tr\left(\hat{G}\hat{u}\right)^2$ yields

$$-\frac{1}{4}Tr(\widehat{G}\widehat{u})^2 = -\frac{1}{4}\sum_{p1,p2,p3,p4} tr\,(G_{p1p2}u_{p2p3}G_{p3p4}u_{p4p1}) =$$
$$= -\sum_p \left\{ \widetilde{A}_{ij}(p)c_{ia}^*(p)c_{ja}(p) + \frac{1}{2}B_{ijab}(p)\left[c_{ia}(p)c_{jb}(-p) + c_{ia}^*(p)c_{jb}^*(-p)\right]\right\} -$$
$$-\sum_p C(p)\left[b^*(p) + b(-p)\right]\cdot\left[b(p) + b^*(-p)\right] -$$
$$\sum_p D_{ia}(p)\left\{\left[b(p) + b^*(-p)\right]c_{ia}^*(p) + \left[b^*(p) + b(-p)\right]c_{ia}(p)\right\},$$

(20.73)

$$\widetilde{A}_{ij}(p) = -\frac{4Z^2}{\beta V}\sum_{p1+p2=p} M_1^{-1}M_2^{-1}(i\omega_1 + \xi_1)(i\omega_2 + \xi_2)n_{1i}n_{1j},$$

$$B_{ijab}(p) = \frac{4Z^2\Delta^2}{\beta V}\sum_{p1+p2=p} M_1^{-1}M_2^{-1}n_{1i}n_{1j}(2n_{1a}n_{1b} - \delta_{ab}),$$

$$C(p) = \frac{Z^2\alpha^2 t_{BF}^2}{\beta V}\sum_{p1+p2=p} M_1^{-1}M_2^{-1}\left[(i\omega_1 + \xi_1)(i\omega_2 + \xi_2) - \Delta^2\right],$$

$$D_{ia}(p) = -D_{ia}(-p) = \frac{4Z^2\alpha\, t_{BF}\Delta}{\beta V} \sum_{p1+p2=p} M_1^{-1} M_2^{-1} (i\omega_1 + \xi_1) n_{1i} n_{1j} =$$

$$= \frac{2i\omega Z^2 \alpha\, t_{BF}\Delta}{\beta V} \sum_{p1+p2=p} M_1^{-1} M_2^{-1} n_{1i} n_{1j}.$$

(20.74)

Using (20.73), (20.74), we can write the quadratic part (20.68) of the functional S_{eff} as follows

$$\sum_p \left(i\omega - \frac{k^2}{2m_B} - \alpha^2 t_B - C(p) \right) b^*(p) b(p) -$$

$$-\frac{1}{2}\sum_p \left(\alpha^2 t_B + C(p)\right)\left[b(p)b(-p) + b^*(p)b^*(-p)\right] -$$

$$-\sum_p \left\{ A_{ij}(p) c^*_{ia}(p) c_{ja}(p) + \frac{1}{2} B_{ijab}(p)\left[c_{ia}(p)c_{jb}(-p) + c^*_{ia}(p)c^*_{jb}(-p)\right]\right\} -$$

$$-\sum_p D_{ia}(p)\left[b(-p) + b^*(p)\right]\cdot\left[c_{ia}(p) - c^*_{ia}(-p)\right]$$

(20.75)

Here,

$$A_{ij}(p) = \tilde{A}_{ij}(p) - g_0^{-1}\delta_{ij} \qquad (20.76)$$

On the transition from (20.68) to (20.75) the formula (20.65) for the condensate density of the Bose–field $b(p)$ was used, and the expression $\lambda_B - 2\alpha^2 t_B - \rho_F Z t_{BF}$ encountered in (20.68) is replaced by $-\alpha^2 t_B$ in accordance with (20.65).

The interaction between the Bose– and Fermi–subsystems influences the structure of the quadratic form (20.75). The quadratic form in $b(p)$ and $b^*(p)$ in (20.75) differs from the quadratic form for the pure

Bose–system by the replacement of $\alpha^2 t_B$ by $\alpha^2 t_B + C(p)$, where $C(p)$ is proportional to t_{BF}^2, the square of the amplitude for scattering of a Bose–particle by a Fermi–particle. The last term in (20.75) describes the interaction between the fields $b(p)$, $b^*(p)$ and $c_{ia}(p), c_{ia}^*(p)$. The coefficient function in this term $D_{ia}(p)$ is proportional to $\omega\delta_{ia}$ at small momenta. It means that only the mode $c_{ii}(p) - c_{ii}^*(-p)$ interacts with Bose–fields $b(p)$, $b^*(p)$ at small momenta. For a pure Fermi–system of the type ^{3}He–B this is a sound mode.

Thus, we arrive at the following conclusion. In the first approximation, the interaction between the Bose–and Fermi–subsystems reduces to interaction between the sound (acoustic) modes of the two subsystems and does not influence the other collective modes of the Fermi–subsystem.

We obtain an equation that determines the spectrum of both acoustic modes. For its derivation, it is easiest to set in (20.75)

$$c_{ia}(p) = c_{ia}(-p) = i\delta_{ia} v(p) = -c_{ia}^*(p) = -c_{ia}^*(-p) \tag{20.77}$$

Then instead of (20.75) we obtain the quadratic form

$$\sum_p \left(i\omega - \frac{k^2}{2m_B} - \alpha^2 t_B - C(p) \right) b^*(p) b(p) -$$
$$-\frac{1}{2}\sum_p \left(\alpha^2 t_B + C(p)\right)\left[b(p)b(-p) + b^*(p)b^*(-p)\right] -$$
$$-\sum_p v^2(p)\left[A_{ii}(p) - B_{ijij}(p)\right] - 2i\sum_p \left[b(p) + b^*(-p)\right] v(p) D_{ii}(p). \tag{20.78}$$

For small $p = (\boldsymbol{k}, \omega)$

$$
\begin{aligned}
&A_{ij}(p)-B_{ijij}(p)= \\
&=-3g_0^{-1}-\frac{4Z^2}{\beta V}\sum_{p1+p2=p} M_1^{-1}M_2^{-1}\left[(i\omega_1+\xi_1)(i\omega_2+\xi_2)+\Delta^2\right]= \\
&=\frac{4Z^2}{\beta V}\sum_{p1+p2=p}\left[\left(\omega_1^2+\xi_1^2+\Delta^2\right)^{-1}-\frac{(i\omega_1+\xi_1)(i\omega_2+\xi_2)+\Delta^2}{\left(\omega_1^2+\xi_1^2+\Delta^2\right)\left(\omega_2^2+\xi_2^2+\Delta^2\right)}\right].
\end{aligned}
\tag{20.79}
$$

If we go from the sum $\sum_p$ to the integral over the neighborhood of the Fermi–sphere and then use the Feynman trick, founded on the identity

$$
\begin{aligned}
&\left[\left(\omega_1^2+\xi_1^2+\Delta^2\right)\left(\omega_2^2+\xi_2^2+\Delta^2\right)\right]^{-1}= \\
&=\int_0^1 d\alpha\left[\alpha\left(\omega_1^2+\xi_1^2+\Delta^2\right)+(1-\alpha)\left(\omega_2^2+\xi_2^2+\Delta^2\right)\right]^{-2},
\end{aligned}
$$

we obtain

$$
\begin{aligned}
&A_{ij}(p)-B_{ijij}(p)=\frac{4Z^2k_F^2}{(2\pi)^4c_F}\int_0^1 d\alpha\int d\Omega d\omega_1 d\xi_1\cdot \\
&\frac{1}{\left(\omega_1^2+\xi_1^2+\Delta^2\right)}-\frac{\omega_1^2+\xi_1^2+\Delta^2-\alpha(1-\alpha)\left(\omega^2+c_F^2(\boldsymbol{n},\boldsymbol{k})^2\right)}{\left[\omega_1^2+\xi_1^2+\Delta^2+\alpha(1-\alpha)\left(\omega^2+c_F^2(\boldsymbol{n},\boldsymbol{k})^2\right)\right]^2}= \\
&=\frac{Z^2k_F^2}{4\pi^3c_F}\int_0^1 d\alpha\int d\Omega\left[\frac{2\alpha(1-\alpha)\left(\omega^2+c_F^2(\boldsymbol{n},\boldsymbol{k})^2\right)}{\Delta^2+\alpha(1-\alpha)\left(\omega^2+c_F^2(\boldsymbol{n},\boldsymbol{k})^2\right)}+\right. \\
&\left.+\ln\left(1+\frac{\alpha(1-\alpha)\left(\omega^2+c_F^2(\boldsymbol{n},\boldsymbol{k})^2\right)}{\Delta^2}\right)\right]\cong\frac{Z^2k_F^2}{2\pi^2c_F}\left(\omega^2+\frac{c_F^2k^2}{3}\right).
\end{aligned}
\tag{20.80}
$$

We also have

$$C(p) \cong C(0) = \frac{2Z^2\alpha^2 t_{BF}^2 k_F^2}{(2\pi)^4 c_F} \int \frac{\left(-2\Delta^2\right) d\Omega d\omega_1 d\xi_1}{\left(\omega_1^2 + \xi_1^2 + \Delta^2\right)^2} = -\frac{Z^2\alpha^2 t_{BF}^2 k_F^2}{\pi^2 c_F},$$

$$D_{ii}(p) = \frac{2i\omega Z^2\alpha\, t_{BF} k_F^2 \Delta}{(2\pi)^4 c_F} \int \frac{d\Omega d\omega_1 d\xi_1}{\left(\omega_1^2 + \xi_1^2 + \Delta^2\right)^2} = \frac{i\omega Z^2\alpha\, t_{BF}^2 k_F^2}{2\pi^2 c_F \Delta}$$

(20.81)

Substitution of (20.80) and (20.81) in (20.78) gives a quadratic form with the matrix

$$\begin{pmatrix} i\omega - \frac{k^2}{2m_B} - \alpha^2 t_B + \frac{Z^2\alpha^2 t_{BF}^2 k_F^2}{\pi^2 c_F}; & -\alpha^2 t_B + \frac{Z^2\alpha^2 t_{BF}^2 k_F^2}{\pi^2 c_F}; & \frac{\omega Z^2\alpha\, t_{BF} k_F^2}{\pi^2 c_F \Delta} \\ -\alpha^2 t_B + \frac{Z^2\alpha^2 t_{BF}^2 k_F^2}{\pi^2 c_F}; & -i\omega - \frac{k^2}{2m_B} - \alpha^2 t_B + \frac{Z^2\alpha^2 t_{BF}^2 k_F^2}{\pi^2 c_F}; & \frac{\omega Z^2\alpha\, t_{BF} k_F^2}{\pi^2 c_F \Delta} \\ -\frac{\omega Z^2\alpha\, t_{BF} k_F^2}{\pi^2 c_F \Delta}; & -\frac{\omega Z^2\alpha\, t_{BF} k_F^2}{\pi^2 c_F \Delta}; & -\frac{Z^2 k_F^2}{\pi^2 c_F \Delta^2}\left(\omega^2 + \frac{c_F^2 k^2}{3}\right) \end{pmatrix}$$

(20.82)

Equating to zero the determinant of this matrix, we arrive after the substitution $i\omega \to E$ at the equation

$$\left[E^2-\left(\frac{k^2}{2m_B}\right)^2-\frac{k^2}{m_B}\alpha^2\left(t_B-\frac{Z^2t_{BF}^2k_F^2}{\pi^2c_F}\right)\right]\cdot\left[E^2-\frac{c_F^2k^2}{3}\right]-$$
$$-E^2k^2\frac{Z^2\alpha^2t_{BF}^2k_F^2}{m_B\pi^2c_F}=0 \tag{20.83}$$

for the spectrum of acoustic modes. The equation for the corresponding velocities has the form

$$u^4-u^2\left(\frac{c_F^2}{3}+\frac{\alpha^2t_B^2}{m_B}\right)+\frac{c_F^2\alpha^2}{3m_B}\left(t_B-\frac{Z^2t_{BF}^2k_F^2}{\pi^2c_F}\right)=0 \tag{20.84}$$

It follows from this equations that the interaction of the Bose– and Fermi–subsystems characterized by the amplitude t_{BF} does not change the sums of the squares of the sound velocities $u_1^2+u_2^2$, but reduces the product $u_1^2u_2^2$ compared with the values for the noninteracting subsystems.

An analogous investigation is possible also for other superfluid phases. We shall consider the *A*–phase of the three–dimensional model and then the *a*– and *b*–phases of the two–dimensional model.

20.6. Bose–Spectrum of a System of the Type ^{3}He–*A*–^{4}He

We now consider a system in which the Fermi–component is in the *A*–phase. The condensate function of the field $c_{ia}(p)$ has the form

$$c_{ia}^{(0)}(p)=(\beta V)^{1/2}\delta_{p0}\delta_{a1}(\delta_{i1}+i\delta_{i2})c \tag{20.85}$$

and the function $b^{(0)}(p)$ is equal to $(\beta V)^{1/2}\alpha\delta_{p0}$ as in the *B*–phase. As in the *B*–phase, the constants α and c are determined from the conditions $\partial_\alpha S_{eff} = 0$, $\partial_c S_{eff} = 0$. The equation for α has the same form (20.64) and the same solution (20.65) as in the *B*–phase. The equation $\partial_c S_{eff} = 0$ has the same form as for pure Fermi–system, namely

$$c\left\{2g_0^{-1} + Z^2(\beta V)^{-1}\sum_p \left(n_1^2 + n_2^2\right)\left[\left(\omega_1^2 + \xi_1^2 + \left(n_1^2 + n_2^2\right)\Delta_0^2\right)\right]\right\} = 0, \tag{20.86}$$

where $\Delta_0 = 2cZ$, $\xi = c_F(k - k_F)$.

For the quadratic part S_{eff} in the fluctuations of the fields $b(p)$, $b^*(p)$, $c_{ia}(p)$, $c^*_{ia}(p)$ around their condensate values the same formula (20.68) holds as for the *B*–phase. However, the expression $-\frac{1}{4}Tr\left(\hat{G}\hat{u}\right)^2$ will differ from the same expression in the *B*–phase, since $\hat{G}$ has a different form ($\hat{u}$ is given by the same formula (20.70) as for the *B*–phase). For the *A*–phase, $\hat{G}$ is the operator that is inverse of

$$\hat{G}^{-1} = \begin{pmatrix} Z^{-1}(i\omega_1 - \xi_1)\delta_{p1p2} & 2c(n_1 + in_2)\sigma_1\delta_{p1+p2,0} \\ -2c(n_1 - in_2)\sigma_1\delta_{p1+p2,0} & Z^{-1}(-i\omega_1 + \xi_1)\delta_{p1p2} \end{pmatrix} \tag{20.87}$$

Inverting $\hat{G}^{-1}$, we obtain

$$\hat{G} = \frac{Z}{M_1}\begin{pmatrix} -(i\omega_1 + \xi_1)\delta_{p1p2} & \Delta_0\sigma_1(n_1 + in_2)\delta_{p1+p2,0} \\ -\Delta_0\sigma_1(n_1 - in_2)\delta_{p1+p2,0} & (i\omega_1 + \xi_1)\delta_{p1p2} \end{pmatrix}, \tag{20.88}$$

where

$$M_1 = \omega_1^2 + \xi_1^2 + \Delta_0^2\left(n_1^2 + n_2^2\right) = \omega_1^2 + \xi_1^2 + \Delta_0^2 \sin^2\theta$$

(θ is the angle between the vector $\boldsymbol{n}$ and the distinguished direction).

Calculation $-\frac{1}{4}Tr\left(\hat{G}\hat{u}\right)^2$ yields:

$$-\frac{1}{4}Tr(\widehat{G}\widehat{u})^2 = -\frac{1}{4}\sum_{p1,p2,p3,p4} tr\,(G_{p1p2}u_{p2p3}G_{p3p4}u_{p4p1}) =$$
$$= -\sum_p \left\{ \widetilde{A}_{ij}(p)c^*_{ia}(p)c_{ja}(p) + \frac{1}{2}B^-_{ijab}(p)c_{ia}(p)c_{jb}(-p) + \right.$$
$$\left. + \frac{1}{2}B^+_{ijab}c^*_{ia}(p)c^*_{jb}(-p) \right\} - \sum_p C(p)\left[b^*(p)+b(-p)\right]\cdot\left[b(p)+b^*(-p)\right] -$$
$$-\sum_p \left\{ D^-_{ia}(p)\left[b^*(p)+b(-p)\right]c_{ia}(p) + D^+_{ia}(p)\left[b(p)+b^*(-p)\right]c^*_{ia}(p) \right\}$$
(20.89)

Here,

$$\widetilde{A}_{ij}(p) = -\frac{4Z^2}{\beta V}\sum_{p1+p2=p} M_1^{-1}M_2^{-1}(i\omega_1+\xi_1)(i\omega_2+\xi_2)n_{1i}n_{1j},$$

$$B^{\pm}_{jab}(p) = \frac{4Z^2}{\beta V}\sum_{p1+p2=p} M_1^{-1}M_2^{-1}n_{1i}n_{1j}(2\delta_{1a}\delta_{1b}-\delta_{ab})(n_1 \pm n_2)^2,$$

$$C(p) = \frac{Z^2\alpha^2 t^2_{BF}}{\beta V}\sum_{p1+p2=p} M_1^{-1}M_2^{-1}\left[(i\omega_1+\xi_1)(-i\omega_2+\xi_2)-\Delta_0^2\left(n_1^2+n_2^2\right)\right],$$

$$D^{\pm}_{ia}(p) = \delta_{a1}\frac{4Z^2\alpha\, t_{BF}\Delta}{\beta V}\sum_{p1+p2=p} M_1^{-1}M_2^{-1}n_i\left(n_1 \pm in_2\right). \qquad (20.90)$$

Using the obtained formulae, we can write the quadratic part of the functional S_{eff} as follows

$$\sum_p \left(i\omega - \frac{k^2}{2m_B} - \alpha^2 t_B - C(p) \right) b^*(p) b(p) - \frac{1}{2} \sum_p \left(\alpha^2 t_B + C(p) \right) \times$$
$$\times \left[b(p) b(-p) + b^*(p) b^*(-p) \right] - \sum_p \{ A_{ij}(p) c^*_{ia}(p) c_{ja}(p) +$$
$$+ \frac{1}{2} B^-_{jab}(p) c_{ia}(p) c_{jb}(-p) + \frac{1}{2} B^+_{jab}(p) c^*_{ia}(p) c^*_{jb}(-p) \Big\} -$$
$$- \sum_p \left[b(p) + b^*(-p) \right] \cdot \left[D^-_{ia}(p) c_{ia}(-p) + D^+_{ia}(p) c^*_{ia}(-p) \right],$$

(20.91)

where

$$A_{ij}(p) = \tilde{A}_{ij}(p) - g_0^{-1} \delta_{ij} \tag{20.92}$$

Let us note, that the functions $A_{ij}(p) B^{\pm}_{ijab}(p), C(p)$ are even on p and $D^{\pm}_{ij}(p)$ is odd on p.

The quadratic form (20.91) obtained for the *A*–phase has a structure similar to that of the form (20.75) for the *B*–phase. In particular, the last term in (20.91) describes the interaction between the fields $b(p)$ and $c_{ia}(p)$. At small momenta $D^{\pm}_{ia}(p) \propto \omega \delta_{a1} (\delta_{i1} + i\delta_{i2})$. This means that the mode corresponding to the variable $c_{11}(p) - ic_{21}(p) - c^*_{11}(-p) + ic^*_{21}(-p)$ interacts with the fields $b(p)$ and $b^*(p)$. For the pure Fermi–system in the *A*–phase this variable corresponds to the acoustic mode of the collective excitations.

We arrive at the same conclusion as for the B–phase: the interaction of the Bose– and Fermi–subsystems reduces in the first approximation to interaction between their acoustic modes and does not influence the other collective modes of the Fermi–subsystem.

We obtain the equation for the "coupled" acoustic modes by proceeding as in the previous section for the B–phase. In the quadratic form (20.91) we set

$$c_{21}(p) = c_{21}^{*}(-p) = u(p) = ic_{11}(p) = -ic_{11}^{*}(-p) \tag{20.93}$$

and set the remaining $c_{ia}(p), c_{ia}^{*}(p)$ equal to zero. Then instead of (20.91) we obtain the quadratic form

$$\begin{aligned}
&\sum_{p}\left(i\omega - \frac{k^2}{2m_B} - \alpha^2 t_B - C(p)\right)b^{*}(p)b(p) - \\
&-\frac{1}{2}\sum_{p}\left(\alpha^2 t_B + C(p)\right)\left[b(p)b(-p) + b^{*}(p)b^{*}(-p)\right] - \\
&-\sum_{p} u^2(p)\left\{-2g_0^{-1} - \frac{4Z^2}{\beta V}\sum_{p1+p2=p} M_1^{-1}M_2^{-1}\times\right. \\
&\times\left.\left[(i\omega_1 + \xi_1)(i\omega_2 + \xi_2) + \Delta_0^2\left(n_1^2 + n_1^2\right)\right]\left(n_1^2 + n_1^2\right)\right\} - \\
&-\sum_{p}\left[b(p) + b^{*}(-p)\right]u(p)\frac{4\omega Z^2 \alpha t_{BF}\Delta}{\beta V}\sum_{p1+p2=p} M_1^{-1}M_2^{-1}\left(n_1^2 + n_1^2\right).
\end{aligned} \tag{20.94}$$

For small $p = (\boldsymbol{k}, \omega)$ we have

$$-2g_0^{-1}-\frac{4Z^2}{\beta V}\sum_{p1+p2=p}M_1^{-1}M_2^{-1}\left[(i\omega_1+\xi_1)(i\omega_2+\xi_2)+\Delta_0^2\left(n_1^2+n_2^2\right)\right]\times$$

$$\times\left(n_1^2+n_2^2\right)=\frac{4Z^2}{\beta V}\sum_{p1+p2=p}\left(n_1^2+n_2^2\right)\left\{M_1^{-1}-M_1^{-1}M_2^{-1}\left[(i\omega_1+\xi_1)(i\omega_2+\xi_2)+\right.\right.$$

$$\left.\left.+\Delta_0^2\left(n_1^2+n_2^2\right)\right]\right\}=\frac{4Z^2k_F^2}{(2\pi)^4c_F}\int_0^1 d\alpha\int d\Omega\left(n_1^2+n_2^2\right)\int d\omega_1 d\xi_1\times$$

$$\times\left\{\frac{1}{\omega_1^2+\xi_1^2+\Delta^2}-\frac{\omega_1^2+\xi_1^2+\Delta^2-\alpha(1-\alpha)\left(\omega^2+c_F^2(\boldsymbol{n},\boldsymbol{k})^2\right)}{\left(\omega_1^2+\xi_1^2+\Delta^2+\alpha(1-\alpha)\left(\omega^2+c_F^2(\boldsymbol{n},\boldsymbol{k})^2\right)\right)^2}\right\}=$$

$$=\frac{Z^2k_F^2}{4\pi^3c_F}\int_0^1 d\alpha\int d\Omega\left(n_1^2+n_2^2\right)\cdot\left\{\frac{2\alpha(1-\alpha)\left(\omega^2+c_F^2(\boldsymbol{n},\boldsymbol{k})^2\right)}{\Delta^2+\alpha(1-\alpha)\left(\omega^2+c_F^2(\boldsymbol{n},\boldsymbol{k})^2\right)}+\right.$$

$$\left.+\ln\left[1+\frac{\alpha(1-\alpha)\left(\omega^2+c_F^2(\boldsymbol{n},\boldsymbol{k})^2\right)}{\Delta^2}\right]\right\}\cong\frac{Z^2k_F^2}{2\pi^2c_F\Delta_0^2}\left(\omega^2+\frac{c_F^2k^2}{3}\right), \tag{20.95}$$

$$C(p)\cong C(0)=\frac{2Z^2\alpha^2t_{BF}^2k_F^2}{(2\pi)^4c_F}\int\frac{\left(-2\Delta^2\right)d\Omega d\omega_1 d\xi_1}{\left(\omega_1^2+\xi_1^2+\Delta^2\right)^2}=-\frac{Z^2\alpha^2t_{BF}^2k_F^2}{\pi^2c_F},$$

$$\frac{4\omega Z^2\alpha\, t_{BF}\Delta_0}{(2\pi)^4c_F}\int\frac{d\Omega d\omega_1 d\xi_1\left(n_1^2+n_2^2\right)}{\left(\omega_1^2+\xi_1^2+\Delta^2\right)^2}=\frac{\omega Z^2\alpha\, t_{BF}^2k_F^2}{\pi^2c_F\Delta_0} \tag{20.96}$$

Here, $\Delta^2=\Delta_0^2\left(n_1^2+n_2^2\right)=\Delta_0^2\sin^2\theta$. Formulae (20.95), (20.96) lead to the same matrix (20.82) and consequently to the same equations for the spectrum (20.83) of the acoustic excitations and for the velocities of the sounds (20.84) as for the *B*–phase.

20.7. Bose–Spectrum of Films of the Types ^{3}He–*a*–^{4}He and ^{3}He–*b*–^{4}He

As was shown by Brusov and Popov[17,19,20], two–dimensional systems of the type ^{3}He can be in two stable superfluid phases *a*– and *b*–. The matrix order parameters in these phases are proportional to

$$\begin{pmatrix} 1 & 0 & 0 \\ i & 0 & 0 \end{pmatrix} (a\text{–phase}), \qquad \begin{pmatrix} 1 & 0 & 0 \\ 0 & 1 & 0 \end{pmatrix} (b\text{–phase}) \tag{20.97}$$

We consider the two possibilities (20.97) for a two–dimensional Fermi–Bose–system as well. We can consider them in an analogous way as it was done in two previous sections for the three–dimensional systems.

We begin with the *a*–phase. Here the expression (20.68) for the quadratic part of S_{eff} with respect to the fluctuations of the fields $b(p)$, $b^*(p)$, $c_{ia}(p)$, $c^*_{ia}(p)$ around their condensate values remains true. The condensate function of the phase has the same form of

$$c^{(0)}_{ia}(p) = (\beta V)^{1/2}\,\delta_{p0}\delta_{a1}(\delta_{i1} + i\delta_{i2})c \tag{20.98}$$

as for the three–dimensional *A*–phase, and the equation for c

$$c\left\{2g_0^{-1} + Z^2(\beta V)^{-1}\sum_p (n_1^2 + n_2^2)\left[(\omega_1^2 + \xi_1^2 + (n_1^2 + n_2^2)\Delta_0^2)\right]\right\} = 0 \tag{20.99}$$

is the same as for the pure Fermi–system. The expressions for $\hat{G}^{-1}$, $\hat{G}$ are the same as (20.87), (20.88), and $M_1 = \omega_1^2 + \xi_1^2 + \Delta^2$. Calculating $-\frac{1}{4}Tr(\hat{G}\hat{u})^2$, we arrive at a quadratic form of the form (20.91). The coefficient functions of the quadratic form differ from (20.90), (20.92) in that the combination $n_1^2 + n_2^2$ encountered in (20.90) is equal to 1 in the two–dimensional case.

Examination of the crossed terms containing both fields $b(p)$, $b^*(p)$ and $c_{ia}(p), c_{ia}^*(p)$ (the last term in (20.91)) shows that in the first approximation the only mode of the Fermi–subsystem "coupled" to the Bose–fields $b(p)$ and $b^*(p)$ is, as for the A–phase, the mode, corresponding to the variable

$$c_{11}(p) - ic_{21}(p) - c_{11}^*(-p) - ic_{21}^*(-p). \tag{20.100}$$

This mode corresponds to acoustic oscillations of pure Fermi–system. To obtain the equation for the "coupled" acoustic modes, let us make a substitution (20.93) and equating to zero the determinant of the quadratic form (20.94). At small $p = (\boldsymbol{k}, \omega)$ the coefficients of the quadratic form in two–dimensional case are

$$\begin{aligned} & -2g_0^{-1} - \frac{4Z^2}{\beta V} \sum_{p1+p2=p} M_1^{-1} M_2^{-1} \left[(i\omega_1 + \xi_1)(i\omega_2 + \xi_2) + \Delta^2 \right] \approx \\ & \approx \frac{Z^2 k_F}{2\pi c_F \Delta_0^2} \left(\omega^2 + \frac{c_F^2 k^2}{2} \right), \end{aligned} \tag{20.101}$$

$$C(p) \cong C(0) = -\frac{Z^2 \alpha^2 t_{BF}^2 k_F}{\pi\, c_F}, \tag{20.102}$$

$$\begin{aligned} & \frac{4\omega Z^2 \alpha\, t_{BF} \Delta_0}{(2\pi)^4 c_F} \sum_{p1+p2=p} M_1^{-1} M_2^{-1} = \\ & = \frac{4\omega Z^2 \alpha\, t_{BF} \Delta_0 k_F}{(2\pi)^3 c_F} \int \frac{d\Omega\, d\omega_1 d\xi_1}{\left(\omega_1^2 + \xi_1^2 + \Delta^2\right)^2} = \frac{\omega Z \alpha\, t_{BF} k_F}{\pi\, c_F \Delta_0}. \end{aligned} \tag{20.103}$$

The difference from the three–dimensional expressions is in replacement of $\left(\omega^2 + c_F^2 k^2/3\right)$ by $\left(\omega^2 + c_F^2 k^2/2\right)$ and k_F^2/π^2 by k_F/π. As a result, the equation for spectrum of acoustic modes (20.83) is replaced by the equation

$$\left[E^2 - \left(\frac{k^2}{2m_B}\right)^2 - \frac{k^2}{m_B}\alpha^2\left(t_B - \frac{Z^2 t_{BF}^2 k_F}{\pi\, c_F}\right)\right]\cdot\left[E^2 - \frac{c_F^2 k^2}{2}\right] - \\ - E^2 k^2 \frac{Z^2 \alpha^2 t_{BF}^2 k_F}{m_B \pi\, c_F} = 0 \qquad (20.104)$$

and the equation for the velocities of the acoustic modes takes the form

$$u^4 - u^2\left(\frac{c_F^2}{2} + \frac{\alpha^2 t_B}{m_B}\right) + \frac{c_F^2 \alpha^2}{2m_B}\left(t_B - \frac{Z^2 t_{BF}^2 k_F}{\pi\, c_F}\right) = 0 \qquad (20.105)$$

Similarly, one can consider the *b*–phase, in which the condensate function of the field $c_{ia}(p)$ has the form

$$c_{ia}^{(0)}(p) = (\beta V)^{1/2}\,\delta_{p0}\delta_{ia} c\,, \qquad (20.106)$$

which is formally identical to (20.59). The analogs of the expressions (20.68)–(20.72), in which it is necessary to understand

$$(\boldsymbol{n}, \boldsymbol{\sigma}) = n_1\sigma_1 + n_2\sigma_2 \qquad (20.107)$$

remains valid. Examination of the crossed terms which contain not only the fields $b(p)$, $b^*(p)$, but also $c_{ia}(p), c_{ia}^*(p)$, shows that only one (acoustic) mode of the fields $c_{ia}(p), c_{ia}^*(p)$ interacts with the fields $b(p)$ and $b^*(p)$. This mode corresponds to the variable

$$c_{ii}(p) - c^*_{ii}(-p) = c_{11}(p) + c_{22}(p) - c^*_{11}(-p) - c^*_{22}(-p)$$

To calculate the spectrum of the "coupled" acoustic modes of the Bose– and Fermi–subsystems, we make the substitution (20.77), which leads to a quadratic form of the type (20.78). The coefficient functions of the quadratic form for the *b*–phase are the same as for the *a*–phase. For the *b*–phase this leads to the same equations (20.104), (20.105) for the spectrum of the acoustic modes and the velocities of sounds as for the *a* –phase.

20.8. Conclusion

We have investigated in the first approximation the Bose–spectrum of model systems of the type ^{3}He–^{4}He in the case when both components of the system (the Bose and the Fermi) are superfluid. The interaction of the Fermi– and Bose–subsystems leads to a coupling of the acoustic modes of the two subsystems, and for the velocities of the sounds we have obtained the biquadratic equation (20.84) for the *A*– and *B*–phases of the three–dimensional system and the equation (20.105) for the *a*– and *b*–phases of the two–dimensional system.

For both the three– and two–dimensional cases, the interaction of the Bose– and Fermi–subsystems does not changes the sums of the squares of the sound velocities, $u_1^2 + u_2^2$, but does reduce the product $u_1^2 u_2^2$ compared with the values in the noninteracting subsystems. In the first approximation, the Fermi–Bose–interaction does not influence the form of the remaining branches of the collective excitations of the Fermi–subsystem (17 branches in the three–dimensional system and 11 in the two–dimensional system), which for the considered model were investigated in details by Alonso and Popov[21] and by Brusov and Popov[22-24].

Chapter XXI

Novel Sound Phenomena in Impure Superfluids

During the last decade, new techniques for producing impure superfluids with unique properties have been developed. This new class of systems includes superfluid helium confined to aerogel, HeII with different impurities (D_2, N_2, Ne, Kr), superfluids in Vycor glasses, and watergel. These systems exhibit very unusual properties including unexpected acoustic features. We discuss the sound properties of these systems and show that sound phenomena in impure superfluids are modified from those in pure superfluids.

We calculate[1,2,21] the coupling between temperature and pressure oscillations for impure superfluids and for superfluid He in aerogel. We show that the coupling between these two kinds of oscillations is governed by terms proportional either to impurity or to aerogel density rather than by thermal expansion coefficient, which is enormously small in pure superfluids. This replacement plays a fundamental role in all sound phenomena in impure superfluids. It enhances the coupling between the two sound modes (first and second sounds) that leads to the existence of such phenomena as the slow mode and heat pulse propagation with the velocity of first sound observed in superfluids in aerogel. This means that it is possible to observe in impure superfluids such unusual sound phenomena as slow "pressure" (density) waves and fast "temperature" (entropy) waves.

The enhancement of the coupling between the two sound modes decreases the threshold values for nonlinear processes as compared to pure superfluids. Sound conversion, which has been observed in pure superfluids only by shock waves should be observed at moderate sound amplitude in impure superfluids. Cerenkov emission of second sound by first sound (which never been observed in pure as well as in impure superfluids) could be observed in impure superfluids. We have shown that the enhanced coupling between first and second sound changes even the nature of the sound modes in impure superfluids. It leads as well to significant shift in fast mode frequency at transition temperature.We also have derived for the first time the nonlinear hydrodynamic equations for superfluid helium in aerogel. These equations are generalizations of McKenna et al.[4] equations for the case of nonlinear hydrodynamics and could be used to study sound propagation phenomena in aerogel–superfluid system, in particular – to study sound conversion phenomena. We get two alternative sets of equations, one of which is a generalization of a traditional set of nonlinear hydrodynamics equations for the case of an aerogel–superfluid system, while second one represents equations which are the analogy of Putterman's hydrodynamic equations for superfluids[17].

21.1. Introduction

For a great deal of time, the only impurities in He II studied experimentally were dissolved ^{3}He atoms. In the last decade new techniques for producing impure superfluids with unique properties have been developed and these new systems have been studied intensively. This new class of systems includes superfluid helium confined to aerogel (Cornell, Lancaster, Manchester, Northwestern), HeII with different impurities (D_2, N_2, Ne, Kr) (Cornell), superfluids in Vycor glasses, and watergel – a frozen water "lattice" in HeII. These systems exhibit very unusual properties including unexpected acoustic features. In the following, we discuss the sound properties of these systems and show that sound phenomena in impure superfluids are modified from those in pure superfluids. We started to discuss these issues in our Rapid Communication[1] and shortly discussed in Ref. 2.

Bulk superfluids support two sound modes: first sound which is ordinary sound corresponding to pressure (density) oscillations at constant entropy density and second sound, representing temperature (entropy) oscillations. We calculate the coupling between temperature (entropy) oscillations and pressure (density oscillations) for impure superfluids (including ^{3}He–^{4}He mixtures) and for superfluid He (both ^{3}He and ^{4}He) in aerogel. We show that the coupling between these two kinds of oscillations is governed either by $c\,\partial\rho/\partial c$ (c is impurity concentration) or by $\sigma\rho_a\rho_s$ (ρ_a is aerogel density), rather than by the thermal expansion coefficient $\partial\rho/\partial T$, which is enormously small in pure superfluids. This replacement plays a fundamental role in sound phenomena, and, in particular, in sound conversion phenomena in impure superfluids and in superfluids in aerogel. It enhances the coupling between the two sound modes that leads to the existence of such phenomena as the slow mode and heat pulse propagation with the velocity of first sound observed in superfluid He in aerogel. This means that it is possible to observe in impure superfluids such unusual sound phenomena as slow "pressure" (density) oscillations and fast "temperature" (entropy) oscillations. The enhancement of the coupling between the sound modes decreases the threshold values for nonlinear processes as compared to pure superfluids. Via this sound conversion, which has been observed in pure superfluids only by shock waves, should be observed at moderate sound amplitudes in impure superfluids. Cerenkov emission of second sound by first sound (which never been observed in pure superfluids) could be observed in impure superfluids.

In this paper we consider novel sound phenomena in impure superfluids, where these phenomena are caused by either impurities (including ^{3}He admixtures in He II) or by the presence of aerogel.

We start from showing the decoupling of first and second sound in pure superfluids. This part also simply demonstrates the general idea of our calculations. In the second part we consider the coupling of sounds in impure superfluids and show that it is determines by a term proportional to $c\,\partial\rho/\partial c$ instead of $\partial\rho/\partial T$ as in pure superfluids. Then we calculate the coupling between two sounds for superfluid He (both ^{3}He and ^{4}He) in aerogel where it is governed by term proportional to $\sigma\rho_a\rho_s$. In the next part we give explanation of some earlier experiments like observation of slow mode in superfluid ^{3}He[3] and HeII in aerogel[4] and in ^{3}He–^{4}He–mixtures[5,6] and propagation of heat pulse with velocity of fast mode in HeII in aerogel[7] which has not been understood that time. In next part we discuss nonlinear hydrodynamics in pure and impure superfluids and sound conversion phenomena, described by it. We show that the enhancement of the coupling between the sound modes, obtained by us in linear approximation, has a dramatic effect on the nonlinear sound conversion processes. It decreases the threshold values for nonlinear processes as compared to pure superfluids, allowing much easier observe sound conversion and observe for the first time Cerenkov emission. We discuss significant frequency shift appears in fast mode frequency in impure superfluids. Impurities change even the nature of sounds as we discuss in this paper.

Then we discuss some experiments on observation of novel sound phenomena as well as sound conversion in different superfluid systems: pure superfluids, impure superfluids and ^{3}He–^{4}He mixtures. In next Chapters we have derived for the first time nonlinear hydrodynamic equations for superfluid helium in aerogel, which are the generalization of the equations obtained by McKenna et al.[4] for the case of nonlinear hydrodynamics. We get two alternative sets of equations, one of which includes an equation for superfluid velocity $\vec{v}_s$, while second one (Putterman's type[17] equations) includes an equation for $\vec{A}=\frac{\rho_n}{\rho\sigma}\vec{w}$, where $\vec{w}=\vec{v}_n-\vec{v}_s$ is relative velocity. In Conclusion we summarize our results.

21.2. Decoupling of First and Second Sound in Pure Superfluids

Bulk superfluids support two sound modes: first (ordinary) sound propagating with a velocity $u_I=\sqrt{\left(\frac{\partial P}{\partial \rho}\right)_\sigma}$ and corresponding to pressure (density) oscillations and second sound propagating with a velocity $u_2=\sqrt{\frac{\sigma^2\rho_s}{\rho_n\left(\partial\sigma/\partial T\right)}}$ and representing temperature (entropy) oscillations.

The coupling of these sound modes turns out to be through the thermal expansion coefficient $\partial\rho/\partial T$, which is always small and is anomalously small for superfluid He.

Sound propagation in a superfluid is described by the following equations obtained by linearizing the two–fluid hydrodynamical equations[8]:

$$\frac{\partial^2 \rho}{\partial t^2}=\Delta P, \qquad \frac{\partial^2 \sigma}{\partial t^2}=\frac{\rho_s}{\rho_n}\sigma^2 \Delta \mathrm{T} \tag{21.1}$$

where σ is the entropy per gram. After substituting $P=P_0+P'$ and $T=T_0+T'$ (where f_0 is equilibrium value and f' is the perturbation due to the sound wave) one gets:

$$\frac{\partial \rho}{\partial P}\frac{\partial^2 P'}{\partial t^2}-\Delta \rho' + \frac{\partial \rho}{\partial \mathrm{T}}\frac{\partial^2 T'}{\partial t^2}=0,$$

$$\frac{\partial \sigma}{\partial P}\frac{\partial^2 P'}{\partial t^2}+\frac{\partial \sigma}{\partial T}\frac{\partial^2 T'}{\partial t^2}-\frac{\sigma^2 \rho_s}{\rho_n}\Delta T'=0. \tag{21.2}$$

Note here, that although both these equations describe the coupling between pressure and temperature, the structure of the equations is quite different. The first equation describes propagating pressure oscillations, which produce attendant temperature oscillations, whereas the second equation describes the reverse. It is an important to keep in mind this distinction when discuss the nature of the slow mode and high velocity thermal waves.

If we consider a plane wave, P' and T' are proportional to $\exp[-i\omega(t-x/u)]$. The secular equation gives us two velocities for sound propagation: u_1 and u_2. We neglect the thermal expansion coefficient $\partial\rho/\partial T$, which is always small and is anomalously small for He II and superfluid ^{3}He. Let us determine the coupling between first and second sounds. First suppose we excite first sound, i.e. pressure oscillations. One gets

$$T' = P'\frac{\partial\sigma}{\partial P}u_1^2 / \frac{\partial\sigma}{\partial T}(u_2^2 - u_1^2) = \alpha P' \tag{21.3}$$

Thus, pressure oscillations are accompanied by temperature oscillations with an amplitude $\alpha P_o'$, which is small because of the size of the thermal expansion coefficient $\frac{\partial\rho}{\partial T}$ ($\left(\frac{\partial\sigma}{\partial P}\right)_T = \frac{1}{\rho^2}\left(\frac{\partial\rho}{\partial T}\right)_P$).

Now consider excited temperature oscillations and the propagation of second sound. We get

$$P' = \frac{\partial\rho}{\partial T}\frac{u_1^2 u_2^2}{u_1^2 - u_2^2}T' \tag{21.4}$$

Thus, P' is small because of the size of the thermal expansion coefficient $\partial\rho/\partial T$. Because of this weak coupling between first and second sound such phenomena as the slow mode as well as heat pulse propagation with the velocity of first sound have never been observed in pure superfluids.

21.3. Sounds Coupling in Impure Superfluids

21.3.1. Superfluids with different impurities, ^3He–^4He mixtures

We will consider dilute mixtures of different impurities (atoms, ions, molecules, electrons, etc.) in superfluids. As an example, the dilute ^3He–^4He mixtures could be considered as weak solutions of ^3He in ^4He because of the 6.4% maximum ^3He concentration (Note that very intensive sound experiments in ^3He–^4He mixtures have been developed by Kharkov group[10,18] from former Soviet Union). The theory of weak solutions was first developed by Khalatnikov[8]. If we suppose that impurities participate just in normal flow we should add to the system of hydrodynamic equations the continuity equation for impurities[8]

$$\frac{\partial}{\partial t}\rho c + div\,\rho c\mathbf{v}_n = 0 \tag{21.5}$$

and write the thermodynamic equality as

$$dE = Tds + \mu d\rho + Zdc + (\mathbf{v}_n - \mathbf{v}_s, d\mathbf{j})$$

including additional variable $c = \dfrac{N_i m_i}{N_i m_i + N_s m_s}$, here $Z = \rho(\mu_i - \mu_s)$, where μ_i is chemical potential of impurity and μ_s is that of the superfluid. From linearized system of hydrodynamic equations after eliminating the velocities $\mathbf{v}_n$ and $\mathbf{v}_s$ we get

$$\ddot{\rho} = \Delta P\,; \quad \frac{\rho_n}{\rho_s}\frac{\dot{\sigma}}{\sigma} = \sigma\Delta T + c\Delta\frac{Z}{\rho}\,; \qquad \frac{\dot{c}}{c} = \frac{\dot{\sigma}}{\sigma}. \tag{21.6}$$

Comparison with equation (21.1) shows that we got additional term $c\Delta Z/\rho$ in the second equation as well as an additional third equation.

Choosing as independent variables P', T' and c' and considering traveling wave $(f \sim \exp[-i\omega(t - x/u)])$ we get a system of three equations for these variables[1].

The secular equation for this system determines the velocities of first and second sounds which have been found[1] in the limit $\partial\rho/\partial T = 0$. For our purpose of investigating of coupling of sounds let us isolate c' from one of the equations and eliminate it from first two equations[1].

For coupling of two sounds we have the following results.

Admixture of second sound at first sound

For admixture of second sound at first sound one gets

$$T' = -P' \frac{\frac{\rho_n}{\rho_s \sigma} \frac{\partial \sigma}{\partial P} u_1^2 + \frac{c}{\rho^2} \frac{\partial \rho}{\partial c} + c \frac{\partial \sigma}{\partial P} \frac{A_1}{B}}{\frac{\rho_n}{\rho_s \sigma} \frac{\partial \sigma}{\partial T} u_1^2 - \sigma + c \frac{\partial \sigma}{\partial c} + c \frac{\partial \sigma}{\partial T} \frac{A_1}{B}} = \alpha P' \tag{21.7}$$

Here,

$$A = \frac{\rho_n}{\rho_s \sigma} \frac{\partial \sigma}{\partial c} u^2 - c \frac{\partial}{\partial c} \frac{Z}{\rho}; \; B = c \frac{\partial \sigma}{\partial c} - \sigma \,, \; A_1 = A(u = u_1) \tag{21.8}$$

Coefficient α determines the amplitude of temperature oscillations which accompany the pressure oscillations in running plane wave. While the coupling between P' and T' oscillations exists at all temperatures, T' oscillations can propagate only below T_c (T_λ for HeII) because above T_c as we mentioned in Introduction they are damped at a distance of order the wavelength, $\lambda \approx \sqrt{\chi/\omega}$ through conductive heat transfer (χ is the thermal conductivity coefficient). So if we excite P' oscillations in impure He we get propagating T' oscillations just below T_c (T_λ).

Admixture of first sound at second sound

For admixture of first sound at second sound one gets

$$P' = -T' \frac{\left(\frac{\partial p}{\partial T} + \tilde{c}\right) u_2^2}{\left(\frac{\partial p}{\partial P} + \tilde{c}\right) u_2^2 - 1} = \left(\frac{\partial p}{\partial T} + \tilde{c}\right) \frac{u_1^2 u_2^2}{u_1^2 - u_2^2} T' = \beta T' \,. \tag{21.9}$$

Here,

$$\tilde{c} = \frac{c}{B} \frac{\partial p}{\partial c} \frac{\partial \sigma}{\partial T} \,.$$

Coefficient β determines the amplitude of pressure oscillations which accompany the temperature oscillations in running plane wave.

It is easy to see that the parameter $\partial\rho/\partial T$ is changed by $\partial\rho/\partial T + \tilde{c}$, which is proportional to $c\,\partial\rho/\partial c$. So, the coupling of first sound to the second sound in impure superfluids is dominated by the term $c\,\partial\rho/\partial c$ instead of $\partial\rho/\partial T$ in pure superfluid.

21.3.2. Sounds coupling in superfluid He in aerogel

Aerogel in superfluids is studied very intensively during last decade. The importance of these systems is connected to the fact that this allows to investigate the influence of impurities on superfluidity. This influence is significant, as it was shown in numerous papers[3–7,9]. Impurities change drastically all properties of the system including superfluid ones, phase diagram, acoustic properties etc. For example, for aerogel–superfluid ^{3}He system there is a critical value of fluid density (pressure), below which there is no superfluidity even at T=0 in opposite to pure superfluid ^{3}He, where at $T < T_C$ superfluidity takes place at all pressures. Phase separation line in ^{3}He–^{4}He mixtures in aerogel is detached from λ–line. So called slow modes (slow pressure (density) waves) have been observed in both HeII[4] and superfluid ^{3}He[3] in aerogel as well as in ^{3}He–^{4}He mixtures[5,6]. In last case simultaneously two slow modes have been observed: one of them terminates at T_C while other at T_λ. First mode attributes to superfluid ^{3}He in large porous and second – to ^{3}He–^{4}He mixture[5,6]. A very interesting coupling of these modes has been observed (for more details see Chapter XX). Also fast temperature waves have been observed[7]. Analysis of sound phenomena in impure superfluids has been done by Brusov et al. [1,2]

To analyze the sounds coupling in superfluid He in aerogel we will use the modified hydrodynamic equations introduced for this case by McKenna[4]

$$\frac{\partial\rho}{\partial t}+\nabla(\rho_n\mathbf{v}_n+\rho_s\mathbf{v}_s)=0\,;\quad \frac{\partial\rho\sigma}{\partial t}+\nabla(\rho\sigma\mathbf{v}_n)=0\,;\frac{\partial\mathbf{v}_s}{\partial t}=-\frac{1}{\rho}\nabla P+\sigma\nabla T\,;$$

$$\rho_{na}\frac{\partial\mathbf{v}_n}{\partial t}=-\frac{\rho_n}{\rho}\nabla P-\nabla P_a-\rho_s\sigma\nabla T\,;\qquad \frac{\partial\rho_a}{\partial t}+div(\rho_a\mathbf{v}_n)=0 \tag{21.10}$$

These differ from the bulk superfluid He equations by the replacement $\rho_n\rightarrow\rho_n+\rho_a=\rho_{na}$ on the left hand side of fourth equation of (10) and the additional restoring force $-\nabla P_a$ due to the aerogel[4].

Performing the same calculations as above we arrive at two equations for P' and T' (excluding P_a' from the system of three equations for P', T' and P_a')[1].

For coupling of two sounds at aerogel–superfluid system we have the following results.

Admixture of second sound at first sound

For admixture of second sound at first sound one gets[1] (taking into account anomalous smallness of $\frac{\partial\rho}{\partial T}$, $\frac{\partial\rho_a}{\partial T}$, $\frac{\partial\sigma}{\partial P}$ and accounting $u_2 << u_1$)

$$T'=-P'\frac{\frac{\sigma\rho_a\rho_s}{\rho\rho_{na}}\left(1-\frac{\rho_n}{\rho_{na}A}\right)}{u_1^2\rho\frac{\partial\sigma}{\partial T}}. \tag{21.11}$$

Here,

$$A=\frac{u^2}{u_a^2}\frac{\rho_a}{\rho_n},\; A_i=A(u=u_i).$$

Admixture of first sound at second sound

For admixture of first sound at second sound one gets[1]

$$P' = -T' \frac{\left[u_2^2 \left(\frac{\partial \rho}{\partial T} + \frac{\rho_n}{\rho_{na} A_2} \frac{\partial \rho_a}{\partial T} \right) + \sigma \frac{\rho_s \rho_a}{\rho_{na}} \left(1 - \frac{\rho_n}{\rho_{na} A_2} \right) \right]}{\frac{u^2}{u_1^2} - \frac{1}{\rho} \left[\rho_s + \frac{\rho_n^2}{\rho_{na}} \left(1 - \frac{\rho_a}{\rho_{na} A_2} \right) \right]} . \qquad (21.12)$$

It is easy to see that the sounds coupling is proportional to a magnitude of $\sigma \rho_a \rho_s$.

Some immediate conclusions follow from our analysis.

1. The coefficients α and β in HeII is much grater than in superfluid ^3He because of the three order of magnitude difference in entropy density in the superfluid regions of both isotopes (in K and mK regions). For example, for superfluid ^3He σ_3=2x10^{-4} J/cm^{3}K at T=0.7 mK, while for HeII σ_4=0.1 J/cm^{3}K at T =1.6 K.

2. The coupling coefficients α and β have a maximum which depends on the values of ρ_s and ρ_a . The values of α and β depend strongly on the sound speed of the aerogel sample u_a.

3. The superfluid density ρ_s increases with decreasing temperature but the entropy decreases: the competition of two these factors produces experimental constraints an intermediate temperature regime is optimum for measurements: far from T_c where ρ_s is small and far from zero temperature, where entropy is equal to zero.

4. From the existence of the large coefficient u_1^2 in denominator of the equation for *T'* and absence of this term in the equation for *P'* it is clear that the admixture of first sound at second sound in superfluid He in aerogel will be easier to observe than the reverse.

The estimations made for HeII and ^3He in aerogel at T=1.6 K and T=0.7 mK consequently give for β coefficients the following results:

HeII: $\beta = 2.5\text{x}10^2$ Pa/mK $\quad$ ^{3}He: $\beta = 2\text{x}10^{-2}$ Pa/mK

while the coefficients α turn out to be a few orders of magnitude less. This is in agreement with above conclusions.

At Figs. 21.1 and 21.2 we show the dependence of α and β parameters for HeII and ^{3}He in aerogel on the aerogel porosity. The dependence of α parameter for HeII on the aerogel porosity for two values of sound velocity in aerogel: 100 m/s and 50 m/s is shown at Fig. 21.3.

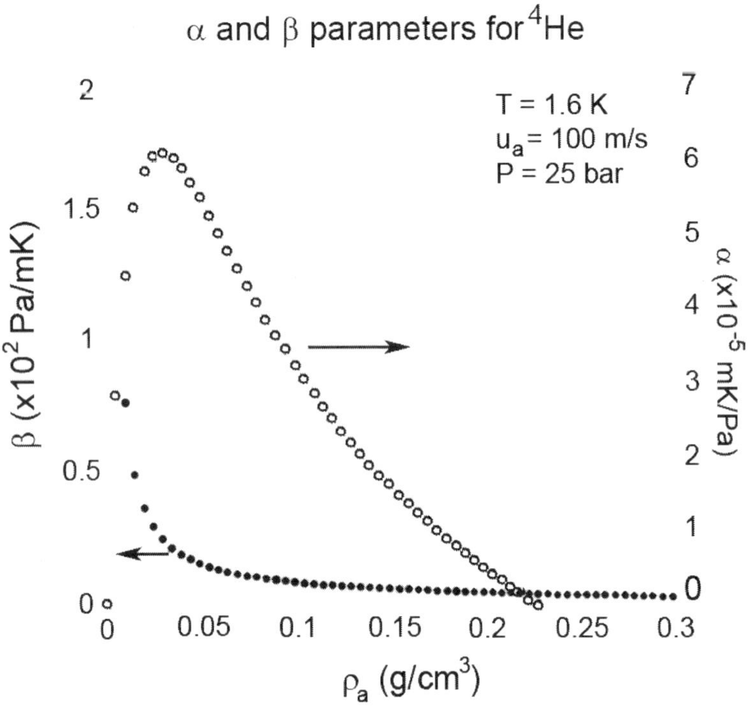

FIG. 21.1. The α (open symbols) and β (filled symbols) coupling parameters for HeII versus aerogel density at pressure P = 25 bar, the temperature T = 1.6 mK and sound velocity in aerogel u_a = 100 m/s.

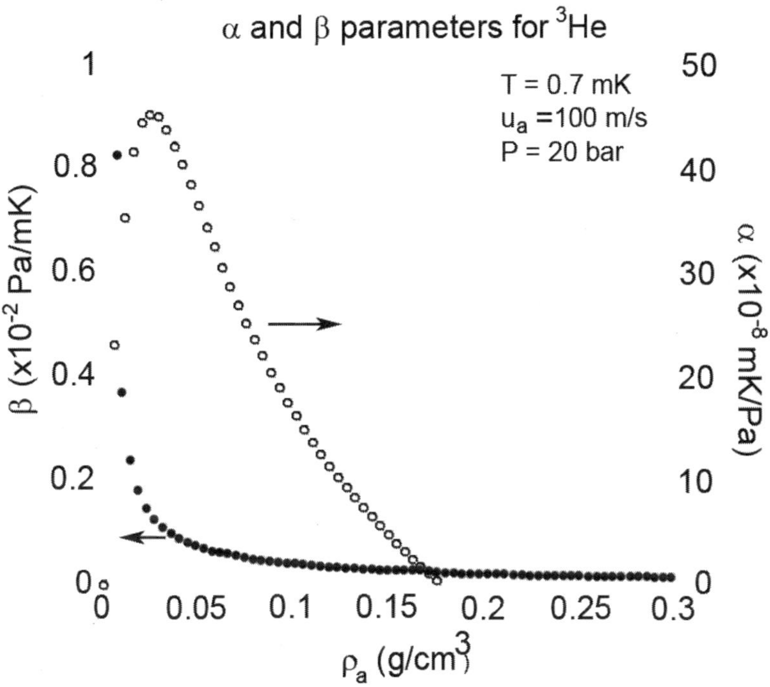

FIG. 21.2. The α (open symbols) and β (filled symbols) coupling parameters for superfluid ^3He versus aerogel density at pressure P=20 bar, the temperature T=0.7 mK and sound velocity in aerogel $u_a = 100$ m/s.

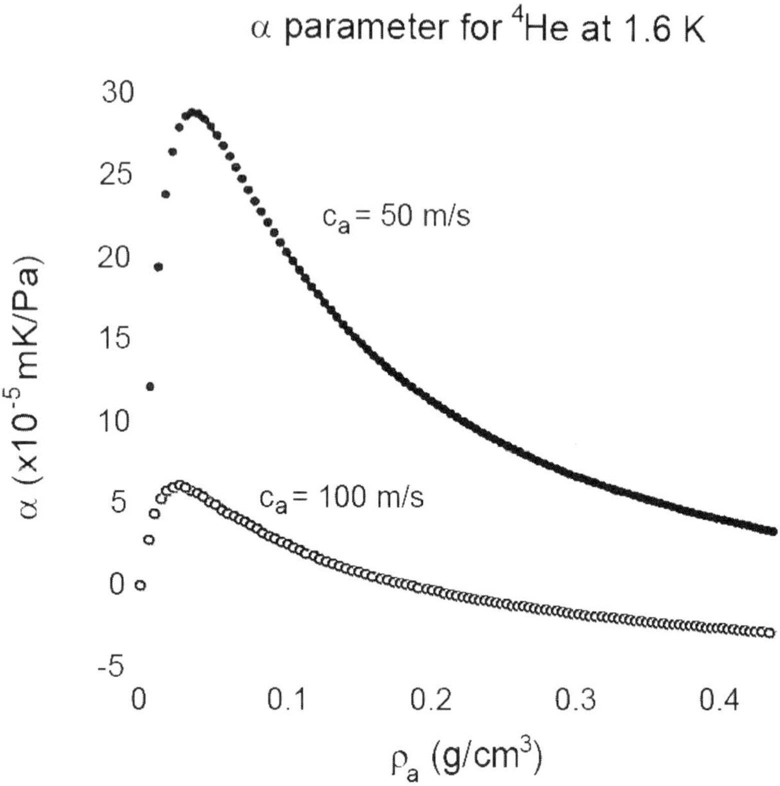

FIG. 21.3. The coupling parameter α of the first to second sound versus aerogel density for HeII at pressure P=25 bar and the temperature T=1.6 K for two values of sound velocity in aerogel: u_a = 100 m/s (open symbols) and u_a = 50 m/s (filled symbols).

21.4. Slow Pressure (Density) Oscillations, Fast Temperature (Entropy) Oscillations

There are some unexplained phenomena observed in superfluid He in aerogel, namely a slow pressure mode with all other attributes of thermal waves[3,4] as well as fast thermal waves[7] mode with all other attributes of pressure waves. These phenomena can be understood in terms of the increased coupling between pressure and temperature oscillations obtained by us above. From our discussion concerning equation (21.2) follows that pressure oscillations only propagate with a speed u_1, while thermal oscillations propagate with a speed u_2. Exciting pressure oscillations cause temperature oscillations in the vicinity of the transducer. If these oscillations have a low frequency, corresponding to the speed of second sound, thermal waves can propagate, while pressure oscillations cannot. These propagating thermal oscillations in turn generate pressure oscillations at the receiving transducer. This is observed in He in aerogel (but not in pure superfluids) because of the significant increasing in coupling between pressure and temperature oscillations. Let us suppose we excite first sound with an amplitude P' and low frequency. The pressure oscillations of transducer excite temperature oscillations with an amplitude $T'=\alpha P'$ following from the equation (21.11). These temperature oscillations propagate below T_c with the velocity of second sound through the cell (note that first sound can not propagate at such low frequency). The temperature waves reached second transducer lead to pressure oscillations (below T_c) with amplitude P'', which is found from the equation of (21.12) $P''=\beta T'=\alpha\beta P'$. This is the so called a slow mode, observed in sound experiments in superfluid ^{3}He in aerogel[3]: pressure oscillations with all other features of a second sound wave (exists only below T_c, small sound velocity etc.). In this case first sound is transformed into second sound at one transducer, second sound propagates through the cell and then second sound is converted into first sound at receiving transducer. Slow mode in HeII in aerogel has been observed by McKenna[4].

Other unusual phenomenon has been observed by Mulders[7] who observed the propagation of heat pulse in HeII in aerogel. The received pulses, detected by bolometer, "contained structure corresponding to the transit times of the fast mode." At that time this was not understood but now we can explain this behavior based on enhancement of the coupling between the two sound modes[1]. Let us suppose we excite by heater second sound with an amplitude T' and high frequency. The temperature oscillations of heater excite pressure oscillations with an amplitude $P'=\beta T'$ following from the equation (21.12). These pressure oscillations propagate with the velocity of first sound through the cell (note that second sound can not propagate at such high frequency). The pressure waves reached bolometer lead to temperature oscillations with amplitude

T'', which is found from the equation (21.11) $T''= \alpha T'=\beta\alpha P'$. This picture can explain the propagation of heat pulse with the velocity of fast mode, observed by Mulders[5]. In this case second sound is transformed into first sound near heater and then first sound is converted into second sound at bolometer.

21.5. Fast Mode Frequency Shift at T_C (T_λ)

Here, we would like to explain the nature of fast mode frequency shift at T_c (T_λ) and show that this shift is much bigger in impure superfluids comparing to bulk ones.

Sound propagation in a superfluid is described by the following equations, obtained by linearizing the two fluid hydrodynamical equations[8]. Velocities of sounds:

$$u_{10}=\sqrt{\left(\frac{\partial P}{\partial \rho}\right)_\sigma} \qquad \text{and} \qquad u_{20}=\sqrt{\frac{\sigma^2\rho_s}{\rho_n\left(\partial\sigma/\partial T\right)}}$$

have been obtained from secular equation upon neglecting by thermal expansion coefficient δ, which is small always, but is enormously small for He II and superfluid ^{3}He.

If we keep δ, we get the following secular equation

$$\begin{vmatrix} \frac{u^2}{u_1^2}-1 & \delta u^2 \\ \alpha u^2 & \frac{\partial\sigma}{\partial T}\left(u^2-u_2^2\right) \end{vmatrix}=0 \tag{21.13}$$

which gives us following expression for velocities of sounds (supposing smallness of δ)

$$u_1^2=u_{10}^2+\gamma u_{20}^2; \qquad u_2^2=u_{20}^2-\gamma u_{20}^2,$$

where $\alpha = \partial\sigma/\partial P$, $\gamma = \alpha\delta u_{10}^2 / \partial\sigma/\partial T$.

1. If we move to T_c (T_λ) upon warming the velocity (frequency) of first sound decreases via the decrease of u_{20} (or ρ_s) (negative shift of first sound frequency at T_c (T_λ)). T–dependence of this decrease is very similar to one of ρ_s.

2. In pure superfluids this shift is quite small (or even negligible) via smallness of δ.

In impure superfluids δ is replaced by bigger values $\sigma\rho_a\rho_s$ or $c\,\partial\rho/\partial c$ that leads to more significant shift in first sound frequency (of order u_{20} (a few hundreds Hz)) as observed in ^{3}He in aerogel[3].

21.6. Difference in Nature of First and Second Sound in Impure Superfluids

Let us show that the presence of impurities changes even the nature of first and second sounds.

In pure superfluids, for first sound $v_n \approx v_s$, while for second sound $j = \rho_n v_n + \rho_s v_s \approx 0$, if one neglects the thermal expansion coefficient δ. To the first order in δ approximate one finds:

$$j = \frac{\delta\rho}{\sigma\rho_n}\frac{u_1^2 u_2^2}{u_1^2 - u_2^2} v_s \tag{21.14}$$

Thus, the mass flow in impure superfluids is non–zero in second sound, because of the replacement of δ by $c\,\partial\rho/\partial c$ or by $\sigma\rho_s\rho_n$.

For the first sound

$$v_n = v_s\left(1+\frac{\delta\rho}{\sigma\rho_s}\frac{u_1^2u_2^2}{u_1^2-u_2^2}\right) \tag{21.15}$$

so $v_n \neq v_s$ in impure superfluids. For example[10], in ^{3}He–^{4}He mixtures at c = 0.2 for the wide temperature region for first sound $v_n = (1.2-1.5)v_s$ and for second sound $v_n = -(0.6-0.8)\frac{\rho_s}{\rho_n}v_s$.

This means that in impure superfluids:

1) normal and superfluid components in first sound do not longer move coherently,
2) mass flow in second sound is nonzero and could be significant.

Thus, the enhanced coupling between first and second sound changes the nature of the sound modes in impure superfluids.

We would like to underline here the difference between the second sound and fourth sound phenomena in aerogel–superfluid system. It is well known that fourth sound is related to the case when normal component of superfluid can not oscillate and only superfluid component can oscillate, producing fourth sound. In aerogel–superfluid system normal component of superfluid is locked to aerogel and they move together. At McKenna equations this fact leads to the replacement $\rho_n \to \rho_n + \rho_a = \rho_{na}$ on the left hand side of fourth equation of (21.10) as it has been mentioned in section 21.3. Thus in aerogel–superfluid system we can not speak about fourth sound.

We should speak either about modified second sound (with properties which are quite different from properties of second sound in bulk superfluid) or about sound which represent the intermediate state between second and fourth sound.

21.7. Sound Conversion Phenomena

Other very interesting and important sound phenomenon in superfluids is sound conversion. We show below that sound conversion in impure superfluids as well as sound phenomena discussed by us above is quite different from the bulk superfluids case.

The sound conversion phenomena have been predicted by Khalatnikov for two strongly inhomogeneous cases: reflection of second sound from the boundary between He II and its vapor and sound reflection from the (phase separated) boundary between ^{3}He–rich phase and ^{4}He–rich phase of ^{3}He–^{4}He mixture[11]. In the first case, second sound in He II under reflection from liquid–vapor interface creates temperature oscillations. This leads to periodical vaporization of liquid and condensation of vapor near the interface. As result, density oscillations appear near the interface which propagate through the vapor like ordinary sound waves. In the second case the transformation of first (second) sound into second (first) sound takes place through the difference of ^{3}He concentration ΔC between the ^{3}He–rich and ^{4}He–rich phases. The sound conversion (SC) effect at the phase boundary is proportional to ΔC and vanishes when $\Delta C \rightarrow 0$. Below we consider sound conversion phenomena for homogeneous superfluids, where these phenomena are caused by either impurities (including ^{3}He admixture in He II) or by the presence of aerogel.

21.7.1. Conservation Laws in Sound Conversion

Let us consider two–phonon processes in linear approximation. Conservation laws for energy and momentum give

$$\omega_1 = \omega_2 \, , \, \boldsymbol{k}_1 = \boldsymbol{k}_2 \tag{21.16}$$

where ω_1, $\boldsymbol{k}_1$ are frequency and momentum of first sound and ω_2, $\boldsymbol{k}_2$ are frequency and momentum of second sound.

The velocities of two sounds are different ($u_1 \neq u_2$), thus conservation laws could not be satisfied.

So, we should consider more complex processes which involve not two but at least three phonons, which are described by a nonlinear hydrodynamic (in linear approximation all the frequencies must be the same and for three phonon processes this violates the energy conservation law).

Conversion of first sound into second sound

Cerenkov emission (1→1′ + 2)

Here, first sound phonon stands for electron and second sound phonon stands for photon, so emission is possible in case $u_1 > u_2$.

Let us write conservation laws for this case

$$\omega_1 = \omega_{1'} + \omega_2, \quad \boldsymbol{k}_1 = \boldsymbol{k}_{1'} + \boldsymbol{k}_2,$$

$$\omega_2 = 2\omega_1(u_2 / u_1) \sin \Theta/2 \tag{21.17}$$

$$k_1 = 2k_2 \sin \Theta/2$$

where Θ is angle between $\boldsymbol{k}_1$ and $\boldsymbol{k}'_1$.

Experiments on Cerenkov emission never been done even in pure superfluids.

Parametric decay of first sound (1→ 2′+ 2)

In case of parametric decay of first sound into two second sounds conservation laws can be written as follows

$$\omega_1 = \omega'_2 + \omega_2, \quad \boldsymbol{k}_1 = \boldsymbol{k}_{2'} + \boldsymbol{k}_2,$$

$$\omega_{2,2'} = \omega_1/2, \ k_2 = \omega_2 / u_2 = \omega_1/2\, u_2 = k_1\, u_1/2\, u_2 >> k_1 \tag{21.18}$$

$$\boldsymbol{k}_2 \approx -\boldsymbol{k}_{2'} \text{ (if } u_2 << u_1).$$

First sound produces second sounds propagating in perpendicular directions to direction of the first sound. In pure HeII experiments on parametric decay of first sound have been done[14] (see section 21.8).

Conversion of second sound into first sound

Emission of first sound by second sound (2→2'+1)

Let us write conservation laws for emission of first sound by second sound

$\boldsymbol{k}_2 = \boldsymbol{k}_{2'} + \boldsymbol{k}_1$, $u_2\, k_2 = u_2\, k_{2'} + u_1\, k_1$

$(k_2 - k_{2'})/\, k_1 = u_1 \,/\, u_2 > 1.$

This violets triangle rule, thus emission of first sound by second sound *is prohibited.*

Parametric decay of second sound (2→1'+1)

For parametric decay of second sound into two first sounds we have

$\boldsymbol{k}_2 = \boldsymbol{k}_{1'} + \boldsymbol{k}_1$, $u_2\, k_2 = u_1\, k_{1'} + u_1\, k_1$

$(k_1 + k_{1'})/\, k_2 = u_2 \,/\, u_1 < 1$

This violates the triangle rule, thus parametric decay of second sound *is prohibited.*

Transformation of two second sounds into first sound (2+2'→1)

In case of transformation of two second sounds into first sound one has

$\boldsymbol{k}_2 + \boldsymbol{k}_{2'} = \boldsymbol{k}_1$, $u_2 k_2 + u_2 k_{2'} = u_1 k_1$

$$(k_2 + k_{2'})/ k_1 = u_1 / u_2 > 1 \tag{21.19}$$

$\omega_1 = 2\, \omega_{2,2'}$,

$k_1 < k_2, k_{2'}$,

$\omega_2 = \omega_{2'}$, $\cos \Theta/2 = u_2/u_1$.

Θ is the angle between $\boldsymbol{k}_2$ and $\boldsymbol{k}_{2'}$.

In pure HeII such experiments have been done[15] (see section 21.8).

So, just three processes are allowed by conservation laws, two of them for conversion of first sound into second sound: parametric decay of first sound and Cerenkov emission of second sound by first sound and just one for conversion of second sound into first sound: transformation of two second sounds into first sound.

Cerenkov emission of second sound by first sound never been observed experimentally even in pure superfluids, while parametric decay of first sound and transformation of two second sounds into first sound have been observed in pure HeII (we shortly describe these experiments below). The experimental conditions for two these experiments follow from momentum conservation law and mean that exciting and excited sounds propagate in approximately perpendicular directions. This condition is different from one for Cerenkov emission.

21.7.2. Sound Conversion in Pure Superfluids

In order to illustrate the importance replacing the thermal expansion coefficient by larger terms $c\,\partial\rho/\partial c$ or $\sigma\rho_a\rho_s$, obtained by us in the linear approximation, let us write the expressions obtained by Pokrovskii and Khalatnikov[12] for the amplitude of emission of second sound by first sound in pure superfluids. Emission amplitude

$$V(\omega,\chi)=\frac{\omega^{3/2}}{2\rho^{1/2}}\frac{\left(\frac{\partial\rho}{\partial T}\right)_p}{\left(\frac{\partial\sigma}{\partial T}\right)^{1/2}}\left(\frac{c_2}{c_1}\sin\frac{\chi}{2}\right)^{1/2}\left[\rho\left(\frac{\partial}{\partial\rho}\ln\frac{\rho c_1^2}{\left(\frac{\partial P}{\partial\sigma}\right)_\rho}\right)_\sigma+\cos\chi\right] \tag{21.20}$$

where χ is the angle between incoming and outgoing first–sound phonons.

Amplitude for parametric decay

$$U(\omega,\theta)=\frac{\rho^{1/2}\omega^{3/2}}{2\rho_s^{1/2}c_1}\left\{-1-\frac{2\rho_s}{\rho}\cos^2\theta-\rho_n\rho c_1^2\frac{\partial\rho_n^{-1}}{\partial P}+\right.$$
$$\left.+\rho_n c_1^2\left[\frac{\partial}{\partial T}\left(\frac{\rho_s}{\rho_n}\right)\frac{\left(\rho^{-2}\frac{\partial\rho}{\partial T}\right)}{\left(\frac{\partial\sigma}{\partial T}\right)}-\frac{\rho_s}{\rho_n}\frac{\partial}{\partial T}\frac{\left(\rho^{-2}\frac{\partial\rho}{\partial T}\right)}{\left(\frac{\partial\sigma}{\partial T}\right)}\right]\right\}. \tag{21.21}$$

Here, Θ is the angle at which the second–sound phonons fly apart, relative to the direction of propagation of the first sound.

Write also the ratio of the threshold first sound amplitude for Cerenkov emission (a_1) and parametric decay (a_2)

$$\frac{|a_1|}{|a_2|}=\left(\frac{\delta_1(\omega)}{\delta_2(\omega)}2\frac{c_2}{c_1}\right)^{1/2}\frac{\rho^2}{\rho_s c_1}\frac{\left(\frac{\partial\sigma}{\partial T}\right)^{1/2}}{\left(\frac{\partial\rho}{\partial T}\right)}\frac{[U]}{[V]} \tag{21.22}$$

Here, $[U]$ and $[V]$ are given by $[\ ]$expressions in (21.20) and (21.21) respectively, $\delta_1(\omega)$ and $\delta_2(\omega)$ are the attenuations of the first and second sounds.

The threshold energy for Cerenkov emission

$$W_1 = c_1\omega|a_1|^2 = \frac{16\delta_1(\omega)\delta_2(\omega)\rho^3 c_2 \dfrac{\partial\sigma}{\partial T}}{\omega^2\left(\dfrac{\partial\rho}{\partial T}\right)^2 |V|^2} \tag{21.23}$$

The replacement of $\partial\rho/\partial T$ by larger terms increases significantly the emission amplitude (which is proportional to $\partial\rho/\partial T$) and changes the amplitude for parametric decay in which the terms proportional to $\partial\rho/\partial T$ are of the same order as the remaining terms. From the expression (21.22) for ratio of threshold amplitude for Cerenkov emission (a_1) and parametric decay (a_2) we find that this replacement also changes the temperature regime where the Cerenkov threshold comes before the decay threshold and vice versa. The expression for the threshold energy for Cerenkov emission shows that in impure superfluids this threshold decreases significantly which makes the observation of Cerenkov emission much more likely. Cerenkov emission which never been observed in pure superfluids could be observed in impure superfluids via decrease the threshold energy for this process. For ^{3}He–^{4}He mixtures we will see this in the next sections: 21.7.3 and 21.10.

21.7.3. Sound Conversion in ^{3}He–^{4}He Mixtures

Now let us illustrate the importance of the replacement of $\partial\rho/\partial T$ by larger terms in nonlinear processes. Below are the results obtained by Lebedev[13] for ^{3}He–^{4}He mixtures in nonlinear approximation in Hamiltonian formalism.

For weak solutions ($\Gamma\gamma >> 1$) one has

$$V(\omega,\chi)=\frac{\omega^{3/2}\Gamma^{1/2}\gamma}{2\rho^{1/2}}-\left(\frac{c_2}{c_1}\sin\frac{\chi}{2}\right)^{1/2}\left[1-\cos\chi-D_\rho\ln\frac{D_\sigma P}{D_\rho P}\right], \tag{21.24}$$

$$U(\omega,\chi)=\frac{\omega^{3/2}}{2^{5/2}\rho^{1/2}c_1}- -\left[1-2\cos\chi-\frac{\rho}{\Gamma}D_\rho\frac{\Gamma}{\rho}+\frac{\gamma}{\rho c_2^2}((D_\sigma-\gamma D_\rho)^2P+2\gamma D_\sigma P)\right]. \tag{21.25}$$

Here,

$$\gamma=-\frac{\sigma}{\rho}\left(\frac{\partial\rho}{\partial\sigma}\right)_{c,P}-\frac{c}{P}\left(\frac{\partial\rho}{\partial\sigma}\right)_{\sigma,P},\ \Gamma=\frac{\rho_s}{\rho_n}.$$

For concentrated solutions ($\Gamma\gamma\approx 1$) one has

$$V(\omega,\chi)=\frac{\omega^{3/2}(D_\rho P)\Gamma^{1/2}\gamma}{2\rho^2c_1^3c_2}-\left(\frac{c_2}{c_1}\sin\frac{\chi}{2}\right)^{1/2}([\gamma^2\rho(D_\rho-\frac{D_\sigma}{\gamma})\frac{1}{\rho_n}+ +(\gamma-1)(1+\gamma(1+2\Gamma))]\cos\chi+ +\frac{1}{\rho c_1^2}(\partial_\rho+\partial_\sigma+\Gamma\gamma\partial_\sigma)^2(\partial_\rho+\partial_\sigma-\gamma^{-1}\partial_\sigma)c), \tag{21.26}$$

$$U(\omega,\chi)=\frac{\omega^{3/2}(D_\rho P)^{1/2}\Gamma\gamma^2}{2^{5/2}\rho^3c_1^4c_2^2}(\partial_\rho+\partial_\sigma+\Gamma\gamma\partial_\sigma)(\partial_\rho+\partial_\sigma-\gamma^{-1}\partial_\sigma)^2c) \tag{21.27}$$

Comparing formulae (21.20)–(21.23) for a pure superfluid with formulae obtained by Lebedev[13] for ^3He–^4He mixture shows that the thermal expansion coefficient is replaced in mixtures by terms proportional to $c\,\partial\rho/\partial c$ in agreement with our results. Note, that nonlinear theory of sound conversion for superfluids in aerogel developed by us in sections 21.11 and 21.12. The parametric decay of first sound into second sound[14] and the transformation of two second sounds into first sound[15] in pure HeII has been observed experimentally. In first case, first sound of amplitude about 50 Pa produced second sound intensity of the order of 10^{-7} μW/cm^2. In second experiment coupling coefficient was about 30 Pa/mK at T=1.3 K. In the next section we describe two these experiments a little bit more detailed.

21.8. Sound Conversion Experiments in Pure Superfluids

Conversion of second sound into first sound

Garrett et al.[15] have observed the conversion of two second sound into first sound.

For two second–sound waves of equal frequency, ω, and equal amplitude, traveling in different directions indicated by their wave vectors $\boldsymbol{k}_2$ and $\boldsymbol{k}_{2'}$, the condition that the phase velocity, c_{ph}, of the component at 2ω be that of first sound is cos $\Theta/2=u_2/u_1$ or $c_{ph}=c_1=2\omega/\left|\boldsymbol{k}_2+\boldsymbol{k}_{2'}\right|$, where Θ is angle between $\boldsymbol{k}_2$ and $\boldsymbol{k}_{2'}$.

This resonance condition leads to the generation of a propagating first–sound wave at a frequency 2ω, whose amplitude grows linearly with x in the region of interaction (x is chosen as the direction which bisects Θ). In order to achieve an interaction region of maximum length, the experiment was performed in a spiral waveguide of rectangular cross section. The waveguide was operated in its lowest nonplane–wave second–sound mode. Used $\omega\approx 650$ Hz. Mode conversion coupling constant p_2 / w_1^2 turns out to be growing and of order 0.1 dynes $cm^{-2}/cm^2 sec^{-2}$ in temperature region 1.25K–1.4K.

Parametric generation of second sound by first sound

This experiment was done[14] on a short cylindrical cell of 50 mm in diameter and 4 mm length. The sides of the cavity were formed by a pair of capacitor transducers. In the short direction, the cell was used as a first sound resonator with the first harmonic frequency about 28 KHz. The frequency of parametrically excited second sound waves was about 14 KHz. The experiment was performed at a constant temperature by generating first sound at pump frequency, which was chosen to be the first harmonic frequency of the cavity. For a given first sound amplitude, the second sound amplitude was measured by the bolometers. The experiment has been done in the vicinity of superfluid transition. It has shown that first sound of amplitude about 50 Pa produced second sound intensity of the order of 10^{-7} $\mu W/cm^2$.

21.9. Some Possible New Sound Experiments in Impure Superfluids

In impure superfluids one can observe such phenomena as the slow pressure mode and fast temperature mode. First mode has been observed in superfluids ^3He and HeII in aerogel. It should be observed in impure superfluids including ^3He–^4He mixtures. Second mode has been observed in HeII in aerogel as heat pulse propagation with the velocity of first sound. It should be observed in impure superfluids including ^3He–^4He mixtures. Sound conversion, which has been observed in pure superfluids only by shock waves, should be observed at moderate sound amplitude in impure superfluids. There are a few possibilities to observe these phenomena in these systems. To observe first sound to second sound conversion one should use the cylindrical cell with one piezoelectric sound transducer as speaker and bolometer (or carbon thermometer) as receiver. To observe second sound to first sound conversion one needs to use Garrett et al.[15] techniques. The simplest experiment is in HeII. Cerenkov emission of second sound by first sound (which never been observed in pure superfluids) could be observed in impure superfluids.

Let us consider the possible experiments in impure HeII. For the impurity we could include atoms or molecules (D_2, N_2, Ne, Kr) used in recent Cornell sound experiements with HeII[16]. Increasing the impurity concentration opens a channel for sound conversion and allows the observation of this phenomenon. Also the same situation takes place for observation of slow mode in superfluid He in aerogel which should not be observe in bulk superfluid He and appears in superfluid He in aerogel and should exhibits itself more clearly in less porous aerogel. In ^{3}He–^{4}He mixtures we could observe this phenomenon for homogenous situation (without phase separation) as well as for phase separated ^{3}He–^{4}He mixtures. Two–fluid hydrodynamic analysis is applied above T_c and three velocities ($\boldsymbol{v}_n$, $\boldsymbol{v}_{s1}$ and $\boldsymbol{v}_{s2}$) analysis should be used below T_c, where both subsystems (spaced separated) are superfluid. For ^{3}He–^{4}He mixtures in aerogel the detachment of the λ–line from the phase separation surface allows new possibilities: we are able to investigate high ^{3}He–concentrations. This has been realized by recent experiments at Cornell, where superfluid ^{3}He in aerogel is studied while adding up to 30% ^{4}He[5,6]. Two slow modes in ^{3}He–^{4}He mixtures have been observed by us recently for the first time[5,6]. In next section we discuss these experiments.

21.10. Coupling of Two Slow Modes in Superfluid ^{3}He–^{4}He Mixture in Aerogel

Lawes et al.[5,6] have studied ^{3}He–^{4}He mixture in aerogel. The main result was the observation for the first time of two slow modes: one of them terminates at T_C while other terminates at T_λ. They attributed first mode (which terminates at T_C) to the slow mode in superfluid ^{3}He and second mode (which terminates at T_λ) to the slow mode in the ^4He reach phase of the dilute ^3He–^4He mixture. Also the frequency of latter mode shifts at temperatures below T_C and this fact has not been understood that time.

Given by Lawes et al.[5] interpretation of two modes has meaning and seems reasonable. In case of ^3He–^4He mixture in aerogel we have two subsystems which are very perfectly separated in space: the ^4He reach phase, which covers strands and superfluid ^3He in large porous. Thus signals coming from these regions seem uncoupled and well defined.

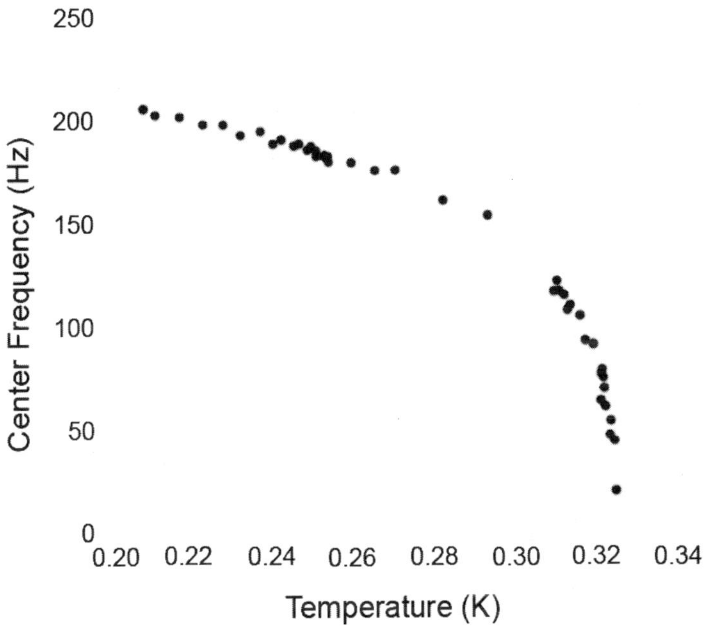

FIG. 21.4. Center frequencies of the mixture slow mode in the 11% ^4He mixture at P=17.2 bar versus temperature.

One more observation supporting this suggestion is that temperatures, at which each of these modes terminates, correspond very precisely to transition temperatures into superfluid phase of pure ^3He in aerogel (T_C)

and of the ^{4}He reach phase in aerogel (T_λ) consequently. The only observation, which remains unclear within this picture[5], is the positive frequency shift of the ^{4}He reach phase slow mode below T_C. Actually it is difficult to understand how can superfluid ^{3}He in large porous of the system have any influence over the ^{4}He reach phase being space separated with the latter subsystem. A couple possible ways to imagine such influence are the proximity effect and the change of the boundary conditions for the ^{4}He reach phase under superfluid phase transition in ^{3}He in large porous.

Here, we would like to suggest more general picture of coupling of two slow modes of ^{3}He–^{4}He mixtures in aerogel. Our model is based on the presence of aerogel in the system as well as on McKenna et al.[4] suggestion concerning joint motion of normal fluid and aerogel (see equations (21.11)). Ishikawa et al.[19] have come to the same conclusion, that the normal fluid is locked to aerogel, studying the impurity scattering effect on the sound propagation in normal liquid ^{3}He in aerogel.

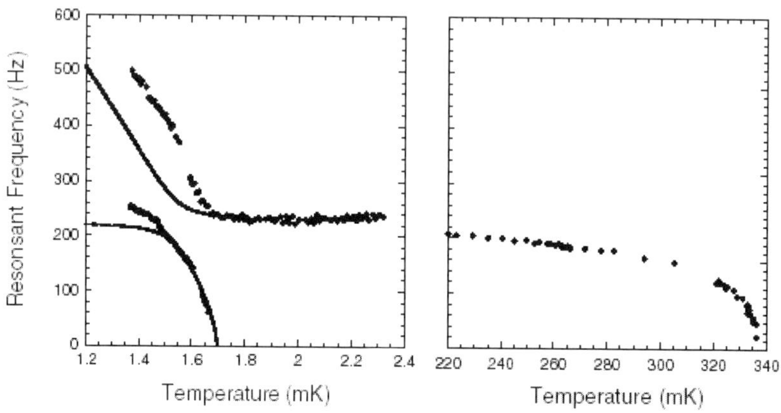

FIG. 21.5. Center frequencies of the mixture slow mode in the 11% ^{4}He mixture at P=17.2 bar versus temperature.

They have gotten for the longitudinal sound velocity the expression

$$u = \left(\rho c_1^2 / (\rho + \rho_a)\right)^{1/2}. \tag{21.28}$$

This means that aerogel moves together with liquid and gives extra inertia to liquid oscillations.

The fact, that the normal fluid is locked to aerogel, allows us to create the following picture. We consider system describing by three densities (three fluid dynamics) $\rho_n, \rho_{s1}, \rho_{s2}$ and three velocities $\vec{v}_n, \vec{v}_{s1}, \vec{v}_{s2}$. Here ρ_n is the normal density (which is inhomogeneous and equal to density of normal component of ^{3}He in large porous and to density of normal component of ^{3}He–^{4}He mixtures around strands) and ρ_{s1}, ρ_{s2} are superfluid densities of ^{3}He and ^{4}He reach phase consequently. The inhomogeneous normal density is locked to aerogel. Thus we have two kind of out–of–phase oscillations (namely these oscillations corresponding to slow modes): ρ_{s1} and ρ_{na} and ρ_{s2} and ρ_{na}. First out–of–phase oscillations of normal component (which includes both pure ^{3}He and mixture normal parts at different regions of system) together with aerogel ρ_{na} and superfluid density of ^{3}He in large porous ρ_{s1} have been attributed by Lawes et al. to "^{3}He" slow mode. Other mode ("a mixture" slow mode in Lawes et al. interpretation) is attributed by us to out–of–phase oscillations of normal component locked to aerogel ρ_{na} and superfluid component of mixture ρ_{s1}. Following Brusov et al.[2] we should mention that out–of–phase oscillations are not precisely anticoherent motion of normal and superfluid components (like second sound). In case of impure superfluids (including superfluids with aerogel) mass flow in second sound is nonzero

$$j = \rho_n v_n + \rho_s v_s \neq 0 \tag{21.29}$$

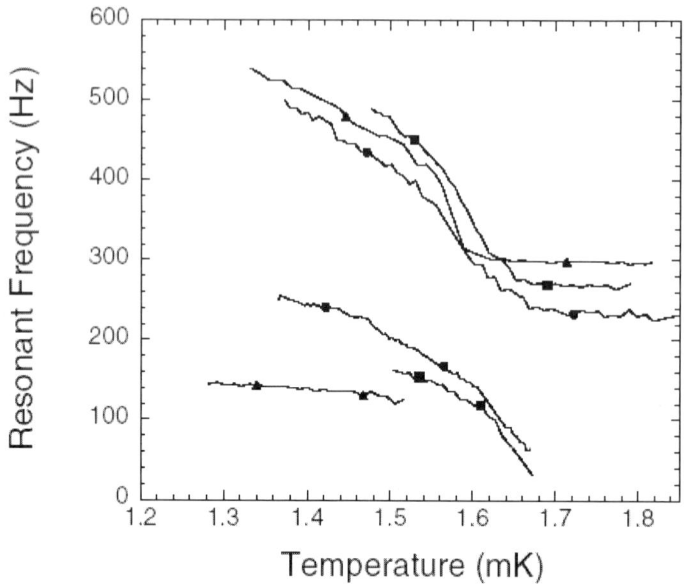

FIG. 21.6. Dependence of resonant frequencies of both slow modes in ^3He-^4He mixture in aerogel, contained 10,5% (circles), 23% (squares), 71% (triangles) of ^4He.

and could be significant in opposite to pure superfluids where it is equal to zero and normal and superfluid components move coherently in first sound

$$\vec{v}_n \approx \vec{v}_s \tag{21.30}$$

The main difference between present interpretation and Lawes et al.[5] one is that we attributed each of observed slow modes to both subsystems (to the ^4He reach phase covering strands ("mixture) and pure ^3He in large porous) while previous interpretation attributed each of observed slow modes to just one particular subsystem (either to the ^4He reach phase or to pure ^3He in large porous). This allows us to explain the coupling between two slow modes following from the presence of aerogel in the system as well as from joint motion of normal fluid and aerogel.

The general theory of behavior of ^3He–^4He mixtures in aerogel will be very complicate. It should take into account that system is strongly inhomogeneous (the ^4He reach phase covers strands and pure ^3He fills the large porous) as well as the theory should be non–local one (the oscillation of aerogel together with normal fluid at chosen point depends on mass of aerogel and normal fluid in other regions of cell). Thus, at least two equations of (21.10) (third and fourth) should be non–local and thus modified. We will consider this theory anywhere also, but here we estimate the unlocked effect: the decoupling of normal component of pure ^3He in large porous from the aerogel.

At very low temperatures (of order T_C) we can imagine that the whole ^4He reach phase is in superfluid state. Thus the only normal component we have is the normal component of pure ^3He in large porous. Because slow mode is manifestation of second sound[1] we can consider the expression for the velocity of second sound

$$u_2 = \sqrt{\frac{\sigma^2 \rho_s}{\rho_n \left(\partial\sigma / \partial T \right)}} \tag{21.31}$$

Here, one can consider ρ_n as the normal component of pure ^3He in large porous while $\rho_s \approx \rho$ is the density of ^4He–reach phase.

From (21.31) one gets

$$du_2 / u_2 = -d\rho_n / 2\rho_n \,. \tag{21.32}$$

Negative sign means that below T_C, where ρ_n decreases, slow mode velocity (and frequency) increases.

In aerogel[3], $d\rho_n / \rho_n \approx 20\% \div 25\%$, thus $du_2 / u_2 \approx 10\% \div 12\%$.

Thus, the estimation of the unlocked effect shows that frequency shift due to decoupling of the superfluid ^3He from the aerogel produces about 10% of the observed value, that means that another effects like the change of pressure gradient ∇P below T_C should be accounted.

21.11. Nonlinear Hydrodynamic Equations for Superfluid Helium in Aerogel

As we mentioned above the first version of hydrodynamic equations for superfluid helium in aerogel has been suggested by McKenna et al.[4], who supposed that the normal fluid is locked to aerogel, and both move together with velocity $\vec{v}_n$. They also accounted the additional restoring force P_a due to aerogel. They have written the set of linear equations (21.10). McKenna et al.[4] equations describe correctly sound velocity of superfluids in aerogel, but description a set of acoustic properties such as sound conversion etc. requires nonlinear hydrodynamic equations. As we also mentioned above sound conversion has been studied first by Khalatnikov et al.[8,11,12] and two of three sound conversion phenomena have been observed experimentally in pure HeII by shock waves[14,15]. Nonlinear hydrodynamic equations and sound conversion phenomena in ^{3}He–^{4}He mixtures have been studied by Lebedev[13].

We have generalized McKenna et al. equations describing superfluids in aerogel for nonlinear hydrodynamics case and derived for the first time the nonlinear hydrodynamic equations for superfluid helium in aerogel. These equations could be used to study sound propagation phenomena in aerogel–superfluid system, in particular – to study sound conversion phenomena.We get two alternative sets of equations, one of which is Putterman's type equations[17] (equation for $\vec{v}_s$ is replaced by equation for

$$\vec{A} = \frac{\rho_n}{\rho\sigma}\vec{w}, \text{ where } \vec{w} = \vec{v}_n - \vec{v}_s)$$

The most general form of hydrodynamic equations for superfluid helium are as follows[8]

$$\frac{\partial\rho}{\partial t}+\nabla J=0,\ \frac{\partial(\rho\sigma)}{\partial t}+div(\rho\sigma\vec{v}_n)=0,\ \frac{\partial J_i}{\partial t}+\frac{\partial\Pi_{ij}}{\partial r_j}=0,$$

$$\frac{\partial\vec{v}_s}{\partial t}+\nabla\left(\frac{v_s^2}{2}+\mu\right)=0 \tag{21.33}$$

Here, μ is chemical potential and $\Pi_{ij}=p\delta_{ij}+\rho v_{ni}v_{nj}+\rho v_{si}v_{sj}$.

First three equations are conservation laws for mass, entropy and momentum and forth is equation for superfluid velocity $\vec{v}_s$.

Supposing $v_n, v_s \le v_2$ (second sound velocity), we get[8] that ρ_s and ρ_n are independent of relative velocity $\vec{w}=\vec{v}_n-\vec{v}_s$.

Writing relation

$$d\mu=-\sigma dT+\frac{1}{\rho}dp-\frac{\rho_n}{\rho}\vec{w}d\vec{w} \tag{21.34}$$

one gets with accuracy of second order of smallness (in w)

$$\mu(P,T,\vec{w})\approx\mu(P,T)-\frac{\rho_n}{2\rho}w^2,\ \ \sigma(P,T,\vec{w})\approx\sigma(P,T)+\frac{w^2}{2}\frac{\partial}{\partial T}\left(\frac{\rho_n}{\rho}\right),$$

$$\rho(P,T,\vec{w})\approx\rho(P,T)+\frac{\rho^2w^2}{2}\frac{\partial}{\partial T}\left(\frac{\rho_n}{\rho}\right). \tag{21.35}$$

Putting these relations into hydrodynamic equations and come to $\vec{w},\vec{v}$ variables instead of $\vec{v}_s,\vec{v}_n$ ones we receive the following nonlinear hydrodynamic equations with accuracy of second order of smallness with respect to relative velocity

$$\frac{\partial}{\partial t}(\rho+\frac{1}{2}w^2\rho^2\frac{\partial}{\partial P}(\frac{\rho_n}{\rho}))+div\vec{J}=0,$$

$$\frac{\partial}{\partial t}[\sigma\rho+\frac{\rho w^2}{2}(\frac{\partial}{\partial T}(\frac{\rho_n}{\rho})+\sigma\rho\frac{\partial}{\partial P}(\frac{\rho_n}{\rho}))]+div\sigma\rho(\vec{v}-\frac{\rho_n}{\rho}\vec{w})=0,$$

$$\frac{\partial J_i}{\partial t}+\frac{\partial}{\partial r_j}(P\delta_{ij}+\rho v_i v_j+\frac{\rho_n\rho_s}{\rho}w_i w_j)=0,$$

$$\frac{\partial}{\partial t}(\vec{v}-\frac{\rho_n}{\rho}\vec{w})+\nabla(\mu+\frac{(\vec{v}-\frac{\rho_n}{\rho}\vec{w})^2}{2}-\frac{\rho_n w^2}{2\rho})=0\,. \tag{21.36}$$

The generalization of hydrodynamic equations (21.33) for superfluid helium in aerogel leads us to following set of equations

$$\rho\frac{\partial\sigma}{\partial t}+J\nabla\sigma=0, \qquad \frac{\partial J_i}{\partial t}+\frac{\partial}{\partial r_j}((P+P_a)\delta_{ij})+\rho_a\dot{v}_{ni}=0,$$

$$\frac{\partial\vec{v}_s}{\partial t}+\nabla\left(\frac{v_s^2}{2}+\mu\right)=0, \quad \frac{\partial\rho_a}{\partial t}+\nabla(\rho_a v_n)=0. \tag{21.37}$$

In terms of $P,T,\vec{w},\vec{v}$ the nonlinear hydrodynamic equations for superfluid helium in aerogel with accuracy of second order with respect to velocities have the following form

$$\frac{\partial}{\partial t}(\rho + \frac{1}{2} w^2 \rho^2 \frac{\partial}{\partial P}(\frac{\rho_n}{\rho})) + div\vec{J} = 0,$$

$$\frac{\partial}{\partial t}[\sigma\rho + \frac{\rho w^2}{2}(\frac{\partial}{\partial T}(\frac{\rho_n}{\rho}) + \sigma\rho\frac{\partial}{\partial P}(\frac{\rho_n}{\rho}))] + div\,\sigma\rho(\vec{v} - \frac{\rho_n}{\rho}\vec{w}) = 0,$$

$$\frac{\partial J_i}{\partial t} + \frac{\partial}{\partial r_j}((P + P_a)\delta_{ij} + \rho v_i v_j + \frac{\rho_n \rho_s}{\rho} w_i w_j) + \rho_a \frac{\partial}{\partial t}(v_i + \frac{\rho_s}{\rho} w_i) = 0,$$

$$\frac{\partial}{\partial t}(\vec{v} - \frac{\rho_n}{\rho}\vec{w}) + \nabla(\mu + \frac{(\vec{v} - \frac{\rho_n}{\rho}\vec{w})^2}{2} - \frac{\rho_n w^2}{2\rho}) = 0,$$

$$\frac{\partial \rho_a}{\partial t} + \nabla(\rho_a \vec{v}) + \nabla(\frac{\rho_a \rho_S}{\rho}\vec{w}) = 0\,. \qquad (21.38)$$

The first three equations are conservation laws for mass, entropy and momentum, forth is equation for superfluid velocity $\vec{v}_s$ and last equation is aerogel mass conservation law.

21.12. Putterman's Type Equations

We also derive Putterman's type equations[17], where equation for $\vec{v}_s$ is replaced by equation for $\vec{A} = \frac{\rho_n}{\rho\sigma}\vec{w}$, where $\vec{w} = \vec{v}_n - \vec{v}_s$.

Conservation law for momentum for superfluid in aerogel can be written as:

$$\frac{\partial J_i}{\partial t} + \frac{\partial}{\partial r_j}((P + P_a)\delta_{ij} + \rho_n v_{ni} v_{nj} + \rho_s v_{si} v_{sj}) + \rho_a \frac{\partial v_{ni}}{\partial t} = 0, \qquad (21.39)$$

Using

$$dp = \rho d\mu + \rho\sigma dT + \frac{1}{2}\rho_n dw^2$$

and the identity

$$\frac{\partial}{\partial r_j}(\rho_n v_{nj} w_i) + \rho v_{sj}\frac{\partial v_{si}}{\partial r_j} + v_{si}\frac{\partial}{\partial r_j}(\rho_s v_{sj} + \rho_n v_{nj}) \equiv$$
$$\equiv \frac{\partial}{\partial r_j}(\rho_n v_{ni} v_{nj} + \rho_s v_{si} v_{sj}) - \rho_n w_j \frac{\partial v_{si}}{\partial r_j}$$

we get instead of (21.39)

$$\frac{\partial}{\partial t}(\rho_n v_n + \rho_s v_s)_i + \frac{\partial}{\partial r_j}(\rho_n v_{nj} w_i) + \rho v_{sj}\frac{\partial v_{si}}{\partial r_j} + v_{si}\frac{\partial}{\partial r_j}(\rho_n v_n + \rho_s v_s) =$$
$$-\rho\frac{\partial \mu}{\partial r_j} - \rho\sigma\frac{\partial T}{\partial r_j} - \frac{\partial}{\partial r_k}P_a\delta_{ik} - \rho_n w_j \frac{\partial v_{nj}}{\partial r_i} + \rho_n w_j\left(\frac{\partial v_{sj}}{\partial r_i} - \frac{\partial v_{si}}{\partial r_j}\right) - \rho_a \frac{\partial v_{ni}}{\partial t} \tag{21.40}$$

From

$$\vec{A} = \frac{\rho_n}{\rho\sigma}\vec{w}$$

one gets the following identity

$$\frac{\partial}{\partial r_j}(\rho_n v_{nj} w_i) \equiv A_i \frac{\partial \rho\sigma v_{nj}}{\partial r_j} + \rho\sigma v_{nj}\frac{\partial A_i}{\partial r_j} \tag{21.41}$$

Multiplying

$$\frac{\partial \vec{v}_s}{\partial t} + (\vec{v}_s \nabla)\vec{v}_s = -\nabla \mu$$

by ρ we have

$$\frac{\partial \rho \vec{v}_s}{\partial t} - \vec{v}_s \frac{\partial \rho}{\partial t} + \rho(\vec{v}_s \nabla)\vec{v}_s = -\rho \nabla \mu . \tag{21.42}$$

Subtracting (21.42) from (21.40) and introducing (21.42) one finds

$$\frac{\partial}{\partial t}\rho_n w_i - A_i \frac{\partial \rho\sigma}{\partial t} + \rho\sigma v_{nj}\frac{\partial A_i}{\partial r_j} =$$
$$= -\rho\sigma \frac{\partial T}{\partial r_i} - \frac{\partial}{\partial r_k} P_a \delta_{ik} - \rho_n w_j \frac{\partial v_{nj}}{\partial r_i} + \rho_n w_j \left(\frac{\partial v_{sj}}{\partial r_i} - \frac{\partial v_{si}}{\partial r_j}\right) - \rho_a \frac{\partial v_{ni}}{\partial t}, \tag{21.43}$$

where the conservation laws for mass and entropy have been also used.

Division of (21.43) by $\rho\sigma$ yields

$$\frac{\partial A_i}{\partial t} + (\vec{v}_n \nabla)A_i =$$
$$= -\frac{\partial T}{\partial r_i} - \frac{1}{\rho\sigma}\frac{\partial}{\partial r_k} P_a \delta_{ik} - \frac{\rho_a}{\rho\sigma}\frac{\partial v_{ni}}{\partial t} - A_j \frac{\partial v_{nj}}{\partial r_i} + [\vec{A} \times (\nabla \times \vec{v}_s)]_i \,, \tag{21.44}$$

where we have used the vector identity

$$B_j\left(\frac{\partial C_j}{\partial r_i} - \frac{\partial C_i}{\partial r_j}\right) \equiv [\vec{B} \times (\nabla \times \vec{C})]_i .$$

When $\nabla \times \vec{v}_s = 0$, we get

$$\frac{\partial A_i}{\partial t}+(\vec{v}_n\cdot\nabla)A_i=-\frac{\partial T}{\partial r_j}-A_i\frac{\partial v_{nj}}{\partial r_i}-\frac{1}{\rho\sigma}\frac{\partial}{\partial r_k}P_a\delta_{ik}-\frac{\rho_a}{\rho\sigma}\dot{v}_{ni}. \tag{21.45}$$

This equation could replace the fourth equation from (21.38): this gives us new form of nonlinear hydrodynamic equations for superfluid helium in aerogel – Putterman's type equations. Using these equations we could analyze sound propagation phenomena in aerogel–superfluid system.

21.13. Conclusion

We have shown that sound phenomena in impure superfluids are quite different from ones in pure superfluids and the cause of this difference is the increase of the coupling between first and second sound modes in impure superfluids. Significant frequency shift appears in fast mode, a slow mode exists in these systems as well as heat pulse propagation with the velocity of first sound. All nonlinear processes are changed drastically. The large increase of the coupling between first and second sound modes in impure superfluids decreases the threshold values for parametric decay of first sound, and the transformation of two second sounds into a first sound, as well as facilitate the observation of Cerenkov emission (which never been observed in pure superfluids). Sound conversion, which has been observed in pure superfluids only by shock waves[14,15] should be observed at moderate sound amplitude in impure superfluids. In addition, the nature of both sound modes in impure superfluids are modified from the pure case. We have derived for the first time the nonlinear hydrodynamic equations for superfluid helium in aerogel. We got two alternative sets of equations, one of which is a generalization of a traditional set of nonlinear hydrodynamics equations for an aerogel–superfluid system, while second one represents Putterman's type equations (equation for $\vec{v}_s$ is replaced by equation for $\vec{A}=\frac{\rho_n}{\rho\sigma}\vec{w}$, where $\vec{w}=\vec{v}_n-\vec{v}_s$). These equations are generalization of McKenna et al. equations for nonlinear hydrodynamics case and could be used to study sound propagation phenomena in aerogel–superfluid system, in particular – to study sound conversion phenomena.

We summarize the difference in sound properties of pure and impure superfluids in Table 21.1 below.

Table 21.1. The difference in sound properties of pure and impure superfluids.

	Pure superfluids	Impure superfluids
Coupling of sounds via	$\partial\rho/\partial T$	$c\,\partial\rho/\partial c$ or $\sigma\rho_a\rho_s$
Sound modes observed and to be observed		
first sound	Yes	Yes
second sound	Yes	Yes
slow pressure waves	No	Yes
fast temperature waves	No	Yes
Sound conversion (1→2, 2→1)	Observed only by shock waves	Should be observed at moderate sound amplitude
Cerenkov emission of second sound by first one	Never been observed, Difficult to observe (amplitude is proportional to $\frac{\partial\rho}{\partial T}$)	Should be observed, Much easier to observe (amplitude is proportional to $c\,\partial\rho/\partial c$ or $\sigma\rho_a\rho_s$)
Nature of sounds:		
First sound	$\vec{V}_n = \vec{V}_s$, normal and superfluid components move coherently	$\vec{V}_n \neq \vec{V}_s$, normal and superfluid components do not move coherently;
Second sound	$j=0$, mass flow is zero	$j\neq 0$, mass flow is nonzero and could be significant
Shift in first sound frequency at T_C (T_λ)	small	significant

Chapter XXII

Path Integral Approach to the Theory of Crystals

In this Chapter we will develop an approach to the microscopic theory of periodic structures in the framework of the functional integration method following to the book by Popov[1]. This approach was suggested by Kapitonov and Popov[2] and was developed by Andrianov, Kapitonov and Popov[3,4].

Our starting point will be a system of electrons and ions with the Columb interaction. The properties of crystals are determined by the collective excitations (phonons). Clearly, a microscopic theory must describe phonons and their interactions starting from the system of electrons and ions. The functional integral method allows us to realize this aim. The main idea is to go from the initial action of electrons and ions to the effective action functional in terms of the electric potential field $\varphi(\boldsymbol{x},\tau)$ .This field has an immediate physical meaning and provides the collective variable we need.

We can find the static field $\varphi_0(\boldsymbol{x})$, corresponding to the crystalline structure from the stationary condition for the effective action functional S_{eff}. If $\varphi_0(\boldsymbol{x})$ is known we can consider small fluctuations in the vicinity of the stationary point of S_{eff}. In order to do this, we have to expand S_{eff} in this neighborhood and to separate the quadratic form of $\varphi(\boldsymbol{x},\tau)-\varphi_0(\boldsymbol{x})$. It is this quadratic form that defines the spectrum of collective excitations. Forms of the third and higher degrees describe the interaction of these excitations.

The model described below is immediately applicable to the description of metallic hydrogen. If we wish to apply this scheme to other metals we have to modify the starting action in order to take into account the finite sizes of ions defined by their filled electron shells.

We begin with the action functional S for a system of electrons and ions:

$$S = \int_0^\beta d\tau d^3x \left(\sum_s \bar{\chi}_{es}(\mathbf{x},\tau)\partial_\tau \chi_s(\mathbf{x},\tau) + \bar{\chi}_i(\mathbf{x},\tau)\partial_\tau \chi_i(\mathbf{x},\tau) \right) - \int_0^\beta H'(\tau) d\tau \tag{22.1}$$

where

$$H'(\tau) = \int d^3x \sum_s (2m)^{-1} \nabla \bar{\chi}_{es}(\mathbf{x},\tau) \nabla \chi_{es}(\mathbf{x},\tau) - \lambda_e \bar{\chi}_{es}(\mathbf{x},\tau)\chi_{es}(\mathbf{x},\tau) +$$
$$+ \int d^3x \sum_s (2M)^{-1} \nabla \bar{\chi}_i(\mathbf{x},\tau) \nabla \chi_i(\mathbf{x},\tau) - \lambda_i \bar{\chi}_i(\mathbf{x},\tau)\chi_i(\mathbf{x},\tau) +$$
$$+ \frac{1}{2}\int d^3x d^3y \frac{e^2}{|\mathbf{x}-\mathbf{y}|} \rho(\mathbf{x},\tau)\rho(\mathbf{y},\tau), \tag{22.2}$$

$$\rho(\mathbf{x},\tau) = \bar{\chi}_i(\mathbf{x},\tau)\chi_i(\mathbf{x},\tau) - \sum_s \bar{\chi}_{es}(\mathbf{x},\tau)\chi_{es}(\mathbf{x},\tau). \tag{22.3}$$

Here $\chi_{es}(\mathbf{x},\tau), \bar{\chi}_{es}(\mathbf{x},\tau)$ is the Grassmannian Fermi–field of electrons, s = ± is the spin index. Let us also regard the $\chi_i(\mathbf{x},\tau), \bar{\chi}_i(\mathbf{x},\tau)$ field as a Fermi–field. We can use the following Fourier expansions of the $\chi_{es}(\mathbf{x},\tau), \bar{\chi}_i(\mathbf{x},\tau)$ fields:

$$\chi_{es}(x) = (\beta V)^{-1/2} \sum_p a_{es}(p) \exp(i(\omega\tau + \boldsymbol{k}\cdot\boldsymbol{x})),$$

$$\chi_i(x) = (\beta V)^{-1/2} \sum_p a_i(p) \exp(i(\omega\tau + \boldsymbol{k}\cdot\boldsymbol{x})), \tag{22.4}$$

where $p = (\mathbf{k},\omega)$; $\omega = (2n+1)\pi T$ are Fermi–frequencies and

$x = (\mathbf{x},\tau)$, $\beta = T^{-1}$.

First of all it is convenient to go from the Coulomb interaction to the interaction via the Bose–field of electric potential $\varphi(\mathbf{x},\tau)$. This method is also useful in plasma theory. In order to do it let us introduce the following Gaussian integral

$$\int d\varphi \exp\left[-(8\pi)^{-1}\int d\tau d^3x(\nabla\varphi(\mathbf{x},\tau))^2\right]. \tag{22.5}$$

into the integral with respect to the fields $\chi_{es}(\mathbf{x},\tau), \overline{\chi}_{es}(\mathbf{x},\tau)$, $\chi_i(\mathbf{x},\tau), \overline{\chi}_i(\mathbf{x},\tau)$. Performing the shift transformation

$$\varphi(\mathbf{x},\tau) \to \varphi(\mathbf{x},\tau) + ie\int\frac{d^3y\rho(\mathbf{y},\tau)}{|\mathbf{x}-\mathbf{y}|}. \tag{22.6}$$

we can cancel the Coulomb interaction term in the action functional (22.1). We then obtain the action functional

$$\begin{aligned}
S(\chi_{es},\overline{\chi}_{es},\chi_i,\overline{\chi}_i,\varphi) &= -(8\pi)^{-1}\int_0^\beta d\tau d^3x(\nabla\varphi(\mathbf{x},\tau))^2 + \\
&+ \int_0^\beta d\tau\int d^3x\Big(\sum_s \overline{\chi}_{es}(\mathbf{x},\tau)\partial_\tau\chi_{es}(\mathbf{x},\tau) - (2m)^{-1}\nabla\overline{\chi}_{es}(\mathbf{x},\tau)\nabla\chi_{es}(\mathbf{x},\tau)\Big) + \\
&+ (\lambda_e + ie\varphi(\mathbf{x},\tau))\overline{\chi}_{es}(\mathbf{x},\tau)\chi_{es}(\mathbf{x},\tau)] + \\
&+ \int_0^\beta d\tau\int d^3x(\overline{\chi}_i(\mathbf{x},\tau)\partial_\tau\chi_i(\mathbf{x},\tau) - (2M)^{-1}\nabla\overline{\chi}_i(\mathbf{x},\tau)\nabla\chi_i(\mathbf{x},\tau)) + \\
&+ (\lambda_i + ie\varphi(\mathbf{x},\tau))\overline{\chi}_i(\mathbf{x},\tau)\chi_i(\mathbf{x},\tau)]
\end{aligned} \tag{22.7}$$

If our systems is in the crystalline state, the ions (whose mass M is much larger than the electron mass m) are located near the sites of some crystalline lattice. In this case it is better to go from the "field" description of ions to the "corpuscular" description. In the language of functional integrals this means that we have to integrate not over the fields $\chi_i(\mathbf{x},\tau), \bar{\chi}_i(\mathbf{x},\tau)$, but over the trajectories of ions oscillating near the lattice points. Here the most suitable tool is the formalism of integration over trajectories in the phase space. The last terms in (22.7) depending on $\chi_i(\mathbf{x},\tau), \bar{\chi}_i(\mathbf{x},\tau)$ must be replaced by the following expression:

$$+\int_0^\beta d\tau \sum_l \left[i p_l(\tau)\partial_\tau q_l(\tau) - (2M)^{-1} p_l^2(\tau) + \lambda_i - ie\varphi(q_l(\tau),\tau)\right]. \tag{22.8}$$

Here $\boldsymbol{l}$ ($\boldsymbol{l} \in L$) is the lattice vector which enumerates the ions. Studying the lattice dynamics we may omit the irrelevant term $\sum_i \lambda_i$ in (22.8).

Now we may integrate over the electron Fermi–fields $\chi_{es}(\mathbf{x},\tau), \bar{\chi}_{es}(\mathbf{x},\tau)$. The integral is Gaussian and we can express it via a regularized determinant of the operator acting on $\chi_{es}(\mathbf{x},\tau)$. As a result, after replacing the ion terms by (22.8), integrating over Fermi–fields, we obtain in place of (22.7) the expression

$$\begin{aligned} S(\boldsymbol{p}_l, \boldsymbol{q}_l, \varphi(\mathbf{x},\tau)) &= -(8\pi)^{-1}\int_0^\beta d\tau d^3x (\nabla\varphi(\mathbf{x},\tau))^2 + \\ &+ 2\ln\det\frac{\left(\partial_\tau + (2m)^{-1}\nabla^2 + \lambda_e + ie\varphi(\mathbf{x},\tau)\right)}{\left(\partial_\tau + (2m)^{-1}\nabla^2 + \lambda_e\right)} + \\ &+ \int_0^\beta d\tau \sum_l \left[i\boldsymbol{p}_l(\tau)\partial_\tau \boldsymbol{q}_l(\tau) - (2M)^{-1}\boldsymbol{p}_l^2(\tau) - ie\varphi(\boldsymbol{q}_l(\tau),\tau)\right]. \end{aligned} \tag{22.9}$$

The factor of two before $\ln\det$ is due to taking electron spin into account. Now we have to integrate $S[\boldsymbol{p}_l,\boldsymbol{q}_l,\varphi]$ with respect to the field $\varphi(\mathbf{x},\tau)$, the ion momenta $\boldsymbol{p}_l(\tau)$ and their coordinates $\boldsymbol{q}_l(\tau)$. It is natural to assume that the main contribution in the integral comes from the field $\varphi(\mathbf{x},\tau)$, the momenta $\boldsymbol{p}_l(\tau)$ and the coordinates $\boldsymbol{q}_l(\tau)$ which are close to some equilibrium values. These values can be found from the equations

$$\frac{\nabla^2}{4\pi}\varphi(\mathbf{x},\tau)+ie\Big\{2\big[\partial_\tau+(2m)^{-1}\nabla^2+\lambda_e+ie\varphi(\mathbf{x},\tau)\big]^{-1}_{\mathbf{x},\tau;\mathbf{x},\tau}-\sum_l\delta(\boldsymbol{x}-\boldsymbol{q}_l(\tau))\Big\}=0,$$

$$i\partial_\tau\boldsymbol{q}_l(\tau)-M^{-1}\boldsymbol{p}_l(\tau)=0, \tag{22.10}$$

$$-i\partial_\tau\boldsymbol{p}_l(\tau)+ie\nabla\varphi(\boldsymbol{q}_l(\tau),\tau)=0\,.$$

First of all we consider the stationary solution of (22.10) of the form

$$\boldsymbol{p}_l=0,\boldsymbol{q}_l=\boldsymbol{l}\in L,\omega=-i\phi(\boldsymbol{x}), \tag{22.11}$$

where $\boldsymbol{l}\in L$ is a lattice vector. It is clear that the second equation is fulfilled and the third equation holds because $\nabla\varphi=0$ in the lattice points. The first equation will hold if $\phi(\boldsymbol{x})$ obeys the following nonlinear equation of the self–consistent field:

$$\nabla^2\phi(\mathbf{x},\tau)=4\pi e\Big\{\,2\big[\partial_\tau+(2m)^{-1}\nabla^2+\lambda_e+ie\phi(\mathbf{x},\tau)\big]^{-1}_{\mathbf{x},\tau;\mathbf{x},\tau}-\sum_l\delta(\boldsymbol{x}-\boldsymbol{l})\Big\} \tag{22.12}$$

We cannot solve (22.12) explicitly. Nevertheless it is not hard to obtain an approximate solution of this equation, if we linearize its right–hand side relative to $\phi(x)$, i.e. take only the first two terms in the functional expansion of In det in $\phi(x)$. Let us write down the linearized equation, moving the terms depending on $\phi(x)$ to the left–hand side:

$$\nabla^2\phi(\mathbf{x})+8\pi e^2\int d\tau' d^3y G_0(\boldsymbol{x}-\boldsymbol{y},\tau-\tau')G_0(\boldsymbol{y}-\boldsymbol{x},\tau'-\tau)\phi(y)= \\ =4\pi e\left\{\rho_0-\sum_{l}\delta(\boldsymbol{x}-\boldsymbol{l})\right\}. \tag{22.13}$$

The constant ρ_0 on the right–hand side

$$\rho_0=2\left[\partial_\tau+(2m)^{-1}\nabla^2+\lambda_e+ie\phi(\mathbf{x},\tau)\right]^{-1}_{\mathbf{x},\tau;\mathbf{x},\tau}=2G_0(0,0) \tag{22.14}$$

is the density of electron gas with chemical potential λ_e. For the system to be electro–neutral, we must take $\rho_0=V_0^{-1}$, where V_0 is the volume of an elementary crystalline cell. G_0 in (22.13) and (22.14) is the Green function of a free electron.

It is not hard to solve (22.13) using the Fourier transform. The solution

$$\phi(\boldsymbol{x})=4\pi e\rho_0\sum_{0\neq \boldsymbol{k}\in L^*}\frac{\exp(i\boldsymbol{k}\boldsymbol{x})}{k^2\varepsilon(\boldsymbol{k})} \tag{22.15}$$

has the form of a sum over nonzero vectors of the inverse lattice L^*.

Here, $\varepsilon(k)$ is defined by the formulae

$$\varepsilon(\boldsymbol{k})=1-\frac{8\pi e^2}{k^2\beta V}\sum_{k_1,\omega_1}G_0(\omega_1,\boldsymbol{k}_1)G_0(\omega_1,\boldsymbol{k}_1-\boldsymbol{k})= \tag{22.16}$$

$$=1-\frac{8\pi e^2}{k^2 V}\sum_{\boldsymbol{k}_1}\frac{n(\boldsymbol{k}_1)-n(\boldsymbol{k}_1-\boldsymbol{k})}{\frac{k_1^2}{2m}-\frac{(\boldsymbol{k}_1-\boldsymbol{k})^2}{2m}}. \tag{22.17}$$

$\varepsilon(\boldsymbol{k})$ is the dielectric function in the random phase approximation (RPA). The solution (22.15) of the linearized equation (22.13) must give a good approximation to an exact solution for systems of large density.

Regarding $\phi(\boldsymbol{x})$ as a known function, we can take into account small deviations of $\varphi(\boldsymbol{x},\tau)$ from $-i\phi(\boldsymbol{x})$, of $\boldsymbol{q}_l(\tau)$ from $\boldsymbol{l}$ and $\boldsymbol{p}_l(\tau)$ from zero. We start by considering the first equation in (22.10) as a equation for $\varphi(\boldsymbol{x},\tau)$ if $\boldsymbol{q}_l(\tau)$ are given functions. Physically it means that we have to obtain the electric potential for a given low of motion of heavy particles. This is the usual way in the so–called adiabatic approach. We denote

$$\varphi(\boldsymbol{x},\tau)=-i\phi(\boldsymbol{x})-i\delta\phi(\boldsymbol{x},\tau). \tag{22.18}$$

Substituting (22.18) into the first equation in (22.10) we obtain the equation

$$\nabla^2(\phi(\mathbf{x})+\delta\phi(\mathbf{x},\tau))=$$
$$=4\pi e\left\{2\left[2\partial_\tau+(2m)^{-1}\nabla^2+\lambda_e+ie(\phi(\mathbf{x})+\delta\phi(\mathbf{x},\tau))\right]^{-1}_{\mathbf{x},\tau;\mathbf{x},\tau}-\sum_{l\in L}\delta(\boldsymbol{x}-\boldsymbol{q}_l(\tau))\right\}. \tag{22.19}$$

Let us subtract (22.12) from (22.19) and linearize relative to $\delta\phi(\boldsymbol{x},\tau)$. We obtain the following linearized equation for $\delta\phi(\boldsymbol{x},\tau)$:

$$\nabla^2\delta\phi(\mathbf{x})+8\pi e^2\int d\tau' d^3y G_0(\boldsymbol{x},\tau,\boldsymbol{y},\tau'|\phi)G_0(\boldsymbol{y},\tau',\boldsymbol{x},\tau|\phi)\delta\phi(\boldsymbol{y},\tau')=$$
$$=4\pi e\sum_{\boldsymbol{l}\in L}\left[\delta(\boldsymbol{x}-\boldsymbol{l})-\delta(\boldsymbol{x}-\boldsymbol{q}_l(\tau))\right]. \tag{22.20}$$

Here $G_0(\boldsymbol{x},\tau,\boldsymbol{y},\tau'|\phi)$ is the Green function of electron in the periodic potential $\phi(\boldsymbol{x})$.

Linearization of the right–hand side of (22.20) relative to $\boldsymbol{q}_l(\tau)-\boldsymbol{l}$ gives

$$\nabla^2\delta\phi(\mathbf{x})+8\pi e^2\int d\tau' d^3y G_0(\boldsymbol{x},\tau,\boldsymbol{y},\tau'|\phi)G_0(\boldsymbol{y},\tau',\boldsymbol{x},\tau|\phi)\delta\phi(\boldsymbol{y},\tau')=$$
$$4\pi e\sum_{\boldsymbol{l}\in L}\left(\boldsymbol{q}_l(\tau)-\boldsymbol{l},\nabla\delta(\boldsymbol{x}-\boldsymbol{l})\right). \tag{22.21}$$

The Green function G can be written as

$$G(\boldsymbol{x},\tau,\boldsymbol{y},\tau'|\phi)=(\beta V)^{-1}\sum_{\substack{\omega,\boldsymbol{p}\in \boldsymbol{B}_1\\ \boldsymbol{k},\boldsymbol{k}_1\in L^*}}G(\omega,\boldsymbol{k},\boldsymbol{k}_1)\exp[i\omega(\tau-\tau')+i\boldsymbol{k}\boldsymbol{x}-i\boldsymbol{k}_1\boldsymbol{y}] \tag{22.22}$$

Taking into account that the periodic potential $\phi(\boldsymbol{x})$ has only Fourier components with $\boldsymbol{q}_l(\tau)-\boldsymbol{l}$, we can write

$$\boldsymbol{k} = \boldsymbol{p} + \boldsymbol{K}\,,\ \boldsymbol{k}_1 = \boldsymbol{p} + \boldsymbol{K}_1\,, \boldsymbol{K}, \boldsymbol{K}_1 \in L^*\,, \boldsymbol{p} \in B_1\,, \tag{22.23}$$

i.e. $\boldsymbol{k}$ and $\boldsymbol{k}_1$ can differ only by an inverse lattice vector, $\boldsymbol{p} \in B_1$ in (22.23) means that $\boldsymbol{p}$ belongs to the first Brillouin zone B_1. As it is well known, B_1 is a polyedron with facets formed by the planes orthogonal to vectors of the lattice points nearest to zero, and intersecting these vectors at their centres. So we can rewrite (22.22) as follows:

$$G(\boldsymbol{x},\tau,\boldsymbol{y},\tau'|\phi) = (\beta V)^{-1} \sum_{\substack{\omega,\boldsymbol{p}\in B_1 \\ \boldsymbol{k},\boldsymbol{k}_1\in L^*}} G(\omega,\boldsymbol{p},\boldsymbol{k},\boldsymbol{k}_1)\times$$
$$\times \exp[i\omega(\tau-\tau') + i\boldsymbol{k}\boldsymbol{x} - i\boldsymbol{k}_1\boldsymbol{y} + i\boldsymbol{p}(\boldsymbol{x}-\boldsymbol{y})]\,. \tag{22.24}$$

Using the substitutions

$$\delta\phi(\boldsymbol{x},\tau) = (\beta V)^{-1} \sum_{\substack{\omega,\boldsymbol{p}\in B_1 \\ \boldsymbol{k}\in L^*}} \delta\phi(\omega,\boldsymbol{p}+\boldsymbol{k}) \exp(i(\omega\tau + (\boldsymbol{p}+\boldsymbol{k},\boldsymbol{x}))), \tag{22.25}$$

$$\nabla\delta(\boldsymbol{x}-\boldsymbol{l}) = i(V)^{-1} \sum_{\boldsymbol{p}\in B_1,\boldsymbol{k}\in L^*} (\boldsymbol{p}+\boldsymbol{k}) \exp(i((\boldsymbol{p}+\boldsymbol{k},\boldsymbol{x}-\boldsymbol{l}))), \tag{22.26}$$

$$\boldsymbol{q}_l(\tau) - \boldsymbol{l} = i\beta^{-1} \sum_{\omega} \delta\boldsymbol{q}_l(\omega) \exp(i\omega\tau), \tag{22.27}$$

we may rewrite (22.20) with the right–hand side (22.21) in terms of Fourier coefficients:

$$-\left(\boldsymbol{p}+\boldsymbol{k}\right)^2\delta\phi\left(\omega,\boldsymbol{p}+\boldsymbol{k}\right)+8\pi e^3\sum_{\boldsymbol{k}\in L^*}\Pi_e\left(\omega,\boldsymbol{p}+\boldsymbol{k},\boldsymbol{p}+\boldsymbol{k}'\right)\delta\phi\left(\omega,\boldsymbol{p}+\boldsymbol{k}'\right)=$$
$$=4\pi ie\sum_{\boldsymbol{l}\in L}\left(\boldsymbol{p}+\boldsymbol{k},\delta\boldsymbol{q}_l\left(\omega\right)\right)\exp\left(-i\boldsymbol{p}\boldsymbol{l}\right).$$

(22.28)

Here Π_e is the so–called electron loop in a periodic field, i.e.

$$\Pi_e\left(\omega,\boldsymbol{p}+\boldsymbol{k},\boldsymbol{p}+\boldsymbol{k}'\right)=\frac{2}{\beta V}\sum_{\omega_1,\boldsymbol{p}_1\in B_1}G\left(\omega_1,\boldsymbol{p}_1,\boldsymbol{k}\right)G\left(\omega+\omega_1,\boldsymbol{p}+\boldsymbol{p}_1,\boldsymbol{k}'\right).$$

(22.29)

Introducing the matrix

$$\varepsilon\left(\omega,\boldsymbol{p}+\boldsymbol{k},\boldsymbol{p}+\boldsymbol{k}'\right)=\delta_{k,k'}-\frac{8\pi e^2}{\left|\boldsymbol{p}+\boldsymbol{k}'\right|}\Pi_e\left(\omega,\boldsymbol{p}+\boldsymbol{k}',\boldsymbol{p}+\boldsymbol{k}\right), \qquad (22.30)$$

we can write the solution of (22.28) as follows:

$$\delta\phi\left(\omega,\boldsymbol{p}+\boldsymbol{k}\right)=4\pi e\sum_{\boldsymbol{l}\in L}\left(\delta\boldsymbol{q}_l\left(\omega\right)\boldsymbol{A}\left(\omega,\boldsymbol{p}+\boldsymbol{k}\right)\right)\exp\left(-i\boldsymbol{p}\boldsymbol{l}\right),$$

(22.31)

where

$$\boldsymbol{A}\left(\omega,\boldsymbol{p}+\boldsymbol{k}\right)=i\sum_{\boldsymbol{k}'\in L^*}\varepsilon^{-1}\left(\omega,\boldsymbol{p}+\boldsymbol{k},\boldsymbol{p}+\boldsymbol{k}'\right)\frac{\boldsymbol{p}+\boldsymbol{k}'}{\left|\boldsymbol{p}+\boldsymbol{k}'\right|^2}. \qquad (22.32)$$

If we know $\phi(\boldsymbol{x})$ and $\delta\phi(\boldsymbol{x},\tau)$ we can expand the functional (22.9) into a power series in the deviation of $\varphi(\boldsymbol{x},\tau)$ from $-i\phi(\boldsymbol{x})-i\delta\phi(\boldsymbol{x},\tau)$ .i.e. into a power series in $\varphi_1(\boldsymbol{x},\tau)$ where

$$\varphi_1(\boldsymbol{x},\tau)=\varphi(\boldsymbol{x},\tau)+i\phi(\boldsymbol{x})+i\delta\phi(\boldsymbol{x},\tau)\,. \tag{22.33}$$

There is no linear term in $\varphi_1(\boldsymbol{x},\tau)$ in this expansion, because the field $-i\phi(\boldsymbol{x})-i\delta\phi(\boldsymbol{x},\tau)$ obeys the "stationary" equation. The terms which do not depend on $\varphi_1(\boldsymbol{x},\tau)$ are followed by the terms of the second order, third order and so on. The coefficient functions of this expansion depend on $\boldsymbol{q}_l(\tau)-\boldsymbol{l}$ and can be expanded into a power series in $\boldsymbol{q}_l(\tau)-\boldsymbol{l}$.

We shall first consider the terms quadratic in $\varphi_1(\boldsymbol{x},\tau)$, $\boldsymbol{q}_l(\tau)-\boldsymbol{l}$. These terms describe noninteracting excitations. The third– and higher–order terms describe the interaction of excitations.

The quadratic form of the variable $\varphi_1(\boldsymbol{x},\tau)$ can be obtained by putting $\boldsymbol{q}_l(\tau)=\boldsymbol{l}$ in the coefficient functions of the above–mentioned expansion. So it is clear that the quadratic form of the variable is nothing but the second variation form of the functional (22.9) relative to the variable $\varphi(\boldsymbol{x},\tau)$ at $\varphi(\boldsymbol{x},\tau)=-i\phi(\boldsymbol{x})$. We can write this form as

$$-(8\pi)^{-1}\int d\tau d^3x(\nabla\varphi_1(\boldsymbol{x},\tau))^2+e^2\int d\tau d\tau' d^3xd^3y\varphi_1(\boldsymbol{x},\tau)\varphi_1(\boldsymbol{y},\tau')\times$$
$$\times G(\boldsymbol{x},\tau;\boldsymbol{y},\tau'|\phi)G(\boldsymbol{y},\tau';\boldsymbol{x},\tau|\phi)\,, \tag{22.34}$$

where $G(\boldsymbol{x},\tau;\boldsymbol{y},\tau'|\phi)$ is the electron Green function in the periodic potential $\phi(\boldsymbol{x})$. This form describes collective excitations of the system of the plasma oscillation type.

We have to add to (22.34) the quadratic form of $\boldsymbol{q}_l(\tau)-\boldsymbol{l}$, $\boldsymbol{p}_l(\tau)$. describing lattice oscillations. In order to calculate this form, we substitute $\varphi(\boldsymbol{x},\tau)=-i\phi(\boldsymbol{x})+i\delta\phi(\boldsymbol{x},\tau)$ into (22.9) and then separate the terms quadratic in $\boldsymbol{q}_l(\tau)-\boldsymbol{l}$, $\boldsymbol{p}_l(\tau)$. First of all we write the last term in (22.9) in the form

$$\begin{aligned}
&-e\int_0^\beta d\tau\sum_l\left[\phi(\boldsymbol{q}_l(\tau))+\delta\phi(\boldsymbol{q}_l(\tau),\tau)\right]=\\
&=-e\int_0^\beta d\tau d^3x\left(\sum_l\delta(\boldsymbol{x}-\boldsymbol{l})\right)(\phi(\boldsymbol{x},\tau)+\delta\phi(\boldsymbol{x},\tau))-\\
&-e\int_0^\beta d\tau d^3x(\phi(\boldsymbol{x},\tau)+\delta\phi(\boldsymbol{x},\tau))\left[\left(\sum_l\delta(\boldsymbol{x}-\boldsymbol{q}_l(\tau))-\delta(\boldsymbol{x}-\boldsymbol{l})\right)\right].
\end{aligned}\tag{22.35}$$

We combine the first term on the right–hand side of (22.35) with the expression

$$\begin{aligned}
&(8\pi)^{-1}\int d\tau d^3x\left[\nabla(\phi(\boldsymbol{x})+\delta\phi(\boldsymbol{x},\tau))\right]^2+\\
&+2\ln\det\left[\partial_\tau+(2m)^{-1}\nabla^2+\lambda_e+(\phi(\boldsymbol{x})+\delta\phi(\boldsymbol{x},\tau))\right]/\left[\partial_\tau+(2m)^{-1}\nabla^2+\lambda_e\right]
\end{aligned}\tag{22.36}$$

We are interested in the terms quadratic in $\boldsymbol{q}_l(\tau)-\boldsymbol{l}$. These terms are given by the quadratic part of the expansion of (22.36) plus the first term on the right–hand side of (22.35). There are no linear terms in $\delta\phi(\boldsymbol{x},\tau)$ due to the equation for $\phi(\boldsymbol{x})$. So we obtain

$$-(8\pi)^{-1}\int d\tau d^3x(\nabla\,\delta\phi(x,\tau))^2 - e^2\int d\tau d\tau' d^3x d^3y\,\delta\phi(x,\tau)\delta\phi(y,\tau)\times$$
$$\times G(x,\tau;y,\tau'|\phi)G(y,\tau';x,\tau|\phi)\,.$$
(22.37)

Now we consider the second term on the right–hand side of (22.35) which can be written as follows:

$$-e\int_0^\beta d\tau\sum_l[\phi(q_l(\tau))-\phi(l)]+e\int_0^\beta d\tau d^3x\,\delta\phi(x,\tau)(q_l(\tau)-l,\nabla(x-l)).$$
(22.38)

In the first term we may make the replacement

$$\phi(q_l(\tau))-\phi(l)\approx\frac{1}{2}(\delta q_l(\tau),\nabla)^2\phi(x)\big|_{x=l}$$

and then rewrite the first term in (22.38) as

$$\frac{e}{2\beta}\sum_{\substack{\omega,k\in L^*\\ l\in L}}(\delta q_l(\omega),\delta q_l(-\omega))_j K_i K_j\phi(K). \tag{22.39}$$

Now we use the equation (22.20) for $\delta\phi(x,\tau)$ and rewrite (22.37) as

$$-\frac{e}{2}\int d\tau d^3x\sum_{l\in L}(\delta q_l(\tau),\nabla\,\delta q_l(x-l))\delta\phi(x,\tau). \tag{22.40}$$

Combining this term with the second added in (22.38) we get

$$-\frac{e}{2}\int d\tau d^3x \sum_{l\in L}(\delta q_l(\tau), \nabla \delta q_l(x-l))\delta\phi(x,\tau)=$$
$$=-\frac{e}{2}\int d\tau \sum_{l\in L}(\delta q_l(\tau), \nabla \delta q_l(x,\tau))\Big|_{x=l}. \tag{22.41}$$

Replacing $\delta\phi(x,\tau)$ in (22.41) by its expression according to (22.25), (22.31) and (22.32), we obtain

$$-\frac{2\pi e^2}{\beta}\sum_{\omega}\sum_{l_1\neq l_1}\left(\delta q_{l_1}(\omega)\right)_i\left(\delta q_{l_2}(\omega)\right)_j a_{ij}(\omega, l_1 - l_2), \tag{22.42}$$

where

$$a_{ij}(\omega, l_1 - l_2) = V^{-1}\sum_{p\in B_1}\exp(i((p, l_1 - l_2)))\times$$
$$\times \sum_{K,K'\in L^*}(p+K)_i(p+K')_j\varepsilon^{-1}(\omega, p+K, p+K')_j|p+K'|^{-2}. \tag{22.43}$$

We thus derive the following quadratic form of $\delta q_l, p_l$:

$$\beta^{-1}\sum_{\omega}\Bigg[\sum_{l_1\neq l_1}\Big(\omega p_l(\omega)\delta q_l(\omega)-(2M)^{-1}p_l(\omega)p_{-l}(-\omega)\Big)_i+$$
$$+\frac{e}{2}\sum_{l}\big(\delta q_{l_1}(\omega)\big)_i\big(\delta q_{l_2}(\omega)\big)_j\sum_{K\in L^*}K_iK_j\phi(K)-$$
$$-2\pi e^2\sum_{l_1,l_2}\big(\delta q_{l_1}(\omega)\big)_i\big(\delta q_{l_2}(\omega)\big)_j a_{ij}(\omega,l_1-l_2)\Bigg]. \tag{22.44}$$

It is this form that describes harmonic oscillations of the lattice.

Let us pass from $\delta q_l, p_l$ to the new (normal) variables:

$$\delta q_l(\omega)=N^{-1/2}\sum_{p\in B_1}q_p(\omega)\exp(ipl),$$

$$p_l(\omega)=N^{-1/2}\sum_{p\in B_1}p_p(\omega)\exp(ipl). \tag{22.45}$$

This is a standard procedure for investigating lattice vibrations. The quadratic form (22.44) becomes simpler in the new variables:

$$\beta^{-1}\sum_{\omega,p\in B_1}\Bigg[\sum_{l_1\neq l_1}\Big(\omega p_p(\omega)q_{-p}(-\omega)-(2M)^{-1}p_p(\omega)p_{-p}(-\omega)\Big)-$$
$$-\frac{1}{2}D_{ij}(\omega,p)\big(q_p(\omega)\big)_i\big(q_{-p}(-\omega)\big)_j\Bigg]. \tag{22.46}$$

It is usual to call the matrix $D_{ij}(\omega, \boldsymbol{p})$ in (22.46) the dynamical matrix. Its eigenvalues determine the spectrum of lattice variations. We can write this matrix explicitly as

$$D_{ij}(\omega, \boldsymbol{p}) = 4\pi e^2 \rho_0 \sum_{\boldsymbol{K},\boldsymbol{K}' \in L^*} \frac{(\boldsymbol{p}+\boldsymbol{K})_i (\boldsymbol{p}+\boldsymbol{K}')_j}{|\boldsymbol{p}+\boldsymbol{K}'|^2} \varepsilon^{-1}(\omega, \boldsymbol{p}+\boldsymbol{K}, \boldsymbol{p}+\boldsymbol{K}') - \\ - \sum_{\boldsymbol{K} \in L^*} \boldsymbol{K}_i \boldsymbol{K}_j \, \phi(\boldsymbol{K}) . \tag{22.47}$$

In the first approximation $\phi(\boldsymbol{k})$ is defined by (22.15):

$$\phi(\boldsymbol{K}) = 4\pi e \rho_0 K^{-2} \varepsilon^{-1}(\boldsymbol{K}). \tag{22.48}$$

The same approximation for ε^{-1} is

$$\varepsilon^{-1}(\omega, \boldsymbol{p}+\boldsymbol{K}, \boldsymbol{p}+\boldsymbol{K}') = \delta_{\boldsymbol{K}\boldsymbol{K}'} \varepsilon^{-1}(\omega, \boldsymbol{p}+\boldsymbol{K}). \tag{22.49}$$

Here,

$$\varepsilon(\omega, \boldsymbol{K}) = 1 - \frac{8\pi e^2}{K^2 \beta V} \sum_{\boldsymbol{K}_1, \omega_1} G_0(\omega_1, \boldsymbol{K}+\boldsymbol{K}_1) G_0(\omega+\omega_1, \boldsymbol{K}_1) = \\ = 1 - \frac{8\pi e^2}{K^2 V} \sum_{\boldsymbol{K}_1} \frac{n(\boldsymbol{K}_1) - n(\boldsymbol{K}+\boldsymbol{K}_1)}{i\omega + \frac{k_1^2}{2m} - \frac{(\boldsymbol{K}+\boldsymbol{K}_1)^2}{2m}} \tag{22.50}$$

is the dielectric function in RPA.

Using approximations (22.48) and (22.49) we may rewrite the dynamical matrix as

$$D_{ij}(\omega,\boldsymbol{p}) = M\omega_p^2 \sum_{\boldsymbol{K}\in L^*} \left[\frac{(\boldsymbol{p}+\boldsymbol{K})_i(\boldsymbol{p}+\boldsymbol{K})_j}{(\boldsymbol{p}+\boldsymbol{K})^2 \varepsilon(\omega,\boldsymbol{p}+\boldsymbol{K})} - \frac{\boldsymbol{K}_i\boldsymbol{K}_j}{K^2\varepsilon(\boldsymbol{K})} \right], \tag{22.51}$$

where

$$\omega_p^2 = \frac{4\pi e^2 \rho}{M} \tag{22.52}$$

is a square of the so–called ion–plasma frequency. Relation (22.51) implies that the dynamical matrix $D_{ij}(\omega,\boldsymbol{p})$ has three real positive eigenvalues

$$D\boldsymbol{e}_{p,\lambda} = M\omega_{p,\lambda}^2 \boldsymbol{e}_{p,\lambda}; \ \lambda = 1,2,3 \tag{22.53}$$

Here $\boldsymbol{e}_{p,\lambda}$ are orthonormal eigenvectors of D_{ij} and $\omega_{p,\lambda}$ are the eigen–frequencies.

We see that the functional integral approach allows us to describe a crystalline state in a system of electrons and ions and to go from the initial variables of functional integration to some new variables (22.45) corresponding to normal oscillations of the crystalline lattice. In the next Chapter we apply this approach to construct the effective interaction between electrons near the Fermi–surface. It is not hard to show that this interaction for the model considered has an attractive character and may lead to superconductivity at sufficiently low temperatures.

Chapter XXIII

Effective Interaction of Electrons Near the Fermi–Surface

As it is well known, the superconductivity phenomenon exists due to attractive interaction between electrons in the narrow energetic layer near the Fermi surface. The effective interaction of electrons consists of two parts: the Coulomb repulsive potential and some attractive potential arising from the phonon exchange. In this Chapter we will derive this effective interaction potential in the framework of the functional integral method following to the book by Popov[1]. Here the procedure of the previous section will be slightly changed. At the first stage we integrate over the electron fields with momenta outside the narrow layer near the Fermi surface. Then we integrate over the electric potential field and also over ion trajectories in the phase space. So we come down to the effective interaction potential between electrons near the Fermi–surface.

The starting action functional is

$$
\begin{aligned}
S(\chi_e, \bar{\chi}_e, \boldsymbol{p}_l, \boldsymbol{q}_l, \varphi(\boldsymbol{x},\tau)) &= -(8\pi)^{-1} \int_0^\beta d\tau d^3x (\nabla \varphi(\boldsymbol{x},\tau))^2 + \\
&+ \int_0^\beta d\tau d^3x \sum_s \bar{\chi}_{es}(\boldsymbol{x},\tau)\left(\partial_\tau + (2m)^{-1}\nabla^2 + \lambda_e + ie\varphi(\mathbf{x},\tau)\right)\chi_{es}(\boldsymbol{x},\tau) + \\
&+ \int_0^\beta d\tau \sum_l \left[i\boldsymbol{p}_l(\tau)\partial_\tau \boldsymbol{q}_l(\tau) - (2M)^{-1}\boldsymbol{p}_l^2(\tau) - ie\varphi(\boldsymbol{q}_l(\tau),\tau)\right].
\end{aligned}
\tag{23.1}
$$

At the first step of functional interaction we extract the slow varying component in the electric potential field $\varphi(x,\tau)$ by performing the following shift transformation:

$$\varphi(x,\tau)=\tilde{\phi}(x)-i\phi(x|q_l(\tau)). \tag{23.2}$$

The equation for $\phi(x|q_l(\tau))$ can be obtained from the variation principle after integration with respect to the field $\chi_{1es}(x,\tau)$, and the fields $\chi_{1es}(x,\tau)$, $\overline{\chi}_{1es}(x,\tau)$ of electrons outside the shell near the Fermi surface. The fields χ_0, χ_1 are defined by the formulac

$$\chi_{0es}(\mathbf{x},\tau)=(\beta V)^{-1/2}\sum_{\substack{k,\omega \\ |k-k_F|<k_0}}\chi_{es}(k,\omega)\exp[i(\omega\tau+k\mathbf{x})],$$

$$\chi_{1es}(\mathbf{x},\tau)=(\beta V)^{-1/2}\sum_{\substack{k,\omega \\ |k-k_F|>k_0}}\chi_{es}(k,\omega)\exp[i(\omega\tau+k\mathbf{x})],$$

$$\chi_{es}(\mathbf{x},\tau)=\chi_{0es}(\mathbf{x},\tau)+\chi_{1es}(\mathbf{x},\tau), \tag{23.3}$$

Upon integrating exp S_0 with respect to the fields

$$\tilde{\varphi}(x,\tau), \chi_{1es}(x,\tau), \overline{\chi}_{1es}(x,\tau)$$

we obtain

$$\begin{aligned}&\int D\tilde{\varphi}D\overline{\chi}_1 D\chi_1 \exp S_0[\chi_0+\chi_1, \overline{\chi}_0+\overline{\chi}_1, p_l, q_l, \tilde{\varphi}-i\phi]=\\&=\exp S_{eff}(\chi_0, \overline{\chi}_0, p_l, q_l, \phi)\,.\end{aligned} \tag{23.4}$$

In the first approximation S_{eff} is a sum of three terms

$$S_{eff} = S_{el}[\chi_0, \bar{\chi}_0] + S_{ph}[\boldsymbol{p}_l, \boldsymbol{q}_l] + S_{el-ph}. \tag{23.5}$$

Here $S_{el}[\chi_0, \bar{\chi}_0]$ is the action functional describing electrons on the shell and their pair Coulomb interaction. It consists of terms of the second and fourth order in $\chi_0, \bar{\chi}_0$. The term $S_{ph}[\boldsymbol{p}_l, \boldsymbol{q}_l]$ describes a phonon system, S_{el-ph} is the electron–phonon interaction.

Formally S_{eff} also contains terms of the sixth and higher degrees on $\chi_0, \bar{\chi}_0$. If the shell is sufficiently narrow, such terms can be neglected.

It is not hard to write down expressions for each term in (23.5). The first term is

$$\begin{aligned} S_{el} &= \sum_{\omega,s} d^3x \bar{\chi}_{0es}(\boldsymbol{x},\omega)\left(i\omega - (2m)^{-1}\nabla^2 + \lambda_e\right)\chi_{0es}(\boldsymbol{x},\omega) - \\ &- \sum_{\omega,s} \int d^3x d^3y \bar{\chi}_{0es}(\boldsymbol{x},\omega)\Sigma_0(\boldsymbol{x},\boldsymbol{y},\omega)\chi_{0es}(\boldsymbol{x},\omega) - \\ &- (2\beta)^{-1}\sum_{\omega} \int d^3x d^3y \frac{e^2}{|\boldsymbol{x}-\boldsymbol{y}|}\varepsilon^{-1}(\boldsymbol{x},\boldsymbol{y},\omega)\rho_0(\boldsymbol{x},\omega)\rho_0(\boldsymbol{y},-\omega)\ , \end{aligned} \tag{23.6}$$

where

$$\rho_0(\boldsymbol{x},\omega) = \sum_{\omega,s} \bar{\chi}_{0es}(\boldsymbol{x},\omega_1)\chi_{0es}(\boldsymbol{x},\omega+\omega_1) \tag{23.7}$$

Here $s = \pm$ is a spin index, $\Sigma_0(\boldsymbol{x},\boldsymbol{y},\omega)$ is the self–energy part, $\varepsilon(\boldsymbol{x},\boldsymbol{y},\omega)$ is a dielectric function.

The slow part of the electric potential field ϕ can be written as

$$\phi = \phi_0(x) + \delta\phi(x|q_l(\tau)). \tag{23.8}$$

In the first approximation we have

$$\phi_0(x) = 4\pi e\rho \sum_{0\neq k\in L^*} \frac{\exp(ikx)}{k^2\varepsilon(k)}. \tag{23.9}$$

Here k stands for the inverse lattice vectors, $\varepsilon(k)$ is a static dielectric function of the electron gas. The term $\delta\phi$ in (23.8) is defined by the formulae

$$\delta\phi(x, \delta q_l(\tau)) = (\beta V)^{-1} \sum_{\substack{\omega, p\in B_1 \\ k\in L^*}} \exp(i(\omega\tau + kx))\delta\phi(\omega, p+k),$$

$$\delta\phi(\omega, p+k) = -4\pi i e N^{1/2}\left(\delta q_l(\omega), \frac{p+k}{(p+k)^2}\varepsilon^{-1}(\omega, p+k)\right), \tag{23.10}$$

$$q_p(\omega) = N^{-1/2}\sum_{l\in L}\delta q_l(\omega)\exp(-ipl).$$

Here, B_1 is the first Brillouin zone in the inverse lattice space.

As a result, we obtain the quadratic part of the phonon action functional S_{ph} in the following form

$$S_{ph}(p_l, q_l) = \beta^{-1}\sum_{\omega, p\in B_l}\Big[\big(\omega p_p(\omega)q_{-p}(-\omega) - (2M)^{-1}p_p(\omega)p_{-p}(-\omega)\big) - \frac{1}{2}D_{ij}(\omega, p)(q_p(\omega))_i(q_{-p}(-\omega))_j\Big]. \tag{23.11}$$

Here, $D_{ij}(\omega, \boldsymbol{p})$ is the dynamical matrix

$$D_{ij}(\omega, \boldsymbol{p}) = M\omega_p^2 \left\{ \frac{p_i p_j}{p^2 \varepsilon(\omega, \boldsymbol{p})} + \sum_{0 \neq \boldsymbol{k} \in L^*} \left[\frac{(\boldsymbol{p}+\boldsymbol{k})_i (\boldsymbol{p}+\boldsymbol{k})_j}{|\boldsymbol{p}+\boldsymbol{k}|^2 \varepsilon(\omega, \boldsymbol{p}+\boldsymbol{k})} - \frac{k_i k_j}{k^2 \varepsilon(\boldsymbol{k})} \right] \right\} \tag{23.12}$$

and

$$\omega_p^2 = \frac{4\pi e^2 \rho}{M} \tag{23.13}$$

is a square of the ion–plasma frequency.

The last term in (23.5) is the electron–phonon interaction, which can be shown to be equal to

$$S_{el-ph}(\boldsymbol{p}_l, \boldsymbol{q}_l) = \beta^{-1} \sum_{\omega} \int d^3x e\, \delta\phi(\boldsymbol{x}, \boldsymbol{q}_l(\omega)) \rho_0(\boldsymbol{x}, -\omega). \tag{23.14}$$

This is just electrostatic energy of electrons near the Fermi–surface in the electric potential due to the displacement of ions from their equilibrium positions.

Now we can obtain the effective interaction between electrons in the shell by integrating with respect to the phonon variables $\boldsymbol{P}_p(\omega), \boldsymbol{q}_p(\omega)$. First we rewrite S_{el-ph} from (23.14) as follows:

$$S_{el-ph}(\boldsymbol{p}_l, \boldsymbol{q}_l) = \beta^{-1} \sum_{\omega, \boldsymbol{p} \in \boldsymbol{B}_1} (\boldsymbol{q}_p(\omega), \boldsymbol{B}(-\boldsymbol{p}, -\omega)), \tag{23.15}$$

where

$$\boldsymbol{B}(-\boldsymbol{p},-\omega)=-iN^{1/2}\frac{4\pi e^2}{\beta V}\sum_{k\in L^*}\frac{\boldsymbol{p}+\boldsymbol{k}}{|\boldsymbol{p}+\boldsymbol{k}|^2\varepsilon(\omega,\boldsymbol{p}+\boldsymbol{k})}\rho_0(-\boldsymbol{p},-\boldsymbol{k},-\omega). \tag{23.16}$$

We denote by $M\omega_\lambda^2(\omega,\boldsymbol{p})$ the eigenvalues and by $e_\lambda(\omega,\boldsymbol{p})$ the eigenvectors of the dynamical matrix. We have

$$D_{\alpha\beta}(\omega,\boldsymbol{p})\boldsymbol{e}_{\lambda\alpha}(\omega,\boldsymbol{p})=M\omega_\lambda^2(\omega,\boldsymbol{p})\boldsymbol{e}_{\lambda\alpha}(\omega,\boldsymbol{p}). \tag{23.17}$$

One can see that the eigenvalues obey the following sum rule:

$$\sum_{\lambda=1}^{3}\omega_\lambda^2(\omega,\boldsymbol{p})=\omega_p^2\varepsilon^{-1}(\omega,\boldsymbol{p})+\omega_p^2\sum_{0\neq k\in L^*}\left(\varepsilon^{-1}(\omega,\boldsymbol{p}+\boldsymbol{k})-\varepsilon^{-1}(\boldsymbol{k})\right) \tag{23.18}$$

We may write the result of the integration with respect to the phonon variables in the following form:

$$\int DpDq\exp\left(S_{ph}+S_{el-ph}\right)=\exp\widetilde{S}_{el-el}, \tag{23.19}$$

where

$$\widetilde{S}_{el-el}=\frac{1}{2\beta}\sum_{\omega,\boldsymbol{p}\in B_1}\sum_{\alpha,\beta}\left(D(\omega,\boldsymbol{p})-M(i\omega)^2\right)^{-1}_{\alpha\beta}B_\alpha(\boldsymbol{p},\omega)B_\beta(-\boldsymbol{p},-\omega), \tag{23.20}$$

$$B_\alpha(\boldsymbol{p},\omega)B_\beta(-\boldsymbol{p},-\omega)=$$
$$=M\omega_p^2\frac{4\pi e^2}{V}\sum_{k,k'\in L^*}\frac{(\boldsymbol{p}+\boldsymbol{k})_\alpha(\boldsymbol{p}+\boldsymbol{k})_\beta\rho_0(\boldsymbol{p}+\boldsymbol{k},\omega)\rho_0(-\boldsymbol{p}-\boldsymbol{k}',-\omega)}{(\boldsymbol{p}+\boldsymbol{k})^2(\boldsymbol{p}+\boldsymbol{k}')^2\varepsilon(\omega,\boldsymbol{p}+\boldsymbol{k})\varepsilon(\omega,\boldsymbol{p}+\boldsymbol{k}')}$$
(23.21)

Using the eigenvector basis of the dynamical matrix we get

$$\left(D(\omega,\boldsymbol{p})-M(i\omega)^2\right)^{-1}_{\alpha\beta}=M^{-1}\left(\omega_\alpha^2(\omega,\boldsymbol{p})-(i\omega)^2\right)^{-1}\delta_{\alpha\beta}\,. \qquad (23.22)$$

Expression (23.20) describes the interaction between electrons near the Fermi surface due to the phonon exchange. A Coulomb repulsive term must be added to (23.20) in order to obtain the full effective interaction of electrons. This term has the following form:

$$S^c_{el-el}=-\frac{1}{2\beta}\sum_\omega\int\frac{d^3p}{(2\pi)^3}\frac{4\pi e^2}{p^2\varepsilon(\omega,\boldsymbol{p})}\rho_0(\boldsymbol{p},\omega)\rho_0(-\boldsymbol{p},-\omega)\,. \qquad (23.23)$$

Here we have replaced the sum over $\boldsymbol{p}$, $\boldsymbol{k}$ by the integral according to the rule:

$$V^{-1}\sum_{p\in B_1,k\in L^*}\to(2\pi)^{-3}\int d^3p\,.$$

It is possible to simplify (23.20) a little by putting $\boldsymbol{k}=\boldsymbol{k}'$ in (23.21). It means that we neglect “Umklapp” processes. This is justified if we are interested only in the interaction of electrons of opposite momenta near the Fermi–surface. Assuming this we may rewrite (23.20) as

$$\tilde{S}_{el-el} = -\frac{1}{2\beta V}\sum_{\omega,\alpha}\int\frac{d^3p}{(2\pi)^3}\frac{4\pi e^2\omega_p^2}{\omega_\alpha^2(\omega,\boldsymbol{p})-(i\omega)^2}\frac{(\boldsymbol{p},\boldsymbol{e}_\alpha(\omega,\boldsymbol{p}))^2}{p^4\varepsilon^2(\omega,\boldsymbol{p})}\rho_0(\boldsymbol{p},\omega)\rho_0(-\boldsymbol{p},-\omega)$$

(23.24)

So we come down to the following effective electron–electron interaction:

$$S_{eff\ el-el} = -\frac{1}{2\beta}\sum_{\omega}\int\frac{d^3p}{(2\pi)^3}\frac{4\pi e^2}{p^2\varepsilon(\omega,\boldsymbol{p})}\left[1-\sum_\alpha\frac{\omega_p^2(\boldsymbol{p},\boldsymbol{e}_\alpha(\omega,\boldsymbol{p}))^2}{(\omega_\alpha^2-(i\omega)^2)p^2\varepsilon(\omega,\boldsymbol{p})}\right]\times$$
$$\times\rho_0(\boldsymbol{p},\omega)\rho_0(-\boldsymbol{p},-\omega).$$

(23.25)

We write down the effective electron–electron potential in the static limit

$$V_{eff}(\boldsymbol{p}) = V^c(\boldsymbol{p})+V^{ph}(\boldsymbol{p}) = \frac{4\pi e^2}{p^2\varepsilon(\omega,\boldsymbol{p})}\left[1-\sum_\alpha\frac{\omega_p^2p_\alpha^2}{\omega_\alpha^2(\boldsymbol{p})p^2\varepsilon(\boldsymbol{p})}\right]. \tag{23.26}$$

In the large density limit $\varepsilon(\boldsymbol{p})\to 1$ and (23.18) goes into

$$\sum_{\alpha=1}^{3}\omega_\alpha^2(\boldsymbol{p}) = \omega_p^2. \tag{23.27}$$

So we may write (23.26) as

$$V_{eff}(\boldsymbol{p}) = \frac{4\pi e^2}{p^2}\left[1-\sum_\alpha\frac{\omega_p^2p_\alpha^2}{\omega_\alpha^2(\boldsymbol{p})p^2}\right]. \tag{23.28}$$

According to (23.27) we have $(\omega_p^2/\omega_\alpha^2(\boldsymbol{p}))>1$, and

$$\sum_\alpha\frac{\omega_p^2p_\alpha^2}{\omega_\alpha^2(\boldsymbol{p})p^2} > \sum_\alpha\frac{p_\alpha^2}{p^2} = 1. \tag{23.29}$$

This implies that the effective electron–electron potential is negative, and that electrons attract each other near the Fermi–surface. The attractive interaction is the reason for the system to become a superconductor at sufficiently low temperatures.

Chapter XXIV

The Path Integral Models of *p*– and *d*–Pairing for Bulk Superconductors

24.1. Models of *p*– and *d*–Pairing

In this Chapter we consider the path integral models of *p*–[1,2,8] and *d*–[3-8] pairing for bulk superconductors. The *p*–pairing model for superconductors is similar to the model, built by us in Chapter V for superfluid ^{3}He. Thus we will just shortly remind the main points of *p*–wave pairing model and will pay the most of attention to case of *d*–wave pairing model. In the method of functional integration, the initial Fermi–system is described by anticommuting functions $\chi_s(\mathbf{x},\tau), \bar{\chi}_s(\mathbf{x},\tau)$ defined in the volume $V = L^3$, which are antiperiodic in time τ with a period $\beta = T^{-1}$. Here s is the spin index. These functions can be expanded into a Fourier series

$$\chi_s(x) = (\beta V)^{-1/2} \sum_p a_s(p) \exp\left(i(\omega\tau + \mathbf{k}\cdot\mathbf{x})\right), \tag{24.1}$$

where $p = (\mathbf{k},\omega)$; $\omega = (2n+1)\pi T$ are Fermi–frequencies and $x = (\mathbf{x},\tau)$.

Let us consider the functional of action for an interacting Fermi–system

$$S = \int_0^\beta d\tau d^3x \sum_s \bar{\chi}_s(\mathbf{x},\tau)\partial_\tau \chi_s(\mathbf{x},\tau) - \int_0^\beta H'(\tau)d\tau \tag{24.2}$$

which corresponds to the following Hamiltonian

$$H'(\tau)=\int d^3x\sum_s(2m)^{-1}\nabla\overline{\chi}_s(\mathbf{x},\tau)\nabla\chi_s(\mathbf{x},\tau)-(\lambda+s\mu_0H)\overline{\chi}_s(\mathbf{x},\tau)\chi_s(\mathbf{x},\tau)$$

$$+\frac{1}{2}\int d^3xd^3yU(\mathbf{x}-\mathbf{y})\sum_{ss'}\overline{\chi}_s(\mathbf{x},\tau)\overline{\chi}_{s'}(\mathbf{y},\tau)\chi_{s'}(\mathbf{y},\tau)\chi_s(\mathbf{x},\tau) \tag{24.3}$$

In order to obtain the effective functional of action, we shall use the method of division of Fermi–fields into "fast" and "slow" fields with subsequent successive integration over these fields. Fast fields χ_{1s} and $\overline{\chi}_{1s}$ are determined by components of expansion (24.1) either with frequencies $|\omega|>\omega_0$, or with momenta $|k-k_F|>k_0$. The remaining component $\chi_{0s}=\chi_s-\chi_{1s}$ of the Fourier expansion define slow fields χ_{0s}. Integrating over fast fields, we obtain

$$\int \exp SD\overline{\chi}_{1s}D\chi_{1s}=\exp\tilde{S}(\chi_{0s},\tilde{\chi}_{0s}). \tag{24.4}$$

The most general form of $\tilde{S}$ is the sum of even–order forms of the fields $\chi_{0s},\overline{\chi}_{0s}$:

$$\tilde{S}=\sum_{n=0}^{\infty}\tilde{S}_{2n} \tag{24.5}$$

Neglecting the insignificant constant $\tilde{S}_0$ and sixth and higher–order terms (this can be done when the layer $|k-k_F|<k_0$ is narrow), we retain

only the second– and fourth–order terms $\widetilde{S}_2$ and $\widetilde{S}_4$ describing noninteracting quasiparticles near the Fermi– surface and their paired interaction respectively:

$$\widetilde{S}_2 \approx \sum_{s,p} Z^{-1}\left[i\omega - c_F(k-k_F) + s\mu H\right]a_s^+(p)a_s(p) \tag{24.6}$$

$\widetilde{S}_4$ is different for different types of pairing, so starting from here we should split our consideration.

24.2. *p*–Pairing

Case of p–pairing has been considered by us[1,2,8] for superfluids in Chapter V. Here we shortly remind the derivation of the effective action functional for this case and later we will discuss the difference between models of p–pairing for superfluids and superconductors. In later case we should take into account the lattice symmetry.

In the case of triplet pairing $\widetilde{S}_4$ is equal to

$$\widetilde{S}_4 = -(\beta V)^{-1} \sum_{p_1+p_2=p_3+p_4} t_0(p_1,p_2,p_3,p_4)a_+^+(p_1)a_-^+(p_2)a_-(p_4)a_+(p_3) -$$
$$-(2\beta V)^{-1} \sum_{p_1+p_2=p_3+p_4} t_1(p_1,p_2,p_3,p_4)[\ 2a_+^+(p_1)a_-^+(p_2)a_-(p_4)a_+(p_3) +$$
$$a_+^+(p_1)a_+^+(p_2)a_+(p_4)a_+(p_3) + a_-^+(p_1)a_-^+(p_2)a_-(p_4)a_-(p_3)\ ], \tag{24.7}$$

where $p=(\mathbf{k},\omega)$ is 4–momentum, $t_0(p_i)$ is symmetric, $t_1(p_i)$ is antisymmetric scattering amplitudes under exchanges $p_1 \leftrightarrow p_2$; $p_3 \leftrightarrow p_4$. In the vicinity of the Fermi–sphere one could put $\omega_i=0$, $\mathbf{k}_i=k_F\,\boldsymbol{n}_i$ (i=1,2,3,4). Amplitudes t_0 and t_1 should depend upon just two invariants, let us say on $(\boldsymbol{n}_1, \boldsymbol{n}_2)$ and $(\boldsymbol{n}_1-\boldsymbol{n}_2, \boldsymbol{n}_3-\boldsymbol{n}_4)$ with t_0 even and t_1 odd with respect to second invariant.

Thus we could write $t_0=f((\boldsymbol{n}_1, \boldsymbol{n}_2), (\boldsymbol{n}_1-\boldsymbol{n}_2, \boldsymbol{n}_3-\boldsymbol{n}_4))$; $t_1=(\boldsymbol{n}_1-\boldsymbol{n}_2, \boldsymbol{n}_3-\boldsymbol{n}_4)$ $g((\boldsymbol{n}_1, \boldsymbol{n}_2), (\boldsymbol{n}_1-\boldsymbol{n}_2, \boldsymbol{n}_3-\boldsymbol{n}_4))$, where f and g are even with respect to second argument. We will consider model with $f=0$, g=const<0.

In order to go over to Bose–fields describing Cooper pairs of quasi–fermions at the Fermi–surface, we introduce the Gaussian integral over Bose–fields c_{ia}, c_{ia}^{+} in the integrand of the integral over Fermi–fields:

$$\int dc_{ia}^{+}dc_{ia}\left[g^{-1}\sum_{p,i,,a}c_{ia}^{+}(p)c_{ia}(p)\right] \tag{24.8}$$

Carrying out a shift in Bose–fields by the quadratic form of slow fields annihilating the fourth–order form, we obtain a Gaussian integral over slow fields. Evaluation of this integral leads to the required effective functional of action

$$S_{\text{eff}} = g^{-1}\sum_{p,i,a} c_{ia}^{+}(p)c_{ia}(p)+\frac{1}{2}\ln\ \det\frac{\hat{M}\left(c_{ia},c_{ia}^{+}\right)}{\hat{M}\left(c_{ia}^{(0)},c_{ia}^{(0)+}\right)} \tag{24.9}$$

where $c_{ia}^{(0)}$ is the condensate value of Bose–fields c_{ia} and $\hat{M}\left(c_{ia},c_{ia}^{+}\right)$ is the 4×4 matrix depending on Bose–fields and parameters of quasi–fermions

$$M_{11}=Z^{-1}[i\omega+\xi-\mu(\mathbf{H\sigma})]\delta_{p_1p_2},$$

$$M_{22}=Z^{-1}[-i\omega+\xi+\mu(\mathbf{H\sigma})]\delta_{p_1p_2},$$

$$M_{12}=(\beta V)^{-1/2}(n_{1i}-n_{2i})c_{ia}(p_1+p_2)\sigma_a,$$

$$M_{21} = -(\beta V)^{-1/2}(n_{1i} - n_{2i})c^{+}_{ia}(p_1 + p_2)\sigma_a . \qquad (24.10)$$

24.3. *d*–Pairing

In the case of singlet pairing $\tilde{S}_4$ is equal to

$$\tilde{S}_4 = -(\beta V)^{-1} \sum_{p_1+p_2=p_3+p_4} t(p_1, p_2, p_3, p_4) a^{+}_{+}(p_1) a^{+}_{-}(p_2) a_{-}(p_4) a_{+}(p_3) . \qquad (24.11)$$

The first version of the model of d–pairing in SC constructed by the method of functional integration was proposed by Brusov and Brusova[3] in 1994 when the idea of d–pairing in HTSC compounds was just put forth. We shall describe below an improved self–consistent model of SC with d–pairing and apply it for analyzing the collective mode spectrum (CMS). It should be noted that a similar model has been constructed by Brusov and Popov for *p*–pairing in superfluid ^{3}He.[1,2,8]

In the case of *d*–pairing, we have[3-8]

$$t(p_1, p_2, p_3, p_4) = V(\hat{k}, \hat{k}') = \sum_{m=-2}^{2} g_m Y_{2m}(\hat{k}) Y^{*}_{2m}(\hat{k}') . \qquad (24.12)$$

Here, $k_1 = k + q/2$, $k_2 = -k + q/2$, $k_3 = k' + q/2$, $k_4 = -k' + q/2$, $Y_{2m}(\hat{k})$ are spherical harmonics with $l = 2$.

In the case of spherical symmetry considered here, we have only one coupling constant g. Less symmetric cases necessitate the introduction of several coupling constants $g_{|m|}$ (the number of such constants is two in the case of a cubic lattice and three in the case of a hexagonal lattice).

In order to go over to Bose–fields describing Cooper pairs of quasi–fermions at the Fermi–surface, we introduce the Gaussian integral over Bose–fields c_{ia}, c^{+}_{ia} in the integrand of the integral over Fermi–fields:

$$\int dc_{ia}^{+} dc_{ia} \left[g^{-1} \sum_{p,i,,a} c_{ia}^{+}(p) c_{ia}(p) \right]. \tag{24.13}$$

In the case of *d*–pairing, c_{ia} are symmetric traceless 3x3 matrices and the number of the degrees of freedom c_{ia} is $2 \cdot (6-1) = 10$ in contrast to *p*–pairing for which we have an arbitrary 3×3 matrix, and the number of degrees of freedom c_{ia} is $2 \cdot 3 \cdot 3 = 18$. It should be noted that the number of degrees of freedom of the order parameter is equal to the number of collective modes in each superconducting phase. Carrying out a shift in Bose–fields by the quadratic form of slow fields annihilating the fourth–order form, we obtain a Gaussian integral over slow fields. Evaluation of this integral leads to the required effective functional of action

$$S_{\text{eff}} = g^{-1} \sum_{p,i,a} c_{ia}^{+}(p) c_{ia}(p) + \frac{1}{2} \ln \ \det \frac{\hat{M}\left(c_{ia}, c_{ia}^{+}\right)}{\hat{M}\left(c_{ia}^{(0)}, c_{ia}^{(0)+}\right)}, \tag{24.14}$$

where $c_{ia}^{(0)}$ is the condensate value of Bose–fields c_{ia} and $\hat{M}\left(c_{ia}, c_{ia}^{+}\right)$ is the 4×4 matrix depending on Bose–fields and parameters of quasi–Fermions.

It was mentioned above that the number of degrees of freedom in the case of *d*–pairing is equal to 10, i.e., we must have five complex canonical variables, which can be naturally chosen in the form

$$c_1 = c_{11} + c_{22},\ c_2 = c_{11} - c_{22},\ c_3 = c_{12} + c_{21},\ c_4 = c_{13} + c_{31},$$
$$c_5 = c_{23} + c_{32}.$$

In the canonical variables, the effective action has the form

$$S_{\text{eff}} = (2g)^{-1}\sum_{p,j} c_j^+(p)c_j(p)(1+2\delta_{j1}) + \frac{1}{2}\ln\ \det\frac{\hat{M}\left(c_j^+,c_j\right)}{\hat{M}\left(c_j^{+(0)},c_j^{(0)}\right)},$$

(24.15)

where

$$M_{11} = Z^{-1}[i\omega + \xi - \mu(\mathbf{H}\boldsymbol{\sigma})]\delta_{p_1 p_2},$$

$$M_{22} = Z^{-1}[-i\omega + \xi + \mu(\mathbf{H}\boldsymbol{\sigma})]\delta_{p_1 p_2},$$

$$M_{12} = M_{21}^* = (\beta V)^{-1/2}\left(\frac{15}{32\pi}\right)^{1/2}\left[c_1\left(1-3\cos^2\theta\right) + \right.$$

$$\left. + \ c_2\sin^2\theta\cos^2\varphi + c_3\sin^2\theta\sin 2\varphi + c_4\sin 2\theta\cos\varphi + c_5\sin 2\theta\sin\varphi\right].$$

(24.16)

This functional determines all the properties of the model superconducting Fermi–system with *d*–pairing[3-8] and in particular the spectrum of collective excitations[9,10].

Chapter XXV

High Temperature Superconductors (HTSC) and Their Physical Properties

25.1. The Discovery of HTSC

HTSC has been discovered by Bednorz and Muller (BM) in 1986.[2] Below we describe shortly their search of HTSC, following their Nobel Prize lecture[3] (see also monograph by Peter Brusov[1]).

The observation of the development of the increase of T_c, shown in Fig. 25.1, 25.2 would naturally lead to the conviction that intermetallic compounds should not be pursued any further. This because since 1973 the highest T_c of 23.3 K could not be raised. But nevertheless, the fact that SC had been observed in several complex oxides evoked BM special interest.

The second oxide after $SrTiO_3$ to exhibit surprisingly high T_c's of 13 K was discovered in the Li–Ti–O system by Johnstone et al.[4] in 1973. Their multiphase samples contained a $Li_{1+x}Ti_{2-x}O_4$ spinel responsible for the high T_c. Owing to the presence of different phases and difficulties in preparation, the general interest remained low, especially as Sleight et al.[5] in 1975 discovered the $BaPb_{1-x}Bi_xO_3$ perovskite also exhibiting a T_c of 13 K. This compound could easily be prepared as a single phase and even thin films for device applications could be grown, a fact that triggered increased activities in the United States and Japan. According to the BCS theory

$$kT_c = 1.13\hbar\omega_D \exp\left\{-1/N(0)V^*\right\} \tag{25.1}$$

both mixed–valent oxides, having a low carrier density $n = 4\times10^{21}/\mathrm{cm}^3$ and a comparatively low density of states per unit cell $N(0)$ at the Fermi–level, should have a large e–ph coupling constant V^*, leading to the high T_c's. Subsequently, attempts were made to raise the T_c in the perovskite by increasing $N(0)$ via changing the Pb:Bi ratio, but the compound underwent a metal–insulator transition with a different structure; thus these attempts failed.

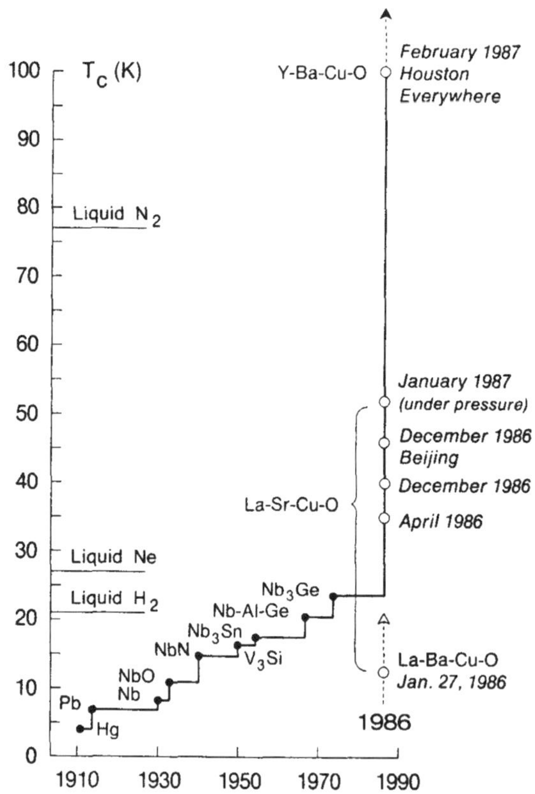

FIG. 25.1. Evolution of the superconductive transition temperature subsequent to the discovery of the phenomenon[53].

BM expected other metallic oxides to exist where even higher T_c's could be reached by increasing $N(0)$ and/or the e–ph coupling. Possibly they could enhance the latter by polaron formation as proposed theoretically by Chakraverty[6] or by the introduction of mixed valencies. The intuitive phase diagram of the coupling constant $\lambda = N(0)V^*$ versus T proposed by Chakraverty[6] for polaronic contributions is shown in Fig. 25.2. There are three phases, a metallic one for small λ, and an insulating bipolaronic one for large λ, with a SC phase between them, i.e., a metal–insulator transition occurs for large λ. For intermediate λ, a HTSC might be expected. The question was, in which systems to look for SC transitions.

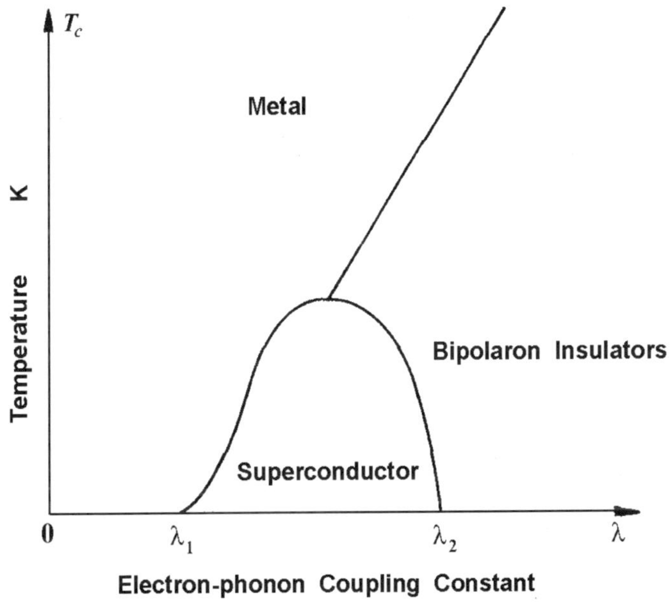

FIG. 25.2. Phase diagram as a function of electron–phonon coupling strength[6].

The guiding idea in developing the concept of the HTSC search was influenced by the Jahn–Teller (JT) polaron model, as studied in a linear chain model for narrow–band intermetallic compounds by Höck et al.

The JT theorem is well known in the chemistry of complex units. A nonlinear molecule or a molecular complex exhibiting an electronic degeneracy will spontaneously distort to remove or reduce this degeneracy. Complexes containing specific transition–metal (TM) central ions with special valency show this effect. In the linear chain model, for small JT distortions with a stabilization energy E_{JT} smaller than the bandwidth of the metal, only a slight perturbation of the traveling electrons is present. With increasing E_{JT}, the tendency to localization is enhanced, and for E_{JT} being of the magnitude of the bandwidth, the formation of JT polarons was proposed.

These composites of an electron and a surrounding lattice distortion with a high effective mass can travel through the lattice as a whole, and a strong e–ph coupling exists. In BM opinion, this model could realize the Chakraverty phase diagram. Based on the experience from studies of isolated JT ions in the perovskite insulators, their assumption was that the model would also apply to the oxide, their field of experience, if oxides could be turned into conductors. BM knew there were many of them. Oxides containing TM ions with partially filled e_g orbitals, like Ni^{3+}, Fe^{4+}, or Cu^{2+}, exhibit a strong JT effect, and BM considered these as possible candidates for new superconductors.

They started the search for HTSC in late summer 1983 with the La–Ni–O system. $LaNiO_3$ is a metallic conductor with the transfer energy of the JT–e_g electrons larger than the JT stabilization energy, and thus the JT distortion of the oxygen octahedra surrounding the Ni^{3+} is suppressed. However, already the preparation of the pure compound brought some surprises, as the material obtained by their standard coprecipitation method and subsequent solid–state reaction turned out to be sensitive not only to the chemicals involved but also to the reaction T. Having overcome all difficulties with the pure compound, BM started to partially substitute the trivalent Ni by trivalent Al to reduce the metallic bandwidth of the Ni ions and make it comparable to the Ni^{3+} JT stabilization energy. With increasing Al concentration, the metallic characteristics of the pure $LaNiO_3$ gradually changed, first giving a general increase in the resistivity and finally with high substitution leading to a semiconducting behavior with a transition to localization at low T. The idea did not seem to work out the way BM had thought, so they considered the introduction of some internal strain within the $LaNiO_3$ lattice to reduce the bandwidth. This was realized by replacing the La^{3+} ion by the smaller Y^{3+} ion, keeping the Ni site unaffected. The resistance behaviour changed in a way they had already recorded in the previous case, and at that point BM started wondering whether the target at which they were aiming really did exist.

Some hopes appeared with new idea, involving another TM element encountered in BM search, namely, copper. In a new series of compounds, partial replacement of the JT Ni^{3+} by the non–JT Cu^{3+} increased the absolute value of the resistance, however, the metallic character of the solid solutions was preserved down to 4 K. But again, they observed no indication of SC. The time to study the literature and reflect on the past had arrived.

It was in late 1985 that the turning point was reached. BM became aware of an article by the French scientists Michel, Er–Rakho, and Raveau, who had investigated a Ba–La–Cu oxide with perovskite structure exhibiting metallic conductivity in the temperature range between 300°C and –100°C. The special interest of that group was the catalytic properties of oxygen–deficient compounds at elevated temperatures. In the Ba–La–Cu oxide with a perovskite–type structure containing Cu in two different valencies, all BM's concept requirements seemed to be fulfilled.

They started preparations for a series of solid solutions, as by varying the Ba/La ratio one would have a sensitive tool to continuously tune the mixed valency of copper. The synthesis had been performed.

When performing the four–point resistivity measurement, the temperature dependence did not seem to be anything special when compared with the dozens of samples measured earlier. During cooling, however, a metallic–like decrease was first observed, followed by an increase at low temperatures, indicating a transition to localization. A sudden resistivity drop of 50% occurred at 11 K. Was this the first indication of SC?

Repeated measurements showed perfect reproducibility and an error could be excluded. Compositions as well as thermal treatment were varied and within two weeks BM were able to shift the onset of the resistivity drop to 30 K. This was an incredibly high value compared to the highest T_c in the Nb_3Ge SC.

BM found strong indications that the Ba–containing La_2CuO_4 was the phase responsible for the SC transition in their samples. Starting from the orthorhombically distorted host lattice, increasing the Ba substitution led to a continuous variation of the lattice toward a tetragonal unit cell. The highest T_c's were obtained with a Ba concentration close to this transition, whereas when the perovskite phase became dominant, the transition was suppressed and the samples showed only metallic characteristics.

The magnetic measurements demonstrated that within samples showing a resistivity drop, a transition from para– to dia–magnetism occurred at slightly lower temperatures, indicating that SC–related shielding currents existed. The diamagnetic transition started below what is presumably the highest T_c in the samples as indicated by theories describing the behavior of percolative SC. In all samples the transition to the diamagnetic state was systematically related to the results of BM's resistivity measurements. The final proof of SC, the presence of the Meissner–Ochsenfeld effect, had been demonstrated. Combining the x–ray analysis, resistivity and susceptibility measurements, it was possible to clearly identify the Ba–doped La_2CuO_4 as the SC compound.

Interesting properties related to the behavior of a spin glass were found. BM then intensively studied the magnetic field and time dependences of the magnetization, before finally starting to realize an obvious idea, namely, to replace La also by other alkaline–earth elements like Sr and Ca. Especially Sr^{2+} had the same ionic radius as La^{3+}. BM began experiments on the new materials which indicated that for the Sr–substituted samples T_c was approaching 40 K and the diamagnetism was even higher. The group of Tanaka at the University of Tokyo, had repeated BM's experiments and could confirm their result (November, 1986). Chu at the University of Houston also convinced that within the Ba–La–Cu–O system SC occurred at 35 K. Colleagues who had not paid any attention to BM's work at all suddenly became alert. By applying hydrostatic pressure to the samples, Chu was able to shift the SC transition from 35 to almost 50 K in 1987[7]. Modification of the original oxides by introducing the smaller Y^{3+} for the larger La^{3+} resulted in a giant jump of T_c to 92 K in multiphase samples.[8-10] At a breathtaking pace, dozens of groups repeated these experiments, and after an effort of only a few days the new SC compound could be isolated and identified. The resistive transition in the new $YBa_2Cu_3O_7$ compound was complete at 92 K, and even more impressive was the fact that the Meissner effect could now be demonstrated without any experimental difficulties with liquid nitrogen as the coolant. Within a few months, the field of SC had experienced a tremendous revival, with an explosive development of T_c's which nobody can predict where it will end.

An early account of the discovery appeared in the September 4, 1987, issue of Science, which was dedicated to science in Europe.

25.2. Physical Properties of HTSC

In this section we will describe very briefly the classes of HTSC synthesized to the moment, and some of their basic properties like structure and phase diagram, isotope effect and value of magnetic flux, normal state properties including the spin–gap phenomena, etc. Most of the other properties concerning HTSC will be described by us in proper parts of consequent Chapters. Here we will consider also the band structure of HTSC and some simple microscopic models like tight–binding as well as the three–band Hubbard model.

25.2.1. Some experimental data

HTSC classes

Up to now there are nine groups of HTSC.

1. $La_{2-x}(Sr,Ba,Ca)_xCuO_4$, which are obtained from La_2CuO_4, or 2–1–4, by doping by Sr, Ba or Ca, T_c=39 K in $La_{1.85}Sr_{0.15}CuO_4$

2. $YBa_2Cu_3O_{6+x}$, or 1–2–3, T_c=92 K at x=0

3. Bi–Sr–Ca–Cu–O, T_c=110 K for $Bi_2Sr_2Ca_2Cu_3O_x$
4. Tl–Ca–Ba–Cu–O, T_c=125 K for $Tl_2Ca_2Ba_2Cu_3O_{10}$
5. $HgBa_2CuO_{4+\delta}$, T_c=156 K
6. $Nd_{2-x}Ce_xCuO_4$, T_c=24 K at x=0.15

The feature of these compounds is that the charge carriers are electrons, while in other compounds — holes.

7. $Ba_{1-x}K_xBiO_{3-\delta}$, T_c=29.5 K at x=0.4

This is the first 3D perovskite superconductor without Cu.

8. The fullerene superconductors:
Cs_3C_{60}, T_c=40 K, $RbCs_2C_{60}$(33 K), K_3C_{60}(19.7 K), Rb_3C_{60}(30 K)

9. Sr_2RuO_4

This compound provides the first example of a layered perovskite material that exhibits SC without Cu. In the 3D analog of Sr_2RuO_4 – $SrRuO_3$ – there is a metallic FM phase. These observation indicate that an odd–parity ($l = 1$) SC state is likely. It has been proposed that an exotic nonunitary SC state similar to the ^{3}He–A_1 phase is realized.

Structure

Despite the apparent complexity there is a striking configurational similarity of undoped parent compounds of the HTSC. All HTSC consist of 2D CuO_2 planes which are sandwiched between intervening atomic layers. These layers are composed mostly out of alkaline–earths, rare–earths, oxygen and halogenides.[52] Depending on the number of CuO_2 planes per unit cell the materials exist in a single–, a double–, or a triple–plane form as in La_2CuO_4, in $YBa_2Cu_3O_6$, or in $Bi_2Sr_2YCu_3O_8$, respectively. The CuO_2 planes are widely accepted to host those electronic excitations which are most relevant to the SC. The intervening layers are viewed as 'inert' charge reservoirs. The actual 3D oxygen coordination of a particular in–plane Cu atom may vary. Planar CuO_4 plaquettes are found, as well as CuO_5 tetrahedra, and CuO_6 octahedra. In the latter so–called apical oxygen atoms are located above and/or below the CuO_2 plane.

HTSC generally have a TG crystal structure, but at low T many of them undergo transformation towards an OR structure in which the CuO_2 planes are buckled.[11] In $YBa_2Cu_3O_6$ the in–plane copper sites have a CuO_5 tetrahedral coordination and form a square lattice with a lattice constant of 3.8 Å . The in–plane Cu–O bond length is 1.9 Å . This may be compared to the distance between the in–plane Cu atom and the apical O which is 2.3 Å and therefore leads to a relatively small overlap. In $YBa_2Cu_3O_6$ additional out–of–plane Cu atoms exist. They lead to the formation of Cu–O chains which are relevant in the doped material.

It is experimentally well established that the parent compounds La_2CuO_4,[12] Nd_2CuO_4 [13] and $YBa_2Cu_3O_6$ [14] are *insulators* and ordered *antiferromagnetically* (AF) below a Neel temperature T_N .

The localized Cu spins provide the magnetic moments for the AF order. The in–plane AF exchange coupling $J_{\parallel}$ is generated by a Cu superexchange and the undoped CuO_2 plane is well described by 2D spin–$\frac{1}{2}$ AF Heisenberg model. Typically $J_{\parallel} \sim 0.1$ eV as determined by inelastic neutron scattering[15,16] and by Raman scattering.[17] The direction of the ordered Cu spins has been found to be perpendicular to the crystal c–axis. In the basal plane the Cu spins are oriented at a 45° angle to the Cu–O bonds. Long–range AF order at finite T requires an inter–plane coupling $J_{\perp}$. The c–axis anisotropy however is extremely large $J_{\perp}/J_{\parallel} \sim 10^5$.[18,19] Therefore, above the Neel temperature, which is $T_N \approx 325$ K for La_2CuO_4 [20] ($T_N \approx 415$ K for $YBa_2Cu_3O_6$), the spin correlations are essentially 2D. The full $Cu3d^9$ magnetic moment is reduced by approximately 40% due quantum spin fluctuations and due to Cu–O covalency.[15,21-24]

The actual 3D spin structure at $T = 0$ can be intricate. In La_2CuO_4 the inter planar frustration of the AF exchange is lifted by a TG to OR transition. This leads to an additional Dzyaloshinski–Moria spin interaction $H_{DM} = \sum_{\rangle ij \langle} \mathbf{D}_{ij} \cdot \mathbf{S}_i \times \mathbf{S}_j$, which influences the 3D spin order.[18,25-27] In fact, a tiny net in–plane FM moment arises which in turn is AF coupled between adjacent planes by $J_{\perp}$.

The phase diagram

Parent compounds can be "doped" which eventually leads to metallic behavior and SC. Doping is achieved either by heterovalent substitution as in $La_{2-x}Sr_xCuO_4$, in $Nd_{2-x}Ce_xCuO_4$, and in $Bi_2Sr_2Ca_{1-x}Y_xCu_2O_8$ or by a variation of the total O content as in $YBa_2Cu_3O_{6+x}$. Doping introduces additional charge carriers into the CuO_2 planes.[28] In $La_{2-x}Sr_xCuO_4$ the formal valency of strontium is Sr^{2+}. To achieve charge neutrality in this case, the in–plane O atoms change from $O2p^6$ into an $O2p^5$ states leaving additional holes in the planes. In $Nd_{2-x}Ce_xCuO_4$ the Nd^{3+} state is replaced by a cerium Ce^{4+} state yielding an excess electron which enters the planes. The sign of the measured Hall coefficient supports the picture of hole carriers in $La_{2-x}Sr_xCuO_4$ and of electron carriers in $Nd_{2-x}Ce_xCuO_4$.[29]

In $YBa_2Cu_3O_{6+x}$ doping amounts to the addition of O atoms in between the out–of–plane copper atoms. This leads to the formation of Cu–O chains. It is believed that these O atoms form O^{2-} states by absorbing electrons from the rest of the material and, to a certain extent, from out of the CuO_2 planes. This process is still controversial. In particular, it introduces the possibility of hole–type carriers in the Cu–O chains, and it is certainly responsible for the nonlinear relation between the dopant concentration and the in–plane carrier density in $YBa_2Cu_3O_{6+x}$.[30]

In Fig. 25.3 we show the phase diagrams of $La_{2-x}Sr_xCuO_4$ and of $Nd_{2-x}Ce_xCuO_4$ (one of $YBa_2Cu_3O_{6+x}$ is similar to 2–1–4) in the T versus x plane. Evidently, depending on T and x the cuprate HTSC can be varied continuously between an AF insulator, a spin glass phase, HTSC and a "normal" metal. Since none of this is accompanied by any major structural changes of the materials, it is quite natural to expect electronic correlations to play a significant role in these systems.[31,32] The notion of overdoped or underdoped cuprates is frequently used to define the region of doping above, or below a so–called "optimal" doping concentration at which the SC T_c is highest. In $YBa_2Cu_3O_{6+x}$, there is a so–called 60K plateau in the T_c versus x curve which is a result of the nonlinear increase of the in–plane carrier density.

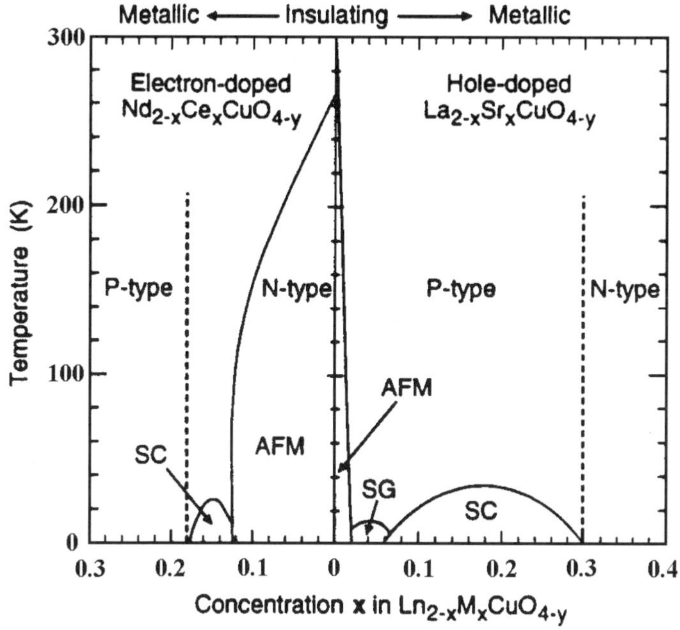

FIG. 25.3. T versus x phase diagram of $La_{2-x}Sr_xCuO_4$ combined with $Nd_{2-x}Ce_xCuO_4$.

Obviously, holes and electrons doped into the AF insulators are very efficient in destroying the AF order. In $La_{2-x}Sr_xCuO_4$ only ~2% of holes are sufficient to achieve this while ~12% of electrons are required in $Nd_{2-x}Ce_xCuO_4$. Therefore, the charge carriers seem to couple strongly to the spin systems. However, there is an apparent asymmetry between the electron and the hole–doped systems regarding the stability of AF as well as that of SC.

Normal–state properties

It is widely believed that understanding the normal–state properties of the HTSC will also shed light on the SC mechanism. The basis for this expectation resides in the unusual normal–state properties of these materials. For example, strong anisotropies are observed, mainly caused by the 2D nature of the problem, and magnetic phases exist close to the SC regions. In addition, there are properties of the HTSC that have raised the possibility of observing deviations from a Fermi–liquid (FL) description of the normal state. For example, we know that in a FL metal the magnetic susceptibility and the Hall coefficient are T independent, the resistivity ρ grows like T^2 at low T and the NMR relaxation $1/T_1 \sim T$. These behaviors have *not* been observed in the HTSC although a FL description is still not ruled out.[33]

Resistivity

Let us briefly describe the normal–state property of the HTSC that is more frequently mentioned as indicative of an unusual normal state, namely, the experimentally observed linear T dependence of the ρ,[34,35] while in conventional low–T SC, it has been experimentally observed that $\rho \approx a + bT^5$, at low T (but larger than T_c). This temperature dependence arises from the scattering of electrons with phonons. At higher temperatures, a linear behavior is expected, and the interpolation between the two regimes is given by the Grüneisen–Bloch formula.[36] The residual resistivity a at $T = 0$ is caused by scattering with magnetic impurities, point or line defects, other electrons, etc. It is important to note that in those materials where the composition can be changed easily by chemical doping, the linear behavior $\rho \sim T$ is observed *only* in a narrow carrier–concentration window near the "optimal" composition, i.e., those corresponding to the highest critical temperatures.[37,38] This detail is not sufficiently remarked upon in the literature. In single crystals of $La_{2-x}Sr_xCuO_4$ contrary to the linear dependence $\rho \sim T$ at $x = 0.15$, under– and overdoped samples follow a different power–law behavior over a wide range. Then, the temperature dependence of the resistivity is more involved than what theorists usually believe. In addition, the electron–doped HTSC do *not* show a linear behavior of the resistivity with temperature, but rather a quadratic dependence[39,40] $\rho \sim T^2$. The T^2 dependence is consistent with e–e scattering in a FL, and thus a luck of universality seems to exist between hole– and electron–doped materials. The behavior of the resistivity with temperature is not the only "anomalous" property of the normal state of the HTSC. Puzzling results have been observed in the optical conductivity $\sigma(\omega)$, Raman scattering, measurements of the Hall coefficient ($R_H \sim T^{-1}$ in YBCO compounds) and several others. While in simple models of weakly interacting electrons it is expected that R_H would be approximately constant, the experimental results for the cuprates show a strong $1/T$ temperature dependence. The presence of a "spin–gap" in YBCO is another normal–state property that deserves considerable attention.

Spin gap

The spin gap phenomenon has attracted a lot of attention. The magnetic susceptibility decreasing with decreasing T and the normal–state maximum in the T dependence of Cu relaxation rates are characteristic anomalies in underdoped 123 materials (as well as in the similarly underdoped but stoichiometric $YBa_2Cu_4O_8$ compound[41] and in $BiSr_2Ca_2Cu_20_8$.[42] This is taken as evidence for a gap to low–lying spin excitations developing in the normal state. The same phenomenon of a spin–pseudogap has also been observed in neutron scattering. When first mentioned in the literature[43] suggestions have been made that the spin gap which appears already significantly above T_c results from precursive SC pair formation.[44] A scenario, however, where the SC transition is viewed as Bose condensation of preformed Cooper pairs is not supported by experiment. Measurements of the work function have shown that the behavior of the chemical potential at and below T_c is not compatible with what is expected for Bose condensation. The nature of the transition is in fact much closer to BCS behavior with modifications arising from the anomalously short coherence length in the HTSC.

Further conclusive evidence against a SC precursor effect as the origin for the anomalous spin gap has come from measurements on Zn substituted 123 O_{6+x}.[45] While the Zn substitution for Cu substantially reduces the T_c, the homogeneous spin susceptibility χ_s and hence the spin gap feature is only weakly modified proving that it is of a different origin.

It is important to stress that the spin–gap behavior disappears for the optimally doped 123 O_7 compound and does not occur for overdoped 1–2–3 materials.[46] In these compounds the gap for charge and spin excitations opens at the same T_c. The existence of apparently two different T for the onsets of the gaps for spin and charge excitations in the underdoped region has been speculated to result from the separation of spin and charge.[47]

However, another possible hint to the origin of the spin gap, that it is not an intrinsic property of the planar spin dynamics but rather results from the AF coupling of layers which is stronger in the oxygen–deficient compounds closest to the composition of the insulator.[48,49] The open question remains whether the intra–bilayer coupling is strong enough to explain the observed spin–pseudogap feature.

A different point of view has been taken by Sokol and Pines[50] who argue that the appearance of the spin gap is not related to the presence of CuO_2–bilayers. Rather it is suggested to reflect a crossover in the spin dynamics from an overdamped to a quantum disordered regime.[51]

The ratio $2\Delta/kT_c$

The value of this ratio has a great interest because it allows to make a conclusion whether HTSC are SC with strong or weak coupling. In $Ba_{1-x}K_xBiO_3$ $2\Delta/kT_c = 3.5 \pm 0.4$ which is consistent with the BCS value of 3.52 for weak coupling SC. In all other HTSC this value is much higher (of order 6–8) that testifies about strong coupling in these materials.

Charge $2e$

The carriers of SC current have charge $2e$. This fact has been established for ceramics as well as for crystal samples by measurement of magnetic flux which turns out to be equal to $h/2e$. The Josephson effect measurements have given the same result.

The carriers in SC state are holes, that follows from the positivity of Hall constant R_H and sign of thermopower. The carrier concentration is quite low: of order $(6–10)\cdot 10^{21}$ cm^{-3} in 2–1–4 and $1.5\cdot 10^{21}$ cm^{-3} in 1–2–3.

Isotope effect

There is zero isotope effect in 1–2–3 under substitution $^{135}Ba \to {}^{138}Ba$ and $^{63}Cu \to {}^{65}Cu$. Under substitution $^{18}O \to {}^{16}O$ $\Delta T_c = 0.3 - 0.5$ K. In 2–1–4 $0.14 < \alpha < 0.35$. In $Bi_2Sr_2Ca_2Cu_3O_x$ for 75 K phase $\Delta T_c = 0.34$ K, for 110 K phase $\Delta T_c = 0.32$ K.

So the isotope effect in HTSC is vanishing or much less than in ordinary SC, that testifies against phonon mediate mechanism of SC.

Only in $Ba_{0.6}K_{0.4}BiO_3$ the substitution $^{18}O \to {}^{16}O$ exhibits $\alpha = 0.42 \pm 0.05$, that is close to ordinary value 0.5.

Spin correlations

Neutron scattering experiments on 1–2–3 and 2–1–4 show that for small doping ($x \le 0.04$ in doped 2–1–4) the AF long–range order is replaced by commensurate short range spin correlations. Experiments have proved that dynamical AF correlations persist even into SC state.

The momentum width of the magnetic Bragg peaks is a direct measure for spin–spin correlation length. In SG state correlation length $\xi \sim 1/\sqrt{x}$ corresponding to the average separation between the holes in the CuO_2 plane as introduced by Sr doping. In the metallic state (for x >5%) ξ becomes considerably larger than the mean spacing $3.8Å/\sqrt{x}$ between carriers. The enhancement over the $1/\sqrt{x}$ behavior has been interpreted to indicate an excellent screening of the donor impurities by the charge carriers in the CuO_2 planes, but it remains puzzling that the correlation length is increasing with hole concentration. In 2–1–4 $\xi = 42Å$ $(x = 0.04)$ and $12Å$ $(x = 0.15)$ while in 1–2–3 $\xi = (9 \pm 0.2)Å$ $(x = 0.5)$ and $3.3Å$ $(x = 0.92)$.

Chapter XXVI

Symmetry of Order Parameter in HTSC

26.1. Introduction

Since the discovery of the HTSC the determination of the pairing state has been a key issue for both experiment and theory. Since the pairing state reflects the fundamental broken symmetries at the PT at T_c, the problem is central to clarifying the microscopic mechanism of superconductivity. Different types of pairing are compatible with different pairing interactions. For example, interaction via AF spin fluctuations only leads to $d_{x^2-y^2}$ pairing, and is never attractive for s–wave pairing. While the interaction via charge fluctuations can lead to either s– or d–wave pairing. The pairing state symmetry also affects key physical properties, such as penetration depth, optical response, sensitivity to impurities, and the macroscopic Ginzburg–Landau level behavior. Even commercial applications of high T_c materials may be affected by pairing symmetry issues. For example a d–wave superconductor will have a finite resistivity at microwave frequencies even if there are no weak links present.

There is now a large body of experimental work indicating a highly anisotropic energy gap $\Delta(k)$ in $YBa_2Cu_3O_{7-\delta}$(YBCO).[1] Furthermore there is also strong evidence that $\Delta(k)$ changes sign, as expected in a predominantly $d_{x^2-y^2}$ pairing state, and unlike some highly anisotropic s–wave states. Several novel SQUID and flux quantization experiments have also strongly supported the $d_{x^2-y^2}$ state, although there are, two prominent exceptions which we discuss below.

The pairing state in the other HTSC has been less thoroughly investigated, however there is good evidence that the Bi, Tl and Hg based cuprates are also predominantly $d_{x^2-y^2}$.[1,2] The situation for the La based materials is less clear cut, since few of the relevant measurements have been done. Only for the n–type Nd materials is there currently good evidence for a finite gap everywhere on the Fermi surface consistent with an isotropic s–wave state.[3]

Much of the recent attention has been focused on the $d_{x^2-y^2}$ pairing state. This state has two key properties that make it a feasible candidate for explaining the behavior of the cuprates. First, it exhibits nodes in the energy gap that lead to an excess of excitations at low T. The presence of such excitations has been firmly established by a host of transport, tunneling, and thermodynamic measurements over the past ten years. Second, this particular d–wave symmetry is implied by a number of possible SC pairing mechanisms, particularly those involving magnetic interactions that are known to be important in the cuprates. However many other pairing symmetries are also possible and have been promoted in the framework of theoretical models and experimental results.

In this Chapter we consider the symmetry classification of the HTSC, following to Annett[4] and Sigrist and Rice.[5] We consider the mixed symmetry states, such as $s+d$, $s+id$ $d_{x^2-y^2}+id_{xy}$ etc., the connection between pairing symmetry and pairing interactions. We describe the ideas of the experimental symmetry probes and analyze the data of these experiments. We develop the path integral model of d–pairing in HTSC (both for bulk systems and CuO_2–planes) and consider the collective properties of superconducting states, calculating the CMS for these systems.

The theoretical part of this chapter is based upon the papers of Annett, Annett *et al.*,[1,4,6-13] Sigrist and Rice (SR),[5] Brusov *et al.*[14-36]

26.1.1. Superconductivity and Broken Symmetry

Superconductivity is a manifestation of broken symmetry in nature. By this one means that the symmetry group, H, of the system's density matrix at T below the transition, is less than the group, G, of the Hamiltonian. The question is to decide which symmetries are broken at T_c in the HTSC. We know definitely that global gauge symmetry is broken at the superconducting phase transition, since Meissner flux expulsion, flux quantization (and hence persistent currents) and the Josephson effect all follow from broken global gauge symmetry alone.[37] Without broken gauge symmetry at most zero resistance could occur. "Spontaneously broken gauge symmetry" means that below T_c the wave function, or density matrix, of the system spontaneously develops a definite phase, ϕ, which can be treated as a thermodynamic variable. There are two caveats to this statement. Firstly, the phase does not commute with the particle number; however this is no problem if one works in an appropriate thermodynamic ensemble. Secondly the quantum mechanical phase itself is not a physical observable. What is observed is the phase rigidity, or macroscopic quantum coherence below T_c, which gives rise to phenomena such as the Meissner and Josephson effects. In conventional s–wave superconductors only gauge symmetry is broken. If the pairing is not conventional then some other symmetries of the Hamiltonian are broken below the transition. Symmetries which might be broken include lattice point and translation group operations and spin rotation symmetries in addition to global gauge symmetry. The aim here is to classify all the possible ways in which the symmetry could be broken, and hence to enumerate all the possible SC.

Since the superconducting transition is one of spontaneously broken symmetry it is possible to construct an order parameter which is zero in the normal state and non–zero in the superconducting state. The starting assumption that we shall make is therefore simply that there is some set of complex numbers, Δ_i, which are zero above T_c and become non–zero below. This is certainly the case in any BCS–type microscopic pairing theory where, for example, the numbers Δ_i may be taken as the values of the gap function $\Delta_{\alpha\beta}(\mathbf{k})$ at various points on the FS, where α and β are spin indices. Indeed one might expect this to be true even if BCS theory does not apply; for example Δ_i could represent a "holon" or holon pair Bose condensate,[38-40] or a Bose–condensate of tightly bound fermion pairs.[41-45] The second key assumption which it is necessary to make is that the numbers Δ_i behave continuously at the phase transition (i.e. that it is a second–order transition). From this it is natural to write the superconductor free energy density (or coarse–grained effective Hamiltonian or effective action) as the following general Ginzburg–Landau type functional[46]:

$$f[\Delta] = \alpha_{ij}\Delta_i^*\Delta_j + \beta_{ijkl}\Delta_i^*\Delta_j^*\Delta_k\Delta_l + K_{ijkl}\partial_i\Delta_j^*\partial_k\Delta_l + \ldots \qquad (26.1)$$

(in the absence of magnetic fields), where the expansion is in powers of Δ_i (which is small close to the transition by the assumption of continuity) and of ∂_i (assuming the order parameter is slowly varying). Eq. (26.1) also uses the fact that the free energy must be invariant under overall gauge transformations $\Delta_i \to \exp(i\phi)\Delta_i$. It will also be assumed that the free energy of Eq. (26.1) is bounded below without requiring higher–order terms in the functional. The free–energy density certainly has the form of Eq. (26.1) if the BCS pairing theory applies, as shown by Gor'kov. However, again this is so general that one might expect it to be true irrespective of the microscopic theory, simply provided that the phase transition is second–order. The existence of the order parameter or validity of Eq. (26.1), certainly would be true for the majority of microscopic models presented to date.

From this point it is possible to apply the powerful methods of group theory to the problem. In particular we can see immediately from Eq. (26.1) that physically distinct superconducting states correspond to different irreducible representations (IR) of the symmetry group. The proof follows simply from considering the symmetry properties of the matrix α_{ij}. The free energy of Eq. (26.1) is invariant under the following symmetry group:

$$G = G_0 \times U(1) \times T \qquad (26.2)$$

where $U(1)$ is the group of global GT, T is time reversal, and G_0 is the group of symmetries, such as the point group and spin rotation symmetries. A typical element of the group acting upon the Δ_i gives

$$\Delta_i \to R_{ij}(g_0)\exp(i\varphi)T^* \Delta_j$$

where $R_{ij}(g_0)$ is a matrix representing element $g_0 \in G_0$, T is $\Delta_i \to \Delta_i^*$ and $n = 0$ or 1. Invariance of the free energy under the group elements implies that α_{ij} is a Hermitian matrix, and that $\alpha_{ij} = R_{ik}(g_0)^+ \alpha_{kl} R_{lj}(g_0)$. The usual methods of group representation theory then imply that the matrix α_{ij} can be block–diagonalized, and there exists a basis for the Δ_i where each block is just a constant α^n times the identity matrix with each block corresponding to an IR of the group G_0. Now consider the T dependence of the matrix α_{ij}. At high T, the superconducting state is unstable and $\Delta_i = 0$ because the matrix α_{ij} is positive definite. As T is lowered, a non–zero order parameter will first occur when α_{ij} ceases to be positive definite, i.e. when one of the numbers α^n first becomes negative. For example, let us assume that the first to become negative is $\alpha^1 = \alpha(T - T_c)$ which becomes negative at T_c (and is assumed an analytic function of T at T_c). The eigenvectors, $\mathbf{x}^{(j)}$ for $j = 1$ to g, corresponding to this block transform according to an IR Γ of dimension g. In practice to find the eigenvectors $\mathbf{x}^{(j)}$ one has to solve a BCS gap equation close to T_c; however, this is unimportant here since we shall only be concerned with their symmetry properties. It follows from these arguments that, sufficiently close to T_c, the complex numbers Δ_i will be of the following form:

$$\Delta_i = \sum_j \eta_j x_i^{(j)}$$

and components along other eigenvectors will be small. Because the thermodynamic state of the system is then entirely determined by the set of g numbers η_i, these can be taken as the order parameter. The number of components of the order parameter is therefore given by the dimensionality of this IR Γ (assuming no accidental degeneracies); if it has g dimensions then the order parameter has g complex components or $n = 2g$ real components. This powerful and simple result means that we can classify all the possible OP, and their dimensionalities, purely by considering the different IR of the symmetry group, and with essentially no input from microscopic theory.

For each of this possible order parameter it is straightforward to construct the most general GL free energy. Firstly, rewriting the free energy (26.1) in terms of the order parameter η_i gives:

$$f[\eta] = \alpha(T - T_c)\eta_i^* \eta_i + \beta_{ijkl}\eta_i^* \eta_j^* \eta_k \eta_l + K_{ijkl}\partial_i \eta_j^* \partial_k \eta_l + \ldots \tag{26.3}$$

from which the quartic and gradient terms can be reduced to a small number of symmetry–distinct terms by a straightforward method.[49] Clearly, the quartic term transforms under group operations in G_0 according to the product representation

$$\Gamma^* \otimes \Gamma^* \otimes \Gamma \otimes \Gamma = \sum_i n_i \Gamma_i \tag{26.4}$$

where Γ^* is the conjugate representation to Γ, and on the right–hand side the product representation has been reduced to a sum of IR Γ_i occurring n_i times. If the identity representation, Γ_1, occurs n_1 times in this sum then the number of independent quartic invariants cannot be greater than n_1. In fact it may be less than n_1, because of additional relationships such as $\beta_{ijkl} - \beta_{jikl} = \beta_{ijlk}$. Similar considerations may be used to determine the number of distinct gradient terms that can occur. Below practical examples of the construction of quartic and gradient invariants will be given in this way when constructing each of the free energies that might be appropriate for the HTSC.

Finally, given the order parameter space, η_i for $i = 1$ to g, and the free energy one needs to construct the possible minima which can occur. Often this is achieved simply by direct calculus. However if the OP has a large number of components (for example, the order parameter for ^{3}He has nine complex components), this method is impractical and it becomes advantageous to use group theory again. This goes back to the original concept of broken symmetry, in that we classify the different subgroups, H, of the full group, G, that leave the GS invariant:

$$R_{ij}^{\Gamma}(h_0)\exp(i\phi)T^h\eta_j = \eta_i \tag{26.5}$$

where group element $h \in H$ is represented by $R_{ij}^{\Gamma}(h_0)\exp(i\phi)T^h$ with $h \in G_0$ and where $R_{ij}^{\Gamma}(h_0)$ is the matrix representing h_0 in Γ. Note that the subgroup H is only unique up to a conjugation gHg^{-1}, since for any minimum, η_i, an equivalent minimum, $R_{ij}(g)\eta_j$, exists for all elements g in G. H is usually denoted the residual symmetry group or the little group.[51] In general it may be proved[48] for quartic free energies such as (26.3) that the GS has some residual symmetries (i.e. that H is non–trivial). Indeed, for a wide class of symmetry groups (but not always), the subgroup H will be as large as possible, corresponding to the fewest possible broken symmetries. Group–theoretically, this means that H is a maximal subgroup of G. The maximal subgroups have a number of important properties. For example, it can be proved that states corresponding to a maximal subgroup will be stationary points of any function invariant under.[51] This implies that they are "inert" in that they are stationary points of each term in the free energy separately, and hence that the order parameter is unchanged by small changes in the coefficients of the quartic terms. The inert states are the most important class of states for ^{3}He, however, non–inert states do also occur for some parameter values.[49,50]

26.1.2. The symmetry group

To proceed with the classification of different superconducting order parameter and free energy minima for the HTSC, we need to know the appropriate symmetry group of the normal state. At least some input from microscopic theory is required to decide what group to use. For example the spin–bag theory[52] and some of the RVB theories[38-40,53-55] of HTSC have a normal state which breaks the full lattice symmetry. For the sake of simplicity, we will not consider these possibilities but merely assume that the normal stale has the conventional lattice and spin symmetries, corresponding to symmetry group:

$$G = SO(3)\times G_c \times U(1)\times T \tag{26.6}$$

where $SO(3)$ is the group of rotations in spin space, G_c is the crystal space group, $U(1)$ is the gauge group and T is time reversal symmetry. Eq. (26.6) assumes that the spin–orbit coupling is negligible.

With the symmetry group (26.6), the previous arguments imply that the possible superconducting order parameter correspond to different IR of

$$G_0 = SO(3)\times G_c \tag{26.7}$$

which are obviously products of representations of $SO(3)$ and those of G_c. First we need to choose an IR for the $SO(3)$ spin group. If we have BCS superconductivity owing to condensation of CP we know that only the $S=0$ and $S=1$ representations are possible, corresponding to singlet and triplet pairing respectively. If some more exotic many–

fermion object were condensing, instead of CP, we might have some other representations, a possibility we shall not consider here. In terms of the BCS pairing gap matrix $\Delta_{\alpha\beta}(\mathbf{k})$ the singlet and triplet components are conveniently separated by the following transformation[56]:

$$\Delta_{\alpha\beta}(\mathbf{k}) = i(\Delta_0(\mathbf{k})I + \mathbf{d}(\mathbf{k})\cdot\quad)\sigma_y \tag{26.8}$$

with Δ_0 corresponding to the 1D singlet representation and the spin vector $\mathbf{d}$ corresponding to the 3D triplet representation. Here σ are the Pauli matrices. Below we shall also use the fact that Δ_0 is even under spatial inversion and $\mathbf{d}$ odd.

For both singlet and triplet states there are order parameter derived from each different IR of the crystal space group, G_c. In general the crystal space group contains both translations as well as point group rotations, glide planes and screw axes. Method of obtaining IR of general space groups have been derived by Koster[57] and a complete listing of the IR of each of the 230 possible space groups has been given by Miller and Love.[58] Because of the subgroup of translational symmetries, each representation is associated with a wave–vector $\mathbf{K}$ in the first BZ. We will assume that $\mathbf{K}$ is zero (i.e. the gamma point in the zone). With this assumption the order parameter is invariant under all primitive lattice translation, while if this were not true the SC would spontaneously break the lattice translation symmetry. In the case $\mathbf{K}=0$ the group representations are especially simple, since they are given by the representations of the point group operations alone. From the crystal structures we only need consider the two cases of orthorhombic or the tetragonal point groups D_{2h} (mmm), or D_{4h} (4/mmm) respectively, each with inversion symmetry. The group character tables for the D_{2h} and D_{4h} groups are given in Ref. 59. From the tables we can see that the orthorhombic (OR) group has only 1D representations, while the tetragonal (TG) group has both 1D and 2D IR.

At this point it is immediately possible to enumerate all the possible order parameters which might occur in the HTSC. Clearly, in the orthorhombic symmetry, since all the spatial representations are 1D, there are only two possibilities: singlet or triplet, giving order parameter with one or three complex components respectively. For the tetragonal crystal with singlet pairing there are two possibilities: either an order parameter with one or two complex components, depending upon whether it belongs to a 1D or 2D representation. Secondly, for triplet pairing in the tetragonal crystal, the order parameter may have three or six complex components, again depending upon whether it transform as a 1D or 2D representation of D_{4h}. There are thus only four possible types of OP, with one, two, three or six complex components.

26.2. Symmetry Classification of HTSC States

We may now proceed with the classification of pairing states corresponding to the IR of the symmetry groups given above. We will follow here to SR paper.[5] In spite of the fact that the triplet pairing is ruled out by NMR Knight shift measurements we will consider both type of pairing: singlet and triplet.

26.2.1. Square lattice

The point group D_4 of the 2D square lattice contains 5 IR Γ_i, four 1D and one 2D. Both the simple (s^+ in Annett notations) and the so–called extended s–wave or s^-– state ($\sim \cos k_x a + \cos k_y a$) are in the trivial representation Γ_1. Therefore no symmetry–related distinction exists between these two, so that a mixing between them is expected in the absence of other symmetries (see below). The d–wave–states belong to either Γ_3 ($\sim k_x^2 - k_y^2$) or to Γ_4 ($\sim k_x k_y$), and therefore are differentiated by the point group symmetry.

Considering p–wave–(spin–triplet)–pairing with spin–orbit coupling we have to reduce the product $\Gamma_5 \otimes (\Gamma_2 \oplus \Gamma_5) = \Gamma_1 \oplus \Gamma_2 \oplus \Gamma_3 \oplus \Gamma_4 \oplus \Gamma_5$, where the orbital basis

$\{k_x, k_y\}$ and the spin triplet basis $\{\hat{\mathbf{z}};\hat{\mathbf{x}},\hat{\mathbf{y}}\}$ are combined in the $x-y-$ plane. The inert states in the absence of spin–orbit coupling are constructed from the Γ_5 representation of the point group. Table 26.1 catalogues the classification of the states in the point group D_4.

Table 26.1. Classification of the point group D_4 and the particle–hole symmetry $\mathbf{K}$. Γ_i are the representations of D_4 and $R_1(1,1)$, $R_2(1,-1)$ the representations of $\mathcal{K}$. $\uparrow$ denotes a unit vector in l–direction ($l = x, y, z$), a) s–wave, extended s–wave, b) d–wave, c) p–wave with spin–orbit coupling; $\Gamma_1 \oplus \Gamma_2 \oplus \Gamma_3 \oplus \Gamma_4$, d) p–wave without spin–orbit coupling: inert phases.

	Γ_i	R_i	$\Delta(\mathbf{k}) \sim \psi(\mathbf{k}), \mathbf{d}(\mathbf{k})$
a)	Γ_1	R_1	$\psi(\mathbf{k}) = \text{const}$
	Γ_1	R_2	$\frac{1}{\sqrt{2}}(k_x^2 + k_y^2) \sim \frac{1}{\sqrt{2}}(\cos k_x a + \cos k_y a)$
b)	Γ_3	R_2	$\psi(\mathbf{k}) = \frac{1}{\sqrt{2}}(k_x^2 - k_y^2) \sim \frac{1}{\sqrt{2}}(\cos k_x a - \cos k_y a)$
	Γ_4	R_1	$k_x k_y \sim \sin k_x a \sin k_y a$
c)	Γ_1	R_2	$\mathbf{d}(\mathbf{k}) = \frac{1}{\sqrt{2}}(\hat{\mathbf{x}}k_x + \hat{\mathbf{y}}k_y)$
	Γ_2	R_2	$\frac{1}{\sqrt{2}}(\hat{\mathbf{x}}k_y - \hat{\mathbf{y}}k_z)$
	Γ_3	R_2	$\frac{1}{\sqrt{2}}(\hat{\mathbf{x}}k_x - \hat{\mathbf{y}}k_y)$
	Γ_4	R_2	$\frac{1}{\sqrt{2}}(\hat{\mathbf{x}}k_y + \hat{\mathbf{y}}k_x)$
	Γ_5	R_2	$\hat{\mathbf{z}}k_x$
			$\hat{\mathbf{z}}k_y$

d)	Γ_3	R_2	$\mathbf{d}(\mathbf{k}) = \frac{1}{\sqrt{2}}(\hat{\mathbf{x}}k_x + \hat{\mathbf{y}}k_y)$
		R_2	$\frac{1}{\sqrt{2}}(\hat{\mathbf{x}}k_z + i\hat{\mathbf{y}}k_y)$
		R_2	$\frac{1}{\sqrt{2}}(\hat{\mathbf{x}} + i\hat{\mathbf{y}})k_x$
		R_2	$\frac{1}{2}(\hat{\mathbf{x}} + i\hat{\mathbf{y}})(k_x + k_y)$
		R_2	$\frac{1}{2}(\hat{\mathbf{x}} + i\hat{\mathbf{y}})(k_x + ik_y)$
		R_2	$\hat{\mathbf{z}}k_x$
		R_2	$\hat{\mathbf{z}}k_y$
		R_2	$\frac{1}{\sqrt{2}}\hat{\mathbf{z}}(k_x + k_y)$
		R_2	$\frac{1}{\sqrt{2}}\hat{\mathbf{z}}(k_x + ik_y)$

26.2.2. Tetragonal lattice

Turning to the tetragonal lattice symmetry D_{4h} (10 IR) SR find that beside the two types of s–wave states a d–wave state is in the trivial representation Γ_1^+ with $\psi(\mathbf{k}) = 6^{-1/2}(2k_z^2 - k_x^2 - k_y^2)$. Therefore the Γ_1^+ – singlet–pairing state is a mixture of these three states. There is a 2–fold degenerate Γ_5^+ representation with states $\{k_x k_z, k_y k_z\}$. The states of Γ_3^+ and Γ_4^+ are the same as SR found in D_4. The p–wave–states correspond to $(\Gamma_2^+ \oplus \Gamma_5^+) \otimes (\Gamma_2^- \oplus \Gamma_5^-) = 2\Gamma_1^- \oplus \Gamma_2^- \oplus \Gamma_3^- \oplus \Gamma_4^- \oplus 2\Gamma_5$ if spin–orbit coupling is considered; otherwise Γ_2^- and Γ_5^- are the basis representations needed to obtain the inert phases (Table 26.2).

Table 26.2. Classification of the point group D_{4h}. $\Gamma_i^{(+,-)}$ are the representations of D_{4h}. a) s–wave, extended s–wave, b) d–wave, c) p–wave with spin–orbit coupling; $2\Gamma_1^- \oplus \Gamma_2^- \oplus \Gamma_3^- \oplus \Gamma_4^- \oplus 2\Gamma_5^-$, d) p–wave without spin–orbit coupling: inert phases.

	Γ_i	$\Delta(\mathbf{k}) \sim \psi(\mathbf{k}), \mathbf{d}(\mathbf{k})$
a)	Γ_1^+	$\psi(\mathbf{k}) = \text{const}$
	Γ_1^+	$\frac{1}{\sqrt{3}}\left(k_x^2 + k_y^2 + k_z^2\right) \sim \frac{1}{\sqrt{3}}\left(\cos k_x a + \cos k_y a + \cos k_z a\right)$
b)	Γ_1^+	$\psi(\mathbf{k}) = \frac{1}{\sqrt{6}}\left(2k_z^2 - k_x^2 - k_y^2\right) \sim \frac{1}{\sqrt{6}}\left(2\cos k_z a - \cos k_x a - \cos k_y a\right)$
	Γ_3^+	$\frac{1}{\sqrt{2}}\left(k_x^2 - k_y^2\right) \sim \frac{1}{\sqrt{2}}\left(\cos k_x a - \cos k_y a\right)$
	Γ_4^+	$k_x k_y \sim \sin k_x a \sin k_y a$
	Γ_5^+	$\left.\begin{matrix} k_y k_z \sim \sin k_y a \sin k_z a \\ k_z k_x \sim \sin k_z a \sin k_x a \end{matrix}\right\}$
c)	Γ_1^-	$\mathbf{d}(\mathbf{k}) = \frac{1}{\sqrt{3}}\left(\hat{\mathbf{x}}k_x + \hat{\mathbf{y}}k_y + \hat{\mathbf{z}}k_z\right)$
		$\frac{1}{\sqrt{6}}\left(2\hat{\mathbf{z}}k_z - \hat{\mathbf{x}}k_x - \hat{\mathbf{y}}k_y\right)$
	Γ_2^-	$\frac{1}{\sqrt{2}}\left(\hat{\mathbf{x}}k_y - \hat{\mathbf{y}}k_x\right)$
	Γ_3^-	$\frac{1}{\sqrt{2}}\left(\hat{\mathbf{x}}k_x - \hat{\mathbf{y}}k_y\right)$
	Γ_4^-	$\frac{1}{\sqrt{2}}\left(\hat{\mathbf{x}}k_y + \hat{\mathbf{y}}k_x\right)$
	Γ_3^-	$\left.\begin{matrix} \hat{\mathbf{z}}k_x \\ \hat{\mathbf{z}}k_y \end{matrix}\right\}$
	Γ_3^-	$\left.\begin{matrix} \hat{\mathbf{x}}k_z \\ \hat{\mathbf{y}}k_z \end{matrix}\right\}$

d)	Γ_2^-	$\mathbf{d}(\mathbf{k}) = \hat{\mathbf{z}}k_z$
		$\frac{1}{\sqrt{2}}(\hat{\mathbf{x}}+i\hat{\mathbf{y}})k_x$
	Γ_5^-	$\frac{1}{\sqrt{2}}(\hat{\mathbf{x}}k_x+\hat{\mathbf{y}}k_y)$
		$\hat{\mathbf{z}}k_z$
		$\frac{1}{\sqrt{2}}\hat{\mathbf{z}}(k_x+k_y)$
		$\frac{1}{\sqrt{2}}(\hat{\mathbf{x}}k_x+i\hat{\mathbf{y}}k_y)$
		$\frac{1}{\sqrt{2}}\hat{\mathbf{z}}(k_x+ik_y)$
		$\frac{1}{\sqrt{2}}(\hat{\mathbf{x}}+i\hat{\mathbf{y}})k_z$
		$\frac{1}{2}(\hat{\mathbf{x}}+i\hat{\mathbf{y}})(k_x+k_y)$
		$\frac{1}{2}(\hat{\mathbf{x}}+i\hat{\mathbf{y}})(k_x+ik_y)$

26.2.3. The orthorhombic lattice

While the tetragonal symmetry is realized in HTSC above T_c (much higher than the onset of the SC phase), the orthorhombic symmetry D_{2h} (8 1D representations) is preferred below T_c.[60] The lower symmetry of the point group D_{2h} yields more states which belong to Γ_1^+: i.e. s^+-, s^-- and d–wave–states, $6^{-1/2}(2k_z^2-k_x^2-k_y^2)$ and $2^{-1/2}(k_x^2-k_y^2)$. A mixing between these four states can occur in the SC phase. Each of the remaining d–wave–states corresponds to a different 1D representation: $\Gamma_2^+ k_x k_y$; $\Gamma_3^+ k_x k_y$; $\Gamma_4^+ k_y k_z$.

The list of the p–wave–states with spin–orbit coupling contains other representations

$3\Gamma_1^- \oplus 2\Gamma_2^- \oplus 2\Gamma_3^- \oplus 2\Gamma_4^- = \left(\Gamma_2^- \oplus \Gamma_3^- \oplus \Gamma_4^-\right) \otimes \left(\Gamma_2^+ \oplus \Gamma_3^+ \oplus \Gamma_4^+\right)$.

Also here a mixture of states belonging to the same representation can occur. The inert phases are simple polar states obtained from Γ_2^-, Γ_3^- and Γ_4^- (Table 26.3).

Table 26.3. Classification of the point group D_{2h}. $\Gamma_i^{(+,-)}$ are the representations of D_{2h}. a) s–wave, extended s–wave, b) d–wave, c) p–wave with spin–orbit coupling; $3\Gamma_1^- \oplus 2\Gamma_2^- \oplus 2\Gamma_3^- \oplus 2\Gamma_4^-$, d) p–wave without spin–orbit coupling: inert phases.

	Γ_i	$\Delta(\mathbf{k}) \sim \psi(\mathbf{k}), \mathbf{d}(\mathbf{k})$
a)	Γ_1^+	$\psi(\mathbf{k}) = \mathrm{const}$
	Γ_1^+	$\frac{1}{\sqrt{3}}\left(k_x^2 + k_y^2 + k_z^2\right) \sim \frac{1}{\sqrt{3}}\left(\cos k_x a + \cos k_y a + \cos k_z a\right)$
b)	Γ_1^+	$\psi(\mathbf{k}) = \frac{1}{\sqrt{6}}\left(2k_z^2 - k_x^2 - k_y^2\right) \sim \frac{1}{\sqrt{6}}\left(2\cos k_z a - \cos k_x a - \cos k_y a\right.$
		$\frac{1}{\sqrt{2}}\left(k_x^2 - k_y^2\right) \sim \frac{1}{\sqrt{2}}\left(\cos k_x a - \cos k_y a\right)$
	Γ_2^+	$k_x k_z \sim \sin k_x a \sin k_z a$
	Γ_3^+	$k_x k_y \sim \sin k_x a \sin k_y a$
	Γ_4^+	$k_y k_z \sim \sin k_y a \sin k_z a$

c)	Γ_1^-	$\mathbf{d}(\mathbf{k})=\frac{1}{\sqrt{3}}\left(\hat{\mathbf{x}}k_x+\hat{\mathbf{y}}k_y+\hat{\mathbf{z}}k_z\right)$
		$\frac{1}{\sqrt{6}}\left(2\hat{\mathbf{z}}k_z-\hat{\mathbf{x}}k_x-\hat{\mathbf{y}}k_y\right)$
		$\frac{1}{\sqrt{2}}\left(\hat{\mathbf{x}}k_x-\hat{\mathbf{y}}k_y\right)$
	Γ_2^-	$\frac{1}{\sqrt{2}}\left(\hat{\mathbf{z}}k_x-\hat{\mathbf{x}}k_z\right)$
		$\frac{1}{\sqrt{2}}\left(\hat{\mathbf{z}}k_x+\hat{\mathbf{x}}k_z\right)$
	Γ_3^-	$\frac{1}{\sqrt{2}}\left(\hat{\mathbf{x}}k_y-\hat{\mathbf{y}}k_x\right)$
		$\frac{1}{\sqrt{2}}\left(\hat{\mathbf{x}}k_y+\hat{\mathbf{y}}k_x\right)$
	Γ_4^-	$\frac{1}{\sqrt{2}}\left(\hat{\mathbf{y}}k_z-\hat{\mathbf{z}}k_y\right)$
		$\frac{1}{\sqrt{2}}\left(\hat{\mathbf{y}}k_z+\hat{\mathbf{z}}k_y\right)$
d)	Γ_2^-	$\mathbf{d}(\mathbf{k})=\hat{\mathbf{y}}k_y$
		$\frac{1}{\sqrt{2}}(\hat{\mathbf{z}}+i\hat{\mathbf{x}})k_y$
	Γ_3^-	$\hat{\mathbf{z}}k_z$
		$\frac{1}{\sqrt{2}}(\hat{\mathbf{x}}+i\hat{\mathbf{y}})k_z$
	Γ_4^-	$\hat{\mathbf{x}}k_x$
		$\frac{1}{\sqrt{2}}(\hat{\mathbf{y}}+i\hat{\mathbf{z}})k_x$

26.2.4. Electron–hole symmetry

Some of the model Hamiltonians used in a 2D square lattice have an additional symmetry. SR have investigated the case of electron–hole symmetry for such Hamiltonians. They construct a transformation $\mathbf{K}$, which leaves H invariant, e.g. the Hubbard Hamiltonian for a half–filled band:

$$H = t\sum_{\sigma}\sum_{\rangle i,j\langle} c_{i\sigma}^{+} c_{j\sigma} + U\sum_{i} n_{i\uparrow} n_{i\downarrow} .$$

Then

$$\mathbf{K}:\ c_{i\sigma}^{+} \to c_{i\sigma} \quad i \in \text{sublattice } A$$

$$c_{j\sigma}^{+} \to -c_{j\sigma} \quad j \in \text{sublattice } B$$

where $\langle i,j \rangle$ denotes a summation only over the nearest neighbours, $\mathbf{K}$ transforms a particle creation operator in a particle annihilation operator and vice versa at the same site, in doing so the operator changes the sign on the sites of the sublattice B. Each site of the sublattice A has only nearest neighbours on the sublattice B and vice versa.

It is obvious that H is invariant under this transformation $\mathbf{K}$, since the $c_{i\sigma}^{+}(c_{i\sigma})$ are fermion operators satisfying anticommutation relations.

The diagonalization of the kinetic energy term in H gives a band energy $\varepsilon(\mathbf{k}) = 2t(\cos k_x a + \cos k_y a)$, with operators

$$c_{\mathbf{k},\sigma}^{+} = \sum_{i \in A} c_{i\sigma}^{+} e^{i\mathbf{k}\cdot\mathbf{R}_i} + \sum_{j \in B} c_{j\sigma}^{+} e^{i\mathbf{k}\cdot\mathbf{R}_j}$$

The property of $\mathbf{K}$ in $\mathbf{k}$–space is easily obtained by applying $\mathbf{K}$ on $c^{+}_{\mathbf{k},\sigma}$ and $c_{\mathbf{k},\sigma}$:

$$\mathbf{K}:\ c^{+}_{\mathbf{k},\sigma} \rightarrow \sum_{i\in A} c_{i\sigma} e^{i\mathbf{k}\cdot\mathbf{R}_i} - \sum_{j\in B} c_{j\sigma} e^{i\mathbf{k}\cdot\mathbf{R}_j}$$

$$= \sum_{i\in A} c_{i\sigma} e^{i\mathbf{k}\cdot\mathbf{R}_i} + \sum_{j\in B} c_{j\sigma} e^{i\mathbf{k}\cdot\mathbf{R}_j} e^{i\mathbf{Q}(\mathbf{R}_j-\mathbf{R}_0)}$$

$$= e^{-i\mathbf{Q}\mathbf{R}_0}\left\{ \sum_{i\in A} c_{i\sigma} e^{i(\mathbf{k}+\mathbf{Q})\mathbf{R}_i} + \sum_{j\in B} c_{j\sigma} e^{i(\mathbf{k}+\mathbf{Q})\mathbf{R}_j} \right\} = c^{\dagger}_{-\mathbf{k}-\mathbf{Q},\sigma}$$

$$\mathbf{K}:\ c_{\mathbf{k},\sigma} \rightarrow c^{+}_{-\mathbf{k}-\mathbf{Q},\sigma}$$

where $\mathbf{Q} = \left(\frac{\pi}{a},\frac{\pi}{a}\right)$ and $\mathbf{R}_0 \in$ sublattice A (we choose $\mathbf{R}_0 = (0,0)$) with $e^{i\mathbf{Q}(\mathbf{R}_j-\mathbf{R}_0)} = -1$ and $e^{i\mathbf{Q}(\mathbf{R}_i-\mathbf{R}_0)} = +1$. Therefore the band energy $\varepsilon(k)$ has the following property:

$$\mathbf{K}\varepsilon(\mathbf{k}) = \varepsilon(-\mathbf{k}-\mathbf{Q}) = -\varepsilon(\mathbf{k})$$

$\mathbf{K}$ connects different areas of the first BZ. Since $\mathbf{K}^2$ is the identity, it generates a group κ of two elements and with the two representations $\mathbf{R}_1(1,1)$ and $\mathbf{R}_2(1,-1)$. Thus an additional classification of the pairing states in the square lattice may be possible. Of course s^{+} state is in the representation $\mathbf{R}_1$. On the other hand the s^{-} state is a $\mathbf{R}_2$–state:

$$\mathbf{K}\psi(\mathbf{k}) = \frac{1}{\sqrt{2}}\left(\cos\left(-k_x - \frac{\pi}{a}\right)a + \cos\left(-k_y - \frac{\pi}{a}\right)a\right) = -\psi(\mathbf{k})$$

This fact allows a distinction between these two states by symmetry. The d–wave–states belong either to $R_2 2^{-1/2}(\cos k_x a - \cos k_y a)$ or to $R_1(\sin k_x a \cdot \sin k_y a)$. All spin–triplet states are in the R_2–representation. So SR examined the group theoretical classification of possible SC states in the HTSC. The key conclusion is that only in the presence of higher symmetry structures and particle–hole symmetry can a distinction be drawn on symmetry grounds between the s^+, s^- states and some of the d–wave states. In the lower symmetry crystal structures that in practice occur, these states belong to the same symmetry classification so that they cannot appear alone but only in mixtures in the SC state. Therefore in general in singlet states the surfaces $\varepsilon(\mathbf{k}) = 0$ and the energy surface $\varepsilon(\mathbf{k}) = 0$ will not coincide but intersect in lines or possibly not at all. The distinction between these states is one of degree and not a symmetry difference. In 26.3.2 we will study the problem of mixing of states belong to different IR.

26.3. Singlet States

26.3.1. The gap functions

We have considered above a complete symmetry classification of possible SC states of HTSC. As we mentioned the triplet pairing is ruled out by NMR Knight shift measurements. Thus we will consider only singlet phases and their mixtures allowed by symmetry.

Table 26.4. Magnitude and phase of the superconducting order parameter as a function of direction in the CuO_2 planes for the primary candidates for the pairing symmetry [Ref. 61].

PAIRING STATE	MAGNITUDE $\lvert\Delta\rvert$ / RELATIVE PHASE ϕ	GAP $\Delta(\mathbf{k})$
isotropic s	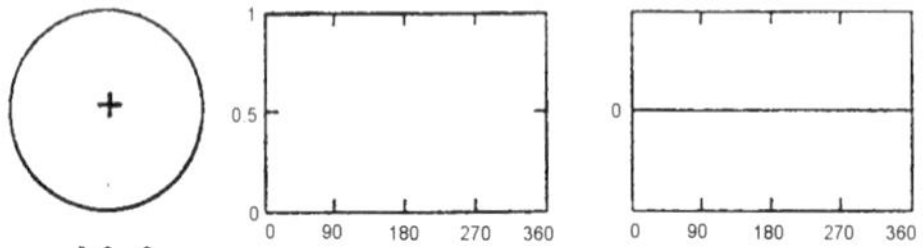	$\Delta(\mathbf{k})=const$
$d_{x^2-y^2}$	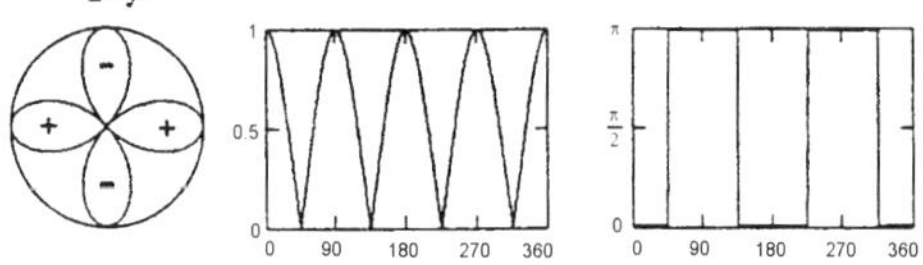	$\Delta(\mathbf{k})=\Delta_0(\cos k_x a-\cos k_y a)$
anisotropic s	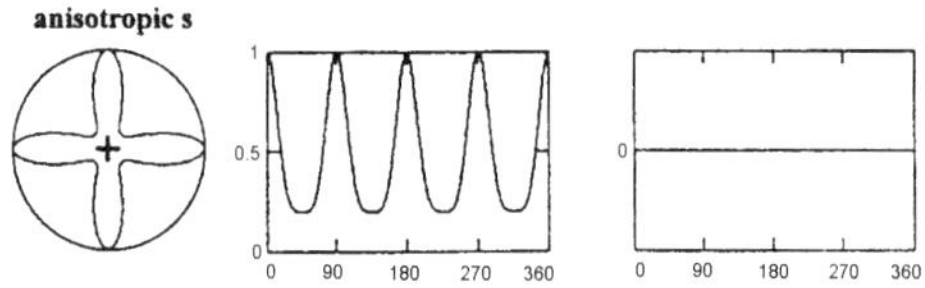	$\Delta(\mathbf{k})=\Delta_0(\cos k_x a-\cos k_y a)^4+\Delta_1$
extended s	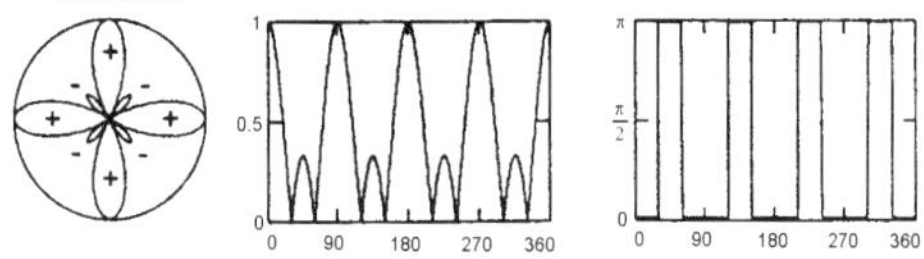	$\Delta(\mathbf{k})=\Delta_0\left[(1+\gamma^2)(\cos k_x a-\cos k_y a)^2-\gamma^2\right]$
$s + i\, d_{x^2-y^2}$	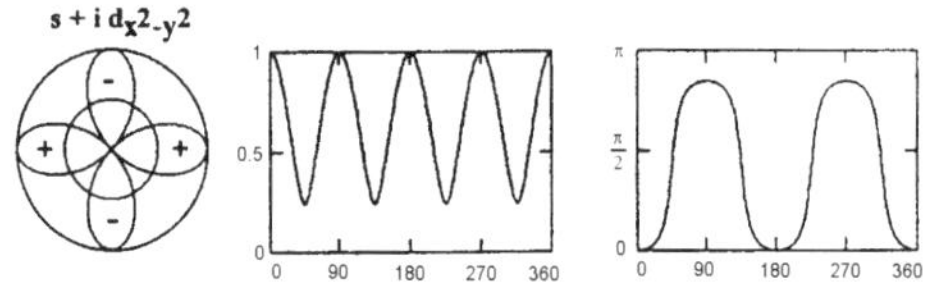	$\Delta(\mathbf{k})=\Delta_0\left[\varepsilon+i(1-\varepsilon)(\cos k_x a-\cos k_y a)\right]$
$d_{x^2-y^2} + i\, d_{xy}$	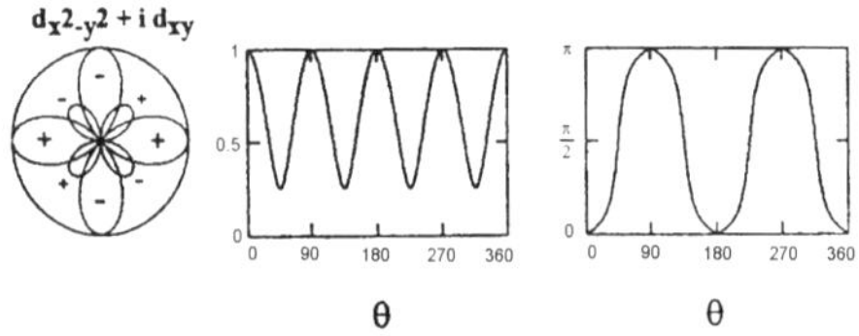	$\Delta(\mathbf{k})=\Delta_0[(1-\varepsilon)(\cos k_x a-\cos k_y a)+ i\varepsilon(2\sin k_x a \sin k_y a)]$

In Table 26.4. we give the gap functions for the singlet states of CuO_2 plane.[61] Here Δ_1 represents the minimum value of the gap which occurs along the (110) directions, $\varepsilon\Delta_0$ is the fraction of s or d_{xy} component mixed in with the $d_{x^2-y^2}$ state and $\varepsilon\Delta_0$ is the minimum energy gap.

26.3.2. Mixing of states of different irreducible representations

Below following to Annett *et al.* paper[1] we will discuss to what extent thermodynamic or related observations enable us to infer constraints on the behavior of the OP under the (exact or approximate) symmetry operations of the crystal Hamiltonian. In particular we will emphasize the conditions under which the pairing state belongs to a single irreducible representation of the symmetry group, and the conditions under which it becomes a "mixed" state involving two or more representations. As we show below, simple thermodynamic considerations place quite strong restrictions on the conditions for such mixed states to occur.

Consider first the group G of exact symmetries of the Hamiltonian for a given crystal. This is the direct product of the gauge group $U(1)$, the crystal lattice translation group T_l and the crystal point group H :

$$G = U(1) \otimes T_l \otimes H \tag{26.9}$$

As remarked above, we will normally assume that ψ transforms as the identity representation of T_l, in which case we can simply take G to be given by $U(1) \otimes H$ however, we may as well keep (26.9) for generality. The SC order parameter is an anomalous Green function $\psi(\mathbf{r}_1, \mathbf{r}_2, \alpha, \beta) \equiv \psi(\mathbf{r}_1, \mathbf{r}_2, \alpha, \beta) \equiv \langle \psi_\alpha(\mathbf{r}_1) \psi_\beta(\mathbf{r}_2) \rangle$. We now expand ψ in the IR $\chi(\mathbf{r}_1 \mathbf{r}_2; \alpha\beta)$ of the group G:

$$\psi(\mathbf{r}_1, \mathbf{r}_2; \alpha, \beta) = \sum_l \sum_{m=0}^{D_1} \psi_{lm} \chi_{lm}(\mathbf{r}_1, \mathbf{r}_2; \alpha, \beta) \tag{26.10}$$

where D_l is the dimension of the l–th IR and the ψ_{lm} are coefficients which are in general complex. Note that it is not assumed, anywhere in this argument, that ψ is constant as a function of the centre of mass variable $R \equiv (r_1 + r_2)/2$.

We now proceed, in the spirit of Ginzburg and Landau, to express the free energy F as a multiple power series in the coefficients ψ_{lm} using the principle that F must be invariant under all operations of G (and moreover must lie real). It is immediately clear that invariance under $U(1)$ implies that all terms containing an odd number of ψ_{lm}'s vanish identically, while the rest must contain an even number of ψ_{lm}'s and ψ^*_{lm}'s. Moreover, the second–order terms must have the form $\sum_l \alpha_l(T)\sum_m |\psi_{lm}|^2$, where the $\alpha_l(T)$ are functions of T which, barring pathology, will be different for different l. At fourth order the generic term is:

$$\tfrac{1}{2}\sum \beta_{l_1m_1l_2m_2l_3m_3l_4m_4}\psi^*_{l_1m_1}\psi^*_{l_2m_2}\psi_{l_3m_3}\psi_{l_4m_4} \tag{26.11}$$

For the tetragonal and orthorhombic symmetry groups relevant to the HTSC the relevant representations and GL expansions up to fourth order are well known.[4] However, such symmetry analyses usually assume that only a single IR, say l, is relevant. Instead, let us examine terms in the GL expansion which couple two or more representations.

The quartic terms can be grouped, for reasons which will become apparent, into what we shall call "mixing" and "non–mixing" terms. The non–mixing terms are those in which for each representation the ψ_{lm}'s enter to even order, and mixing terms are the rest. The non–mixing terms are thus the terms such as $|\psi_{l_1m_1}|^2|\psi_{l_2m_2}|^2$ and $(\psi^*_{l_1m_1})^2(\psi_{l_2m_2})^2 + \text{c.c.}$, while the mixing terms are of the form $\psi^*_{l_1m_1}\psi^*_{l_2m_2}\psi^2_{l_1m_1}$. The usefulness of this separation into mixing and non–mixing terms lies in the fact that in all orthorhombic and TG SC, such as HTSC, simple symmetry arguments imply that the mixing terms are absent.[4] Thus the general form of the GL expansion up to fourth order is:

$$F(T)=\sum_{lm}\left(\alpha_l(T)|\psi_{lm}|^2+\tfrac{1}{2}\beta_{l,m_1,m_2,m_3,m_4}\psi^*_{lm_1}\psi^*_{lm_2}\psi_{lm_3}\psi_{lm_4}\right)$$

$$+\sum_{l_1,l_2,m_1,m_2,m_3,m_4}\tfrac{1}{2}\beta_{l_1l_2m_1m_2m_3m_4}\psi^*_{l_1m_1}\psi_{l_1m_2}\psi^*_{l_2m_3}\psi_{l_2m_4}$$

$$+\sum_{l_1,l_2,m_1,m_2,m_3,m_4}\tfrac{1}{2}k_{l_1l_2m_1m_2m_3m_4}\psi^*_{l_1m_1}\psi^*_{l_1m_2}\psi_{l_2m_3}\psi_{l_2m_4}$$

plus the terms involving three or four distinct representations, l. For the purposes of the present argument, we may neglect the sixth– and higher–order terms, which do not affect the results qualitatively. In fact there are no mixing terms to *any* order[4] unless three or four distinct representations are mixed.

At a sufficiently high T all the $\alpha_l(T)$ are positive, and minimisation of the free energy is achieved by setting all ψ_{lm} equal to zero, i.e. the system is in the normal phase. As T falls, there comes a point, $T_c^{l_0}$, where one α_l corresponding (e.g.) to $l=l_0$, becomes negative while (in the absence of pathological coincidence) all other α_l remain positive. We first consider the case (actually not very likely to be relevant to the HTSC) where the IR l_0 is multidimensional. In this case, for T just below $T_c^{l_0}$, some or all of the corresponding ψ_{l_0m} will be nonzero; the weights with which the various m are represented will be controlled primarily by the fourth–order terms, and may in principle depend on T through the β's (or through the omitted higher–order terms). Thus it is possible that the "configuration" (i.e. the relative weight of the various m) undergoes either a continuous or a discontinuous change below $T_c^{l_0}$; an example of the latter is the A–B transition in superfluid ^{3}He.

We turn to the case, probably more relevant to the HTSC, that the IR l_0 is one–dimensional. The crucial question is: what is the condition that below $T_c^{l_0}$ some ψ_{lm} corresponding to *other* values of l are nonzero? It is clear that if the mixing terms are nonzero this can happen in a continuous way (even if all other α_l are positive for all T). But let us consider the case that the mixing terms vanish, corresponding to the HTSC. For simplicity of notation, we shall specialize to the case where there is only one relevant IR besides l_0 and it is moreover also one–dimensional (the generalizations are straightforward). In this case we note that we can always minimize the free energy by choosing $\arg\left(\psi_{l_0 m_0}\psi_{lm}\right)$ to be either 0 or π; having done this, and changing the notation for convenience ($l_0 \to l$, etc.), we can write:

$$F(T) = F_0(T) + \alpha_1(T)|\psi_1|^2 + \alpha_2(T)|\psi_2|^2 +$$

$$\frac{1}{2}\beta_1(T)|\psi_1|^4 + \frac{1}{2}\beta_2(T)|\psi_2|^4 + \kappa(T)|\psi_1|^2 \cdot |\psi_2|^2 \qquad (26.12)$$

For purposes of illustration we shall choose the T dependences of the coefficients to have the simple GL form of:

$$\alpha_1(T) = \alpha_1\left(T - T_{c_1}\right),\ \alpha_2(T) = \alpha_2\left(T - T_{c_2}\right),\ T_{c_2} \le T_{c_1}\ \text{(and possibly}$$
$$T_{c_2} < 0),\ \beta_1(T) = b_1,\ \beta_2(T) = b_2,\ \kappa(T) = \kappa \qquad (26.13)$$

where b_1, b_2, and κ are constants, and where stability requires that $\beta_1, \beta_2 > 0$, $\kappa > -\sqrt{\beta_1\beta_2}$. (Note, that in BCS–type theories these conditions are automatically satisfied). The qualitative results are independent of this ansatz.

The phase diagram of a system with a free energy of the form (26.13) is discussed in detail by Imry.[62] For our purposes it is sufficient to know the following: for T above the "upper" transition T_{c_1}, it is clear that $\psi_1 = \psi_2 = 0$ (normal phase). For T just below T_{c_1}, the free energy is minimized by the following choice:

$$\psi_1 = (\alpha_1/\beta_1)\left(T_{c_1} - T\right)^{1/2}, \ \psi_2 = 0 \tag{26.14}$$

At lower T there are various possibilities, depending on the ratios of the parameters.[62] However, it is clear that if ψ_2 is ever to obtain a finite value, one of two things must happen: either it must jump discontinuously from zero to this value, which evidently corresponds to a first–order PT with an actual discontinuity in the value of various physical quantities, or there must be a second second–order PT at T^* given by:

$$T^* - T_{c_2} = \lambda\left(T^* - T_{c_1}\right), \quad \lambda = \kappa\alpha_1/\alpha_2\beta_1 \tag{26.15}$$

This second possibility requires $\lambda \le T_{c_2}/T_{c_1}$. In this case it is straightforward to show that the (positive) specific–heat discontinuity at T^* is given by the following expression:

$$\Delta c_v^* = \left(a_2^2/b_2\right)\frac{(1-\lambda)^2}{1-\kappa^2/b_1b_2} \ge \left(a_2^2/b_2\right)\frac{\left[\left(T_{c_1} - T_{c_2}\right)/T_{c_1}\right]^2}{1-\kappa^2/b_1b_2} \tag{26.16}$$

Thus, assuming that the ratio $\left(a_2^2/b_2\right)/\left(a_1^2/b_1\right)$ is not pathologically small, the anomaly at T^* is comparable to that at $T_c\left(\equiv T_{c_1}\right)$ unless T_{c_2} is extremely close to T_{c_1} (and quite likely even then). A second quantity of interest is the change in slope at T^* of the "total" order parameter $|\psi|^2 = |\psi_1|^2 + |\psi_2|^2$; crudely speaking, many physical quantities such as the mean–square energy gap are likely to be roughly proportional to this. We find for the *relative* change in slope δ^* at T^* the following expression:

$$\delta^* = \frac{(1-\lambda)(1-\lambda a_2/a_1)(a_2 b_1/a_1 b_2)}{1-\kappa^2/b_1 b_2} \tag{26.17}$$

In simple BCS–type theories it turns out that $a_j \propto T_{cj}^{-1}$, and thus (in view of the constraint on λ) the quantity δ^* is positive in such theories and of order one except possibly for T_{c_2} very close to T_{c_1}.

It is clear that the above results, derived for the explicit form (26.12) of the free energy, should generalize qualitatively to more realistic "non–mixing" forms, with the sole caveat that if the T^* is very low compared to T_c the anomalies δc_v^* and δ^* are likely to be correspondingly reduced. With this caveat, therefore, we draw the following very important conclusion: if the order parameter is a superposition of two functions from different IR of the crystal symmetry group and symmetry considerations forbid "mixing" of these functions in the free energy (as in the cuprates), then there must inevitably be a second PT at some T below T_c. Unless the relevant IR have the transition temperatures which are very close, both the entropy and other physical quantities will undergo either discontinuities or substantial changes in slope. The almost complete absence to date, of any suggestion of such phenomena in the HTSC is a very strong argument against this type of superposition.

For completeness we should briefly discuss what happens when, owing perhaps to some small breaking of the symmetry (such as may be constituted, for some at least of the HTSC by the orthorhombic crystalline anisotropy) the mixing terms are small but not zero. We first note that any "quadratic" mixing, that is, any term in the free energy of the following form:

$$\gamma(\psi_1\psi_2^* + c.c) \tag{26.18}$$

can always be eliminated by a re–diagonalization of the quadratic terms; however, the quartic terms will then contain terms of the form $\psi_1^3\psi_2$ and $\psi_2^3\psi_1$, whose coefficients will be proportional to $\gamma/|\alpha_1-\alpha_2|$ when this is small. We note that in most cases this coefficients will not be strongly T – dependent; in particular, in BCS theory, because of the special form of the $\alpha_j(T)$, $|\alpha_1-\alpha_2|$ is approximately $T\left(T_{c_2}^{-1}-T_{c_1}^{-1}\right)$. There may in addition be "direct" fourth–order mixing terms: in the following, let T^* be the T at which a second second–order phase transition would have occurred in the absence of a mixing term. (If the original second transition was first order, it would be little affected by a small amount of mixing). For small mixing, the term in $\psi_1\psi_2^3$ will have little effect for $T \geq T^*$, so we shall neglect it. Suppose now that the coefficient of $\psi_1^3\psi_2$ is of order of magnitude ξ. Then for $T<T_{c_1}$ and not too close to T^*, the effect of the mixing terms is that ψ_2 is non–zero, increasing as $T\left(T_{c_1}-T\right)^{3/2}$:

$$\psi_2 \sim \xi\psi_1^3/\alpha_2'(T) \sim \xi(\alpha_1/\beta_1)^{3/2}\left(T_{c_1}-T\right)^{3/2}/\alpha_2'(T) \tag{26.19}$$

where $\alpha_2'(T) \equiv \alpha_2(T)(1-\lambda)$.

Well below T^*, on the other hand, the mixing will have only a small effect and ψ_2 will increase approximately as $\left[-\left(\alpha_2'(T)/\beta_2\right)\cdot\left(1-\kappa^2/\beta_1\beta_2\right)^{-1}\right]^{1/2}$. We may obtain an estimate of the order of magnitude ΔT of the crossover region by equating the value of the first expression at $T^*+\Delta T$ to that of the second at $T^*-\Delta T$; because $\alpha_2'(T) \equiv \alpha_2(1-\lambda)\left(T-T^*\right)$, this gives:

$$\Delta T = \left(\frac{\xi}{\beta_2}\right)^{2/3}\left(\frac{\alpha_1\beta_2}{\alpha_2\beta_1}\right)\left(\frac{T_{c_1}-T^*}{1-\lambda}\right)\left(1-\kappa^2/\beta_1\beta_2\right)^{1/3} \tag{26.20}$$

Thus, for ξ/β_2 non–zero but <<1 (and λ not too close to 1, etc.) the vestige of the second second–order transition persists in the form of a sharp kink in the thermodynamic properties near T^*, as we should expect intuitively; the absence of observation of such behavior may be used to put limits on ξ in any specific case of interest.

Finally, what about the quartic terms in the GL expansion which involve three or four distinct representations? In this case the "mixing" terms need not vanish. For example, mixing terms such as $\psi^*_{l_1m_1}\psi^*_{l_2m_2}\psi_{l_3m_3}\psi_{l_4m_4}$ are symmetry–allowed in a TG crystal with $l_1 = A_{1g}$, $l_2 = A_{2g}$, $l_3 = B_{1g}$, $l_4 = B_{2g}$. However, mixing terms of this kind do not lead to a continuous mixing of representations of the GS, and thus do not change the qualitative observations given above, regardless of how many α_l 's go negative. Again a second PT below $T_c^{l_0}$ is required if a mixed symmetry state is to exist at low T.

26.3.3. Orthorhombicity and twins

As is well known, the CuO_2 planes in most of HTSC, the exception of the Tl– and Hg–based compounds, are not exactly TG but have an orthorhombic structure; in addition to the slight difference (usually < 2%) of the a– and b–crystal axes, the "buckling" of the planes usually picks out a special axis. In the case of YBCO (both "1237" and "1248") a more substantial anisotropy of the crystal lattice as a whole is induced by the presence of chains in the b–direction; although one's immediate instinct is that most of the "action" as regards to SC is likely to be in the CuO_2 planes and the chains should therefore be a relatively minor perturbation, the fact that the penetration depth is observed[63,64] to be appreciably anisotropic (~50%) in the ab–plane shows that this "perturbation" cannot necessarily be neglected.

The presence of the orthorhombic anisotropy means that the relevant symmetry group is no longer the group of the square but a subgroup of it. However, it is important to appreciate that this subgroup is not the same for all HTSC. In the case of YBCO, the *a*– and *b*–crystal axes become inequivalent, but each remain a twofold axis and a mirror plane: thus $\hat{I}_{axis}$ remains a symmetry operation while $\hat{R}_{\pi/2}$ and $\hat{I}_{\pi/4}$ are no longer so. The effect is to permit mixing of what were in a square lattice the s^+ and $d_{x^2-y^2}$ states, and likewise the pair s^- and d_{xy}; however, there can for example still be no mixing of $d_{x^2-y^2}$ and s^-. On the other hand, in LSCO and BSCCO it is the two orthogonal 45° axes which become inequivalent, while remaining mirror planes; thus $\hat{I}_{\pi/4}$ is still a good symmetry operation but $\hat{R}_{\pi/2}$ and $\hat{I}_{axis}$ are not, and the members of the pairs (s^+, d_{xy}) and $(s^-, d_{x^2-y^2})$ can mix. These considerations have important consequences for the structure of the gap function (a "local" quantity which is defined separately in each twin domain, see below): while in YBCO the nodes of the gap, if it is "$d_{x^2-y^2}$ like" (cf. below) may now occur at an angle different from 45°, in LSCO and BSCCO it must still occur at exactly 45°.

In many of the experiments conducted on the orthorhombic cuprates to determine the symmetry of the order parameter, the samples have been heavily twinned; untwinned samples are the exception. In heavily twinned samples (such as thin films) the order parameter will be effectively TG in symmetry, since the order parameter domains average over many twin domains.

26.3.4. Multilayer structures

Up to now, we have assumed that all pairing states take place within a single plane, i.e. that in the expression $\psi(\mathbf{r}_1, \mathbf{r}_2; \alpha, \beta) \equiv \langle \psi_\alpha(\mathbf{r}_1) \psi_\beta(\mathbf{r}_2) \rangle$ for the OP r_1 and r_2 lie in the same CuO_2 layer. But many of the cuprates, including YBCO, possess double or in some cases triple $Cu0_2$ planes; and that even for single–layer materials such as Tl 2201, it is not obvious that there cannot be pairing between electrons in different layers.

While for single layer materials the suggestion about intra–layer pairing seems to be natural for double– or triple–layer materials such as YBCO or BSCCO 2223 it is less obvious *a priori* that the hypothesis of "exclusively inter–layer" pairing can be excluded. For example, Anderson considered the possibility of inter–layer pairing of holons in RVB theory. Let us focus for definiteness on a bilayer material such as YBCO. If the order parameter is symmetric with respect to interchange of the "layer" indices of r_1 and r_2 (a possibility which actually seems pathological if there is to be no intra–layer pairing) then it is clear that the intra–layer pairing given above analysis goes through unchanged. If on the other hand it is antisymmetric (and we assume spin singlet even–frequency pairing), then the symmetry with respect to inversion within the *ab*–plane must be odd rather than even, and we have to deal with a set of IR different from s^+, s^-, $d_{x^2-y^2}$, d_{xy}. We will not explore this possibility further here, since (a) in view of the general qualitative similarity between the superconducting behavior of various classes of cuprates, it seems extremely unlikely that the pairing state is radically different in one– and two–layer materials, and (b) a state with odd parity in the ab–plane seems difficult to reconcile with the existence of a reproducible (*a*– or *b*–direction) Josephson effect with ordinary s–wave SC. Thus we conclude that within a single CuO_2 plane the OP is finite and belongs to the set s^+, s^-, $d_{x^2-y^2}$, d_{xy}. Given this state of affairs, it is still a nontrivial question how the order parameter behaves (a) under reflection in the symmetry plane "spacing" the two layers (e.g. in the case of YBCO, the plane containing the Y atoms), and (b) under translation up the *c*–axis from one unit cell to the next. With regard to (b), the "natural" assumption is that the order parameter is identical both in magnitude and (in the absence of superflow) in phase from one unit cell to the next, (i.e. behaves according to the identity representation of the translation group T_l), and any assumption different from this would seem to complicate the picture gratuitously without helping in any way to resolve the apparent experimental inconsistencies. We will therefore make the "natural" assumption from now on. (The same comment applies to the behaviour of the order parameter within the *ab*–plane).

Finally, what is the behaviour of the order parameter under reflection in the "spacing" plane? Evidently the IR are even or odd under this operation, and thus are "non–mixing". Thus, by the above arguments, the absence of more than one phase transition in the bilayer materials is strong evidence in favour of one and only one IR being realized. Now if it is the odd representation that occurs, it is difficult, if not impossible to understand the existence of an (*a*– or *b*–direction) Josephson effect with ordinary SC. Thus we conclude that the symmetry under reflection is even, and it is then clear that we can for many (though not all) purposes neglect the bilayer structure entirely, i.e. treat each bilayer as effectively a single layer.

It is clearly possible to give a similar discussion of trilayer structures, but since few if any of the crucial experiments have been done on trilayer materials, there seems no point in doing so here.

26.4. Pairing Symmetry and Pairing Interactions

The increasingly strong experimental evidence for a $d_{x^2-y^2}$ pairing implies that the physical properties and phenomenology of the superconducting state will differ considerably from those of conventional isotropic s–wave SC. The d–wave state also places strong constraints on the possible microscopic mechanisms for superconductivity.

In order to analyze this problem Annett (A)[11] tried to answer the next logical question: given that there is a $d_{x^2-y^2}$ pairing state what experiments can probe the underlying interactions giving rise to it? In other words, are there reasonably clear cut experiments which can eliminate possible pairing mechanisms leading to the d–wave state?

Annett discussed two straightforward pairing mechanisms which would lead to such a state. These are a simple generic model of pairing due to exchange of AF spin fluctuations, and a similar model of pairing due to exchange of charge fluctuations, such as phonons or plasmons. Also he discussed possible experiments which could distinguish between these two generic pairing mechanisms.

26.4.1. Two scenarios for d–wave pairing

Assuming that $d_{x^2-y^2}$ pairing is indeed present, then the next key question is to ask what pairing interaction leads to such a state. Assuming a BCS gap equation is valid near T_c, then it is quite straightforward to see that there are at least two quite distinct scenarios which lead to $d_{x^2-y^2}$ pairing in the cuprates. There are: AF spin fluctuation exchange, with an interaction peaked near $\mathbf{Q} = (\pi, \pi)$, and phonon (or plasmon) scattering with an interaction peaked near $\mathbf{Q} = 0$. In the latter case it is also necessary to invoke a strong on–site Coulomb repulsion to suppress s–wave pairing compared to d–wave.

Assuming for the sake of discussion that a simple BCS gap equation is valid:

$$\Delta(\mathbf{k}) = \sum_{\mathbf{k}'} V(\mathbf{k},\mathbf{k}')\Delta(\mathbf{k}')\frac{\tanh\beta\varepsilon_{\mathbf{k}'}/2}{\varepsilon_{\mathbf{k}'}} \tag{26.21}$$

it is then simple to examine the pairing states derived from different model interactions $V(\mathbf{k},\mathbf{k}')$. The simplest AF fluctuation exchange model would be for $V(\mathbf{k},\mathbf{k}')$ which is negative (due to the magnetic scattering vertex) and peaked near (π, π). For definiteness taking the simplest such function:

$$V(\mathbf{k},\mathbf{k}') = -\frac{V_0}{4}\left(2 - \cos(k_x - k_x') - \cos(k_y - k_y')\right) - \mu^* \tag{26.22}$$

and the FS of $La_{2-x}Sr_xCuO_4$ gives a strong attraction in the $d_{x^2-y^2}$ channel. This is because the negative sign in $V(\mathbf{k},\mathbf{k}')$ cancels with the sign change in the $d_{x^2-y^2}$ gap function giving a net positive eigenvalue for the gap equation. In contrast, for the s–wave channel there is always a net pair breaking interaction. Here μ^* is the Coulomb pseudopotential due to on–site repulsions. It is assumed to be **k** independent and always positive.

Table 26.5 shows values of the highest eigenvalues of the linearized gap equation, Eq. (26.21), in each main pairing channel for this interaction. The results are expressed as an effective coupling, λ_{eff}, so that $k_B T_c \approx \hbar\Omega_D \exp(-1/\lambda_{eff})$ for channels with an attractive interaction. In Table 26.5 $\lambda_0 = N(0)V_0$. The values given assume a FS appropriate for $La_{2-x}Sr_xCuO_4$ and the simple model interaction of Eq. (26.22). Using model interactions $V(\mathbf{k},\mathbf{k}')$ which are more sharply peaked near (π,π), such as the Montoux Pines interaction,[159] leads to qualitatively similar values of λ_{eff}. This is because the dominant contributions to the gap equation come from the regions of Fermi surface near the saddle points at $(\pi,0)$ $(0,\pi)$ etc. The most important parts of $V(\mathbf{k},\mathbf{k}')$ are thus those near the saddle point nesting vector of $\mathbf{Q} = (\pi,\pi)$, and whether V is strongly peaked there or not is less important.

The second simple scenario for the effective interaction is an e–ph (or electron–plasmon) coupling. Here the d–wave state can be stabilized if the interaction is peaked at small $\mathbf{Q}$. The simplest possible such form is:

$$V(\mathbf{k},\mathbf{k}') = +\frac{V_0}{4}\left(2 + \cos(k_x - k_x') + \cos(k_y - k_y')\right) - \mu^*, \tag{26.23}$$

where again μ^* is the Coulomb pseudopotential. Again Table 26.5 lists the values of the highest eigenvalue λ_{eff}. One can see that the attraction is strongest in both the s–wave and $d_{x^2-y^2}$ channels. If μ^* is negligible then s–wave pairing is more favorable, but when μ^* becomes larger than $0.2\lambda_0$ then the $d_{x^2-y^2}$ state will have the highest T_c. Again these results are qualitatively unchanged with interactions more sharply peaked around $\mathbf{Q} = \mathbf{k} - \mathbf{k}' = (0,0)$ than the simple cosine form of Eq. (26.23). For example the extreme small $\mathbf{Q}$ picture[65] also leads to stable d–wave pairing if μ^* is sufficiently large.

Of course these two scenarios are not unique, and many other possible pairing interactions $V(\mathbf{k},\mathbf{k}',\omega)$ could give rise to d–wave pairing in a BCS or Eliashberg framework. D–wave pairing could also arise from other mechanisms which are well outside a BCS / Eliashberg framework. There include bipolaron models, in which the bipolaron has an internal $d_{x^2-y^2}$ pairing state, and large positive U HM, which may have a $d_{x^2-y^2}$ paired GS in 2D. The main point of Table 26.5 is not the specific models for the interactions, but the general trends. The key point is that the AF scenario only even leads to $d_{x^2-y^2}$ pairing, and is never attractive for s–wave pairs. While, on the other hand, the charge fluctuation scenario can lead to either s– or d–wave pairing, depending on the specific model parameters.

Table 26.5. Eigenvalues of the linearized BCS gap equation for the two scenarios discussed in the text: magnetic scattering broadly peaked near (π,π) and charge scattering broadly peaked near $(0,0)$.

IR	Pairing State	Magnetic λ_{eff}	Charge λ_{eff}
A_{1g}	s–wave	$-0.50\lambda_0 - \mu^*$	$+0.50\lambda_0 - \mu^*$
A_{2g}	$g_{xy(x^2-y^2)}$	0	0
B_{1g}	$d_{x^2-y^2}$	$+0.31\lambda_0$	$+0.31\lambda_0$
B_{2g}	d_{xy}	0	0
E_u	(P_x, P_y)	$+0.09\lambda_0$	$+0.09\lambda_0$

26.4.2. Tests of the pairing interaction

None of the experiments discussed above can easily distinguish between these two generic d–wave pairing models. This is because whatever the underlying interactions they both lead to the same gap function $\Delta(\mathbf{k})$ and excitation spectrum $E(\mathbf{k})$. They also have the same macroscopic symmetry of the order parameter and would thus be indistinguishable in any macroscopic quantum coherence experiment. What is needed are experiments which can distinguish between different the underlying interactions which lead to the d–wave state.

Since both model effective interactions lead to the same bulk d–wave order parameter it is necessary to look at other types of experiments in order to distinguish them. One possibility is to look at SC fluctuation effects. In the AF spin fluctuation scenario only d–wave pairing is stable (ignoring the weak p–wave coupling) and so the relevant Ginzburg–Landau free energy per unit volume is simply that of a single $O(2)$ OP:

$$f(T) = f_0(T) + \frac{\hbar^2}{2m_d^*}|\nabla\psi_d|^2 + \alpha_d(T - T_{cd})|\psi_d|^2 + \frac{1}{2}\beta_d(T)|\psi_d|^4 \quad (26.24)$$

On the other hand in the charge fluctuation scenario, both s– and d–wave pairing states are stable, only with different T_c values. The appropriate Ginzburg–Landau theory is thus[1,66]

$$f(T) = f_0(T) + \frac{\hbar^2}{2m_d^*}|\nabla\psi_d|^2 + \frac{\hbar^2}{2m_s^*}|\nabla\psi_s^2|^2 + \frac{\hbar^2}{2m_{sd}^*}\left(\psi_d^*\left(\nabla_x^2 - \nabla_y^2\right)\psi_s + c.c\right)$$

$$+\alpha_s(T - T_{cd})|\psi_s|^2 + \alpha_d(T - T_{cd})|\psi_d|^2$$

$$+\frac{1}{2}\beta_s(T)|\psi_s|^4 + \frac{1}{2}\beta_d(T)|\psi_d|^4 + \kappa(T)|\psi_s|^2 \cdot |\psi_d|^2 . \quad (26.25)$$

If the s–wave and d–wave T_c values are close enough, then the fluctuations should show evidence of an underlying 4 component order parameter, rather than the usual 2 component one. Measurements of critical fluctuations in specific heat and penetration depth of $YBa_2Cu_3O_7$ appear to be consistent with a conventional $O(2)$ order parameter. This is to be expected with either form of the free energy, since only one of the two order parameter ψ_d or ψ_s becomes critical at T_c. However Gaussian fluctuations well away from T_c should show some evidence for the second order parameter on T scales of order $T_{cd} - T_{cs}$ about the critical point.

It should also be possible to enhance the effects of the s–wave component of the order parameter if one can tune experimental parameters to make T_{cd} and T_{cs} approach one another. For example, if the interpretation of the photoemission results[67] is that overdoped systems become s–wave, then we can expect that there is a crossover as a function of doping and that T_{cs} eventually exceeds T_{cd}. There will then be a critical doping for which $T_{cd} = T_{cs}$, at which point the system has an order parameter with four real components. The critical behavior at this point would be that of an anisotropic $O(4)$ model, rather than the usual $O(2)$ critical behavior.

Another method of tuning T_{cd} and T_{cs} is with irradiation, or with impurities. Since d–wave pairing states are strongly sensitive to disorder it is expected that T_{cd} will decrease linearly with radiation dose, or with impurity concentration.[68,69] Indeed just this behavior is found experimentally for both irradiation and impurity doping.[1,70-72] A linear decrease in T_{cd} is expected for both the AF spin fluctuation and the charge fluctuation pairing scenarios. However they differ qualitatively in what happens at large radiation doses. In the charge fluctuation scenario T_{cd} decreases rapidly, while T_{cs} remains rather constant, because of Anderson's theorem. Thus at after some critical dose T_{cs} exceeds T_{cd} and the system is an s–wave SC. On the other hand, in the AF spin fluctuation scenario, the pairing interaction is strongly *pair breaking* for s–wave CP, and so s–wave pairing is never stable. One therefore expects that after increasing radiation doses T_{cd} gradually decreases to zero, after which the system is a normal metal. These two possibilities are shown schematically in Fig. 26.1. Thus by an appropriate irradiation experiment one should be able to distinguish whether spin–fluctuations or phonons provide the underlying pairing interaction in the cuprates. The irradiation experiments[70] give results which closely resemble Fig. 26.1a, showing a d–wave to normal metal transition as a function of radiation dose. This can thus be viewed as strong evidence in favor of the AF spin fluctuation scenario. The same behavior was seen in the impurity doping experiments.[71] On the other hand, if the effect of overdoping with oxygen also serves primarily to increase disorder, then the results of Kelley *et al.*[67] would appear to favor Fig. 26.1b over Fig. 26.1a.

There are some caveats to any attempt to eliminate possible pairing mechanisms using these arguments. Firstly, both irradiation and doping with impurities may change the underlying electronic structure in unforeseen and complicated ways, which would invalidate the assumption that the FS remains roughly the same at all doses and only the disorder is increased. This caveat is especially true for overdoping. Secondly, even s–wave SC can be destroyed by sufficiently strong disorder if the system undergoes a transition to an Anderson or Mott insulator phase. The experimental claim that the highly dosed state is a normal metal, not an insulator, is thus crucial.[70] Thirdly, it is also necessary to assume that the pairing interaction itself, $V(\mathbf{k},\mathbf{k}')$, is not destroyed by the radiation. This, in turn, depends on whether quasi long range order is necessary for the relevant interactions or not. For example the nearest neighbor phonon mediated pairing interaction implied by Eq. (26.23) should be quite insensitive to the disorder, as should short range spin fluctuations. However quasi long range order may be disrupted by irradiation, and so mechanisms which depend strongly on nesting, vHs, or small $\mathbf{Q}$ phonon pairing[65] may be also strongly influenced by disorder, even for s–wave pairing states. However presumably in these cases although the T_c would cease to be strongly enhanced in the disordered system, T_c would remain non–zero at a typical value of a low T_c SC in the disordered state.

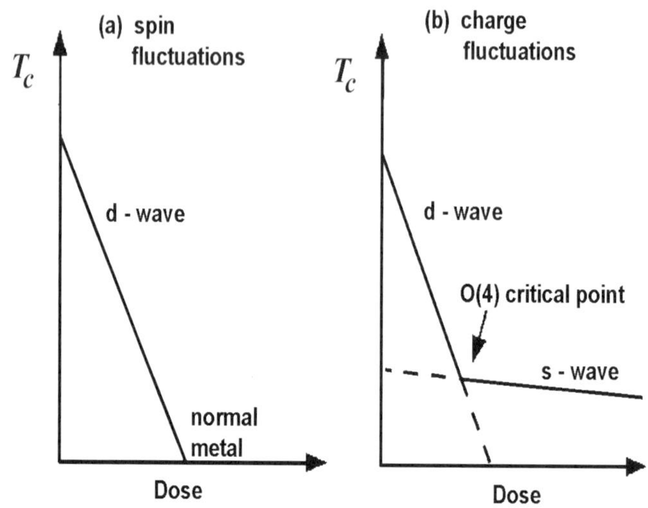

a) b)

FIG. 26.1. Schematic of effects of irradiation on a d–wave superconductor, assuming either (a) spin or (b) charge fluctuations provide the pairing interaction.

Further tests are necessary to confirm whether irradiation, impurities or overdoping correspond to a d–wave to normal metal transition, or whether there is d–wave to s–wave transition followed by an s–wave to normal metal transition. Perhaps it would be possible to irradiate the same samples used in the SQUID and flux quantization experiments[61,73-76] and thus observe a change in macroscopic symmetry *in the same sample* before and after irradiation. Alternatively it would be interesting to know whether the fluctuation effects near T_c change qualitatively as a function of irradiation or doping. Again this must happen if there is a d–wave to s–wave cross–over as shown in Fig. 26.1b.

Conclusion

It can now be considered a well established fact that the pairing state in most of the cuprate systems corresponds to $d_{x^2-y^2}$ superconductivity. While this fact alone eliminates many possible pairing mechanisms, it still leaves many possibilities. A. have attempted to highlight two simple BCS type of models which give rise to a d–wave pairing state. These correspond to exchange of AF spin fluctuations or of small **Q** phonons. Experiments which can distinguish between these two generic possibilities include investigation of Gaussian fluctuation effects outside the true critical regime, doping studies in which the pairing state may change as a function of doping, and irradiation and impurity experiments. In particular the irradiation experiments[70] appear to show a d–wave to normal metal transition (Fig. 26.1a) rather than a d–wave to s–wave transition (Fig. 26.1b). This provides strong support to the AF spin fluctuation scenario, and is difficult to explain with a phonon pairing model.

26.4.3. Influence of electron–phonon interaction on $d_{x^2-y^2}$ – pairing

Bulut and Scalapino (BS)[77] studied the relationship of the $d_{x^2-y^2}$ symmetry to the pairing mechanism by examining how three types of e–ph interactions affect $d_{x^2-y^2}$ pairing.

As we discussed above (as well as it was discussed by a number of authors[78-80]), a $d_{x^2-y^2}$ gap naturally arises from the exchange of AF spin fluctuations. However, the physical picture that emerges from these calculations is more general and shows that a $d_{x^2-y^2}$ gap will occur for a nearly half–filled band when there is an effective singlet interaction which is repulsive for onsite pairing and attractive for near–neighbor pairing. This spatial structure of the interaction means that its Fourier transform becomes more positive as the momentum transfer increases toward large values. This is easily understood from the BCS gap equation:

$$\Delta_{\mathbf{p}} = -\sum_{\mathbf{p}'} V(\mathbf{p}-\mathbf{p}')\Delta_{\mathbf{p}'}/2E_{\mathbf{p}'} \tag{26.26}$$

Near half–filling of the 2D system, the phase space is such that the important scattering processes take electrons from $(\mathbf{p}\uparrow, -\mathbf{p}\downarrow)$ with $\mathbf{p}$ near a corner of the FS, say, near $(\pi, 0)$, to $(\mathbf{p}'\uparrow, -\mathbf{p}'\downarrow)$ with $\mathbf{p}'$ near $(0,\pi)$ or $(0,-\pi)$. Since the interaction $V(\mathbf{p}-\mathbf{p}')$ is positive, the relative phase of the states $(\mathbf{p}\uparrow, -\mathbf{p}\downarrow)$ making up the bound CP changes sign as $\mathbf{p}$ goes from $(\pi,0)$ to $(0,\pi)$ or $(0,-\pi)$, leading to a gap with $d_{x^2-y^2}$ symmetry. This is the case within a RPA in which the interaction is mediated by the exchange of AF spin fluctuations. It has also been found by Monte Carlo calculations[81,82] that for the Hubbard model the pairing interaction is attractive in the $d_{x^2-y^2}$ channel. In these cases the interaction is positive at all momentum transfers becoming larger in the region near (π,π) associated with the short–range AF correlations.

In order to explore the effect of e–ph interactions, BS begin with a 2D HM on a square lattice:

$$H = -t\sum_{\langle i,j\rangle,s}\left(c_{is}^{\dagger}c_{js} + c_{js}^{\dagger}c_{is}\right) + U\sum_{i} n_{i\uparrow}n_{i\downarrow} \tag{26.27}$$

Here t is a near–neighbor hopping and U is the on–site Coulomb interaction. BS take a simple phenomenological RPA form[83] of the singlet pairing interaction associated with the exchange of spin fluctuations:

$$V_{SF}(p'-p) = \frac{3}{2}\bar{U}^2\chi(p'-p) \tag{26.28}$$

with $\chi(q) = \chi_0(q)/[1-\bar{U}\chi_0(q)]$. Here $p = (\mathbf{p}, i\omega_n)$, $\bar{U}$ is a renormalized Coulomb interaction, and χ_0 is the spin susceptibility:

$$\chi_0(\mathbf{q},\omega)=\frac{1}{N}\sum_{\mathbf{p}}\frac{f(\varepsilon_{\mathbf{p+q}})-f(\varepsilon_{\mathbf{p}})}{\omega-(\varepsilon_{\mathbf{p+q}}-\varepsilon_{\mathbf{p}})+i0^+} \tag{26.29}$$

with $\varepsilon_{\mathbf{p}}=-2t(\cos p_x+\cos p_y)-\mu$. Now, in addition to V_{SF}, we following BS will examine three model e–ph interactions. The first is a simple on–site Holstein coupling of the following form:

$$V_1=\sum_i g x_i n_i \tag{26.30}$$

with x_i the atomic displacement at site $\hat{i}$ and $n_i=n_{i\uparrow}+n_{i\downarrow}$ the on–site electron density. One could imagine this type of coupling arising from the interaction with an apical oxygen O(4). In Eq. (26.30) the coupling is linear in the atomic displacement rather than quadratic. This is possible since O(4) breaks the reflection symmetry with respect to a single CuO_2 layer. The second e–ph interaction can be viewed as arising from the in–plane breathing motion[84] of an O(2) oxygen:

$$V_2=\sum_i g\left[x_i(n_i-n_{i+x})+y_i(n_i-n_{i+y})\right] \tag{26.31}$$

Here, x_i describes the displacement of the O(2) along the $\mathbf{x}$ axis between the Cu sites at $\hat{i}$ and $\hat{i}+\hat{\mathbf{x}}$ and y_i the $\mathbf{y}$–axis displacement of an O(2) along the $\mathbf{y}$ axis between the Cu sites at $\hat{i}$ and $\hat{i}+\hat{\mathbf{y}}$. The third interaction involves an axial $\mathbf{z}$ motion of a buckled O(2) atom:

$$V_3=\sum_i g\left[z_i^x(n_i+n_{i+x})+z_i^y(n_i+n_{i+y})\right] \tag{26.32}$$

Here, z_i^x is for an O(2) between the $\hat{i}$ and $\hat{i}+\hat{\mathbf{x}}$ sites and z_i^y is for an O(2) between the $\hat{i}$ and $\hat{i}+\hat{\mathbf{y}}$ sites. Linear coupling of this type is possible for buckled Cu–O–Cu bonds. The e–ph interactions considered here are diagonal in the electron number representation. Note that it would also be interesting to study the effects of phonon modes where the e–ph coupling is not diagonal in the electron number representation.

Assuming for the discussion that the lattice coordinate can be described as a local harmonic oscillator with frequency ω_0 (which just as g is of course different for the different modes), the effective e–e interaction mediated by the exchange of these phonons is:

$$V_{ph} = -2|g(\mathbf{q})|^2 \omega_0 / \left(\omega_m^2 + \omega_0^2\right), \tag{26.33}$$

where ω_m is the Matsubara frequency $2m\pi T$. Here for the local interaction, Eq. (26.30),

$$|g(\mathbf{q})|^2 = |g|^2 / 2M\omega_0 \tag{26.34}$$

while for the breathing mode, Eq. (26.31),

$$|g(\mathbf{q})|^2 = |g|^2 \left(\sin^2 q_x/2 + \sin^2 q_y/2\right) / 2M\omega_0 \tag{26.35}$$

and for the axial mode, Eq. (26.32),

$$|g(\mathbf{q})|^2 = -|g|^2 \left(\cos^2 q_x/2 + \cos^2 q_y/2\right) / M\omega_0 \tag{26.36}$$

with M the O ion mass.

In order to see how these interactions affect the pairing BS examine the leading eigenvalue λ and eigenfunction $\phi(p)$ of the Bethe–Salpeter equation, neglecting self–energy contributions:

$$\lambda\phi(p) = -\frac{T}{N}\sum_{p'}\left[V_{SF}(p-p') + V_{ph}(p-p')\right] \times G(p')G(-p')\phi(p'), \tag{26.37}$$

where $G(p)$ is the single–particle Green's function given as:

$$G(p) = \left(i\omega_n - \varepsilon_p\right)^{-1} \tag{26.38}$$

From now on, we will measure energies in units of t. The chemical potential has been chosen so that the site occupation $\left(n_{i\uparrow} + n_{i\downarrow}\right) = 0.875$ and an effective Coulomb interaction $\overline{U} = 2$ has been taken. We will also take $\omega_0 = 0.25$ and $|g|^2/M\omega_0 = 1$ corresponding to an e–ph coupling strength $|g|^2 N(0)/M\omega_0^2 = 0.8$, where $N(0)$ is the electron DOS at μ_F.

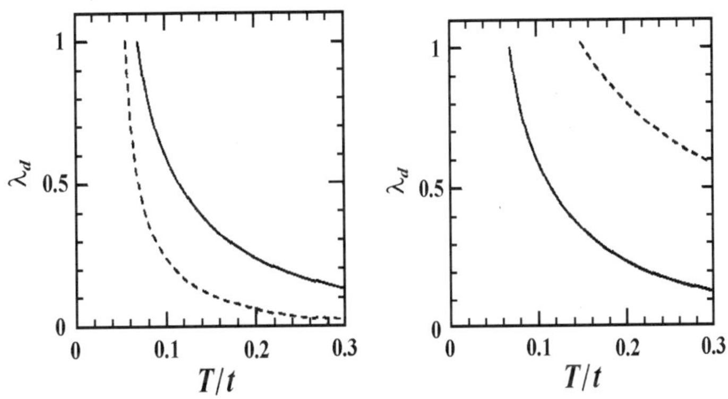

FIG. 26.2. The $d_{x^2-y^2}$ eigenvalue of Eq. (26.37) vs T in units of the hopping t. The solid curve gives the eigenvalue for just the spin–fluctuation interaction V_{SF} and the dashed curve shown the effect when the e–ph interaction V_2 associated with the breathing mode, Eq. (26.31) and (26.35), is added to V_{SF}. The coupling constants are given in the text.

FIG. 26.3. Same as Fig. 26.2 except for the axial O election–phonon interaction V_3. Eqs. (26.32) and (26.36).

BS find that the leading eigenvalue in the even frequency singlet channel has $d_{x^2-y^2}$ symmetry and the T dependence of the eigenvalue $\lambda_{x^2-y^2}(T)$ is shown in Figs. 26.2 and 26.3 for the e–ph interactions given by Eqs. (26.31) and (26.32), respectively. The solid line in each figure shows the eigenvalue in the absence of the phonon–mediated interaction ($g=0$), while the dashed curve shows the effect of including the phonon–mediated term. It is clear that the breathing mode interaction, Eqs. (26.31) and (26.35), suppresses $d_{x^2-y^2}$ pairing[85] while the axial O(2) mode of Eqs. (26.32) and (26.36) enhances the $d_{x^2-y^2}$ pairing, raising T_c. The local interaction, Eqs. (26.30) and (26.34), is orthogonal to the $d_{x^2-y^2}$ gap and hence does not affect the $d_{x^2-y^2}$ eigenvalue when self–energy effects are neglected. Including it in the self–energy will act to suppress T_c due to the wave function renormalization.

To understand the behavior shown in Figs. 26.2 and 26.3 we note that the strength of the coupling to the axial mode, Eq. (26.36), decreases as $\mathbf{q}$ approaches (π,π). Because the phonon–mediated interaction, Eq. (26.33), is negative, decreasing the magnitude of the coupling $|g(\mathbf{q})|^2$ acts to make the interaction *more positive* as the momentum transfer increases. As discussed above, this is the criteria for a $d_{x^2-y^2}$ gap to form when the system is near half–filling. Clearly it will be interesting to examine the isotope effect within models in which the strength of the e–ph coupling decreases at large momentum transfers.

Thus BS conclude that a $d_{x^2-y^2}$ gap implies that the pairing interaction becomes more positive for large momentum transfers. This is clearly the case for the spin–fluctuation interaction, Eq. (26.28), but as shown it can also occur for the attractive phonon–mediated interaction if

$|g(\mathbf{q})|^2$ *decreases* at large momentum transfers. This form of coupling would also give rise to an e–ph coupling constant λ which could be large compared to the effective coupling constant λ_{tr} entering transport processes since the transport coupling constant λ_{tr} weights large momenta transfers more heavily.[86,217] For small momentum transfers **q**, a scattering of $(\mathbf{p}\uparrow, -\mathbf{p}\downarrow)$ to $(\mathbf{p}+\mathbf{q}\uparrow, -\mathbf{p}-\mathbf{q}\downarrow)$ with $\mathbf{p}$ near a corner of the FS connects regions which have the same sign of the $d_{x^2-y^2}$ gap so that according to Eq. (26.26) an attractive e–ph interaction [negative $V(q)$] enhances $\Delta_\mathbf{p}$. Another way to see that this latter case is similar to the spin–fluctuation interaction is to add U onto the phonon interaction, giving $U+V_{ph}$. For the $d_{x^2-y^2}$ channel, a constant has no effect, but if it is larger than the magnitude of the phonon–mediated interaction, Eq. (26.33), then as $|g(\mathbf{q})|^2$ *de*creases, the total interaction $U+V_{ph}$ is positive and increases as $\mathbf{q}$ becomes large, just as V_{SF}. In order to obtain more quantitative information on the role of the e–ph interaction, it would be useful to have BS calculations[87,88] of the $d_{x^2-y^2}$ e–ph coupling constant for the ν mode:

$$\lambda^{\nu}_{d_{x^2-y^2}} = \frac{2\sum_{k,k'} g(k)g(k')\left(\left|M^{\nu}_{kk'}\right|^2/\omega_{k-k'}\right)\delta(\varepsilon_k)\delta(\varepsilon_{k'})}{\sum_k g^2(k)\delta(\varepsilon_k)} \qquad (26.39)$$

Here, $g(k)\sim(\cos k_x - \cos k_y)$, ε_k is the band energy with k including the band index, and $M^{\nu}_{kk'}$ is the e–ph matrix element for the $\nu\, th$ phonon mode.

Note, that two other papers are relevant to BS's work.[189]

26.5. Experimental Symmetry Probes

Having classified the many different ways in which the superconducting symmetry breaking may occur, it remains to be seen which experimental measurements may distinguish between the different states, and what relevant experimental evidence actually exists in the HTSC. In this section we, following Annett paper,[4] will discuss a number of experiments which could help narrow down the range of possibilities. We will concentrate in particular on those experiments which are fundamental probes of the symmetry properties of the superconducting state. By "fundamental probes of symmetry" we mean that the outcome of the particular experiment can be decided purely by symmetry arguments, expressed in terms of the superconducting order parameter. The experiment thus determines the superconducting symmetry independently of the actual microscopic basis of the superconductivity. The experiments of this kind that we will discuss are: the Josephson effect, splitting of the transition and strain anomalies, fluctuations and critical behavior, CM and exotic vortices etc. Only in section 26.5.7 we discuss other experiments that can shed light on the symmetry, but for which it is necessary to have a microscopic theory to describe the SC. This latter group of experiments includes low–T properties of thermodynamic and transport properties, such as specific heat, thermal conductivity, penetration depth, NMR relaxation and Knight shift, and other probes of gap nodal structure. For each experiment we will give a brief description of the current experimental situation; however, this will be far from a comprehensive review of all the relevant experimental literature. Similar discussions of how to identify unconventional superconducting phases have been given in Refs. 89–93.

26.5.1. Josephson effects

General principles

By far the most widely known test for exotic superconductivity is the existance, or non–existance of the Josephson effect in a junction between a conventional and an unconventional superconductor. There have been many observations during the last years of non–vanishing Josephson effects between $YBa_2Cu_30_7$ and the familiar "low T" SC such as niobium,[94,95] and this is often sited as evidence against an unconventional symmetry state in $YBa_2Cu_30_7$.[96] In order to be absolutely clear about what the experiment does and does not tell us, let us examine this situation from a fundamental symmetry viewpoint.

Given two superconductors, each with broken gauge symmetry, then one expects to see a Josephson effect whenever the two order parameter phases are coupled. If the order parameters are η_i and ψ (assuming for simplicity that one has a scalar order parameter, as expected in Nb, for example) then a Josephson effect will occur whenever there are non–zero terms in the free energy of the following form:

$$a_{ij...kl.}(\psi^*)^n(\psi)^m(\eta_i^*\eta_j^*...)(\eta_k\eta_l...)+c.c. \tag{26.40}$$

with $n \neq m$ and where $c.c.$ denotes the complex conjugate. In that case a $d.c.$ voltage bias, V, across the junction, which causes the phase of the order parameter ψ to rotate relative to η_i like $\exp(ie^* Vt/\hbar)$ will produce an oscillatory current like $\sin[c^* V(m-n)t/\hbar+\delta]$ in the junction.[37] Here e^* is the effective charge which couples the condensate to the EMF ($2e$ for BCS superconductivity) and which for simplicity we have assumed the same for both SC.[97,98,100] The fundamental symmetry issue is thus whether or not the coupling (26.40) is allowed or whether it must vanish identically for some group–theoretic reason.

In the absence of spin–orbit coupling it is simple to see that if the niobium has a singlet pairing state, and the Y–Ba–Cu–O has a triplet state then the leading order coupling:

$$a_\mu \psi^* \eta_\mu + c.c. \tag{26.41}$$

must be zero.[101] This is simply because there is no preferred direction in spin space for the coupling constant a_μ. Put more formally: under global spin rotations in SO(3) the product $\psi^m \eta_\mu$ transforms according to $\mathbf{0} \otimes \mathbf{1} = \mathbf{1}$ and so cannot be an invariant transforming like $\mathbf{0}$ (here $\mathbf{0}$ and $\mathbf{1}$ are the $S=0$ and $S=1$ representations of SO(3) respectively). Clearly the type of argument is identical to the one used to find the possible Ginzburg–Landau quartic and gradient invariants.

The fact that (26.41) is identically zero is the content of the frequently made statement that the observation of Josephson coupling with niobium rules out triplet pairing in the copper–oxide SC. Notice however how restrictive the conditions are. As soon as the spin–orbit coupling is non–zero (however small, and in either the junction, the Y–Ba–Cu–O or in the Nb), then the Josephson current will be non–zero because there are preferred directions in spin–space, such as normal to the junction plane.[102,103] A magnetic interface would also have the same effect. The only statement which can then be made about the pairing is to estimate the magnitude of the coupling, which requires microscopic theory. In BCS theory it is generally expected[102,103] that the singlet–triplet coupling is smaller than singlet–singlet coupling by approximately a/ξ_0 where a is a lattice constant and ξ_0 is a coherence length, a ratio which is of order unity (~0.2) for Y–Ba–Cu–O.

In fact, even in the absence of spin–orbit coupling or magnetic fields there would be non–zero higher order couplings of the following form:

$$a\psi^* \left(\boldsymbol{\eta}^* \times \boldsymbol{\eta}\right) \cdot \boldsymbol{\eta} + c.c. \tag{26.42}$$

However, in weak–coupling BCS theory they would be small by some powers of Δ/ε_F, where Δ is the maximum gap ($\sim T_c$) and ε_F is the Fermi energy. They would be significant in an extreme strong coupling limit where Δ is comparable to ε_F.

As well as selection rules based upon spin symmetry, there are angular momentum considerations that might lead to a vanishing Josephson current, first investigated by Akhtyamov.[104] For general junctions this does not lead to any special symmetry selection rules, since the junction breaks any rotational symmetries. However, if the junction could be made in such a way as to preserve a twofold or fourfold symmetry axis of the crystal, or a mirror plane, then the Josephson current can vanish rigorously for certain types of superconducting state.[105] For example, suppose that the Josephson junction is fabricated by epitaxial growth or some similar technique on an $a-b$ face of Y–Ba–Cu–O normal to the c axis. If the interface preserves the C_2 (or approximate C_4) symmetry of the c axis, then it is still possible to classify SC states of the entire S–I–S system according to their transformation properties under C_2 (or C_4). For example, for the singlet states of orthorhombic symmetry the ${}^1A_{1g}$ and ${}^1B_{1g}$ states are invariant under this C_2 operation, while the ${}^1B_{2g}$ and ${}^1B_{3g}$ states change sign. Thus, if the Y–Ba–Cu–O were in a ${}^1B_{2g}$ state, then the Josephson coupling (26.40) to niobium (presumably a ${}^1A_{1g}$ superconductor) must vanish at all powers of m and n with $m-n$ odd. The first non–zero couplings would be for $n-m=2$, which would result in a Josephson current with frequency $2e^*V/h$ instead of the conventional e^*V/h. Clearly, selection rules such as this can occur for junctions fabricated on a variety of different crystallographic symmetry axes or planes. A thorough listing of the directions in which the leading order Josephson effect must vanish has been given by Gor'kov[91] for the TG singlet and triplet states with strong spin orbit coupling. It is hoped that advances in surface preparation techniques will make this oriented junction experiment a possibility in the near future, since it provides a clear and unambiguous test of the SC state symmetry. Anderson has argued that there is no evidence to date of Josephson effect in planar junctions, and hence that this constitutes evidence for an unconventional order parameter. Finally, there are similar predictions of tests of unconventional superconductivity using the proximity effect.[106,107]

A third type of Josephson coupling selection rule is worth mentioning here. If the superconductor order parameter changed sign between successive CuO_2 planes, then it is possible that the order parameter vanishes on planes in real space. A movable tunneling probe, such as a scanning tunneling microscope, could then directly observe these zeros by measuring Josephson tunneling as a function of position in the $a-c$ or $b-c$ surface planes. Similar real–space zeros could exist on or between twinning planes[108] and could be observed by the same technique.

Macroscopic symmetry tests

Macroscopic quantum coherence phenomena provide a completely different route to explore the OP symmetry. Because these phenomena do not rely on particular microscopic aspects of the theory they may be somewhat more robust. Nevertheless there are still some complication in interpretation, as we discuss below.[12]

The Josephson tunneling between two SC leads to a coupling energy which, in general is given by:[1]

$$E_j = -J\cos(\phi_1 - \phi_2 + \alpha)$$

Here, ϕ_1 and ϕ_2 are the phases of the Ginzburg–Landau order parameter of the two superconductors far from the junction. The real constants $J(\geq 0)$ and α reflect both the junction geometry and the pairing states of the two superconductors. If both superconductors superconductors and the junction respect time reversal symmetry and $J \neq 0$ then α is either 0 or π. The latter case is termed a π junction.

The Josephson effect is useful as a probe of the pairing state symmetry, in two ways. Firstly J (and hence $I_c = 2\pi J/\varphi_0$) will vanish by symmetry for certain unconventional pairing states *provided the junction is an ideal plane aligned with a crystallographic symmetry axis.* AGL have termed this key point PRINCIPLE A.[1]

The second key principle, PRINCIPLE B, relies on the fact that for unconventional superconductors there will be π Junctions for certain crystallographic orientations. Since the value of α depends on a choice of gauge for the superconducting phases ϕ_1 and ϕ_2, this is not directly observable in a single junction. However for superconducting loops containing two or more junctions the net phase for the loop, $\sum_i a_i$, is gauge invariant and physically observable. For example, assuming inductive energies are negligible such loops will have free energy minima at half integral flux $\Phi = (n + 1/2)\varphi_0$.

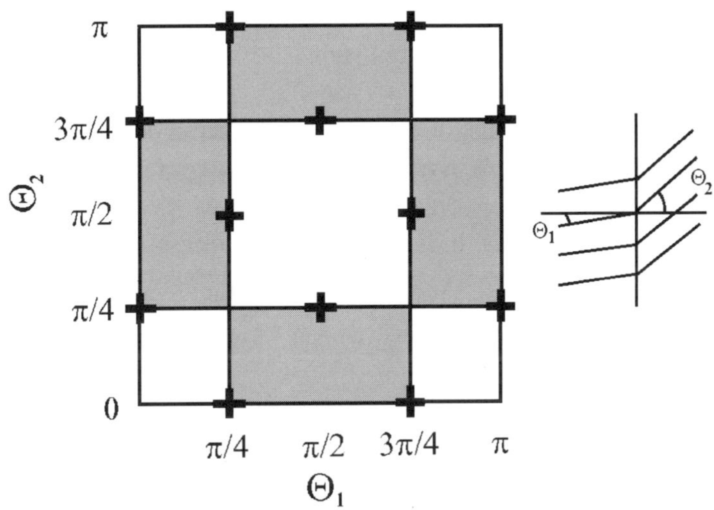

FIG. 26.4. $d_{x^2-y^2}$ Josephson coupling in an a–b plane grain boundary junction.

In order to make use of principle B one must relate the junction parameters J and α to the two superconducting gap functions $\Delta_1(\mathbf{k})$ and $\Delta_2(\mathbf{k})$ and the junction orientation. This is non–trivial, and in principle requires a detailed microscopic calculation of order parameter and electronic states in the junction region. Simplifying by assuming that the tunneling is dominated by the single $\mathbf{k}$ vector normal to the junction leads to the following estimate:

$$E_J = -A\Delta_1(\mathbf{k}_1)\Delta_2^*(\mathbf{k}_2) + c.c. \tag{26.43}$$

where $\mathbf{k}_1$ and $\mathbf{k}_2$ are the vectors on the FS of superconductors 1 and 2 in the direction of the junction normal. For example, in a a–b plane grain boundary junction of two $d_{x^2-y^2}$ SC this yields:

$$E_j = -J_0 \cos(2\theta_1)\cos(2\theta_2)\cos(\phi_1 - \phi_2) \tag{26.44}$$

where θ_1 and θ_2 are the angles of the two crystallographic a–b axes relative to the junction.[109] Fig. 15.4 shows (as shaded regions) the angles where π–junction behavior occurs according to Eq. (26.44). The figure also shows (as crosses) the orientations where zero coupling must occur according to principle A. Even if the microscopic approximation Eq. (26.43) is not accurate, changes from normal to π–junction behavior must occur at the crosses. In fact experimental measurements of I_c as a function of θ_2 for $\theta_1 = 0$ show a much more rapid variation with angle than Eq. (26.44) suggests,[110] however one which is consistent with these principle A zeros.

We shall only very briefly summarize the key experimental results, since more detailed reviews are available.[1,111] The original Wollman *et al.*[76] experiment demonstrates a π phase shift in SQUID loops connecting a and b axis faces of YBCO with Nb. This is consistent with Eq. (26.43) for a mainly $d_{x^2-y^2}$ gap function. There is also evidence for time reversal symmetry[112] showing that $\alpha_a + \alpha_b = \pi$ to a precision of at least 5%, and hence $\Delta(\mathbf{k})$ is real. The tricrystal experiments of Tsuei *et al.*[73-75] are also consistent with the π–junctions predicted by Eq. (26.44) in both YBCO and in $Tl_2Ba_2CuO_{6+\delta}$, a single plane TG SC. Taken together these and other experiments[1] provide a strong confirmation of a predominantly $d_{x^2-y^2}$ gap in the HTSC. While, in principle, π–junctions may occur even in s–wave SC (e.g. due to magnetic impurities) it would be highly unlikely that these would occur in all of these experiments, and not in any of the corresponding control experiments. The measurements could be consistent with either pure $d_{x^2-y^2}$ or mixed s+d states, although the observation of time reversal invariance severely limits the possible s+id type mixtures.

Only two phase sensitive experiments are, taken at face value, not consistent with the pure d–wave predictions. These are the observations of *c*–axis Josephson effects with Pb,[113] and the observation of non–zero Josephson coupling between a YBCO film and a hexagonal region of the film with crystal axis rotated by 45°.[114] These experiments may provide evidence for a mixed s/d pairing state. Alternatively it may be that the interpretation of the experiments is more complex than originally thought.

Sun *et al.*[113] Lguchi and Wen[115] and Kleiner *et al.*[116] measured a Josephson effect between Pb the c–axis face of YBCO thin films and crystals. Kleiner *et al.* found that the product $I_c R_N$ was largest ($1meV$) for untwinned single crystals, decreasing on twinned samples and being smallest ($10\mu eV$) for films. The non–zero I_c for the untwinned crystals is not at all surprising, since the YBCO is orthorhombic and $d_{x^2-y^2}$ and s^+ pairing states become mixed. The difficulty is in explaining the observations on twinned crystals and films. If we assume the YBCO gap function to be of the mixed s/d form then Δ_s/Δ_d will have a definite sign determined by the OR Fermi surface. There are then two possible behaviors in twinned crystals: Either (i) Δ_d and Δ_s can both change sign across twin boundaries, or (ii) Δ_d and Δ_s can both keep the same sign on either side of a twin boundary. Averaging over a macroscopic number of twins in case (i) the system has a finite average d–wave order parameter and no s–wave component, while in case (ii) it is the s–wave component which survives averaging. Effectively the averaging over twins restores a macroscopic TG symmetry and either a pure s or pure d–wave state. The results of Wollman *et al.* on heavily twinned samples show we must have case (i) and the twin averaged OP is pure d–wave.[115] However then we would expect $I_c = 0$ for c–axis tunneling in heavily twinned samples, in contradiction to the observations of a finite (if small) I_c.

How can one reconcile these apparent contradictions? Ideally one would like c–axis tunneling experiments on TG samples, which would eliminate the issue of twinning altogether. Possibly this idea of a macroscopic TG "twin–averaged" crystal is not applicable. For example such averaging assumes an equal density of both type of twin, which may not be the case if macroscopic strains are present during the crystal growth. The $I_c R_N$ values observed by Kleiner *et al.*[116] did decrease with number of twins, but more slowly than a simple $1/\sqrt{N}$ statistical averaging would imply. It is possible the effective value of N in the averaging may be smaller than expected if the tunneling is dominated by surface defects (such as step edges, dislocations etc.) rather than the regions of flat ideal YBCO surface. It is also possible that the order parameter is deformed near the surface in order to eliminate the "frustration" implicit in the YBCO–Pb junction. For example a surface mixing of s+d could explain the Sun *et al.*[113] and Kleiner *et al.*[116] measurements without requiring a two bulk phase transitions.[66] Of course a bulk mixing of s and d is also possible, but note that this would have to be a mixing even in the effective TG twin–averaged system, and thus would require a second phase transition.

The other experiment which apparently contradicts a pure $d_{x^2-y^2}$ pairing state involves the observation of a non–zero Josephson current, between a YBCO oriented thin film and a hexagonal region rotated by -45°.[114] For a $d_{x^2-y^2}$ order parameter principle A implies that the I_c would be zero on two of the hexagon faces. The other faces occur in pairs having opposite signs for E_J implying a total I_c of zero for the complete hexagon. Furthermore, removing faces one at a time, I_c was found to be simply proportional to the number of remaining faces. Taken at face value this observation would appear to imply an *isotropic* s–wave gap. However this cannot be the correct explanation. Isotropic s–wave would be inconsistent with the other phase sensitive measurements and with the microscopic measurements of the energy gap, such as optical,

Raman and $\lambda(T)$.[1] An I_c independent of junction orientation also contradicts other similar thin film grain boundary measurements, such as the tricrystal ring experiments of Tsuei *et al.*[73-75] and the systematic measurements of I_c as a function of miscut angle by Ivanov *et al.*[110] An effect which might indeed influence the Lin and Chaudhari type experiments has been suggested by Millis.[117] He argued that half integer fluxoids would spontaneously nucleate in the hexagon in order to make I_c non–zero. Indeed Kirtley *et al.*[118] did find such fluxoids in similar hexagons, although they were not distributed in exactly the way Millis predicted.

Conclusion

It is clear that a large body of experiments now provide strong evidence for a predominantly $d_{x^2-y^2}$ order parameter in the cuprates. Whether this is a pure d–wave state or has an admixture of s or d_{xy} is less clear at this time (for d_{xy} admixture in magnetic field see 15.5.2). Imaginary mixed states such as s+id or d+id are very strongly constrained by photoemission, T dependence of the penetration depth, and the time reversal symmetries seen in certain SQUID interference patterns. Photoemission also now provides some quite stringent bounds on real s+d mixtures. Such mixed states might explain the Sun *et al.*[113] and Lin and Chaudhari experiments, which are otherwise hard to reconcile with a pure d–wave state. On the other hand it is possible that both of these experiments are complicated by other effects. It would be very helpful to repeat the Sun *et al.*[113] type experiments using TG systems or BSCO where s/d mixing is not permitted without a second phase transition.

26.5.2. Magnetic induction of $d_{x^2-y^2}+id_{xy}$ order in HTSC

Laughlin[120] proposed that the phase transition in $Bi_2Sr_2CaCu_2O_8$ observed by Krishana *et al.*[119] is the development of a small d_{xy}

superconducting order parameter phased by $\pi/2$ with respect to the principal $d_{x^2-y^2}$ one to produce a minimum energy gap Δ. Krishana *et al.*[119] have reported a phase transition in $Bi_2Sr_2CaCu_2O_8$ induced by a magnetic field and characterized by a kink in the thermal conductivity as a function of field strength, followed by a flat plateau. The high–field state is also SC. They argued from the existence of this plateau that heat transport by quasiparticles was zero in the new state and that this probably indicated the development of an energy gap. The transition has the peculiarity of being easily induced by small fields. Krishana *et al.*[119] report the empirical relation $T_c \propto \sqrt{B}$, although over the limit field range of $0.6T < B < 5T$, and also that the transition sharpens as T_c is reduced.

Laughlin proposed that the new high–field state is the parity and time–reversal symmetry violating $d_{x^2-y^2} + id_{xy}$ superconducting state proposed long ago by him,[120] which has many properties in common with quantum Hall states, including particularly chiral edge modes and exactly quantized boundary currents. The essential point of his argument is that the state must have a magnetic moment in order to account for the experiment, and this is possible only if it violates both parity and time–reversal symmetry. The development of $s + id$ order, for example, or high–momentum Cooper pairing are both ruled out for this reason, as is a restructuring of the vortex lattice.

The $d_{x^2-y^2} + id_{xy}$ state differs fundamentally from $s + id$ conventional s–wave states in not being continuously deformable to a Fermi sea on a sample with edges, although it can be so deformed on a torus. This is the property underling Wiegmann's concept of a "topological superconductivity".[121]

The $d_{x^2-y^2} + id_{xy}$ state is, however, continuously deformable into a doubly occupied Landau level.

26.5.3. Transition splitting, spontaneous strain and magnetism

Transition splitting

A conclusive experimental proof that the order parameter of the SC phase has more than a single component would be the observation of multiple phases and/or splitting of the transition under symmetry breaking perturbations. This phenomenon is well known in ^3He, where a number of different phases may occur as a function of pressure and magnetic field.[122] A splitting of the phase transition was also observed in UPt_3,[123,124] adding weight to previous evidence for unconventional SC in that material.[90,91,125] For simplicity, we will only discuss two generic types of splitting of the phase transition that might occur for the HTSC, namely splitting due to orthorhombic distortions, or due to magnetic fields.

The simplest case of phase transition splitting occurs for the two component E representation states under a weak orthorhombic distortion or strain. In this case, a perturbation which lowers the symmetry from D_4 to D_2 would split the D_4 E representation into two one–dimensional representations:

$$E \to B_2 + B_3 \tag{26.45}$$

of D_2. The two components of the E OP η_1, and η_2 are then no longer degenerate and can be associated with slightly different T_c say T_{c1} and T_{c2}. At the higher, say T_{c1}, the order parameter η_1 becomes non–zero. On lowering the T to T_{c2} there may or may not be a second PT at which the component η_2 becomes non–zero. There will be a second transition if the stable phase of the E representation is $(1,i)$ or $(1,1)$ type, while there will not be a second transition if the $(1,0)$ phase is stable (assuming the D_2 axes are along x and y).

A systematic way to investigate the transition splittings that occur under external symmetry breaking perturbations is to construct the GL functionals which include coupling of the order parameter to the external field. For example if ε_{ij} is the 2D strain tensor relative to the D_4 structure, then it couples to an E representation order parameter as:

$$\Delta f_s = K_s \varepsilon_{ij} \eta_i^* \eta_j \tag{26.46}$$

to leading order. If the orthorhombic distortion occurs so that the D_2 twofold axes are the TG x and y axes then $\varepsilon_{ij} = \lambda \sigma_z$, while if they are the TG $<11>$ and $<1\bar{1}>$ axes then $\varepsilon_{ij} = \lambda \sigma_z$. The phase diagram can then be determined by minimization of the GL functional in the presence of the strain field ε_{ij}. The various splittings that occur for the E phases due to the slight orthorhombic distortion have been enumerated in detail.

Transition splittings may also occur in the presence of magnetic fields. For example for triplet phases without spin orbit coupling, the magnetic field may couple to the order parameter by the following SO(3) symmetry breaking terms in the Ginzburg–Landau functional[122]

$$\Delta f_H = K_H i \varepsilon_{\gamma\mu\nu} H_\gamma \eta_{\mu i}^* \eta_{\nu i} + K'_H H_\mu H_\nu \eta_{\mu i}^* \eta_{\nu i} \,. \tag{26.47}$$

Since the linear term vanishes for unitary triplet phases but is non–zero for non–unitary phases, the presence of a magnetic field is likely to stabilize a non–unitary phase, at least close to T_c. This is well known in ^{3}He where the non–unitary ε and $A_1(\gamma)$ phases occur in a field.[122] Since there are analogous triplet phases in Tables 15.2, 15.3, the same may be possible for the HTSC. In addition to Eq. (26.47), there may also be coupling of a magnetic field to the spatial indices of the order parameter for E representation states (either directly or via the covariant derivatives $\partial_i = \nabla_i - ie^* A_i/\hbar c$), leading to splittings reviewed by Joynt[92] and Gor'kov.[91] These authors also discuss other related effects, such as angular dependence of H_{c2}, which could also be used to determine experimentally whether or not a SC has an unconventional order parameter. Possible splitting of the phase transition due to disorder has been proposed in Ref. 126. Splitting due to softening of CM has been proposed by Hirashima.[127] Finally, a splitting due to a phenomenological two–component order parameter has been proposed by Das, He and Choy.[128,129]

Whether or not splittings of the phase transition actually occur in the HTSC remains an open question for now. Some authors of early papers report observations of a split transition,[130-132] but latter Inderhees's studies on largely untwinned single crystals find no evidence for a splitting. It is clear now that features seen for some twinned crystals[133] are due to a superconducting transition occurring on the twinning boundaries.[134]

Spontaneous strain and magnetism

There is one further consequence of the coupling terms (26.46) and (26.47) which could be used to establish whether or not unconventional SC occurs. That is the spontaneous generation of strain or magnetic fields if the SC transition itself breaks the symmetry related to the field. For example in the E representation states the $(1,0)$ type phases spontaneously break the D_4 symmetry to D_2. The coupling (26.46) then immediately leads to a non–zero OR strain ε_{ij} in the lattice. Potentially, this spontaneous strain distortion could be observed by structural probes such as X–ray scattering. Observations of this type were made by Horn *et al.*[135] who saw such a strain anomaly at T_c in $YBa_2Cu_3O_7$. However, other authors found no evidence for changes at T_c (although they had fewer T points close to T_c).[136,137] In any case, since $YBa_2Cu_3O_7$ is OR above and below T_c, a spontaneous OR strain distortion is not symmetry forbidden in any superconducting state, conventional or unconventional, through second–order couplings to the strain.[138] Indeed, lattice distortions are possible at T_c even in conventional SC.[139] The observation of the strain would thus only be fully conclusive of an unconventional order parameter in a TG superconductors if it were observed to become OR below T_c. The analogous magnetic process would be the observation of a FM moment below T_c for a non–unitary triplet phase as dictated by the first term in Eq. (26.47).

26.5.4. Critical phenomena and Gaussian fluctuations

Critical phenomena

In the SC phase transition, just like any other phase transition, the critical behavior is universal, and is expected to depend only upon the dimensionality of the system and the symmetry group of the Ginzburg–Landau free energy. Thus each distinct type of order parameter and corresponding Ginzburg–Landau free energy gives rise to a unique set of critical exponents, amplitude ratios and so on.[140] By observing the critical behaviour it should therefore be possible to distinguish the various different types of superconducting order parameter described above. For example, the conventional single complex order parameter has the usual free energy of:

$$f[\eta] = \alpha(T - T_c)|\eta|^2 + \beta|\eta|^4 + \frac{\hbar^2}{2m_{ij}}\partial_i\eta \cdot \partial_j\eta \tag{26.48a}$$

which is equivalent to the XY model, or $O(n)$ Heisenberg model with $n = 2$. The heat capacity should therefore diverge with an exponent $\alpha = -0.02 \pm 0.03$ at T_c (assuming 3D), while the correlation length of the order parameter should diverge with an exponent $\nu = 0.670 \pm 0.006$.[140] On the other hand, if the superconductor had an E representation order parameter with two complex components and the free energy of:

$$f[\eta_1,\eta_2]=\alpha(T-Tc)\eta_i^*\eta_i+\beta_1(\eta_i^*\eta_i)^2+\beta_2|\eta_i\eta_i|^2+\beta_3(|\eta_1|^4+$$

$$|\eta_2|^4)+\frac{\hbar^2}{2m_1'}(|\partial_x\eta_1|^2+|\partial_y\eta_2|^2)+\frac{\hbar^2}{2m_1''}(|\partial_y\eta_1|^2+$$

$$|\partial_x\eta_2|^2)+\frac{\hbar^2}{2m_2}(|\partial_x\eta_1|^2+|\partial_x\eta_2|^2)+\frac{\hbar^2}{2m_3'}(\partial_x\eta_2^*\partial_y\eta_1+$$

$$\text{c.c.})+\frac{\hbar^2}{2m_3''}(\partial_x\eta_1^*\partial_y\eta_2+\text{c.c.}) \tag{26.48b}$$

then according to Toledano, Michel, Toledano and Brezin[141] the critical exponents would be the same as the $O(4)$ Heisenberg model:[141] $\alpha=-\varepsilon^2/6$ and $\nu=1/2+\varepsilon/8+7\varepsilon^2/96$ (in the ε expansion to order ε^2) giving $\alpha\sim-0.167$, $\nu\sim0.698$ for the physical case of $4-d=\varepsilon=1$. (These exponents[142] are for a free energy which does not include the anisotropic gradient terms in the free energy of Eq. (26.48b). It is therefore only correct for this case if either $1/m_1''=1/m_3'=1/m_3''=0$ or these variables are irrelevant in the renormalization group equations.) The critical properties have not been calculated for either of the triplet order parameter. It might be the case that the fixed points are the same as for the related isotropic HsM $O(6)$ and $O(12)$ (see Ref. 140); however, this is not guaranteed. For example, similar free energies with "cubic" anisotropy terms can give rise to new types of fixed points and different exponents.[142]

Is it practical to observe critical fluctuations in the HTSC? An estimate of the range of T around T_c, where the behaviour is critical is given by the Ginzburg criterion.[140,143] This gives the width of the critical region as the range of T for which $C_{f1}(T)>\Delta C$, where $C_{f1}(T)$ is the contribution to the heat capacity due to fluctuations and ΔC is the MF discontinuity in heat capacity at the transition. Using the conventional Ginzburg–Landau free energy for a single complex order parameter this gives the range[133]

$$t_G \approx \frac{1}{32\pi^2}\left(\frac{k_B}{\Delta C \xi(0)^3}\right)^2$$

in 3D, where $\Delta C = (\alpha T_c)^2/2\beta$, $t \equiv |T - T_c|/T_c$, k_B is Boltzmann's constant, $\xi(0)^3 = \xi_x(0)\xi_y(0)\xi_z(0)$ and $\xi_i(0) = \left(\hbar^2/2m_{ii}\alpha T_c\right)^{1/2}$. For $YBa_2Cu_3O_7$ the parameters have been estimated experimentally, to give $\Delta C \approx 35 - 45$ $\mathrm{mJcm^{-3}K^{-1}}$ [133] and $\xi_x(0) = \xi_y(0) \approx 14\text{Å}$, $\xi_z(0) \approx 2.5\text{Å}$. These numbers give t_G of order 2×10^{-3}, or a critical region of approximately 0.2 K above and below T_c. It should prove possible to perform experiments with this T resolution on the HTSC, provided that the samples can be made sufficiently uniform.

Even if the critical region is too narrow to be observed in zero–field it is possible to widen the critical region by applying a magnetic field. Lee and Shenoy[144] showed that in the limit $2h >> t_H$ (where $h = H/(T_c dH_{c2}/dT)$ and $t_H \equiv |T - T_{c2}(H)|/T_{c2}(H)$) that the fluctuation contribution to the specific heat is enhanced by a factor of h/t_H, thus widening the critical region. Salamon *et al.*[145] showed that their field–dependent heat capacity data obeyed a scaling relation consistent with a critical correlation length exponent of $\nu \sim 0.75$. This exponent is not consistent with the usual $O(2)$ GL theory which gives $\nu \sim 0.67$. Whether the exponent is consistent with any of the other possible GL theories is not known, however it would be consistent with an $O(n)$ model with $n \sim 6$.[146-148]

Gaussian fluctuations

Even outside the critical region many physical properties of SC are affected by fluctuations. For the HTSC fluctuation effects have been observed in the conductivity,[149-151] heat capacity,[133,146,152-155] magnetic susceptibility,[156,157] and thermopower.[158] These can be detected within a few degrees above T_c for the heat capacity and thermopower and can apparently be observed over a range of at least a hundred degrees from T_c for the conductivity and susceptibility. A review of all these effects is beyond the scope of this book, and so we will only discuss cases where the fluctuations have been used to test the nature of the order parameter and its GL theory.

While most fluctuation quantities are non–universal outside the critical region, Inderhees *et al.*[133] observed that within the $O(2)$ Ginzburg–Landau theory there is at least one universal quantity, namely the amplitude ratio C_+/C_- where the Gaussian fluctuation contribution to the heat capacity is $C_{\pm}t^{-1/2}$ above and below the transition respectively (in 3D). In the $O(2)$ Ginzhurg–Landau theory this dimensionless quantity has the numerical value of $1/\sqrt{2}$. Thus an experimental measurement of C_+/C_- can show whether or not the GL theory is indeed the conventional one. The original experimental data of Inderhees *et al.*[133] gave an amplitude ratio 2.8 ± 0.8 and 2.5 ± 0.3 on two different crystals, inconsistent with the expected $1/\sqrt{2}$. Since a single complex order parameter was incompatible with the data Annett, Randeria and Renn[7] calculated the (nonuniversal) possible ranges of C_+/C_- for the other allowed order parameters, and concluded that either the two–, three– or six–complex component order parameter could be consistent with the experiment. More recent experimental data on untwinned samples shows a sharper transition and a smaller amplitude ratio. The new data appears to agree with the $1/\sqrt{2}$ expected for the conventional Ginzburg–Landau free energy (26.48a), but could also be explained by any of the alternative allowed (26.48b) etc., and so cannot constrain the possibilities. The explanation for the discrepancy between the two sets of data may be rounding due to inhomogeneities or the effects of the winning planes.[134] An anomalously high amplitude ratio was also observed in a re–analysis of 1967 data on Ti–Mo alloys.[161]

While there are no universal quantities for the fluctuation conductivity, susceptibility or thermopower it is still possible to calculate the magnitudes of each effect in a Gaussian approximation and to compare these predictions with the data. In several of the fits to the fluctuation conductivity for $YBa_2Cu_3O_7$ the fluctuation conductivity appears to be approximately a factor of two or three larger than expected in the 3D–regime,[149-151,162,163] which may be due to a higher order parameter dimensionality. Similar analysis of the fluctuation diamagnetism for $YBa_2Cu_3O_7$ showed that it was a factor of 2–3 larger than predicted by the usual Ginzburg–Landau theory (26.48a),[6] however an alternative fit suggested that the data was consistent with conventional Ginzburg–Landau theory. The principal difference between these two fits was that in the former the normal state background was assumed linear, while in the latter the normal state background was curved. This illustrates a severe problem with drawing conclusions about the fluctuations from this type of analysis: one has to be able to separate unambiguously fluctuations from normal state effects in order for the analysis to be meaningful.

There appear to be hints from several pieces of data that the critical and Gaussian fluctuation properties of $YBa_2Cu_3O_7$ are not consistent with a conventional single complex order parameter. However, it remains to be seen whether these discrepancies are confirmed by subsequent experiments on better samples, or whether they can be attributed to materials problems, or other difficulties in the theory or in the data analysis.

26.5.5. Collective modes

Among many different methods which could help to identify the type of pairing and the order parameter in HTSC there is one which already has been successfully applied to the study of the superfluid phases of He^3– where one has the triplet p–pairing – namely investigation of the collective mode spectra (CMS) in these systems.

The order parameter and Ginzburg–Landau free energy of HTSC not only completely characterize their equilibrium thermodynamic properties close to T_c, but they also describe their non–equilibrium behavior for weak long–wave length departures from equilibrium when supplemented by phenomenological kinetic equations of the form[140,164]:

$$\frac{d\delta\eta_i(\mathbf{k})}{dt} = -\Gamma(\mathbf{k})\frac{\partial F[\eta]}{\partial\delta\eta_i^*(-\mathbf{k})} \tag{26.49}$$

where $\delta\eta_i$ is the departure of η_i from its equilibrium value, $\delta\eta_i(\mathbf{k})$ is its Fourier transform, and $F[\eta]$ is the total free–energy functional. Equations of this form can indeed be derived on the basis of microscopic BCS theory[165]. The eigenvalues of Eq. (26.49) define a set of CM of the SC in which the order parameter varies slowly in space and time (analogous to second sound in superfluid ^{4}He). Experimental observations of these CM of the SC state may, in principle, be used to determine the order parameter, since these modes are determined only by the Ginzburg–Landau theory and by Eq. (26.49), at least close to T_c. In particular, from Eq. (26.49), the number of CM close to T_c is simply related to the number of components of the order parameter, if it has g complex components then there are $2g = n$ CM. Above T_c these n modes are all degenerate and have imaginary frequencies which approach zero in the limit $T \to T_c$, $\mathbf{k} \to 0$. Below T_c some or all of the modes will acquire finite excitation energies in the $\mathbf{k} \to 0$ limit. It is possible to decide on symmetry grounds alone which modes are Gd (Gd) modes (which remain gapless at $\mathbf{k} = 0$), because there is at least one Gd mode for each continuous symmetry which is broken at T_c.[37,164] A complete discussion of the CM for all the various SC states is obviously an enormous task beyond the scope of this book. The calculation of whole collective mode spectrum for all SC states arisen in symmetry classification of HTSC in case of p–wave[25,31] and d–wave[14-23] pairing has been done by Brusov *et al.* The results for d–wave pairing will be discussed in 15.8, but here we shall just give a few representative examples. The discussion is limited to the situation close to T_c since it is only there that Eq. (26.49) is valid. Well below T_c the modes may change in character, for example by becoming propagating instead of diffusive, or may simply cease to be meaningful elementary excitations.

Firstly, as is well known, the conventional single–complex–component order parameter gives rise to two collective modes, of which one corresponds to phase variations of the order parameter and the other to amplitude variations.[37,166] The amplitude mode has a gap, while the phase mode (Bogoliubov sound) would be a Gd–mode because of the breaking of the continuous $U(1)$ gauge symmetry. As Anderson has pointed out, this mode ceases to be gapless when coupling to the long–ranged electromagnetic fields is taken into account, an example of the Higgs mechanism.[37,166-168]

As a second example of collective modes, consider the three complex–component order parameter of a triplet state without spin–orbit coupling in an OR crystal. Here, there are therefore six collective modes. Assuming that the SC GS is of the $(0,0,1)$ type then there are obviously two degenerate Gd modes corresponding to changes to the order parameter like $\delta\eta_i = (\varepsilon,0,0)$or $(0,\varepsilon,0)$, which infinitesimally rotate the order parameter direction in spin space (spin modes).The remaining modes have gaps and they are an amplitude mode $\delta\eta_i = (0,0,\varepsilon)$ a phase mode $\delta\eta_i = (0,0,i\varepsilon)$, and two modes, which mix in components of the other $(1,i,0)$ type GS: $\delta\eta_i = (i\varepsilon,0,0)$ and $(0,i\varepsilon,0)$. The gaps for the latter two excitations vanish if the Ginzburg–Landau parameter $\beta_2 \to 0$. If the GS were of the $(1,i,0)$ type there would also be two analogous spin rotation Gd modes.

Similarly the two–complex–component order parameter E representation states gives rise to four CM, of which none are Gd since no continuous symmetries are broken except $U(1)$. If the microscopic symmetry of the FS were close to cylindrical D_∞ instead of TG D_4 then the β_3 coefficient in the Ginzburg–Landau theory vanishes and there is a

Gd mode for the $(1,0)$ and $(1,1)$ type GS phases corresponding to OP rotations in *a–b* plane. There is no corresponding Gd mode for the $(1,i)$ type phase.

Finally, for each of the six–complex–component order parameter states there are twelve CM. Clearly for these phases which break symmetries in the continuous spin rotation group there will be two Gd spin modes. A complete investigation of all the remaining non–Gd modes would be rather involved, however in view of the close similarity between many of these phases and phases of superfluid ^{3}He it should be possible to take over many of the results of the extensive work on superfluid ^{3}He done by Brusov *et al.*[14-36]

For the p–pairing in bulk SC one has: a B–phase (Γ_1 rep. OR and TG, SOC), an A–phase (Γ_3^-, Γ_4^-, OR; Γ_2^-, Γ_5^-, TG) and an A_1–phase (Γ_5^-, TG) and a 2D–phase (Γ_5^-, TG). The whole CES is well known for A– and B–phases[35,36] and it's calculated for A_1– and 2D–phase.[381]

For the triplet pairing in 2D–systems Brusov and Popov[35,36] have calculated the whole CMS for two SC phases: a – with order parameter $\sim\begin{pmatrix}100\\ i00\end{pmatrix}$ and b – with order parameter $\sim\begin{pmatrix}100\\ 010\end{pmatrix}$, where the matrix denotes the non–zero amplitudes connecting three spin states to the two orbit states. The spectrum in each of these phases consists of 12 modes: Gd– and nonphonon (nph) ones. The analogy of *a*– and *b*–phases appears in Sigrist and Rice classification in the Γ_1 representation (*b*–phases, case with SOC) or Γ_5 one (*a*– and *b*–phase, case without SOC). The CMS in bulk HTSC and CuO_2 planes under d – pairing has been calculated by Brusov, Brusova, Brusov.[32-34]

The experimental observation of these various collective modes would clearly help to identify the type of pairing and the symmetries of the superconducting GS. However, so far as we are aware, no experimental results on the HTSC to date have been interpreted in terms of evidence for collective modes of one type or another. One of the possible reasons for this is that the experimental aspects of collective mode spectrum are more complicate in HTSC comparing with superfluid

He^3 and HFSC. This follows from the fact that the frequencies of ultrasound as well as microwaves used in experiments on exciting of the collective modes in SC Fermi–systems are proportional to the gap in Fermi–spectrum which is proportional to T_c. It is easy to estimate that while one needs the ultrasound with frequencies of order 5–10 MHz in superfluid He^3 (we speak about Ginzburg–Landau region) where T_c ~1–3 mK, in HFSC, where T_c~1K, one needs frequencies of order a few GHz and in HTSC, where T_c~100K, of order 100 GHz. (Note, however, that there is not principle limitations on the frequencies of ultrasound / microwaves and approaching T_c one can decrease the requested frequencies significantly). By this reason the collective mode observation in HFSC and HTSC is a more difficult task but at least in HFSC the experiments with microwaves at 20 GHz have been done at Northwestern University (USA). On the other hand none of the modes couple directly to light (except the phase mode which is the same for each possible case and simply corresponds to the London penetration depth), and so they cannot be observed simply in spectroscopy. Certainly some of them may couple to phonons through interactions of the type of Eq. (26.47), and so they may in principle he observed as peaks in phonon linewidth as a function of T (which is analogous to the way they are observed in liquid ^{3}He). Maybe the anomalous phonon linewidth observed by Cooper *et al.*[214] is due to this phenomenon. Those which correspond to spin rotation modes may in principle be observed in neutron scattering, although probably the cross–section is too small by a factor Δ/ε_f in BCS theory. Another experiment that might detect the spin modes has been proposed by Gor'kov,[169] involving observation of microwave transmission through thin films. We will return to the problem of collective excitations in HTSC in 26.8.

26.5.6. Exotic vortices

Another important consequence of an unusual order parameter in the HTSC would be the possibility of unusual types of vortices and other defects. For a multicomponent order parameter it is possible to classify the different types of topological defects that can occur by homotopy group theory.[37,89] Since this is such a vast subject we will concentrate only on the most common type of defect in superconductors, namely the vortex. Even here there are a rich range of possibilities, as the case of ^{3}He illustrates.[170] In particular the vortices may have normal cores or superconducting cores, and may even spontaneously break the axial symmetry about the vortex. We will only consider some of the simplest cases that could occur for the various HTSC order parameter.

First, let us review the case of the vortex in a conventional superconductors from the point of view of group theory. First of all we shall assume that the superconductor is isotropic. In that case if a vortex is present the vortex core should reduce the symmetry group to one of axial symmetry alone. If the vortex is a conventional one, then the phase of the order parameter $\eta(\mathbf{r})$ changes by 2π around the vortex. Clearly, if we have axial symmetry, then under rotation by an angle φ about the axis, the order parameter will change phase by $\exp(i\varphi)$. The symmetry group of operations which leaves the order parameter invariant at the vortex can therefore be described as:

$$D_\infty(E)=\{E,\exp(-i\varphi)C_\varphi,\exp(i\pi C_2')\} \tag{26.50}$$

where C_φ represents rotation by an angle φ about the vortex, and C_2' rotation by π about any line normal to the vortex. Now the operations in this group are not compatible with the symmetry of a conventional "s–wave" order parameter (which would be invariant under $D_\infty\{E,C_\varphi,C_2'\}$) and hence the amplitude of the "s–wave" order parameter $\eta(\mathbf{r})$ must vanish on the vortex core. This is of course the reason that vortices in conventional superconductors have normal cores. Notice in fact that we have been overly restrictive in the assumption of an isotropic superconductor and all that would have been necessary to prove the point was a twofold axis C_2 along the vortex. With just a twofold axis then the vortex state must be invariant under $\exp(i\pi)C_2$, while an "s–wave" order parameter is invariant under C_2 alone. This shows that even taking into account the reduced symmetry in orthorhombic crystals vortices of "s–wave" states must have normal cores (at least for vortices along the symmetry directions).

Consider now a somewhat more complicated case, namely a "d–wave" ${}^1E_g$ state in a TG crystal. Let us assume first that the vortex is along the C_4 symmetry axis of the crystal. Remembering that there are three possible phases for this order parameter $(1,0),(1,1),(1,i)$, one might ask the following: is it possible for a vortex in one of these phases to have a superconducting core belonging to one of the other phases? In fact the answer is still no, at least if even the twofold symmetry of the C_4 axis still holds. This follows because all of the three ${}^1E_g$ phases are invariant under the operation $\exp(i\pi)C_2$. The additional π phase change due to the phase winding means that the vortex is invariant under C_2, which is incompatible with any of the ${}^1E_g$ states, hence proving the point. A similar argument also shows that vortices in the $a-b$ plane would also be expected to have normal cores.

The paragraphs above show that simple crystal symmetry arguments can be used to severely restrict the cases when vortices may have superconducting instead of normal cores. Of course this type of argument breaks down for an important class of exotic vortices, those that spontaneously break the axial symmetry. Examples of vortices which break the axial symmetry have been proposed for ^{3}He, as well as for UPt_3.[170] In the latter case the crystal symmetry is hexagonal, and the superconductor is thought to occur in the E representation of the hexagonal group. Schenstrom *et al.*[170] showed that, if the bulk is in a $(1,i)$ type phase and the vortex is along the c axis, then the vortex may spontaneously break the C_6 symmetry to C_3. In that case the argument used in the previous paragraph becomes invalid (since it required the C_2 symmetry) and it is possible for the vortex core to be SC in the $(1,-i)$ phase. The hexagonal E representation is so similar to the E

representation of the TG group that presumably the same type of vortex may be expected here.

Experimentally, it may be difficult to ascertain whether the HTSC have conventional vortices or not. Certainly the properties of the flux lattice are remarkable enough, with 'giant flux creep' and the possibility of a melted flux liquid.[171-173] These properties, however, are on the length scale of the penetration depth, which is two orders of magnitude larger than the coherence length which determines the core size. Presumably the flux flow contribution to transport properties, as well as the behaviour in flux decoration experiments, would be the same whatever the actual nature of the cores. Of course if a transition between two types of vortex core state took place, as is proposed for ^{3}He then that might be observable by acoustic attenuation or by nuclear magnetic resonance (NMR), for example.[170] One possible indirect way of deciding whether the cores are normal or superconducting would be from experimental estimates of the vortex pinning energies, which would presumably be different if the vortex core is SC instead of normal. Finally, a direct and elegant method of determining whether the vortex cores are normal or superconducting would be by direct observation of the local quasiparticle density of states by scanning tunneling microscopy.[174,175]

Other interesting possibilities have been suggested for properties of vortex cores and other defects, which might have application to the HTSC. For example the multicomponent order parameter generally have degenerate GS, which would be separated by domain walls. Burlachkov and Kopnin have proposed that the linear term in the low–T specific heat could be explained if these domain walls are normal and carry a finite DOS. Another possibility is that completely new states of matter, including a four–fermion condensate, may occur in the cores of vortices in both superfluids and SC.

26.5.7. Probes of the gap function

AGL[1] discussed the evidence that the excitation gap $|\Delta(\mathbf{k})|$ has a zero on the Fermi–surface. Photoemission experiments show that this zero is on or close to the location expected for $d_{x^2-y^2}$ pairing.[176,177] We also emphasized evidence that $\Delta(\mathbf{k})$ has a sign change on the FS. A sign change is present for $\Delta(\mathbf{k}) \sim \cos k_x - \cos k_y$ but is not for highly anisotropic s–wave gaps such as $\Delta(\mathbf{k}) \sim \left|\cos k_x - \cos k_y\right|$. The evidence for a sign change comes from the effects of impurities on the T dependence of the penetration depth $\lambda(T)$. The prediction for resonant impurity scattering in a d–wave superconductors is that $\lambda(T)$ is linear in T: $\lambda(T) = \lambda(0) + aT + ...$ in clean samples, but that there is a cross–over to a T^2 power law in the presence of impurities. The two regimes are characterized by a crossover T, T^*, which depends on the impurity concentration. The predicted behavior for anisotropic s–wave systems where $\Delta(\mathbf{k})$ does not change sign is qualitatively different.[178,179] The experiments of Bonn *et al.*[180] agree very well with the d–wave predictions and disagree with the anisotropic s–wave scenario. Evidence for a d–wave sign change in $\Delta(\mathbf{k})$ also comes from the polarization dependence of the Raman scattering.

To what extent do these measurements rule out mixed states such as s+d or s+id? Suppose that the gap was of the following form:

$$\Delta(\mathbf{k}) = \Delta_s + \Delta_d \cos 2\theta \tag{26.51}$$

with θ the angle of $\mathbf{k}$ around the Fermi surface (measured with respect to the zone corner S in YBCO, X in BSCO). If Δ_s/Δ_d is real and less than unity, then the gap function has a node, but it is shifted away from θ =45°. On the other hand if Δ_s/Δ_d is imaginary then the excitation gap $|\Delta(\mathbf{k})|$ has a minimum $\Delta_{\min} = |\Delta_s|$ at 45°. For BSCO the most recent photoemission measurements of Ding *et al.*[622] imply that the minimum excitation gap occurs either at 45° or no more than 5° either side of 45°. This limits any real admixture to $|\Delta_s/\Delta_d| < 0.2$. If Δ_s/Δ_d is were imaginary, Ding *et al.*'s[176] data implies $|\Delta_s/\Delta_d| < 0.15$. A similar bound is also found for YBCO from the data of Schabel *et al.* A more stringent bound on $\Delta_{\min}$ can be estimated from the penetration depth data of Bonn *et al.*[180] In their cleanest samples $\lambda(T)$ is linear down to at least $T^* \sim 3$K, suggesting $|\Delta_s/\Delta_d| < 0.03$ for s+id states. Similar bounds would also apply to $d_{xy} + id_{x^2-y^2}$ mixtures.

These bounds disagree with the large s+id mixing found by Li and Joynt in their analysis of NMR data.[181] Several factors could account for this discrepancy. Firstly the NMR data, are not in the asymptotic low T regime comparable to the $\lambda(T)$ data. The fits may therefore be influenced by the T dependent AF spin fluctuations. Secondly Li and Joynt[181] only considered the simplest $\cos 2\theta$ type $d_{x^2-y^2}$ state, and did not include higher cubic harmonics which can change the prefactors in the low T power laws.

Finally, it is possible that the pairing state symmetry changes with doping. The photoemission data of Kelley *et al.*[67] show a gap opening up on the ΓX line in BSCO for highly overdoped samples. Assuming these results are not influenced by the incommensurate modulations which affected the Ding *et al.*[182,215] data for ΓX then this would be clear evidence for either a mixed symmetry state or even a pure s–wave state in these samples. The $n-$ type Nd–based compounds appear to have a non–zero $\Delta_{\min}$.[3] Again this could either be due to a mixed state such as s+id, or d+id or simply pure s–wave pairing.

26.5.8. Distinction of a scalar from a tensor order parameter

The authors of Ref. 89 have suggested some tests to distinguish a scalar from the tensor order parameter. They obtained the following results.

1) For magnetic field H parallel to axis of unit cell in the single crystal every tensor order parameter (except a scalar) can lead to the occurrence of a transverse magnetic field along with a longitudinal component of the resulting current density.

2) The product of penetration length λ_i, and coherence length ξ_i is an invariant along crystallographic axis only for the scalar order parameter. The absence of this invariance at any T implies the existence of a

multicomponent order parameter. 3) For muiticomponent OP more than one intermediate upper critical field in a given direction must exists.

26.6. Experimental Evidence for $d_{x^2-y^2}$–Pairing

At SCES–93 in San–Diego Pines[183] presented a "collection" of experiments (both for "clean" and "dirty" samples) supporting d–wave pairing. At the same meeting, Leggett, Scalapino and others in their talks gave further reasoning in support of this idea. Namely at SCES–93 d–wave pairing has become one of the main alternative pairing being completely formed in Grenoble at M^2S–HTSC IV. Below we describe the "Pines's[183] set of experiments".

26.6.1. "Clean samples"

Because they have been carried out on unusually clean samples, the experiments listed below provide strong direct evidence for the signature of spin–fluctuation–induced superconductivity, the $d_{x^2-y^2}$ pairing state.

Nuclear magnetic resonance in $YBa_2Cu_3O_7$

For s–wave isotropic or anisotropic pairing, one expects that both nuclear magnetic relaxation rates and the Knight shift [~ $\chi_0(T)$] will fall of exponentially at low T, while the presence of a line of nodes in the $d_{x^2-y^2}$ pairing state leads to $T_1^{-1} \sim T^3$ and $\chi_0(T) \sim T$ for $T \le T_c/2$ in systems like $YBa_2Cu_3O_7$ and $Bi_2Sr_2CaCu_2O_8$ in which the maximum value of the energy gap, Δ_0, has been observed to open up rapidly below T_c.

Martindale *et al.*[184] found a significantly different T dependence for the three nuclear magnetic relaxation rates they have measured for similar samples in low magnetic fields: ${}^{63}T_1^{-1}$ and ${}^{17}T_1^{-1}$ in a field perpendicular to the Cu–O planes, and ${}^{63}T_1^{-1}$ for a field parallel to the planes. Although all three exhibit approximately T^3 behavior for

$T \leq T_c/2$, it is only when one takes AF correlations into account and employs a FS appropriate to the quasiparticle spectrum derived from ARPES experiments (a next neighbor hopping parameter, $t' \sim -0.45$ that of the NN hopping parameter t), that one is able to obtain a *quantitative* account of all three relaxation rates.[185] Barrett *et al.*[186] found that the chain ^{62}Cu Knight shift varies linearly with T at low T in accord with $d_{x^2-y^2}$ pairing; they also find the planar Knight shirt is linear in T for $T_c/5 \leq T \leq T_c/2$, falling off less rapidly below $T_c/5$, an effect which is likely due to imperfections in their sample (see below).

Penetration depth in $YBa_2Cu_3O_7$

Hardy *et al.*[187] measured the penetration depth, $\lambda(T)$, as a function of T at quite low T in an unusually clean single crystal; they found the ratio $\left[\lambda^2(0)/\lambda^2(T)\right]$ varies linearly with T for $(T_c/10) \leq T \leq (T_c/2)$, as expected for $d_{x^2-y^2}$ pairing, but in striking contradiction to the gapped, exponential behavior predicted by isotropic s–wave pairing or the anisotropic s–wave pairing state of Chakravarty *et al.*[188]

SQUID experiments on SNS tunnel junctions

Wollman *et al.*[76] carried out a quite difficult experiment, suggested independently by Sigrist and Rice[109] and by Leggett, which can determine directly the symmetry of the pairing state. The spatial anisotropy of the phase of the order parameter in single crystals of YBCO was determined from the magnetic field modulation of YBCO–Pb DC SQUIDS. Their experimental results rule out s–wave pairing, and taken as a whole "give rather strong evidence for a phase shift of π that is predicted for the $d_{x^2-y^2}$ pairing state".[76]

Let us consider this experiment in detail.[61] The resistance versus applied flux for YBCO–Pb SQUID has been measured. In order to extract the intrinsic phase shift δ_{ab} inside the YBCO crystal which is interesting, it is necessary either to reduce the asymmetry until it is of little significance or to extrapolate the results to the zero–current limit. Authors have taken both approaches. Samples that are nearly symmetrical and require little extrapolation, and those in which they could monitor the resistance modulation all the way down to zero bias current, exhibit a minimum resistance at a flux of $(1/2)\Phi_0$. In other samples, it is necessary to extrapolate the phase of the R versus Φ currents to zero bias current by plotting the value of applied flux at which the resistance is a minimum (corresponding to the maximum critical current) versus the bias current. This procedure is valid only in the noise–rounded regime near the critical current. It is verified by the experimental data, which do in fact exhibit quasilinear behavior in this regime, and also by the results of the edge SQUIDs, which always extrapolate to zero flux as required independent of the pairing state. For a corner SQUID, *s*–wave pairing gives a zero–current intercept at $\Phi = 0$; the $d_{x^2-y^2}$ state would yield an intercept at $\Phi = \Phi_0/2$.

In Fig. 26.5(a), we show measurements and linear extrapolations for seven corner SQUIDs, each cooled slowly in zero field. The slopes, a measure of the asymmetry, vary over a wide range. The intercepts are distributed in a range centered at $0.5\Phi_0$ with a spread of about $\pm 0.1\Phi_0$. Thus, although the pure *d*–wave signature is not strictly indicated by each sample, each one plotted here – and every other corner SQUID which has been studied – does exhibit a significant phase shift of order π consistent with the $d_{x^2-y^2}$ state.

Raman scattering experiments on $Bi_2Sr_2CaCu_2O_8$

Deveraux *et al.*[190] investigated the sensitivity of the polarization dependence of the electronic Raman spectra to the symmetry of the pairing state, and carried out new measurements on the Bi 2:2:1:2 system for which an almost perfect surface can be obtained. They found that measurements of the electronic Raman effect on the cuprates provide a large amount of symmetry–dependent information, all of which agree with predictions of $d_{x^2-y^2}$ pairing.

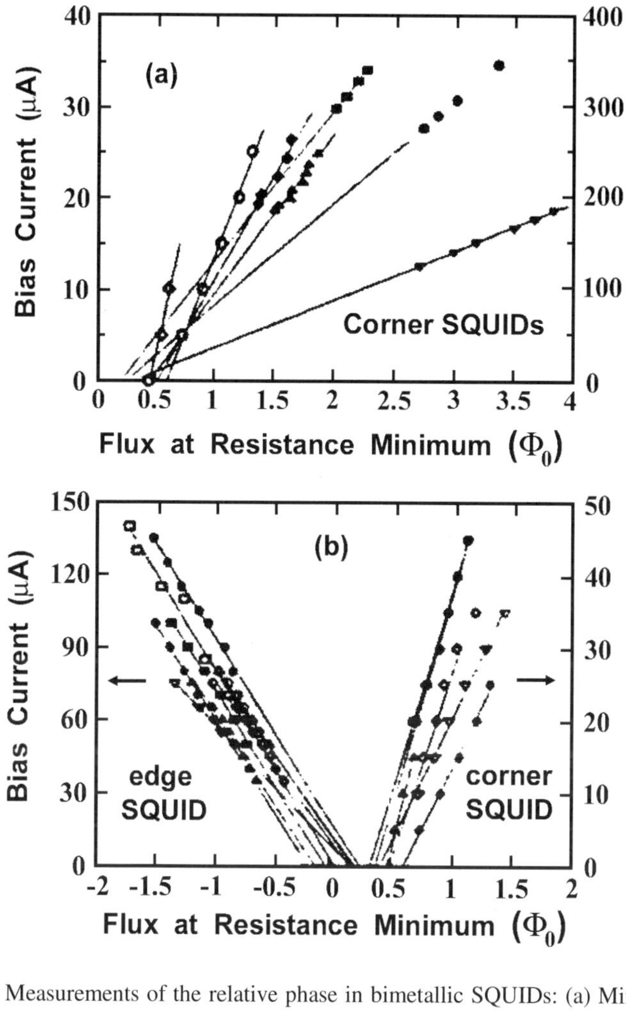

FIG. 26.5. Measurements of the relative phase in bimetallic SQUIDs: (a) Minima in the SQUID modulation curves extrapolated to zero bias current for seven different samples with varying amounts of asymmetry. An intercept of zero would occur for s–wave pairing; the observed intercept of order $\Phi_0/2$ is indicative of the $d_{x^2-y^2}$ symmetry. The distribution is due to magnetic vortices trapped near the SQUID. (b) Comparison of a corner SQUID and edge SQUID on the same crystal cooled down multiple times. As required independent of the pairing state, the edge SQUID extrapolates to zero, but the corner SQUID shows clear evidence for a sign change in the OP between orthogonal directions in the crystal [Ref. 76].

Tunneling and ARPES experiments

Coffey and Coffey[191] showed that in a StC $d_{x^2-y^2}$ superconductors the spontaneous decay of quasiparticles gives rise to a feature at $2\Delta_0$ in S–I–N tunneling, at $3\Delta_0$ in S–I–S tunneling, and a dip feature in ARPES data. Coffey reviewed the considerable body of experimental evidence for the presence of this feature. Its presence has now been further confirmed in the StC calculations of Monthoux and Scalapino,[192] and in the S–I–N tunneling experiments by Liu *et al.* on high quality single crystals of $Bi_2Sr_2CaCu_2O_8$, who found a distinct feature at $2\Delta_0 = 6.4kT_c$ and a tunneling DOS which agrees well with $d_{x^2-y^2}$ pairing predictions. Moreover, in their ARPES experiments on $Bi_2Sr_2CaCu_2O_8$, Shen *et al.*[177] found a marked spatial variation of the gap parameter which is fully consistent with $d_{x^2-y^2}$ pairing.

Paramagnetic Meissner effect (Wohlleben effect)

A number of granular HTSC exhibit a PM Meissner effect,[193] christened the Wohlleben effect by Sigrist and Rice[109] who explained it, and a concomitant unusual microwave absorption, as the natural consequence of the existence of a Josephson network of coupled $d_{x^2-y^2}$ paired regions in these systems.

Thermal conductivity and microwave surface impedance of $YBa_2Cu_3O_7$

Yu et al.[194] measured the thermal conductivity of an untwinned single crystal of $YBa_2Cu_3O_7$ in the $a-b$ plane. They found a large enhancement of the thermal conductivity in the SC state, which they show arises from the strong suppression of the power law quasiparticle scattering rate due to spin fluctuations in a d–wave superconductor. A comparable rapid rise in the microwave surface resistance was seen by Bonn et al.[195], who also found a linear T–dependence below 30K.

26.6.2. Impurities

Monthoux and Pines (MP)[196] in their strong coupling calculations for $YBa_2Cu_3O_7$ considered the influence of potential impurity scattering on T_c. They found that isotropic potential scattering, even in the unitary limit, has a small influence on T_c, essentially because it represents a comparatively small add–on to the contribution to the self–energy arising from the spin–fluctuation scattering responsible for the resistivity: the latter does reduce T_c substantially. MP also were able to explain the remarkable difference between the influence of planar Ni and Zn impurities on the SC of $YBa_2Cu_3O_7$. Experimentally it has been shown[197] that both its low–frequency magnetic properties and T_c are quite sensitive to the presence of Zn impurities, while being surprisingly insensitive to Ni impurities; the latter thus act primarily as potential scatterers. MP's calculations show that Ni acts as a sub–unitary scatterer, while the much more substantial influence of Zn on T_c and other properties, which is a direct consequence of the fact that it changes the local magnetic order and hence the spin–fluctuation–induced interaction responsible for SC, means that it acts as a super–unitary scatterer. An alternative explanation, that $YBa_2Cu_3O_7$ is an s–wave superconductor in which spin–exchange scattering from local moments associated with Zn impurities acts to reduce T_c, has been shown by Walstedt *et al.*[198] to fail by more than an order of magnitude.

Impurities and imperfections also markedly change the low (and sometimes not so low) T properties of $d_{x^2-y^2}$ SC.[199-201] The reason is that at low T, where the quasiparticle excitations of interest are all near the point nodes of the 2D system (line nodes for 3D), imperfections act to change the clean–limit quasiparticle DOS, $N_s(T) \sim T$, to a constant value, $N_{\text{imp}}(0)$ as $T \to 0$.[200,201] Hirschfeld and Goldenfeld (HG)[200] have shown that in the unitary limit, at low T $N_{\text{imp}}(T)$ varies quadratically with T, crossing over to the expected linear behavior at a temperature

$T^* \sim n_0 T_c$, where $n_0 = \left[N_{\text{imp}}(0)/N(0)\right]$ varies as $\sqrt{N_f}$ where N_f is the impurity density.

As might be expected, all other low T properties of a "dirty d–wave" SC are altered: indeed the nodal quasiparticles behave like a normal FL with an effective mass, $m_{imp} \sim n_0 m^*$ where m^* is a suitably averaged normal state quasiparticle mass. One therefore expects that below T^*, $\left(T_1^{-1}\right) \sim T$, $c_v \sim T$, etc. Moreover, as HG have shown, below T^* one gets a shift in the low T penetration depth, $\lambda(T)$ from the expected linear in T behavior to quadratic behavior, while $\lambda_{imp}(0)$ is shifted from its clean limit value, $\lambda(0)$, by an amount $n_0\lambda(0)$. We would also predict that below T^*, T_1^{-1} will follow the Korringa relation, since in this T range FL corrections are likely negligible.

Let us now consider the experiments on deliberately or inadvertently imperfect samples of HTSC, which find a natural explanation with dirty d–wave superconductivity.

NMP experiments

Ishida et al.[197] found that Ni and Zn also influence the low T NMR behavior of $YBa_2Cu_3O_7$ quite differently. A small concentration, $N_{Zn} \cong 1\,\%$ brings about a remarkable change in the Knight shift; below $T^* \sim 35$K, they found a nearly T independent Knight shift which is a substantial fraction (~ 0.3) of its normal state value. These results are consistent with the HG theory: since Zn acts as a super–unitary scatterer,[196] it is not implausible that $n_0 \sim 3\sqrt{N_{Zn}} \sim 0.3$, $T^* \sim 30$K, with correspondingly $\sqrt{2}$ larger values for $N_{Zn} = 2\%$.

Ishida et al.[197] also emphasized that their results on nuclear magnetic relaxation times are likewise in good agreement with one's expectations for a dirty d–wave superconductivity. They found that on Zn–doped samples of $YBa_2Cu_3O_7$ ${}^{63}T_1^{-1} \sim T$ below some T^* [197] while similar behavior was seen for ${}^{63}T_1^{-1}$ measurements on $Bi_2Sr_2CaCu_2O_8$,[197] a result which has also been obtained by Takigawa. These latter results suggest that even in the "best" samples of the latter material, there exist a sufficient number of imperfections (which act as unitary scatterers) to yield $T^* \sim T_c/5$, corresponding to $n_0 \sim 0.2$.

Penetration depth in $YBa_2Cu_3O_7$ and $Bi_2Sr_2CaCu_2O_8$

Lee *et al.*[202,203] in experiments on the penetration depth, $\lambda(T)$, of quite pure thin films of $YBa_2Cu_3O_{7-\delta}$, find evidence for the dirty d–wave crossover. They found a T^* ~ 25 K, with a $\lambda(T)$ at higher T $(T^* \leq T \leq T_c)$ which agrees extremely well with the linear behavior which Hardy *et al.*[60] find extends down to 4 K in their single crystal samples. Their work thus supplies a significant bridge between thin film and single crystal results. It demonstrates that even in the best thin films, there is a concentration of imperfections sufficiently large to lead to $n_0 \sim (T^*/T_c) \geq 0.2$. When combined with the NMR results on T_1^{-1} for Bi 2:2:1:2, this result provides a plausible explanation for the failure of Ma *et al.*[205] to observe a linear T dependence for the $\lambda(T)$ at low T in experiments on single crystals of this material, or thin films of $YBa_2Cu_3O_7$.

Inelastic neutron scattering

Mason et al.[206] carried out inelastic neutron scattering experiments on a sample of $La_{1.86}Sr_{0.14}CuO_4$ which they found displays a linear specific heat at low T which is some 20% of its value in the normal state. The corresponding value of n_0 ~ 0.2 is large for their measured density (0.07%) of magnetic impurities; it suggests that the concentration of imperfections which act as unitary scatterers is at least an order to magnitude larger. If the reduction in T_c produced by these imperfections scales with that observed for Zn–doped $YBa_2Cu_3O_7$, these act to reduce T_c by some 3 K, a result consistent with the measured T_c ~ 35 K. Taken together, these results make a strong case for dirty d–wave superconductivity in this sample. Dirty d–wave superconductivity also provides a plausible explanation for the finding of Mason et al.[206] that SC has a relatively small effect on the strength of the low–frequency peak at the incommensurate wave vector $\mathbf{Q}_\delta$; they found that at 4.3 K:

$$\frac{\chi''(\mathbf{Q}_\delta,\omega)}{\omega}=\frac{\chi_{\mathbf{Q}_\delta}}{\omega_{SF}} \tag{26.52}$$

is only reduced to some 60% of its value above T_c, a result which would be inexplicable were the system a dirty s–wave superconductor with an impurity concentration of $\leq 1\%$. Finally, the failure of Mason et al.[206] to find any appreciable changes in the superconducting state in the much weaker scattering intensity measured at a wave vector, $\mathbf{Q}_\gamma$ which corresponds to node–node scattering across the Fermi–surface, is easily understood within the dirty d–wave context. Magnetic impurities and imperfections mask the kind of significant changes in the low–frequency magnetic properties (Knight shift and T_1^{-1}) seen in clean samples. It would thus seem that the small concentration of imperfections and impurities which, in the present state of the art, necessarily are produced during the growing of crystals large enough to carry out neutron scattering experiments, render it extremely difficult to use this technique to test predictions for clean $d_{x^2-y^2}$ superconductors.

26.7. Irradiation Studies

Irradiation studies are possibly a more clear cut test of sensitivity of the SC to disorder, since unlike doping studies there is no danger of substantially altering the electronic structure at the same time as introducing disorder. Basov *et al.*[207] measured the evolution of the infrared optical conductivity in a $YBa_2Cu_3O_{7-\delta}$ crystal both before and after irradiation with low–energy He ions. The irradiation suppressed T_c from 93.5K to 80K. Basov *et al.*[207] observed that after irradiation the optical conductivity was finite at all frequencies, and even developed a

Drude–like low frequency peak. Detailed calculations of the far infrared conductivity were performed for a $d_{x^2-y^2}$ superconductors with both elastic and inelastic scattering.[208,209] These calculations show that the qualitative evolution of the spectra with increasing disorder, including the Drude–like peak, could be understood naturally in a d–wave picture. In contrast the optical conductivity of disordered isotropic s–wave is qualitatively different.[207] It is particularly noteworth that, at least within the Eliashberg formalism of the calculations, there is a reasonable *quantitative* agreement between the decrease in T_c under irradiation, and the observed spectral shape.[208] In other words, there is a consistent choice of parameters such that the same defect scattering rate leads to both the observed T_c decrease and the spectral shape. Carbotte et al.[208] emphasize that the calculated spectral shape would lie essentially the same for any gap function $\Delta(\mathbf{k})$ with a node, including extended s–wave states, but would be inconsistent with states in which $\Delta(\mathbf{k})$ does not change sign (such as $|k_x^2 - k_y^2|$). Thus, we conclude that the d–wave pairing state accounts qualitatively (and possibly even quantitatively) for the experimentally observed features of the far infrared conductivity. Perhaps the main question about the theoretical calculations for this comparison is the use of the Eliashberg theory: it is by no means obvious that this is applicable, given that there is presently no understanding of the microscopic mechanism, and hence no justification to assume Migdal's theorem applies. Nevertheless one believes that the comparison is probably qualitatively correct, in the sense that it is hard to see how a more detailed theory could restore agreement between the s–wave calculations and the experimental observations.

A very elegant technique to explore the effects of impurities on the cuprates is to use electron irradiation to displace atoms from the crystal lattice. These studies[210] were performed on high quality, detwinned single crystals, and the electron energy was chosen so as to selectively remove the planar oxygens O(2,3); a careful study was undertaken to demonstrate that this had in fact occurred. The resulting defects do not contribute a Curie–Weiss tail to the normal state susceptibility, neither do they affect the plasma frequency nor the carrier concentration. Thus, the resulting samples contained non–magnetic defects in the CuO_2 planes, without any change in carrier concentration or plasma frequency, at least to the extent that could be determined experimentally. The a–axis resistivity, measured as a function of irradiation dose, indicates clearly that for 4.1% displacement of planar oxygens, the $T_c = 0$, or at least less than 12K, the lowest T attained in the experiment. If we follow the authors, and assume that T_c actually is zero, or would have become zero, at the appropriate extrapolated irradiation dose, then it appears that these data are in sharp contrast to the expectation based upon a s–wave state and non–magnetic scattering, viz. an originally anisotropic s wave state would become isotropised by the impurity scattering, leading to a non–zero value of T_c above some critical concentration of impurities. We conclude two things: first, the pairing state must have nodes, but could be either d–wave or an extended s–wave state (with 8 nodes). The latter state is ruled out by (e.g.) the photoemission data on gap anisotropy. Secondly, the d–wave state may have a small admixture of s, but not so much that the gap function symmetry reverts to s–wave before the SC is destroyed. Thus, this experiment places an upper bound on any non–zero value for the gap function integrated over the Fermi–surface.

The results of electron irradiation experiments agree qualitatively with the earlier study of Sun *et al.*[211] who also found that T_c gets driven to zero in both Pr–doped and ion–beam damaged $YBa_2Cu_3O_{7-\delta}$ single crystals. These authors did not attempt to determine if the resulting impurity scattering was magnetic or not. In both experiments, taken in isolation from the body of other data described here, it would be conceivable that the pairing state is s–wave, and that there is a small magnetic component to the scattering: a quantitative comparison of the T_c dependence and the experimentally implied upper bounds on magnetic scattering would be needed to explore this possibility. A potentially serious issue for the $d_{x^2-y^2}$ scenario, which is still not resolved, is the large quantitative discrepancy between the observed T_c

in the experiments of Sun et *al.* and calculations based on Eliahsberg theory combined with the observed residual resistance at T_c.[69,211]

There are three logically distinct points at which any of the above arguments could fail. Firstly, the applicability of Abrikosov–Gor'kov theory to the cuprates is certainly not obvious *a priori.* It should be emphasised that the Abrikosov–Gor'kov theory is in essence a variational calculation, which compares the free energy of one possible kind of pairing state, namely that in which pairs form in plane–wave states, despite the fact that these are no longer eigenstates of the single–particle Hamiltonian in the presence of impurities, with the normal–state free energy. While it is clear that some obvious alternatives, e.g. pairing in the basis of single–particle eigenstates, do worse than the Abrikosov–Gor'kov ansatz it is not entirely obvious that there is no possible alternative which might do better in the rather unusual environment of the HTSC, and this must remain a caveat with respect to conclusions drawn on this basis.

Secondly, as mentioned earlier, Eliashberg theory itself is dubious in the cuprates since it is unlikely that Migdal's theorem is valid if the pairing mechanism Is either purely electronic (e.g. spin fluctuation exchange) or involves the lattice (e.g. bipolaron theorics).

Thirdly, even assuming that the above two points can be dealt with, It is far from clear that all the conclusions of Abrikosov–Gor'kov theory are then strictly valid in a disordered 2D system. In particular Wenger and Nersesyan[212,213] have shown that the vertex corrections are large for 2D d – wave SC and hence Abrikosov–Gor'kov theory may fail for the quasiparticle states near the gap–node. They also found[212,213] a density of quasiparticle, states near the d–wave gap node of the form $\rho(\omega) \sim |\omega|^{\alpha}$ as $\omega \to 0$ with $\alpha < 1$ which is qualitatively different from the d–wave Abrikosov–Gor'kov result $\rho(\omega)$ ~ const. On the other hand, numerical calculations[216] of the DOS in a disordered 2D d–wave superconductor were qualitatively similar to the Abrikosov–Gor'kov predictions (see. Ref. 216, Fig. 8).

26.8. List of Abridgements for Chapters XXV and XXVI

1D – one dimensional
2D – two dimensional
3D – three dimensional
AF – antiferromagnet(ic)
BCS – Bardeen–Cooper–Schrieffer
BE(C) – Bose–Einstein (condensation)
BSC – bipolaronic SC
BS – band structure
BZ – Brillouin zone
CDW – charge density wave
CI – Coulomb interaction
CM(S) – collective mode (spectrum)
CO – charge ordered
CP – Cooper pair(s)
CR – Coulomb repulsion
CS(T) – Chern–Simons (term)
CS(L)S – chiral spin (liquid) state
DOS – density of states
DM – diamagnetic
ED – exact diagonalization
e–e – electron–electron
EHM – extended Hubbard model
EM(F) – electromagnetic (field(s))
e–ph – electron–phonon
FD – Fermi–Dirac
FE – Fermi–energy
FL – Fermi–liquid
FM – ferromagnet(ic)
FS – Fermi–surface
GF – gauge field(s)
GI – gauge invariance
GL – Ginzburg–Landau
GS – ground state

GT – gauge transformation
HF(A) – Hartree–Fock (approximation)
HsM – Heisenberg model
HTSC – high temperature superconductor(s) (superconductivity)
HM – Hubbard model
IR – irreducible representation
JT – Jahn–Teller
KT – Kosterlitz–Thouless
LR RVB – long range RVB
ME – Meissner effect
MF(A) – mean field (approximation)
MFT – mean field theory
NFL – non Fermi liquid
(O)DLRO – (off) diagonal long–range order
OP – order parameter
OR – orthorhombic
PM – paramagnet(ic)
PrT – perturbation theory
PS – phase separation
PSC – polaronic SC
PT – phase transition
QFT – quantum field theory
(Q)MC – (quantum) Monte Carlo
QED – quantum electrodynamics
(R)VB – (resonating) valence bond
RPA – random phase approximation
SC(E)S – strongly correlated (electron) system
SC – superconductor(s), superconductivity
SR RVB – short range RVB
SL – spin liquid
SDW – spin density wave
StC(A) – strong coupling (approximation)
SL(S) – spin liquid (state)
TG – tetragonal
T – temperature

T_c – transition temperature

vHs – van Hove singularity

WC(A) – weak coupling (approximation)

Chapter XXVII

D–Pairing in HTSC

27.1. Introduction

In this Chapter we start to study the collective excitations in unconventional superconductors. As we noted in Introduction, up to now study of the collective excitations in unconventional superconductors (USC) carries exotic character via a few reasons. First of all while there were some evidences of nontrivial type of pairing in some superconductors (HFSC, HTSC etc.) there was not superconductor in which unconventional pairing has been established exactly. Secondly, there was not found strong evidences of existing of the collective excitations in superconductors.

The situation has changed drastically within last few years removing study of the collective excitations in USC into real plane. In light of recent experiments this topic becomes very important. First of all an amplitude mode (with frequency of order 2Δ) has been observed in films of ordinary superconductors. Secondly, now the type of pairing is established for many superconductors. We have **s–pairing** in ordinary superconductors and electron–type HTSC; **p–pairing** in pure ^3He; ^3He in aerogel, Sr_2RuO_4 (HTSC), UPt_3 (HFSC) and **d–pairing** in hole–type HTSC, organic superconductors, some HFSC (UPd_2Al_3, $CePd_2Si_2$, $CeIn_3$, $CeNi_2Ge_2$ etc.).

Recently Northwestern University (John Ketterson's group) has presented results of a microwave surface impedance study of the heavy fermion superconductor UBe_{13}. They clearly have observed an absorption peak whose frequency – and temperature–dependence scales with the BCS gap function $\Delta(T)$. This was the first direct observation of the resonant absorption into a collective mode, with energy approximately proportional to the superconducting gap. This discovery opens a new page in study of the collective excitations in unconventional superconductors.

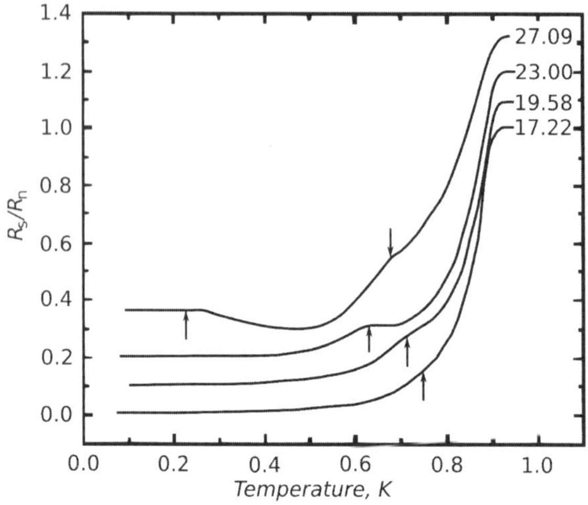

FIG. 27.1. Temperature dependence of surface impedance in heavy fermion superconductor UBe_{13} [Ref. 16].

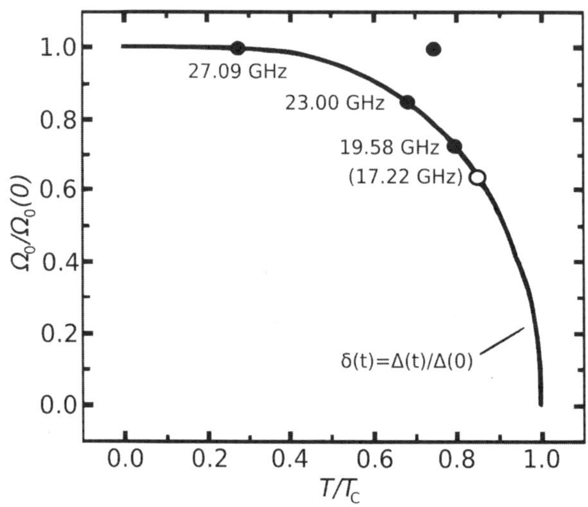

FIG. 27.2. Normalized frequency of collective mode in heavy fermion superconductor UBe_{13} [Ref. 16].

27.2. Bulk HTSC Under d–Pairing

We shall use the effective functional of action obtained above for analyzing the collective mode spectrum in HTSC.[1-15] In the first approximation, the spectrum is determined by the quadratic form of the effective functional of action obtained as a result of the shift $c_{ia}(p) \to c_{ia}(p) + c_{ia}^{(0)}(p)$ in Bose–fields by a condensate wave function $c_{ia}^{(0)}(p)$, whose form is determined by the concrete superconducting phase. The spectrum can be found from the equation, $\det Q = 0$, where Q is the matrix of the quadratic form. Let us consider the results of calculation of the collective mode spectrum for superconducting phases appearing in the symmetry classification of HTSC. We considered the following states: $d_{x^2-y^2}$, d_{xy}, d_{xz}, d_{yz}, $d_{3z^2-r^2}$. For each superconducting phase, five high–frequency modes were determined as well as five Goldstone– (quasi–Goldstone) modes whose energies are either equal to zero or small ($\leq 0.1\ \Delta_0$).

First, consider $d_{x^2-y^2}$ state, $\begin{pmatrix} 1 & 0 & 0 \\ 0 & -1 & 0 \\ 0 & 0 & 0 \end{pmatrix}$ with order parameter $\Delta_0(T)\,(0; \sin^2\theta \cos 2\varphi; 0; 0; 0)$.

The gap equation in this case has the following form:

$$g^{-1} + \frac{\alpha^2 Z^2}{2\beta V} \sum_p \frac{\sin^4\Theta \cos^2 2\phi}{\omega^2 + \xi^2 + \Delta_0^2 \sin^4\Theta \cos^2 2\phi} = 0 \tag{27.1}$$

and the gap $\Delta^2(T) = \Delta_0^2 \sin^4\Theta \cos^2 2\phi]$. (27.2)

As we mentioned above, the spectrum is determined by the quadratic form of the effective functional of action which coefficients are proportional to the sums of the products of Green's functions of quasifermions. At low temperatures ($T_c - T \sim T_c$) we can go from a summation to an integration by the following rule:

$$\frac{1}{\beta V}\sum_p \to \frac{1}{(2\pi)^4}\frac{k_F^2}{c_F}\int d\omega d\xi d\Omega . \tag{27.3}$$

To evaluate these integrals it is useful to use the Feynman equality:

$$\left[\left(\omega_1^2+\xi_1^2+\Delta^2\right)\left(\omega_2^2+\xi_2^2+\Delta^2\right)\right]^{-1}=$$

$$\int d\lambda\left[\alpha\left(\omega_1^2+\xi_1^2+\Delta^2\right)+(1-\lambda)\left(\omega_2^2+\xi_2^2+\Delta^2\right)\right]^{-2} \tag{27.4}$$

It is easy to evaluate the integrals with respect to variables ω and ξ and then with respect to parameter α and the angular variables. After calculating all integrals except over the angular variables and equating the determinant of the resulting quadratic form to zero one gets the following set of ten equations, which determine the whole spectrum of the collective modes for $d_{x^2-y^2}$–state:

$$\int_0^1 dx\int d\varphi\{\frac{\sqrt{\omega^2+4(1-x^2)^2\cos^2 2\phi}}{\omega}\ln\frac{\sqrt{\omega^2+4(1-x^2)^2\cos^2 2\phi}+\omega}{\sqrt{\omega^2+4(1-x^2)^2\cos^2 2\phi}-\omega}$$

$$(1-3x^2)^2+[(1-3x^2)^2-\frac{3}{8}((1-3x^2)^2+3(1-x^2)^2\cos^2 2\phi)]\times$$

$$\times\ln(1-x^2)^2\cos^2 2\phi\}=0,$$

$$\int_0^1 dx \int d\varphi \{ \frac{\omega}{\sqrt{\omega^2 + 4(1-x^2)^2 \cos^2 2\phi}} \ln \frac{\sqrt{\omega^2 + 4(1-x^2)^2 \cos^2 2\phi} + \omega}{\sqrt{\omega^2 + 4(1-x^2)^2 \cos^2 2\phi} - \omega}$$

$$(1-3x^2)^2 + [(1-3x^2)^2 - \frac{3}{8}((1-3x^2)^2 + 3(1-x^2)^2 \cos^2 2\phi)] \times$$

$$\times \ln(1-x^2)^2 \cos^2 2\phi\} = 0,$$

$$\int_0^1 dx \int d\varphi \{ \frac{\sqrt{\omega^2 + 4(1-x^2)^2 \cos^2 2\phi}}{\omega} \ln \frac{\sqrt{\omega^2 + 4(1-x^2)^2 \cos^2 2\phi} + \omega}{\sqrt{\omega^2 + 4(1-x^2)^2 \cos^2 2\phi} - \omega}$$

$$(1-x^2)^2 \cos^2 2\phi + [(1-x^2)^2 \cos^2 2\phi - 2(1-x^2)^2 x^2 \cos^2 \phi] \times$$

$$\times \ln(1-x^2)^2 \cos^2 2\phi\} = 0,$$

$$\int_0^1 dx \int d\varphi \{ \frac{\omega}{\sqrt{\omega^2 + 4(1-x^2)^2 \cos^2 2\phi}} \ln \frac{\sqrt{\omega^2 + 4(1-x^2)^2 \cos^2 2\phi} + \omega}{\sqrt{\omega^2 + 4(1-x^2)^2 \cos^2 2\phi} - \omega}$$

$$(1-x^2)^2 \cos^2 2\phi + [(1-x^2)^2 \cos^2 2\phi - 2(1-x^2)^2 x^2 \cos^2 \phi)] \times$$

$$\times \ln(1-x^2)^2 \cos^2 2\phi\} = 0,$$

$$\int_0^1 dx \int d\varphi \{ \frac{\sqrt{\omega^2 + 4(1-x^2)^2 \cos^2 2\phi}}{\omega} \ln \frac{\sqrt{\omega^2 + 4(1-x^2)^2 \cos^2 2\phi} + \omega}{\sqrt{\omega^2 + 4(1-x^2)^2 \cos^2 2\phi} - \omega}$$

$$4(1-x^2)^2 x^2 \cos^2 \phi + [4(1-x^2)^2 x^2 \cos^2 \phi - 2(1-x^2)^2 x^2 \cos^2 \phi] \times$$

$$\times \ln(1-x^2)^2 \cos^2 2\phi\} = 0,$$

$$\int_0^1 dx \int d\varphi \{ \frac{\omega}{\sqrt{\omega^2 + 4(1-x^2)^2 \cos^2 2\phi}} \ln \frac{\sqrt{\omega^2 + 4(1-x^2)^2 \cos^2 2\phi} + \omega}{\sqrt{\omega^2 + 4(1-x^2)^2 \cos^2 2\phi} - \omega}$$

$$4(1-x^2)^2 x^2 \cos^2 \phi + [4(1-x^2)^2 x^2 \cos^2 \phi - 2(1-x^2)^2 x^2 \cos^2 \phi)] \times$$

$$\times \ln(1-x^2)^2 \cos^2 2\phi\} = 0,$$

$$\int_0^1 dx \int d\varphi \{ \frac{\sqrt{\omega^2 + 4(1-x^2)^2 \cos^2 2\phi}}{\omega} \ln \frac{\sqrt{\omega^2 + 4(1-x^2)^2 \cos^2 2\phi} + \omega}{\sqrt{\omega^2 + 4(1-x^2)^2 \cos^2 2\phi} - \omega}$$

$$4(1-x^2)^2 x^2 \sin^2 \phi + [4(1-x^2)^2 x^2 \sin^2 \phi - 2(1-x^2)^2 x^2 \cos^2 \phi] \times$$

$$\times \ln(1-x^2)^2 \cos^2 2\phi\} = 0,$$

$$\int_0^1 dx \int d\varphi \{ \frac{\omega}{\sqrt{\omega^2+4(1-x^2)^2\cos^2 2\phi}} \ln \frac{\sqrt{\omega^2+4(1-x^2)^2\cos^2 2\phi}+\omega}{\sqrt{\omega^2+4(1-x^2)^2\cos^2 2\phi}-\omega}$$
$$4(1-x^2)^2 x^2 \sin^2\phi + [4(1-x^2)^2 x^2 \sin^2\phi - 2(1-x^2)^2 x^2 \cos^2\phi)]\times$$
$$\times \ln(1-x^2)^2 \cos^2 2\phi\} = 0,$$
$$\int_0^1 dx \int d\varphi \{ \frac{\sqrt{\omega^2+4(1-x^2)^2\cos^2 2\phi}}{\omega} \ln \frac{\sqrt{\omega^2+4(1-x^2)^2\cos^2 2\phi}+\omega}{\sqrt{\omega^2+4(1-x^2)^2\cos^2 2\phi}-\omega}$$
$$(1-x^2)^2 \sin^2\phi + [(1-x^2)^2 \sin^2\phi - 2(1-x^2)^2 x^2 \cos^2\phi]\times$$
$$\times \ln(1-x^2)^2 \cos^2 2\phi\} = 0,$$
$$\int_0^1 dx \int d\varphi \{ \frac{\omega}{\sqrt{\omega^2+4(1-x^2)^2\cos^2 2\phi}} \ln \frac{\sqrt{\omega^2+4(1-x^2)^2\cos^2 2\phi}+\omega}{\sqrt{\omega^2+4(1-x^2)^2\cos^2 2\phi}-\omega}$$
$$(1-x^2)^2 \sin^2\phi + [(1-x^2)^2 \sin^2\phi - 2(1-x^2)^2 x^2 \cos^2\phi)]\times$$
$$\times \ln(1-x^2)^2 \cos^2 2\phi\} = 0. \tag{27.5}$$

Now, let us consider d_{xy} state

$$\begin{pmatrix} 0 & 1 & 0 \\ 1 & 0 & 0 \\ 0 & 0 & 0 \end{pmatrix}$$

with order parameter $\Delta_0(T)\,(0;0;i\sin^2\theta\sin 2\varphi;0;0)$.

The gap equation has the following form:

$$g^{-1} + \frac{\alpha^2 Z^2}{2\beta V} \sum_p \frac{\sin^4\Theta \sin^2 2\phi}{\omega^2+\xi^2+\Delta_0^2 \sin^4\Theta \sin^2 2\phi} = 0 \tag{27.6}$$

and gap $\Delta^2(T) = \Delta_0^2 \sin^4 \Theta \sin^2 2\phi$. (27.7)

After making the same calculations as in case of $d_{x^2-y^2}$–state (evaluation of the integrals in the quadratic form with respect to variables ω and ξ and then with respect to parameter α; and equating the determinant of the resulting quadratic form to zero one gets the following set of ten equations, which determine the whole spectrum of the collective modes for d_{xy} –state:

$$\int_0^1 dx \int d\varphi \{ \frac{\sqrt{\omega^2 + 4(1-x^2)^2 \sin^2 2\phi}}{\omega} \ln \frac{\sqrt{\omega^2 + 4(1-x^2)^2 \sin^2 2\phi} + \omega}{\sqrt{\omega^2 + 4(1-x^2)^2 \sin^2 2\phi} - \omega}$$

$$(1-3x^2)^2 + [(1-3x^2)^2 - \frac{3}{8}((1-3x^2)^2 + 3(1-x^2)^2 \cos^2 2\phi)] \times$$

$$\times \ln(1-x^2)^2 \sin^2 2\phi \} = 0,$$

$$\int_0^1 dx \int d\varphi \{ \frac{\omega}{\sqrt{\omega^2 + 4(1-x^2)^2 \sin^2 2\phi}} \ln \frac{\sqrt{\omega^2 + 4(1-x^2)^2 \sin^2 2\phi} + \omega}{\sqrt{\omega^2 + 4(1-x^2)^2 \sin^2 2\phi} - \omega}$$

$$(1-3x^2)^2 + [(1-3x^2)^2 - \frac{3}{8}((1-3x^2)^2 + 3(1-x^2)^2 \cos^2 2\phi)] \times$$

$$\times \ln(1-x^2)^2 \sin^2 2\phi \} = 0,$$

$$\int_0^1 dx \int d\varphi \{ \frac{\sqrt{\omega^2 + 4(1-x^2)^2 \sin^2 2\phi}}{\omega} \ln \frac{\sqrt{\omega^2 + 4(1-x^2)^2 \sin^2 2\phi} + \omega}{\sqrt{\omega^2 + 4(1-x^2)^2 \sin^2 2\phi} - \omega}$$

$$(1-x^2)^2 \cos^2 2\phi + [(1-x^2)^2 \cos^2 2\phi - 2(1-x^2)^2 x^2 \cos^2 \phi] \times$$

$$\times \ln(1-x^2)^2 \sin^2 2\phi \} = 0,$$

$$\int_0^1 dx \int d\varphi \{ \frac{\omega}{\sqrt{\omega^2 + 4(1-x^2)^2 \sin^2 2\phi}} \ln \frac{\sqrt{\omega^2 + 4(1-x^2)^2 \sin^2 2\phi} + \omega}{\sqrt{\omega^2 + 4(1-x^2)^2 \sin^2 2\phi} - \omega}$$

$$(1-x^2)^2 \cos^2 2\phi + [(1-x^2)^2 \cos^2 2\phi - 2(1-x^2)^2 x^2 \cos^2 \phi)] \times$$

$$\times \ln(1-x^2)^2 \sin^2 2\phi \} = 0,$$

$$\int_0^1 dx \int d\varphi \{ \frac{\sqrt{\omega^2 + 4(1-x^2)^2 \sin^2 2\phi}}{\omega} \ln \frac{\sqrt{\omega^2 + 4(1-x^2)^2 \sin^2 2\phi} + \omega}{\sqrt{\omega^2 + 4(1-x^2)^2 \sin^2 2\phi} - \omega}$$

$$4(1-x^2)^2 x^2 \cos^2 \phi + [4(1-x^2)^2 x^2 \cos^2 \phi - 2(1-x^2)^2 x^2 \cos^2 \phi] \times$$

$$\times \ln(1-x^2)^2 \sin^2 2\phi \} = 0,$$

$$\int_0^1 dx \int d\varphi \{ \frac{\omega}{\sqrt{\omega^2 + 4(1-x^2)^2 \sin^2 2\phi}} \ln \frac{\sqrt{\omega^2 + 4(1-x^2)^2 \sin^2 2\phi} + \omega}{\sqrt{\omega^2 + 4(1-x^2)^2 \sin^2 2\phi} - \omega}$$

$$4(1-x^2)^2 x^2 \cos^2 \phi + [4(1-x^2)^2 x^2 \cos^2 \phi - 2(1-x^2)^2 x^2 \cos^2 \phi)] \times$$

$$\times \ln(1-x^2)^2 \sin^2 2\varphi \} = 0$$

$$\int_0^1 dx \int d\varphi \{ \frac{\sqrt{\omega^2 + 4(1-x^2)^2 \sin^2 2\phi}}{\omega} \ln \frac{\sqrt{\omega^2 + 4(1-x^2)^2 \sin^2 2\phi} + \omega}{\sqrt{\omega^2 + 4(1-x^2)^2 \sin^2 2\phi} - \omega}$$

$$4(1-x^2)^2 x^2 \sin^2 \phi + [4(1-x^2)^2 x^2 \sin^2 \phi - 2(1-x^2)^2 x^2 \cos^2 \phi] \times$$

$$\times \ln(1-x^2)^2 \sin^2 2\phi \} = 0,$$

$$\int_0^1 dx \int d\varphi \{ \frac{\omega}{\sqrt{\omega^2 + 4(1-x^2)^2 \sin^2 2\phi}} \ln \frac{\sqrt{\omega^2 + 4(1-x^2)^2 \sin^2 2\phi} + \omega}{\sqrt{\omega^2 + 4(1-x^2)^2 \sin^2 2\phi} - \omega}$$

$$4(1-x^2)^2 x^2 \sin^2 \phi + [4(1-x^2)^2 x^2 \sin^2 \phi - 2(1-x^2)^2 x^2 \cos^2 \phi)] \times$$

$$\times \ln(1-x^2)^2 \sin^2 2\phi \} = 0,$$

$$\int_0^1 dx \int d\varphi \{ \frac{\sqrt{\omega^2 + 4(1-x^2)^2 \sin^2 2\phi}}{\omega} \ln \frac{\sqrt{\omega^2 + 4(1-x^2)^2 \sin^2 2\phi} + \omega}{\sqrt{\omega^2 + 4(1-x^2)^2 \sin^2 2\phi} - \omega}$$

$$(1-x^2)^2 \sin^2 \phi + [(1-x^2)^2 \sin^2 \phi - 2(1-x^2)^2 x^2 \cos^2 \phi] \times$$

$$\times \ln(1-x^2)^2 \sin^2 2\phi \} = 0,$$

$$\int_0^1 dx \int d\varphi \{ \frac{\omega}{\sqrt{\omega^2 + 4(1-x^2)^2 \sin^2 2\phi}} \ln \frac{\sqrt{\omega^2 + 4(1-x^2)^2 \sin^2 2\phi} + \omega}{\sqrt{\omega^2 + 4(1-x^2)^2 \sin^2 2\phi} - \omega}$$

$$(1-x^2)^2 \sin^2 \phi + [(1-x^2)^2 \sin^2 \phi - 2(1-x^2)^2 x^2 \cos^2 \phi)] \times$$

$$\times \ln(1-x^2)^2 \sin^2 2\phi \} = 0.$$

(27.8)

It is easy to see, that equations for the collective mode spectrum for d_{xy} state can be obtained from the ones for $d_{x^2-y^2}$–state by replacement of $\cos^2 2\phi$ by $\sin^2 2\phi$. Brusov et al.[6-15] have obtained as well equations for d_{xz}, d_{yz}, $d_{3z^2-r^2}$ states. Solving the equations for $d_{x^2-y^2}$, d_{xy}, d_{xz}, d_{yz}, $d_{3z^2-r^2}$ states numerically we obtain the complete collective–mode spectrum for all five phases.

For the high–frequency modes the following results were obtained:

$$\boldsymbol{d}_{3z^2-r^2}: \quad \begin{aligned} E_1 &= \Delta_0(2.0 - i\,1.65) \\ E_{2,3} &= \Delta_0(1.85 - i\,0.69) \\ E_{4,5} &= \Delta_0(1.64 - i\,0.50) \end{aligned}$$

$$\boldsymbol{d}_{x^2-y^2};\, \boldsymbol{d}_{xy}: \quad \begin{aligned} E_1 &= \Delta_0(1.88 - i\,0.79) \\ E_2 &= \Delta_0(1.66 - i\,0.50) \\ E_3 &= \Delta_0(1.14 - i\,0.68) \\ E_4 &= \Delta_0(1.13 - i\,0.71) \\ E_5 &= \Delta_0(1.10 - i\,0.65) \end{aligned}$$

$$\boldsymbol{d}_{xz};\, \boldsymbol{d}_{yz}: \quad \begin{aligned} E_1 &= \Delta_0(1.76 - i\,1.1) \\ E_2 &= \Delta_0(1.70 - i\,0.48) \\ E_3 &= \Delta_0(1.14 - i\,0.68) \\ E_4 &= \Delta_0(1.13 - i\,0.73) \\ E_5 &= \Delta_0(1.04 - i\,0.83) \end{aligned} \tag{27.9}$$

Let us discuss the results obtained. For each superconducting phase, five high–frequency modes were determined as well as five Goldstone (quasi–Goldstone) modes, whose energies are either equal to zero or small ($\leq 0.1\ \Delta_0$). The spectra in $d_{x^2-y^2}$ and d_{xy} states as well as in d_{xz}, d_{yz} turn out to be identical. In $d_{x^2-y^2}$, d_{xy}, d_{xz}, d_{yz}, states all modes are nondegenerated while in $d_{3z^2-r^2}$ state two high frequency modes are twice degenerated. The energies (frequencies) of all high–frequency modes turn out to be complex and their imaginary parts, $\operatorname{Im} E_i$, describe the damping of the collective excitations via decay of Cooper pairs into fermions. Their energies (frequencies) in $d_{x^2-y^2}$ and d_{xy} states are ranged between $1.1\ \Delta_0(T)$ and $1.88\ \Delta_0(T)$, in d_{xz}, d_{yz} states – between $1.04\ \Delta_0(T)$ and $1.76\ \Delta_0(T)$ and in $d_{3z^2-r^2}$ state between $1.64\ \Delta_0(T)$ and $2.0\,\Delta_0(T)$.

We should note here that the high–frequency collective modes in superconductors are damped much strongly in case of *d*–pairing than in case of *p*–pairing. In d–wave superconductors $\operatorname{Im} E_i$ are ranged from 30% to 65%. It is much higher than in case of p–wave superconductors, where $\operatorname{Im} E_i$ are mainly ranged from 2.5% to 16% (just for some modes at some phases $\operatorname{Im} E_i$ is higher). This fact is connected with the nodal structure of energy gap. As a rule one has points of nodes under *p*–pairing and lines of nodes under *d*–pairing.

In the next Chapter we will see that similar difference in attenuation of the collective modes via the different topology of nodes takes place as well for pure and mixed d–wave superconductors.

The results on high–frequency modes can be useful in determining the order parameter and the type of pairing in HTSC as well as for interpreting the ultrasound and microwave absorption experiments with these systems[1-15].

Chapter XXVIII

How To Distinguish The Mixture of Two D–Wave States from Pure D–Wave State of HTSC

28.1. The Mixture of Two d–Wave States

Recent experiments[1] and theoretical considerations[2,3] show that in high–temperature superconductors the mixture of different d–wave states is realized. Paul Brusov and Peter Brusov[4] have calculated for the first time the collective mode spectrum in a mixed $d_{x2-y2} + id_{xy}$ state of high–temperature superconductors. They used the model of d–pairing for superconductive and superfluid fermi–systems (HTSC, HFSC etc.) created by Brusov and Brusova within path integration technique earlier[5-7]. They have shown that in spite of the fact that spectra in both states d_{x2-y2} and d_{xy} are identical[8-10] the spectrum in mixture $d_{x2-y2} + id_{xy}$ state turns out to be quite different from them. Thus the probe of the spectrum in ultrasound and/or microwave absorption experiments could be used to distinguish the mixture of two d–wave states from pure d–wave states.

As we mentioned above the most scientists believe that there is a *d*–wave pairing in oxide superconductors. At the same time the different ideas concerning extended s–wave pairing, mixture of *s*– and *d*–states, as well as of different *d*–states still discuss actively. One of the cause of such a situation is the uncertainty in answer the question: do we have exact zero gap along some chosen lines in momentum space (like the case of d_{x2-y2}) or gap is anisotropic but nonzero everywhere (except maybe some points). Existing experiments (tunneling etc.) do not give the certain answer this question while the answer is quite principle. From other side there are some experiments[1] which could be explained[3] under suggestion about realization in high–temperature superconductors of a mixed states, like $d_{x2-y2}+id_{xy}$. Annett *et al.*[2] considered the possibility of mixture of different d–wave states in high–temperature superconductors and came to conclusion that $d_{x2-y2}+id_{xy}$ is the most likely state.

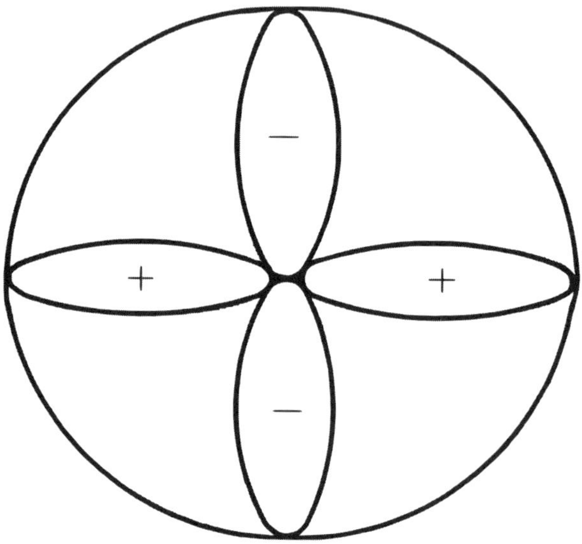

FIG. 28.1. Gap symmetry in pure *d*–wave state (d_{xy} state).

Pavel Brusov and Peter Brusov[4] suggested one of the possible ways to distinguish the mixture of two *d*–states from pure *d*–states. For this they considered the mixed $d_{x2-y2}+id_{xy}$ state and calculate the spectrum of collective modes in this state. The comparison of this spectrum with the spectrum of a pure d–wave states of high–temperature superconductors shows that they are significantly different and could be the probe of the symmetry of the order parameter in high–temperature superconductors.

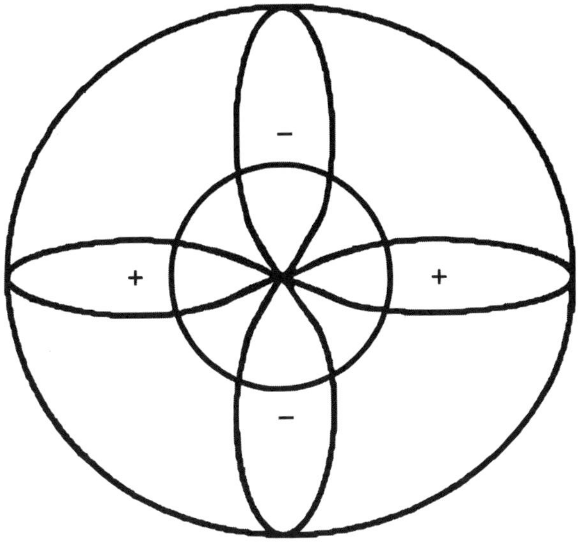

FIG. 28.2. Gap symmetry in the mixed s–wave and d– wave states ($s+id_{xy}$).

28.2. Equations for Collective Modes Spectrum in a Mixed d–Wave State of Unconventional Superconductors

28.2.1. Model for the mixed state

We have generalized here Brusov et al.[4] consideration for the case of arbitrary admixture of d_{xy} state. We consider the mixed $(1-\gamma)d_{x^2-y^2} + i\gamma d_{xy}$ state in high temperature superconductors and derive a full set of equations for collective modes spectrum in mixed d–wave state with arbitrary admixture of d_{xy} state.

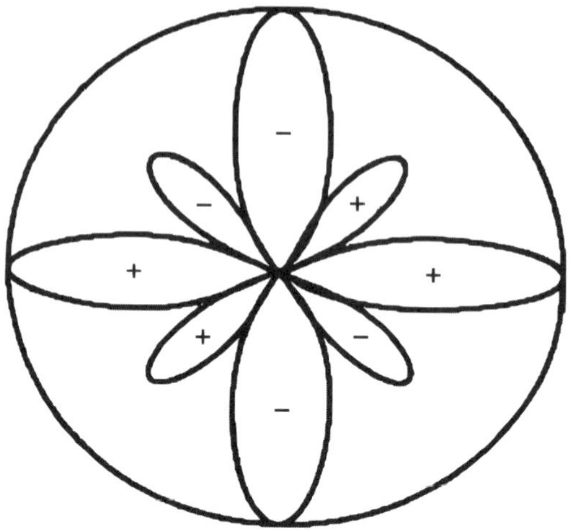

FIG. 28.3. Gap symmetry in the mixed $(1-\gamma)d_{x^2-y^2} + i\gamma d_{xy}$ state.

We have used the model of d–pairing in high–temperature superconductors and HFSC, created by Brusov et al.[5,11]

It is described by the effective functional of action:

$$S_{\text{eff}} = g^{-1} \sum_{p,i,a} c_{ia}^{+}(p)c_{ia}(p) + \frac{1}{2} \ln \ \det \frac{\hat{M}\left(c_{ia}, c_{ia}^{+}\right)}{\hat{M}\left(c_{ia}^{(0)}, c_{ia}^{(0)+}\right)}, \qquad (28.1)$$

where $c_{ia}^{(0)}$ is the condensate value of Bose–fields c_{ia} (symmetric traceless matrix) and $\hat{M}\left(c_{ia}, c_{ia}^{+}\right)$ is the 4×4 matrix depending on Bose–fields and parameters of quasi–fermions.

The number of degrees of freedom in the case of d–pairing is equal to 10, i.e., one must have five complex canonical variables, which can be naturally chosen in the following form:

$$c_1 = c_{11} + c_{22},\ c_2 = c_{11} - c_{22},\ c_3 = c_{12} + c_{21},$$

$$c_4 = c_{13} + c_{31},\ c_5 = c_{23} + c_{32}.$$

In the canonical variables, the effective action has the following form:

$$S_{\text{eff}} = (2g)^{-1} \sum_{p,j} c_j^{+}(p) c_j(p) \left(1 + 2\delta_{j1}\right) + \frac{1}{2} \ln \det \frac{\hat{M}\left(c_j^{+}, c_j\right)}{\hat{M}\left(c_j^{+(0)}, c_j^{(0)}\right)}, \tag{28.2}$$

where

$$M_{11} = Z^{-1}[i\omega + \xi - \mu(\mathbf{H}\boldsymbol{\sigma})]\delta_{p_1 p_2},$$

$$M_{22} = Z^{-1}[-i\omega + \xi + \mu(\mathbf{H}\boldsymbol{\sigma})]\delta_{p_1 p_2},$$

$$M_{12} = M_{21}^{*} = (\beta V)^{-1/2} \left(\frac{15}{32\pi}\right)^{1/2} \left[c_1\left(1 - 3\cos^2\theta\right) + \right.$$

$$+ c_2 \sin^2\theta \cos 2\varphi + c_3 \sin^2\theta \sin 2\varphi +$$

$$\left. + c_4 \sin 2\theta \cos\varphi + c_5 \sin 2\theta \sin\varphi \right], \tag{28.3}$$

where $p = (\mathbf{k}, \omega)$; $\omega = (2n+1)\pi T$ are Fermi–frequencies and $x = (\mathbf{x}, \tau)$, ξ is the kinetic energy with respect to Fermi–level, μ–magnetic moment of quasifermion, $\boldsymbol{H}$–magnetic field, $\boldsymbol{\sigma}=(\sigma_1, \sigma_2, \sigma_3)$ – Pauli–matrices.

This functional determines all the properties of the model superconducting Fermi–system with *d*–pairing.

Let us consider the mixed $(1-\gamma)d_{x^2-y^2} + i\gamma d_{xy}$ state of high–temperature superconductors. The order parameter in this state takes the following form:

$$\Delta_0(T)\,[\,(1-\gamma)\begin{pmatrix}1 & 0 & 0\\ 0 & -1 & 0\\ 0 & 0 & 0\end{pmatrix}+i\gamma\begin{pmatrix}0 & 1 & 0\\ 1 & 0 & 0\\ 0 & 0 & 0\end{pmatrix}\,] \tag{28.4}$$

or in canonical variables:

$$\Delta_0(T)\,(0;(1-\gamma)\sin^2\theta\cos 2\varphi; i\gamma\sin^2\theta\sin 2\varphi;0;0) \tag{28.5}$$

The gap equation has the following form:

$$g^{-1}+\frac{\alpha^2 Z^2}{2\beta V}\sum_p \frac{\sin^4\Theta[\gamma^2+\cos^2 2\phi(1-2\gamma)]}{\omega^2+\xi^2+\Delta_0^2\sin^4\Theta[\gamma^2+\cos^2 2\phi(1-2\gamma)]}=0\,, \tag{28.6}$$

where

$\Delta_0 = 2cZ\alpha$, $\alpha=(15/32\pi)^{1/2}$ and gap

$$\Delta^2(T)=\Delta_0^2\sin^4\Theta[\gamma^2+\cos^2 2\phi(1-2\gamma)]\,. \tag{28.7}$$

For a limited case of γ=0 one gets d_{x2-y2} state with order parameter $(0;\sin^2\theta\cos 2\varphi;0;0;0)$.

The gap equation in this case has the following form:

$$g^{-1}+\frac{\alpha^2 Z^2}{2\beta V}\sum_p \frac{\sin^4\Theta\cos^2 2\phi}{\omega^2+\xi^2+\Delta_0^2\sin^4\Theta\cos^2 2\phi}=0 \tag{28.8}$$

and gap $\Delta^2(T)=\Delta_0^2\sin^4\Theta\cos^2 2\phi]\,.$ (28.9)

For limited case γ=1 one gets d_{xy} state with order parameter $\Delta_0(T)$ $(0;0;i\sin^2\theta\sin 2\varphi;0;0)$.

The gap equation has the following form:

$$g^{-1}+\frac{\alpha^2 Z^2}{2\beta V}\sum_p \frac{\sin^4\Theta\sin^2 2\phi}{\omega^2+\xi^2+\Delta_0^2\sin^4\Theta\sin^2 2\phi}=0 \qquad (28.10)$$

and gap $\Delta^2(T)=\Delta_0^2\sin^4\Theta\sin^2 2\phi$ (28.11)

Brusov et al.[8] case of equal admixtures of d_{x2-y2} and d_{xy} states in our consideration corresponds to the case γ=1/2. The order parameter takes the following form:

$$\Delta_0(T)\,(0;\sin^2\theta\cos 2\varphi;i\sin^2\theta\sin 2\varphi;0;0) \qquad (28.12)$$

The gap equation has the following form:

$$g^{-1}+\frac{\alpha^2 Z^2}{2\beta V}\sum_p \frac{\sin^4\Theta}{\omega^2+\xi^2+\Delta_0^2\sin^4\Theta}=0 \qquad (28.13)$$

and gap $\Delta(T)=\Delta_0(T)\sin^2\theta$.

28.2.2. Equations for collective modes spectrum in a mixed d–wave state at arbitrary admixture of d_{xy} state

The spectrum of collective excitations in the first approximations is determined by the quadratic part of S_{eff}, obtained after shift $c_j \to c_j+c_j^o$ where c_j^o are the condensate values of c_j, which take the following form[5,11]

$$c_j^0(p) = (\beta V)^{1/2} c\delta_{p0} b_j^0 \text{ and } b_2^0 = 2(1-\gamma), \quad b_3^0 = 2i\gamma$$

with all remaining components of b_j^0 equal to zero.

Excluding terms involving g^{-1} by gap equation, one obtains the following form for the quadratic part of S_h

$$S_H = \frac{\alpha^2 Z^2}{8\beta V} \sum_p \frac{[c^0 Y^*][c^{+0} Y]}{\omega^2 + \xi^2 + [c^0 Y^*][c^{+0} Y]} \sum_j (1 + 2\delta j 1) c_j^+(p) c_j(p)$$

$$+ Z^2/4\beta V \sum_{p1+p2=p} \frac{1}{M_1 M_2} \{(i\omega_1 + \xi_1)(i\omega_2 + \xi_2)([c^+(p)Y(p_2)]$$

$$[c(p)Y^*(p_1)] + [c^+(p)Y(p_1)][c(p)Y^*(p_2)]) - \Delta^2[c^+(p)Y(-p_1)]$$

$$[c^+(-p)Y(-p_2)] - \Delta^{+2}[c(p)Y^*(-p_1)][c(-p)Y^*(-p_2)]]\}. \qquad (28.14)$$

Here,

$$[cY^*] = c_1(1 - 3\cos^2\theta) + c_2 \sin^2\theta \cos 2\varphi + c_3 \sin^2\theta \sin 2\varphi +$$

$$+ c_4 \sin 2\theta \cos\varphi + c_5 \sin 2\theta \sin\tilde{\varphi}.$$

The coefficients of the quadratic form are proportional to the sums of the products of Green's functions of quasifermions. At low temperatures ($T_c - T \sim T_c$) we can go from a summation to an integration by the following rule

$$\frac{1}{\beta V}\sum_{p} \rightarrow \frac{1}{(2\pi)^4}\frac{k_F^2}{c_F}\int d\omega d\xi d\Omega . \tag{28.15}$$

To evaluate these integrals it is useful to use the Feynman equality

$$\left[\left(\omega_1^2+\xi_1^2+\Delta^2\right)\left(\omega_2^2+\xi_2^2+\Delta^2\right)\right]^{-1} =$$

$$\int d\lambda\left[\alpha\left(\omega_1^2+\xi_1^2+\Delta^2\right)+\left(1-\lambda\right)\left(\omega_2^2+\xi_2^2+\Delta^2\right)\right]^{-2}. \tag{28.16}$$

It is easy to evaluate the integrals with respect to variables ω and ξ and then with respect to parameter α and the angular variables.

After calculating all integrals except over the angular variables and equating the determinant of the resulting quadratic form to zero one gets the following set of ten equations, which determine the whole spectrum of the collective modes for $(1-\gamma)d_{x^2-y^2} + i\gamma d_{xy}$ state at arbitrary γ

$$\int_0^1 dx\int d\varphi\{\ \frac{\sqrt{\omega^2+4(1-x^2)^2[\gamma^2+\cos^2 2\phi(1-2\gamma)]}}{\omega}\times$$

$$\times\ln\frac{\sqrt{\omega^2+4(1-x^2)^2[\gamma^2+\cos^2 2\phi(1-2\gamma)]}+\omega}{\sqrt{\omega^2+4(1-x^2)^2[\gamma^2+\cos^2 2\phi(1-2\gamma)]}-\omega}(1-3x^2)^2+[(1-3x^2)^2-$$

$$-\frac{3}{8}((1-3x^2)^2+3(1-x^2)^2\cos^2 2\phi)]\ln(1-x^2)^2[\gamma^2+\cos^2 2\phi(1-2\gamma)]\ \}=0$$

$$\int_0^1 dx \int d\varphi \{ \frac{\omega}{\sqrt{\omega^2 + 4(1-x^2)^2[\gamma^2 + \cos^2 2\phi(1-2\gamma)]}} \cdot (1-3x^2)^2 \times$$
$$\times \ln \frac{\sqrt{\omega^2 + 4(1-x^2)^2[\gamma^2 + \cos^2 2\phi(1-2\gamma)]} + \omega}{\sqrt{\omega^2 + 4(1-x^2)^2[\gamma^2 + \cos^2 2\phi(1-2\gamma)]} - \omega} + [(1-3x^2)^2 -$$
$$- \frac{3}{8}((1-3x^2)^2 + 3(1-x^2)^2 \cos^2 2\phi)] \ln(1-x^2)^2[\gamma^2 + \cos^2 2\phi(1-2\gamma)]\} = 0$$

$$\int_0^1 dx \int d\varphi \{ \frac{\sqrt{\omega^2 + 4(1-x^2)^2[\gamma^2 + \cos^2 2\phi(1-2\gamma)]}}{\omega} \cdot (1-x^2)^2 \cos^2 2\phi \times$$
$$\times \ln \frac{\sqrt{\omega^2 + 4(1-x^2)^2[\gamma^2 + \cos^2 2\phi(1-2\gamma)]} + \omega}{\sqrt{\omega^2 + 4(1-x^2)^2[\gamma^2 + \cos^2 2\phi(1-2\gamma)]} - \omega} + [(1-x^2)^2 \cos^2 2\phi -$$
$$- 2(1-x^2)^2 x^2 \cos^2 \phi] \ln(1-x^2)^2[\gamma^2 + \cos^2 2\phi(1-2\gamma)]\} = 0$$

$$\int_0^1 dx \int d\varphi \{ \frac{\omega}{\sqrt{\omega^2 + 4(1-x^2)^2[\gamma^2 + \cos^2 2\phi(1-2\gamma)]}} \cdot (1-x^2)^2 \cos^2 2\phi \times$$
$$\times \ln \frac{\sqrt{\omega^2 + 4(1-x^2)^2[\gamma^2 + \cos^2 2\phi(1-2\gamma)]} + \omega}{\sqrt{\omega^2 + 4(1-x^2)^2[\gamma^2 + \cos^2 2\phi(1-2\gamma)]} - \omega} + [(1-x^2)^2 \cos^2 2\phi -$$
$$- 2(1-x^2)^2 x^2 \cos^2 \phi)] \ln(1-x^2)^2[\gamma^2 + \cos^2 2\phi(1-2\gamma)]\} = 0$$

$$\int_0^1 dx \int d\varphi \{ \frac{\sqrt{\omega^2 + 4(1-x^2)^2[\gamma^2 + \cos^2 2\phi(1-2\gamma)]}}{\omega} \cdot 4(1-x^2)^2 x^2 \cos^2 \phi \times$$
$$\times \ln \frac{\sqrt{\omega^2 + 4(1-x^2)^2[\gamma^2 + \cos^2 2\phi(1-2\gamma)]} + \omega}{\sqrt{\omega^2 + 4(1-x^2)^2[\gamma^2 + \cos^2 2\phi(1-2\gamma)]} - \omega} + [4(1-x^2)^2 \times$$
$$\times x^2 \cos^2 \phi - 2(1-x^2)^2 x^2 \cos^2 \phi] \ln(1-x^2)^2[\gamma^2 + \cos^2 2\phi(1-2\gamma)]\} = 0$$

$$\int_0^1 dx \int d\varphi \{ \frac{\omega}{\sqrt{\omega^2 + 4(1-x^2)^2[\gamma^2 + \cos^2 2\phi(1-2\gamma)]}} \cdot 4(1-x^2)^2 x^2 \cos^2 \phi \times$$

$$\times \ln \frac{\sqrt{\omega^2 + 4(1-x^2)^2[\gamma^2 + \cos^2 2\phi(1-2\gamma)]} + \omega}{\sqrt{\omega^2 + 4(1-x^2)^2[\gamma^2 + \cos^2 2\phi(1-2\gamma)]} - \omega} + [4(1-x^2)^2 x^2 \cos^2 \phi -$$

$$-2(1-x^2)^2 x^2 \cos^2 \phi)] \ln(1-x^2)^2[\gamma^2 + \cos^2 2\phi(1-2\gamma)]\} = 0$$

$$\int_0^1 dx \int d\varphi \{ \frac{\sqrt{\omega^2 + 4(1-x^2)^2[\gamma^2 + \cos^2 2\phi(1-2\gamma)]}}{\omega} \cdot 4(1-x^2)^2 x^2 \sin^2 \phi \times$$

$$\times \ln \frac{\sqrt{\omega^2 + 4(1-x^2)^2[\gamma^2 + \cos^2 2\phi(1-2\gamma)]} + \omega}{\sqrt{\omega^2 + 4(1-x^2)^2[\gamma^2 + \cos^2 2\phi(1-2\gamma)]} - \omega} + [4(1-x^2)^2 x^2 \sin^2 \phi -$$

$$-2(1-x^2)^2 x^2 \cos^2 \phi] \ln(1-x^2)^2[\gamma^2 + \cos^2 2\phi(1-2\gamma)]\} = 0$$

$$\int_0^1 dx \int d\varphi \{ \frac{\omega}{\sqrt{\omega^2 + 4(1-x^2)^2[\gamma^2 + \cos^2 2\phi(1-2\gamma)]}} \cdot 4(1-x^2)^2 x^2 \sin^2 \phi \times$$

$$\times \ln \frac{\sqrt{\omega^2 + 4(1-x^2)^2[\gamma^2 + \cos^2 2\phi(1-2\gamma)]} + \omega}{\sqrt{\omega^2 + 4(1-x^2)^2[\gamma^2 + \cos^2 2\phi(1-2\gamma)]} - \omega} + [4(1-x^2)^2 x^2 \sin^2 \phi -$$

$$-2(1-x^2)^2 x^2 \cos^2 \phi)] \ln(1-x^2)^2[\gamma^2 + \cos^2 2\phi(1-2\gamma)]\} = 0$$

$$\int_0^1 dx \int d\varphi \{ \frac{\sqrt{\omega^2 + 4(1-x^2)^2[\gamma^2 + \cos^2 2\phi(1-2\gamma)]}}{\omega} \cdot (1-x^2)^2 \sin^2 \phi \times$$

$$\times \ln \frac{\sqrt{\omega^2 + 4(1-x^2)^2[\gamma^2 + \cos^2 2\phi(1-2\gamma)]} + \omega}{\sqrt{\omega^2 + 4(1-x^2)^2[\gamma^2 + \cos^2 2\phi(1-2\gamma)]} - \omega} + [(1-x^2)^2 \sin^2 \phi -$$

$$-2(1-x^2)^2 x^2 \cos^2 \phi] \ln(1-x^2)^2[\gamma^2 + \cos^2 2\phi(1-2\gamma)]\} = 0$$

$$\int_0^1 dx \int d\varphi \{ \frac{\omega}{\sqrt{\omega^2 + 4(1-x^2)^2[\gamma^2 + \cos^2 2\phi(1-2\gamma)]}} \cdot (1-x^2)^2 \sin^2\phi \times$$
$$\times \ln \frac{\sqrt{\omega^2 + 4(1-x^2)^2[\gamma^2 + \cos^2 2\phi(1-2\gamma)]} + \omega}{\sqrt{\omega^2 + 4(1-x^2)^2[\gamma^2 + \cos^2 2\phi(1-2\gamma)]} - \omega} + [(1-x^2)^2 \sin^2\phi -$$
$$-2(1-x^2)^2 x^2 \cos^2\phi)] \ln(1-x^2)^2[\gamma^2 + \cos^2 2\phi(1-2\gamma)]\} = 0. \tag{28.17}$$

The following substitutions have been used: $\cos\theta = x$, $\omega = \omega/\Delta_0$.

These equations determine the whole spectrum of collective modes in mixed $(1-\gamma)d_{x^2-y^2} + i\gamma d_{xy}$ state of high temperature superconductors (HTSC) with arbitrary admixture of d_{xy} state. Knowledge of the collective mode spectrum could be used for interpretation of the sound attenuation and microwave absorption data as well as for identification of the type of pairing and order parameter in unconventional superconductors. In particular, they allow to estimate the extent of admixture of a d_{xy}–state in a possible mixed state.

The most interesting case is the case of small γ: we suppose that dominant state is $d_{x^2-y^2}$–state and admixture of d_{xy} state is small, say 5–10%. In these case we could expand all expressions in powers of small γ and obtain the corrections to the spectrum of pure $d_{x^2-y^2}$–state, which has been found before[4,5,11].

28.2.3. Equations for collective modes spectrum in a mixed d–wave state at an equal admixtures of $d_{x^2-y^2}$ and d_{xy} states

Brusov et al.[4] have supposed the equal admixtures of $d_{x^2-y^2}$– and d_{xy} – states – in our consideration this corresponds to the case $\gamma = 1/2$ – and derived the following equations:

$i=1$

$$\int_0^1 dx \int d\varphi \{ \frac{\sqrt{\omega^2+4f}}{\omega} \ln \frac{\sqrt{\omega^2+4f}+\omega}{\sqrt{\omega^2+4f}-\omega} g_1 + (g_1 - \frac{3}{2} f_1) \ln f \} = 0$$

$$\int_0^1 dx \int d\varphi \{ \frac{\omega}{\sqrt{\omega^2+4f}} \ln \frac{\sqrt{\omega^2+4f}+\omega}{\sqrt{\omega^2+4f}-\omega} g_1 + (g_1 - \frac{3}{2} f_1) \ln f \} = 0 \qquad (28.18)$$

$i = 2,3,4,5$

$$\int_0^1 dx \int d\varphi \{ \frac{\sqrt{\omega^2+4f}}{\omega} \ln \frac{\sqrt{\omega^2+4f}+\omega}{\sqrt{\omega^2+4f}-\omega} g_i + (g_i - \frac{1}{2} g \ ) \ln f \} = 0$$

$$\int_0^1 dx \int d\varphi \{ \frac{\omega}{\sqrt{\omega^2+4f}} \ln \frac{\sqrt{\omega^2+4f}+\omega}{\sqrt{\omega^2+4f}-\omega} g_i + (g_i - \frac{1}{2} g \ ) \ln f \} = 0 \ . \qquad (28.19)$$

Here,

$$g_1 = (1-3x^2)^2; \ g_2 = (1-x^2)^2 \cos^2 2\varphi \ ; \ g_3 = g = 4(1-x^2)x^2 \cos^2 \varphi ;$$
$$g_4 = 4(1-x^2)x^2 \sin^2 \varphi \ ; \ g_5 = (1-x^2)^2 \sin^2 \varphi \ ;$$
$$f_1 = (1/4)\left[(1-3x^2)^2 + 3(1-x^2)^2 \cos^2 2\varphi \right]; \ f = (1-x^2)^2 . \qquad (28.20)$$

They have solved[4, 11] these equations numerically and have found five high frequency modes in each state obtained from the second equations while the first ones appear to give either Goldstone modes or modes with vanishing energies (of order 0.03 $\Delta_0(T)$ – 0.08 $\Delta_0(T)$).

Below we give their results for high frequency modes (E_i is the energy (frequency) of i–th branch).

$$E_{1,2} = \Delta_0(T)(1.93 - i\ 0.41); \tag{28.21}$$

$$E_3 = \Delta_0(T)(1.62 - i\ 0.75); \tag{28.22}$$

$$E_{4,5} = \Delta_0(T)(1.59 - i\ 0.83) \tag{28.23}$$

Comparison of these results with spectrum of pure $d_{x^2-y^2}$– and d_{xy} –states, obtained by Brusov et al.[4,11]:

$$E_1 = \Delta_0(T)(1.88 - i\ 0.79); \tag{28.24}$$

$$E_2 = \Delta_0(T)(1.66 - i\ 0.50); \tag{28.25}$$

$$E_3 = \Delta_0(T)(1.40 - i\ 0.68); \tag{28.26}$$

$$E_4 = \Delta_0(T)(1.13 - i\ 0.71); \tag{28.27}$$

$$E_5 = \Delta_0(T)(1.10 - i\ 0.65) \tag{28.28}$$

led them to conclusion, that in spite of the fact that spectra in both pure states, $d_{x^2-y^2}$ and d_{xy}, turn out to be identical, spectrum in mixed $d_{x^2-y^2}$ – and d_{xy} –state is quite different from that in pure states. In pure states all modes are nondegenerated while in mixed state two high frequency modes are twice degenerated. The energies (frequencies) of high frequency modes are ranged between 1.1 $\Delta_0(T)$ and 1.88 $\Delta_0(T)$ while in mixed state between 1.59 $\Delta_0(T)$ and 1.93 $\Delta_0(T)$ and the collective modes have higher frequencies. Also, damping of collective modes in pure d–states is more than in mixed state (Im E_i is from 30% to 65% in pure states and from 20% to 50% in mixed state). It could be easy understood, because in pure states the gap vanishes along chosen lines while in mixed state it vanishes just at two points (poles) [4].

The difference of spectrum of collective excitations in pure d –wave states and in mixed state give us a possibility to probe the state symmetry by ultrasound and/or microwave absorption experiments. Note, that while these experiments could require high frequencies (of order of tens GHz as in case of Feller et al. experiment[16]) there is no principle restrictions for ultrasound (microwave) frequencies: because frequencies of collective modes are proportional to $\Delta_0(T)$, which vanishes at T_c, thus, one could use in principle any frequency approaching to T_c.

Note, that case of $d_{x^2-y^2} + id_{xy}$ – state has been considered as well by Balatsky et al. [17], who studied one of the possible collective mode in this state. They consider a superconducting state with mixed–symmetry order parameter components, e.g., $d+is$ and $d_{x^2-y^2}+id_{xy}$ and argued for the existence of a new orbital magnetization mode which corresponds to oscillations of relative phase φ between two components around an equilibrium value of $\varphi = \pi / 2$. It is similar to the clapping –mode in superfluid ^{3}He–A. They estimated the frequency of this mode $\omega_0(B,T)$ depending on the field and temperature for the specific case of magnetic field induced d_{xy}–state. This mode is tunable with a magnetic field with $\omega_0(B,T) \propto B\Delta_0$, where Δ_0 is the magnitude of the d–wave order parameter. As well they estimated the velocity $s(B,T)$ of this mode.

28.2.4. $d_{x^2-y^2}$ – state of high temperature superconductors with a small admixture of d_{xy} – state

Obtained above equations (28.17) determine the whole spectrum of collective modes in mixed $d_{x^2-y^2} + i\varepsilon d_{xy}$ state of high temperature superconductors (HTSC) with arbitrary admixture of d_{xy} state. Knowledge of the collective mode spectrum could be used for interpretation of the sound attenuation and microwave absorption data as well as for identification of the type of pairing and order parameter in unconventional superconductors. In particular, they allow to estimate the extent of admixture of a d_{xy} state in a possible mixed state.

The most interesting case however is the case of small γ: we suppose that dominant state is $d_{x^2-y^2}$ state and admixture of d_{xy} state is small, say 3–10%. In these case we could expand all expressions in powers of small γ and obtain the corrections to the equations for spectrum of pure $d_{x^2-y^2}$ state, which has been found before[11-13].

Let us consider the case of small admixture of d_{xy} state (small ε): we suppose that dominant state is $d_{x^2-y^2}$ state and admixture of d_{xy} state is small, about 3–10%. In these case we could expand all expressions in powers of small ε and obtain the corrections to the equations for spectrum of pure $d_{x^2-y^2}$ state, which have been found before[11-13].

Using the following notations:

$$a = (1-x^2)^2 \cos^2 2\phi;$$
$$b = (1-x^2)^2 \sin^2 2\phi; \tag{28.29}$$
$$a_1 = a + \omega^2,$$

we get the following expressions:

$$\sqrt{\omega^2 + 4(1-x^2)^2[\cos^2 2\phi + \gamma^2 \sin^2 2\phi]} \approx \sqrt{a_1}\left(1 + \gamma^2 \frac{b}{2a_1}\right) \tag{28.30}$$

$$\ln f \approx \ln a + \gamma^2 \frac{b}{a} \tag{28.31}$$

$$\frac{\sqrt{\omega^2 + 4(1-x^2)^2[\cos^2 2\phi + \gamma^2 \sin^2 2\phi]} + \omega}{\sqrt{\omega^2 + 4(1-x^2)^2[\cos^2 2\phi + \gamma^2 \sin^2 2\phi]} - \omega} \approx A + \gamma^2 B, \tag{28.32}$$

where

$$A = \frac{(\sqrt{a_1} + \omega)^2}{a} \;;$$

$$B = \frac{b}{a^2} \frac{\sqrt{a_1} + \omega}{\sqrt{a_1}} [a + \sqrt{a_1} (\omega + \sqrt{a_1})] \tag{28.33}$$

$$\begin{aligned} &\ln \frac{\sqrt{\omega^2 + 4(1-x^2)^2[\cos^2 2\phi + \gamma^2 \sin^2 2\phi]} + \omega}{\sqrt{\omega^2 + 4(1-x^2)^2[\cos^2 2\phi + \gamma^2 \sin^2 2\phi]} - \omega} \approx \\ &\approx \ln \frac{\sqrt{\omega^2 + 4(1-x^2)^2 \cos^2 2\phi} + \omega}{\sqrt{\omega^2 + 4(1-x^2)^2 \cos^2 2\phi} - \omega} + \gamma^2 C, \end{aligned} \tag{28.34}$$

where

$$C = \frac{b(\sqrt{a_1} + \omega)}{\sqrt{a}} + \frac{b}{a} \; . \tag{28.35}$$

For other expressions one gets with the accuracy of order γ^2:

$$\begin{aligned} &\frac{\sqrt{\omega^2 + 4(1-x^2)^2[\cos^2 2\phi + \gamma^2 \sin^2 2\phi]}}{\omega} \times \\ &\times \ln \frac{\sqrt{\omega^2 + 4(1-x^2)^2[\cos^2 2\phi + \gamma^2 \sin^2 2\phi]} + \omega}{\sqrt{\omega^2 + 4(1-x^2)^2[\cos^2 2\phi + \gamma^2 \sin^2 2\phi]} - \omega} \approx \\ &\frac{\sqrt{\omega^2 + 4(1-x^2)^2 \cos^2 2\phi}}{\omega} \ln \frac{\sqrt{\omega^2 + 4(1-x^2)^2 \cos^2 2\phi} + \omega}{\sqrt{\omega^2 + 4(1-x^2)^2 \cos^2 2\phi} - \omega} + \\ &+ \frac{\gamma^2}{\omega} \sqrt{a_1} \left(C + \frac{b}{2a_1} \ln \frac{\sqrt{\omega^2 + 4(1-x^2)^2 \cos^2 2\phi} + \omega}{\sqrt{\omega^2 + 4(1-x^2)^2 \cos^2 2\phi} - \omega} \right); \end{aligned} \tag{28.36}$$

$$\frac{\omega}{\sqrt{\omega^2+4(1-x^2)^2[\cos^2 2\phi+\gamma^2\sin^2 2\phi]}}\times$$

$$\times\ln\frac{\sqrt{\omega^2+4(1-x^2)^2[\cos^2 2\phi+\gamma^2\sin^2 2\phi]}+\omega}{\sqrt{\omega^2+4(1-x^2)^2[\cos^2 2\phi+\gamma^2\sin^2 2\phi]}-\omega}\approx$$

$$\frac{\omega}{\sqrt{\omega^2+4(1-x^2)^2\cos^2 2\phi+}}\ln\frac{\sqrt{\omega^2+4(1-x^2)^2\cos^2 2\phi}+\omega}{\sqrt{\omega^2+4(1-x^2)^2\cos^2 2\phi}-\omega}+$$

$$+\gamma^2\frac{\omega}{\sqrt{a_1}}\left(C-\frac{b}{2a_1}\ln\frac{\sqrt{\omega^2+4(1-x^2)^2\cos^2 2\phi}+\omega}{\sqrt{\omega^2+4(1-x^2)^2\cos^2 2\phi}-\omega}\right). \tag{28.37}$$

Putting all expressions (28.29)–(28.37) into (28.24) one gets the whole set of equations for collective mode spectrum of mixed $d_{x^2-y^2}+i\varepsilon d_{xy}$ state with small admixture of d_{xy} state to $d_{x^2-y^2}$ state.

28.2.5. Conclusion

In order to solve one of the problem of unconventional superconductivity – the exact form of the order parameter – we consider the mixed $(1-\gamma)d_{x^2-y^2}+i\gamma d_{xy}$ state in high temperature superconductors and derive a full set of equations for collective modes spectrum in a mixed d–wave state with an arbitrary admixture of d_{xy} state. Brusov et al. [4,11-14] have solved these equations for the case of equal admixtures $d_{x^2-y^2}$– and d_{xy}–states. They have shown that the difference of spectrum of collective excitations in pure d – wave states and in mixed state give a possibility to probe the state symmetry by ultrasound and/or microwave absorption experiments.

The most interesting case is the case of small γ: we suppose that dominant state is $d_{x^2-y^2}$–state and admixture of d_{xy}–state is small, say 3–10%. In this case we expand all expressions in powers of small γ and obtain the corrections to the spectrum of pure $d_{x^2-y^2}$–state, which has been found before.

The results obtained could be useful for identification of the type of pairing and determination of the exact form of the order parameter in unconventional superconductors. In particular, they allow to estimate the extent of admixture of a d_{xy} state in a possible mixed state. Derived equations allow to calculate the whole collective mode spectrum, which could be used for interpretation of the sound attenuation and microwave absorption data.

The results obtained could also allow answer three very important questions:

1) does the gap disappear along some chosen lines?

2) do we have a pure or mixed *d*–wave state in high–temperature superconductors?

3) how large is the admixture of d_{xy} state in a possible mixed state?

Chapter XXIX

P–Wave Superconductors

29.1. Introduction

P–pairing is realized not only in pure superfluid ^3He, superfluid ^3He in aerogel, but as well in unconventional superconductors: Sr_2RuO_4 (HTSC) and UPt_3 (HFSC). The possibility of the triplet p–wave pairing in Sr_2RuO_4 was first suggested by Rice and Sigrist[1] from the analogy to superfluid ^3He. Now there is considerable experimental evidence of the unconventional triplet superconductivity in Sr_2RuO_4. The absence of the Hebel–Slichter peak in $1/T_1$, and substantial reduction of T_C by nonmagnetic impurities indicate that Sr_2RuO_4 is at least not a conventional s–wave superconductor. The muon–spin resonant experiments suggest a pairing state with broken time reversal symmetry. More strong evidence of the triplet pairing has been found by ^{17}O NMR experiments in which the temperature–independent Knight shift was observed for the magnetic field parallel to the RuO_2–plane[2].

In the symmetry classification of the possible pairing–states in Sr_2RuO_4 it has been shown that the quasi–two–dimensional electronic structure in this superconductor leads to five possible p–wave states stabilized by the absence of gap nodes. Among this states, the pairing–state compatible with all above experiments is $k_x + ik_y$ state, which is the two–dimensional analog of ^3He–A.

In this Chapter we study the collective mode spectrum in bulk superconducting phases which are realized in HTSC as well as in HFSC under *p*–pairing.

29.2. Bulk *p*–Wave Superconductivity

The first results for case of *p*–pairing have been obtained by us[3,4] for A–, B–, A_1–, 2D– and polar states, which relate to superfluid phases of ^3He, where first three states have been observed. Here we consider additional possible superconducting states, which could be realized in HTSC as well as in HFSC under *p*–pairing. Below we give a summary of the obtained results[5,8]. Remind that spectrum of collective excitation in each superconducting (superfluid) state consists of 18 modes, among which there are high frequency modes as well as Goldstone ones.

A–phase

Pairbreaking modes $E = \Delta_0(T)(1.96 - i\ 0.31)$ (3 modes);

Clapping modes $E = \Delta_0(T)(1.17 - i\ 0.13)$ (6 modes);

Sound modes $E^2 = c_F^2 k^2/3$ (3 modes);

Orbital waves $E^2 = c_F^2 k_{||}^2$ (6 modes).

B–phase

Real squashing modes $E^2 = 12\Delta^2/5$ (5 modes);

Squashing modes $E^2 = 8\Delta^2/5$ (5 modes);

Pairbreaking modes $E^2 = 4\Delta^2$ (4 modes);

Sound modes $E^2 = c_F^2 k^2/3$ (1 mode);

Spin waves:

longitudinal $E^2 = c_F^2 k^2/5$ (1 mode);

transverse $E^2 = 2c_F^2 k^2/5$ (2 modes).

2D–phase

Goldstone modes $E=0$ (6 modes);

Pairbreaking modes $E = \Delta_0(T)(1.96 - i\ 0.31)$ (2 modes);

Clapping modes $E = \Delta_0(T)(1.17 - i\,0.13)$ (4 modes);

Quasi–Goldstone modes $E=2\mu H$ (2 modes);

Quasi–pairbreaking modes

$$E_0^2 = \Delta_0^2(T)(1.96 - i\,0.31)^2 + 4\mu^2 H^2 \text{ (2 modes);}$$

$$E^2 = \Delta_0^2(T)(0.518)^2 + 4\mu^2 H^2 \text{ (1 modes);}$$

$$E^2 = \Delta_0^2(T)(0.495)^2 + 4\mu^2 H^2 \text{ (1 modes).}$$

A_1–phase

Pairbreaking modes $E = \Delta_0(T)(1.96 - i\,0.31)$ (1 mode);

Clapping modes $E = \Delta_0(T)(1.17 - i\,0.13)$ (2 modes);

Quasi–Goldstone modes $E=2\mu H$ (8 modes);

Goldstone modes $E=0$ (1 mode).

Six other modes have an imaginary spectrum (this fact is connected with the instability of the A_1–phase with respect to small perturbations).

In the polar phase with gap $\sim \begin{pmatrix} 000 \\ 000 \\ 001 \end{pmatrix}$ with gap in single–particle spectrum $\Delta^2 = \Delta_0^2 \cos^2\theta$ we obtain the following set of equations for collective mode spectrum:

$$\int_0^1 dx(1-x^2)[(1-\frac{4\Delta^2}{q^2})J-2]=0 \text{ (6 modes)},$$

$$\int_0^1 dx(1-x^2)[J-2]=0 \text{ (6 modes)},$$

$$\int_0^1 dxx^2(1+\frac{4\Delta^2}{q^2})J=0 \text{ (3 modes)},$$

$$\int_0^1 dxx^2 J=0 \text{ (3 modes)}. \tag{29.1}$$

Here,

$$J=\frac{1}{\sqrt{1+4\Delta^2/q^2}}\ln\frac{1+\sqrt{1+4\Delta^2/q^2}}{1-\sqrt{1+4\Delta^2/q^2}},$$

$$x=\cos\theta,\ q^2=\omega^2+c_F^2(\boldsymbol{k}\cdot\boldsymbol{n})^2.$$

Quantity J depends on the gap in the single particle spectrum Δ, which depends on the angle θ (as it will be seen below, in general case it depends on the angles θ and φ).

Putting $\boldsymbol{k}$=0 and solving above equations numerically we have found the root $E=\Delta_0(T)(1.20-i\ 1.75)$ for the second equation and E=0 for the third equation. We did not find any roots for the first and fourth equations.

Thus, for the polar phase we have found six highly damped modes with energy (frequency) $E = \Delta_0(T)(1.20 - i\,1.75)$ and three Goldstone modes. The presence of highly damped modes is associated with the fact that in polar phase the gap disappears along the equator in contrast to axial and planar phases, where the gap disappears in two points (poles) and thus in these two phases the collective modes attenuate moderately and can be observed as resonances in sound absorption experiments.

For the following three states $\frac{1}{\sqrt{2}}\begin{pmatrix} 1 & 0 & 0 \\ 0 & -1 & 0 \\ 0 & 0 & 0 \end{pmatrix}$, $\frac{1}{\sqrt{2}}\begin{pmatrix} 0 & 1 & 0 \\ 1 & 0 & 0 \\ 0 & 0 & 0 \end{pmatrix}$, and $\frac{1}{\sqrt{2}}\begin{pmatrix} 0 & -1 & 0 \\ 1 & 0 & 0 \\ 0 & 0 & 0 \end{pmatrix}$, with gap in single particle spectrum $\Delta^2 = \Delta_0^2 \sin^2\theta$ spectra are identical and can be determined from the following set of equations:

$$\int_0^1 dx(1-x^2)[1+\frac{4\Delta^2}{q^2}]J = 0 \text{ (2 modes)},$$

$$\int_0^1 dx(1-x^2)[1+\frac{6\Delta^2}{q^2}]J = 0 \text{ (3 modes)},$$

$$\int_0^1 dx(1-x^2)[1+\frac{8\Delta^2}{q^2}]J = 0 \text{ (1 mode)},$$

$$\int_0^1 dx(2-x^2)(1+\frac{4\Delta^2}{q^2})[J-1] = 0 \text{ (1 mode)},$$

$$\int_0^1 dx 2x^2 (1+\frac{6\Delta^2}{q^2})[J-1]=0 \text{ (2 modes)},$$

$$\int_0^1 dx(1-x^2)[1-\frac{2\Delta^2}{q^2}]J=0 \text{ (2 modes)},$$

$$\int_0^1 dx(1-x^2)J=0 \text{ (3 modes)},$$

$$\int_0^1 dx(1-x^2)[1-\frac{4\Delta^2}{q^2}]J=0 \text{ (1 mode)},$$

$$\int_0^1 dx x^2[(1-\frac{2\Delta^2}{q^2})J-1]=0 \text{ (2 modes)},$$

$$\int_0^1 dx x^2[J-1]=0 \text{ (1 mode)}. \qquad (29.2)$$

The numerical solution of these equations at $\boldsymbol{k}$=0 leads to the following spectrum for high–frequency modes:

$$E=\Delta_0(T)(1.83-i\,0.06) \text{ (1 mode)};$$

$$E=\Delta_0(T)(1.58-i\,0.04) \text{ (2 modes)};$$

$$E=\Delta_0(T)(1.33-i\,0.10) \text{ (1 mode)};$$

$$E = \Delta_0(T)(1.33 - i\,0.08) \text{ (2 modes);}$$

$$E = \Delta_0(T)(1.28 - i\,0.04) \text{ (2 modes);}$$

$$E = \Delta_0(T)(1.09 - i\,0.22) \text{ (3 modes);}$$

$$E = \Delta_0(T)(0.71 - i\,0.05) \text{ (3 modes);}$$

$$E = \Delta_0(T)(0.33 - i\,0.34) \text{ (1 mode);}$$

$$E = \Delta_0(T)(0.23 - i\,0.71) \text{ (2 modes).} \qquad (29.3)$$

The last two modes have imaginary parts of the same order of magnitude as the real ones. This means that they are damped very strongly and could not be treated as resonances.

For the phase $\begin{pmatrix} 1 & 1 & 0 \\ i & i & 0 \\ 0 & 0 & 0 \end{pmatrix}$, with gap $\Delta^2 = \Delta_0^2 \sin^2\theta$ we obtain the following set of equations for the collective mode spectrum:

$$\int_0^1 dx x^2 (1 + \frac{2\Delta^2}{q^2})[J - 1] = 0 \text{ (6 modes),}$$

$$\int_0^1 dx (1 - x^2)(1 + \frac{2\Delta^2}{q^2}) J = 0 \text{ (4 modes),}$$

$$\int_0^1 dx(1-x^2)(1+\frac{\Delta^2}{q^2})J = 0 \text{ (4 modes)},$$

$$\int_0^1 dx(1-x^2)(1+\frac{3\Delta^2}{q^2})J = 0 \text{ (4 modes)}. \tag{29.4}$$

The numerical solution of these equations at $\boldsymbol{k}$=0 leads to the following spectrum of high frequency modes

$$E = \Delta_0(T)(0.66 - i\ 0.02);$$

$$E = \Delta_0(T)(0.64 - i\ 0.02);$$

$$E = \Delta_0(T)(0.46 - i\ 0.04);$$

$$E = \Delta_0(T)(0.36 - i\ 0.04). \tag{29.5}$$

For the phase $\frac{1}{\sqrt{6}}\begin{pmatrix} -1 & 0 & 0 \\ 0 & -1 & 0 \\ 0 & 0 & 2 \end{pmatrix}$, with the gap $\Delta^2 = \Delta_0^2(1+3\cos^2\theta)$ we obtained the following set of equations for the collective mode spectrum:

$$\int_0^1 dx[(1+\frac{n\Delta^2}{q^2}+\frac{m\Delta^2}{q^2}\frac{1-x^2}{3})J(1-x^2)-\frac{4}{3}+\frac{16\pi}{27\sqrt{3}}]=0 \tag{29.6}$$

At n=4, m=1 we get the equation, describing the modes corresponding to the variables:

$u_{11}-u_{22}$, $u_{12}+u_{21}$, where $c_{ij}=u_{ij}+v_{ij}i$ are the Bose–fields.

At $n=4$, $m=0$ – to the variables $u_{12}-u_{21}$.

At $n=0$, $m=-1$– to the variables $v_{11}-v_{22}$, $v_{12}+v_{21}$.

At $n=0$, $m=0$ – to the variables $v_{12}-v_{21}$.

$$\int_0^1 dx[(1+\frac{n\Delta^2}{q^2}+\frac{m\Delta^2}{q^2}\frac{2}{3}[A+(2A-1)x^2])J(1-x^2)-\frac{4}{3}+\frac{16\pi}{27\sqrt{3}}]$$
$$\times\int_0^1 dx[(1+\frac{n\Delta^2}{q^2}+\frac{mA\Delta^2}{q^2}\frac{2x^2}{3})J2x^2+\frac{2}{3}-\frac{8\pi}{27\sqrt{3}}]$$
$$-2[\int_0^1 dxJ\frac{\Delta^2}{q^2}\frac{2}{3}(1-x^2)x^2]^2=0 \quad (29.7)$$

At $A=1;$ $n=4$, $m=1$ we get the equations, describing the modes corresponding to the variables $u_{11}+u_{22}$, u_{33}.

At $A=1$; $n=0$, $m=-1$– to the variables $v_{11}+v_{22}$, v_{33}.

At $A=0$; $n=4$, $m=1$– to the variables (u_{23}, u_{32}), (u_{13}, u_{31}).

At $A=0$; $n=0$, $m=-1$– to the variables (v_{23}, v_{32}), (v_{13}, v_{31}).

For the phase $\begin{pmatrix} 0 & 0 & 0 \\ 0 & 0 & 0 \\ 1 & 1 & 0 \end{pmatrix}$, with the gap $\Delta^2=\Delta_0^2\cos^2\theta$ we get the following equation for the spectrum:

$$\int_0^1 dx\int_0^{2\pi} d\varphi(A-x^2)(1\pm A\sin\varphi)[(1+\frac{n\Delta^2}{q^2})J+A-1]=0 \quad (29.8)$$

At $A=0;$ $n=0$ we get the equations, describing the modes corresponding to the variables u_{31}, u_{32}, v_{33}.

At $A=0$; $n=4$ – to the variables u_{33}, v_{31}, v_{32}.

At $A = 1$; $n = 0$ – to the variables $u_{11}\pm u_{21}$, $u_{12}\pm u_{22}$, $v_{13}\pm v_{23}$.

At $A = 1$; $n = 4$ – to the variables $v_{11}\pm v_{21}$, $v_{12}\pm v_{22}$, $u_{13}\pm u_{23}$.

For the phase $\begin{pmatrix} 0 & 0 & 1 \\ 0 & 0 & 1 \\ 0 & 0 & 0 \end{pmatrix}$, with the gap $\Delta^2 = \Delta_0^2 \sin^2\theta$ we obtained the following two equations for the spectrum:

$$\int_0^1 dx \int_0^{2\pi} d\varphi x^2 [(1+\frac{n\Delta^2}{q^2})J - 1] = 0 \tag{29.9}$$

$$\int_0^1 dx \int_0^{2\pi} d\varphi (1-x^2)(1\pm\sin\phi)(1+\frac{n\Delta^2}{q^2})J = 0 . \tag{29.10}$$

The first equation describes the modes corresponding to variables u_{31}, u_{32}, v_{33} for $n = 0$ and to variables $u_{33,}$ v_{31}, v_{32} for $n = 4$.

The second equation describes the modes corresponding to the variables $u_{11}\pm u_{21}$, $u_{12}\pm u_{22}$, $v_{13}\pm v_{23}$ for $n = 0$ and to variables $v_{11}\pm v_{21}$, $v_{12}\pm v_{22}$, $u_{13}\pm u_{23}$ for $n = 4$.

For the phases with order parameters $\frac{1}{\sqrt{2}}\begin{pmatrix} 0 & 0 & 1 \\ 0 & 0 & 0 \\ \pm 1 & 0 & 0 \end{pmatrix}$ and $\frac{1}{\sqrt{2}}\begin{pmatrix} 0 & 0 & 1 \\ 0 & 0 & 0 \\ 0 & \pm 1 & 0 \end{pmatrix}$ with the gaps $\Delta^2 = \Delta_0^2 \sin^2\theta \cos^2\varphi$ and $\Delta^2 = \Delta_0^2 \sin^2\theta \sin^2\varphi$ respectively the spectrum turns out to be identical and for its calculation we get the following two equations, the first of which:

$$\int_0^1 dx \int_0^{\frac{\pi}{2}} d\varphi [(1+\frac{2\Delta^2}{q^2}+\frac{2\Delta_0^2}{q^2}[A(1-x^2)\cos^2\varphi + Bx^2])J(1-x^2)\cos^2\varphi +$$

$$\frac{1}{2}[(1-x^2)\cos^2\varphi - x^2]\ln(1-(1-x^2)\cos^2\varphi)]\times \int_0^1 dx \int_0^{\frac{\pi}{2}} d\varphi [(1+\frac{2\Delta^2}{q^2}+$$

$$\frac{2\Delta_0^2}{q^2}[B(1-x^2)\cos^2\varphi + Ax^2])Jx^2 - \frac{1}{2}[(1-x^2)\cos^2\varphi - x^2]\times$$

$$\times \ln(1-(1-x^2)\cos^2\varphi)] - [\int_0^1 dx \int_0^{\frac{\pi}{2}} d\varphi \frac{4\Delta_0^2}{q^2} x^2(1-x^2)\cos^2\varphi J]^2 = 0$$

(29.11)

leads to an equation describing the modes corresponding to variables v_{11}, v_{33} for A=1, B=0; to variables u_{11}, u_{33} for A=–1, B=0; to the variables v_{13}, v_{31} for A=0, B=1; and to variables u_{13}, u_{31} for A=0, B =–1.

The second equation takes the following form:

$$\int_0^1 dx \int_0^{\pi/2} d\varphi \{(1+\frac{2\Delta^2}{q^2}N+\frac{2\Delta_0^2}{q^2}x^2P)J[(1-x^2)((z-y)\cos^2\varphi + y)+$$

$$x^2(1-z-y)]+((1-x^2)\cos^2\varphi - x^2)(2z+y-1)+$$

$$+(3(1-x^2)\sin^2\varphi - 1)y\frac{1}{2}\ln(1-(1-x^2)\cos^2\varphi)\} = 0,$$

(29.12)

For $y = 1, z = 0$ and $N = 0, P = 0$ we get the equation, describing the modes corresponding to the variable u_{22},

For $N = 2, P = 0$ – to variable v_{22},

For $N = 0, P = 1$– to variable u_{21},

For $N = 2, P = -1$ – to variable v_{21},

For $N = 1$, P $= -1$ – to variable u_{23},

For $N = 1$, P $= 1$ – to variable v_{23}.

For y=0, z=1 и N=0, P=0 we get the equation, describing the modes corresponding to variable u_{12},

For N=2, P=0 – to the variables v_{12}.

For $y = 0$, $z = 0$ и $N = 0$, $P = 0$ we get the equation, describing the modes corresponding to the variable u_{32},

For $N = 2$, $P = 0$ – to the variable v_{32}.

For the phase $\begin{pmatrix} 0 & 0 & 1 \\ 0 & 0 & 0 \\ 1 & 0 & 0 \end{pmatrix}$ we obtain the following high–frequency modes:

$$E = \Delta_0(T)(1.80 - i\,0.09);$$

$$E = \Delta_0(T)(0.55 - i\,0.80). \tag{29.13}$$

The last mode has imaginary part of the same order of magnitude as real one. This means that it is damped very strongly and could not be treated as resonance.

Tewordt[9] analyzed the order parameter collective mode spectrum in Sr_2RuO_4 supposed the p–pairing in this superconductor. He has considered two possible superconducting states with the order parameters $\hat{d} = \Delta_0 \hat{z}(k_x + ik_y)$ and $\hat{d} = \frac{\Delta_0}{2}\hat{z}(k_x + k_y)$, arisen in the symmetry classification.

Note, that the first phase is the analog to the A–phase of the superfluid ^{3}He. In this phase Tewordt[9] has found the mode $E = 2\Delta_0$, while for the second phase he has found the mode $E = \sqrt{3}\Delta_0$.

Both modes are coupled to the charge density fluctuations, but this coupling is small via smallness of the value $dN(E)/dE$, which is the measure of the electron–hole asymmetry on the Fermi–surface. Comparing the results by Tewordt[9] ($E = 2\Delta_0$) with ours[5-8], we should note that for high frequency modes at the phase, which is the analog to the A–phase of the superfluid ^{3}He, we get the frequency (energy) $E = \Delta_0(T)(1.96 - i0.31)$, which is more exact value of the frequency. This is connected to the fact, that Tewordt[9] did not calculate the imaginary parts of the collective mode frequencies, which presence renormalizes the real parts of the collective mode frequencies via dispersion relations.

Brusov et al.[2,5-8] have considered as well the second superconducting phase, studied by Tewordt, with the order parameter $\begin{pmatrix} 0 & 0 & 1 \\ 0 & 0 & 1 \\ 0 & 0 & 0 \end{pmatrix}$, for the spectrum of which the authors have got the equations, shown above (however, the solution of these equations have not been obtained).

Considering the pairing amplitude different in the xy–plane and in a plane perpendicular to it, Tewordt[9] obtained a set of quasi–Godstone modes with the frequencies $\omega^2 = \Delta_0^2 \ln(T_c/T_{cj})$, where $T_{cj} < T_c$ is the temperature of the superconducting transition, corresponding to the pairing in the x–y – plane.

Because Brusov et al.[2,5-8] have considered both amplitude to be equal, they obtained pure Godstone modes ($T_c = T_{cj}$ and hence, $\omega = 0$) instead of the quasi–Godstone modes. It should be noted, however, that the number of superconducting phases considered by Brusov et al.[2,5-8] in the case of p–pairing, is much larger than those analyzed by Tewordt[9].

Higashitani et al.[10] studied the effect of the order parameter collective modes on electromagnetic response in Sr_2RuO_4 considering a spin triplet superconductor with $(k_x \pm ik_y)$ orbital symmetry. They obtained a clapping mode with frequency $\omega = \sqrt{2}\Delta$. This result has been obtained by Brusov et al.[2,5-8] twenty years earlier.

Chapter XXX

Two Dimensional P– and D–Wave Superconductivity

30.1. Two–Dimensional Models of p– and d–Pairing in USC

The existence of CuO_2 planes[1] – the common structural factor of HTSC – suggests we consider 2D models. Brusov and Popov[2,3] developed a 2D–model of *p*–pairing using a path integration technique. Brusov et al.[4-9] developed a 2D model of *d*–pairing within the same technique. The models involve hydrodynamic action functionals, obtained by path integration over "fast" and "slow" Fermi–fields. These functionals determine all properties of 2D–superconductors (for example, of CuO_2 planes of HTSC) and, in particular, the spectrum of collective excitations.

30.2. *p*–Pairing

To develop the model of *p*–pairing in the 2D–case let us start with the 3D scheme considered above.

The main distinctions are as follows:

a) The orbital moment $\boldsymbol{l}$ ($\boldsymbol{l}$=1) should be perpendicular to the plane and can have only two projections on the $\hat{z}$–axis: ±1. Because the *p*–pairing is a triplet, the total spin of the pair is equal to 1, so in the case of 2D *p*–pairing one has $3\times2\times2=12$ degrees of freedom. Thus we can describe the superconductive state in this case by complex 2×3 matrices $c_{ia}(p)$. The number of degrees of freedom is equal to the number of the CM in each phase. Note that in the 3D case this number is equal to 18.

b) $\mathbf{x}$ will be a 2D vector and square "volume" will be $S=L^2$ (instead of $V=L^3$ as in 3D case).

30.2.1. Two–dimensional *p*–wave superconducting states

Effective action in case of two–dimensional *p*–wave superconductivity takes a form (see the case of two–dimensional superfluidity of ^{3}He in Chapter XIX)

$$S_{eff} = -\beta V \frac{16\pi^2 T_C \Delta T}{7\zeta(3)} F \tag{30.1}$$

where

$$F = -trAA^+ + \nu\ trA^+AP + (trA^+A)^2 + trAA^+AA^+ + trAA^+A^*A^T -$$
$$- trAA^T A^* A^+ - (1/2)trAA^T trA^* A^+,$$

$$\nu = 7\zeta(3)\mu^2 H^2 / 4\pi^2 T_C \Delta T \tag{30.2}$$

Equation (30.2) is identical in form with that arising in the three–dimensional system. The difference is that the matrix A with elements a_{ia} for the two–dimensional system is a 2 x 3 matrix. The matrix P is the projector on the third axis:

$$P = \begin{pmatrix} 0 & 0 & 0 \\ 0 & 0 & 0 \\ 0 & 0 & 1 \end{pmatrix}$$

Minimizing F, we obtain the following equation for the condensate matrix A:

$$-A + \nu\ AP + 2(trAA^+)A + 2AA^+A + 2A^*A^TA - 2AA^TA^* -$$
$$-\left(trAA^T\right)A^* = 0. \tag{30.3}$$

This equation has several solutions, corresponding to the different superfluid phases. We consider the possibilities:

$$A_1 = \frac{1}{2}\begin{pmatrix} 1 & 0 & 0 \\ i & 0 & 0 \end{pmatrix},\ A_2 = \frac{1}{2}\begin{pmatrix} 1 & 0 & 0 \\ 0 & 1 & 0 \end{pmatrix},\ A_3 = \frac{1}{4}\begin{pmatrix} 1 & i & 0 \\ i & -1 & 0 \end{pmatrix},$$

$$A_5 = \frac{1}{\sqrt{3}}\begin{pmatrix} 0 & 0 & 0 \\ 0 & 1 & 0 \end{pmatrix},\ A_6 = \frac{1}{2}\begin{pmatrix} 1 & 0 & 0 \\ 0 & -1 & 0 \end{pmatrix},\ A_7 = \frac{1}{2}\begin{pmatrix} 0 & \pm 1 & 0 \\ 1 & 0 & 0 \end{pmatrix},$$

$$A_8 = \left(\frac{1-\nu}{3}\right)^{1/2}\begin{pmatrix} 0 & 0 & 1 \\ 0 & 0 & 0 \end{pmatrix},\ A_9 = \left(\frac{1-\nu}{4}\right)^{1/2}\begin{pmatrix} 0 & 0 & 1 \\ 0 & 0 & i \end{pmatrix}. \tag{30.4}$$

The corresponding values of F are equal to:

$$F_1 = -\frac{1}{4},\ F_2 = -\frac{1}{4},\ F_3 = -\frac{1}{8},\ F_4 = -\frac{1}{6},\ F_5 = -\frac{1}{6},$$
$$F_6 = -\frac{1}{4},\ F_7 = -\frac{1}{4},\ F_8 = -(1-\nu)^2/6,\ F_9 = -(1-\nu)^2/4. \tag{30.5}$$

For the first eight phases, the quantity F does not depend on H. The minimum value of $F = -1/4$ is reached for phases with matrices A_1 and A_2 as well as for A_6 and A_7 and A_9 (last state has minimum energy in zero magnetic field (ν=0)). As we know from two–dimensional superfluidity of ^{3}He the first two phases are called the *a*– and *b*–phases. Brusov and Popov[2,3] have proved that the phases *a*– and *b*– are stabile relative to small perturbations. They have calculated the collective mode spectrum for two these phases, while Brusov et al.[4-9] have calculated the collective mode spectrum for A_6 and A_7 states.

30.2.2. The collective mode spectrum

Below we show results obtained by Brusov and Popov[2,3] and Brusov et al.[4-9] for collective mode spectrum for different two–dimensional superconducting states under p–pairing.

a **–phase** with order parameter $\frac{1}{2}\begin{pmatrix} 1 & 0 & 0 \\ i & 0 & 0 \end{pmatrix}$:

$$E^2 = \frac{c_F^2 k^2}{2}\left(1 - \frac{5c_F^2 k^2}{96\Delta^2}\right), \text{ (3 modes)}$$

$$E^2 = 2\Delta^2 + c_F^2 k^2/2, \text{ (6 modes)} \tag{30.6}$$

$$E^2 = 4\Delta^2 + (0.500 + i0.433)c_F^2 k^2. \text{ (3 modes)}$$

b **–phase** with order parameter $\frac{1}{2}\begin{pmatrix} 1 & 0 & 0 \\ 0 & 1 & 0 \end{pmatrix}$:

$$E^2 = \frac{c_F^2 k^2}{2}\left(1 - \frac{5c_F^2 k^2}{48\Delta^2}\right), \text{ (2 modes)}$$

$$E^2 = \frac{3c_F^2 k^2}{4}\left(1 - \frac{c_F^2 k^2}{72\Delta^2}\right), \text{ (1 mode)}$$

$$E^2 = \frac{c_F^2 k^2}{4}\left(1 - \frac{c_F^2 k^2}{48\Delta^2}\right), \text{ (1 mode)}$$

$$E^2 = 2\Delta^2 + c_F^2 k^2/2, \text{ (4 modes)}$$

$$E^2 = 4\Delta^2 + (0.500 - i0.433)c_F^2 k^2, \text{ (2 mode)}$$

$$E^2 = 4\Delta^2 + (0.152 - i0.218)c_F^2 k^2, \text{ (1 mode)}$$

$$E^2 = 4\Delta^2 + (0.849 - i0.216)c_F^2 k^2. \text{ (1 mode)} \quad (30.7)$$

It is interesting to note the existence in *a*– and *b*–phases of so–called two–dimensional sound with velocity $v_2 = c_F/\sqrt{2}$ (just remind that in bulk systems the three–dimensional sound with velocity $v_3 = c_F/\sqrt{3}$ is well known). After Brusov et al. [2,3] this result has been reproduced by a number of authors (Nagai[10], Tewordt[11] etc.).

phase $\frac{1}{2}\begin{pmatrix} 0 & 0 & 1 \\ 0 & 0 & i \end{pmatrix}$:

$$E^2 = 0, \text{ (3 modes);}$$

$$E^2 = 2\Delta^2, \text{ (6 modes);}$$

$$E^2 = 4\Delta^2. \text{ (3 modes)} \quad (30.8)$$

phase $\frac{1}{2}\begin{pmatrix} 0 & \pm 1 & 0 \\ 1 & 0 & 0 \end{pmatrix}$:

$$E^2 = 0, \text{ (4 modes);}$$

$E^2 = 2\Delta^2$, (4 modes);

$E^2 = 4\Delta^2$. (4 modes) (30.9)

phase $\frac{1}{2}\begin{pmatrix} 1 & 0 & 0 \\ 0 & -1 & 0 \end{pmatrix}$:

$E^2 = 0$, (4 modes);

$E^2 = 2\Delta^2$, (4 modes);

$E^2 = 4\Delta^2$. (4 modes) (30.10)

30.3. Two–Dimensional d–Wave Superconductivity

30.3.1. 2D–model of d–pairing in CuO_2 planes of HTSC

As we mentioned above the existence of CuO_2 planes — the common structural factor of HTSC — suggests we consider 2D models. For 2D quantum antiferromagnet (AF) it was shown that only the *d*–channel provides an attractive interaction between fermions. The *d*–pairing arises also in symmetry classifications of CuO_2 planes HTSC.

Brusov and Brusova (BB)[4,5,8] and BBB[9] develop a 2D model of *d*–pairing in the CuO_2 planes of HTSC using a path integration technique. The model involves a hydrodynamic action functional, obtained by path integration over "fast" and "slow" Fermi–fields. This functional determines all properties of the CuO_2 planes and, in particular, the spectrum of collective excitations.

To develop the model of *d*–pairing in the 2D–case one starts with the 3D scheme considered above.

The main distinctions are as follows:

(a) The orbital moment $\vec{l}\left(\left|\vec{l}\right|=2\right)$ should be perpendicular to the plane and can have only two projections on the $\hat{z}$– axis: ±2. Because the *d*–pairing is a singlet the total spin of the pair is equal zero, so in the case of 2D *d*–pairing one has $1\times2\times2=4$ degrees of freedom. Thus we can describe the superconductive state in this case by complex symmetric traceless 2×2 matrices $c_{ia}(p)$, which have the same number of degrees of freedom ($2\times2\times2-2-2=4$). This number is equal to the number of the CM in each phase. Note that in the 3D case this number is equal to 10.

(b) The pairing potential t is given by:

$$t=v\left(\hat{k},\hat{k}'\right)=\sum_{m=-2,2} g_m Y_{2m}\left(\hat{k}\right)Y_{2m}^{*}\left(\hat{k}'\right) \tag{30.11}$$

In the case of circular symmetry $g_2=g_{-2}=g$ and we have one coupling constant g, while less symmetric cases require both constants g_2 and g_{-2}. We will consider the circularly symmetric case where:

$$v\left(\hat{k},\hat{k}'\right)=g\left[Y_{2-2}\left(\hat{k}\right)Y_{2-2}^{*}\left(\hat{k}'\right)+Y_{22}\left(\hat{k}\right)Y_{22}^{*}\left(\hat{k}'\right)\right] \tag{30.12}$$

(c) **x** will be a 2D–vector and square "volume" will be $S=L^2$ (instead of $V=L^3$ as in 3D case).

Taking into account these distinctions we will describe our Fermi–system by the anticommuting functions $\chi_s(\mathbf{x},\tau),\overline{\chi}_s(\mathbf{x},\tau)$, defined in the square volume $S=L^2$ and antiperiodic in "time" τ with period $\beta=T^{-1}$.

After the procedure of path integrating over slow and fast Fermi–fields which is a very similar to 3D one we get the effective action functional which takes formally the same form as in 3D case.

In the case of 2D d–pairing, the number of degrees of freedom is equal to 4. In other words we should have two complex canonical variables. From non–diagonal elements of $\hat{M}$ matrix it is easy to see that the following canonical variables should be chosen:

$$c_1 = c_{11} - c_{22},\ c_2 = c_{12} + c_{21}. \tag{30.13}$$

For the conjugate variables one has:

$$c_1^+ = c_{11}^+ - c_{22}^+,\ c_2^+ = c_{12}^+ + c_{21}^+. \tag{30.14}$$

Let us transform the effective action functional S_{eff} to these new variables.

One has:

$$S_{\text{eff}} = (2g)^{-1} \sum_{p,j} c_j^+(p) c_j(p) + \frac{1}{2} \ln \det \frac{\hat{M}\left(c_j^+, c_j\right)}{\hat{M}\left(c_j^{+(0)}, c_j^{(0)}\right)} \tag{30.15}$$

where

$$M_{11} = Z^{-1}[i\omega - \xi + \mu(\mathbf{H}\boldsymbol{\sigma})]\delta_{p_1 p_2}$$

$$M_{22} = Z^{-1}[-i\omega + \xi + \mu(\mathbf{H}\boldsymbol{\sigma})]\delta_{p_1 p_2} \tag{30.16}$$

$$M_{12} = M_{21}^+ = \sigma_0 \alpha (\beta S)^{-1/2} (c_1 \cos 2\varphi + c_2 \sin 2\varphi).$$

The functional $S_{\rm eff}$ determines all properties of model system–superconducting CuO_2 planes. In particular it determines the collective–mode spectrum.

30.3.2. The collective mode spectrum

Two SC states arise in the symmetry classification of CuO_2 planes with OP which are proportional to $\begin{pmatrix} 1 & 0 \\ 0 & -1 \end{pmatrix}$ and $\begin{pmatrix} 0 & 1 \\ 1 & 0 \end{pmatrix}$ respectively. In the former phase the gap is proportional to $Y_{22}+Y_{2-2}\sim \sin^2\theta|\cos 2\varphi| \sim|\cos 2\varphi|$ while in the later one is proportional to $-i(Y_{22}-Y_{2-2})\sim \sin^2\theta|\sin 2\varphi|\sim|\sin 2\varphi|$. For 2D case we put $\theta=\pi/2$ and $\sin\theta=1$.

Brusov and Brusova[4,5] and Brusov, Brusova and Brusov[9] have calculated the collective– mode spectrum for both of these states. In the first approximation the collective excitations spectrum is determined by the quadratic part of $S_{\rm eff}$, obtained by the shift $c_j(p)\rightarrow c_j^{(0)}+c_j(p)$ in $S_{\rm eff}$. Here $c_j^{(0)}$ are the condensate values of the canonical Bose–fields $c_j(p)$.

The spectra in both phases turns out to be identical. Brusov and Brusova[4,5] found two high frequency modes in each phase with following energies (frequencies):

$$E_1=\Delta_0(1.42-i0.65),$$
$$E_2=\Delta_0(1.74-i0.41). \tag{30.17}$$

Note that the energies of both modes turn out to be complex. This results from the *d*–pairing, or in other words, via the disappearance of a gap in the chosen directions. In this case the Bose–excitations decay into fermions. This leads to a damping of the collective modes. The value of imaginary part of energy is 23% for the second mode and 46 % for first one. Thus both modes should be regarded as resonances and the second mode is better defined than the first.

The other two modes are Goldstone or low–energy modes (with energy $\leq 0.1\Delta_0$).

30.3.3. Lattice symmetry and collective mode spectrum

The above calculations of the collective mode spectrum are self–consistent not completely because Brusov, Brusova and Brusov work within spherical symmetry approximation but use the order parameters obtained with taking the lattice symmetry into account. As it was mentioned above the taking the lattice symmetry into account requires a few coupling constants using instead of one. The number of collective modes in superconducting state will change too (remind that it is equal to 10 in spherical symmetry case). If the simple irreducible representation (IR) occurs then the number of collective modes is equal to twice number of irreducible representation dimensionality. For orthorhombic (OR) symmetry and singlet pairing all irreducible representations are 1D, so in each superconducting state there are two modes corresponding to phase and amplitude variations. Amplitude mode is high frequency with $E \approx 2\Delta$.

Among irreducible representations of tetragonal (TG) symmetry (we consider singlet pairing) there are 1D as 2D. So in addition to the states with two collective modes of conventional superconductors there are states having four collective modes, none of which are Goldstone. As we mentioned above for cylindrical Fermi–surface (D_∞) there is Goldstone mode in $(1,0)$ and $(1,1)$ states but not in $(1,i)$ state.

Because it looks like that there is a mixture of different irreducible representations (corresponding, for example, to *s*– and *d*–wave states or to two different *d*–wave states: $d_{x^2-y^2}$ and d_{xy}) it will be interesting to investigate the collective mode spectrum in this case for different admixture values of *s*–wave state (d_{xy}–state). Considered by us particular case of $d_{x^2-y^2}+id_{xy}$ state (see 30.3.3) shows that such consideration leads to very interesting results.

Chapter XXXI

Collective Modes in the Heavy–Fermion Superconductors

31.1. Physical Properties of Heavy–Fermion Superconductors

Heavy–fermion superconductors were the first superconductors, which have demonstrated the unconventional pairing in charged systems. In this section we describe shortly their properties following to Ref.1.

Heavy–fermion superconductors are intermetalic compounds consisting from *f*–electron metal and *s*–, *p*– or *d*–electron metals. At room temperatures, the *f*–electrons behave as localized spins, the conduction electrons are the *s*–, *p*– or *d*–electrons and have quite ordinary effective masses. Under cooling the *f*–electrons begin to couple coherently to the conduction electrons, resulting in very large effective masses for the resulting hybridized carriers. The effective mass m^* is equal to $180m_0$ for UPt_3, $25m_0$ for URu_2Si_2, $300m_0$ for UBe_{13}, $1100m_0$ for $CeCu_6$. These compounds have some features, that distinguish them from the ordinary metals. The electronic heat capacities are 10^2–10^3 times larger than that observed in most metals; the Pauli susceptibility at low temperatures is about 10^2 times larger. Both of these indicate a very large effective mass for the conduction electrons.

The resistivity and the sound attenuation both continue to change rapidly with temperature down to very low temperatures (<10K) rather than staying constant as in most metals. At low temperatures the resistivity is of the form $\rho = \rho_0 + AT^2$, where A is large, indicating that electron–electron scattering is important. Most of heavy–fermion superconductors (except UPt_3) show at low temperatures a peak in the resistivity, similar to the peak seen in Kondo systems.

The susceptibility at high temperatures follows a Curie–Weiss low, as would be expected for a collection of local moments (i.e. localized *f*–electrons). At low temperatures the susceptibility is large (and very weakly temperature–dependant), indicating strongly interacting conduction electrons.

In addition to a large electronics specific heat (common to all heavy–fermion superconductors) UPt_3 has a $T^3 \log T$ term in a specific heat. A similar term has been observed in normal liquid ^{3}He and has been attributed to spin fluctuations. In general, a $T^3 \log T$ term is a feature of an interacting Fermi–liquid and can come from both spin and density fluctuations. This term in the specific heat, together with the T^2 term in the resistivity, has led to attempts to use Fermi–liquid theory to account for the low–temperature properties of the heavy–fermion superconductors. However, the anisotropy of these systems due to the crystal lattice makes this complicated.

Some heavy–fermion superconductors (UBe_{13}, UPt_3 and URu_2Si_2 etc) show a weak antiferromagnetic ordering followed by a subsequent superconducting transition. The Neel temperature, T_N, is about 10 times higher than the superconducting transition temperature T_C.

T_N is equal to 5K for UPt_3, 17.5 K for URu_2Si_2, 8.8 K for UBe_{13}; T_C is equal to 0.54K for UPt_3, 1.4 K for URu_2Si_2, 0.85 K for UBe_{13}.

The heat–capacity jump at the transition and the large critical field slopes dH_{C2}/dT at T_C in heavy–fermion superconductors indicate that the heavy fermions are responsible for the superconductivity. The properties in the superconducting state are unusual. The heat capacity and ultrasound attenuation follow power low temperature dependence, indicating the presence of line or point nodes in the gap. In addition, UPt_3 shows a variety of properties that make it a very strong candidate for unconventional superconductivity.

The upper critical field in cubic crystal UBe_{13} at zero temperature is of order 8 T with a slope of $dH_{C2}/dT = 40T/K$ at T_C. The upper critical field in hexagonal crystal UPt_3 shows a number of interesting features. Two $H_{C2}(T)$ curves (for $H\| c$ and $H \perp c$) cross at a temperature of 200 *mK*, implying that the anisotropy of the upper critical field is not just a result of the symmetry of the Fermi–surface. Such behavior may be explained by odd–parity pairing with strong spin–orbit coupling. Measurements revealed a kink in the $H_{C2}(T)$ curve for fields in the basal plane.

The upper critical field in a tetragonal crystal URu_2Si_2 is highly anisotropic, having upper critical field of 2 T for $H\| c$ and 8 T for $H\| a$. There is no crossover in $H_{C2}(T)$ as in case of UPt_3.

31.2. Bulk Heavy–Fermion Superconductors Under d–Pairing

The model of d–pairing for superconductive and superfluid fermi–systems (HTSC, HFSC etc.) created by Brusov and Brusova within path integration technique[2] has been applied to investigation of the collective mode spectrum in HTSC in Chapter XXVII. We have considered five states in HTSC ($d_{x^2-y^2}$, $d_{3z^2-r^2}$, d_{xy},d_{xz},d_{yz}).

Here, we consider three states in heavy–fermion superconductors (HFSC) ($d\gamma$, Y_{2-1}, $\sin^2\theta$). We have found[3-5] five high frequency modes in each state as well as five Goldstone–like ones. Results could be used to interpret microwave and ultrasonic absorption experiments as well as to identify the superconductive states through these experiments, because the nature of the paring in heavy–fermion superconductors has not been settled for all compounds. The conventional BCS pairing is in disagreement with the nonexponential temperature dependence of the most of thermodynamic quantities, such as the heat capacity etc. Two popular candidates are *p*–and *d*–pairing (see examples in Introduction). *P*–pairing has been considered by us above and here we consider the *d*–pairing in heavy–fermion superconductors. Brusov–Brusova model[2] for

d–pairing in superconductors turns out to be especially suitable for the investigation of the collective properties of superconductors.The collective excitations in two of three possible phases have been studied earlier[6].Here we calculate the collective mode spectrum in all three phases of heavy–fermion superconductors. We consider here the following states in heavy–fermion superconductors:
dγ, Y_{2-1} and $\sin^2\theta$:

1) **dγ–phase** with the order parameter:

$$c_{ia}^{(0)}(p) = c(\beta V)^{1/2}\delta_{p0}\begin{pmatrix} \exp(4\pi i/3) & 0 & 0 \\ 0 & \exp(2\pi i/3) & 0 \\ 0 & 0 & 1 \end{pmatrix} \tag{31.1}$$

and single particle gap

$$\Delta(T) = \Delta_0(T)\left[\exp(4\pi i/3)k_x^2 + \exp(2\pi i/3)k_y^2 + k_z^2\right]. \tag{31.2}$$

2) **$\mathbf{Y_{2-1}}$–phase** with the order parameter:

$$c_{ia}^{(0)}(p) = c(\beta V)^{1/2}\delta_{p0}\begin{pmatrix} 0 & -i & 1 \\ -i & 0 & 0 \\ 1 & 0 & 0 \end{pmatrix} \tag{31.3}$$

and single particle gap

$$\Delta(T) = \Delta_0(T)\sin 2\theta \cdot \exp(-i\varphi) \tag{31.4}$$

3) **sin²θ –phase** with the order parameter:

$$c_{ia}^{(0)}(p) = c(\beta V)^{1/2}\delta_{p0}\begin{pmatrix} 1 & i & 0 \\ i & -1 & 0 \\ 0 & 0 & 0 \end{pmatrix} \qquad (31.5)$$

and single particle gap $\Delta(T) = \Delta_0(T)\sin^2\theta$.

In the first approximation, the spectrum is determined by the quadratic form of the effective functional of action obtained as a result of the shift $c_{ia}(p) \to c_{ia}(p) + c_{ia}^{(0)}(p)$ in Bose–fields by a condensate wave function $c_{ia}^{(0)}(p)$, whose form is determined by the superconducting phase. The spectrum can be found from the equation $\det Q = 0$, where Q is the matrix of the quadratic form.

After evaluating all integrals (except those over angular variables) and equating the determinant of the quadratic form to zero, we arrive at the following set of equations determining the complete spectrum of collective modes in heavy–fermion superconductors under *d*–pairing (index *i* labels the branches of collective modes pertaining to the same phase, which is labeled by the index *k*):

1) *k*=1, *i*=1

$$\int_0^1 dx \int_0^{2\pi} d\varphi \left\{ \frac{\sqrt{\omega^2 + 4f_1}}{\omega} \ln F_1 g_1 + (g_1 - 2f_1)\ln f_1 \right\} = 0$$

$$\int_0^1 dx \int_0^{2\pi} d\varphi \left\{ \frac{\omega}{\sqrt{\omega^2 + 4f_1}} \ln F_1 g_1 + (g_1 - 2f_1)\ln f_1 \right\} = 0$$

k=1, *i*=2,3,4,5

$$\int_0^1 dx \int_0^{2\pi} d\varphi \left\{ \frac{\sqrt{\omega^2 + 4f_1}}{\omega} \ln F_1 g_i + \left(g_i - \frac{2}{3} f_i \right) \ln f_1 \right\} = 0$$

$$\int_0^1 dx \int_0^{2\pi} d\varphi \left\{ \frac{\omega}{\sqrt{\omega^2 + 4f_1}} \ln F_1 g_i + \left(g_i - \frac{2}{3} f_1 \right) \ln f_i \right\} = 0 \tag{31.6}$$

2) k=2,3; i=1

$$\int_0^1 dx \int_0^{2\pi} d\varphi \left\{ \frac{\sqrt{\omega^2 + 4f_k}}{\omega} \ln F_k g_1 + \left(g_1 - \frac{3}{2} f_1 \right) \ln f_k \right\} = 0$$

$$\int_0^1 dx \int_0^{2\pi} d\varphi \left\{ \frac{\omega}{\sqrt{\omega^2 + 4f_k}} \ln F_k g_1 + \left(g_1 - \frac{3}{2} f_i \right) \ln f_1 \right\} = 0$$

k=2,3; i=2,3,4,5

$$\int_0^1 dx \int_0^{2\pi} d\varphi \left\{ \frac{\sqrt{\omega^2 + 4f_2}}{\omega} \ln F_k g_i + \left(g_i - \frac{1}{2} g_2 \right) \ln f_2 \right\} = 0$$

$$\int_0^1 dx \int_0^{2\pi} d\varphi \left\{ \frac{\omega}{\sqrt{\omega^2 + 4f_2}} \ln F_k g_i + \left(g_i - \frac{3}{2} g_2 \right) \ln f_2 \right\} = 0. \tag{31.7}$$

Here,

$$\ln \frac{\sqrt{\omega^2 + 4f_k} + \omega}{\sqrt{\omega^2 + 4f_k} - \omega} = F_k, \; g_1 = \left(1 - 3x^2\right)^2, \; g_2 = \left(1 - x^2\right)^2 \cos^2 2\varphi,$$

$$g_3 = 4(1-x^2)x^2\cos^2\varphi,\ g_4 = 4(1-x^2)x^2\sin^2\varphi,$$

$$g_5 = (1-x^2)^2\sin^2\varphi,\ f_1 = \frac{1}{4}\left[(1-3x^2)^2 + 3(1-x^2)^2\cos^2 2\varphi\right],$$

$$f_2 = 4(1-x^2)x^2,\ f_3 = (1-x^2)^2,\ \cos\theta = x,\ \omega = \omega/\Delta_0. \tag{31.8}$$

Solving these equations numerically we find ten collective modes in each phase: five high–frequency–modes (derived from the second equations) as well as five Goldstone–(quasi–Goldstone) modes with vanishing (of order $(0.03 \div 0.08)\Delta_0(T)$) (obtained from the first equations). Below we give the results for high–frequency–mode spectra (E_i is the energy (frequency) of i–branch):

a) **dγ–state**

$E_1 = \Delta_0(1.66 - i\,0.5)$, $E_4 = \Delta_0(1.21 - i\,0.60)$,

$E_2 = \Delta_0(1.45 - i\,0.48)$, $E_5 = \Delta_0(1.19 - i\,0.60)$,

$E_3 = \Delta_0(1.24 - i\,0.64)$. (31.9)

It should be noted that the last three modes (3–5) are near degenerated.

b) **Y_{2-1} and $\sin^2\theta$ states**

The spectra of these two phases turn out to be identical:

$E_{1,2} = \Delta_0(1.93 - i\,0.41)$;

$E_3 = \Delta_0(1.62 - i\,0.75)$;

$E_{4,5} = \Delta_0(1.59 - i\,0.83)$. (31.10)

The collective mode spectrum in each state of heavy–fermion superconductors (similar to the case of HTSC, considered by us in Chapter XXVII) consists from ten modes: five high frequency modes and five Goldstone (or Goldstone–like) modes with vanishing energies at zero momenta of excitations. All energies of high frequency modes turn out to be complex and their imaginary parts Im E_i describe the damping of collective modes via decay of Cooper pairs into fermions.

31.3. Conclusion

We calculated the entire collective mode spectrum for the three superconducting phases of heavy–fermion superconductors: dγ, $Y_{2-1,}$ as well as phase with a gap proportional to $\sin^2\theta$. The case of low temperature ($T_C - T \propto T_C$) and spherical symmetry have been studied. Equations describing the collective excitation spectrum have been obtained for arbitrary momenta of collective excitations and the energies (frequencies) of the collective modes have been calculated numerically for zero momenta of collective excitations $\boldsymbol{k}$=0.

We have two notices concerning the accounting of the lattice symmetry:

1. While our model has been done in the spherical symmetry approximation (we use only one coupling constant g), we were accounting the lattice symmetry in indirect way: we use the phases with the order parameters obtained in the symmetry classification of heavy–fermion superconductors.
2. The taking into account the lattice symmetry in self–consistent way requires a few coupling constants g_m. Their number in the general case is equal to five (number of spherical harmonics with l=2), but it reduces to two for the cubic symmetry case (one for the 2D or dγ–representation and the other for 3D or dε–representation) and to three for the case of hexagonal symmetry: $g_{|m|}$ $(m = 0,\pm1,\pm2)$.

Let us discuss the results obtained. First of all we should note that the energies (frequencies) of all modes turn out to be complex and their imaginary parts, $\operatorname{Im} E_i$, describe the damping of the collective excitations via decay of Cooper pairs into fermions. These imaginary parts, $\mathrm{Im}\,E_i$, lie from 20% to 50% of the real parts $\mathrm{Re}\,E_i$. This means that collective modes in the case of *d*–pairing are more strongly damped than in case of *p*–pairing, where imaginary parts, $\mathrm{Im}\,E_i$, range from 8% to 15% of the real parts $\mathrm{Re}\,E_i$. This arises from the different topology of the gap nodes, which are points in the latter case, and a combination of points and lines in the former one. Note, that a similar situation sometimes occurs for *p*–pairing (for example, in the case of the polar phase of ^{3}He the damping is stronger than in axial phase via the fact that there is a line of nodes).

Let us compare the results, obtained by Brusov et al.[3-5] by path integral method with that obtained by Hirashima and Namaizawa by the kinetic equation method[6]. The damping of collective modes was not calculated in Ref. 6. This is a shortcoming of the kinetic equation method, compared with the path integral technique; the first yields only the real parts of the collective mode energies. Accounting the imaginary parts of the collective mode energies leads to a shift of the real parts as well via the dispersion relations. Thus we could compare results by Brusov et al.[3-5], described here with results of Ref. 6 for real parts of energies $\mathrm{Re}\,E_i$ only. Brusov et al.[3-5] found five high frequency modes in each phase. In dγ–phase the energies (frequencies) of these modes lie in the interval $(0.9 \div 1.66)\Delta_0(T)$. In Ref. 6, five modes were found in the interval $(0.9 \div 1.87)\Delta_0(T)$ as well as two low–lying modes with energies $E \approx 0.32\Delta_0(T)$. In the Y_{2-1}–phase Brusov et al. have determined the energies lying in the interval $(0.59 \div 1.93)\Delta_0(T)$, while the energies of six modes of Ref. 6 lie in the interval $(0.22 \div 1.57)\Delta_0(T)$. In both papers some Goldstone and low–lying modes have been found. Note that the spectrum of the third phase with a gap proportional to $\sin^2\theta$ has been calculated by Brusov et al.[3-5] for the first time and the spectrum turns out to be identical to one of the Y_{2-1}–phase. (This third phase has not been considered in Ref. 6). In conclusion, we would like to note that the results on the collective mode spectrum could be used to interpret the experimental data on ultrasound and microwave absorption (microwave experiments at $\propto 20GHz$ have been done at Northwestern University) as well as to identify the type of pairing and the order parameter in heavy–fermion superconductors.

Chapter XXXII

Other Applications of the Theory of Collective Excitations

32.1. Relativistic Analogs of ^3He

More than twenty years ago very interesting and unexpected connections between ^3He theory and gauge field theory particularly the theory of quantum chiral anomalies were discovered[1,2]. It turns out that there exist such effects as the chiral anomaly and the null–charge phenomenon in the A–phase of ^3He. This superfluid phase demonstrates the mechanism of arising chiral fermions and gauge fields, and also the mechanism of generation of the W–boson mass different from the Higgs mechanism. These remarkable properties of ^3He–A arise due to singular points in two poles of the Fermi–sphere $(\theta = 0, \pi)$ in which the gap function $\Delta = \Delta_0 \sin\theta$ of the Fermi–spectrum disappears.

Another interesting possibility is a search of analogs of ^3He in models of relativistic field theory. These are models where an effective action functional resembling that of ^3He–model arises after introduction auxiliary fields and integration with respect to Fermi–fields. In such models the phenomenon of Bose–condensation of auxiliary fields is possible. It should be mentioned that the idea of Bose–condensation in quantum field theory is not new. This idea is employed in the Weinberg–Salam theory of electroweak interactions (here, Bose–condensation of the scalar Higgs–field takes place) and also in quantum chromodynamics.

There exist models of quantum field theory with a more complicated form of Bose–condensate which is analogous to that in ^3He. As an example let us consider the theory with four–fermionic interaction which has the following Euclidean action

$$S = \int dx \overline{\psi}(x)(i\hat{p} - M)\psi(x) \pm \frac{1}{2} g^2 \int dx j_\mu^a(x) j_\mu^a(x) \qquad (32.1)$$

Here, $\psi(x)$ is an isotopic spinor field, $i\hat{p} = \gamma_\mu \partial_\mu$ is the Dirac operator, $j_\mu^a(x) = \overline{\psi}(x)\gamma_\mu \tau_a \psi(x)$ are current componens, in which γ_μ are Dirac matrices and τ_a are isotopic matrices.
Using the formalism of functional integral

$$\int \exp S D\overline{\psi} D\psi \qquad (32.2)$$

let us insert the following Gaussian integral

$$\int \prod_{\mu,a} DB_\mu^a(x) \exp\left(-\frac{1}{2} \int dx B_\mu^a(x) B_\mu^a(x) \right), \qquad (32.3)$$

with respect to an auxiliary vector isovector Bose–fields $B_\mu^a(x)$ into the functional integral (32.2). Then after the shift transformation

$$B_\mu^a(x) \to B_\mu^a(x) + \varepsilon g j_\mu^a(x), \qquad (32.4)$$

where $\varepsilon = 1$ for the "+" sign in (32.1) and $\varepsilon = i$ for the "–" sign we obtain an action of the following form:

$$\tilde{S} = \int dx \left[-\frac{1}{2} B_\mu^a(x) B_\mu^a(x) + \overline{\psi}(x)(i\hat{p} - M)\psi(x) + \varepsilon g j_\mu^a(x) B_\mu^a(x) \right]. \qquad (32.5)$$

After integration over Fermi–fields $\psi(x), \overline{\psi}(x)$ we arrive at the effective action

$$S_{eff} = -\frac{1}{2}\int dx B_\mu^a(x) B_\mu^a(x) + \ln\det \hat{M}\hat{M}_0^{-1} \tag{32.6}$$

where

$$\hat{M} = \gamma_\mu \partial_\mu - M - \varepsilon g \tau_a \gamma^\mu B_\mu^a(x). \tag{32.7}$$

One can consider such an effective action as an analog of the effective action for the ^{3}He–model, where the fields $B_\mu^a(x)$ play a role of fields c_{ia}, describing Cooper pairs. We can apply methods elaborated for the ^{3}He–theory to the effective action (32.6). The first problem is to look for possible condensate values of $B_\mu^a(x)$. In the simplest case the condensate value of $B_\mu^a(x)$ does not depend on *x*, and we have $S_{eff} = -\int dx V(B)$, where $V(B)$ is an effective potential of the following form:

$$V(B) = \frac{1}{2} B_\mu^a(x) B_\mu^a(x) - \int \frac{d^D p}{(2\pi)^D} \ln\det\left[\left(i\hat{p} - M - \varepsilon g \hat{B}\right)\left(i\hat{p} - M\right)^{-1}\right] \tag{32.8}$$

Here, $\hat{B} = \tau_a \gamma_\mu B_\mu^a$, *D* is a dimension of an Euclidean space, "det" means a determinant of a finite–dimensional matrix. The integral in (32.8) converges at *D*=2. For *D*=3 and *D*=4 it should be regularized.

The case *D*=2 and the isotopic group *G*=SU(2) is investigated by Bogdanova and Popov[3]. We have in this case

$$\det\left(i\hat{p} - M - \varepsilon g \hat{B}\right) = \left(p^2 + M^2 - \varepsilon^2 g^2 B_\mu^a B_\mu^a\right)^2 + 4\varepsilon^2 g^2 p_\mu p_\nu B_\mu^a B_\nu^a - $$
$$- 4g^4 [B_1, B_2]^2, \quad [B_1, B_2]^2 = B_1^a B_1^a B_2^b B_2^b - B_1^a B_2^a B_1^b B_2^b. \tag{32.9}$$

The effective potential $V(B)$ depends on two invariants

$$x = B_\mu^a B_\mu^a, \quad y = 1 - 4[B_1, B_2]^2 / (B_\mu^a B_\mu^a)^2 \tag{32.10}$$

It was shown that a nontrivial minimum of $V(B)$ can exist only in the case of $\varepsilon = 1$ and of the pure imaginary M (so we have to change M^2 by $-M^2$ in (32.9)). The necessary condition for theory to be stable is $g^2 < \pi$. There exists also a critical value $g_C^2 = 1.74$. If $g^2 < g_C^2$, there exists a nontrivial minimum of $V(B)$ with y=0, and we have

$$g^2 B_\mu^a B_\mu^a = \Delta^2 \delta_{\mu\nu}. \tag{32.11}$$

This state was called the symmetric phase of the system, and it may be considered as an analog of the *B*–phase of ^{3}He. At $g_C^2 < g^2 < \pi$ the system can exist only in the nonsymmetric phase with $g^2 B_\mu^a B_\mu^a \neq \Delta^2 \delta_{\mu\nu}$.

The determinant (32.9) depends on p^2 for the symmetric phase (32.11). The equation $\det(i\hat{p} - M - \varepsilon g \hat{B}) = 0$ for the Fermi–spectrum gives us the isotropic spectrum. So the isotropy is restored which was broken arising of anomalous average $\langle B_\mu^a \rangle \neq 0$. The analogous situation takes place in ^{3}He–*B*, where the Fermi–spectrum is also isotropic.

Symmetric phases (32.11) are possible also at *D*=3 and *D*=4. But it turns out that the Euclidean rotation group must be a subgroup of the isotopic group $(O(D) \subset G)$.

At the conclusion let us dwell on the renormalization problem. From the standard point of view only the two–dimensional theory with four–fermionic interaction is renormalizable. But it turns out that the theory is renormalizable as a theory of an auxiliary field B_μ^a also for *D*=3. Let us write down S_{eff} (32.6) in the following form

$$S_{eff} = -\left(2g^2\right)^{-1} \int dx A_\mu^a(x) A_\mu^a(x) + \sum_{n=2}^{\infty} S_n \tag{32.12}$$

where $A_\mu^a = gB_\mu^a$, and S_n is the n–th term of the expansion lndet in powers of $A_\mu^a = gB_\mu^a$, which can be interpreted as a one–loop diagram with n tails. If D=3, only S_2 contains divergences. Let us take the quadratic part of S_{eff}

$$Q = -\frac{1}{2} \int dp A_\mu^a(p) A_\nu^b(-p) C_{ab}^{\mu\nu}(p) \ , \tag{32.13}$$

where

$$C_{ab}^{\mu\nu} = g^{-2}(p)\delta_{ab}\delta_{\mu\nu} + (2\pi)^{-3} \int d^3 p_1 tr \tau_a \gamma_\mu \left(i\hat{p} + i\hat{p}_1\right)^{-1} \tau_b \gamma_\nu \left(i\hat{p}_1\right)^{-1} =$$
$$= g_R^{-2} \delta_{ab}\delta_{\mu\nu} + (2\pi)^{-3} \int d^3 p_1 tr \tau_a \gamma_\mu \left[\left(i\hat{p} + i\hat{p}_1\right)^{-1} - \left(i\hat{p}_1\right)^{-1}\right] \tau_b \gamma_\nu \left(i\hat{p}_1\right)^{-1}, \tag{32.14}$$

where

$$g_R^{-2} = g^{-2} + (2\pi)^{-3} \int d^3 p_1 tr \tau_a \gamma_\mu \left(i\hat{p}_1\right)^{-1} \tau_b \gamma_\nu \left(i\hat{p}_1\right)^{-1}, \tag{32.15}$$

and there are no summation over repeated indices a, μ in (32.15). Now the integral in the right hand side of (32.14) converges and it is proportional to $|p|$. As a result the propagator of A_μ^a–field behaves as $|p|^{-1}$ large p and 3D–theory becomes renormalizable.

A more complicated situation is in the four–dimensional theory (D=4). But here renormalizability also is not excluded.

References

Chapter I

1. Bogoliubov N.N. (1947), Izv. Akad. Nauk SSSR, **11**, 77.
2. Bogoliubov N.N., D.V. Shirkov (1980), Introduction to the theory of quantized fields, Wiley & Sons Inc.
3. Bogoliubov N.N., V.V. Tolmachev and D.V. Shirkov (1959), New Method in Superconductivity Theory. New York, Consultants Bureau.
4. Bogoliubov N.N. (1961), Quasiaverages in Problems of Statistical Mechanics, JINR Preprint D–781 (in Russian).
5. Bogoliubov N.N., D.N. Zubarev and Yu. A. Tserkovnikov (1960), JETP, **39**,120.
6. Bogoliubov N.N., Jr. (1984), Method of Investigation of Model Hamiltonian, Nauka, Moscow (in Russian).
7. Bogoliubov N.N., Jr, Sadovnikov B.I. (1975), Some Problems of Statistical Mechanics, Moscow, Nauka.
8. Bogoliubov N.N., Jr, I.G. Brankov, V.A. Zagrebnov, A.M. Kurbatov, Tonchev (1981), Method of Approximation Hamiltonians. The Academia of Sciences of Bulgaria, Sofia.
9. Abrikosov A.A., L.P. Gor'kov and I.E. Dzyaloshinskii (1975), Methods of Quantum Field Theory in Statistical Mechanics, Dover, New York.
10. Berezin P.A. (1968), Method of Second Quantization. Academic Press, N.Y.

Chapter II

1. Bogoliubov N.N., V.V. Tolmachev and D.V. Shirkov (1959), New Method in Superconductivity theory, New York, Consultants Bureau.
2. Andrianov V.A., V.N. Popov (1976), Sov. J. Theor. and Math. Phys. **28**, 340.
3. Abrikosov A.A., L.P. Gor'kov and I.E. Dzyaloshinskii (1975), Methods of Quantum Field Theory in Statistical Mechanics, Dover, New York.
4. Lahdau L.D. (1956), JETP **30**, 1058.
5. Landau L.D. (1957), JETP **32**, 59.
6. Svidzinski A.V. (1971), Sov. J. Theor. and Math. Phys. **9**, 273.
7. Kapitonov V.S., V.N. Popov (1976), Sov. J. Theor. and Math. Phys. **6**, 146.

Chapter III

1. Sarma B.K., et al. (1992), Physical acoustics, **XX**, 107.
2. Shivaram B.S., et al. (1982), Phys. Rev. Letts. **49**, 1646.
3. Brusov P.N. and Popov V.N. (1980), Sov. Phys. JETP **51**, 117.
4. Movshovich R., et al. (1988), Phys. Rev. Letts. **61**, 1732.
5. Movshovich R., et al. (1990), Phys. Rev. Letts. **64**, 431.
6. Pippard A.B. (1953), Proc. Royal Society **A216**, 547.
7. Maki K. (1964), Physica, **1**, 127.
8. Maki K. (1967), Phys. Rev. **156**, 437.
9. Maki K. and Cyrot M. (1967), Phys. Rev. **156**, 433.

Chapter IV

1. Brusov P.N., V.N. Popov (1980), Sov. Phys. JETP, **51**(6), 1217.
2. Brusov P.N., V.N. Popov (1980), Sov. Phys. JETP, **51**(1), 117.
3. Brusov P.N., V.N. Popov (1980), Sov. Phys. JETP, **52**(5), 945.
4. Dobbs E.R., R. Ling, J. Saunders, W. Wojtanowski (1989), In: Proc. of Symposium of quantum fluids and solids 1989, Florida.
5. Brusov P.N. and Brusov P.P. (2009), J. Low Temp. Phys., **155**, 200.
6. Brusov P.N. and P.P. Brusov (2009), Phys. Lett. **373A**, 1972.
7. Koch V.E, Wolfle P. (1981) Phys. Rev. Letts. **46**, 489.
8. Combescot R. (1982), J. Low Temp. Phys. **49**, 295.
9. Pitaevski L.P. (1959), JETP, **37**, 1794.
10. Anderson P.W., P. Morel (1961), Phys. Rev. **123**, 1911.
11. Balian R., V.R.Werthamer, (1963), Phys. Rev. **131**, 110.
12. Osheroff D.D., R.C. Richardson, D.M. Lee (1972), Phys. Rev. Letts, **28**, 885.
13. Khalatnikov I.M. (1965), Theory of superfluidity, New York, Benjamin.
14. Wheatley J.C. (1975), Rev. Mod. Phys. **47**, 415.
15. Maris H.T., W.E. Massey (1970), Phys. Rev. Letts, **25**, 220.
16. Wheatley J.C. (1978), In: Progr. in Low. Temp. Phys. Ed. Brewer D.F., Amsterdam, North Holland, **7A**, 1.
17. Leggett A.J. (1975), Rev. of Modern Phys. **47**, 331.
18. Wolfle P. (1976), Phys. Rev. Letts. **37**, 1279.
19. Giannetta R.W., A. Ahonen, E. Polturak, J. Saunders, E.K. Zeise, R.C. Richardson, D.M. Lee (1980) Phys. Rev. Letts, **45**, 262.
20. Mast D.B., B.K. Sarma, Y.R. Owers–Bradley, J.D. Galder, J.B. Ketterson, W.P. Halperin (1980), Phys. Rev. Lett. **45**, 256.
21. Shivaram B.S., et al. (1982), Phys. Rev. Letts. **49**, 1646.
22. Avenel O., E. Varogaux, H. Ebisawa (1980), Phys. Rev. Letts. **45**, 1952.

23. Daniels M.E., E.R. Dobbs, J. Saunders, P.L. Ward (1983), Phys. Rev. **B27**, 6988.
24. Ling R., W. Wojtanowski, J. Saunders, E.R. Dobbs, (1990), J. Low Temp. Phys. **78**, 187.

Chapter V

1. Alonso V, V.N. Popov (1977), Sov. Phys. JETP, **46**, 760.
2. Brusov P.N., V.N. Popov (1980), Sov. Phys. JETP, **51**(6), 1217.
3. Brusov P.N., V.N. Popov (1980), Sov. Phys. JETP, **51**(l), 117.
4. Brusov P.N., V.N. Popov (1980), Sov. Phys. JETP, **52**(5), 945.
5. Brusov P.N., V.N. Popov (1984), The Collective Excitations in Superfluid Quantum Fluids (Rostov–on–Don State University Press, Rostov–on–Don, USSR).
6. Lahdau L.D. (1956), JETP **30**, 1058.
7. Landau L.D. (1957), JETP **32**, 59.
8. Kleinert H. (1978), preprint FUB–HEP N 14–78.
9. Wo1fle P. (1977), Physica, 90B, 96–106.

Chapter VI

1. Brusov P.N., V.N. Popov (1980), Sov. Phys. JETP, **51**(6), 1217.
2. Shivaram B.S., et al. (1982), Phys. Rev. Letts. **49**, 1646.
3. Brusov P.N., V.N. Popov (1984), The Collective Excitations in Superfluid Quantum Fluids (Rostov–on–Don State University Press, Rostov–on–Don, USSR).
4. Movshovich R., et al. (1988), Phys. Rev. Letts. **61**, 1732.
5. Zhao Z., S. Adenwalla, P.N. Brusov, J.B. Ketterson and B.K. Sarma (1990), Phys. Rev. Lett., **65**, 2688.
6. Fujita T., M. Nakahara, T. Ohmi, et al. (1980), Progr. Theor. Phys. **64**, 396.
7. Brusov P.N. and M.V. Lomakov (1991), Sov. Phys. JETP, **73**(3), 470.
8. Wo1fle P. (1977), Physica **90B**, 96.
9. Andrianov V.A., V.N. Popov (1976), Soviet J. Theor. and Math. Phys. **28**, 340.
10. Vdovin Yu. A. (1962), In: Proc. of MEPI, Moscow, p. 94.
11. Nagai K. (1974), Progr. Theor. Phys. **54**, 1.
12. Halperin W.P. and E. Varoquaux: (1990) in Helium Three, Eds. W. P. Halperin and L. P. Pitaevskii , Elsevier, Amsterdam, p. 353.
13. Brusov P.N., et al. (1990), Physica **165–166**, 685.
14. Volovik G.E. (1984), JETP Letts. **39**, 365.
15. Smith H., W.F. Brinkman and S. Engelsberg (1977), Phys. Rev. B **15**, 199.
16. Lee D.M. and R.C. Richardson (1978), In: Physics of Liquid and Solid Helium, Eds. K.H. Benneman and J.B. Ketterson (Wiley, New York).

17. Shivaram B.S., et al. (1986), J. Low Temp. Phys. **63**, 57.
18. Volovik G.E. (1984), Sov. Phys. Usp. **27**, 363.
19. Volovik G.E., M.V. Khazan (1983), Sov. Phys. JETP, **60**, 276.
20. Fishman R.S. and J.A. Sauls (1988), Phys. Rev. B **38**, 2526.
21. Giannetta R.W., A. Ahonen, E. Polturak, J. Saunders, E.K. Zeise, R.C. Richardson, and D.M. Lee (1980), Phys. Rev. Lett. **45**, 262.
22. Mast D.B., Bimal K. Sarma, J.R. Owers–Bradley, I.D. Calder, J.B. Ketterson, and W.P. Halperin (1980), Phys. Rev. Lett. **45**, 266.
23. Koch V.E. and P. Wolfle (1981), Phys. Rev. Letts. **46**, 486.
24. Wolfle P. (1978), In Progress in Low Temperature Physics, Ed. by D.F. Brewer, North–Holland, Amsterdam, Vol. **7**.
25. Wheatley J.C. (1978), In: Progress in Low Temperature Physics, edited by D.F. Brewer (North–Holland, Amsterdam) Vol. **7A**, p. 1.
26. Alvesalo T.A., T. Haavasoja, M.T. Manninen, and A.T. Soinne (1980), Phys. Rev. Lett. **44**, 1076.
27. Wolfle P. (1978), In: Physics of Ultralow Temperatures, edited by T. Sugawara et al. (The Physical Society of Japan).
28. Rainer D. and J.W. Serene (1976), Phys. Rev. **B 13**, 4745.
29. Baldo M., G. Giansiracusa, U. Lombardo, R. Pucci, and G. Petronio (1978), Phys. Lett. **65A**, 418.
30. Sauls J.A. and J.W. Serene (1981), Phys. Rev. **B 23**, 4798.
31. Tewordt L. and N. Schopohl (1979), J. Low Temp. Phys. **37**, 421.
32. Avenel O., E. Varoquaux and H. Ebisawa, private communication.

Chapter VII

1. Brusov P.N. (1991), Phys. Rev. B, **43**, 12888.
2. Brusov P.N. and V.N. Popov (1980), Zh. Eksp. Teor. Fiz. **79**, 1871 [Sov. Phys. JETP 52, 945].
3. Brusov P.N. and V.N. Popov (1980), Zh. Eksp. Teor. Fiz. **78**, 234 [Sov. Phys. JETP **51**, 117].
4. Brusov P.N. and V.N. Popov (1980), Zh. Eksp. Teor. Fiz. **78**, 2419 (Sov. Phys. JETP **51**, 1217).
5. Dobbs E.R., R. Ling, J. Saunders and W. Wojtanowski (1989), In: Symposium on Quantum Fluids and Solids (University of Gainesville, Florida), Proceedings of the Symposium on Quantum Fluids and Solids, AIP Conf. Proc. No. 194, Ed. by Gary G. Ihas and Uasumasa Takano (AIP, New York).
6. Ling R., W. Wojtanowski, J. Saunders and E.R. Dobbs (1990), J. Low Temp. Phys. **78**, 187.
7. Koch E. and P. Wolfe (1978), J. Phys. (Paris) Colloq. Suppl. 8, **39**, C6–6.

8. Koch V.E. and P. Wolfle (1981), Phys. Rev. Letts. **46**, 486.
9. Brusov P.N. and V.N. Popov (1988), The Superfluidity and Collective Properties of Quantum Liquids (Nauka, Moscow), p. 215.
10. Combescot R. (1982), J. Low Temp. Phys. **49**, 295.
11. Alonso V. and V.N. Popov (1977), Zh. Eksp. Teor. Fiz. **73**, 1445 [Sov. Phys. JETP **46**, 760].
12. Volovik G.E., M.V.Khazan (1983), Zh. Eksp. Teor. Fiz. **85**, 948 [Sov. Phys. JETP **58**, 551 (1983)].
13. Wolfe P. (1977), Physica **90B**, 96.
14. Brusov P.N. (1991), Simple explanation of Ling et al. experiment on the cl–mode–frequency measurement, J. Low Temp. Phys. **82**, 31.
15. G.E. Volovik (1986), Pis'ma Zh. Eksp. Teor. Fiz. **43**, 535 [JETP Lett. 43, 693].
16. Brusov P.N., M.Y. Nasten'ka, T. V. Filatova–Novoselova, M.V. Lomakov and V.N. Popov (1989), In: Symposium on Quantum Fluids and Solids, p. 111.
17. Brusov P.N., et al. (1990), Physica **B 165–166**, 625.
18. Nasten'ka M.Y. and P.N. Brusov (1989), Phys. Lett. **136A**, 321.
19. Gurgenishvili G.E. and G.A. Kharadze (1986), Superfluid Phases of Liquid He3 (Martseba, Tbilisi), p. 180.
20. Tewordt L. and N. Schopohl (1979), J. Low Temp. Phys. **34**, 489.
21. Schopohl N., W. Marquardt and L. Tewordt (1985), J. Low Temp. Phys. **59**, 469.
22. Brusov P.N. (1991), J. Low Temp. Phys. **82**, 31.
23. Halperin W.P. and E. Varoquaux: (1990), In: Helium Three, Eds. W.P. Halperin and L.P. Pitaevskii , Elsevier, Amsterdam, p. 353.
24. Ebisawa H. and K. Maki (1974), Progr. Theor. Phys. **51**, 337.
25. Tewordt L., et al. (1975), J. Low Temp. Phys. **21**, 645.
26. Tewordt L., et al. (1977), J. Low Temp. Phys. **29**, 119.
27. Tewordt L., et al. (1979), J. Low Temp. Phys. **34**, 489; J. Low Temp. Phys. **37**, 421.
28. Wolfle P. (1978), In: Physics of Ultralow Temperatures, edited by T. Sugawara et al. (The Physical Society of Japan, Tokyo).
29. Leggett A.J. and S. Takagi (1976), Phys. Rev. Letts. **36**, 1379.
30. Leggett A.J. and S. Takagi (1978), Ann. Phys. (New York) **110**, 353.
31. Volovik G.E. (1984), Sov. Phys. Usp. **27**, 363.
32. Hall H.E. and Hook (1986), in Progress in Low Temperature Physics, edited by D. F. Brewer (North–Holland, Amsterdam), Vol. IX, p. 143.
33. Einzel D. and P. Wolfle (1978), J. Low Temp. Phys. **32**, 19.
34. Wolfle P. (1976), Phys. Rev. Letts. **37**, 1279.
35. Ling et al. (1989), Europhys. Letts. **10**, 323.
36. Volovik G.E., M.V. Khazan (1982), Zh. Eksp. Teor. Fiz. **82**, 1498 [Sov. Phys. JETP **55**, 867].
37. Dombre T. and R. Combescot (1981), Physica **B 107**, 51.

38. Brinkman W.F. and M.C. Cross (1978), In: Progress in Low Temperature Physics, edited by D.F. Brewer (North–Holland, Amsterdam) Vol. VIIA, p. 105.
39. Leggett A.J. and S. Takagi (1977), Ann. Phys. (New York) **106**, 79.
40. Combescot R. (1975), Phys. Rev. Letts. **35**, 1646.
41. Cross M.C. (1975), J. Low Temp. Phys. **21**, 525.
42. Leggett A.J. (1975), Rev. Mod Phys.**47**, 331.
43. Combescot R. (1975), Phys. Rev. B **18**, 3139.
44. Brusov P.N. and M. Bukshpun (1990), Physica **B 165–166**, 685.
45. Lawson D.T., et al. (1975), Phys. Rev. Letts. **34**, 121.
46. Roach P.R., et al. (1975), Phys. Rev. Letts. **34**, 715.
47. Paulson D.N., et al. (1976), Phys. Rev. Letts. **36**,1322.
48. Paulson D.N., et al. (1976), Phys. Rev. Letts. **37**, 599.
49. Kleinberg R.L. (1979), J. Low Temp. Phys. **35**, 489.
50. Salomaa M.M. and G.E. Volovik (1987), Rev. Mod. Phys. **59**, 533.
51. Ketterson J.B., et al. (1975), In: Quantum Statistics and the Many Body Problem, eds. S.B.Trickey, W.P. Kirk and J.W. Dufty (Plenum Press, New York), p. 35.
52. Cross M.C. and P.W. Anderson (1975), In: Low Temp. Phys., LT14, eds. M. Krusius and M. Vuorio, v.1 (North Holland, Amsterdam), p. 29.
53. Reppy J. D. and P.L. Gammel (1987), Jpn. J. Appl. Phys. **26** (3), 117.
54. Lawson D.T., et al. (1974), J. Low Temp. Phys. **15**, 169.
55. Bhattacharyya P., et al. (1977), Phys. Rev. Letts. **39**, 1290.
56. Cross M.C. and M. Liu (1978), J. Phys.C **11**, 1795.
57. Fetter A.L. (1978), Phys. Rev. Letts. **40**, 1656.
58. Fetter A.L. (1979), Phys. Rev. B. **20**, 303.
59. Lin–Liu Y.R., et al. (1978), J. Phys. Lett. (Paris) **39**, L381.
60. Saslow W.M. and C.–R.Hu (1978), J. Phys. Lett. (Paris) **39**, L379.
61. Fetter A.L. and M.R.Williams (1981), Phys. Rev. **B 23**, 2186.
62. Kleinberg R.L. (1979), J. Low Temp. Phys. **35**, 489.
63. Bates D.M., et al. (1984), Phys. Rev. Letts. **53**, 1574.
64. Bates D.M., et al. (1986), J. Low Temp. Phys. **62**, 177.
65. Bozler H.M. and C.M. Gould (1987), Can. J. Phys., **65**, 1476.
66. Serene J.W. (1974), PhD Thesis (Cornell University).
67. Hook J.R., et al. (1986), Phys. Rev. Letts. **57**, 1749.
68. Hook J.R., et al. (1987), Can. J. Phys. **65**, 1486.
69. Wheatley J.C. (1978), In: Progress in Low Temperature Physics, edited by D.F. Brewer (North–Holland, Amsterdam), Vol. **VIIA**, p. 1.
70. Krusius M., et al. (1978), J. Low Temp. Phys. **33**, 255.
71. Dow R.C.M. (1984), PhD Thesis (Lancaster University).
72. Leggett A.J. (1975), Rev. Modern Phys. **47**, 331.

Chapter VIII

1. Gould C. (1992), Physica B **178**, 266.
2. Mullins et al. (1994), Phys. Rev. Lett. **72**, 4177.
3. Brusov P.N., J.B. Ketterson, P.P. Brusov and N.P. Kulik (2000), Identification of ^{3}He–A by ultrasound experiments, Physica B, **284–288**, p. 264.
4. Sprague D.T., et al. (1995), Phys. Rev. Lett. **75**, 661.
5. Sprague D.T., et al. (1996), Phys. Rev. Lett. **77**, 456.
6. Mermin N.D. and C. Stare (1973), Phys. Rev. Lett. **30**, 1135.

Chapter IX

1. Brusov P.N. and V.N. Popov (1980), Zh. Eksp. Teor. Fiz. **78**, 234 [Sov. Phys. JETP **51**, 117].
2. Brusov P.N. and V.N. Popov (1988), The Superfluidity and Collective Properties of Quantum Liquids (Nauka, Moscow), p. 215.
3. Andrianov V.A. and V.N. Popov (1976), Theor. and Math. Phys. (Sov.) **28**, 340.
4. Combescot M. and R. Combescot (1976), Phys. Rev. Letts, **37**, 388.
5. Alonso V. and V.N. Popov (1977), JETP **77**, 1445.

Chapter X

1. Brusov P.N. and V.N. Popov (1980), Zh. Eksp. Teor. Fiz. **78**, 2419 (Sov. Phys. JETP **51**, 1217).
2. Brusov P.N. (1981), Influence of spin–orbit interaction and magnetic fields on superfluid Phases of He3, In: Problems of Quantum field theory and Statistical physics, eds. P.P. Kulish & V.N. Popov, **101**.
3. Leggett A.J. and S. Takagi (1978), Annlals of Phys. **106**, 79.
4. Sonin E.B. (1970), JETP, **59**, 1416
5. Brusov P.N. and V.N. Popov (1980), Zh. Eksp. Teor. Fiz. **78**, 234 [Sov. Phys. JETP **51**, 117].

Chapter XI

1. Khalatnikov I.M. (1971), Superfluidity theory, Moscow, Nauka.
2. Landau L.D., E.M. Lifshitz, (1959) Theoretical Physics, v. IX, Electrodynamics of continuous media, Moscow, Nauka.
3. Pitaevsky L.P. (1956), JETP, **31**, 536.
4. Putterman S.L. (1978), Hydrodynamics of superfluid liquid.

5. Combas P. (1950), Theorie und Losungen Methoden des Mehrteilchen problems der Wellenmechanic, Basel.
6. Delrieu T.M. (1974), J. de Physique Letts, **35**, 189.
7. Maki K. (1975), Phys.Letts **56**, 101.
8. Fomin I.A., C.I. Pethick, I.W. Serene (1976), Phys. Rev. Letts, **40**, 1144.
9. Brusov P.N., V.N. Popov (1982), Influence of the electric field on collective excitations in superfluid A– and B–Phases of He3, Preprint of V.A. Steklov Math Institute, Leningrad, P–11–82.
10. Brusov P.N., E.D. Gutlianskii, V.N. Popov (1982), Effect of electric field on spectrum and hydrodynamics of superfluid helium. In Problems of Quantum field theory and Statistical physics, ads. P.P. Kulish & V.N. Popov, **120**.
11. Brusov P.N. (1983), Bose–spectrum of superfluid He3 fillms in electric field, Soviet Journal of Low temperature physics, **9** (9) 477.
12. Brusov P.N., B.D. Gutlianskii, V.N. Popov (1983), Energy spectrum and hydrodynamics of superfluid HeII in a strong electric field, Phys. Lett., **A 97** (1/2).
13. Brusov P.N., V.N. Popov (1983), Collective excitations in A– and B–phases of He3 in electric field, Soviet Journal of Theoretical and Mathematical Physics, **57**(2).
14. Brusov P.N., M.V. Lomakov, V.N. Popov (1991), Effect of electric field on collective excitations in He3–B, Sov. Phys. JETP, **72**(4), 649.

Chapter XII

1. Brusov P.N. and V.N. Popov (1984), Phys. Rev. **B30**, 4060.
2. P.N. Brusov (1985), Gap distortion and collective excitations in B–phase of He3, Sov. Phys. JETP, **88** (4).
3. P.N. Brusov, V.N. Popov (1984), Gap distortion and collective modes in He3–B, Physics B+C, Suppl. VIII, Proceedings of LT–17.
4. Vollhardt D., K. Maki (1978), J. Low Temp.Phys. **31**, 457.
5. Fetter A.L. (1975), Phys Letts. **54A**, 63.
6. Fetter A.L. (1976), J. Low.Temp.Phys. **23**, 245.
7. Sauls J.A., J.W. Serene (1984), Proc. of LT–17 Eds. U. Eckern, A. Schmidt, W. Weber, H. Wuhl, North–Holland, Amsterdam, 775.
8. Shivaram B.S. et al (1983), Phys. Rev. Letts **50**, 1070.
9. Brinkrnan W.F., M.C. Cross, (1978), in Progr. in Low. Temp. Phys. Ed. Brewer D.F., Amsterdam, North Holland, **7A**, 105.
10. Gondadze V.P., G.E. Gurgenishvili, G.A. Kharadze (1981), Sov. J. Low. Temp. Phys. **7**, 821.
11. Mineev V.P. (1983), Sov. J. Uspehi Phys. Nauk **139**, 303.
12. Kleinert H. (1980), Phys. Letts, **78A**, 155.
13. Kleinert H. (1980), J. Low. Temp. Phys. **39**, 451.

14. Hakonen P.J., O.T. Ikkala, S.T. Islander, O.V. Lounasmaa, G.E. Volovik (1983), Low. Temp. Phys. **53**, 425.
15. Hakonen P.J., M. Krusius, M.M. Salomaa, J.Y. Simola, Yu. M. Bunkov, V.P. Mineev, G.E. Volovik (1983), Phys. Rev. Letts, **53**, 1362.
16. Volovik G.E., V.P.Mineev (1983), JETP Letts. **37**, 103.
17. Hakonen P.J., M. Krusius, G. Maniashvili, J.T. Simola (1984), in Proceedings of LT–**17**, 49.
18. Brusov P.N. (1985), The influence of superflow on collective excitations in He3–B, Journal of Low temperature physics, **58**, #3/4, 265.
19. Wheatley J.C. (1978), In: Progress in Low Temperature Physics, edited by D.F. Brewer (North–Holland, Amsterdam, (1978), Vol. **VIIA**, p. 1.
20. Brusov P.N. and V.N. Popov (1980), Zh. Eksp. Teor. Fiz. **79**, 1871 [Sov. Phys. JETP 52, 945].
21. Schopohl N., L. Tewordt (1984), In: Proc. of LT–17, 777.
22. Brusov P.N. and V.N. Popov (1980), Zh. Eksp. Teor. Fiz. **78**, 2419 (Sov. Phys. JETP **51**, 1217).
23. Shivaram B.S., et al. (1983), Phys. Rev. Letts **50**, 1070.
24. Schopohl N., M. Warnke and L. Tewordt (1983), Phys. Rev. Letts. **50**, 1070.
25. Vdovin Yu. A. (1962), In: Proc. of MEPI Moscow, p. 94.
26. Daniels M.E., E.R. Dobbs, J. Saunders, P.L. Ward (1983), Phys. Rev. **B27**, 6988.
27. Davis J.P., H. Choi, J. Pollanen and W.P. Halperin, (2006), Phys. Rev. Letts. **97**, 115301.
28. Sauls J.A. and J.W. Serene (1981), Phys. Rev. **B 23**, 4798.
29. Halperin W.P. and E. Varoquaux, (1990), In: Helium Three, edited by W.P. Halperin and L.P. Pitaevskii (Elsevier Science Publishers, Amsterdam).
30. Rainer D. and J.W. Serene (1976), Phys. Rev. **B 13**, 4745.
31. Hamot P.J., H.H. Hensley and W.P. Halperin (1989), J. Low Temp. Phys. **77**, 429.
32. Greywall D.S. (1986), Phys. Rev. **B 33**, 7520.
33. Kalbfeld S., D.M. Kucera, and J.B. Ketterson (1993), Phys. Rev. Lett. **71**, 2264.
34. Lee Y., T.M. Haard, W.P. Halperin and J.A. Sauls (1999), Nature (London) **400**, 431.
35. Moores G.F. and J.A. Sauls (1993), J. Low Temp. Phys. **91**, 13.
36. Fraenkel P.N., R. Keolian, and J.D. Reppy (1989), Phys. Rev. Lett. **62**, 1126.
37. Sauls J.A., cond–mat/9910260.
38. Meisel M.W., O. Avenel and E. Varoquaux (1987), Jpn. J. Appl. Phys. **26**, Suppl. 26–3–1, 183.
39. Masuhara N., B.C. Watson, and M.W. Meisel (2000), Phys. Rev. Lett. **85**, 2537.
40. Brusov P.N., V.N.Popov (1988), Superfluidity and collective properties of quantum liquids, Nauka publishing, Moscow, 216 p.
41. Koch V.E, P. Wolfle (1981), Phys. Rev. Letts. **46**, 489.
42. Movshovich R., et al. (1988), Phys. Rev. Letts. **61**, 1732.
43. Fomin I.A., C.I. Pethick, I.W. Serene (1976), Phys. Rev. Letts, **40**, 1144.

44. Brusov P.N. (1981), Influence of spin–orbit interaction and magnetic fields on superfluid Phases of He3, In: Problems of Quantum field theory and Statistical physics, eds. P.P. Kulish & V.N. Popov, **101**.
45. Paulson D.N., et al. (1975), In: Low Temp. Phys. – LT14, Eds. M. Krusius and M. Vuorio, Vol. 5 (North–Holland, Amsterdam), p. 417.
46. Schopohl N., L. Tewordt (1984), In: Proc. of LT–17, 777.
47. Dahl D.A. (1977), J. Low Temp. Phys. **27**, 139.
48. Panov V.I. and A.A. Sobyanin (1984), In: Proc. of LT–17 (Eds. V. Eckern et al.), E16.
49. Wheatley J.C. (1975), Rev. Mod. Phys. **47**, 415.

Chapter XIII

1. Paulson D.N., R.M. Johnson and J.C. Wheatley (1973), Phys. Rev. Lett. **30**, 829.
2. Roach, B.M. Abraham, M. Kuchnir and J.B. Ketterson, (1975), Phys. Rev. Lett. **34**, 711.
3. Movshovich R., E. Varoquaux, N. Kim and D.M. Lee (1998), Phys. Rev. Lett. **61**, 1732.
4. Adenwala S., Z. Zhao, J.B. Ketterson and B.K. Sarma (1989), Phys. Rev. Lett. **63**, 1811.
5. Avenel O., L. Piche, W.M. Saslow, E. Varoquaux and R. Combescot (1981), Phys. Rev. Lett. **47**, 803.
6. Shivaram B.S., et al. (1982), Phys. Rev. Lett. **49**, 1646.
7. Fujita T., et al. (1980), Progr. Theor. Phys. **64**, 396.
8. Brusov P.N., M.V. Lomakov (1991), The collective excitations in a planar 2D–phase of superfluid He3, Sov. Phys. JETP, **73**(3), 470; Physica B, **165–166**, 635 (1990).
9. Kleinert H. (1980), Phys. Lett. 78A, 155.
10. Brusov P.N. (1985), J. Low Temp. Phys. **58**, 265.
11. Nasten'ka M.Y. and P.N. Brusov (1989), Phys. Lett. **136A**, 321.
12. Bates D.M., et al. (1984), Phys. Rev. Lett. **53**, 1574.
13. Adenwala S., Z. Zhao, P.N. Brusov, et al. (1990), Phys. Rev. Lett. **65**, 2688.
14. Zhao Z., Adenwalla S., P.N. Brusov, et al. (1991), Sov. Phys. JETP Letts. **53**(11), 591.

Chapter XIV

1. Brusov P.N., V.N. Popov (1988), Collective properties of quantum fluids, Moscow, Nauka Publishing, 215 p.

2. Brusov P.N. (1991), Phys. Rev.**B 43**, 12888.
3. Brusov P.N., V.N. Popov (1984), Phys. Rev. **B 30**, 4060.
4. Kalbfeld S., D.M. Kucera and J.B. Ketterson (1993), Phys. Rev. B, **48**, 4160.
5. Kalbfeld S., D.M. Kucera and J.B. Ketterson (1993), Phys. Rev. Letts.**71**, 2264.
6. Moores G.F. and J.A. Sauls (1993), J. Low Temp. Phys. **91**, 13.
7. Aoki Y., Y. Wada, A. Ogino, M. Saitoh, R. Nomura and Y. Okuda (2005), J. Low Temp. Phys. **138**, 783.
8. Brusov P.N., V.N. Popov (1980), ZhETP, **78**, 234.
9. Shivaram B.S., M.V. Meisel, B.K. Sarma, D.B. Mast, W.P. Halperin, J.B. Ketterson (1982), Phys. Rev. Letts. **49**, 1646.
10. Fujita T., M. Nakahara, T. Ohmi et al. (1980), Progr. Theor. Phys. **64**, 396.
11. Fisher M.E. and M.N. Barber (1972), Phys. Rev. Lett. **28**, 1516.
12. Ginzburg V.L. and A.A. Sobyanin (1982), J. Low Temp. Phys. **49**, 507.
13. Brusov Peter, Jeevak Parpia, Paul Brusov and Gavin Lawes (2001), Phys. Rev. B Rapid Commun., **63**, 140507.
14. Brusov Peter and Pavel Brusov (2005) Phys. Rev. **B 72**, 134503.
15. Brusov Peter and Pavel Brusov (2009), J. of Phys. (UK): Conference Series **150** 032012.

Chapter XV

1. Brusov P.N. and V.N. Popov (1988), Collective properties of quantum fluids, Moscow, Nauka Publishing, 215 p.
2. Alonso V. and V.N. Popov (1977), Sov. Phys. JETP **46**, 760.
3. T. Fujita, M. Makahara, T. Ohmi et al. (1980), Progr. Theor. Phys. **64**, 396.
4. Z. Zhao, S. Adenwalla, P.N. Brusov, J.B. Ketterson and B.K. Sarma (1990), Phys. Rev. Lett., **65**, 2688.
5. Brusov Peter and Pavel Brusov (2005), Phys. Rev. **B 72**, 134503.
6. Brusov Peter and Pavel Brusov (2009), J. of Phys. (UK): Conference Series **150** 032012.
7. Brusov P.N. and V.N. Popov (1980), Zh. Eksp. Teor. Fiz. 79, 1871 [Sov. Phys. JETP 52, 945].
8. D.S. Hiroshima and H. Namizawa (1985), Progr. Theor. Phys. **74**, 400.
9. P. Wolfle, (1976) Phys. Rev. Lett., **37**, 1279.
10. Brusov P.N. and M.V. Lomakov (1991), Zh. Eksp. Teor. Fiz. **100**, 849 [Sov. Phys. JETP 73, 470].
11. Peter Brusov (1991), J. Low Temp. Phys. **82**, 31.
12. Brusov P.N. and V.N. Popov (1980), Zh. Eksp. Teor. Fiz. **78**, 2419 (Sov. Phys. JETP **51**, 1217).
13. Shivaram B.S., et al. (1982), Phys. Rev. Letts **49**, 1646.

Chapter XVI

1. Brusov Peter, Jeevak Parpia, Paul Brusov and Gavin Lawes (2001), Phys. Rev. B Rapid Commun., **63**, 140507.
2. Brusov Peter and Pavel Brusov (2005), Phys. Rev. **B 72**, 134503.
3. Brusov Peter (2000), Journal of low temperature physics, **119**, 291.
4. Nasten'ka M.Y. and P.N. Brusov (1989) Phys. Lett. **A 136**, 321.
5. Brusov P.N., Phys. (1991), Rev. **B 43**, 12888.
6. Wolfle P., Physica (1977), **90B**, 96.
7. Brusov P.N., V.N. Popov (1980), Sov. Phys. JETP, **52**, 945.
8. Ling R., W. Wojtanowski, J. Saunders and E.R. Dobbs (1990), J. Low Temp. Phys. **78**, 187.
9. Brusov P.N. (1990), J. Low Temp. Phys. **83**.
10. Combescot M., R. Combescot (1976), Phys. Rev. Lett., **37**, 388.
11. Brusov P.N., V.N. Popov, Sov. Phys. (1980), JETP **51**, 117.
12. Alonso V. and V.N. Popov (1977), Sov. Phys. JETP **46**, 760.
13. Brusov P.N. and M.V. Lomakov (1991), Sov. Phys. JETP **73**, 470.
14. Fujita T., M. Makahara, T. Ohmi et al. (1980), Progr. Theor. Phys. **64**, 396.
15. Zhao Z., S. Adenwalla, P.N. Brusov, J.B. Ketterson and B.K. Sarma (1990), Phys. Rev. Lett., **65**, 2688.
16. Brusov P.N. and V.N. Popov (1980), Sov. Phys. JETP **52**, 714.
17. Hiroshima D.S. and H. Namizawa (1985), Progr. Theor. Phys. **74**, 400.
18. Wolfle P. (1976), Phys. Rev. Lett., **37**, 1279.
19. Brusov P.N. (1991), J. Low Temp. Phys. **82**, 31.
20. Brusov P.N. and V.N. Popov (1980), Sov. Phys. JETP **51**, 117.
21. Brusov P.N. and V.N. Popov (1980), Sov. Phys. JETP **51**, 1217.
22. Mast D. B., et al. (1980), Phys. Rev. Lett., **45**, 256.
23. Brusov P.N., V.N. Popov (1988), Collective properties of quantum fluids, Moscow, Nauka, 215 p.
24. Davis J.P., J. Pollanen, H. Choi, J.A. Sauls and W.P. Halperin (2008) Nature Physics **4**, 571.
25. Davis J.P., H. Choi, J. Pollanen and W.P. Halperin (2008) Phys. Rev. Lett., **100**, 015301.
26. Brusov Peter and Pavel Brusov (2009) Phys. Lett. A, **373**, 1972.
27. Brusov Peter and Pavel Brusov (2009) J. Low Temp. Phys. **155**, 200.

Chapter XVII

1. Brusov P.N., V.N. Popov (1988), Collective properties of quantum fluids, Moscow, Nauka, 215 p.

2. Brusov P.N., M.V. Lomakov, N.P. Brusova (1994), The collective mode spectrum in polar phase of superfluid He3, Physica B, **194–196**, 835.
3. Brusov P.N., M.V. Lomakov, N.P. Brusova (1995), The collective mode spectrum in polar phase of superfluid He3, Sov. J. Low Temp. Phys. **21**, 113.
4. Hiroshima D.S. and H. Namizawa (1985) Progr. Theor. Phys. **74**, 400.

Chapter XVIII

1. Brusov P.N., V.N. Popov (1988), Collective properties of quantum fluids, Moscow, Nauka, 215 p.
2. Brusov P.N., M.V. Lomakov (1991), Collective excitations in the A_1 phase of superfluid He3, Soviet Journal of Low temperature physics, **17**, #4, 227.
3. Brusov P.N., M.V. Lomakov (1991), The collective excitations in He3–A_1, Journal of Low temperature physics, **85**, 91.
4. Brusov P.N. and V.N. Popov (1980), Zh. Eksp. Teor. Fiz. **79**, 1871 [Sov. Phys. JETP **52**, 945].
5. Brusov P.N. (1991), Phys. Rev. B, **43**, 12888.
6. Brusov P.N., M.V. Lomakov (1991), Sov. Phys. JETP, **73**(3), 470.
7. Brusov P.N., M.V. Lomakov (1990), Physica B, **165–166**, 635.

Chapter XIX

1. Kosterlitz J.M., D.J. Thouless (1972), J. Phys, **5**, L124.
2. Berezinsky V.L. (1971), JETP **61**, 1144.
3. Popov V.N. (1972), Sov. J. Theor. Math. Phys., **11**, 236.
4. Popov V.N. (1972), Sov. J. Theor. Math. Phys., **11**, 356.
5. Bogoliubov N.N. (1961), Quasiaverages in Problems of Statistical Mechanics, JINR Preprint D–781 (in Russian).
6. Pitaevski L.P. (1961), JETP, **40**, 646.
7. Bycling E. (1965), Annals of Phys. 32, 367.
8. Wiegel F.W. (1973), Physica **65**, 321.
9. Lieb E.H., W. Liniger (1973), Phys. Rev. **130**, 1605.
10. Yang C.N., C.P.Yang (1969), J. Math. Phys. **10**, 1115.
11. Sonin E.B. (1970), JETP, **59**, 1416.
12. Kurihar S. (1983), J. Phys. Soc. Jap., **52**, 1311.
13. Stein D.L., M.C. Cross (1979), Phys. Rev. Letts., **42**, 504.
14. Rainer D., J.N. Serene (1984), Proc. Of LT–17 Eds. U. Eckern, A. Schmidt, W. Weber, H. Wuhl, North–Holland, Amsterdam, 775.
15. Polyakov A.M. (1975), Phys. Letts. **59B**, 79.

16. Brusov P.N. and V.N. Popov (1980), Zh. Eksp. Teor. Fiz. **79**, 1871 [Sov. Phys. JETP 52, 945].
17. Brusov P.N. and V.N. Popov (1980), Zh. Eksp. Teor. Fiz. **78**, 234 [Sov. Phys. JETP **51**,117].
18. Brusov P.N. and V.N. Popov (1980), Zh. Eksp. Teor. Fiz. **78**, 2419 [Sov. Phys. JETP **51**, 1217].
19. Sachrajda A., F.F. Harris–Lowe et al. (1985), Phys .Rev. Letts, **55**, 1602.
20. Brusov, P.N., V.N. Popov, (1981), JETP, **80**, 1565.
21. Popov V.N. (1983), Functional integrals in quantum field theory and statistical physics, Reidel.
22. Korshunov S.E. (1985) Zh. Eksp. Teor. Fiz. **89**, 531.
23. Davis J.C., A. Amar, J.P. Pekola and R.E. Packard (1988), Phys. Rev. Lett. **60**, 302.
24. Brusov P.N. and V.N. Popov (1982), Superfluidity and Bose excitations in He3 films Phys. Lett., **87A** (9), 472.
25. Fujita T., M. Nakahara and T. Ohmi et al. (1980), Progr. Theor. Phys. **64**, 396.
26. Wolfle P. (1978), Progr. in Low Temp. Phys. **7B**, 191.

Chapter XX

1. Khalatnikov M. (1965), Theory of Superfluidity, Benjamin, NY.
2. Bardeen J., G. Baym, D. Pines (1966), Phys. Rev. Letts., **17**, 372.
3. Baym G., C.I. Petnik (1978), The Physics of Liquid and Solid Helium, **4**, 123.
4. Landau J., J.T. Tough, N.R. Brubaker et al. (1970), Phys. Rev. **A2**, 2472; Phys. Rev. Lett. **23**, 283 (1969)
5. Hoffberg M.B. (1972) Phys. Rev. **A5**, 1986.
6. Volovik G.E., V.P. Mineev, I.M. Khalatnikov (1975) JETP **69**, 675.
7. Andrianov V.A., V.N. Popov (1973). JETP, **64**, 1861.
8. Bashkin E.P. (1980), JETP **78**, 360.
9. Andreyev A.F. (1966), **50**, 1415.
10. Zinovieva K.N., S.T. Boldurev (1969), JETP, **56**, 1089.
11. Andrianov V.A., V.N. Popov (1976), Theor. and Math. Phys. (Sov.) **28**, 340.
12. Andrianov V.A., V.N. Popov (1976), Vestn. Leningrad Univ. **22**, 11.
13. Abrikosov A.A., L.P. Gor'kov and I.E. Dzyaloshinskii (1975), Methods of Quantum Field Theory in Statistical Mechanics, Dover, New York.
14. Popov V.N. (1972), Sov. J. Theor. Math. Phys., **11**, 236.
15. Popov V.N. (1972), Sov. J. Theor. Math. Phys., **11**, 356.
16. Brusov P.N., V.N. Popov (1984), Phys. Lett. **30B**, 4060.
17. Brusov P.N., V.N. Popov (1982), Phys. Rev. **87A**, 472.
18. Brusov P.N., V.N. Popov (1984), Phys. Lett. **101A**, 154.
19. Brusov P.N., V.N. Popov (1981), JETP, **80**, 1565.

20. Brusov, P.N. (1981), In: Problems in quantum field theory and statistical physics eds. P.P. Kulish, V.N. Popov, Nauka, Leningrad, **101**, 28.
21. Alonso V., V.N. Popov (1977), JETP **77**, 1445–1459.
22. Brusov P.N., V.N. Popov (1980), Sov. Phys. JETP, **51**(6), 1217.
23. Brusov P.N., V.N. Popov (1980), Sov. Phys. JETP, **51**(l), 117.
24. Brusov P.N., V.N. Popov (1980), Sov. Phys. JETP, **52**(5), 945.
25. Brusov P.N., V.N. Popov (1982), Bose–spectrum of superfluid systems of He3–He4, Soviet Journal of Theoretical and Mathematical Physics, **53**, #3.
26 Brusov P.N., V.N. Popov (1984), Coupling of the acoustic modes in superfluid mixtures of He3–He4, Europhysice conf. abstracts, **8A**, P. Pl–PO44.

Chapter XXI

1. Brusov Peter, Jeevak Parpia, Paul Brusov and Gavin Lawes (2001), Phys. Rev. B Rapid Commun., **63**, 140507.
2. Brusov Peter, Paul Brusov, Gavin Lawes et al. (2003), Novel sound phenomena in impure superfluids, Physica B **329–333**, 333.
3. Golov A., D.A. Geller and J.M. Parpia (1999), Phys. Rev. Letts. **82**, 3492.
4. McKenna M.J., T. Slawecki and J.D. Maynard (1991), Phys. Rev. Letts. **66**, 1878.
5. Lawes G., E. Nazaretski, N. Mulders, P.N. Brusov, J. Parpia (2002), J. Low Temp. Phys., **126**, 691.
6. Lawes G., E. Nazaretski, N. Mulders, P.N. Brusov, J. Parpia (2002), Distribution of ^{4}He atoms in ^{3}He–^{4}He mixtures in aerogel, JLTP, **129**, 49.
7. Mulders N., R. Mehrotra, L.S. Goldner and G. Ahlers (1991), Phys. Rev. Letts. **67**, 695.
8. Khalatnikov M. (1965), Theory of Superfluidity, Benjamin, NY.
9. Thuneberg E.V., et al. (1998), Phys. Rev. Lett., **80**, 2861.
10. Eselson B., M.I. Kaganov, E. Ya. Rudavskii and I.A. Serbin (1974) Sov. Uspekhi Fizicheskih Nauk, **112**, 591.
11. Arkhipov R.G. and I.M. Khalatnikov (1958), Soviet Phys. JETP, **6**, 583.
12. Pokrovsky V.L. and I.M. Khalatnikov (1976), Sov. Phys. JETP, **44**, 1036.
13. Lebedev V. (1977), Sov. Phys. JETP, **45**, 1169.
14. Rinberg D., V. Cherapanov and V. Steinberg (1996), Phys. Rev. Lett. **76**, 2105.
15. Garrett S., et al. (1978), Phys. Rev. Lett., **41**,413.
16. Kiselev S.I, V.V. Khmelenko and D.M. Lee (2000), Low temperature physics, **26**, 641.
17. Putterman S.J. (1974), Superfluid hydrodynamics, North–Holland Publishing.
18. Dikina L.S., E. Ya. Rudavskii and I.A. Serbin (1970), ZETP, **58**,813.
19. Ishikawa T., M. Yamamoto, S. Higashitani and K. Nagai (2001), Journal of the Physical Society of Japan, **70**, 3486

20. Lawes G., E. Nazaretski, N. Mulders, P.N. Brusov, J. Parpia (2002) J. Low Temp. Phys., **129**, 49.
21. Brusov Peter, Pavel Brusov (2006), Int. J. Modern Phys. **B 20**, 355.

Chapter XXII

1. Popov V.N. (1991), Functional integrals and Collective Excitations, Cambridge University Press.
2. Kapitonov V.S., V.N. Popov (1981) Zap. Nauchn. Sem. LOMI, **101**, 77.
3. Andrianov V.A., V.S. Kapitonov and V.N. Popov (1982), Zap. Nauchn. Sem. LOMI, **120**, 3.
4. Andrianov V.A., V.S. Kapitonov and V.N. Popov (1982), Zap. Nauchn. Sem. LOMI, **131**, 3.

Chapter XXIII

1. Popov V.N. (1991), Functional integrals and Collective Excitations, Cambridge University Press.

Chapter XXIV

1. Alonso V, V.N. Popov (1977), JETP **77**, 1445.
2. Brusov P.N., V.N. Popov (1980), Sov. Phys. JETP, **51**(1), 117.
3. Brusov P.N., N.P. Brusova (1994), Physica **B 194–196**, 1479.
4. Brusov P.N. (1999), Mechanisms of High Temperature Superconductivity, Rostov State University Publshing, v.I,II, 1389 p.
5. Brusov P.N. and N.P. Brusova (1996), The path integral model of d–pairing for HTSC, heavy fermion superconductors and superfluids. J. Low. Temp. Phys. **103**, 251.
6. Brusov P.N., N.P. Brusova and P.P. Brusov (1996), The path integral model of d–pairing superconductors. Czechoslovak J. of Physics **46**, suppl. 2, 1041.
7. Brusov P.N., N.P. Brusova, P.P. Brusov (1996), Soviet Low Temperature Physics, **22**, 506.
8. Brusov P.N., P.P. Brusov (2001), Collective properties of superconductors with nontrivial pairing, Sov. Phys. JETP, **119,** 913–930 [Zh.Eksp.Theor. Phys. **92**, 795].
9. Brusov Peter, Paul Brusov and Chong Lee (2004), Collective properties of unconventional superconductors, Int. J. of Mod. Phys. B, **18**, 867.

10. Brusov P.N. and P.P. Brusov (2006.), Order parameter collective modes in unconventional superconductors, In: Proceedings NATO Advanced Research Workshop, Electron correlation in new materials and nanosystems, Edited by K. Scharnberg and S. Kruchinin, Springer.

Chapter XXV

1. Brusov P.N. (1999), Mechanisms of High Temperature Superconductivity, Rostov State University Publshing, v. I, II, 1389 p.
2. Bednortz J.G. and K.A. Muller (1986), Z. Phys. B – Condensed Matter **64**, 189.
3. Bednortz J.G. and K.A. Muller (1987), Perovskite–Type Oxides – the new approach to high Tc superconductivity, Nobel lecture, Stockholm.
4. Johnston D.C., H. Prakash and W.H. Zachariasen, Mater. Res. Bull. **8**, 777 (1973).
5. Sleight A.W., J.L. Gillson and P.E. Bierstedt, Solid State Commun. **17**, 27 (1975).
6. Chakraverty B.K. (1979), J. Phys. (Paris), Lett. **40**. L99; J. Phys. (Paris) **42**, 1351 (1979).
7. Wu M.K., J.R. Ashburn, C.J. Torg, P.H. Hor, R.L. Meng, L. Gao, Z.J. Huang, Y.Q. Wang and C.W. Chu (1987), Phys. Rev. Lett. **58**, 908.
8. Chu C.W., et al. (1987), Science **235**, 567.
9. Chu C.W., J. Bechtold, L. Gao, P.H. Hor, Z.J. Huang, R.L. Meng, Y.Y. Sun, Y.Q. Wang and Y.Y. Xue (1988), Phys. Rev. Lett. **60**, 941.
10. Chu CW., P.H. Hor, P.L. Meng, L. Gao, Z.J. Huang and Y.Q. Wang (1987), Phys. Rev. Lett. **58**, 405.
11. Jorgensen J.D., M.A. Beno, D.G. Hinks, L. Sonderholm, K.J. Volin, R.L. Hitterman, J.D. Grace, I.K. Schuller, C.U. Segre, K. Zhang and M.S. Kleefisch (1987), Phys. Rev. **B 36**, 3608.
12. Shirane G., Y. Endoh, R.J. Birgeneau, M.A. Kastner, Y. Hidaka, M. Oda, M. Suzuki and T. Murakami (1987), Phys. Rev. Lett. **59**, 1613.
13. Matsuda M., K. Yamada, K. Kakurai, H. Kadowaki, T.R. Thurston, Y. Endoh, Y. Hidaka, R.J. Birgenau, M.A. Kastner, P.M. Gehring, A.H. Moudden and G. Shirane (1990), Phys. Rev. **B 42**, 10098.
14. Tranquada J.M., D.E. Cox, W. Konnmann, A.H. Moudden, G, Shirane, M. Suenage, P. Zailiker, D. Vaknin and S.K. Sinha (1988), Phys. Rev. Lett. **60**, 156.
15. Birgeneau R J. and G. Shirane (1989), In: Physical Properties of High Temperature Superconductors, vol. 1, ed. D.M. Ginsberg (World Scientific, Singapore), Ch. 4, p. 151.
16. Rossat–Mignod J., L.P. Regnault, C. Vettier, P. Burlet, J.Y. Henry and G. Lapertot (1991), Physica **B 169**, 58.
17. Lyons K.B., P.A. Fleury, R.R. Singh and P.E. Sulewski (1990), Proc. SPIE **1336**, 66, In: Proc. NATO Advanced Research Workshop on Dynamics of Magnetic

Fluctuations in High–Temperature Superconductors, G. Reiter, P. Horsch and G. Psaltakis, eds. (Plenum Press, New York, 1991), p. 159.

18. Cheong S.–W., Z. Fisk, J.O. Willis, S.E. Brown, J.D. Thompson. J.P. Remeika, A.S. Cooper, R.M. Aikin, D. Schiferi and G. Griiner (1988), Solid State Commun. **65**, 111.
19. Thio T., T.R. Thurston, N.W. Preyer, P.J. Picone, M.A. Kastner, H.P. Jenssen, D.R. Gabbe, C.Y. Chen, R.J. Birgenau and A. Aharony (1988), Phys. Rev. **B 38**, 905.
20. Vaknin D., S.K. Sinha, DE. Moncton, D. C. Johnson, J. Newsam, C.R. Safinya and H.E. King, Jr. (1987), Phys. Rev. Lett. **58**, 2802.
21. Chakravarty S. (1989), In: High Temperature Superconductivity, Proc. Los Alamos Symp., K.S. Bedell, D. Coffey, D.E. Meltzer, D. Pines, J.R. Schrieffer, eds. (Addison–Wesley, Reading, MA, Advanced Book Program, 1990), p. 136.
22. Endoh Y., K. Yamada, R.J. Birgenau, D.R. Gabbe, H.P. Jenssen, M.A. Kastner, C.J. Peters. P.J. Picone, T.R, Thurston, J.M. Tranquanda, G. Shirane, Y. Hikada, M. Oda, Y. Enomoto, M. Suzuki and T. Murakami (1988), Phys. Rev. **B 37**, 7443.
23. Rossat–Mignod, J., P. Burler, M.J. Jurges, J.Y. Henry and C. Vettier (1988), Physica **C 152**, 19.
24. Tranquada J.M., D.E. Cox, W. Konnmann, A.H. Moudden, G, Shirane, M. Suenage, P. Zailiker, D. Vaknin and S.K. Sinha (1988), Phys. Rev. Lett. **60**, 156.
25. Dzyaloshinskii I. (1958), J. Phys. Chem. Solids **4**, 241.
26. Moriya T. (1985), In: Spin Fluctuations in Itinerant Electron Magnetism, Springer Series in Solid–State Science, vol. 56 (Springer, Berlin, Heidelberg); Phys. Rev. **20**, 91 (1960).
27. Thio T., T.R. Thurston, N.W. Preyer, P.J. Picone, M.A. Kastner, H.P. Jenssen, D.R. Gabbe, C.Y. Chen, R.J. Birgenau and A. Aharony (1988), Phys. Rev. **B 38**, 905.
28. Tokura Y. (1992), In: Physics of High–Temperature Superconductors, Proc. Toshiba Int. School of Superconductivity (ITS2), Kyoto, Japan, eds. S. Maekawa, M. Sato, Springer Series in Solid–State Science, vol. 106 (Springer, Berlin, Heidelberg), p. 191.
29. Ong N.P. (1989), In: Physical Properties of High Temperature Superconductors II, ed. D.M. Ginsberg (World Scientific, Singapore).
30. Cava R.J., A.W. Hewat, E.A. Hewat, B. Batlogg, M. Marezio, K.M. Rabe, J.J. Krajewski, W.F. Peck, Jr. and L.W. Rupp (1990), Physica **C 165**, 419.
31. Anderson P.W. (1988), In: Frontiers and Borderlines in Many–Particle Physics, eds. J.R. Schrieffer and R.A. Broglia (North–Holland, Amsterdam).
32. Anderson P.W. (1987), Science **235**, 1196.
33. Levin K., J.H. Kim, J.P. Lu and Q.Si (1991), Physica **C 175**, 449; In: Electronic Structure and Mechanisms for High Temperature Superconductivity, eds. J. Ashkenazi and G. Vezzoli (Plenum. New York); Preprint, 1992.
34. Batlogg B. (1989), In: High Temperature Superconductivity, Proc. Los Alamos Symp., eds. K.S. Bedell, D. Coffey, D.E. Meltzer, D. Pines, J.R. Schrieffer (Addison–Wesley, Reading , MA, Advanced Book Program, 1990), p. 37.

35. Batlogg B. (1991), Phys. Today **44**, No. 6, p. 44; No. 6, 55 (1991).
36. Burns G. (1992), High Temperature Superconductivity (Academic, New York).
37. Batlogg B., H. Takagi, H.L. Kao and J. Kwo (1992), In: Electronic Properties of High–Tc Superconductors, The Normal and the Superconducting State, eds. Kuzmany et al. (Springer–Verlag, Berlin/Vienna/New York).
38. Takagi H., B. Batlogg, H.L. Kao, J. Kwo, R.J. Cava, J.J. Krajewski and W.F. Peck, Jr. (1992), Phys. Rev. Lett. **69**, 2975.
39. Hidaka Y. and M. Suzuki (1989), Nature **338**, 635.
40. Tsuei C.C., A. Gupta and G. Koren (1989), Physica **C 161**, 415.
41. Poilblanc D., T. Ziman, H.J. Schulz and E. Dagotto (1993), Phys. Rev. **B 47**, 14267.
42. Tranquada J.M., P.M. Gehring., G. Shirane, S. Shamoto and M. Sato (1992), Phys. Rev. **B 46**, 5561.
43. Kane C.L., P.A. Lee, T.K. Ng, B. Chakraborty and N. Read (1990), Phys. Rev. **B 41**, 2653.
44. Hayden S.M., T.E. Mason, G. Aeppli, H.A. Mook, S.–W. Cheong and Z. Fisk (1992), In: Workshop on Phase Separation in Cuprate Superconductors, Erice, Italy, May 6–7, 1992, ed. K.A. Muller (World Scientific, Singapore).
45. Ong N.P., T.W. Jing, T.R. Chien, Z.Z. Wang, T.V. Ramankrishnan, J.M. Tarascon and K. Remschnig (1991), Physica **C 185–189**, 34.
46. Matsuyama H., et al. (1989), Physica **C 160**, 567.
47. Poilblanc D., T. Ziman, H.J. Schulz and E. Dagotto (1993), Phys. Rev. **B 47**, 14267.
48. Knoll P., S. Lo, P. Murugaraj, E. Schonherr and M. Cardona (1993), In: Electronic Properties of High–Tc Superconductors, Kirchberg, Tyrol, 1992, H. Kuzmany, M. Mehring and J. Fink, eds. Springer Series in Solid–State Science, Vol. 113 (Springer, Berlin, Heidelberg), p. 286.
49. Luttinger J.M. (1963), J. Math. Phys. **4**, 1154.
50. Sokol A. and D. Pines (1994), Phys. Rev. Lett. **71**, 2813.
51. Denteneer, P.J. and J.M.J. van Leeuwen (1993), Europhys. Lett. **22**, 413.
52. Tokura Y. (1992), In: Physics of High–Temperature Superconductors, Proc. Toshiba Int. School of Superconductivity (ITS2), Kyoto, Japan, eds. S. Maekawa, M. Sato, Springer Series in Solid–State Science, vol. 106 (Springer, Berlin, Heidelberg), p. 191.
53. Muller K.A. and J.G. Bednorz (1987), Science **237**, 1139.

Chapter XXVI

1. Annett J.F., N.D. Goldenfeld and A.J. Leggett (1996), In: Physical Properties of High Temperature Superconductors V, ed. D.M. Ginsberg (World Scientific, Singapore).
2. Panagopulous C., et al. (1996), Phys. Rev. **B 53**, R2999.

3. Anlage S.M., et al. (1994), Phys. Rev **B 50**, 523; J. Superconductivity **7**, 453 (1994).
4. Annett J.F. (1990), Adv. Phys. **39**, 83.
5. Sigrist M. and T.M. Rice (1994), Rev. Mod. Phys. **67**, 491.
6. Annett J.F., N. Golgenfeld and S.R. Renn (1989), Phys. Rev. **B 40**, 13303.
7. Annett J.F., M. Randeria and S.R. Renn (1988), Phys. Rev. **B 38**, 4660.
8. Annett J.F. and R.M. Martin (1990), Phys. Rev. **B 42**, 3929.
9. Annett J., N. Goldenfeld, and S.R. Renn (1990), In: Physical Properties of High Temperature Superconductors II, ed. D.M. Ginsberg (1991), World Scientific, Singapore, p. 571; Phys. Rev. **B 43**, 2778.
10. Annett J.F. and N. Goldenfeld (1992), J. Low Temp. Phys. **89**, 197.
11. Annett J F. (1997), In: Fluctuation Phenomena in HTSC.
12. Annett J.F., N. Goldenfeld and A. Leggett (1966), J. Low Temp. Phys. **105**, 473
13. Annett J.F., N. Goldenfeld and S.R. Ren (1989), Phys Rev. **B 39**, 708.
14. Brusov P.N., N.P. Brusova (1994), The model of d–pairing for HTSC, Heavy fermion systems and superfluids. Physica B **194–196**, 1479–80.
15. Brusov P.N., M.V. Lomakov, N.P. Brusova (1994), The collective excitations in HTSC under d–pairing (3D–CASE), Physica C, **235–240**, 2259–2260.
16. Brusov P.N. and N.P. Brusova (1996), The path integral model of d–pairing for HTSC, heavy fermion superconductors and superfluids. J. Low. Temp. Phys. **103**, 251.
17. Brusov P.N., N.P. Brusova and P.P. Brusov (1996), The path integral model of d–pairing superconductors, Czechoslovak J. of Physics **46**, suppl. 2, 1041.
18. Brusov P.N., N.P. Brusova, P.P. Brusov (1996), Low temperature physics, **22**, 506.
19. Brusov P.N., N.P. Brusova and P.P. Brusov (1997), The collective excitations of the order parameter in HTSC and heavy fermion superconductors (HFSC) under d–pairing, J. Low Temp. Phys. **108**, 143.
20. Brusov P.N., N.P. Brusova, P.P. Brusov, N.N. Harabaev (1997), The collective modes in HTSC and heavy fermion superconductors (HFSC) under d–pairing, Physica C, **282–287**, p. 1881–1882.
21. Brusov P.N. (1999), The mechanisms of HTSC, RSU publishing, v. **1**, **2**, 1389 p., (in English).
22. Brusov P.N., P.P. Brusov and V.P. Sachenko (2000), Collective excitations in unconventional superconductors, Proceedings of Samara Technical University, v. 10, p.1–26.
23. Brusov P.N., P.P. Brusov (2000), Collective excitations in unconventional superconductors, Proceedings of SPIE Vol. 4058.
24. Brusov P.N., P.P. Brusov (2000), Collective excitations of unconventional superconductors, In Proceedings of International conference on low temp. physics LT–32, Kazan', Russia, p. 43–45.

25. Brusov P.N., P.P. Brusov (2001), "Collective properties of superconductors with nontrivial pairing", Sov. Phys. JETP, **119,** 913–930 [Zh. Exp. Theor. Phys. **92**, 795].
26. Brusov P.N., P.P. Brusov (2006), In: Proceedings NATO Advanced Research Workshop, Electron correlation in new materials and nanosystems, Edited by K. Scharnberg and S. Kruchinin, Springer.
27. Brusov P.N., M.Y. Nasten'ka, T.V. Filatova–Novoselova (1982), The determination of type of pairing and order parameter for HFS and HTSC by sound and microwave absorption experiments. Phys. Lett. **142A**, 179.
28. Brusov Peter, Paul Brusov and Chong Lee (2004), Collective properties of unconventional superconductors, Int. J. of Mod. Phys. B, **18**, 867.
29. Brusov Paul and Peter Brusov (2000), A method of distinction of a mixed $d_{x^2-y^2} + d_{xy}$ state of HTSC from the pure d–wave state, Physica B, **281&282**, p. 949–950.
30. Brusov Peter, Paul Brusov, Pinaki Majumdar and Natali Orehova (2003), Ultrasound attenuation and collective modes in mixed mixed $d_{x^2-y^2} + d_{xy}$ state of unconventional superconductors, Brazilian Journal of Physics, **33**, 729–732.
31. Brusov P.N., M.V. Lomakov, N.P. Brusova, T.V. Filatova–Novoselova (1994), The Spectroscopy of Collective Modes in HTSC for P–Pairing (3D–Case), Physica B **194–196**, 1477–1478.
32. Brusov P.N. and N.P. Brusova (1995), The model of d–pairing in CuO_2 planes of HTSC and the the collective modes. J. Low Temp. Phys. **101**, 1003.
33. Brusov P.N., N.P. Brusova (1994), The collective excitations in Cuo2 planes of HTSC under d–pairing, Physica C, **235–240**.
34. Brusov P.N., N.P. Brusova, P.P. Brusov, N.N. Harabaev (1997), The path integral model of d–pairing in CuO2 planes of HTSC and the collective modes, Physica C, **282–287**, p. 1833–1834.
35. Brusov P.N., et al. (1990), The microscopic theory of collective excitations in superfluid He3, Northwestern University preprint.
36. Brusov P.N. and V.N. Popov (1988), Collective properties of quantum fluids, Moscow, Nauka Publishing, 215 p.
37. Anderson P.W. (1984), Basic Notions of Condensed Matter Physics (Melno–Park, CA; Benjamin–Cumings).
38. Anderson P.W. (1987), Science **235**, 1196 .
39. Kalmeyer V. and R.B. Laughlin (1987), Phys. Rev. Lett., **59**, 2095.
40. Kivelson S.A., D.S. Rokhsar and J.P. Sethna (1987) Phys. Rev. **B 35**, 8865.
41. Alexandrov A.S., J. Ranninger and S. Robaszkiewicz (1986), Phys. Rev. **B 33**, 4526, Phys. Rev. Lett. **56**, 949 (1986).
42. Alexandrov A.S. (1988), Phys. Rev. **B 38**, 925.
43. Randeria M., J. Duan and L. Shieh (1989), Phys. Rev. Lett. **62**, 981; Phys. Rev. **B 41**, 327 (1990).

44. Schmitt–Rink S., C.M. Varma and A.E. Ruckenstein (1989), Phys. Rev. Lett. **63**, 445.
45. Sobyanin A.A. (1988), Physica **C 153–155**, 1203.
46. Landau L.D. and E.M. Lifshitz (1980), Statistical Physics, Part 2 (Oxford, Pergamon).
47. Michel L. (1980), Rev. Mod. Phys., **52**, 591.
48. Abud M. and G. Sartori (1983), Ann. Phys., **150**, 307.
49. Barton G. and M.A. Moore (1974), J. Phys. **C 7**, 2989; 4220.
50. Bhattacharya S., et al. (1988), Phys. Rev. **B 37**, 5901.
51. Michel L., (1980), Rev. Mod. Phys., **52**, 591.
52. Schrieffer J.R., X.G. Wen and S.C. Zhang (1988), Phys. Rev. Lett. **60**, 944.
53. Wen X.G. and A. Zee (1989), Phys. Rev. Lett. **62**, 1937; 2873.
54. Wen X.G., F. Wilczek and A. Zee (1989), Phys. Rev. **B 39**, 11413.
55. Zou Z., B. Doucot and B.S. Shastry (1989), Phys. Rev. **B 39**, 11434.
56. Anderson P.W. and W.F. Brinkman (1974), The Helium Liquids: Proceedings of the 15th Scottish Universities Summer School in Physics, eds. J.G.M. Armitage and I.E. Farquhar (New York, Academic).
57. Koster G.F. (1957), Solid St. Phys. **5**, 173.
58. Miller S.C. and W.F. Love (1967), Tables of Irreducible Representations of Space Groups and Co–representations of Magnetic Space Groups (Boulder, Colorado, Pruett Press).
59. Tinkham M. (1964), Group Theory and Quantum Mechanics (New York, McGraw–Hill).
60. Paul D., et al. (1987), Phys. Rev. Lett. **58**, 1976.
61. Harlingen D. Van (1995), Rev. Mod. Phys. **67**, 515.
62. Imry Y. (1975), J. Phys. **C 8**, 567.
63. Basov D.N., et al. (1995), Phys. Rev. Lett. **74**, 598.
64. Tallon J.L., et al. (1995), Phys. Rev. Lett. **74**, 1008.
65. Varelogiannis G., A. Perali, E. Cappelluti and L. Pietronero (1996), Phys. Rev **B 54**,1.
66. Bahcall S. (1996), Phys. Rev. Lett. **76**, 3634.
67. Kelley R.J., et. al. (1996), Science **271**, 1255.
68. Abrikosov A.A. (1995), Physica **C 244**, 243.
69. Radtke R.J., K. Levin, H.–B. Schuttler and M.R. Norman (1993), Phys. Rev. **B 48**, 653.
70. Giapintzakis J., D.M. Ginsberg, M.A. Kirk and S. Ockers (1994), Phys. Rev. **B 50**, 15967.
71. Peng N. and W.Y. Liang (1994), Physica **C 233**, 61.
72. Westerholt K. and B. van Hedt (1994), J. Low Temp. Phys. **95**, 123.
73. Kirtley J.R., et al. (1996), Phys. Rev. Lett. **76**, 1336.
74. Tsuei C.C., et al. (1996), Science **271**, 329.

75. Tsuei C.C., J.R. Kirtley, C.C. Chi, L.S. Yu–Jahnes, A. Gupta, T. Shaw, J.Z. Sun and M.B. Ketchen (1994), Phys. Rev. Lett. **73**, 593.
76. Wollman D.A., D.J. Van Harlingen, W.C. Lee, D.M. Ginsberg and A.J. Leggett (1993), Phys. Rev. Lett. **71**, 2134.
77. Bulut N. and D.J. Scalapino (1994), Phys. Rev. **B 54**, 14971.
78. Levi B.G. (1993), Phys. Today **46** (5), 17.
79. Scalapino D.J. (1995), Phys. Rep. **250**, 329.
80. Schrieffer J.R. (1994), Solid State Commun. **92**, 129.
81. Bulut N., D.J. Scalapino and S.R. White (1993), Phys. Rev. **B 47**, 2742; 6157.
82. White S.R., D.J. Scalapino, R.L. Sugar, N. Bickers and R. Scalettar (1989), Phys. Rev. **B 39**, 839.
83. Bulut N., D.J. Scalapino and S.R. White (1994), Phys. Rev. **B 50**, 9623.
84. Sokoloff J.B. (1970), Phys. Rev. **B 1**, 1144.
85. Schuttler H.–B. and C.–H. Pao (1995), Phys. Rev. Lett. **75**, 4504.
86. Crespi V.H. and M.L. Cohen (1992), Solid State Commim. **81**, 187.
87. Andersen O.K., A.I. Liechtenstein, O. Rodriguez, I.I. Mazin, O. Jepsen, V.P. Antropov, O. Gunnarsson and S. Goplan (1991), Physica **C 185–189**, 147.
88. Krakauer H., W.E. Pickett and R.E. Cohen (1993), Phys. Rev. **B 47**, 1002.
89. Brandt H.R. and M.S. Doria (1988), Phys. Rev. **B 37**, 9722; Preprint, 1989.
90. Fulde P., J. Keller and G. Zwicknagel (1988), Solid State Phys. **41**, 1.
91. Gor'kov L. P. (1987), Sov. Sci. Rev. **9A**, 1.
92. Joynt R. (1988), Supercond. Sci. Technol. **1**, 210; Phys. Scripta **36**, 175 (1987).
93. Rainer D. (1988), Phys. Scripta **23**, 106.
94. Barone A., et al. (1987), Novel Superconductivity, Eds. S.A. Wolf and V.Z. Kresin (New York, Plenum).
95. Niemeyer J., M. Dietrich and C. Politis (1987), Z. Phys. **B 67**, 155.
96. Little W.A. (1988), Science **242**, 1390.
97. Esteve D., et al. (1987), Europhys. Lett. **3**, 1237.
98. Gammel P.L., D.J. Bishop, G.J. Dolan et al. (1987), Phys. Rev. Lett. **59**, 107.
99. Vedeneev S.I., I.P. Kazakov, S.N. Maksimovskii and V.A. Stepanov (1988), JETP Lett. **47**, 585.
100. Witt T.J. (1988), Phys. Rev. Lett. **61**, 1423.
101. Pals J.A., W. van Haeringen and M.H. van Maaren (1977), Phys. Rev. **B 15**, 2592.
102. Fenton E.W. (1985), Solid St. Commun. **54**, 709; Physica **B 135**, 60.
103. Sauls J.A., Z. Zou and P.W. Anderson (1986), Solid State Commun. **60**, 347.
104. Akhytyamov O.S. (1966), Pis'ma Zh. Eksp. Teor. Fiz. **3**, 284 [Sov. Phys. JETP Lett. **3**, 183].
105. Geshkenbein V.B. and A.I. Larkin (1986), Pis'ma Zh. Eksp. Teor. Fiz., **43**, 306 [JETP Lett., 43, 396].
106. Millis A.J. (1985), Physica **B 135**, 69.
107. Millis A.J., D. Rainer and J.A. Sauls (1987), Novel Superconductivity, Eds. S.A. Wolf and V.Z. Kresin (New York, Plenum).

108. Danilov Y.A. and V.L. Safonov (1988), Physica **C 153–155**, 683.
109. Sigrist M, and T.M. Rice (1993), J. Phys. Soc. Japan **61**, 4283.
110. Ivanov Z.G., et al. (1991), Appl. Phys. Lett. **59**, 3030.
111. Iguchi I. (1995), In: Coherence in High Tc Superconductors, Workshop Herzyla, Israel May 1–3, 1995 (World Scientific, Singapore).
112. Mathai A., et al. (1995), Phys. Rev. Lett. **74**, 4523.
113. Sun A.G., D.A. Gajewski, M.B. Maple and R.C. Dynes (1994), Phys. Rev. Lett. **72**, 2267.
114. Chaudhari P. and S.Y. Lin (1994), Phys. Rev. Lett. **72**, 1084.
115. Iguchi I. and Z. Wen (1994), Phys. Rev. **B 49**, 12388.
116. Kleiner R., et al. (1996), Phys. Rev. Lett. **76**, 2161.
117. Millis A.J. (1994), Phys. Rev. **B 49**, 15408.
118. Kirtley J.R., et.al. (1987), Phys. Rev. **B 35**, 8846; **B 51**, 12057 (1995).
119. Krishana K., et al. (1997), Science **277**, 83.
120. Laughlin R.B. (1998), Phys. Rev. Lett. **80**, 5188.
121. Wiegmann P.B. (1992), Progr. Theor. Phys. **107**, 243.
122. Jones D.R.T., A. Love and M.A. Moore (1976), J. Phys. **C 9**, 743.
123. Fisher R.A., et al. (1989), Phys. Rev. Lett. **62**, 1411.
124. Hasselbach K., L. Taillefer and J. Floquet (1989), Phys. Rev. Lett., **63**, 93.
125. Hess D. W., T.A. Tokuyasu and J.A. Sauls (1989), J. Phys. Condens. Matter. **1**, 8135.
126. Volovik G.E. and D.E. Khmel'nitskii (1985) Pis'ma Zh. Eksp. Teor. Fiz. **40**, 469 [JETP Lett. **40**, 1299].
127. Hirashima D.S. (1988), Physica **C 153–155**, 667.
128. Das M.P., H.X. He and T.C. Choy (1988), Int. J. Mod. Phys. **B 2**, 1513.
129. Das M.P., T.C. Choy and H. He (1988), Int. J. Mod. Phys. **B 1**, 549.
130. Butera R.A. (1988), Phys. Rev. **B 37**, 5909.
131. Ishikawa M., et al. (1988), Solid State Commun. **66**, 201.
132. Ishikawa M., et al. (1988), Physica **C 153–I55**, 1089.
133. Inderhees S.E., et al. (1988), Phys. Rev. Lett. **60**, 1178.
134. Sobyanin A.A. and A.A. Stratonnikov (1988), Physica **C 153–155**, 1681.
135. Horn P.M., et al. (1987), Phys. Rev. Lett. **59**, 2772.
136. Caponi J.J., C. Chaillout, A.W. Hewat, P. Lejay, M. Marezio, N. Nguyen, B. Raveau, J.I. Soubeyroux, J.L. Tholence, and R. Tournier (1987), Europhys. Lett. **3**, 1301.
137. Day P., et al. (1987), J. Phys. **C 20**, L429.
138. Joynt R. and T.M. Rice (1985), Phys. Rev. **B 32**, 6074.
139. Millis A.J. and K.M. Rabe (1988), Phys. Rev. **B 38**, 8908.
140. Ma S.K. (1976), Modern Theory of Critical Phenomena (Reading, Massachusetts, W.A. Benjamin).
141. Toledano J.C., et al. (1985), Phys. Rev. **B 31**, 7171.
142. Millev Y.T. and D.I. Uzonov (1989), Phys. Lett. **A 138**, 523.

143. Ginzburg V.L. (1960), Soviet Phys. Solid State **2**, 1824.
144. Lee P.A. and S.R. Shenoy (1972), Phys. Rev. Lett. **28**, 1025.
145. Rammer J. (1987), Phys. Rev. **B 36**, 5665.
146. Douglass D.H. (1989), Phys. Rev. **B 39**, 4748.
147. Ginsberg D.M., et al. (1988), Physica **C 153–155**, 1082.
148. Salamon M.B., et al. (1988), Phys. Rev. **B 38**, 885.
149. Freitas P.P., et al. (1987), Phys. Rev. **B 36**, 833.
150. Goldenfeld N., et al. (1988), Solid St. Commun. **65**, 465.
151. Hagen S., et al. (1988), Phys. Rev. **B 38**, 7137.
152. Fisher R.A., et al. (1988), Physica **C 153–155**, 1092.
153. Fossheim K., et al. (1988), Int. J. Mod. Phys. **B 1**, 1171.
154. Ginsberg D.M., et al. (1988), Physica **C 153–155**, 1082.
155. Salamon M.B., et al. (1988), Phys. Rev. **B 38**, 885.
156. Kanoda K. (1988), J. Phys. Soc. Japan, **57**, 1544.
157. Lee W.C., R.A. Klemm and D.C. Johnston (1989), Phys. Rev. Lett. **63**, 1012.
158. Howson M.A., et al. (1989), J. Phys. Cond. Matter, 1, 465.
159. Monthoux P., A. Balatsky and D. Pines (1991), Phys. Rev. Lett. **67**, 3348; Phys. Rev. **B 46**, 14803 (1992).
160. Sharifi F., et al. (1989), Physica, **C 161**, 555.
161. Douglass D.H. and V.G. Zarifis (1989), Phys. Rev. **B 40**, 11303.
162. Aronov A.G., et al. (1989), Phys. Rev. Lett. **62**, 965.
163. Matsuda Y., et al. (1988), Solid State Commun. **68**, 103.
164. Forster D. (1975), Hydrodynamic Fluctuations, Broken Symmetry, and Correlation Functions (Frontiers in physics, 47, W.A. Benjamin, Reading, Massachusetts).
165. Abrahams E. and T. Tsuneto (1966), Phys. Rev. **152**, 416.
166. Popov V.N. (1987), Functional Integrals and Collective Excitations (Cambridge University Press, Cambridge).
167. Anderson P.W. (1963), Phys. Rev. **130**, 439.
168. Anderson P.W. (1958), Phys. Rev. **110**, 827.
169. Gor'kov L.P. (1986), Pis'ma Zh. Eksp. Theor. Fiz.. 44, 537 [JETP Lett. 44, 693].
170. Schenstrom A., et al. (1989), Phys. Rev. Lett. **62**, 332.
171. Malozemoff A.P. (1989), Physical Properties of High Temperature Superconductors I, ed. D.M. Ginsberg (Singapore, World Scientific).
172. Moore M.A. (1989), Rev. **B 39**, 136.
173. Nelson D.R. (1995), Nature **375**, 356.
174. Hess H.F., et al. (1989), Phys. Rev. Lett. **62**, 214.
175. Shore J.D., et al. (1989), Phys. Rev. Lett. **62**, 3089.
176. Ding H. (1994), Phys. Rev. **B 50**, 1333.
177. Shen Z.–X., et al. (1993), Phys. Rev. Lett. **70**, 1553.
178. Borkowski L.S. and P.J. Hirschfeld (1993), Phys. Rev **B 49**, 15404.
179. Fehrenbacher R. and M.R. Norman (1994), Phys. Rev. **B 50**, 3495.
180. Bonn D.A., S. Kamal, K. Zhang, et al. (1994), Phys. Rev. **B 50**, 4051.

181. Li Q.P. and R. Joynt (1993), Phys. Rev. **B 47**, 530.
182. Norman M.R., et al. (1995), Phys. Rev. **B 52**, 15107.
183. Pines D. (1994), Physica **B 199–200**, 300.
184. Martindale J.A., S.E. Barrett, K.E. O'Hara, C.P. Slichter, W.C. Lee and D.M. Ginsberg (1993), Phys. Rev. **B 47**, 9155.
185. Thelen D., D. Pines and J.P. Lu (1993), Phys. Rev. **B 47**, 9151.
186. Barrett S.E., J.A. Martindale, D.J. Durand, C.P. Pennington, C.P. Slichter, T.A. Priedmann, J.P. Rice and D.M. Ginsberg (1991), Phys. Rev. Lett. **66**, 108.
187. Hardy W.N., D.A. Bonn, D.C. Morgan, R. Liang and K. Zhang (1993), Phys. Rev. Lett. **70**, 3999.
188. Chakravarty S., A. Sudbo, P.W. Anderson and S. Strong (1993), Science **251**, 337.
189. Dahm T., D. Manske, D. Fay and L. Tewordt (1996), Phys. Rev. **B 54**, 12006; Nazarenko, A. and E. Dagotto, Phys. Rev. **B 53**, R2987 (1996).
190. Devereaux T.P., D. Einzel, B. Stadlober, R. Hackl, D.H. Leach and J.J. Neumeier (1994), Phys. Rev. Lett. **72**, 396.
191. Coffey D. and L. Coffey (1993), Phys. Rev. Lett. **70**, 1529.
192. Monthoux P. and D.J. Scalapino (1994), Phys. Rev. Lett. **72**, 1874.
193. Braunisch W., et al. (1993), Phys. Rev. **B 48**, 403.
194. Yu R.C., M.B. Salamon and W.C. Lee (1992), Phys. Rev. Lett. **69**, 1431.
195. Bonn D.A., et al. (1992), Phys. Rev. Lett. **68**, 2390.
196. Monthoux P. and D. Pines (1992), Phys. Rev. Lett. **69**, 961; Phys. Rev. **B 47**, 6069 (1993); **B 49**, 4261 (1994).
197. Ishida K., et al. (1991), Physica **C 115**, 185; J. Phys. Soc. Japan **62**, 2803 (1993); Preprint, 1993.
198. Walstedt R.E., et al. (1989), Phys. Rev. **B 34**, 1011; Preprint, 1993.
199. Gor'kov L.P. (1958), Sov. Phys. JETP **7**, 505; **40**, 1155 (1985).
200. Hirschfeld P. and N. Goldenfeld (1993), Phys. Rev. **B 48**, 4219.
201. Hotta T. (1993), J. Phys. Soc. Japan **62**, 274.
202. Lee J., K. Paget, and T.R. Lemberger (1994), Phys. Rev. **B 50**, 3337.
203. Lee J.Y. and T.R. Lemberger (1993), Appl. Phys. Lett. **62**, 2419.
204. Hardy W.N., D.A. Bonn, D.C. Morgan, R. Liang and K. Zhang (1993), Phys. Rev. Lett. **70**, 3999.
205. Ma Z., et al. (1993), Phys. Rev. Lett. **71**, 781.
206. Mason T.E., G. Aeppli, S.M. Hayden, A.P. Ramirez and H.A. Mook (1993), Phys. Rev. Lett. **71**, 919.
207. Basov D.N., et al. (1994), Phys. Rev. **B 49**, 12165.
208. Carbotte J.P., C. Jiang, D.N. Basov and T. Timusk (1995), Phys. Rev. **B 51**, 11798.
209. Graf M.J., M. Palumbo, D. Rainer and J.A. Sauls (1995), Phys. Rev. **B 52**, 10588.
210. Giapintzakis J., D.M. Ginsberg, M.A. Kirk and S. Ockers (1994), Phys. Rev. **B 50**, 15967.
211. Sun A.G., et al. (1994), Phys. Rev. **B 50**, 3266.
212. Nersesyan A.A., et al. (1995), Nucl. Phys. **B 438**, 561.

213. Nersesyan A.A., et al. (1994), Phys. Rev. Lett. **72**, 2628.
214. Cooper S.L., et al. (1988), Phys. Rev. **B 38**, 11934.
215. Ding H. (1994), Phys. Rev. **B 50,** 1333.
216. Xiang T. and J.C. Wheatley (1995), Phys.Rev. **B 51**, 11721.
217. Zang F.C. (1989), Phys. Rev. **B 39**, 7375 ; Phys. Rev. **B 40**, 5219 (1989).

Chapter XXVII

1. Brusov P.N., N.P. Brusova (1994), The model of d–pairing for HTSC, Heavy fermion systems and superfluids, Physica **B 194–196**, 1479.
2. Brusov P.N., M.V. Lomakov, N.P. Brusova (1994), The collective excitations in HTSC under d–pairing (3D–CASE), Physica C, **235–240**, 2259.
3. Brusov P.N. and N.P. Brusova (1996), The path integral model of d–pairing for HTSC, heavy fermion superconductors and superfluids. J. Low. Temp. Phys. **103**, 251.
4. Brusov P.N., N.P. Brusova and P.P. Brusov (1996), The path integral model of d–pairing superconductors. Czechoslovak J. of Physics **46**, suppl. 2, 1041.
5. Brusov P.N., N.P. Brusova, P.P. Brusov (1996), Soviet Low temperature physics, **22**, 506.
6. Brusov P.N., N.P. Brusova and P.P. Brusov (1997), The collective excitations of the order parameter in HTSC and heavy fermion superconductors (HFSC) under d–pairing. J. Low Temp. Phys. **108**, 143.
7. Brusov P.N., N.P. Brusova, P.P. Brusov, N.N. Harabaev (1997), The collective modes in HTSC and heavy fermion superconductors (HFSC) under d–pairing, Physica C, **282–287**, 1881–1882.
8. Brusov P.N. (1999), The mechanisms of HTSC, RSU publishing, v. **1**, **2**, 1389 p., (in English).
9. Brusov P.N., P.P. Brusov and V.P. Sachenko (2000), Collective excitations in unconventional superconductors, Proceedings of Samara Technical University, v.10, p. 1–26.
10. Brusov P.N., P.P. Brusov (2000), Proceedings of SPIE Vol. 4058.
11. Brusov P.N., P.P. Brusov (2000), In: Proceedings of International conference on low temp. physics LT–32, Kazan', Russia, p. 43–45.
12. Brusov P.N., P.P. Brusov (2001), Collective properties of superconductors with nontrivial pairing, Sov. Phys. JETP **92**, 795 [Zh. Exp. Theor. Phys. **119**, 913].
13. Brusov P.N. and P.P. Brusov (2006), Order parameter collective modes in unconventional superconductors, In: Proceedings NATO Advanced Research Workshop, Electron correlation in new materials and nanosystems, Edited by K. Scharnberg and S. Kruchinin, Springer.

14. Brusov P.N., M.Y. Nasten'ka, T.V.Filatova–Novoselova (1982), The determination of type of pairing and order parameter for HFS and HTSC by sound and microwave absorption experiments. Phys. Lett. **142A**, 179.
15. Brusov Peter, Paul Brusov and Chong Lee (2004), Collective properties of unconventional superconductors, Int. J. of Mod. Phys. B, **18**, 867.
16. Feller J.R., C.C. Tsai, J.B. Ketterson, et al. (2002), Phys. Rev. Lett. **88**, 247005.

Chapter XXVIII

1. Krishana, K., et al. (1997), Science **277**, 83.
2. Annett J.F., N.D. Goldenfeld and A.J. Leggett (1996), In: Physical Properties of High Temperature Superconductors V, ed. D.M. Ginsberg (World Scientific, Singapore).
3. Laughlin R.B. (1998), Phys. Rev. Lett. **80**, 5188.
4. Brusov Paul and Peter Brusov (2000), Physica B, **281&282**, 949.
5. Brusov P.N., N.P. Brusova (1994), Physica **B 194–196**, 1479.
6. Brusov P.N., N.P.Brusova (1996), J. Low Temp. Phys., **101**, 1003.
7. Brusov P.N., N.P. Brusova, P.P. Brusov (1996), Czechoslovak Journal of Physics, **46**, suppl. s2, 1041.
8. Brusov P.N., N.P. Brusova and P.P. Brusov (1997), J. Low Temp. Phys. **108**, 143; Physica **B 259–261**, 496 (1999).
9. Brusov P.N., N.P. Brusova, P.P. Brusov, N.N. Harabaev (1997), Physica **C 282–287**, 1881.
10. Brusov, P.N., Brusova N.P., Brusov P.P. (1996), Fiz. Nizk. Temp. **22**, 506 [Low Temp. Phys., **22**, 389].
11. Brusov P.N. (1999), Mechanisms of High Temperature Superconductivity, v.1, 2; Rostov State University Publishing, 1389 p.
12. Brusov P.N., P.P. Brusov (2001), "Collective properties of superconductors with nontrivial pairing", Sov. Phys. JETP, **119**, 913–930 [Zh. Eksp. Theor. Phys. v. 92, p. 795].
13. Brusov P.N., N.P. Brusova and P.P. Brusov (1997), The collective excitations of the order parameter in HTSC and heavy fermion superconductors (HFSC) under d–pairing. J. Low Temp. Phys. **108**, 143.
14. Brusov Peter, Paul Brusov, Pinaki Majumdar and Natali Orehova (2003), "Ultrasound attenuation and collective modes in mixed mixed $d_{x^2-y^2} + d_{xy}$ state of unconventional superconductors" , Brazilian Journal of Physics, **33**, 729–732.
15. Brusov Peter, Paul Brusov and Chong Lee (2004), Collective properties of unconventional superconductors, Int. J. Mod. Phys. B, v. 18, 867–882.
16. Feller J.R., C.C. Tsai, J.B. Ketterson et al. (2002), Phys. Rev. Lett. **88**, 247005.
17. Balatsky A.V., P. Kumar, and J.R. Schrieffer (2000), Phys. Rev. Lett. **84**, 4445.

Chapter XXIX

1. Maeno Y., T.M. Rice, M. Sigrist (2001), The intriguing superconductivity of strontium ruthenate, Physics Today **54**, 42.
2. Brusov P.N. (1999), Mechanisms of High Temperature Superconductivity, v.1, 2; Rostov State University Publishing, p. 1384.
3. Brusov P.N., V.N. Popov (1984), The Collective Excitations in Superfluid Quantum Fluids (Rostov–on–Don State University Press, Rostov–on–Don, USSR).
4. Brusov P.N. and V.N. Popov (1988), The Superfluidity and Collective Properties of Quantum Liquids (Nauka, Moscow), p. 215.
5. Brusov P.N., M.V. Lomakov, N.P. Brusova, T.V. Filatova–Novoselova (1994), The Spectroscopy of Collective Modes in HTSC for P–Pairing (3D–Case), Physica B **194–196**, 1477.
6. Brusov P.N., P.P. Brusov (2000), "Collective excitations in unconventional superconductors", Proceedings of SPIE Vol. 4058.
7. Brusov Peter, Paul Brusov and Chong Lee (2004), Collective properties of unconventional superconductors, Int. J. of Mod. Phys. B, **18**, 867.
8. Brusov P.N. and P.P. Brusov (2006), Order parameter collective modes in unconventional superconductors, In: Proceedings NATO Advanced Research Workshop, Electron correlation in new materials and nanosystems, Edited by K. Scharnberg and S. Kruchinin, Springer.
9. Tewordt L. (1999), Phys. Rev. Letts., **83**, 1007.
10. Higashitani S. and K. Nagai (2000), Phys. Rev. **B 62**, 3042.

Chapter XXX

1. Brusov P.N. (1999), Mechanisms of High Temperature Superconductivity", v. 1, 2; Rostov State University Publishing, p. 1384.
2. Brusov P.N., V.N. Popov (1981), Superfluidity and Bose–excitations in He3 films Sov. Phys. JETP, **53**(4), 804–810.
3. Brusov P.N., V.N. Popov (1982), Superfluidity and Bose–excitations in He3 films Phys. Lett., **87A**, #9, 472.
4. Brusov P.N. and N.P. Brusova (1995), The model of d–pairing in CuO_2 planes of HTSC and the the collective modes. J. Low Temp. Phys. **101**, 1003.
5. Brusov P.N., N.P. Brusova (1994), The collective excitations in CuO_2 planes of HTSC under d–pairing, Physica C, **235–240**.
6. Brusov P.N., N.P. Brusova, P.P. Brusov, N.N. Harabaev (1997), "The path integral model of d–pairing in CuO2 planes of HTSC and the collective modes", Physica C, **282–287**, p.1833–1834.
7. Brusov Peter, Paul Brusov and Chong Lee (2004), Collective properties of unconventional superconductors, Int. J. of Mod. Phys. **B 18**, 867–882.

8. Brusov, P.N., N.P. Brusova (1994), Physica **B 194–196**, 1479.
9. Brusov, P.N., N.P. Brusova, P.P. Brusov (1996), Czechoslovak Journal of Physics, **46**, suppl. s2, 1041.
10. Tewordt L. (1999), Phys. Rev. Letts. **83**, 1007.
11. Higashitani S. and K. Nagai (2000), Phys. Rev. **B 62**, 3042.

Chapter XXXI

1. Sarma B.K., et al. (1992), In: Physical Acoustics, **XX**, 107 (Academic Press, Inc).
2. Brusov P.N. and N.P. Brusova (1994), The model of d–pairing for HTSC, Heavy fermion systems and superfluids. Physica B **194–196**, 1479–80.
3. Brusov P.N., M.Y. Nasten'ka, T.V. Filatova–Novoselova (1982), The determination of type of pairing and order parameter for HFS and HTSC by sound and microwave absorption experiments, Phys. Lett. **142A**, 179.
4. Brusov P.N., N.P. Brusova, P.P. Brusov, N.N. Harabaev (1997), The collective modes in HTSC and heavy fermion superconductors (HFSC) under d–pairing, Physica **C 282–287**, 1881–1882.
5. Brusov Peter, Paul Brusov and Chong Lee (2004), Collective properties of unconventional superconductors, Int. J. of Mod. Phys. **B 18**, 867–882.
6. Hirashima D.S. and M. Namizawa (1988), J. Low Temp. Phys. **73**, 137.

Chapter XXXII

1. Volovik G.E. (1986), JETP Lett. **43**, 535.
2. Balatsky A.V., G.E. Volovik and V.A. Konushev (1986), JETP, **90**, 2038.
3. Bogdanova N.E. and V.N. Popov (1981), Sov. J. Theor. and Math. Phys., **46**, 325.

About Authors

PETER NIKITOVICH BRUSOV

was born in Rostov–on–Don, Russia, on 23. 09. 1949. He graduated from Rostov State University in 1972 (physics); Ph.D. (physics) – 1980, Leningrad Polytechnical Institute; Dr. of physical & mathematical sciences, Professor – 1988, Moscow / Dubna, Joint Institute for nuclear research. Now he is a Professor, Financial Academy, Moscow, Russia.

POSITIONS HELD

1. 1977–1980 Postgraduate student, Leningrad Polytechnical Institute, (advisor Prof. V. N. Popov)
2. 1980–1988 Senior scientist, Physical Research Institute, Rostov State University
3. 1983–1984 Visiting scientist, Helsinki University of Technology (c/o Prof. O. Lounasmaa and Prof. M. Krusius), Finland
4. 1988–1989, 1995 Visiting Professor, Northwestern University (c/o Prof. J. B. Ketterson), USA
5. 1988–up to now Head of the Department of HTSC, Physical Research Institute, Professor, Rostov State University
6. 2000–2001 Visiting Professor, Cornell University
7. 2004 Visiting Professor, Houston University, USA
8. 2005 Visiting Professor, Osaka City University, Japan
9. 2004–up to now, Professor, Financial Academy, Moscow, Russia

PUBLICATIONS

Total number: above 130, among them four books:

1. The collective excitations in superfluid quantum liquids, Rostov State University publishing, 200 p. p., 1984.

2. The superfluidity and collective properties of quantum liquids, Moscow, Nauka, 216 p. p., 1988.
3. Mechanisms of HTSC, Rostov State University publishing, v. 1, 2; 1389 p. p., 1999 (in English).
4. Collective excitations in unconventional superconductors and superfluids, World Scientific Publishing, Singapore, 800 p., 2009.

TOPICS OF INVESTIGATIONS

1. The superfluidity and collective excitations in superfluid He^3 and He^3–He^4 mixtures.
2. The superfluidity and collective excitations in He^3–films.
3. The boundary state in superfluid He^3–B.
4. Sound conversion in impure superfluids.
5. Nonlinear sound phenomena in superfluid He^3 in aerogel.
6. The type of pairing and the order parameter in HTSC and HFSC.
7. The d–pairing in HTSC and HFSC, mixtures of states ($d + id$, $d + s$, etc.) in HTSC.
8. The collective excitations in HTSC and HFSC under s–, p– and d–pairing.

RESULTS

The path integration technique

The path integration technique has been developed (initially with V. Popov) for investigation of the collective excitations in superfluid and superconducting Fermi–systems with nontrivial pairing.

Superfluidity and collective excitations in superfluid He^3

The application of path integration technique has allowed P. Brusov to create the microscopic theory of collective excitations (CE) in superfluid He^3. Within this theory the whole spectrum of CE has been calculated in A–, B–, A_1–, 2D and polar phases at k=0. The frequencies of all collective modes in A–, A_1–, 2D– and polar phases where gap disappears in some directions or along lines turn out to be complex. The presence of imagine part of frequency leads to renormalization of real one via the dispersion relations. The experiment in He^3–A on the clapping mode frequency measurement (J. Low Temp. Phys. **78**, 187, (1990)) is in excellent agreement with Brusov et al. old predictions, while 6% difference between kinetic equation method prediction and these data remains even under taking into account the Fermi–liquid and higher pairing corrections.

The dispersion induced (DI) splitting of all 18 collective modes has been calculated in A–, B–, 2D–phases. This splitting has been observed in He^3–B at Northwestern University (Phys. Rev. Lett. **49**, 1646 (1982)) and the data are in good agreement with predictions. In A–phase the prediction of DI splitting shows the necessity of search of "thin" structure of sound absorption spectrum in He^3–A.

The stability of Goldstone modes has been investigated (it requires a calculation of the corrections of order k^4). The linear Zeeman effect for cl– and pb–modes in He^3–A has been predicted. The gap distortion effect has been discovered and it was shown that it plays a significant role in dependence of collective modes spectrum on external pertubations. In particular the resonant absorption of ultrasound at the absorption edge observed by Daniels, et al. in 1983 should be observed not only in magnetic field, but also in electric field, under superflow and some other pertubations.

The influence of superflow on collective modes in He^3–B has been investigated and 3–fold splitting of collective modes has been predicted. This splitting has been observed for sq–mode and this gives us the good method of measurement of superfluid velocity.

The electric field induced splitting of collective modes in He^3–B has been predicted. It should be observed in fields of order $1.5 \div 2.7 \cdot 10^6$ v/cm.

The collective excitations in superfluid He^3–He^4 mixtures has been studied and the coupling of the acoustic modes of both systems has been obtained.

The theory of superfluidity and collective excitations in the He^3 films has been developed. It was shown that two superfluid phases (we called them *a*– and *b*–phases) are favorable energetically and whole spectrum of collective modes (12 modes in each phase) has been calculated for both phases. In 1985 Sachrajda et al. (Phys. Rev. Lett., **55**, 1602 (1985)) have discovered 2D superfluidity in He^3 films. They have obtained a superfluid He^3–film with thickness less than the coherence length.

The boundary state in superfluid He^3–B has been studied with Pavel Brusov.
We have proved that boundary state of ^{3}He–B is deformed *B*–phase, predicted by Brusov and Popov more then twenty five years ago for case of presence of external perturbations like magnetic and electric fields. Influence of wall or, generally speaking, of confined geometry does not lead to existence of a new phase near the boundary, as it was supposed many yeas ago and seemed up to now, but like other external perturbations (magnetic and electric fields, etc.) wall deforms the order parameter of *B*–phase and this deformation leads to very important consequences. In particular, frequencies of collective modes in the vicinity of boundary are shifted up to 20%.

Peter Brusov have introduced the very fruitful idea of gap distortion, which led to discovery of a lot of very interesting effects in collective mode theory in superfluid He^3.

The following effects, predicted by Peter Brusov, has been then observed experimentally:

- dispersion induced splitting of real squashing–mode in He^3–B
- superflow induced splitting of squashing–mode in He^3–B
- two–dimensional superfluidity in He^3–films
- the exact value of the clapping mode frequency in He^3–A
- existence of two slow–modes in He^3–He^4 solutions in aerogel.

Superconductors (SC) with nontrivial pairing

The model of *d*–pairing in SC and superfluid liquids has been developed by path integration technique. This model has been applied to investigation of the collective mode spectrum in HTSC and HFSC. All superconductive states arisen in symmetry classification of these systems have been considered: five superconductive states for HTSC ($d_{x^2-y^2}$, $d_{3z^2-r^2}$, d_{xy}, d_{xz}, d_{yz}) and three superconductive states for HFSC (d_γ, Y_{2-1}, $\sin^2\theta$). There are ten collective modes in each state (in spherical symmetry approximation): five of them are high frequency and five others seem to be Goldstone – like ones. These spectra could be used to interpret the ultrasound and microwave absorption experiments as well as to identify the type of pairing and the order parameter in HTSC.

The 2D model of *d*–pairing for CuO_2–planes – the mutual structural factor of most HTSC – has been created and the collective mode spectrum for two superconductive states have been calculated.

Peter Brusov and Pavel Brusov have calculated for the first time the collective mode spectrum in a mixed state $d_{x^2-y^2}+id_{xy}$ of HTSC. They have shown that in spite of the fact that spectra in both states $d_{x^2-y^2}$ and d_{xy} are identical the spectrum in mixture $d_{x^2-y^2}+id_{xy}$ state turns out to be quite different from them. Thus the probe of the spectrum in ultrasound and/or microwave absorption experiments could be used to distinguish the mixture of two *d*–wave states from pure *d*–wave state.

PAVEL PETROVICH BRUSOV

was born in Leningrad, Russia, on 01. 12. 1979.
He graduated from Rostov State University in 2000 (physics), MSc Degree.
MSc Thesis: "A method of distinction of the mixed $d_{x^2-y^2} + d_{xy}$ state of high-temperature superconductors from a pure d-wave state"
Ph.D. (physics) – 2004,
Ph.D. Thesis "Investigation of the collective modes of the order parameter in superconducting and superfluid Fermi–systems under p– and d–pairing".
Now he is a senior scientist of Physical Research Institute Rostov, South

Federal University, and leading scientist of Company "Kordon Ltd", Rostov–on–Don, Russia.

POSITIONS HELD

1. 2000–2004 Postgraduate student, Superconducting Devices Group, Department of Physics, University of Strathclyde, Glasgow, UK.
2. 2004–2007 Department of Physics Case Western Reserve University
3. 2007–up to now, senior scientist of Physical Research Institute Rostov, South Federal University, and leading scientist of Company "Kordon Ltd", Rostov–on–Don, Russia

PUBLICATIONS

Total number: above 40, among them one monograph:
Collective excitations in unconventional superfluids and superconductors, World Scientific Publishing, 2009, 800 p.

Awards and Grants

1. US grants for Dark Matter Search (2004–2005)
   CDMS Phase–II and Xenon dark matter search
2. UK EPSRC (United Kingdom Engineering and Physical (2000–2003) Sciences Research Council) research studentship
   "HTS SQUID Technology for the 21st Century"
3. ORS (Overseas Research Scholarship) award (2001/2002; 2002/2003)
4. Soros undergraduate and postgraduate student award (1996–2000)
5. RFBR (Russian Fund for Basic Research) research grants:
   – "Theoretical study of collective excitations in superfluids under p–pairing" (1997–1999)
   – Collective properties of unconventional superconductors" (1998–2000).

TOPICS OF INVESTIGATION

Superfluidity and collective excitations in superfluid ^{3}He

1. A method developed to identify ^{3}He–A by ultrasound experiments allows us to distinguish an axial phase from an axiplanar one while other methods do not solve the issue.
2. The theory of torsional oscillator with superfluid ^{3}He in aerogel has been developed at Cornell University.
3. We consider the conversion of first sound into second sound and the reverse in different impure superfluid systems. The coupling between temperature (entropy) oscillations and pressure (density) oscillations for impure superfluids (including He^3–He^4 mixtures) and for superfluid He in aerogel and show that presence of impurity or aerogel plays a fundamental role in the sound conversion phenomena. It enhances the coupling strengths of the two sounds and decreases the threshold values for non–linear processes in homogeneous superfluids. Some phenomena such as the slow mode of superfluid helium in aerogel, and heat pulse propagation in He–II in aerogel can be understood as double sound conversion phenomena.
4. The boundary state in superfluid He^3–B has been studied with Peter Brusov. We have proved that boundary state of ^{3}He–B is deformed *B*–phase, predicted by Brusov and Popov many years ago.

Superconductors with nontrivial pairing

2D and 3D models of p– and d–pairing for superfluids and superconductors (SC) have been developed using path integration technique. Within these models the collective excitations in different USC (high temperature superconductors, heavy fermion superconductors etc.) under p– and d–pairing were calculated by Peter Brusov and Pavel Brusov.

Pavel Brusov investigated some recent ideas concerning the realization in HTSC of mixtures of different states. In particular, he was attracted to the study of mixture of $d_{x^2-y^2}$ and d_{xy} states in HTSC. The results obtained are a powerful tool for interpreting the sound and microwave absorption data as well as for identification of the actual type of pairing and the order parameter in unconventional SC.

Xenon Dark Matter Search Experiment

XENON is a next–generation Dark Matter Direct Detection experiment, which uses liquid xenon as a sensitive detector medium to search for Weakly Interacting Massive Particles (WIMPs) postulated by Super Symmetry (SUSY) and believed to have a mass in the neighborhood of 100 GeV.

Design, development and use of fully automated vacuum system for krypton purification of xenon.

A potentially big problem for liquid xenon for the low background experiment such as Xenon Dark Matter Search Experiment lies in the fact that commercially available xenon gas is contaminated with radioactive ^{85}Kr. Pavel Brusov has taken a part in the testing of the system and development of a software driven control FieldPoint unit that allowed to build a fully functional automated krypton purification system for xenon gas with the projected reduction in the Kr concentraion in Xenon from 5 ppm down to ppt level.

The project will allow to take the system to underground experimental site in Gran Sasso to be installed for purification of significant quantities of xenon involved in the development of a large–scale experiment.

CDMS Phase–II Experiment

The Cryogenic Dark Matter Search (CDMS) experiment uses super–sensitive devices aiming to detect Weakly Interacting Massive Particles (WIMPs). The experiment is located 2341 feet underground in the Soudan mine in northern Minnesota.

Testing and installation of CDMS Z–sensitive detectors.

CDMS II uses five towers of silicon and germanium detectors, called ZIPs, to sense the WIMPs that stream through space. To perform the cryo–testing we made use of out test facility that provides a 100 square feet Class 100 Clean room, Oxford Kelvinox 400 dilution refrigerator capable of cooling down a stack of 5 detectors down to 20 mK for the maximum sensitivity, 12–channel electronic readout, DAQ system and a

14–node computing cluster for processing and alanysing the data for routine evaluation of the ZIP detectors.

Thermal modelling of the Soudan 5–tower "Icebox" cryostat.

Pavel Brusov participated and in 2005 led the development and optimisation of a thermal model that is currently used successfully as a helpful tool in predicting the amount of heat load associated with the detectors, supports and black–body radiation that given the cooling power of the Icebox allows to predict with good accuracy the temperature distribution in the complex Soudan CDMS cryosystem. The latest efforts were aimed towards the development of a new SuperIcebox suitable for a 1–tonn scale CDMS experiment beyond 2008.

Fabrication, characterisation and optimisation of superconducting tungsten thin films for fabrication of CDMS detectors.

Pavel Brusov was in charge of the leading role in the study of superconducting and normal properties of tungsten thin films studied at temperatures down to 15 mK. The athermal–phonon sensors of the ZIP detectors require tungsten thin films that are uniform to better than 10% in the transition temperature, Tc, at an operating temperature of 85 mK. At present this requires a separate dilution fridge measurement run in order to map out the uniformity of Tc and thickness of the grown films, there consequent selective iron implantation and another cool down that allows to verify the uniformity afterwards. The goal of Pavel Brusov studies was to establish correlation between room temperature and superconducting properties of the films and a number of processing parameters during their fabrication and post–processing.

Highly–balanced second–order gradiometer systems

In order to be able to measure small magnetic fields in an unshielded environment, in presence of substantial modern world electromagnetic background interference, one has a choice of either utilising very expensive commercially available magnetic shielding or implement several attracting background interference reduction mechanisms. For these purposes, Pavel Brusov developed a second–order gradiometer capable of rejecting uniform fields and first–order field gradients. Pavel

Brusov's study involved supplementing the second–order gradiometer with additional reference sensors measuring the background fields in the orthogonal *x*, *y* and *z* directions, so that the signals can be combined according to several novel background rejection techniques to achieve the required degree of cancellation.

Pavel Brusov demonstrated a simple prototype system comprising of two long–baseline first–order gradiometers configured as a planar second–order gradiometer with three reference flux–gate magnetometers used to improve the balance. The signals were combined synthetically on a computer and adaptively–balanced using linear regression techniques to remove the interference either in the time–domain or the frequency–domain. The latter is particularly successful and provides sufficient resolution for detection of extremely small magnetic fields in an unshielded environment such as, e.g. a signal comparable in size to a human magneto cardiogram (MCG), and some very small signals originating in non–destructive evaluation (NDE) studies.